Craftsman

2022 NATIONAL ELECTRICAL ESTIMATOR

By Mark C. Tyler

Edited by Richard Pray

Includes inside the back cover:

Includes Free Estimating Software Download

Inside the back cover of this book you'll find a software download certificate. To access the download, follow the instructions printed there. The download includes the National Estimator, an easy to-use estimating program with all the cost estimates in this book. The software will run on PCs using Windows XP, Vista, 7, 8, or 10 operating systems.

Quarterly price updates on the Web are free and automatic all during 2022. You'll be prompted when it's time to collect the next update. A connection to the Web is required.

Access all of Craftsman's most popular costbooks – subscribe to National Estimator Cloud. https://craftsman-book.com/national-estimator-cloud

Craftsman

Craftsman Book Company

6058 Corte del Cedro, Carlsbad, CA 92011

Acknowledgments

The author wishes to thank the following individuals and companies for providing materials and information used in this book.

George H. Booth, Vice President Sales — Graybar Electric Company, Inc.

Steve Koundouriotis — P-W Western, Inc.

Don Geibel — Walker Division of Butler Manufacturing Company.

The tables on pages 439 and 440 are reprinted with permission from NFPA 70®-2017, the *National Electrical Code*®, Copyright 2016, National Fire Protection Association, Quincy, MA 02169. This reprinted material is not the complete and official position of the National Fire Protection Association on the referenced subject, which is represented only by the standard in its entirety.

Cover design: *Jennifer Johnson*

ISBN 978-1-57218-373-5
Published October 2021 for the year 2022.

Contents

How to Use This Book

This manual is a guide to the cost of installing electrical work in buildings. It lists costs to the electrical subcontractor for a wide variety of electrical work.

Before using any estimate in this book, you should understand one important point about estimating electrical construction costs. Estimating is an art, not a science. There's no estimate that fits all work. The manhour estimates in this book will be accurate for many jobs, but remember that no two jobs are identical. And no two crews complete all tasks in exactly the same amount of time. That's why electrical cost estimating requires exercising good judgment. Every estimate has to be custom-made for the specific job, crew and contractor. No estimating reference, computerized cost estimating system or estimating service can take into consideration all the variables that make each job unique.

This book isn't meant to replace well-informed decisions. But when supplemented with an estimator's professional evaluation, the figures in this manual will be a good aid in developing a reliable cost of electrical systems.

National Estimator '22 Inside the back cover of this book you'll find a software download certificate. To access the download, follow the instructions printed there. The download includes the National Estimator, an easy-to-use estimating program with all the cost estimates in this book. The software will run on PCs using Windows XP, Vista, 7, 8, or 10 operating systems. When the National Estimator program has been installed, click Help on the menu bar to see a list of topics that will get you up and running. Or go online to www.craftsman-book.com, click on "Support," then click on "Tutorials" to view an interactive video for National Estimator.

Labor Costs

The labor costs listed in this manual will apply to most jobs where the hourly wage in effect is the same or similar to the following rates:

Journeyman Electrician

Base Wage..$31.28 per hr.

Taxable Fringe Benefits at 5.59%.........$1.75 per hr.

Taxes & Insurance at 19.53%...............$6.45 per hr.

Non-taxable Fringe Benefits at 4.94%...$1.55 per hr.

Total Labor Cost...................................$41.03 per hr.

The total hourly cost includes the basic wage, taxable fringe benefits (vacation pay), workers' compensation insurance, liability insurance, taxes (state and federal unemployment, Social Security and Medicare), and typical nontaxable fringe benefits such as medical insurance.

If your hourly labor cost is much lower or higher, costs of installation can be expected to be proportionately lower or higher than the installation costs listed in this book. If your total hourly labor cost is 25 percent less, for example, reduce the labor figures in the cost tables by 25 percent to find your local cost.

The Craft@Hrs column shows the recommended crew and manhours per unit for installation. For example, L2 in the Craft@Hrs column means that we recommend a crew of two electricians. L1 means that a crew of one electrician is recommended. Costs in the Labor Cost column are the result of multiplying the manhours per unit by the rate of $41.03 per hour.

For example, if the Craft@Hrs column shows L2@.250, the Labor Cost column will show $10.30. That's .250 manhours multiplied by $41.03 per manhour and rounded to the nearest ten cents.

Divide the manhours per unit into 8 to find the number of units one electrician can install in one 8-hour day: 8 divided by .250 equals 32 units per day. Multiply that amount by the number of crew members to find the number of units the crew is likely to install in an 8-hour day. For example, if the crew is two electricians, multiply 32 by 2 to find that the crew can be expected to install 64 units in an 8-hour day.

Some tasks require less labor under certain conditions. For example, when conduit is run in groups, less labor is required for each 100 linear feet. It's the estimator's responsibility to identify conditions likely to require more or less labor than the standard for the type of work being estimated.

This book lists both the labor cost per installed unit and the manhours required for installation. Manhours are listed in hundredths of an hour rather than minutes, making it easier to calculate units.

Material Costs

Material prices in this book are based on actual costs in late 2021, with projections to mid-2022 based on the editor's judgment. Prices are neither "retail" nor "wholesale" costs. Instead, they're intended to reflect typical costs to electrical contractors who buy at electrical supply houses in mid-2022. Volume purchases may cost less because many dealers offer quantity discounts to good customers. Expect prices to vary with location, terms demanded, services offered, and competitive conditions.

Prices in this manual are not representative of shelf prices for electrical materials at big box building material retailers – and for good reason. Most electrical contractors don't buy from big box retailers. They buy from specialized electrical material dealers who offer the selection, service and terms that electrical contractors expect. Big box retailers stock limited quantities, no more than a few hundred electrical SKUs, specialize in commodity-grade merchandise and are generally not set up to meet the needs of professional electrical contractors.

Material costs in this book include normal waste. If waste of materials or breakage is expected to exceed 3 to 5 percent of the materials used on the job, include a separate allowance for excessive waste.

Material delivery cost to the job site isn't included in this book. When delivery cost is significant and can be identified, add that cost to these figures.

Please note that the cost of some electrical materials is highly volatile. For example, copper wire prices have been known to fluctuate 10 percent or more in one month. There's no reliable way to forecast price movements like this. If you're bidding on a project that has a quantity of copper products, you may want to add a qualification to your bid proposal which would allow you to pass on a pricing increase (or decrease), based upon the actual materials pricing at the time of purchase. This way, you can use the current price quoted at the time of your bid, but still leave the door open to any major pricing fluctuations.

Add Sales Tax

No state or local sales tax is included in material prices listed here. Sales tax varies from area to area and may not be applicable on purchases for some types of projects. Add at the appropriate rate when sales tax is charged on materials bought for the job.

Add Overhead and Profit

To complete the estimate, add your overhead and expected profit. Many contractors add an additional 10 to 15 percent for profit to yield an acceptable return on the money invested in the business. But no profit percentage fits all jobs and all contractors. Profit should be based on the current market in each user's local area.

For some electrical contractors, overhead may add as little as 10 percent to the labor and material cost. But routinely adding 10 percent for overhead is poor estimating practice. Overhead should be based on each user's built-in costs. It's the estimator's responsibility to identify all overhead costs and include them in the estimate, either as a lump sum or as a percentage of the total labor and material cost.

Other Costs to Add

A few other costs are excluded from the figures in this manual: electrical building permits, special hoisting costs, freight costs not absorbed by the supplier, utility company charges for installation and service, special insurance and bonds, power equipment other than small tools, mobilization to remote sites, demobilization, nonproductive labor, and nonworking supervisors. If these costs are significant and can be determined, add them to your estimate. If not, you should exclude them and specify clearly that they're not a part of your bid.

All Tables Assume "Good" Conditions

This means that there are few or no unusual conditions to delay production. Conditions are good when work is performed during usual working hours in relatively clean surroundings and in readily accessible areas not over 12 feet above the finish floor. The temperature is between 50 and 85 degrees F. Electricians are working no more than 8 hours a day, 5 days a week.

Good conditions require that all tools and materials be available on the job site when needed. Tools, including power tools, are assumed to be in good working order. Where power tools are appropriate, it's assumed that temporary power is provided. Add the cost of temporary power when it's furnished at your expense.

Proper supervision makes a big difference in labor productivity. The tables assume there is adequate supervision but make no allowance for nonproductive labor — supervisors who direct but do no installation. If you plan to have nonproductive supervision on the job, add that cost to the figures in this manual.

Conditions are seldom "good" when the work area is confined, or when a short construction schedule makes it necessary for many trades to work at the same time. The usual result will be stacks of material obstructing the work space and several tradesmen competing for access at the point of installation.

If the conditions on the job you're estimating aren't expected to be "good," adjust the labor figures in this book as appropriate. Occasionally, larger jobs can be done faster because specialized equipment or crews can be used to good advantage. This will usually reduce the installation cost. More often, conditions are less than "good." In that case, labor costs will be higher.

There's no accepted way to decide how much "bad" conditions will increase the labor hours needed. But it's accepted estimating practice to assign a cost multiplier of more than 1.0 to a job that can be expected to require more than the usual amount of labor per unit installed. For example, if conditions are less than "good" only in minor respects, you might multiply labor costs by 1.10. If conditions are very poor, a multiplier of 1.50 or more may be appropriate.

Other Factors That Affect Productivity

This book's tables assume that the crew used for the job is the smallest crew appropriate for the work at hand. Usually this means that the crew is one journeyman electrician.

Most experts on the productivity of construction trades agree that the smallest crew that can do the job is usually the most efficient. For example, it's foolish to have two men working together setting duplex receptacles — one handing tools and material to the other as needed. Only one of them would be working at any given time. It's more productive to use two one-man crews, each working independently.

Of course, there are exceptions. Sometimes a crew of one takes twice as long as a crew of two. When pulling feeder cable or setting floor-standing switchboards or motor control centers, more help usually cuts the labor cost per installed unit. Some jobs simply can't be done by a crew of one.

When work is done on a scaffold, someone should be on the ground to chase parts and equipment and prepare lighting fixtures for hanging. It wastes manpower to have an electrician leave the scaffold and return when parts or tools are needed. Scaffold installers should install one fixture while the "grunt" below prepares the next. Conduit should be prefabricated on the ground from measurements taken by the electricians on the scaffold. The assistant should bend the conduit and hand it up to the installer.

These labor savings are obvious to anyone who's done this type of work, and are assumed in this book's labor tables.

The Electrician

This book's labor hours are typical of what a trained and motivated journeyman electrician with 5 years of experience will do on most jobs. It's assumed that the installer can read and follow plans and specifications and has the ability to lay out the work to code.

It shouldn't make any difference whether the work is in a hospital, a grocery store, a wood mill or a small convenience store. An experienced journeyman electrician should be able to handle the work at the rates shown here even though the materials and code requirements differ. But you'll have to make allowances if your installers are only familiar with residential work, and the job at hand is something else.

Improving Estimating Accuracy & Profits

It's been said that electrical estimators learn by making mistakes. The best estimators are the ones who've made the most mistakes. Once you've made every mistake possible, you're a real expert.

I can't subscribe 100 percent to that theory, but I know that there are plenty of pitfalls for unsuspecting electrical estimators. This section is intended to suggest ways to spot potential problems before they become major losses. It'll also recommend steps you can take to increase the profit on most jobs.

Labor Productivity

Improving output even slightly can result in major cost savings. Cutting only a minute or two off the installation time for each duplex receptacle or handy box can reduce the labor cost by several hundred to a thousand dollars a job. Getting better productivity from your electricians should be a primary concern for every electrical contractor.

Assuming your electricians are experienced, well-trained, and have all the tools and materials they need to complete the work, the most significant increase in productivity will probably be through motivation.

The best form of motivation for most electricians is to encourage pride in the work they do. Every alert supervisor knows the value of recognizing a job well done. Acknowledging good work builds confidence and encourages extra effort in the future.

Labor Availability

Labor in each locale may not always be readily available. Prior to bidding any project, make an evaluation of the available work force. You may need to make staffing or salary adjustments for the duration of that project. Your work force evaluation will help you prepare for adding another workman, or adjusting a current employee's salary and benefits to compete with rates in your area.

Handling Inspections

The on-site supervisor or foreman should be responsible for dealing with all inspectors. Don't let others circumvent the supervisor's or foreman's authority.

An inspector's only job is to see that the installation complies with the code. They aren't supervisors and don't direct the work. They can and do interpret the code and sometimes make mistakes. Encourage the foreman or supervisor to take issue promptly with a questionable interpretation. Ask the inspector to cite a specific code as his reference. If the inspector insists that his interpretation is correct, and if you believe it's wrong, call the building official to initiate an appeal. Your trade association or the National Electrical Contractors' Association may also be able to persuasively argue in your favor.

Some inspectors have a reputation for being impossible to deal with. Aggressive enforcement of questionable code interpretations can severely hurt project productivity. Following the code carefully will keep you out of most compliance arguments. Every electrician and electrical supervisor must know the code. Code classes are taught at continuation schools in many communities. You can take code classes to both understand how the code is applied and to remain current on code changes.

Mobilization and Demobilization

Many electrical subcontractors have job shacks and lockup boxes that can be moved onto the job for storing tools and materials. Some larger firms have trailers that can be moved from job to job. No matter what type of on-site storage you use, setting up takes time. The bigger the job, the more time will probably be needed.

Usually the first step is getting permission to set up your storage area on the site. Sometimes storage space is at a premium. Some city projects literally have no storage space until parts of the building are completed and can then be used. Occasionally tools and equipment will have to be stored off site. This can require daily mobilization and demobilization, which increases your labor cost substantially. Be sure your estimate includes these costs.

Demobilization usually takes less time and costs less than mobilization. Removing the surplus material, tools and equipment can be done by helpers or material handlers rather than electricians.

One important item in mobilization is temporary electrical service. Be sure you know who pays for installation of temporary power and who pays for power used on site during construction. It's common for the electrical contractor to cover the cost of electrical distribution and service. Installation is usually done by your electricians and will have to pass inspection.

Most communities require temporary electrical permits prior to starting work. Before applying for the permit, contact the electric utility provider and request a meeting with whoever coordinates extensions of service — usually the planner. Before your meeting, determine what size service you need. The planner will tell you what voltage is available and where the point of connection will be. Don't end this meeting with the planner until you've covered every requirement and procedure imposed by the electric utility.

Job Cleanup

Trash and debris that obstructs access to (and on) the job site can make good production next to impossible. That alone should be encouragement to regularly dispose of accumulated waste. Most specifications require that subcontractors remove unused materials, cartons, wrappers and discarded equipment. On many jobs, the general contractor has the right to backcharge subs for removal of their discards if they don't clean the site themselves.

Encourage your crews to do their cleanup while installation is in progress. For example, each time a fixture is removed from a carton, the tradesman should collapse the carton and throw it on the discard pile. It takes slightly more time to dispose of trash this way, but cleanup is less likely to be forgotten.

Some contractors and subcontractors have a reputation for running a dirty job. You've probably seen sites that are so cluttered that you can't understand how anyone could work efficiently. Of course, as the electrical contractor, you can't dictate to the general contractor or the other subcontractors. But the work habits of others affect your productivity, and consequently, your profit.

I believe that if accumulated debris is slowing progress on the job, it's within your rights to discuss it with the general and the other subs. Request a meeting, right in the middle of the clutter. That alone may do the trick.

If you don't insist on a clean site, the fire department probably will. A clean job is more efficient and safer. A cluttered job costs everyone time and money.

Production

No matter how simple and quick you anticipate them to be, most jobs will have some production problems. Every job is unique. Every job brings together skilled tradesmen with varying preferences and habits. Some have never worked together before. Yet each must coordinate the work he does with those who precede him and those who follow. It's normal to expect that some adjustments will be needed before cooperation becomes routine.

Of course, the general contractor is the key to cooperation among the trades. A general who schedules trades properly will have fewer problems and will help all subcontractors earn the profit they're hoping for. This isn't automatic. And some general contractors never learn how to schedule properly. From an estimating prospective, it's more expensive to work for a contractor who has scheduling problems than it is to work for a contractor who's efficient at job coordination. If you anticipate production problems like this on a job, your estimate should reflect it.

Good supervision helps avoid most production problems. Try to schedule material deliveries in a timely manner. Have the right tools on hand when needed. Keep crews as small as possible. Don't work your crews more than 40 hours a week unless absolutely necessary. Too many bodies and too many hours will erode production.

If you're using a larger crew, don't have everyone work at the same time. Instead, break the crew into two units and encourage friendly competition between the two. Offer a reward for the winning crew.

Corrections

This book's tables assume that little or no time is spent making corrections after the work is done. Electrical contractors should have very few callbacks.

If you're called back often to replace faulty materials or correct defective workmanship, one of four things is happening. First, you could be working for some very particular contractors or owners, or handling some very sensitive work. In that case, callbacks could be part of the job and should be included in each estimate. Second, you could be installing substandard materials. Third, your electricians could be doing haphazard work. Finally, your installation procedure could be omitting fixture and circuit tests that could locate problems before the owner finds them.

When qualified electricians install quality materials, the risk of a callback is small. Occasionally a ballast will fail after 10 or 20 hours in use. And sometimes an owner's negligence will damage a circuit or switch. When this happens, accept the service work as routine. Complete it promptly at no extra charge. Consider it cheap advertising — a chance to establish your reputation with the owner. You could turn the service call into some extra work later.

Your Type of Work

Most electrical contractors prefer to handle specific types of work. Only a few have the capital, equipment and skills needed to handle the largest jobs. Most will do residential wiring because that's the most plentiful work available. Some prefer private work with as little government interference as possible. Others bid only government jobs.

The most profitable electrical contractors specialize in one type of work or customer. The electrical construction field is too broad to do everything well. Select an area that you feel comfortable with, and concentrate on doing it as well or better than anyone else. Of course, some of the older and larger electrical shops will do almost any type of work. But nearly every electrical contractor prefers some class of job over all others — and would take only that work if there was enough available to stay busy.

Observe the electrical contractors in your area. Notice the companies that seem to be busiest and most profitable. See what class of customers they service or what type of work they do most. It's probably easier to follow the success of another contractor who's found a winning formula than it is to invent a new formula yourself.

Specialization lets you hire electricians who are specialists, too. That tends to improve productivity, keep costs down, and improve profits — as long as you're handling work that's within your specialty.

Coordination is easier and the profits will usually be higher if you work for a limited number of general contractors. Some contractors seem to be masters at putting a project together. These same contractors probably pay promptly and treat their subs fairly. That makes your job easier and tends to fatten your bottom line. If you've found several contractors who make life more pleasant for you, keep them supplied with competitive bids that'll bring more work into both your shop and theirs.

Most electrical contractors don't bid government work. It's a specialty that requires specific knowledge: complying with detailed general conditions, observing regulations, anticipating inspection criteria and following administrative procedures. And every branch of federal, state and local government has its own requirements. Those who've mastered the procedures usually do quite well when work is plentiful. But government work is a tide that rises and falls just like that of general construction.

Bid Shopping

Many contractors prefer projects that require subcontractor listings. The general contractor must list the subcontractors he plans to use, and has to use the subs he lists. When listing of subs isn't required, in some cases the general contractor shops for lower subcontract bids right up to the time work begins. Even if the general has to list his subs in the contract with the owner, he'll still usually have a month or two to shop bids after the contract is awarded.

When a general contractor uses your bid to land a job, it's normal to expect that your company will get the contract. Giving all your competition a second look at the job is in no one's interest but the general contractor's. It's a waste of time to bid for general contractors who shop their sub bids. Nor is it good practice to undercut another electrical contractor whose estimate was used by the winning general contractor. Support the effort of reputable subcontractors who promote subcontractor listing at bid time.

Need More Help?

This book is concerned primarily with labor and material costs for electrical construction. You'll find only limited information here on how to compile an estimate. If you need a detailed explanation on how to make a material take-off and complete the bid, another book by this publisher may be helpful. You can read about and purchase *Estimating Electrical Construction Revised* using the order form bound into the back of this book.

Section 1: Conduit and Fittings

Every electrical estimator should be familiar with the *National Electrical Code®*. Nearly all inspection authorities follow *NEC®* recommendations on what is and what is not good electrical construction practice. Most inspection authorities accept electrical materials that comply with *NEC* standards. But some cities and counties have special requirements that supplement the current *NEC*. Others are still following an older edition of the *NEC*. The *NEC* is revised every three years to incorporate changes deemed necessary to keep the code up-to-date.

Be aware of the version of the *NEC* that applies at each job you're estimating, and stay current on special requirements that the inspection authority may impose.

Job specifications usually state that all work must comply with the *NEC*. But on many jobs the *NEC* sets only the minimum standard. Job specifications may prohibit what the *NEC* permits. For example, job specs might require specific installation methods or mandate specification grade fixtures.

The *National Electrical Code* classifies all enclosed channels intended to carry electrical conductors as "raceway." This includes conduit, busway and wireway. The most common raceway is electrical conduit. The code identifies the size and number of conductors that can be run through each size of conduit.

Conduit is intended to serve two purposes. First, it's a protective shield for the conductor it carries. It reduces the chance of accidental damage to the wire or insulation. Second, it protects people and property from accidental contact with the conductors. A ground or short is both a safety and a fire hazard.

Conduit is generally required in commercial and industrial buildings, hospitals, hotels, office buildings, stores and underground facilities. It's not generally used in wiring homes and apartments.

Several types of electrical conduit have been approved for electrical construction. Each is designed for a specific purpose or use. All conduit used in electrical construction as a raceway for conductors must bear a label issued by the Underwriter's Laboratories. The UL label indicates that the product has been approved for use under the *National Electrical Code*.

The *NEC* permits a maximum of four bends totaling 360 degrees between terminations in a run of conduit. Exposed conduit should be installed horizontal or vertical and should run parallel to building members. Concealed conduit should be run in the shortest direct line to reduce the length of run. Long runs waste materials, require excessive labor and, if long enough, can reduce the voltage available at the load end.

Electrical Metallic Tubing

EMT is also known as **thin wall** or **steel tube**. EMT conduit is nonferrous steel tubing sold in 10-foot lengths. Unlike water pipe, the ends aren't threaded. The conduit has a corrosion-resistant coating inside and outside. This coating may be hot-dipped galvanizing, electroplating, or some other material. The conduit sizes are ½", ¾", 1", 1¼", 1½", 2", 2½", 3", 3½" and 4".

Many types of EMT fittings are available. There are elbows, compression, set screw, indent and drive-on fittings which may be made of steel or die cast. Couplings and connectors are sold separately and not included in the price of the conduit. Various types of connectors may be purchased with or without insulated throats. The locknuts for the connectors are included in the cost of the connector.

Couplings are available for joining EMT to rigid metal conduit and to flexible conduit. These couplings are available in compression, set screw and drive-on type and are made of steel or die cast.

EMT conduit is sold without couplings. You have to figure the number of couplings needed and price them separately. To figure the number needed, allow one coupling for each 10 feet of conduit. Then add one coupling for each factory-made elbow.

EMT should be bent with a special conduit bender. The bender has a shoe that fits over and around about half of the conduit to keep the conduit from collapsing as it bends. With a bender it's easy to produce smooth, consistent bends up to 90 degrees. Hand benders are used on sizes from ½" to 1¼" . EMT bending machines are available for all sizes of conduit. There are manual, hydraulic and electrically driven machines.

Offsets are made to take EMT conduit around obstructions, and when needed, to align the conduit at a box or cabinet. You can make offsets with a hand bender on sizes up to 1¼". Offsets in EMT conduit over 1¼" should be made with a machine.

In smaller sizes, conduit can be cut with a tubing cutter. Cut larger diameters with a hacksaw or by machine. Cut ends must be reamed to remove the burrs made while cutting. Burrs can damage insulation when wire is pulled through the conduit. Ream with a pocket knife or pliers on smaller sizes and with a metal file or pipe reamer on larger sizes.

EMT must be supported so it doesn't deflect on longer runs. Straps and nailers are the most common way of supporting EMT. Straps usually have one or two holes for securing to the building. Most inspection authorities won't let you support EMT on plumber's perforated metal tape. Straps come in thin steel, heavy duty steel or malleable types. There are special straps made of spring steel for supporting small sizes of EMT to hanger rods or drop ceiling wires.

EMT conduit should be supported at least every 10 feet with a strap or hanger and within 18 inches of every junction box or cabinet.

Other supports include beam clamps for attaching conduit to structural steel members and straps for mounting EMT on steel channel strut. These two-piece straps or clamps are inserted into the strut and bolted together to hold the conduit in place.

EMT can be installed inside or outside, in concrete or masonry, exposed or concealed in walls, floors or ceilings. But be sure to use the correct fittings in wet locations. EMT is not approved for most types of hazardous locations. Some specs limit the use of EMT to dry areas and don't allow placement in masonry or concrete. Conduit placed in concrete floor slab is generally placed below the reinforcing bar curtain or between curtains when two curtains are used. Tie the conduit to the rebar to prevent shifting as the concrete is placed.

Where conduit is turned up above the surface of the concrete, the radius of the turn must be concealed. Part of it can be concealed in a wall, but none should be visible after the building finish has been installed.

As with all types of conduit, EMT should be installed with a minimum of damage to the structure. Keep it clear of heating, ventilating and air conditioning ducts, fire sprinkler systems, plumbing lines, access doors, etc. When necessary, the installer will have to make offsets and bends so the conduit fits into devices, electrical boxes and cabinets.

Flexible Metal Conduit

There are several types of flex conduit: standard wall steel flex, reduced wall steel flex, and aluminum flex. It comes in diameters from 3/8" to 4" and is coiled in rolls of 100 feet in the small sizes and 25 feet in the larger sizes. Flex is usually used in concealed locations but never underground or in concrete. It's cut with a special flex cutter, a hand hacksaw, or with a power cutter such as a portable band saw. The inside cut edge must be reamed to remove cutting burrs which would damage insulation when wire is pulled through conduit.

Flex connectors are available with set screw, screw-in, clamp type, straight, or angled connectors. They're made of steel or die cast. Insulated connectors are also available. Die cast flex couplings are available for joining flex to flex, flex to EMT, or flex to threaded conduit. Support flex with conduit straps or nailers.

Most inspection authorities require that a bonding conductor be installed when electrical wiring is run in flex. Bonding ensures that there's electrical continuity in the flex from one end to the other.

Some specifications restrict the use of flex to short connections to equipment that is subject to vibration (such as motors and machinery) and for built-ins, recessed lighting, and lay-in lighting fixtures.

Flex conduit is popular in remodeling work where wiring in raceway has to be run through an existing cavity wall or in a ceiling cavity. With a little effort, your installer can fish the flex from point to point without opening the wall or ceiling.

Polyvinyl Chloride Conduit

PVC conduit is approved by the *NEC* for many types of applications. But there are some situations where it cannot replace metallic conduit. It's not approved for hazardous locations or in return air plenums. Check with the inspection authority for other restrictions. The standard length is 10 feet and sizes range from ½" to 6". Schedule 40 PVC is the standard weight. Schedule 80 has a heavier wall. PVC can be installed directly underground, concrete encased underground, exposed, in concrete walls, and in unit masonry.

One coupling is furnished with each length of conduit and is usually attached to the conduit. PVC must be bent with a special hot box which heats the conduit until it becomes pliable. Once heated to the right temperature, the tube is bent and then allowed to cool. PVC fittings fit both Schedule 40 and 80 conduit. Couplings, terminal adapters, female adapters, expansion fittings, end bells, caps, conduit bodies, pull boxes, outlet boxes and elbows require a special cement. The glue is air-drying and comes in half-pints, pints, quarts, and gallon containers. The smaller containers have a brush attached to the cap for applying the cement to the conduit or fittings. PVC conduit can join other types of conduit if you use the right fittings to tie the two types together.

PVC is nonconductive. That makes a bonding conductor necessary to ensure electrical continuity from the device to the service panel. You probably won't need a bonding conductor when PVC is used as communications conduit or in some application that doesn't include electrical wiring. When installed exposed, PVC requires extra support to keep it from sagging.

Some job specs restrict use of PVC to specific locations. One common restriction is to limit PVC to underground installations encased in a concrete envelope. Many specifications restrict its use to certain applications.

PVC conduit can be cut with a hand hacksaw, a wood crosscut saw, or with a power cutting machine. The inside cut edge should be reamed to remove the cutting burr. Use a pocket knife or a file.

Power and communications duct is usually called **P&C duct**. It's made of PVC in 25-foot lengths and in diameters from 1" to 6". There are two types of P&C duct. One is called **EB** for encased burial. The other is **DB** for direct burial. Fittings for P&C duct include couplings (one is furnished with each length), end bells, caps and plugs, terminal adapters, female adapters, elbows, and expansion fittings. The elbows are available in various shapes and with either long or short radii. Fittings can be used either on type EB or DB. Use a special cement to weld the fittings to the conduit.

Bend P&C duct with a hot box. It can be cut with the same tools as PVC conduit. The inside cut edge must be reamed to remove the cutting burr.

P&C duct is used for underground systems only, never above ground.

ABS underground duct is used and installed the same as PVC P&C duct. It requires a special ABS cement to weld the fittings to the conduit. The job specifications or the utility company may require either P&C, ABS or PVC duct, depending on the specific use.

Galvanized Rigid Conduit

GRS or **RSC** (for rigid steel conduit) is made with nonferrous metal and has a corrosion-resistant coating on the inside. The outer coating is either hot-dipped galvanizing or electroplate. It comes in diameters from ½" to 6" and in 10-foot lengths with a thread on each end. A coupling is furnished on one end of each length. GRS can be cut with a hand hacksaw, a pipe cutter, or with a cutting machine. The inner cut edge must be reamed to remove the burr. Use a pipe reamer or a file.

After the pipe has been cut and reamed, it can be threaded. Use a hand die for threading on a small job. Where there's more cutting and threading to be done, use a threading machine. Several types are available. Small portable electric threading tools cut sizes up to 2". Larger threading machines can cut, ream and thread conduit diame-ters up to 6". Another good choice for GRS up to 6" is a threading set that uses a tripod vise stand and a threading head that clamps to the pipe in the vise stand. The threading head is turned with a universal joint connected to a power vise. Another set uses a tripod vise stand to hold the conduit. The threading head clamped on the conduit is turned with a reduction gear assembly powered by an electric drill. This rig works well on diameters over 2".

Use enough cutting oil to keep the die cool and lubricated during thread cutting. Cutting oil comes in clear or dark and in small cans, gallons and barrels. Use an oil can to keep a film of oil ahead of the dies. Commercial oiling units hold about a gallon of cutting oil and recirculate oil back to the cutting teeth as oil drips into the catch basin. Most threading machines have automatic oilers that filter the oil as it's reused.

Elbows are available for all sizes of GRS. Long radius bends are available for the larger sizes. Some specifications require concentric bends for all exposed conduit installed parallel on a common hanging assembly or trapeze.

GRS fittings include couplings, locknuts, bushings, one-hole straps, two-hole straps, heavy duty two-hole straps, expansion fittings, threadless compression couplings, threadless set-screw couplings, threadless compression connectors, threadless set-screw connectors, three-piece union-type couplings, strut clamps, beam clamps, hanger clamps, condulets, split couplings, caps, and plugs.

Galvanized rigid conduit is bent about the same way as EMT except that the bender is made for bending rigid conduit. Hand benders are used on conduit up to 1". There are hand benders for 1¼" and 1½" rigid steel conduit, but it takes a lot of effort to make the bend. Power benders can be used on all sizes of conduit, even the ½".

There are three common types of rigid steel benders: one-shot benders create a single standard radius arc. Segment benders must be moved along the conduit as each few degrees of bend are made. The electric sidewinder bender has up to three bending shoes in place ready to bend any of three sizes of conduit. The sidewinder saves labor on larger rigid conduit jobs.

Supports for rigid conduit must be no more than 10 feet apart from support to support and within 18 inches of junction boxes or cabinets.

Trapeze hangers are often used to carry multiple runs of GRS conduit. Trapeze hangers can be made from strut, angle iron, or channel iron. The trapeze is supported from the structural frame of the building with threaded rod — usually either 3/8" or ½" diameter. The upper part of the rod is attached to beam clamps or concrete anchors. The lower portion of the rod is run through the trapeze and is secured with double nuts and flat washers.

Like other hangers, trapezes have to be placed within 10 feet of each other and should be sized to support the total weight of the conduit and all cable. Trapeze hangers can be stacked one over the other with conduit clamped on each one.

IMC Conduit

Intermediate metal conduit (IMC) has a thinner wall than GRS. It comes in the same sizes and uses the same fittings as GRS. The same tools can be used for cutting, threading, and bending. It's made about the same way as GRS, comes in 10-foot lengths and is galvanized for corrosion resistance. The difference is that IMC is lighter and easier to install than GRS. Some specifications restrict its use to specific applications.

PVC Coated Conduit

Both GRS and IMC conduit come with a PVC coating for use in highly corrosive locations. Aluminum tubing also comes with a PVC coating, but applications are restricted to specific uses. The PVC coating is either 10, 20 or 40 mils thick, and is bonded directly to the conduit wall. Most fittings made for use with GRS are available with a PVC coating.

To thread PVC coated conduit, the PVC coating must be cut back away from the end to be threaded. When PVC coated conduit is put in a vise, be sure the coating is protected from the vise jaws. Also be careful when you're bending PVC coated conduit not to damage the coating. If the coating is damaged, patching material is available to restore the surface. The material comes in a spray can. Apply several thin layers to repair worn spots.

Conduit Take-Off

Here's how to calculate conduit quantities. First, scan the specs that cover conduit and conduit installation. Absorb all the information that relates to conduit. Then review the drawings for anything about conduit. The symbol list may include the engineer's design notations. Notes on the drawings or in the specs may set specific minimum conduit sizes. It's common for an engineer to require a minimum size conduit in the home run to the panel or cabinets or to specify a minimum size of ¾" throughout the job. It's also common practice to limit the maximum size of EMT to 2". Ignoring a note like that can be expensive.

For your quantity take-off, use any ruled 8½" by 11" tablet. Draw a pencil line down the left side of the sheet about an inch from the edge. Begin by looking for the smallest diameter of EMT. Write "EMT" at the top left of your take-off sheet. On the next line down, to the left of the vertical line, list the smallest EMT size found in the project — probably ½". To the right of the vertical line and on the same horizontal line as the size, you're going to list lengths of EMT of that diameter. Then you'll go to the next larger diameter, listing quantities until all EMT on the plans has been covered.

Check the plan scale before you start measuring conduit. If the plan has been reduced photographically to save paper, the scale will be inaccurate. Once you're sure of the correct scale, select the appropriate map measure or rule to compute conduit lengths.

Measure the length of each run of ½" EMT. Add enough conduit to include the run down to the wall switch, receptacle or panel. Write down the calculated length. As each run is listed on your take-off sheet, put a check mark on the plan over the line you just measured. Use an erasable color pencil and let each color stand for a particular conduit type. For example, red might be for GRS conduit. Follow the same color code on all estimates to avoid mistakes.

If there are more than two or three plan sheets, it's good practice to calculate the length of ½" EMT on each plan sheet and list that number separately on your take-off form. When you've finished taking off ½" EMT on the first plan sheet, list that quantity, and at the top of the column write in the plan sheet number. Then draw a vertical line to the right of that column and start accumulating lengths from the next plan sheet. As each plan page is taken off, enter the total and write the plan sheet number at the top of the column. Figure 1-1 shows what your take-off might look like if conduit and fittings are found on plan sheets E3 to E11.

When all of the smallest-diameter EMT has been listed, go on to the next larger size. Follow the same procedure.

After listing all EMT, begin with the fittings. Below the last horizontal line used for conduit, and to the left of the vertical line, write the word "Connectors." Below that, list all sizes of connectors needed for the job, again working from the smallest size to the largest. Don't bother to list the couplings. They'll be figured later from the total conduit length — one for each 10 feet and one for each elbow.

Count each connector needed for each conduit run on each plan sheet. Enter the total on your take-off form. When all connectors are counted, count EMT elbows from 1¼" to the largest size needed.

Follow this system for all estimates and for each item on every estimate. Keep it simple and uniform to avoid mistakes and omissions. When finished, your conduit and fitting take-off form might look like Figure 1-1. The right column is the sum of the columns to the left.

Work Sheet					Estimate No.: M351					
					Conduit / Fittings					
	E3	E4	E5	E6	E7	E8	E9	E10	E11	Total
½" EMT	550	420	200	90	290	130	190	320		2190
¾"	20		30	20	80					150
1"			3		5		50			58
1¼"			30							30
1½"									90	90
2"					4				16	20
½" Conn	76	52	124	47	48	16	14	18		395
¾"	4		26	4	19	2				55
1"			4		5	2	2			13
1¼"			2							2
1½"									4	4
2"					2				4	6
1¼" Elb			2							2
1½"									3	3
2"									3	3
½" PVC			310	380	50					740
¾"			120	100	220	50				490
1"			40		320	40				400
1¼"						180				180
1½"				60					75	135
2"				10	25			70	75	180
4"								150		150
½" FA			45	30	4					79
¾"			4	4	12	2				22
1"			2		17	2				21
1¼"										0
1½"				4					2	6
2"				2	2				2	6
½" TA			5							5
¾"					4					4
1"					1					1
½" Elb			50	30	4					84
¾"			2	2	16	2				22

Figure 1-1

Many jobs limit the use of EMT to dry locations. So your EMT take-off will probably start with the lighting plans or the lighting portion of the plan.

Taking Off the Wire

Next, compute the quantity of wire needed. Head up another take-off form with the word "Wire" at the top. Put a vertical line down the left side of the page about an inch from the left edge. In this margin, list wire sizes from the smallest to the largest. To the right of the vertical line you'll list lengths for each wire gauge, on each plan sheet.

Start by measuring the length of ½" EMT with two #12 wires. Multiply by 2 to find the wire length. Then measure the length of ½" EMT with three #12 wires and multiply by 3. Keep following this procedure until the wire needed in all EMT has been computed. But watch for changes in the wire size on long runs. Sometimes the engineer will decide that a larger wire size is needed in the first portion of a run to reduce the voltage drop at the end of the line. This is common where the last device or fixture on a circuit is a long way from the panel.

Follow the same procedure for all conduit and wire. Record all of the measurements on the work sheets. Don't worry about waste of conduit or wire at this point. We'll include an allowance for waste after the totals are added and before figures are transferred to the pricing sheets.

Sometimes the specifications or a note on the plans will allow the use of aluminum feeder wire over a certain size, providing the ampacity of the wire is maintained and the conduit size is increased to accommodate the larger wire size. Be sure to observe these restrictions.

Taking Off Other Conduit

Some specifications permit the use of aluminum conduit in certain locations. The aluminum conduit is made in the same sizes as GRS. The fittings are identical except that they're made of aluminum instead of steel. Most specs prohibit the use of dissimilar metals in a conduit run and don't allow placing of aluminum conduit in concrete. Aluminum conduit saves time because it's lighter and easier to handle. But large wire sizes may be a little more difficult to pull in aluminum conduit. The insulation of the wire, the length of the conduit run, and the pulling lubricant used have an effect on pulling resistance.

When taking off the underground conduit, start a separate work sheet for trenching, surface cutting, breaking, and patching. List all excavation for underground pull boxes, handholes, manholes, poles, and light pole bases. Be sure the trenches are big enough for the number of duct they have to carry. If the specifications require concrete or sand encasement around underground duct, calculate the amount of concrete or sand as you compute measurements for each trench.

Be systematic. Follow the same procedure consistently on every take-off. If there are other estimators in your office, be sure they are using the same procedures. Being consistent reduces errors, minimizes omissions, and makes the work easier for others to check.

We've covered all common conduit. But some other types are used occasionally for special purposes:

Fiber duct is a paper and creosote duct. Type 1 is intended for concrete encasement and Type 2 is used for direct burial. Sizes range from 2" to 5". Lengths can be 5, 8 or 10 feet. End fittings are tapered. Ends that have been cut must be tapered with a duct lathe.

Transite duct is cement asbestos duct. Type 1 is for concrete encasement and Type 2 is for direct burial. Sizes range from 2" to 6". It's made in 5, 8 and 10-foot lengths. Transite is harder to cut and must have tapered ends for fittings.

Soapstone duct is made from a soapstone-like material in sizes from 2" to 4".

Wrought iron pipe comes in sizes from 2" to 4". It's used only for certain types of underground communications lines and has to be threaded on each end to accept fittings.

Clay conduit comes in sizes from 2" to 4". It's used for underground communication runs only.

These types of conduit are seldom specified today. You'll see them used only when an old duct line has to be extended. It may be hard to find a fitting that will join an existing duct system made with one type of duct to a new run of duct made from some other material. Sometimes an oversize plastic coupling can be used. In some cases an inside plastic coupling can be inserted into the old conduit. Then new conduit can be joined to start the new run.

Before extending an old underground duct system, check the old conduit with a mandrel to be sure the line is clean and clear. Old fiber duct that's been under water for a long time will swell, making the inside diameter too small to pull new cable.

Silicon-bronze conduit comes in sizes from ½" to 4". It's threaded like GRS and uses similar fittings, except that fittings are silicon-bronze also. It's used in extremely corrosive locations. This type of conduit will be available from your dealer on special request only. It's harder to bend, but can be bent with standard rigid bending tools. It threads very well with the standard threading tools and cutting oil.

Liquid-tight flexible metal conduit comes in sizes from ½" to 4". It's used to extend conduit to electrical equipment in damp or wet locations. Special fittings are available for connecting electrical systems and devices with this conduit. Your dealer probably stocks a limited supply of liquid-tight flex and will quote prices on request. The conduit can be cut with a hacksaw. Be sure to remove the cutting burr. Special connectors with grips are available to support the conduit and prevent any pulling strain.

Liquid-tight flexible non-metallic conduit comes in sizes from ½" to 1½". It's used in place of flexible metal conduit in concealed locations. Special fittings are available for making connections. Your dealer may have a limited supply in stock.

Flexible metallic tubing is available only in sizes from ⅜" to ¾". Special fittings are available for making connections. The tubing can be bent by hand and is cut with a hacksaw. The cutting burr must be removed before connectors are installed.

Other UL-approved raceways for electrical systems are covered in other sections of this book. See the sections on surface metal raceway, underfloor ducts, header ducts, cable tray, and wireway.

Using the Conduit Tables

The labor tables that follow are for conduit runs that average 50 feet. You'll note that there is no modification in the tables for shorter runs or longer runs of conduit. I agree that it takes more time per linear foot to install a 5-foot run of conduit than it does to install a 95-foot run of conduit. But I don't recommend that you tally shorter runs and longer runs separately and then compute labor separately for each. There's an easier way.

On most jobs the conduit runs average 50 feet. There will usually be about as many runs under 50 feet as there are runs over 50 feet. It's safe then, to use a 50-foot run as our benchmark. As long as the conduit runs on a job average close to 50 feet, there's no need to modify the figures in these tables. If conduit runs average well over 50 feet, consider reducing the cost per linear foot slightly. Increase the cost slightly if conduit runs average less than 50 feet.

The labor costs that follow include the labor needed to bore holes in wood stud walls. Where holes have to be cut through concrete or unit masonry, add these costs separately.

Typical conduit bending is included in the tables that follow. Usually you will have a bend or offset about every 20 feet. Labor needed to make bends and offsets is minor when installing the smaller sizes of conduit.

Concealed conduit is installed where it will be inaccessible once the structure or finish of the building is completed. ***Exposed conduit*** is attached to the surface where access is possible even after the building is completed. It's usually faster to run concealed conduit through wall and ceiling cavities that will be covered later by finish materials. Installing conduit on surfaces that won't be covered later usually takes more time.

If only a small percentage of the conduit is to be installed exposed, the cost difference will be minor and probably can be ignored. But if most of the job is exposed, add about 20 percent to the labor cost.

The conduit tables that follow assume that electricians are working from ladders and lifts up to 12 feet above the floor. Add to the labor cost for heights beyond 12 feet. If a large quantity of conduit has to be installed at 18 feet above the floor, for example, add 15 percent to the labor cost.

If there are conduit runs over 20 feet above the floor, check your labor contract for a ***high time clause***. Some agreements require that electricians be paid time and one-half for heights from 20 to 50 feet and double time for heights beyond 50 feet. If high time must be paid, be sure the extra cost is covered in your bid.

Job Size Modifiers

It's seldom necessary to estimate lower productivity just because the job is small. If you're figuring a very small job with only four or five conduit runs, each with only a strap or two, you might want to use a higher hourly labor rate. On any other job that takes from two days to several years, you can use the labor units in the tables that follow. Of course, you'll still have to modify the figures for other than "good" conditions. And if you have long runs of feeder conduit with parallel runs on a common trapeze, you can reduce those labor units by as much as 40 percent.

Pitfalls

The most common error when estimating conduit is failing to read the plans and specs. Read carefully! Your profit depends on it. It's easy to miss a little note where the designer sets the minimum size for conduit at ¾" and 1" for all home runs to the panel. Look for a note on the plans that requires stub ups to ceiling cavities from power and lighting panels. The designer may require one ¾" conduit run for each three spare circuit breakers in a panel.

It's common for rigid conduit to be installed in a concrete floor slab. Where GRS is stubbed up out of the concrete for a wall switch, it's easier and cheaper to use EMT for the wall extension. The *NEC* permits making that extension in EMT. But some specs don't! Others require that a junction box be used to separate the two types of conduit. Failing to catch that note can be an expensive mistake.

You'll find all sorts of restrictions in specs and notes on the plans. That's why it's so important to read the plans and specs carefully. It's elementary, but it's so often overlooked.

Waste of Material

There will always be some waste on a job. Rounding off the conduit and wire needed to the next even 100 feet will usually allow enough extra material to cover all waste. But there are some cases where you can anticipate a waste problem. For example, suppose there will be 2 feet of waste for every 20 feet of conduit installed because of an unusual lighting pattern. Or suppose a row of junction boxes is spaced at 9 feet. Then a 10 percent waste allowance may be called for. That's almost certainly true if your job is installing the lighting only. There may be no chance to use waste materials on another part of the job.

Allowances

Be sure to make allowances for the vertical portion of every conduit run that stubs up or down in a wall. The floor plan doesn't show the 4 or 5 feet needed to run from the slab to the wall switch or panel. Even worse, if the job is a warehouse, the stub up to a switch or panel may be 15 to 20 feet. That's a wide miss! Watch for stub up.

Material	Craft@Hrs	Unit	Material Cost	Labor Cost	Installed Cost
EMT conduit in floor slab or multiple runs on a trapeze					
1/2"	L1@3.25	CLF	36.70	133.00	169.70
3/4"	L1@3.50	CLF	69.20	144.00	213.20
1"	L1@4.00	CLF	116.00	164.00	280.00
1-1/4"	L1@4.50	CLF	176.00	185.00	361.00
1-1/2"	L1@5.50	CLF	217.00	226.00	443.00
2"	L1@7.00	CLF	266.00	287.00	553.00
2-1/2"	L1@9.00	CLF	433.00	369.00	802.00
3"	L1@10.0	CLF	531.00	410.00	941.00
3-1/2"	L1@11.0	CLF	772.00	451.00	1,223.00
4"	L1@12.0	CLF	784.00	492.00	1,276.00
EMT conduit in concealed areas, walls and closed ceilings					
1/2"	L1@3.50	CLF	36.70	144.00	180.70
3/4"	L1@3.75	CLF	69.20	154.00	223.20
1"	L1@4.25	CLF	116.00	174.00	290.00
1-1/4"	L1@5.00	CLF	176.00	205.00	381.00
1-1/2"	L1@6.00	CLF	217.00	246.00	463.00
2"	L1@8.00	CLF	266.00	328.00	594.00
2-1/2"	L1@10.0	CLF	433.00	410.00	843.00
3"	L1@12.0	CLF	531.00	492.00	1,023.00
3-1/2"	L1@14.0	CLF	772.00	574.00	1,346.00
4"	L1@16.0	CLF	784.00	656.00	1,440.00
EMT conduit installed in exposed areas					
1/2"	L1@3.75	CLF	36.70	154.00	190.70
3/4"	L1@4.00	CLF	69.20	164.00	233.20
1"	L1@4.50	CLF	116.00	185.00	301.00
1-1/4"	L1@6.00	CLF	176.00	246.00	422.00
1-1/2"	L1@8.00	CLF	217.00	328.00	545.00
2"	L1@10.0	CLF	266.00	410.00	676.00
2-1/2"	L1@12.0	CLF	433.00	492.00	925.00
3"	L1@14.0	CLF	531.00	574.00	1,105.00
3-1/2"	L1@16.0	CLF	772.00	656.00	1,428.00
4"	L1@18.0	CLF	784.00	739.00	1,523.00

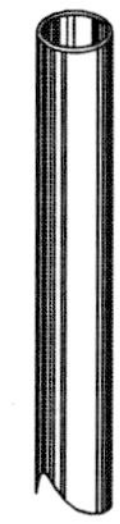

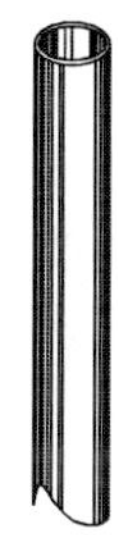

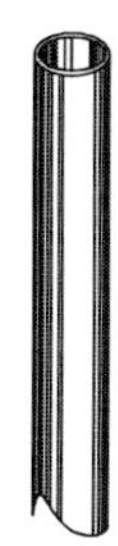

Use these figures to estimate the cost of EMT conduit installed in a building under the conditions described on pages 5 and 6. Costs listed are for each 100 linear feet installed. The crew is one electrician working at a labor cost of $41.03 per manhour. These costs include typical bending, boring out wood studs and joists (in concealed locations only), layout, material handling, and normal waste. Add for connectors, couplings, straps, boxes, wire, sales tax, delivery, supervision, mobilization, demobilization, cleanup, overhead and profit. Note: Conduit runs are assumed to be 50' long. Shorter runs will take more labor and longer runs will take less labor per linear foot.

EMT Hand Benders are on page 27.

EMT Fittings

Material	Craft@Hrs	Unit	Material Cost	Labor Cost	Installed Cost

EMT 45 degree elbows

Material	Craft@Hrs	Unit	Material Cost	Labor Cost	Installed Cost
1"	L1@0.06	Ea	5.31	2.46	7.77
1-1/4"	L1@0.08	Ea	6.68	3.28	9.96
1-1/2"	L1@0.08	Ea	11.30	3.28	14.58
2"	L1@0.10	Ea	14.30	4.10	18.40
2-1/2"	L1@0.15	Ea	34.90	6.15	41.05
3"	L1@0.20	Ea	52.10	8.21	60.31
3-1/2"	L1@0.20	Ea	69.20	8.21	77.41
4"	L1@0.25	Ea	82.00	10.30	92.30

EMT 90 degree elbows

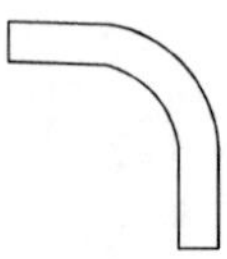

Material	Craft@Hrs	Unit	Material Cost	Labor Cost	Installed Cost
1"	L1@0.08	Ea	6.80	3.28	10.08
1-1/4"	L1@0.10	Ea	8.46	4.10	12.56
1-1/2"	L1@0.10	Ea	9.74	4.10	13.84
2"	L1@0.15	Ea	14.30	6.15	20.45
2-/2"	L1@0.15	Ea	34.90	6.15	41.05
3"	L1@0.20	Ea	52.10	8.21	60.31
3-1/2"	L1@0.20	Ea	69.20	8.21	77.41
4"	L1@0.25	Ea	82.00	10.30	92.30

Use these figures to estimate the cost of EMT elbows installed on EMT conduit in a building under the conditions described on pages 5 and 6. Costs listed are for each elbow installed. The crew is one electrician working at a labor cost of $41.03 per manhour. These costs are for factory-made elbows and include layout, material handling, and normal waste. Add for field bending, couplings and connectors at the end of the run, sales tax, delivery, supervision, mobilization, demobilization, cleanup, overhead and profit. Note: Material costs assume purchase of full box quantities.

Conduit weight per 100 feet (in pounds)

Diameter	EMT steel	ENT plastic	PVC 40	Rigid steel	Intermediate rigid steel	Rigid aluminum
1/2"	30	11	18	79	57	30
3/4"	46	14	23	105	78	40
1"	66	20	35	153	112	59
1-1/4"	96	—	48	201	114	80
1-1/2"	112	—	57	249	176	96
2"	142	—	76	334	230	129
2-1/2"	230	—	125	527	393	205
3"	270	—	164	690	483	268
3-1/2"	350	—	198	831	561	321
4"	400	—	234	982	625	382
5"	—	—	317	1344	—	522
6"	—	—	412	1770	—	678

Indent EMT connectors

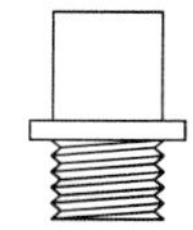

Material	Craft@Hrs	Unit	Material Cost	Labor Cost	Installed Cost
1/2"	L1@0.05	Ea	.46	2.05	2.51
3/4"	L1@0.06	Ea	.82	2.46	3.28

Die cast set screw EMT connectors

Material	Craft@Hrs	Unit	Material Cost	Labor Cost	Installed Cost
1/2"	L1@0.05	Ea	.28	2.05	2.33
3/4"	L1@0.06	Ea	.46	2.46	2.92
1"	L1@0.08	Ea	.86	3.28	4.14
1-1/4"	L1@0.10	Ea	1.52	4.10	5.62
1-1/2"	L1@0.10	Ea	2.07	4.10	6.17
2"	L1@0.15	Ea	2.76	6.15	8.91
2-1/2"	L1@0.15	Ea	6.25	6.15	12.40
3"	L1@0.20	Ea	7.59	8.21	15.80
3-1/2"	L1@0.20	Ea	9.01	8.21	17.22
4"	L1@0.25	Ea	11.30	10.30	21.60

Insulated die cast set screw EMT connectors

Material	Craft@Hrs	Unit	Material Cost	Labor Cost	Installed Cost
1/2"	L1@0.05	Ea	.46	2.05	2.51
3/4"	L1@0.06	Ea	.70	2.46	3.16
1"	L1@0.08	Ea	1.26	3.28	4.54
1-1/4"	L1@0.10	Ea	2.52	4.10	6.62
1-1/2"	L1@0.10	Ea	3.07	4.10	7.17
2"	L1@0.15	Ea	4.12	6.15	10.27
2-1/2"	L1@0.15	Ea	11.20	6.15	17.35
3"	L1@0.20	Ea	13.10	8.21	21.31
3-1/2"	L1@0.20	Ea	16.70	8.21	24.91
4"	L1@0.25	Ea	18.30	10.30	28.60

Steel set screw EMT connectors

Material	Craft@Hrs	Unit	Material Cost	Labor Cost	Installed Cost
1/2"	L1@0.05	Ea	.69	2.05	2.74
3/4"	L1@0.06	Ea	1.12	2.46	3.58
1"	L1@0.08	Ea	1.96	3.28	5.24
1-1/4"	L1@0.10	Ea	4.09	4.10	8.19
1-1/2"	L1@0.10	Ea	5.96	4.10	10.06
2"	L1@0.15	Ea	8.45	6.15	14.60
2-1/2"	L1@0.15	Ea	27.80	6.15	33.95
3"	L1@0.20	Ea	32.70	8.21	40.91
3-1/2"	L1@0.20	Ea	42.90	8.21	51.11
4"	L1@0.25	Ea	49.30	10.30	59.60

Use these figures to estimate the cost of EMT connectors installed on EMT conduit under the conditions described on pages 5 and 6. Costs listed are for each connector or expanded elbow installed. The crew is one electrician working at a labor cost of $41.03 per manhour. These costs include the connector locknut, removing the knockout, layout, material handling, and normal waste. Add for insulated bushings, sales tax, delivery, supervision, mobilization, demobilization, cleanup, overhead and profit. Note: Material costs assume purchase of full box quantities.

Indenter tools are on page 22.

EMT Connectors

Insulated steel set screw EMT connectors

Material	Craft@Hrs	Unit	Material Cost	Labor Cost	Installed Cost
1/2"	L1@0.05	Ea	.92	2.05	2.97
3/4"	L1@0.06	Ea	1.48	2.46	3.94
1"	L1@0.08	Ea	2.46	3.28	5.74
1-1/4"	L1@0.10	Ea	4.93	4.10	9.03
1-1/2"	L1@0.10	Ea	7.23	4.10	11.33
2"	L1@0.15	Ea	10.50	6.15	16.65
2-1/2"	L1@0.15	Ea	47.00	6.15	53.15
3"	L1@0.20	Ea	58.40	8.21	66.61
3-1/2"	L1@0.20	Ea	78.40	8.21	86.61
4"	L1@0.25	Ea	85.70	10.30	96.00

Die cast compression EMT connectors, raintight

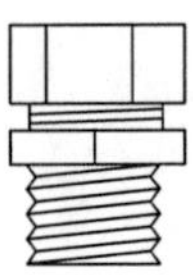

Material	Craft@Hrs	Unit	Material Cost	Labor Cost	Installed Cost
1/2"	L1@0.05	Ea	.46	2.05	2.51
3/4"	L1@0.06	Ea	.82	2.46	3.28
1"	L1@0.08	Ea	1.32	3.28	4.60
1-1/4"	L1@0.10	Ea	2.17	4.10	6.27
1-1/2"	L1@0.10	Ea	2.85	4.10	6.95
2"	L1@0.15	Ea	4.53	6.15	10.68
2-1/2"	L1@0.15	Ea	9.68	6.15	15.83
3"	L1@0.20	Ea	11.80	8.21	20.01
3-1/2"	L1@0.20	Ea	15.70	8.21	23.91
4"	L1@0.25	Ea	18.20	10.30	28.50

Insulated die cast compression EMT connectors, raintight

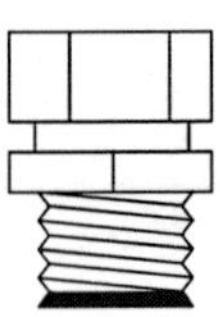

Material	Craft@Hrs	Unit	Material Cost	Labor Cost	Installed Cost
1/2"	L1@0.05	Ea	.61	2.05	2.66
3/4"	L1@0.06	Ea	1.05	2.46	3.51
1"	L1@0.08	Ea	1.65	3.28	4.93
1-1/4"	L1@0.10	Ea	3.06	4.10	7.16
1-1/2"	L1@0.10	Ea	3.78	4.10	7.88
2"	L1@0.15	Ea	5.57	6.15	11.72
2-1/2"	L1@0.15	Ea	16.50	6.15	22.65
3"	L1@0.20	Ea	19.40	8.21	27.61
3-1/2"	L1@0.20	Ea	24.10	8.21	32.31
4"	L1@0.25	Ea	28.20	10.30	38.50

Use these figures to estimate the cost of EMT connectors installed on EMT conduit under the conditions described on pages 5 and 6. Costs listed are for each connector installed. The crew is one electrician working at a labor cost of $41.03 per manhour. These costs include the connector locknut, removing the knockout, layout, material handling, and normal waste. Add for insulated bushings, sales tax, delivery, supervision, mobilization, demobilization, cleanup, overhead and profit. Note: Material costs assume purchase of full box quantities.

Steel compression EMT connectors, raintight

Material	Craft@Hrs	Unit	Material Cost	Labor Cost	Installed Cost
1/2"	L1@0.05	Ea	.08	2.05	2.13
3/4"	L1@0.06	Ea	.13	2.46	2.59
1"	L1@0.08	Ea	.19	3.28	3.47
1-1/4"	L1@0.10	Ea	.40	4.10	4.50
1-1/2"	L1@0.10	Ea	.58	4.10	4.68
2"	L1@0.15	Ea	.82	6.15	6.97
2-1/2"	L1@0.15	Ea	3.97	6.15	10.12
3"	L1@0.20	Ea	5.51	8.21	13.72
3-1/2"	L1@0.20	Ea	8.33	8.21	16.54
4"	L1@0.25	Ea	8.52	10.30	18.82

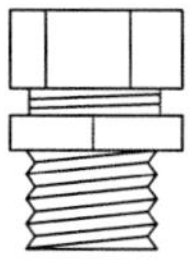

Insulated steel compression EMT connectors, raintight

Material	Craft@Hrs	Unit	Material Cost	Labor Cost	Installed Cost
1/2"	L1@0.05	Ea	.09	2.05	2.14
3/4"	L1@0.06	Ea	.14	2.46	2.60
1"	L1@0.08	Ea	.25	3.28	3.53
1-1/4"	L1@0.10	Ea	.53	4.10	4.63
1-1/2"	L1@0.10	Ea	.75	4.10	4.85
2"	L1@0.15	Ea	1.07	6.15	7.22
2-1/2"	L1@0.15	Ea	6.68	6.15	12.83
3"	L1@0.20	Ea	8.63	8.21	16.84
3-1/2"	L1@0.20	Ea	12.60	8.21	20.81
4"	L1@0.25	Ea	13.00	10.30	23.30

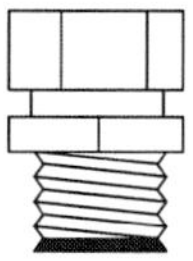

Die cast indent offset EMT connectors

Material	Craft@Hrs	Unit	Material Cost	Labor Cost	Installed Cost
1/2"	L1@0.10	Ea	1.41	4.10	5.51
3/4"	L1@0.10	Ea	1.95	4.10	6.05

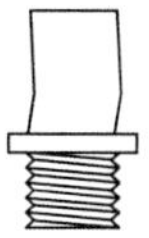

Die cast set screw offset EMT connectors

Material	Craft@Hrs	Unit	Material Cost	Labor Cost	Installed Cost
1/2"	L1@0.10	Ea	1.78	4.10	5.88
3/4"	L1@0.10	Ea	2.58	4.10	6.68
1"	L1@0.15	Ea	3.71	6.15	9.86

Use these figures to estimate the cost of EMT connectors installed on EMT conduit under the conditions described on pages 5 and 6. Costs listed are for each connector installed. The crew is one electrician working at a labor cost of $41.03 per manhour. These costs include the connector locknut, removing the knockout, layout, material handling, and normal waste. Add for insulated bushings, sales tax, delivery, supervision, mobilization, demobilization, cleanup, overhead and profit. Note: Material costs assume purchase of full box quantities.

Indenter tools are on page 22.

EMT Connectors and Couplings

Material	Craft@Hrs	Unit	Material Cost	Labor Cost	Installed Cost

Steel compression offset EMT connectors, raintight

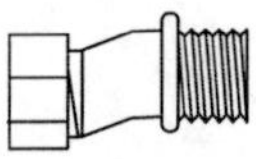

Material	Craft@Hrs	Unit	Material Cost	Labor Cost	Installed Cost
1/2"	L1@0.10	Ea	2.72	4.10	6.82
3/4"	L1@0.10	Ea	3.94	4.10	8.04
1"	L1@0.15	Ea	4.37	6.15	10.52

Indent EMT couplings

Material	Craft@Hrs	Unit	Material Cost	Labor Cost	Installed Cost
1/2"	L1@0.05	Ea	.47	2.05	2.52
3/4"	L1@0.06	Ea	.92	2.46	3.38

Die cast set screw EMT couplings

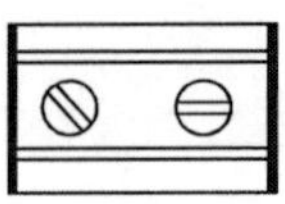

Material	Craft@Hrs	Unit	Material Cost	Labor Cost	Installed Cost
1/2"	L1@0.05	Ea	.30	2.05	2.35
3/4"	L1@0.06	Ea	.47	2.46	2.93
1"	L1@0.08	Ea	.81	3.28	4.09
1-1/4"	L1@0.10	Ea	1.40	4.10	5.50
1-1/2"	L1@0.10	Ea	2.14	4.10	6.24
2"	L1@0.15	Ea	2.85	6.15	9.00
2-1/2"	L1@0.15	Ea	5.47	6.15	11.62
3"	L1@0.20	Ea	6.24	8.21	14.45
3-1/2"	L1@0.20	Ea	7.22	8.21	15.43
4"	L1@0.25	Ea	8.82	10.30	19.12

Indenter tools

Material	Craft@Hrs	Unit	Material Cost	Labor Cost	Installed Cost
With jaws for 1/2" EMT	—	Ea	25.60	—	25.60
With jaws for 3/4" EMT	—	Ea	35.60	—	35.60
Replacement points, 1/2" EMT	—	Ea	1.93	—	1.93
Replacement points, 3/4" EMT	—	Ea	2.00	—	2.00

Use these figures to estimate the cost of EMT connectors and couplings installed on EMT conduit under the conditions described on pages 5 and 6. Costs listed are for each coupling or connector installed. The crew is one electrician working at a labor cost of $41.03 per manhour. These costs include the connector or coupling, layout, material handling, and normal waste. Add for conduit, sales tax, delivery, supervision, mobilization, demobilization, cleanup, overhead and profit. Note: Drive-on EMT fittings are rated as raintight and are also concrete tight. They are threaded with a standard electrical pipe thread and can be adapted easily to rigid conduit or other threaded fittings. Material costs assume purchase of full box quantities.

Set screw steel EMT couplings

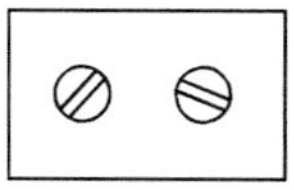

Material	Craft@Hrs	Unit	Material Cost	Labor Cost	Installed Cost
1/2"	L1@0.05	Ea	.18	2.05	2.23
3/4"	L1@0.06	Ea	.22	2.46	2.68
1"	L1@0.08	Ea	.34	3.28	3.62
1-1/4"	L1@0.10	Ea	.70	4.10	4.80
1-1/2"	L1@0.10	Ea	1.05	4.10	5.15
2"	L1@0.15	Ea	1.38	6.15	7.53
2-1/2"	L1@0.15	Ea	3.01	6.15	9.16
3"	L1@0.20	Ea	3.36	8.21	11.57
3-1/2"	L1@0.20	Ea	4.12	8.21	12.33
4"	L1@0.25	Ea	4.50	10.30	14.80

Die cast compression EMT couplings, raintight

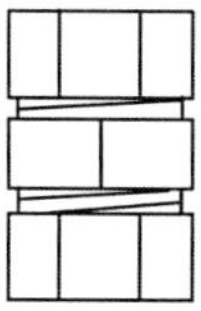

Material	Craft@Hrs	Unit	Material Cost	Labor Cost	Installed Cost
1/2"	L1@0.05	Ea	.26	2.05	2.31
3/4"	L1@0.06	Ea	.33	2.46	2.79
1"	L1@0.08	Ea	.54	3.28	3.82
1-1/4"	L1@0.10	Ea	.98	4.10	5.08
1-1/2"	L1@0.10	Ea	1.52	4.10	5.62
2"	L1@0.15	Ea	1.86	6.15	8.01
2-1/2"	L1@0.15	Ea	7.22	6.15	13.37
3"	L1@0.20	Ea	7.72	8.21	15.93
3-1/2"	L1@0.20	Ea	9.44	8.21	17.65
4"	L1@0.25	Ea	9.86	10.30	20.16

Steel compression EMT couplings, raintight

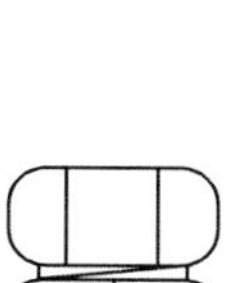

Material	Craft@Hrs	Unit	Material Cost	Labor Cost	Installed Cost
1/2"	L1@0.05	Ea	.26	2.05	2.31
3/4"	L1@0.06	Ea	.36	2.46	2.82
1"	L1@0.08	Ea	.56	3.28	3.84
1-1/4"	L1@0.10	Ea	1.00	4.10	5.10
1-1/2"	L1@0.10	Ea	1.45	4.10	5.55
2"	L1@0.15	Ea	2.00	6.15	8.15
2-1/2"	L1@0.15	Ea	8.20	6.15	14.35
3"	L1@0.20	Ea	10.50	8.21	18.71
3-1/2"	L1@0.20	Ea	15.10	8.21	23.31
4"	L1@0.25	Ea	15.50	10.30	25.80

Use these figures to estimate the cost of EMT couplings installed on EMT conduit under the conditions described on pages 5 and 6. Costs listed are for each coupling installed. The crew is one electrician working at a labor cost of $41.03 per manhour. These costs include the coupling, layout, material handling, and normal waste. Add for conduit, sales tax, delivery, supervision, mobilization, demobilization, cleanup, overhead and profit. Note: Compression fittings are raintight and can be used in concrete. Material costs assume purchase of full box quantities.

EMT Couplings and Straps

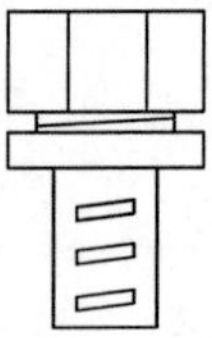

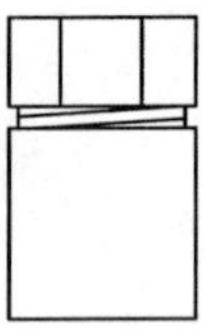

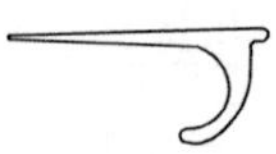

Material	Craft@Hrs	Unit	Material Cost	Labor Cost	Installed Cost
Die cast EMT to flex couplings					
1/2"	L1@0.05	Ea	1.31	2.05	3.36
3/4"	L1@0.05	Ea	1.75	2.05	3.80
1"	L1@0.06	Ea	2.44	2.46	4.90
Steel EMT to GRS compression couplings, raintight					
1/2"	L1@0.05	Ea	2.01	2.05	4.06
3/4"	L1@0.06	Ea	2.85	2.46	5.31
1"	L1@0.08	Ea	4.33	3.28	7.61
1-1/4"	L1@0.10	Ea	7.53	4.10	11.63
1-1/2"	L1@0.10	Ea	9.24	4.10	13.34
2"	L1@0.15	Ea	18.30	6.15	24.45
Steel EMT nail straps					
1/2"	L1@0.02	Ea	.07	.82	.89
3/4"	L1@0.03	Ea	.07	1.23	1.30
1"	L1@0.05	Ea	.09	2.05	2.14
Steel one hole EMT straps					
1/2"	L1@0.03	Ea	.03	1.23	1.26
3/4"	L1@0.04	Ea	.05	1.64	1.69
1"	L1@0.05	Ea	.08	2.05	2.13
1-1/4"	L1@0.06	Ea	.15	2.46	2.61
1-1/2"	L1@0.06	Ea	.22	2.46	2.68
2"	L1@0.10	Ea	.27	4.10	4.37
2-1/2"	L1@0.10	Ea	1.02	4.10	5.12
3"	L1@0.15	Ea	1.24	6.15	7.39
3-1/2"	L1@0.15	Ea	1.95	6.15	8.10
4"	L1@0.15	Ea	2.46	6.15	8.61

Use these figures to estimate the cost of EMT couplings and EMT straps installed on EMT conduit under the conditions described on pages 5 and 6. Costs listed are for each coupling and strap installed. The crew is one electrician working at a labor cost of $41.03 per manhour. These costs include cutting the EMT conduit, layout, material handling, and normal waste. Add the cost of conduit, sales tax, delivery, supervision, mobilization, demobilization, cleanup, overhead and profit. Note: Material costs assume purchase of full box quantities.

Material	Craft@Hrs	Unit	Material Cost	Labor Cost	Installed Cost
One hole heavy duty steel EMT straps					
1/2"	L1@0.03	Ea	.14	1.23	1.37
3/4"	L1@0.04	Ea	.17	1.64	1.81
1"	L1@0.05	Ea	.31	2.05	2.36
1-1/4"	L1@0.06	Ea	.41	2.46	2.87
1-1/2"	L1@0.06	Ea	.61	2.46	3.07
2"	L1@0.10	Ea	.94	4.10	5.04

Material	Craft@Hrs	Unit	Material Cost	Labor Cost	Installed Cost
One hole malleable EMT straps					
1/2"	L1@0.03	Ea	.19	1.23	1.42
3/4"	L1@0.04	Ea	.28	1.64	1.92
1"	L1@0.05	Ea	.40	2.05	2.45
1-1/4"	L1@0.06	Ea	.78	2.46	3.24
1-1/2"	L1@0.06	Ea	.91	2.46	3.37
2"	L1@0.10	Ea	1.77	4.10	5.87
2-1/2"	L1@0.10	Ea	3.81	4.10	7.91
3"	L1@0.15	Ea	4.82	6.15	10.97
3-1/2"	L1@0.15	Ea	6.28	6.15	12.43
4"	L1@0.15	Ea	13.90	6.15	20.05

Material	Craft@Hrs	Unit	Material Cost	Labor Cost	Installed Cost
Two hole steel EMT straps					
1/2"	L1@0.03	Ea	.12	1.23	1.35
3/4"	L1@0.04	Ea	.17	1.64	1.81
1"	L1@0.05	Ea	.27	2.05	2.32
1-1/4"	L1@0.06	Ea	.39	2.46	2.85
1-1/2"	L1@0.06	Ea	.46	2.46	2.92
2"	L1@0.10	Ea	.77	4.10	4.87
2-1/2"	L1@0.10	Ea	1.15	4.10	5.25
3"	L1@0.15	Ea	1.34	6.15	7.49
3-1/2"	L1@0.15	Ea	1.40	6.15	7.55
4"	L1@0.15	Ea	1.66	6.15	7.81

Use these figures to estimate the cost of EMT straps installed on EMT conduit under the conditions described on pages 5 and 6. Costs listed are for each strap installed. The crew is one electrician working at a labor cost of $41.03 per manhour. These costs include cutting the EMT conduit, layout, material handling, and normal waste. Add the cost of conduit, screws or nails to hold the straps, sales tax, delivery, supervision, mobilization, demobilization, cleanup, overhead and profit. Note: Material costs assume purchase of full box quantities.

EMT Straps, Hangers and Clips

Material	Craft@Hrs	Unit	Material Cost	Labor Cost	Installed Cost
Two hole heavy duty steel EMT straps					
1"	L1@0.05	Ea	.12	2.05	2.17
1-1/4"	L1@0.06	Ea	.17	2.46	2.63
1-1/2"	L1@0.06	Ea	.24	2.46	2.70
2"	L1@0.10	Ea	.40	4.10	4.50
2-1/2"	L1@0.10	Ea	.46	4.10	4.56
3"	L1@0.10	Ea	.66	4.10	4.76
3-1/2"	L1@0.15	Ea	.99	6.15	7.14
4"	L1@0.15	Ea	1.68	6.15	7.83

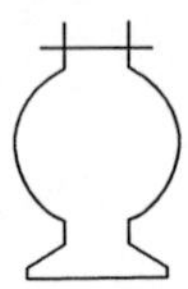

Steel EMT conduit hangers with bolt					
1/2"	L1@0.03	Ea	.31	1.23	1.54
3/4"	L1@0.04	Ea	.34	1.64	1.98
1"	L1@0.05	Ea	.40	2.05	2.45
1-1/4"	L1@0.06	Ea	.48	2.46	2.94
1-1/2"	L1@0.06	Ea	.59	2.46	3.05
2"	L1@0.10	Ea	.74	4.10	4.84
2-1/2"	L1@0.10	Ea	.84	4.10	4.94
3"	L1@0.15	Ea	1.11	6.15	7.26
3-1/2"	L1@0.15	Ea	1.29	6.15	7.44
4"	L1@0.15	Ea	3.46	6.15	9.61

Beam clamp EMT conduit hanger assembly					
1/2"	L1@0.05	Ea	.59	2.05	2.64
3/4"	L1@0.06	Ea	.68	2.46	3.14
1"	L1@0.08	Ea	.79	3.28	4.07
1-1/4"	L1@0.10	Ea	.96	4.10	5.06
1-1/2"	L1@0.10	Ea	1.19	4.10	5.29
2"	L1@0.15	Ea	1.47	6.15	7.62

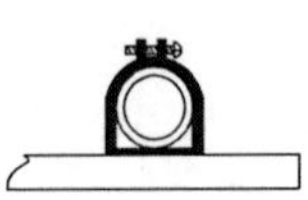

EMT Strut Clamp					
1/2"	L1@0.06	Ea	.44	2.46	2.90
3/4"	L1@0.08	Ea	.45	3.28	3.73
1"	L1@0.10	Ea	.52	4.10	4.62
1-1/4"	L1@0.10	Ea	.58	4.10	4.68
1-1/2"	L1@0.10	Ea	.71	4.10	4.81
2"	L1@0.15	Ea	.74	6.15	6.89

Use these figures to estimate the cost of EMT straps, hangers and clips installed on EMT conduit under the conditions described on pages 5 and 6. Costs listed are for each strap, hanger or clip installed. The crew is one electrician working at a labor cost of $41.03 per manhour. These costs include cutting the EMT conduit, layout, material handling, and normal waste. Add the cost of conduit, screws or nails to hold the straps, sales tax, delivery, supervision, mobilization, demobilization, cleanup, overhead and profit. Note: Material costs assume purchase of full box quantities.

EMT Clips, Adapters, Elbows, Caps and Benders

EMT clips for rod, wire, or steel flange

Material	Craft@Hrs	Unit	Material Cost	Labor Cost	Installed Cost
1/2"	L1@0.04	Ea	.15	1.64	1.79
3/4"	L1@0.05	Ea	.16	2.05	2.21
1"	L1@0.06	Ea	.18	2.46	2.64
1-1/4"	L1@0.08	Ea	.22	3.28	3.50

EMT split adapters

Material	Craft@Hrs	Unit	Material Cost	Labor Cost	Installed Cost
1/2"	L1@0.08	Ea	1.25	3.28	4.53
3/4"	L1@0.10	Ea	1.13	4.10	5.23
1"	L1@0.15	Ea	1.55	6.15	7.70

Die cast 90 degree EMT elbows

Material	Craft@Hrs	Unit	Material Cost	Labor Cost	Installed Cost
1/2"	L1@0.10	Ea	2.43	4.10	6.53
3/4"	L1@0.10	Ea	3.80	4.10	7.90
1"	L1@0.15	Ea	5.26	6.15	11.41
1-1/4"	L1@0.15	Ea	26.20	6.15	32.35
1-1/2"	L1@0.15	Ea	34.10	6.15	40.25

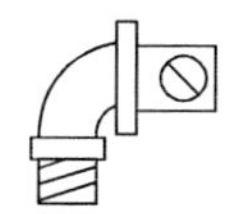

90 degree EMT short elbows

Material	Craft@Hrs	Unit	Material Cost	Labor Cost	Installed Cost
1/2"	L1@0.10	Ea	2.18	4.10	6.28
3/4"	L1@0.10	Ea	3.02	4.10	7.12
1"	L1@0.15	Ea	5.31	6.15	11.46
1-1/4"	L1@0.15	Ea	21.00	6.15	27.15

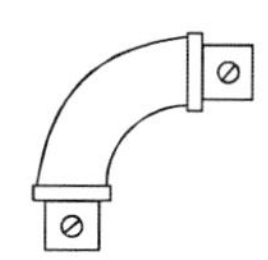

Slip-fitter EMT entrance caps

Material	Craft@Hrs	Unit	Material Cost	Labor Cost	Installed Cost
1/2"	L1@0.10	Ea	3.58	4.10	7.68
3/4"	L1@0.10	Ea	4.18	4.10	8.28
1"	L1@0.15	Ea	4.93	6.15	11.08
1-1/4"	L1@0.15	Ea	5.51	6.15	11.66

EMT hand benders

Material	Craft@Hrs	Unit	Material Cost	Labor Cost	Installed Cost
1/2"	—	Ea	17.80	—	17.80
3/4"	—	Ea	38.50	—	38.50
1"	—	Ea	43.00	—	43.00

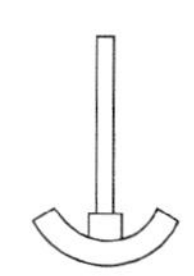

Use these figures to estimate the cost of items shown above installed on EMT conduit under the conditions described on pages 5 and 6. Costs listed are for each item installed. The crew is one electrician working at a labor cost of $41.03 per manhour. These costs include the connector locknut, removing the knockout when required, layout, material handling, and normal waste. Add for conduit boxes, insulated bushings, sales tax, delivery, supervision, mobilization, demobilization, cleanup, overhead and profit. Note: Material costs assume purchase of full box quantities.

Flexible Conduit

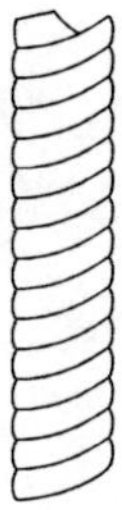

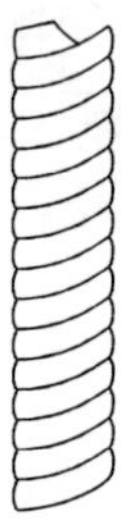

Material	Craft@Hrs	Unit	Material Cost	Labor Cost	Installed Cost
Flex steel conduit					
3/8"	L1@2.50	CLF	19.60	103.00	122.60
1/2"	L1@2.75	CLF	17.40	113.00	130.40
3/4"	L1@3.00	CLF	23.70	123.00	146.70
1"	L1@3.25	CLF	43.30	133.00	176.30
1-1/4"	L1@3.50	CLF	55.70	144.00	199.70
1-1/2"	L1@3.75	CLF	91.20	154.00	245.20
2"	L1@4.00	CLF	111.00	164.00	275.00
2-1/2"	L1@4.25	CLF	134.00	174.00	308.00
3"	L1@4.50	CLF	235.00	185.00	420.00
3-1/2"	L1@4.75	CLF	334.00	195.00	529.00
4"	L1@5.00	CLF	303.00	205.00	508.00
Flex aluminum conduit					
3/8"	L1@2.25	CLF	21.50	92.30	113.80
1/2"	L1@2.50	CLF	17.80	103.00	120.80
3/4"	L1@2.75	CLF	24.50	113.00	137.50
1"	L1@3.00	CLF	46.00	123.00	169.00
1-1/4"	L1@3.25	CLF	61.20	133.00	194.20
1-1/2"	L1@3.50	CLF	110.00	144.00	254.00
2"	L1@3.75	CLF	117.00	154.00	271.00
2-1/2"	L1@4.00	CLF	187.00	164.00	351.00
3"	L1@4.25	CLF	306.00	174.00	480.00
3-1/2"	L1@4.50	CLF	354.00	185.00	539.00
4"	L1@4.75	CLF	395.00	195.00	590.00

Use these figures to estimate the cost of flexible conduit installed in a building, and for equipment hookup under the conditions described on pages 5 and 6. Costs listed are for each 100 linear feet installed. The crew is one electrician working at a labor cost of $41.03 per manhour. These costs include boring or notching wood studs and joists (in concealed locations), cutting flex conduit, layout, material handling, and normal waste. Add for connectors, couplings, straps, boxes, wire, bonding wire, sales tax, delivery, supervision, mobilization, demobilization, cleanup, overhead and profit. Note: Conduit runs are assumed to be 25' long. Labor costs per linear foot will be higher on shorter runs and lower on longer runs.

Material	Craft@Hrs	Unit	Material Cost	Labor Cost	Installed Cost
Die cast screw-in flex connectors					
3/8"	L1@0.03	Ea	.15	1.23	1.38
1/2"	L1@0.03	Ea	.20	1.23	1.43
3/4"	L1@0.05	Ea	.34	2.05	2.39
1"	L1@0.06	Ea	.76	2.46	3.22
1-1/4"	L1@0.10	Ea	1.46	4.10	5.56
1-1/2"	L1@0.10	Ea	2.03	4.10	6.13
2"	L1@0.15	Ea	2.46	6.15	8.61

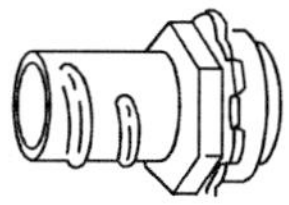

Material	Craft@Hrs	Unit	Material Cost	Labor Cost	Installed Cost
Insulated die cast screw-in flex connectors					
3/8"	L1@0.03	Ea	.30	1.23	1.53
1/2"	L1@0.03	Ea	.35	1.23	1.58
3/4"	L1@0.05	Ea	.46	2.05	2.51
1"	L1@0.06	Ea	.98	2.46	3.44
1-1/4"	L1@0.10	Ea	2.27	4.10	6.37
1-1/2"	L1@0.10	Ea	2.69	4.10	6.79
2"	L1@0.15	Ea	3.10	6.15	9.25

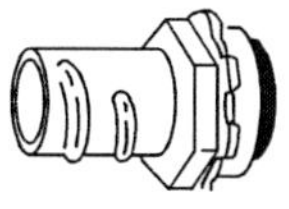

Material	Craft@Hrs	Unit	Material Cost	Labor Cost	Installed Cost
Die cast squeeze flex connectors					
3/8"	L1@0.05	Ea	.60	2.05	2.65
1/2"	L1@0.05	Ea	.69	2.05	2.74
3/4"	L1@0.06	Ea	.77	2.46	3.23
1"	L1@0.08	Ea	1.52	3.28	4.80
1-1/4"	L1@0.10	Ea	3.31	4.10	7.41
1-1/2"	L1@0.10	Ea	6.25	4.10	10.35
2"	L1@0.15	Ea	10.60	6.15	16.75

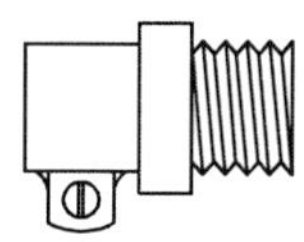

Use these figures to estimate the cost of flexible conduit connectors installed on flex conduit under the conditions described on pages 5 and 6. Costs listed are for each connector installed. The crew is one electrician working at a labor cost of $41.03 per manhour. These costs include the locknut, removing the knockout, layout, material handling, and normal waste. Add for connectors, couplings, straps, boxes, wire, bonding wire, sales tax, delivery, supervision, mobilization, demobilization, cleanup, overhead and profit. Note: Material costs assume purchase of full boxes.

Squeeze Flexible Conduit Connectors

Insulated die cast squeeze flex connectors

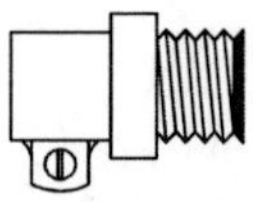

Material	Craft@Hrs	Unit	Material Cost	Labor Cost	Installed Cost
1/2"	L1@0.05	Ea	1.05	2.05	3.10
3/4"	L1@0.06	Ea	1.06	2.46	3.52
1"	L1@0.08	Ea	1.30	3.28	4.58
1-1/4"	L1@0.10	Ea	3.14	4.10	7.24
1-1/2"	L1@0.10	Ea	5.15	4.10	9.25
2"	L1@0.15	Ea	7.84	6.15	13.99

Malleable squeeze flex connectors

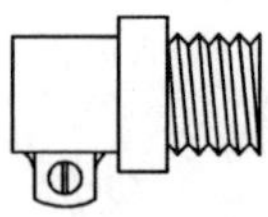

Material	Craft@Hrs	Unit	Material Cost	Labor Cost	Installed Cost
3/8"	L1@0.05	Ea	.58	2.05	2.63
1/2"	L1@0.05	Ea	.86	2.05	2.91
3/4"	L1@0.06	Ea	.70	2.46	3.16
1"	L1@0.08	Ea	2.55	3.28	5.83
1-1/4"	L1@0.10	Ea	4.01	4.10	8.11
1-1/2"	L1@0.10	Ea	5.47	4.10	9.57
2"	L1@0.15	Ea	7.65	6.15	13.80
2-1/2"	L1@0.15	Ea	15.00	6.15	21.15
3"	L1@0.20	Ea	20.70	8.21	28.91

Insulated malleable squeeze flex connectors

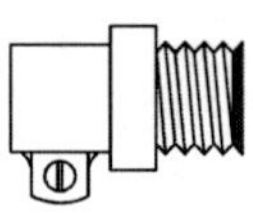

Material	Craft@Hrs	Unit	Material Cost	Labor Cost	Installed Cost
3/8"	L1@0.05	Ea	1.19	2.05	3.24
1/2"	L1@0.05	Ea	1.26	2.05	3.31
3/4"	L1@0.06	Ea	1.41	2.46	3.87
1"	L1@0.08	Ea	2.55	3.28	5.83
1-1/4"	L1@0.10	Ea	5.47	4.10	9.57
1-1/2"	L1@0.10	Ea	8.02	4.10	12.12
2"	L1@0.15	Ea	12.10	6.15	18.25
2-1/2"	L1@0.15	Ea	23.70	6.15	29.85
3"	L1@0.20	Ea	31.20	8.21	39.41
3-1/2"	L1@0.25	Ea	86.30	10.30	96.60
4"	L1@0.25	Ea	102.00	10.30	112.30

Use these figures to estimate the cost of flexible conduit connectors installed on flex conduit under the conditions described on pages 5 and 6. Costs listed are for each connector installed. The crew is one electrician working at a labor cost of $41.03 per manhour. These costs include the locknut, removing the knockout, layout, material handling, and normal waste. Add for conduit boxes, insulated bushings, sales tax, delivery, supervision, mobilization, demobilization, cleanup, overhead and profit. Note: Material costs assume purchase of full boxes.

Material	Craft@Hrs	Unit	Material Cost	Labor Cost	Installed Cost
45 degree die cast flex connectors					
3/8"	L1@0.05	Ea	.26	2.05	2.31
1/2"	L1@0.05	Ea	.33	2.05	2.38
3/4"	L1@0.06	Ea	.92	2.46	3.38

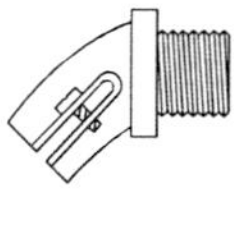

Material	Craft@Hrs	Unit	Material Cost	Labor Cost	Installed Cost
45 degree malleable flex connectors					
3/8"	L1@0.05	Ea	.78	2.05	2.83
1/2"	L1@0.06	Ea	.84	2.46	3.30

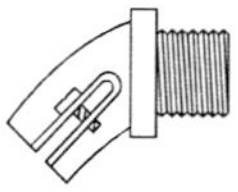

Material	Craft@Hrs	Unit	Material Cost	Labor Cost	Installed Cost
90 degree die cast two screw flex connectors					
3/8"	L1@0.05	Ea	.23	2.05	2.28
1/2"	L1@0.05	Ea	.32	2.05	2.37
3/4"	L1@0.06	Ea	.84	2.46	3.30
1"	L1@0.08	Ea	1.76	3.28	5.04
1-1/4"	L1@0.10	Ea	2.67	4.10	6.77
1-1/2"	L1@0.15	Ea	6.42	6.15	12.57
2"	L1@0.20	Ea	17.40	8.21	25.61
2-1/2"	L1@0.25	Ea	21.60	10.30	31.90
3"	L1@0.25	Ea	29.00	10.30	39.30

Material	Craft@Hrs	Unit	Material Cost	Labor Cost	Installed Cost
Insulated 90 degree die cast two screw flex connectors					
3/8"	L1@0.05	Ea	.32	2.05	2.37
1/2"	L1@0.05	Ea	.42	2.05	2.47
3/4"	L1@0.06	Ea	.98	2.46	3.44
1"	L1@0.08	Ea	2.00	3.28	5.28
1-1/4"	L1@0.10	Ea	3.61	4.10	7.71
1-1/2"	L1@0.15	Ea	6.99	6.15	13.14
2"	L1@0.20	Ea	18.20	8.21	26.41
2-1/2"	L1@0.25	Ea	22.40	10.30	32.70
3"	L1@0.25	Ea	29.60	10.30	39.90
3-1/2"	L1@0.30	Ea	67.40	12.30	79.70
4"	L1@0.30	Ea	80.20	12.30	92.50

Use these figures to estimate the cost of flexible conduit connectors installed on flex conduit under the conditions described on pages 5 and 6. Costs listed are for each connector installed. The crew is one electrician working at a labor cost of $41.03 per manhour. These costs include the locknut, removing the knockout, layout, material handling, and normal waste. Add for conduit boxes, insulated bushings, sales tax, delivery, supervision, mobilization, demobilization, cleanup, overhead and profit. Note: Material costs assume purchase of full boxes.

Flexible Conduit Connectors and Couplings

90 degree malleable squeeze flex connectors

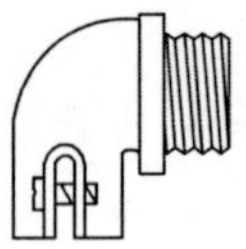

Material	Craft@Hrs	Unit	Material Cost	Labor Cost	Installed Cost
3/8"	L1@0.05	Ea	.53	2.05	2.58
1/2"	L1@0.05	Ea	1.03	2.05	3.08
3/4"	L1@0.06	Ea	1.47	2.46	3.93
1"	L1@0.08	Ea	1.96	3.28	5.24
1-1/4"	L1@0.10	Ea	4.29	4.10	8.39
1-1/2"	L1@0.15	Ea	8.33	6.15	14.48
2"	L1@0.20	Ea	12.10	8.21	20.31
2-1/2"	L1@0.25	Ea	14.00	10.30	24.30
3"	L1@0.25	Ea	37.80	10.30	48.10
3-1/2	L1@0.30	Ea	57.40	12.30	69.70
4"	L1@0.30	Ea	140.00	12.30	152.30

Insulated 90 degree malleable squeeze flex connectors

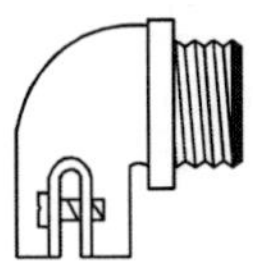

3/8"	L1@0.05	Ea	.72	2.05	2.77
1/2"	L1@0.05	Ea	1.20	2.05	3.25
3/4"	L1@0.06	Ea	1.52	2.46	3.98
1"	L1@0.08	Ea	2.11	3.28	5.39
1-1/4"	L1@0.10	Ea	4.88	4.10	8.98
1-1/2"	L1@0.15	Ea	8.63	6.15	14.78
2"	L1@0.20	Ea	13.00	8.21	21.21
2-1/2"	L1@0.25	Ea	15.20	10.30	25.50
3"	L1@0.25	Ea	38.40	10.30	48.70
3-1/2"	L1@0.30	Ea	58.40	12.30	70.70
4"	L1@0.30	Ea	146.00	12.30	158.30

Die cast screw-in flex couplings

1/2"	L1@0.03	Ea	.96	1.23	2.19
3/4"	L1@0.05	Ea	1.57	2.05	3.62
1"	L1@0.06	Ea	2.72	2.46	5.18

Die cast screw-in flex to EMT couplings

3/8"	L1@0.05	Ea	1.03	2.05	3.08
1/2"	L1@0.06	Ea	1.59	2.46	4.05
3/4"	L1@0.08	Ea	2.22	3.28	5.50
1"	L1@0.10	Ea	3.11	4.10	7.21

Die cast set screw flex to rigid couplings

1/2"	L1@0.05	Ea	2.17	2.05	4.22
3/4"	L1@0.06	Ea	2.77	2.46	5.23

Use these figures to estimate the cost of flexible conduit connectors and couplings installed on flex conduit under the conditions described on pages 5 and 6. Costs listed are for each connector or coupling installed. The crew is one electrician working at a labor cost of $41.03 per manhour. These costs include the locknut, removing the knockout, layout, material handling, and normal waste. Add for conduit boxes, insulated bushings, sales tax, delivery, supervision, mobilization, demobilization, cleanup, overhead and profit. Note: Material costs assume purchase of full boxes.

Type EF or Type LT flex steel conduit

Material	Craft@Hrs	Unit	Material Cost	Labor Cost	Installed Cost
3/8"	L1@4.00	CLF	108.00	164.00	272.00
1/2"	L1@4.00	CLF	128.00	164.00	292.00
3/4"	L1@4.50	CLF	172.00	185.00	357.00
1"	L1@5.00	CLF	260.00	205.00	465.00
1-1/4"	L1@6.00	CLF	356.00	246.00	602.00
1-1/2"	L1@7.00	CLF	484.00	287.00	771.00
2"	L1@9.00	CLF	606.00	369.00	975.00
2-1/2"	L1@11.0	CLF	1,570.00	451.00	2,021.00
3"	L1@15.0	CLF	2,160.00	615.00	2,775.00
4"	L1@17.0	CLF	3,110.00	698.00	3,808.00

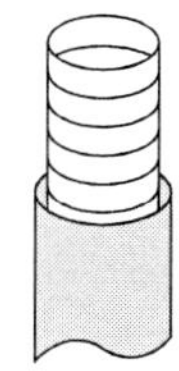

Type UA liquid-tight flex conduit

Material	Craft@Hrs	Unit	Material Cost	Labor Cost	Installed Cost
3/8"	L1@4.00	CLF	253.00	164.00	417.00
1/2"	L1@4.00	CLF	298.00	164.00	462.00
3/4"	L1@4.50	CLF	411.00	185.00	596.00
1"	L1@5.00	CLF	641.00	205.00	846.00
1-1/4"	L1@6.00	CLF	913.00	246.00	1,159.00
1-1/2"	L1@7.00	CLF	952.00	287.00	1,239.00
2"	L1@9.00	CLF	1,190.00	369.00	1,559.00
2-1/2"	L1@11.0	CLF	2,170.00	451.00	2,621.00
3"	L1@15.0	CLF	3,080.00	615.00	3,695.00
4"	L1@17.0	CLF	4,960.00	698.00	5,658.00

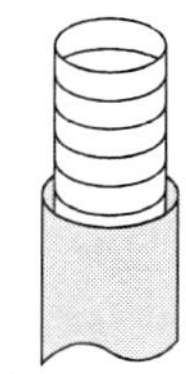

Type OR liquid-tight flex conduit

Material	Craft@Hrs	Unit	Material Cost	Labor Cost	Installed Cost
3/8"	L1@4.00	CLF	245.00	164.00	409.00
1/2"	L1@4.00	CLF	283.00	164.00	447.00
3/4"	L1@4.50	CLF	430.00	185.00	615.00
1"	L1@5.00	CLF	590.00	205.00	795.00
1-1/4"	L1@6.00	CLF	774.00	246.00	1,020.00
1-1/2"	L1@7.00	CLF	1,090.00	287.00	1,377.00
2"	L1@9.00	CLF	1,350.00	369.00	1,719.00
2-1/2"	L1@11.0	CLF	2,670.00	451.00	3,121.00
3"	L1@15.0	CLF	3,630.00	615.00	4,245.00
4"	L1@17.0	CLF	5,130.00	698.00	5,828.00

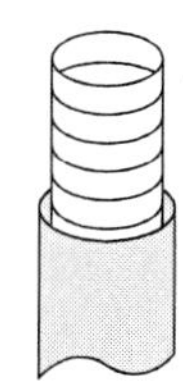

Construction grade liquid-tight flex conduit

Material	Craft@Hrs	Unit	Material Cost	Labor Cost	Installed Cost
3/8"	L1@4.00	CLF	91.70	164.00	255.70
1/2"	L1@4.00	CLF	147.00	164.00	311.00
3/4"	L1@4.50	CLF	208.00	185.00	393.00
1"	L1@5.00	CLF	317.00	205.00	522.00
1-1/4"	L1@6.00	CLF	449.00	246.00	695.00
1-1/2"	L1@7.00	CLF	516.00	287.00	803.00
2"	L1@9.00	CLF	631.00	369.00	1,000.00

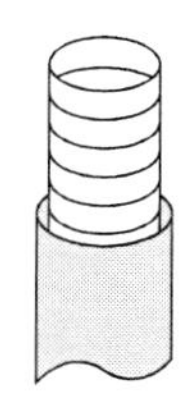

Use these figures to estimate the cost of liquid-tight flex conduit installed as part of equipment hookup under the conditions described on pages 5 and 6. Costs listed are for each linear foot installed. The crew is one electrician working at a labor cost of $41.03 per manhour. These costs include cutting conduit, layout, material handling, and normal waste. Add for connectors, boxes, straps, wire, bonding wire, sales tax, delivery, supervision, mobilization, demobilization, cleanup, overhead and profit. Note: Conduit runs are assumed to be 25' long. Labor costs per linear foot will be higher on shorter runs and lower on longer runs.

Liquid-tight Flexible Conduit and Connectors

Type HC liquid-tight extra flex conduit

Material	Craft@Hrs	Unit	Material Cost	Labor Cost	Installed Cost
3/8"	L1@4.00	CLF	317.00	164.00	481.00
1/2"	L1@4.00	CLF	331.00	164.00	495.00
3/4"	L1@4.50	CLF	461.00	185.00	646.00
1"	L1@5.00	CLF	672.00	205.00	877.00
1-1/4"	L1@6.00	CLF	911.00	246.00	1,157.00
1-1/2"	L1@7.00	CLF	1,250.00	287.00	1,537.00
2"	L1@9.00	CLF	1,540.00	369.00	1,909.00
2-1/2"	L1@11.0	CLF	2,860.00	451.00	3,311.00
3"	L1@15.0	CLF	3,990.00	615.00	4,605.00
4"	L1@17.0	CLF	5,830.00	698.00	6,528.00

Type CN-P liquid-tight flex non-metallic conduit

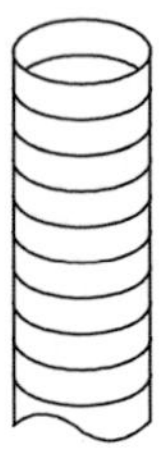

Material	Craft@Hrs	Unit	Material Cost	Labor Cost	Installed Cost
3/8"	L1@3.50	CLF	375.00	144.00	519.00
1/2"	L1@3.75	CLF	485.00	154.00	639.00
3/4"	L1@4.00	CLF	690.00	164.00	854.00
1"	L1@4.50	CLF	885.00	185.00	1,070.00
1-1/4"	L1@5.00	CLF	1,160.00	205.00	1,365.00
1-1/2"	L1@5.50	CLF	1,940.00	226.00	2,166.00
2"	L1@6.00	CLF	2,690.00	246.00	2,936.00

Malleable liquid-tight flex connectors

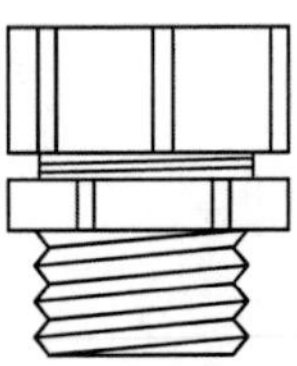

Material	Craft@Hrs	Unit	Material Cost	Labor Cost	Installed Cost
3/8"	L1@0.10	Ea	6.21	4.10	10.31
1/2"	L1@0.10	Ea	6.21	4.10	10.31
3/4"	L1@0.10	Ea	8.84	4.10	12.94
1"	L1@0.15	Ea	13.00	6.15	19.15
1-1/4"	L1@0.20	Ea	22.30	8.21	30.51
1-1/2"	L1@0.20	Ea	31.80	8.21	40.01
2"	L1@0.25	Ea	58.50	10.30	68.80
2-1/2"	L1@0.25	Ea	267.00	10.30	277.30
3"	L1@0.30	Ea	304.00	12.30	316.30
4"	L1@0.30	Ea	394.00	12.30	406.30

Insulated malleable liquid-tight flex connectors

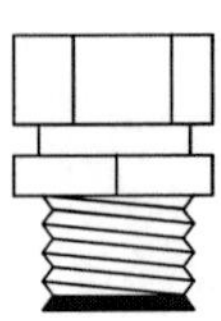

Material	Craft@Hrs	Unit	Material Cost	Labor Cost	Installed Cost
3/8"	L1@0.10	Ea	7.80	4.10	11.90
1/2"	L1@0.10	Ea	7.80	4.10	11.90
3/4"	L1@0.10	Ea	11.40	4.10	15.50
1"	L1@0.15	Ea	17.50	6.15	23.65
1-1/4"	L1@0.20	Ea	28.30	8.21	36.51
1-1/2"	L1@0.20	Ea	41.20	8.21	49.41
2"	L1@0.25	Ea	77.10	10.30	87.40
2-1/2"	L1@0.25	Ea	423.00	10.30	433.30
3"	L1@0.30	Ea	473.00	12.30	485.30
4"	L1@0.30	Ea	563.00	12.30	575.30

Use these figures to estimate the cost of liquid-tight flex conduit (two top tables) and connectors (two bottom tables) installed with equipment hookup under the conditions described on pages 5 and 6. Costs listed are for each 100 linear feet of conduit and each connector installed. The crew is one electrician working at a labor cost of $41.03 per manhour. These costs include cutting conduit, removal of the knockout for the connector, layout, material handling, and normal waste. Add for straps, boxes, wire, bonding wire, sales tax, delivery, supervision, mobilization, demobilization, cleanup, overhead and profit. Note: Connector costs assume the purchase of full box quantities.

Die cast liquid-tight flex connectors

Material	Craft@Hrs	Unit	Material Cost	Labor Cost	Installed Cost
3/8"	L1@0.10	Ea	6.38	4.10	10.48
1/2"	L1@0.10	Ea	6.28	4.10	10.38
3/4"	L1@0.10	Ea	8.94	4.10	13.04
1"	L1@0.15	Ea	13.20	6.15	19.35
1-1/4"	L1@0.20	Ea	22.80	8.21	31.01
1-1/2"	L1@0.20	Ea	31.80	8.21	40.01
2"	L1@0.25	Ea	58.50	10.30	68.80
2-1/2"	L1@0.25	Ea	276.00	10.30	286.30
3"	L1@0.30	Ea	307.00	12.30	319.30
4"	L1@0.30	Ea	403.00	12.30	415.30

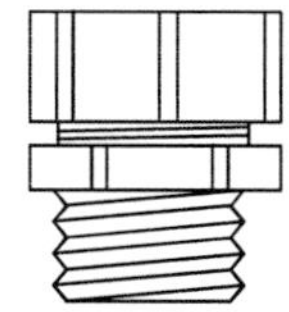

Insulated die cast liquid-tight flex connectors

Material	Craft@Hrs	Unit	Material Cost	Labor Cost	Installed Cost
3/8"	L1@0.10	Ea	6.83	4.10	10.93
1/2"	L1@0.10	Ea	6.83	4.10	10.93
3/4"	L1@0.10	Ea	10.00	4.10	14.10
1"	L1@0.15	Ea	15.40	6.15	21.55
1-1/4"	L1@0.20	Ea	24.30	8.21	32.51
1-1/2"	L1@0.20	Ea	34.50	8.21	42.71
2"	L1@0.25	Ea	63.80	10.30	74.10
2-1/2"	L1@0.25	Ea	329.00	10.30	339.30
3"	L1@0.30	Ea	364.00	12.30	376.30
4"	L1@0.30	Ea	429.00	12.30	441.30

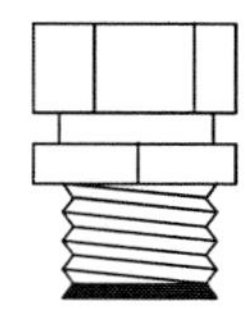

45 degree malleable liquid-tight flex connectors

Material	Craft@Hrs	Unit	Material Cost	Labor Cost	Installed Cost
3/8"	L1@0.12	Ea	4.03	4.92	8.95
1/2"	L1@0.12	Ea	3.96	4.92	8.88
3/4"	L1@0.15	Ea	6.09	6.15	12.24
1"	L1@0.15	Ea	11.90	6.15	18.05
1-1/4"	L1@0.20	Ea	20.30	8.21	28.51
1-1/2"	L1@0.20	Ea	26.90	8.21	35.11
2"	L1@0.25	Ea	35.40	10.30	45.70
2-1/2"	L1@0.25	Ea	172.00	10.30	182.30
3"	L1@0.30	Ea	185.00	12.30	197.30
4"	L1@0.30	Ea	230.00	12.30	242.30

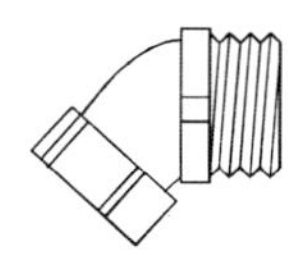

Insulated 45 degree malleable liquid-tight flex connectors

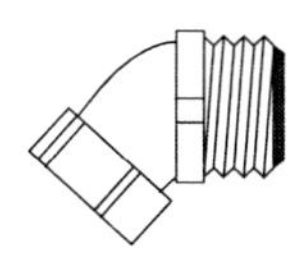

Material	Craft@Hrs	Unit	Material Cost	Labor Cost	Installed Cost
3/8"	L1@0.12	Ea	8.11	4.92	13.03
1/2"	L1@0.12	Ea	13.00	4.92	17.92
3/4"	L1@0.15	Ea	19.50	6.15	25.65

Use these figures to estimate the cost of liquid-tight flex connectors installed on liquid-tight flex conduit under the conditions described on pages 5 and 6. Costs listed are for each connector installed. The crew is one electrician working at a labor cost of $41.03 per manhour. These costs include locknuts and removal of the knockout, layout, material handling, and normal waste. Add for conduit, insulating bushings, sales tax, delivery, supervision, mobilization, demobilization, cleanup, overhead and profit. Note: Material costs assume purchase of full box quantities.

Liquid-tight Flex Connectors and Couplings

Insulated 45 degree malleable liquid-tight flex connectors

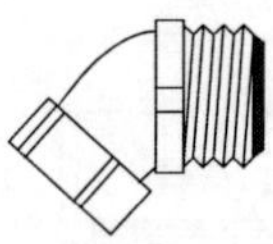

Material	Craft@Hrs	Unit	Material Cost	Labor Cost	Installed Cost
1"	L1@0.15	Ea	37.80	6.15	43.95
1-1/4"	L1@0.20	Ea	58.50	8.21	66.71
1-1/2"	L1@0.20	Ea	71.30	8.21	79.51
2"	L1@0.25	Ea	107.00	10.30	117.30
2-1/2"	L1@0.25	Ea	52.90	10.30	63.20
3"	L1@0.30	Ea	63.40	12.30	75.70
4"	L1@0.30	Ea	73.10	12.30	85.40

90 degree malleable liquid-tight flex connectors

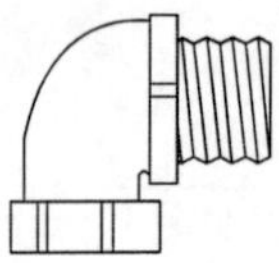

Material	Craft@Hrs	Unit	Material Cost	Labor Cost	Installed Cost
3/8"	L1@0.15	Ea	9.93	6.15	16.08
1/2"	L1@0.15	Ea	9.74	6.15	15.89
3/4"	L1@0.15	Ea	14.80	6.15	20.95
1"	L1@0.20	Ea	30.40	8.21	38.61
1-1/4"	L1@0.25	Ea	45.70	10.30	56.00
1-1/2"	L1@0.25	Ea	55.30	10.30	65.60
2"	L1@0.30	Ea	80.70	12.30	93.00
2-1/2"	L1@0.30	Ea	364.00	12.30	376.30
3"	L1@0.40	Ea	444.00	16.40	460.40
4"	L1@0.40	Ea	655.00	16.40	671.40

Insulated 90 degree malleable liquid-tight flex connectors

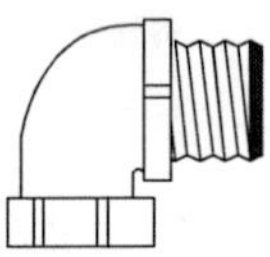

Material	Craft@Hrs	Unit	Material Cost	Labor Cost	Installed Cost
3/8"	L1@0.15	Ea	13.00	6.15	19.15
1/2"	L1@0.15	Ea	13.00	6.15	19.15
3/4"	L1@0.15	Ea	19.50	6.15	25.65
1"	L1@0.20	Ea	37.30	8.21	45.51
1-1/4"	L1@0.25	Ea	56.50	10.30	66.80
1-1/2"	L1@0.25	Ea	68.90	10.30	79.20
2"	L1@0.30	Ea	104.00	12.30	116.30
2-1/2"	L1@0.30	Ea	524.00	12.30	536.30
3"	L1@0.40	Ea	630.00	16.40	646.40
4"	L1@0.40	Ea	821.00	16.40	837.40

Malleable liquid-tight flex to rigid combination couplings

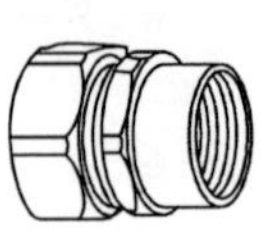

Material	Craft@Hrs	Unit	Material Cost	Labor Cost	Installed Cost
1/2"	L1@0.15	Ea	8.69	6.15	14.84
3/4"	L1@0.15	Ea	8.69	6.15	14.84
1"	L1@0.20	Ea	12.30	8.21	20.51
1-1/4"	L1@0.25	Ea	40.30	10.30	50.60
1-1/2"	L1@0.25	Ea	72.00	10.30	82.30
2"	L1@0.30	Ea	98.80	12.30	111.10
2-1/2"	L1@0.30	Ea	461.00	12.30	473.30
3"	L1@0.40	Ea	509.00	16.40	525.40
4"	L1@0.40	Ea	630.00	16.40	646.40

Use these figures to estimate the cost of liquid-tight flex connectors installed on liquid-tight flex conduit under the conditions described on pages 5 and 6. Costs listed are for each connector installed. The crew is one electrician working at a labor cost of $41.03 per manhour. These costs include locknuts and removal of the knockout, layout, material handling, and normal waste. Add for conduit, insulating bushings, sales tax, delivery, supervision, mobilization, demobilization, cleanup, overhead and profit. Note: Material costs assume purchase of full box quantities.

Material	Craft@Hrs	Unit	Material Cost	Labor Cost	Installed Cost

Schedule 40 PVC conduit, 10' lengths with coupling

Material	Craft@Hrs	Unit	Material Cost	Labor Cost	Installed Cost
1/2"	L1@3.10	CLF	29.80	127.00	156.80
3/4"	L1@3.20	CLF	34.80	131.00	165.80
1"	L1@3.30	CLF	53.00	135.00	188.00
1-1/4"	L1@3.40	CLF	76.40	140.00	216.40
1-1/2"	L1@3.45	CLF	86.70	142.00	228.70
2"	L1@3.50	CLF	106.00	144.00	250.00
2-1/2"	L2@3.60	CLF	178.00	148.00	326.00
3"	L2@3.75	CLF	213.00	154.00	367.00
4"	L2@4.00	CLF	297.00	164.00	461.00
5"	L2@4.25	CLF	444.00	174.00	618.00
6"	L2@4.50	CLF	526.00	185.00	711.00

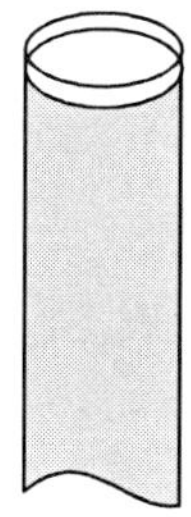

Schedule 80 heavy wall PVC conduit, 10' lengths with coupling

Material	Craft@Hrs	Unit	Material Cost	Labor Cost	Installed Cost
1/2"	L1@3.20	CLF	55.70	131.00	186.70
3/4"	L1@3.30	CLF	75.90	135.00	210.90
1"	L1@3.40	CLF	95.80	140.00	235.80
1-1/4"	L1@3.50	CLF	126.00	144.00	270.00
1-1/2"	L1@3.60	CLF	159.00	148.00	307.00
2"	L1@3.70	CLF	195.00	152.00	347.00
2-1/2"	L2@3.90	CLF	304.00	160.00	464.00
3"	L2@4.00	CLF	374.00	164.00	538.00
4"	L2@4.50	CLF	581.00	185.00	766.00
5"	L2@5.00	CLF	780.00	205.00	985.00
6”	L2@6.00	CLF	1,100.00	246.00	1,346.00

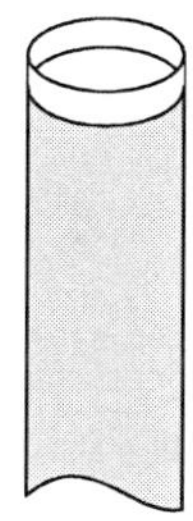

30 degree Schedule 40 PVC elbows

Material	Craft@Hrs	Unit	Material Cost	Labor Cost	Installed Cost
1/2"	L1@0.05	Ea	3.42	2.05	5.47
3/4"	L1@0.06	Ea	3.53	2.46	5.99
1"	L1@0.08	Ea	4.23	3.28	7.51
1-1/4"	L1@0.10	Ea	6.05	4.10	10.15
1-1/2"	L1@0.10	Ea	8.33	4.10	12.43
2"	L1@0.15	Ea	12.10	6.15	18.25
2-1/2"	L2@0.15	Ea	23.00	6.15	29.15
3"	L2@0.20	Ea	39.40	8.21	47.61
4"	L2@0.25	Ea	65.30	10.30	75.60
5"	L2@0.30	Ea	105.00	12.30	117.30
6"	L2@0.50	Ea	123.00	20.50	143.50

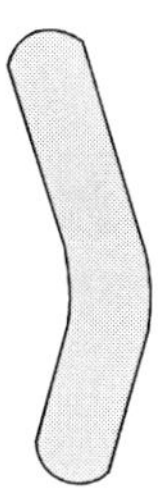

Use these figures to estimate the cost of PVC conduit and elbows installed underground or in a building under the conditions described on pages 5 and 6. Costs listed are for 100 linear feet of conduit installed or for each elbow installed. The crew is one electrician for diameters to 2" and two electricians for 2-1/2" and larger conduit. The labor cost is $41.03 per manhour. These costs include making up joints with cement (glue), layout, material handling, and normal waste. Add for bends, connectors, end bell, spacers, wire, trenching, encasement, sales tax, delivery, supervision, mobilization, demobilization, cleanup, overhead and profit. Conduit runs are assumed to be 50' long. Shorter runs will take more labor and longer runs will take less labor per linear foot.

PVC Elbows and Couplings

45 degree Schedule 40 PVC elbows

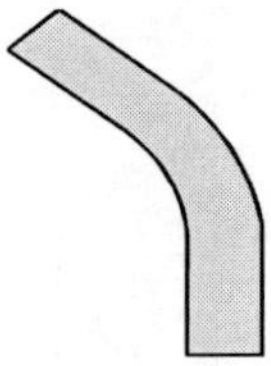

Material	Craft@Hrs	Unit	Material Cost	Labor Cost	Installed Cost
1/2"	L1@0.05	Ea	2.53	2.05	4.58
3/4"	L1@0.06	Ea	2.69	2.46	5.15
1"	L1@0.08	Ea	4.22	3.28	7.50
1-1/4"	L1@0.10	Ea	5.96	4.10	10.06
1-1/2"	L1@0.10	Ea	8.12	4.10	12.22
2"	L1@0.15	Ea	11.20	6.15	17.35
2-1/2"	L1@0.15	Ea	19.40	6.15	25.55
3"	L1@0.20	Ea	27.10	8.21	35.31
4"	L1@0.25	Ea	59.60	10.30	69.90
5"	L1@0.35	Ea	57.40	14.40	71.80
6"	L1@0.50	Ea	97.40	20.50	117.90

90 degree Schedule 40 PVC elbows

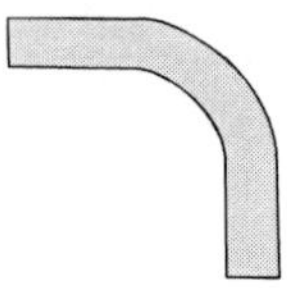

Material	Craft@Hrs	Unit	Material Cost	Labor Cost	Installed Cost
1/2"	L1@0.05	Ea	2.57	2.05	4.62
3/4"	L1@0.06	Ea	2.93	2.46	5.39
1"	L1@0.08	Ea	4.93	3.28	8.21
1-1/4"	L1@0.10	Ea	6.53	4.10	10.63
1-1/2"	L1@0.10	Ea	8.69	4.10	12.79
2"	L1@0.15	Ea	9.17	6.15	15.32
2-1/2"	L1@0.15	Ea	20.60	6.15	26.75
3"	L1@0.20	Ea	36.80	8.21	45.01
4"	L1@0.25	Ea	62.80	10.30	73.10
5"	L1@0.35	Ea	111.00	14.40	125.40
6"	L1@0.50	Ea	186.00	20.50	206.50

90 degree Schedule 80 PVC elbows

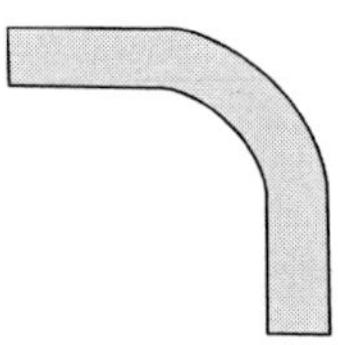

Material	Craft@Hrs	Unit	Material Cost	Labor Cost	Installed Cost
1/2"	L1@0.06	Ea	2.85	2.46	5.31
3/4"	L1@0.08	Ea	3.11	3.28	6.39
1"	L1@0.10	Ea	4.65	4.10	8.75
1-1/4"	L1@0.15	Ea	6.28	6.15	12.43
1-1/2"	L1@0.15	Ea	9.43	6.15	15.58
2"	L1@0.20	Ea	10.70	8.21	18.91
2-1/2"	L1@0.20	Ea	23.70	8.21	31.91
3"	L1@0.25	Ea	64.90	10.30	75.20
4"	L1@0.30	Ea	96.10	12.30	108.40

Schedule 40 PVC couplings

Material	Craft@Hrs	Unit	Material Cost	Labor Cost	Installed Cost
1/2"	L1@0.02	Ea	.67	.82	1.49
3/4"	L1@0.03	Ea	.79	1.23	2.02
1"	L1@0.05	Ea	1.25	2.05	3.30
1-1/4"	L1@0.06	Ea	1.66	2.46	4.12
1-1/2"	L1@0.08	Ea	2.30	3.28	5.58

Use these figures to estimate the cost of PVC elbows and couplings installed on PVC conduit under the conditions described on pages 5 and 6. Costs listed are for each elbow or coupling installed. The crew is one electrician working at a labor cost of $41.03 per manhour. These costs include applying cement (glue), layout, material handling, and normal waste. Add for conduit, couplings, connectors, end bells, spacers, sales tax, delivery, supervision, mobilization, demobilization, cleanup, overhead and profit. Material costs assume purchase of full box quantities.

PVC Couplings, Adapters and Expansion Couplings

Schedule 40 PVC couplings

Material	Craft@Hrs	Unit	Material Cost	Labor Cost	Installed Cost
2"	L1@0.10	Ea	3.01	4.10	7.11
2-1/2"	L1@0.10	Ea	5.31	4.10	9.41
3"	L1@0.15	Ea	8.77	6.15	14.92
4"	L1@0.15	Ea	13.60	6.15	19.75
5"	L1@0.20	Ea	34.50	8.21	42.71
6"	L1@0.25	Ea	43.90	10.30	54.20

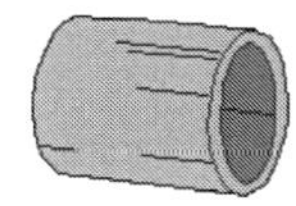

Type FA female PVC adapters

Material	Craft@Hrs	Unit	Material Cost	Labor Cost	Installed Cost
1/2"	L1@0.05	Ea	1.03	2.05	3.08
3/4"	L1@0.06	Ea	1.66	2.46	4.12
1"	L1@0.08	Ea	2.24	3.28	5.52
1-1/4"	L1@0.10	Ea	2.97	4.10	7.07
1-1/2"	L1@0.10	Ea	3.19	4.10	7.29
2"	L1@0.15	Ea	4.33	6.15	10.48
2-1/2"	L1@0.15	Ea	9.56	6.15	15.71
3"	L1@0.20	Ea	11.90	8.21	20.11
4"	L1@0.25	Ea	15.90	10.30	26.20
5"	L1@0.30	Ea	39.90	12.30	52.20
6"	L1@0.40	Ea	52.40	16.40	68.80

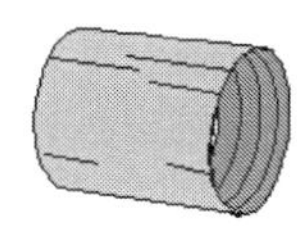

Type TA terminal PVC adapters

Material	Craft@Hrs	Unit	Material Cost	Labor Cost	Installed Cost
1/2"	L1@0.05	Ea	.91	2.05	2.96
3/4"	L1@0.06	Ea	1.57	2.46	4.03
1"	L1@0.08	Ea	1.95	3.28	5.23
1-1/4"	L1@0.10	Ea	2.50	4.10	6.60
1-1/2"	L1@0.10	Ea	3.01	4.10	7.11
2"	L1@0.15	Ea	4.36	6.15	10.51
2-1/2"	L1@0.15	Ea	7.41	6.15	13.56
3"	L1@0.20	Ea	10.80	8.21	19.01
4"	L1@0.25	Ea	18.60	10.30	28.90
5"	L1@0.30	Ea	39.90	12.30	52.20
6"	L1@0.40	Ea	52.40	16.40	68.80

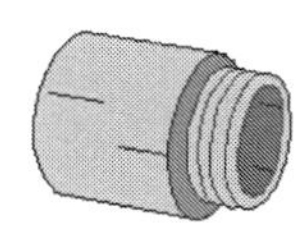

2" range expansion PVC couplings

Material	Craft@Hrs	Unit	Material Cost	Labor Cost	Installed Cost
1/2"	L1@0.15	Ea	36.00	6.15	42.15
3/4"	L1@0.20	Ea	36.70	8.21	44.91
1"	L1@0.25	Ea	52.40	10.30	62.70
1-1/4"	L1@0.30	Ea	77.80	12.30	90.10
1-1/2"	L1@0.30	Ea	98.50	12.30	110.80
2"	L1@0.40	Ea	121.00	16.40	137.40

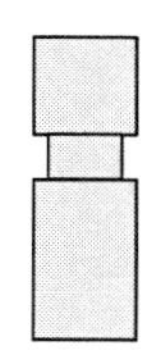

Use these figures to estimate the cost of PVC fittings installed on PVC conduit under the conditions described on pages 5 and 6. Costs listed are for each fitting installed. The crew is one electrician working at a labor cost of $41.03 per manhour. These costs include applying cement (glue), removal of knockouts, layout, material handling, and normal waste. Add for conduit, couplings, connectors, end bells, spacers, sales tax, delivery, supervision, mobilization, demobilization, cleanup, overhead and profit. Note: Material costs assume purchase of full box quantities.

PVC Expansion Couplings, End Bells, Caps and Plugs

6" range expansion PVC couplings

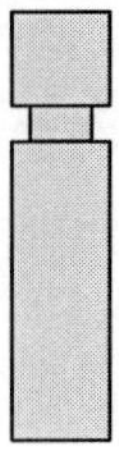

Material	Craft@Hrs	Unit	Material Cost	Labor Cost	Installed Cost
1/2"	L1@0.15	Ea	71.50	6.15	77.65
3/4"	L1@0.20	Ea	73.00	8.21	81.21
1"	L1@0.25	Ea	77.20	10.30	87.50
1-1/4"	L1@0.30	Ea	78.80	12.30	91.10
1-1/2"	L1@0.30	Ea	83.30	12.30	95.60
2"	L1@0.40	Ea	90.30	16.40	106.70
2-1/2"	L1@0.40	Ea	92.50	16.40	108.90
3"	L1@0.50	Ea	105.00	20.50	125.50
4"	L1@0.60	Ea	155.00	24.60	179.60
5"	L1@0.70	Ea	194.00	28.70	222.70
6"	L1@0.75	Ea	245.00	30.80	275.80

PVC end bells

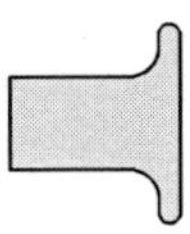

1"	L1@0.10	Ea	12.70	4.10	16.80
1-1/4"	L1@0.15	Ea	15.70	6.15	21.85
1-1/2"	L1@0.15	Ea	15.80	6.15	21.95
2"	L1@0.20	Ea	23.50	8.21	31.71
2-1/2"	L1@0.20	Ea	25.70	8.21	33.91
3"	L1@0.25	Ea	27.20	10.30	37.50
4"	L1@0.30	Ea	32.60	12.30	44.90
5"	L1@0.35	Ea	51.30	14.40	65.70
6"	L1@0.40	Ea	56.10	16.40	72.50

PVC caps and plugs

1/2" caps	L1@0.05	Ea	3.96	2.05	6.01
3/4" caps	L1@0.06	Ea	4.87	2.46	7.33
1" caps	L1@0.08	Ea	5.19	3.28	8.47
1-1/4" caps	L1@0.10	Ea	7.09	4.10	11.19
1-1/2" plugs	L1@0.10	Ea	8.41	4.10	12.51
2" plugs	L1@0.10	Ea	9.10	4.10	13.20
2-1/2" plugs	L1@0.10	Ea	9.78	4.10	13.88
3" plugs	L1@0.15	Ea	7.16	6.15	13.31
4" plugs	L1@0.15	Ea	15.00	6.15	21.15
5" plugs	L1@0.20	Ea	67.20	8.21	75.41
6" plugs	L1@0.20	Ea	112.00	8.21	120.21

Use these figures to estimate the cost of PVC fittings installed on PVC conduit under the conditions described on pages 5 and 6. Costs listed are for each fitting installed. The crew is one electrician working at a labor cost of $41.03 per manhour. These costs include applying cement (glue), removal of knockouts, layout, material handling, and normal waste. Add for conduit, locknuts, insulated bushings, sales tax, delivery, supervision, mobilization, demobilization, cleanup, overhead and profit. Note: Material costs assume purchase of full box quantities.

PVC reducing bushings

Material	Craft@Hrs	Unit	Material Cost	Labor Cost	Installed Cost
3/4" to 1/2"	L1@0.03	Ea	4.80	1.23	6.03
1" to 1/2"	L1@0.03	Ea	5.29	1.23	6.52
1" to 3/4"	L1@0.03	Ea	5.50	1.23	6.73
1-1/4" to 3/4"	L1@0.05	Ea	5.77	2.05	7.82
1-1/4" to 1"	L1@0.05	Ea	5.86	2.05	7.91
1-1/2" to 1"	L1@0.10	Ea	6.13	4.10	10.23
1-1/2" to 1-1/4"	L1@0.10	Ea	6.53	4.10	10.63
2" to 1-1/4"	L1@0.15	Ea	6.79	6.15	12.94
2-1/2" to 2"	L1@0.15	Ea	7.60	6.15	13.75
3" to 2"	L1@0.20	Ea	22.80	8.21	31.01
4" to 3"	L1@0.25	Ea	26.90	10.30	37.20

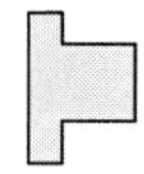

Type C PVC conduit bodies

Material	Craft@Hrs	Unit	Material Cost	Labor Cost	Installed Cost
C 1/2"	L1@0.10	Ea	22.80	4.10	26.90
C 3/4"	L1@0.10	Ea	27.90	4.10	32.00
C 1"	L1@0.15	Ea	29.30	6.15	35.45
C 1-1/4"	L1@0.15	Ea	47.40	6.15	53.55
C 1-1/2"	L1@0.20	Ea	62.50	8.21	70.71
C 2"	L1@0.25	Ea	88.50	10.30	98.80

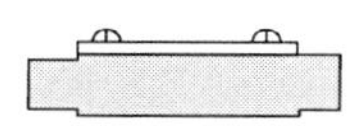

Type E PVC conduit bodies

Material	Craft@Hrs	Unit	Material Cost	Labor Cost	Installed Cost
E 1/2"	L1@0.10	Ea	18.30	4.10	22.40
E 3/4"	L1@0.10	Ea	27.10	4.10	31.20
E 1"	L1@0.15	Ea	32.00	6.15	38.15
E 1-1/4"	L1@0.15	Ea	39.60	6.15	45.75
E 1-1/2"	L1@0.20	Ea	47.40	8.21	55.61
E 2"	L1@0.25	Ea	84.00	10.30	94.30

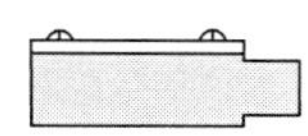

Type LB PVC conduit bodies

Material	Craft@Hrs	Unit	Material Cost	Labor Cost	Installed Cost
LB 1/2"	L1@0.10	Ea	17.50	4.10	21.60
LB 3/4"	L1@0.10	Ea	22.80	4.10	26.90
LB 1"	L1@0.15	Ea	25.00	6.15	31.15
LB 1-1/4"	L1@0.15	Ea	37.80	6.15	43.95

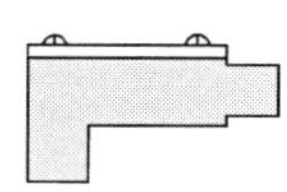

Use these figures to estimate the cost of PVC fittings installed on PVC conduit under the conditions described on pages 5 and 6. Costs listed are for each fitting installed. The crew is one electrician working at a labor cost of $41.03 per manhour. These costs include applying cement (glue), removal of knockouts, layout, material handling, and normal waste. Add for conduit, locknuts, insulated bushings, sales tax, delivery, supervision, mobilization, demobilization, cleanup, overhead and profit. Note: Material costs assume purchase of full box quantities.

PVC Conduit Bodies and Service Entrance Caps

Type LB PVC conduit bodies

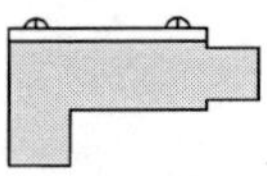

Material	Craft@Hrs	Unit	Material Cost	Labor Cost	Installed Cost
LB 1-1/2"	L1@0.20	Ea	45.60	8.21	53.81
LB 2"	L1@0.25	Ea	80.70	10.30	91.00
LB 2-1/2"	L1@0.30	Ea	294.00	12.30	306.30
LB 3"	L1@0.30	Ea	301.00	12.30	313.30
LB 4"	L1@0.40	Ea	329.00	16.40	345.40

Type LL PVC conduit bodies

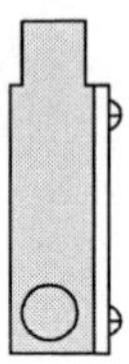

Material	Craft@Hrs	Unit	Material Cost	Labor Cost	Installed Cost
LL 1/2"	L1@0.10	Ea	18.10	4.10	22.20
LL 3/4"	L1@0.10	Ea	27.10	4.10	31.20
LL 1"	L1@0.15	Ea	27.90	6.15	34.05
LL 1-1/4"	L1@0.15	Ea	40.30	6.15	46.45
LL 1-1/2"	L1@0.20	Ea	47.40	8.21	55.61
LL 2"	L1@0.25	Ea	82.20	10.30	92.50

Type LR PVC conduit bodies

Material	Craft@Hrs	Unit	Material Cost	Labor Cost	Installed Cost
LR 1/2"	L1@0.10	Ea	18.10	4.10	22.20
LR 3/4"	L1@0.10	Ea	27.10	4.10	31.20
LR 1"	L1@0.15	Ea	27.90	6.15	34.05
LR 1-1/4"	L1@0.15	Ea	40.30	6.15	46.45
LR 1-1/2"	L1@0.20	Ea	47.40	8.21	55.61
LR 2"	L1@0.25	Ea	82.20	10.30	92.50

Type T PVC conduit bodies

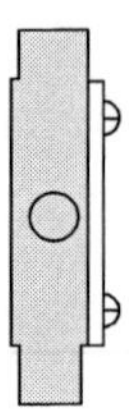

Material	Craft@Hrs	Unit	Material Cost	Labor Cost	Installed Cost
T 1/2"	L1@0.10	Ea	22.80	4.10	26.90
T 3/4"	L1@0.15	Ea	27.90	6.15	34.05
T 1"	L1@0.15	Ea	29.30	6.15	35.45
T 1-1/4"	L1@0.20	Ea	48.80	8.21	57.01
T 1-1/2"	L1@0.25	Ea	62.50	10.30	72.80
T 2"	L1@0.30	Ea	88.50	12.30	100.80

PVC slip-fitter entrance caps

Material	Craft@Hrs	Unit	Material Cost	Labor Cost	Installed Cost
3/4"	L1@0.15	Ea	20.40	6.15	26.55
1"	L1@0.15	Ea	27.20	6.15	33.35
1-1/4"	L1@0.25	Ea	33.80	10.30	44.10
1-1/2"	L1@0.30	Ea	40.80	12.30	53.10
2"	L1@0.50	Ea	68.60	20.50	89.10
2-1/2"	L1@0.60	Ea	353.00	24.60	377.60
3"	L1@0.60	Ea	370.00	24.60	394.60
4"	L1@0.75	Ea	1,030.00	30.80	1,060.80

Use these figures to estimate the cost of PVC fittings installed on PVC conduit under the conditions described on pages 5 and 6. Costs listed are for each fitting installed. The crew is one electrician working at a labor cost of $41.03 per manhour. These costs include applying cement (glue), removal of knockouts, layout, material handling, and normal waste. Add for conduit, locknuts, insulated bushings, sales tax, delivery, supervision, mobilization, demobilization, cleanup, overhead and profit. Note: Material costs assume purchase of full box quantities.

Type FS PVC boxes

Material	Craft@Hrs	Unit	Material Cost	Labor Cost	Installed Cost
FS1 1/2"	L1@0.20	Ea	40.00	8.21	48.21
FS2 3/4"	L1@0.20	Ea	40.00	8.21	48.21
FS3 1"	L1@0.25	Ea	40.00	10.30	50.30
FSC1 1/2"	L1@0.25	Ea	43.70	10.30	54.00
FSC2 3/4"	L1@0.25	Ea	43.70	10.30	54.00
FSC3 1"	L1@0.30	Ea	43.70	12.30	56.00
FSS1 1/2"	L1@0.25	Ea	43.70	10.30	54.00
FSS2 3/4"	L1@0.25	Ea	43.70	10.30	54.00
FSS3 1"	L1@0.30	Ea	43.70	12.30	56.00
FCSS1 1/2"	L1@0.30	Ea	44.70	12.30	57.00
FCSS2 3/4"	L1@0.30	Ea	44.70	12.30	57.00
FCSS3 1"	L1@0.35	Ea	44.70	14.40	59.10

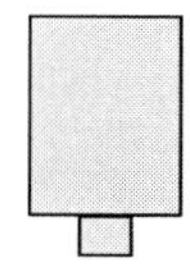

Type FS, WP PVC box covers

1 gang blank	L1@0.10	Ea	8.27	4.10	12.37
1 gang single outlet	L1@0.10	Ea	10.80	4.10	14.90
1 gang duplex outlet	L1@0.10	Ea	17.40	4.10	21.50
1 gang single switch	L1@0.10	Ea	17.40	4.10	21.50
1 gang GFCI	L1@0.10	Ea	17.40	4.10	21.50

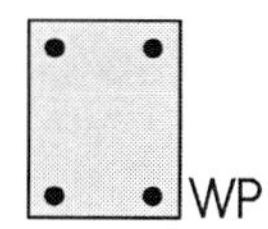

PVC junction boxes

4" x 4" x 2"	L1@0.25	Ea	55.00	10.30	65.30
4" x 4" x 4"	L1@0.25	Ea	90.20	10.30	100.50
4" x 4" x 6"	L1@0.30	Ea	104.00	12.30	116.30
5" x 5" x 2"	L1@0.30	Ea	108.00	12.30	120.30
6" x 6" x 4"	L1@0.35	Ea	111.00	14.40	125.40
6" x 6" x 6"	L1@0.40	Ea	129.00	16.40	145.40
8" x 8" x 4"	L1@0.40	Ea	210.00	16.40	226.40
8" x 8" x 7"	L1@0.50	Ea	309.00	20.50	329.50
12" x 12" x 4"	L1@0.70	Ea	323.00	28.70	351.70
12" x 12" x 6"	L1@0.75	Ea	329.00	30.80	359.80

30 degree sweeping PVC elbows

2" 24" radius	L1@0.15	Ea	78.80	6.15	84.95
2" 36" radius	L1@0.20	Ea	88.80	8.21	97.01
2" 48" radius	L1@0.25	Ea	98.70	10.30	109.00
3" 24" radius	L1@0.20	Ea	157.00	8.21	165.21
3" 36" radius	L1@0.25	Ea	169.00	10.30	179.30
3" 48" radius	L1@0.30	Ea	181.00	12.30	193.30
4" 24" radius	L1@0.25	Ea	228.00	10.30	238.30
4" 36" radius	L1@0.30	Ea	269.00	12.30	281.30
4" 48" radius	L1@0.40	Ea	308.00	16.40	324.40

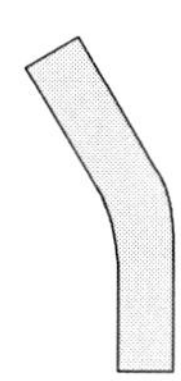

Use these figures to estimate the cost of PVC fittings installed on PVC conduit under the conditions described on pages 5 and 6. Costs listed are for each fitting installed. The crew is one electrician working at a labor cost of $41.03 per manhour. These costs include applying cement (glue), removal of knockouts, layout, material handling, and normal waste. Add for conduit, locknuts, insulated bushings, sales tax, delivery, supervision, mobilization, demobilization, cleanup, overhead and profit. Note: Material costs assume purchase of full box quantities.

PVC Elbows

30 degree sweeping PVC elbows

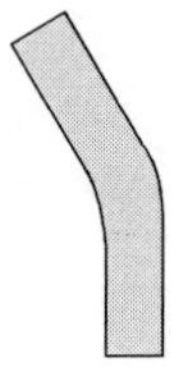

Material		Craft@Hrs	Unit	Material Cost	Labor Cost	Installed Cost
5"	36" radius	L1@0.40	Ea	403.00	16.40	419.40
5"	48" radius	L1@0.50	Ea	464.00	20.50	484.50
6"	36" radius	L1@0.75	Ea	653.00	30.80	683.80
6"	48" radius	L1@1.00	Ea	702.00	41.00	743.00

45 degree sweeping PVC elbows

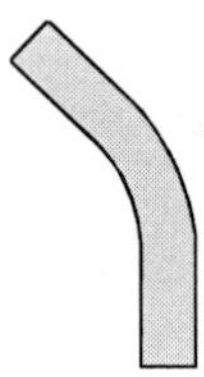

Material		Craft@Hrs	Unit	Material Cost	Labor Cost	Installed Cost
2"	24" radius	L1@0.15	Ea	78.80	6.15	84.95
2"	30" radius	L1@0.15	Ea	84.00	6.15	90.15
2"	36" radius	L1@0.20	Ea	88.80	8.21	97.01
2"	48" radius	L1@0.30	Ea	98.70	12.30	111.00
2-1/2"	30" radius	L1@0.20	Ea	118.00	8.21	126.21
2-1/2"	36" radius	L1@0.25	Ea	128.00	10.30	138.30
2-1/2"	48" radius	L1@0.30	Ea	140.00	12.30	152.30
3"	24" radius	L1@0.25	Ea	157.00	10.30	167.30
3"	30" radius	L1@0.30	Ea	164.00	12.30	176.30
3"	36" radius	L1@0.30	Ea	169.00	12.30	181.30
3"	48" radius	L1@0.40	Ea	181.00	16.40	197.40
4"	24" radius	L1@0.30	Ea	228.00	12.30	240.30
4"	30" radius	L1@0.30	Ea	249.00	12.30	261.30
4"	36" radius	L1@0.35	Ea	269.00	14.40	283.40
4"	48" radius	L1@0.40	Ea	308.00	16.40	324.40
5"	30" radius	L1@0.35	Ea	367.00	14.40	381.40
5"	36" radius	L1@0.40	Ea	403.00	16.40	419.40
5"	48" radius	L1@0.50	Ea	463.00	20.50	483.50
6"	36" radius	L1@0.75	Ea	653.00	30.80	683.80
6"	48" radius	L1@1.00	Ea	702.00	41.00	743.00

90 degree sweeping PVC elbows

Material		Craft@Hrs	Unit	Material Cost	Labor Cost	Installed Cost
2"	24" radius	L1@0.20	Ea	84.00	8.21	92.21
2"	30" radius	L1@0.25	Ea	88.80	10.30	99.10
2"	36" radius	L1@0.30	Ea	98.70	12.30	111.00
2"	48" radius	L1@0.35	Ea	104.00	14.40	118.40
2-1/2"	30" radius	L1@0.30	Ea	91.00	12.30	103.30
2-1/2"	36" radius	L1@0.35	Ea	97.40	14.40	111.80
2-1/2"	48" radius	L1@0.40	Ea	105.00	16.40	121.40

Use these figures and the table at the top of the next page to estimate the cost of PVC sweeps installed on PVC conduit under the conditions described on pages 5 and 6. Costs listed are for each sweep installed. The crew is one electrician working at a labor cost of $41.03 per manhour. These costs include applying cement (glue), layout, material handling, and normal waste. Add for couplings, connectors, end bells, spacers, sales tax, delivery, supervision, mobilization, demobilization, cleanup, overhead and profit. Note: Material costs assume purchase of full packages.

90 degree sweeping PVC elbows

Material	Craft@Hrs	Unit	Material Cost	Labor Cost	Installed Cost
3" 24" radius	L2@0.30	Ea	157.00	12.30	169.30
3" 30" radius	L2@0.35	Ea	162.00	14.40	176.40
3" 36" radius	L2@0.40	Ea	169.00	16.40	185.40
3" 48" radius	L2@0.50	Ea	174.00	20.50	194.50
4" 24" radius	L2@0.35	Ea	228.00	14.40	242.40
4" 30" radius	L2@0.40	Ea	249.00	16.40	265.40
4" 36" radius	L2@0.45	Ea	269.00	18.50	287.50
4" 48" radius	L2@0.55	Ea	308.00	22.60	330.60
5" 30" radius	L2@0.45	Ea	367.00	18.50	385.50
5" 36" radius	L2@0.50	Ea	403.00	20.50	423.50
5" 48" radius	L2@0.60	Ea	463.00	24.60	487.60
6" 36" radius	L2@0.60	Ea	653.00	24.60	677.60
6" 48" radius	L2@0.75	Ea	702.00	30.80	732.80

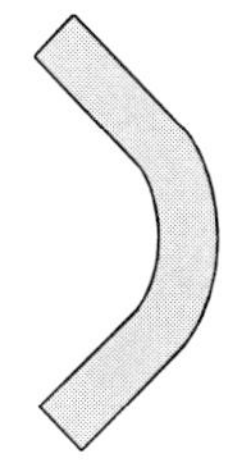

Type EB power and communication duct

Material	Craft@Hrs	Unit	Material Cost	Labor Cost	Installed Cost
2"	L2@3.30	CLF	327.00	135.00	462.00
3"	L2@3.50	CLF	477.00	144.00	621.00
4"	L2@4.00	CLF	778.00	164.00	942.00
5"	L2@4.50	CLF	1,180.00	185.00	1,365.00
6"	L2@5.00	CLF	1,690.00	205.00	1,895.00

Type DB power and communication duct

Material	Craft@Hrs	Unit	Material Cost	Labor Cost	Installed Cost
2"	L2@3.30	CLF	367.00	135.00	502.00
4"	L2@4.00	CLF	1,080.00	164.00	1,244.00
5"	L2@4.50	CLF	1,180.00	185.00	1,365.00
6"	L2@5.00	CLF	1,810.00	205.00	2,015.00

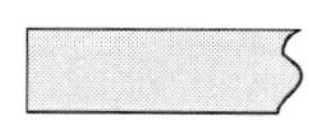

Type EB or DB power and communication duct couplings

Material	Craft@Hrs	Unit	Material Cost	Labor Cost	Installed Cost
2"	L1@0.05	Ea	4.88	2.05	6.93
3"	L1@0.10	Ea	11.20	4.10	15.30
4"	L1@0.10	Ea	17.50	4.10	21.60
5"	L1@0.15	Ea	32.20	6.15	38.35
6"	L1@0.15	Ea	98.50	6.15	104.65

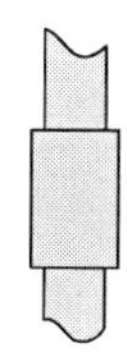

45 degree Type EB or DB power and communication duct elbows

Material	Craft@Hrs	Unit	Material Cost	Labor Cost	Installed Cost
2" 24" radius	L1@0.15	Ea	33.40	6.15	39.55
3" 36" radius	L1@0.30	Ea	46.30	12.30	58.60
3" 48" radius	L1@0.40	Ea	75.90	16.40	92.30
4" 36" radius	L1@0.40	Ea	60.90	16.40	77.30
4" 48" radius	L1@0.75	Ea	85.50	30.80	116.30
5" 48" radius	L1@0.50	Ea	110.00	20.50	130.50

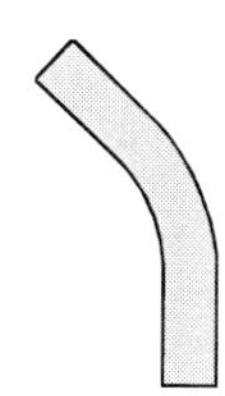

Use these figures to estimate the cost of PVC elbows (top table) and power and communication duct couplings and elbows (bottom tables). The footnote on the previous page applies to PVC sweep elbows. P&C duct is installed underground under the conditions described on pages 5 and 6. Costs listed are for each 100 linear feet installed. The crew is two electricians working at a labor cost of $41.03 per manhour. These costs include one coupling, applying cement (glue), multiple runs in the same trench, layout, material handling, and normal waste. Add for trenching, encasement, spacers and chairs, single duct runs, sales tax, delivery, supervision, mobilization, demobilization, cleanup, overhead and profit. Note: Encased burial requires spacers or chairs every 5 feet. Costs for spacers, chairs, encasement and trenching are listed elsewhere in this manual.

Power & Communication Duct Couplings, Elbows & Adapters

90 degree Type EB or DB power and communication duct elbows

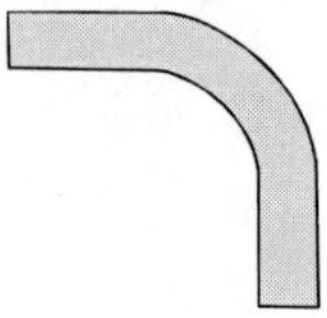

Material	Craft@Hrs	Unit	Material Cost	Labor Cost	Installed Cost
2" 18" radius	L1@0.20	Ea	26.90	8.21	35.11
2" 24" radius	L1@0.30	Ea	34.20	12.30	46.50
2" 36" radius	L1@0.35	Ea	46.30	14.40	60.70
3" 24" radius	L1@0.30	Ea	46.30	12.30	58.60
3" 36" radius	L1@0.40	Ea	48.80	16.40	65.20
3" 48" radius	L1@0.50	Ea	51.70	20.50	72.20
4" 24" radius	L1@0.35	Ea	39.20	14.40	53.60
4" 36" radius	L1@0.45	Ea	55.40	18.50	73.90
4" 48" radius	L1@0.55	Ea	154.00	22.60	176.60
5" 36" radius	L1@0.50	Ea	131.00	20.50	151.50
5" 48" radius	L1@0.60	Ea	172.00	24.60	196.60
6" 48" radius	L1@0.75	Ea	227.00	30.80	257.80

5 degree power and communication bend couplings

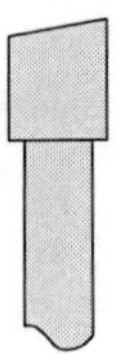

Material	Craft@Hrs	Unit	Material Cost	Labor Cost	Installed Cost
2"	L1@0.05	Ea	40.30	2.05	42.35
3"	L1@0.10	Ea	50.70	4.10	54.80
4"	L1@0.10	Ea	60.20	4.10	64.30
5"	L1@0.15	Ea	65.30	6.15	71.45
6"	L1@0.15	Ea	67.50	6.15	73.65

Power and communication duct plugs

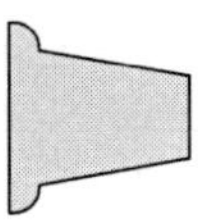

Material	Craft@Hrs	Unit	Material Cost	Labor Cost	Installed Cost
2"	L1@0.05	Ea	7.08	2.05	9.13
3"	L1@0.10	Ea	10.40	4.10	14.50
4"	L1@0.10	Ea	11.80	4.10	15.90
5"	L1@0.15	Ea	15.90	6.15	22.05
6"	L1@0.15	Ea	20.10	6.15	26.25

Type FA female power and communication duct adapters

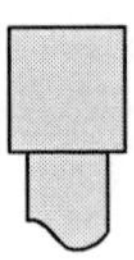

Material	Craft@Hrs	Unit	Material Cost	Labor Cost	Installed Cost
2"	L1@0.20	Ea	4.33	8.21	12.54
3"	L1@0.25	Ea	11.90	10.30	22.20
4"	L1@0.30	Ea	15.90	12.30	28.20
5"	L1@0.40	Ea	39.90	16.40	56.30
6"	L1@0.50	Ea	52.40	20.50	72.90

Type TA terminal power and communication adapters

Material	Craft@Hrs	Unit	Material Cost	Labor Cost	Installed Cost
2"	L1@0.20	Ea	4.36	8.21	12.57
3"	L1@0.25	Ea	10.80	10.30	21.10
4"	L1@0.30	Ea	18.60	12.30	30.90
5"	L1@0.40	Ea	36.70	16.40	53.10
6"	L1@0.50	Ea	43.90	20.50	64.40

Use these figures to estimate the cost of PVC fittings installed on PVC power and communication duct under the conditions described on pages 5 and 6. Costs listed are for each fitting installed. The crew is one electrician working at a labor cost of $41.03 per manhour. These costs include cutting and fitting, applying cement (glue), layout, material handling, and normal waste. Add for extra couplings, sales tax, delivery, supervision, mobilization, demobilization, cleanup, overhead and profit. Note: Material costs are based on purchase of full packages. All of these fittings can be used either on type EB or type DB duct.

Material	Craft@Hrs	Unit	Material Cost	Labor Cost	Installed Cost

Power and communication duct end bells

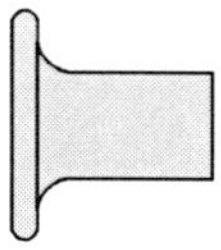

2"	L1@0.15	Ea	22.70	6.15	28.85
3"	L1@0.20	Ea	27.10	8.21	35.31
4"	L1@0.25	Ea	32.60	10.30	42.90
5"	L1@0.30	Ea	51.30	12.30	63.60
6"	L1@0.50	Ea	56.10	20.50	76.60

Base type plastic duct spacers

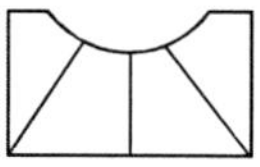

2" 1-1/2" separation	L1@0.05	Ea	4.33	2.05	6.38
3" 1-1/2" separation	L1@0.05	Ea	4.70	2.05	6.75
4" 1-1/2" separation	L1@0.05	Ea	5.19	2.05	7.24
5" 1-1/2" separation	L1@0.05	Ea	5.61	2.05	7.66
6" 1-1/2" separation	L1@0.05	Ea	9.03	2.05	11.08
2" 2" separation	L1@0.05	Ea	4.70	2.05	6.75
3" 2" separation	L1@0.05	Ea	5.19	2.05	7.24
4" 2" separation	L1@0.05	Ea	5.54	2.05	7.59
5" 2" separation	L1@0.05	Ea	5.77	2.05	7.82
6" 2" separation	L1@0.05	Ea	9.61	2.05	11.66
2" 3" separation	L1@0.05	Ea	5.07	2.05	7.12
3" 3" separation	L1@0.05	Ea	5.62	2.05	7.67
4" 3" separation	L1@0.05	Ea	5.95	2.05	8.00
5" 3" separation	L1@0.05	Ea	7.30	2.05	9.35
6" 3" separation	L1@0.05	Ea	10.20	2.05	12.25

Intermediate type plastic duct spacers

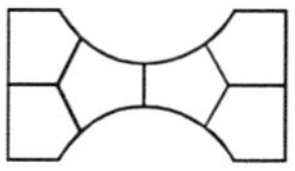

2" 1-1/2" separation	L1@0.05	Ea	4.33	2.05	6.38
3" 1-1/2" separation	L1@0.05	Ea	4.70	2.05	6.75
4" 1-1/2" separation	L1@0.05	Ea	5.19	2.05	7.24
5" 1-1/2" separation	L1@0.05	Ea	5.61	2.05	7.66
6" 1-1/2" separation	L1@0.05	Ea	9.03	2.05	11.08
2" 2" separation	L1@0.05	Ea	4.70	2.05	6.75
3" 2" separation	L1@0.05	Ea	5.19	2.05	7.24
4" 2" separation	L1@0.05	Ea	5.54	2.05	7.59
5" 2" separation	L1@0.05	Ea	5.77	2.05	7.82
6" 2" separation	L1@0.05	Ea	9.61	2.05	11.66
2" 3" separation	L1@0.05	Ea	5.07	2.05	7.12
3" 3" separation	L1@0.05	Ea	5.62	2.05	7.67
4" 3" separation	L1@0.05	Ea	5.95	2.05	8.00
5" 3" separation	L1@0.05	Ea	7.30	2.05	9.35
6" 3" separation	L1@0.05	Ea	10.20	2.05	12.25

Use these figures to estimate the cost of P&C end bell or plastic spacer installed with duct systems under the conditions described on pages 5 and 6. Costs listed are for each end bell or spacer installed. The crew is one electrician working at a labor cost of $41.03 per manhour. These costs include ganging spacers, tying duct to the spacer, layout, material handling, and normal waste. Add for duct, other fittings, sales tax, delivery, supervision, mobilization, demobilization, cleanup, overhead and profit. Note: Material costs are based on purchase of full packages. Tie wire should never be tied completely around the duct, it should be tied in a figure 8 pattern through open spaces in the side of the spacer and over the top part of the duct. Running wire completely around the duct will cause the wire to pick up an induction field from the current passing through the conductor, generating heat which will weaken the insulation.

ENT Conduit and Fittings

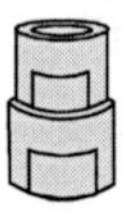

Material	Craft@Hrs	Unit	Material Cost	Labor Cost	Installed Cost
ENT conduit, non-metallic tubing					
1/2"	L1@2.15	CLF	148.00	88.20	236.20
3/4"	L1@2.25	CLF	204.00	92.30	296.30
1"	L1@2.50	CLF	327.00	103.00	430.00
ENT connectors					
1/2"	L1@0.03	Ea	4.07	1.23	5.30
3/4"	L1@0.04	Ea	6.25	1.64	7.89
1"	L1@0.05	Ea	9.43	2.05	11.48
ENT couplings					
1/2"	L1@0.03	Ea	2.86	1.23	4.09
3/4"	L1@0.04	Ea	3.77	1.64	5.41
1"	L1@0.05	Ea	6.61	2.05	8.66
ENT male adapters					
1/2"	L1@0.03	Ea	2.98	1.23	4.21
3/4"	L1@0.04	Ea	4.08	1.64	5.72
1"	L1@0.05	Ea	6.80	2.05	8.85

Use these figures to estimate the cost of ENT conduit and fittings installed under the conditions described on pages 5 and 6. Costs listed are for each 100 linear feet installed and for each fitting installed. The crew is one electrician working at a labor cost of $41.03 per manhour. These costs include cutting and fitting, applying cement (glue), layout, material handling, and normal waste. Add for extra couplings, sales tax, delivery, supervision, mobilization, demobilization, cleanup, overhead and profit. Note: Material costs are based on purchase of full packages.

Galvanized Rigid Steel (GRS) Conduit and Elbows

Standard wall galvanized rigid steel conduit

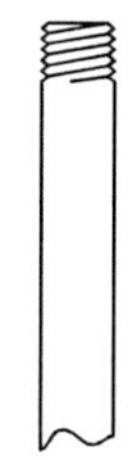

Material	Craft@Hrs	Unit	Material Cost	Labor Cost	Installed Cost
1/2"	L1@4.00	CLF	238.00	164.00	402.00
3/4"	L1@4.50	CLF	248.00	185.00	433.00
1"	L1@5.00	CLF	389.00	205.00	594.00
1-1/4"	L1@7.00	CLF	404.00	287.00	691.00
1-1/2"	L1@8.00	CLF	654.00	328.00	982.00
2"	L1@10.0	CLF	787.00	410.00	1,197.00
2-1/2"	L1@12.0	CLF	1,570.00	492.00	2,062.00
3"	L1@14.0	CLF	1,730.00	574.00	2,304.00
3-1/2"	L1@16.0	CLF	2,640.00	656.00	3,296.00
4"	L1@18.0	CLF	2,390.00	739.00	3,129.00
5"	L1@25.0	CLF	4,810.00	1,030.00	5,840.00
6"	L1@30.0	CLF	5,700.00	1,230.00	6,930.00

45 degree galvanized rigid steel elbows

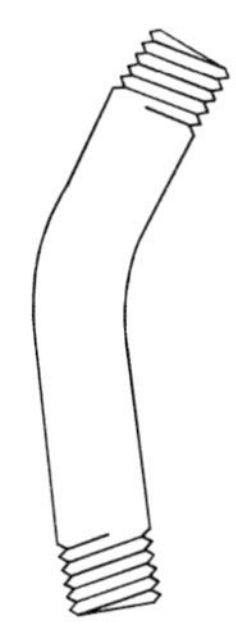

Material	Craft@Hrs	Unit	Material Cost	Labor Cost	Installed Cost
1/2"	L1@0.10	Ea	20.70	4.10	24.80
3/4"	L1@0.10	Ea	21.40	4.10	25.50
1"	L1@0.12	Ea	33.20	4.92	38.12
1-1/4"	L1@0.15	Ea	45.40	6.15	51.55
1-1/2"	L1@0.15	Ea	55.80	6.15	61.95
2"	L1@0.20	Ea	81.10	8.21	89.31
2-1/2"	L1@0.25	Ea	151.00	10.30	161.30
3"	L1@0.25	Ea	208.00	10.30	218.30
3-1/2"	L1@0.30	Ea	333.00	12.30	345.30
4"	L1@0.30	Ea	374.00	12.30	386.30
5"	L1@0.50	Ea	1,040.00	20.50	1,060.50
6"	L1@1.00	Ea	1,570.00	41.00	1,611.00

90 degree galvanized rigid steel elbows

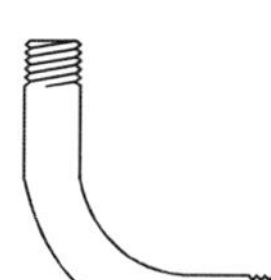

Material	Craft@Hrs	Unit	Material Cost	Labor Cost	Installed Cost
1/2"	L1@0.10	Ea	12.80	4.10	16.90
3/4"	L1@0.10	Ea	14.10	4.10	18.20
1"	L1@0.12	Ea	21.50	4.92	26.42
1-1/4"	L1@0.15	Ea	26.50	6.15	32.65
1-1/2"	L1@0.15	Ea	30.80	6.15	36.95
2"	L1@0.20	Ea	45.30	8.21	53.51
2-1/2"	L1@0.25	Ea	110.00	10.30	120.30
3"	L1@0.25	Ea	162.00	10.30	172.30
3-1/2"	L1@0.30	Ea	221.00	12.30	233.30
4"	L1@0.30	Ea	431.00	12.30	443.30
5"	L1@0.50	Ea	547.00	20.50	567.50
6"	L1@1.00	Ea	750.00	41.00	791.00

Use these figures to estimate the cost of GRS conduit and elbows installed in buildings under the conditions described on pages 5 and 6. Costs listed are for each 100 linear feet of conduit or for each elbow installed. The crew is one electrician working at a labor cost of $41.03 per manhour. These costs include one coupling on each length of conduit, threading, cutting, straps, layout, material handling, and normal waste. Add for other fittings, boxes, wires, sales tax, delivery, supervision, mobilization, demobilization, cleanup, overhead and profit. Note: Couplings are not included with elbows. The elbows listed are factory made and have a standard radius. Conduit runs are assumed to be 50' long. Installation costs per linear foot will be less on longer runs and more on shorter runs.

GRS Hand Benders are on page 52.

Galvanized Rigid Steel Large Radius Elbows

Material		Craft@Hrs	Unit	Material Cost	Labor Cost	Installed Cost
90 degree galvanized rigid steel large radius elbows						
1"	12" radius	L1@0.10	Ea	40.50	4.10	44.60
1-1/4"	12" radius	L1@0.15	Ea	47.50	6.15	53.65
1-1/2"	12" radius	L1@0.15	Ea	56.30	6.15	62.45
2"	12" radius	L1@0.20	Ea	72.90	8.21	81.11
2-1/2"	12" radius	L2@0.20	Ea	99.90	8.21	108.11
1"	15" radius	L1@0.10	Ea	42.10	4.10	46.20
1-1/4"	15" radius	L1@0.15	Ea	47.90	6.15	54.05
1-1/2"	15" radius	L1@0.15	Ea	58.90	6.15	65.05
2"	15" radius	L1@0.20	Ea	67.90	8.21	76.11
2-1/2"	15" radius	L2@0.20	Ea	91.70	8.21	99.91
1"	18" radius	L1@0.10	Ea	44.30	4.10	48.40
1-1/4"	18" radius	L1@0.15	Ea	48.70	6.15	54.85
1-1/2"	18" radius	L1@0.15	Ea	58.10	6.15	64.25
2"	18" radius	L1@0.20	Ea	70.70	8.21	78.91
2-1/2"	18" radius	L2@0.20	Ea	94.30	8.21	102.51
3"	18" radius	L2@0.25	Ea	120.00	10.30	130.30
3-1/2"	18" radius	L2@0.25	Ea	137.00	10.30	147.30
4"	18" radius	L2@0.30	Ea	150.00	12.30	162.30
1"	24" radius	L1@0.10	Ea	47.90	4.10	52.00
1-1/4"	24" radius	L1@0.15	Ea	50.60	6.15	56.75
1-1/2"	24" radius	L1@0.15	Ea	61.00	6.15	67.15
2"	24" radius	L1@0.20	Ea	75.00	8.21	83.21
2-1/2"	24" radius	L2@0.20	Ea	99.90	8.21	108.11
3"	24" radius	L2@0.25	Ea	146.00	10.30	156.30
3-1/2"	24" radius	L2@0.25	Ea	196.00	10.30	206.30
4"	24" radius	L2@0.30	Ea	223.00	12.30	235.30
1"	30" radius	L1@0.15	Ea	118.00	6.15	124.15
1-1/4"	30" radius	L1@0.20	Ea	123.00	8.21	131.21
1-1/2"	30" radius	L1@0.20	Ea	157.00	8.21	165.21
2"	30" radius	L1@0.25	Ea	192.00	10.30	202.30
2-1/2"	30" radius	L2@0.25	Ea	148.00	10.30	158.30
3"	30" radius	L2@0.30	Ea	192.00	12.30	204.30
3-1/2"	30" radius	L2@0.30	Ea	227.00	12.30	239.30
4"	30" radius	L2@0.35	Ea	282.00	14.40	296.40
5"	30" radius	L2@0.50	Ea	395.00	20.50	415.50
1"	36" radius	L1@0.20	Ea	92.70	8.21	100.91
1-1/4"	36" radius	L1@0.25	Ea	143.00	10.30	153.30
1-1/2"	36" radius	L1@0.25	Ea	171.00	10.30	181.30
2"	36" radius	L1@0.30	Ea	223.00	12.30	235.30
2-1/2"	36" radius	L2@0.30	Ea	457.00	12.30	469.30
3"	36" radius	L2@0.35	Ea	529.00	14.40	543.40

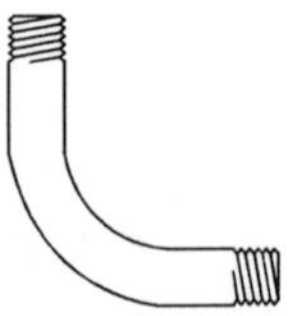

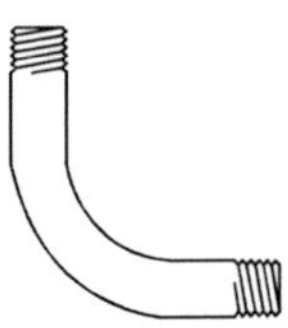

Use these figures to estimate the cost of large radius GRS elbows installed on GRS conduit under the conditions described on pages 5 and 6. Costs listed are for each elbow installed. The crew is one electrician for size to 2" and two electricians for sizes over 2". The labor cost is $41.03 per manhour. These costs include layout, material handling, and normal waste. Add for other GRS fittings, conduit, field bending, sales tax, delivery, supervision, mobilization, demobilization, cleanup, overhead and profit. Note: All elbows are assumed to be factory made.

Material		Craft@Hrs	Unit	Material Cost	Labor Cost	Installed Cost
90 degree galvanized rigid steel large radius elbows						
3-1/2"	36" radius	L2@0.35	Ea	242.00	14.40	256.40
4"	36" radius	L2@0.40	Ea	264.00	16.40	280.40
5"	36" radius	L2@0.60	Ea	491.00	24.60	515.60
6"	36" radius	L2@1.00	Ea	549.00	41.00	590.00
1"	42" radius	L1@0.25	Ea	126.00	10.30	136.30
1-1/4"	42" radius	L1@0.30	Ea	153.00	12.30	165.30
1-1/2"	42" radius	L1@0.30	Ea	171.00	12.30	183.30
2"	42" radius	L1@0.35	Ea	227.00	14.40	241.40
2-1/2"	42" radius	L2@0.35	Ea	310.00	14.40	324.40
3"	42" radius	L2@0.40	Ea	403.00	16.40	419.40
3-1/2"	42" radius	L2@0.40	Ea	538.00	16.40	554.40
4"	42" radius	L2@0.50	Ea	369.00	20.50	389.50
5"	42" radius	L2@0.75	Ea	715.00	30.80	745.80
6"	42" radius	L2@1.25	Ea	750.00	51.30	801.30
1"	48" radius	L1@0.30	Ea	148.00	12.30	160.30
1-1/4"	48" radius	L1@0.35	Ea	171.00	14.40	185.40
1-1/2"	48" radius	L1@0.35	Ea	186.00	14.40	200.40
2"	48" radius	L1@0.40	Ea	284.00	16.40	300.40
2-1/2"	48" radius	L2@0.40	Ea	359.00	16.40	375.40
3"	48" radius	L2@0.50	Ea	554.00	20.50	574.50
3-1/2"	48" radius	L2@0.50	Ea	690.00	20.50	710.50
4"	48" radius	L2@0.70	Ea	866.00	28.70	894.70
5"	48" radius	L2@1.00	Ea	1,160.00	41.00	1,201.00
6"	48" radius	L2@1.50	Ea	1,200.00	61.50	1,261.50
Galvanized rigid steel couplings						
1/2"		L1@0.05	Ea	1.24	2.05	3.29
3/4"		L1@0.06	Ea	1.50	2.46	3.96
1"		L1@0.08	Ea	2.21	3.28	5.49
1-1/4"		L1@0.10	Ea	3.77	4.10	7.87
1-1/2"		L1@0.10	Ea	4.75	4.10	8.85
2"		L1@0.15	Ea	6.26	6.15	12.41
2-1/2"		L2@0.15	Ea	14.60	6.15	20.75
3"		L2@0.20	Ea	19.20	8.21	27.41
3-1/2"		L2@0.20	Ea	25.70	8.21	33.91
4"		L2@0.25	Ea	68.10	10.30	78.40
5"		L2@0.30	Ea	144.00	12.30	156.30
6"		L2@0.50	Ea	220.00	20.50	240.50

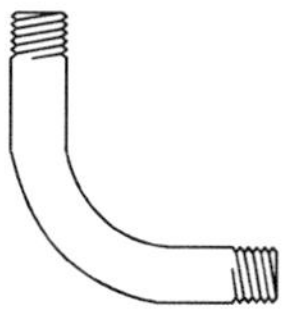

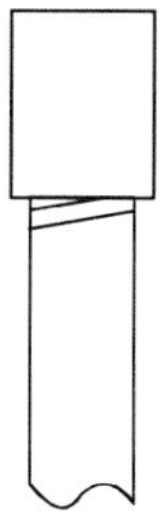

Use these figures to estimate the cost of large radius GRS elbows and couplings installed on GRS conduit under the conditions described on pages 5 and 6. Costs listed are for each elbow or coupling installed. The crew is one electrician for sizes to 2" and two electricians for sizes over 2". The labor cost is $41.03 per manhour. These costs include layout, material handling, and normal waste. Add for other GRS fittings, conduit, field bending, sales tax, delivery, supervision, mobilization, demobilization, cleanup, overhead and profit. Note: All elbows are assumed to be factory made.

GRS Terminations, Intermediate Metal Conduit (IMC) and Elbows

Galvanized rigid steel conduit terminations

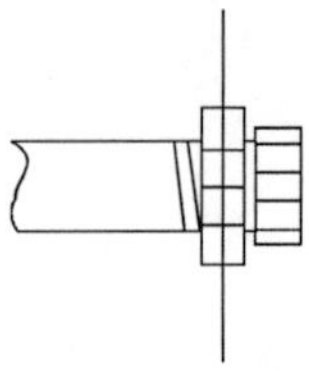

Material	Craft@Hrs	Unit	Material Cost	Labor Cost	Installed Cost
1/2"	L1@0.05	Ea	1.48	2.05	3.53
3/4"	L1@0.06	Ea	2.03	2.46	4.49
1"	L1@0.08	Ea	3.18	3.28	6.46
1-1/4"	L1@0.10	Ea	3.96	4.10	8.06
1-1/2"	L1@0.10	Ea	6.00	4.10	10.10
2"	L1@0.15	Ea	8.26	6.15	14.41
2-1/2"	L2@0.15	Ea	20.00	6.15	26.15
3"	L2@0.20	Ea	25.60	8.21	33.81
3-1/2"	L2@0.20	Ea	50.80	8.21	59.01
4"	L2@0.25	Ea	58.50	10.30	68.80
5"	L2@0.30	Ea	124.00	12.30	136.30
6"	L2@0.50	Ea	235.00	20.50	255.50

Intermediate metal conduit

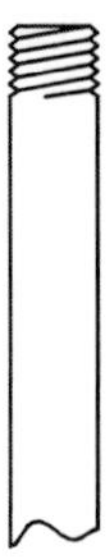

Material	Craft@Hrs	Unit	Material Cost	Labor Cost	Installed Cost
1/2"	L1@3.75	CLF	98.70	154.00	252.70
3/4"	L1@4.00	CLF	104.00	164.00	268.00
1"	L1@4.50	CLF	167.00	185.00	352.00
1-1/4"	L1@6.50	CLF	201.00	267.00	468.00
1-1/2"	L1@7.25	CLF	257.00	297.00	554.00
2"	L1@9.00	CLF	343.00	369.00	712.00
2-1/2"	L2@11.0	CLF	703.00	451.00	1,154.00
3"	L2@13.0	CLF	843.00	533.00	1,376.00
3-1/2"	L2@15.0	CLF	973.00	615.00	1,588.00
4"	L2@17.0	CLF	1,150.00	698.00	1,848.00

45 degree intermediate metal conduit elbows

Material	Craft@Hrs	Unit	Material Cost	Labor Cost	Installed Cost
1/2"	L1@0.10	Ea	14.00	4.10	18.10
3/4"	L1@0.10	Ea	17.10	4.10	21.20
1"	L1@0.10	Ea	26.50	4.10	30.60
1-1/4"	L1@0.15	Ea	40.30	6.15	46.45
1-1/2"	L1@0.15	Ea	44.20	6.15	50.35
2"	L1@0.20	Ea	63.80	8.21	72.01
2-1/2"	L2@0.20	Ea	112.00	8.21	120.21
3"	L2@0.25	Ea	171.00	10.30	181.30
3-1/2"	L2@0.25	Ea	257.00	10.30	267.30
4"	L2@0.30	Ea	303.00	12.30	315.30

Galvanized rigid steel hand benders

Material	Craft@Hrs	Unit	Material Cost	Labor Cost	Installed Cost
1/2"	--	Ea	49.10	--	49.10
3/4"	--	Ea	77.30	--	77.30
1"	--	Ea	105.00	--	105.00
1-1/4"	--	Ea	133.00	--	133.00

Use these figures to estimate the cost of GRS terminations, intermediate metal conduit and IMC elbows installed under the conditions described on pages 5 and 6. Costs listed are for each fitting or 100 linear feet installed. The crew is one electrician for GRS terminations and IMC to 2" and two electricians for GRS or IMC over 2". The labor cost is $41.03 per manhour. These costs include removing the knockout, field bending of the IMC and one coupling for each 10' length, layout, material handling, and normal waste. Add for straps and other fittings, sales tax, delivery, supervision, mobilization, demobilization, cleanup, overhead and profit. Note: Material cost is based on purchase of full packages. Conduit runs are assumed to be 50' long. Installation costs per linear foot will be less on longer runs and more on shorter runs.

IMC Elbows, Couplings and Running Thread

90 degree intermediate metal conduit elbows

Material	Craft@Hrs	Unit	Material Cost	Labor Cost	Installed Cost
1/2"	L1@0.10	Ea	15.90	4.10	20.00
3/4"	L1@0.10	Ea	19.20	4.10	23.30
1"	L1@0.10	Ea	25.60	4.10	29.70
1-1/4"	L1@0.15	Ea	40.30	6.15	46.45
1-1/2"	L1@0.15	Ea	42.50	6.15	48.65
2"	L1@0.20	Ea	64.70	8.21	72.91
2-1/2"	L1@0.20	Ea	110.00	8.21	118.21
3"	L1@0.25	Ea	175.00	10.30	185.30
3-1/2"	L1@0.25	Ea	263.00	10.30	273.30
4"	L1@0.30	Ea	310.00	12.30	322.30

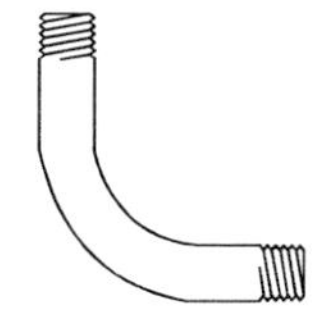

Rigid steel couplings (used on IMC)

Material	Craft@Hrs	Unit	Material Cost	Labor Cost	Installed Cost
1/2"	L1@0.05	Ea	1.24	2.05	3.29
3/4"	L1@0.06	Ea	1.50	2.46	3.96
1"	L1@0.08	Ea	2.21	3.28	5.49
1-1/4"	L1@0.10	Ea	3.77	4.10	7.87
1-1/2"	L1@0.10	Ea	4.75	4.10	8.85
2"	L1@0.15	Ea	6.26	6.15	12.41
2-1/2"	L1@0.15	Ea	14.60	6.15	20.75
3"	L1@0.20	Ea	19.20	8.21	27.41
3-1/2"	L1@0.20	Ea	25.60	8.21	33.81
4"	L1@0.25	Ea	68.10	10.30	78.40
5"	L1@0.30	Ea	146.00	12.30	158.30
6"	L1@0.50	Ea	220.00	20.50	240.50

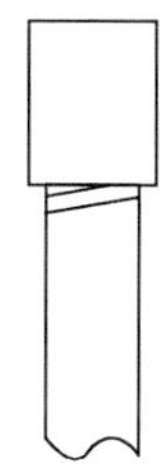

Steel running thread in 36" lengths

Material	Craft@Hrs	Unit	Material Cost	Labor Cost	Installed Cost
1/2"	L1@0.15	Ea	23.30	6.15	29.45
3/4"	L1@0.15	Ea	25.70	6.15	31.85
1"	L1@0.20	Ea	43.40	8.21	51.61
1-1/4"	L1@0.20	Ea	49.30	8.21	57.51
1-1/2"	L1@0.25	Ea	53.70	10.30	64.00
2"	L1@0.25	Ea	72.40	10.30	82.70
2-1/2"	L1@0.30	Ea	114.00	12.30	126.30
3"	L1@0.30	Ea	143.00	12.30	155.30
3-1/2"	L1@0.35	Ea	174.00	14.40	188.40
4"	L1@0.40	Ea	206.00	16.40	222.40
5"	L1@0.50	Ea	477.00	20.50	497.50
6"	L1@0.75	Ea	482.00	30.80	512.80

Use these figures to estimate the cost of elbows, couplings and running thread installed on intermediate metal conduit under the conditions described on pages 5 and 6. Costs listed are for each fitting installed. The crew is one electrician working at a labor cost of $41.03 per manhour. These costs include cutting, removal of the knockout, layout, material handling, and normal waste. Add for elbow couplings, terminations, sales tax, delivery, supervision, mobilization, demobilization, cleanup, overhead and profit. Note: Elbows and running thread are factory made. Job specifications may prohibit the use of running thread.

Galvanized Steel Locknuts and Plastic or Insulated Bushings

Galvanized steel locknuts

Material	Craft@Hrs	Unit	Material Cost	Labor Cost	Installed Cost
1/2"	L1@0.02	Ea	.28	.82	1.10
3/4"	L1@0.02	Ea	.45	.82	1.27
1"	L1@0.02	Ea	.77	.82	1.59
1-1/4"	L1@0.03	Ea	1.00	1.23	2.23
1-1/2"	L1@0.03	Ea	1.45	1.23	2.68
2"	L1@0.05	Ea	2.14	2.05	4.19
2-1/2"	L1@0.05	Ea	5.28	2.05	7.33
3"	L1@0.07	Ea	6.73	2.87	9.60
3-1/2"	L1@0.07	Ea	13.00	2.87	15.87
4"	L1@0.09	Ea	14.20	3.69	17.89
5"	L1@0.10	Ea	28.80	4.10	32.90
6"	L1@0.20	Ea	63.10	8.21	71.31

Plastic bushings

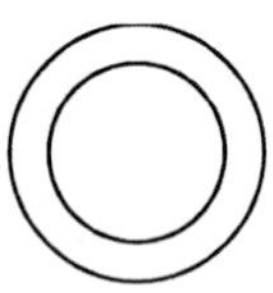

Material	Craft@Hrs	Unit	Material Cost	Labor Cost	Installed Cost
1/2"	L1@0.02	Ea	.35	.82	1.17
3/4"	L1@0.02	Ea	.64	.82	1.46
1"	L1@0.03	Ea	1.03	1.23	2.26
1-1/4"	L1@0.04	Ea	1.49	1.64	3.13
1-1/2"	L1@0.04	Ea	2.03	1.64	3.67
2"	L1@0.05	Ea	3.77	2.05	5.82
2-1/2"	L1@0.05	Ea	8.90	2.05	10.95
3"	L1@0.07	Ea	8.91	2.87	11.78
3-1/2"	L1@0.07	Ea	12.50	2.87	15.37
4"	L1@0.09	Ea	13.40	3.69	17.09
5"	L1@0.10	Ea	25.50	4.10	29.60
6"	L1@0.20	Ea	44.80	8.21	53.01

Insulated ground bushings

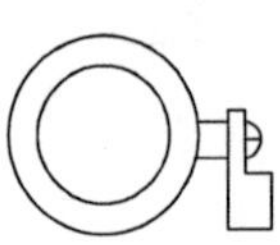

Material	Craft@Hrs	Unit	Material Cost	Labor Cost	Installed Cost
1/2"	L1@0.10	Ea	7.41	4.10	11.51
3/4"	L1@0.10	Ea	9.51	4.10	13.61
1"	L1@0.10	Ea	10.60	4.10	14.70
1-1/4"	L1@0.15	Ea	14.50	6.15	20.65
1-1/2"	L1@0.15	Ea	16.00	6.15	22.15
2"	L1@0.20	Ea	10.20	8.21	18.41
2-1/2"	L1@0.20	Ea	38.10	8.21	46.31
3"	L1@0.25	Ea	49.50	10.30	59.80
3-1/2"	L1@0.25	Ea	61.10	10.30	71.40
4"	L1@0.30	Ea	75.30	12.30	87.60
5"	L1@0.40	Ea	103.00	16.40	119.40
6"	L1@0.50	Ea	182.00	20.50	202.50

Use these figures to estimate the cost of locknuts and bushings installed on GRS or IMC conduit under the conditions described on pages 5 and 6. Costs listed are for each locknut or bushing installed. The crew is one electrician working at a labor cost of $41.03 per manhour. These costs include removal of the knockout, layout, material handling, and normal waste. Add for conduit, sales tax, delivery, supervision, mobilization, demobilization, cleanup, overhead and profit. Note: Material costs assume purchase of full box quantities. The locknuts are steel for sizes up to 2" and malleable for sizes over 2". On conduit terminations at boxes or cabinets, one locknut is used inside the box and one locknut is used outside the box. A bushing is used at the end of each conduit run to protect the wire. An insulated ground bushing is used when connecting a ground wire to the conduit system.

Material	Craft@Hrs	Unit	Material Cost	Labor Cost	Installed Cost
Galvanized rigid steel nipples					
1/2" x close	L1@0.05	Ea	1.74	2.05	3.79
1/2" x 1-1/2"	L1@0.05	Ea	1.87	2.05	3.92
1/2" x 2"	L1@0.05	Ea	2.04	2.05	4.09
1/2" x 2-1/2"	L1@0.05	Ea	2.16	2.05	4.21
1/2" x 3"	L1@0.05	Ea	2.52	2.05	4.57
1/2" x 3-1/2"	L1@0.05	Ea	2.94	2.05	4.99
1/2" x 4"	L1@0.05	Ea	3.88	2.05	5.93
1/2" x 5"	L1@0.05	Ea	6.07	2.05	8.12
1/2" x 6"	L1@0.05	Ea	7.24	2.05	9.29
1/2" x 8"	L1@0.05	Ea	12.60	2.05	14.65
1/2" x 10"	L1@0.05	Ea	14.50	2.05	16.55
1/2" x 12"	L1@0.05	Ea	16.70	2.05	18.75
3/4" x close	L1@0.06	Ea	3.71	2.46	6.17
3/4" x 2"	L1@0.06	Ea	4.42	2.46	6.88
3/4" x 2-1/2"	L1@0.06	Ea	4.87	2.46	7.33
3/4" x 3"	L1@0.06	Ea	5.35	2.46	7.81
3/4" x 3-1/2"	L1@0.06	Ea	5.60	2.46	8.06
3/4" x 4"	L1@0.06	Ea	6.32	2.46	8.78
3/4" x 5"	L1@0.06	Ea	7.25	2.46	9.71
3/4" x 6"	L1@0.06	Ea	8.47	2.46	10.93
3/4" x 8"	L1@0.06	Ea	13.90	2.46	16.36
3/4" x 10"	L1@0.06	Ea	16.70	2.46	19.16
3/4" x 12"	L1@0.06	Ea	18.80	2.46	21.26
1" x close	L1@0.08	Ea	5.61	3.28	8.89
1" x 2"	L1@0.08	Ea	6.13	3.28	9.41
1" x 2-1/2"	L1@0.08	Ea	6.67	3.28	9.95
1" x 3"	L1@0.08	Ea	7.43	3.28	10.71
1" x 3-1/2"	L1@0.08	Ea	8.47	3.28	11.75
1" x 4"	L1@0.08	Ea	9.31	3.28	12.59
1" x 5"	L1@0.08	Ea	10.60	3.28	13.88
1" x 6"	L1@0.08	Ea	11.50	3.28	14.78
1" x 8"	L1@0.08	Ea	18.20	3.28	21.48
1" x 10"	L1@0.08	Ea	23.60	3.28	26.88
1" x 12"	L1@0.08	Ea	26.70	3.28	29.98

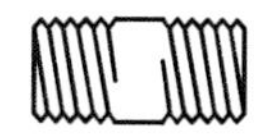
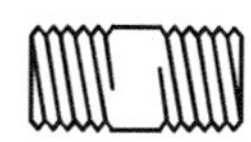
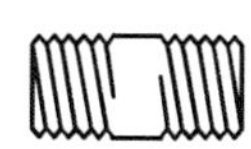

Use these figures to estimate the cost of nipples installed on GRS conduit under the conditions described on pages 5 and 6. Costs listed are for each nipple installed. The crew is one electrician at a labor cost of $41.03 per manhour. These costs include removal of the knockout, layout, material handling, and normal waste. Add for terminations, couplings, sales tax, delivery, supervision, mobilization, demobilization, cleanup, overhead and profit. Note: Nipples are factory made, not field made. In many cases a coupling will be needed with a nipple.

Galvanized Rigid Steel Nipples

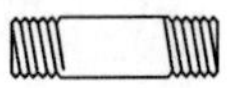
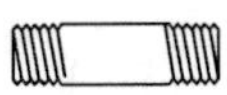
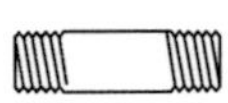
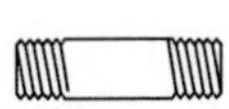

Material	Craft@Hrs	Unit	Material Cost	Labor Cost	Installed Cost
Galvanized rigid steel nipples (continued)					
1-1/4" x close	L1@0.10	Ea	7.49	4.10	11.59
1-1/4" x 2"	L1@0.10	Ea	8.36	4.10	12.46
1-1/4" x 2-1/2"	L1@0.10	Ea	8.80	4.10	12.90
1-1/4" x 3"	L1@0.10	Ea	9.49	4.10	13.59
1-1/4" x 3-1/2"	L1@0.10	Ea	10.80	4.10	14.90
1-1/4" x 4"	L1@0.10	Ea	11.40	4.10	15.50
1-1/4" x 5"	L1@0.10	Ea	13.20	4.10	17.30
1-1/4" x 6"	L1@0.10	Ea	14.80	4.10	18.90
1-1/4" x 8"	L1@0.10	Ea	24.50	4.10	28.60
1-1/4" x 10"	L1@0.10	Ea	30.80	4.10	34.90
1-1/4" x 12"	L1@0.10	Ea	35.50	4.10	39.60
1-1/2" x close	L1@0.10	Ea	8.99	4.10	13.09
1-1/2" x 2"	L1@0.10	Ea	9.52	4.10	13.62
1-1/2" x 2-1/2"	L1@0.10	Ea	10.70	4.10	14.80
1-1/2" x 3"	L1@0.10	Ea	14.60	4.10	18.70
1-1/2" x 3-1/2"	L1@0.10	Ea	16.80	4.10	20.90
1-1/2" x 4"	L1@0.10	Ea	18.50	4.10	22.60
1-1/2" x 5"	L1@0.10	Ea	20.90	4.10	25.00
1-1/2" x 6"	L1@0.10	Ea	25.70	4.10	29.80
1-1/2" x 8"	L1@0.10	Ea	38.80	4.10	42.90
1-1/2" x 10"	L1@0.10	Ea	40.30	4.10	44.40
1-1/2" x 12"	L1@0.10	Ea	43.50	4.10	47.60
2" x close	L1@0.15	Ea	10.70	6.15	16.85
2" x 2-1/2"	L1@0.15	Ea	12.60	6.15	18.75
2" x 3"	L1@0.15	Ea	14.50	6.15	20.65
2" x 3-1/2"	L1@0.15	Ea	16.60	6.15	22.75
2" x 4"	L1@0.15	Ea	18.50	6.15	24.65
2" x 5"	L1@0.15	Ea	21.50	6.15	27.65
2" x 6"	L1@0.15	Ea	24.50	6.15	30.65
2" x 8"	L1@0.15	Ea	35.30	6.15	41.45
2" x 10"	L1@0.15	Ea	42.50	6.15	48.65
2" x 12"	L1@0.15	Ea	48.10	6.15	54.25
2-1/2" x close	L1@0.15	Ea	30.00	6.15	36.15
2-1/2" x 3"	L1@0.15	Ea	30.30	6.15	36.45
2-1/2" x 3-1/2"	L1@0.15	Ea	35.40	6.15	41.55
2-1/2" x 4"	L1@0.15	Ea	37.40	6.15	43.55
2-1/2" x 5"	L1@0.15	Ea	44.40	6.15	50.55
2-1/2" x 6"	L1@0.15	Ea	50.10	6.15	56.25

Use these figures to estimate the cost of nipples installed on GRS conduit under the conditions described on pages 5 and 6. Costs listed are for each nipple installed. The crew is one electrician at a labor cost of $41.03 per manhour. These costs include removal of the knockout, layout, material handling, and normal waste. Add for terminations, couplings, sales tax, delivery, supervision, mobilization, demobilization, cleanup, overhead and profit. Note: Nipples are factory made, not field made. In many cases a coupling will be needed with a nipple.

Galvanized rigid steel nipples (continued)

Material	Craft@Hrs	Unit	Material Cost	Labor Cost	Installed Cost
2-1/2" x 8"	L1@0.15	Ea	66.00	6.15	72.15
2-1/2" x 10"	L1@0.15	Ea	76.70	6.15	82.85
2-1/2" x 12"	L1@0.15	Ea	89.10	6.15	95.25

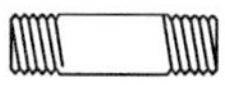

Material	Craft@Hrs	Unit	Material Cost	Labor Cost	Installed Cost
3" x close	L1@0.20	Ea	35.20	8.21	43.41
3" x 3"	L1@0.20	Ea	37.30	8.21	45.51
3" x 3-1/2"	L1@0.20	Ea	41.80	8.21	50.01
3" x 4"	L1@0.20	Ea	45.20	8.21	53.41
3" x 5"	L1@0.20	Ea	52.70	8.21	60.91
3" x 6"	L1@0.20	Ea	60.30	8.21	68.51
3" x 8"	L1@0.20	Ea	89.90	8.21	98.11
3" x 10"	L1@0.20	Ea	108.00	8.21	116.21
3" x 12"	L1@0.20	Ea	112.00	8.21	120.21

Material	Craft@Hrs	Unit	Material Cost	Labor Cost	Installed Cost
3-1/2" x close	L1@0.25	Ea	43.00	10.30	53.30
3-1/2" x 4"	L1@0.25	Ea	55.30	10.30	65.60
3-1/2" x 5"	L1@0.25	Ea	62.50	10.30	72.80
3-1/2" x 6"	L1@0.25	Ea	71.40	10.30	81.70
3-1/2" x 8"	L1@0.25	Ea	89.90	10.30	100.20
3-1/2" x 10"	L1@0.25	Ea	108.00	10.30	118.30
3-1/2" x 12"	L1@0.25	Ea	126.00	10.30	136.30

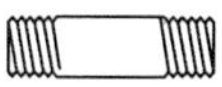

Material	Craft@Hrs	Unit	Material Cost	Labor Cost	Installed Cost
4" x close	L1@0.25	Ea	51.00	10.30	61.30
4" x 4"	L1@0.25	Ea	62.00	10.30	72.30
4" x 5"	L1@0.25	Ea	72.80	10.30	83.10
4" x 6"	L1@0.25	Ea	81.10	10.30	91.40
4" x 8"	L1@0.25	Ea	101.00	10.30	111.30
4" x 10"	L1@0.25	Ea	125.00	10.30	135.30
4" x 12"	L1@0.25	Ea	148.00	10.30	158.30

Material	Craft@Hrs	Unit	Material Cost	Labor Cost	Installed Cost
5" x close	L1@0.40	Ea	37.70	16.40	54.10
5" x 5"	L1@0.40	Ea	96.00	16.40	112.40
5" x 6"	L1@0.40	Ea	105.00	16.40	121.40
5" x 8"	L1@0.40	Ea	257.00	16.40	273.40
5" x 10"	L1@0.40	Ea	145.00	16.40	161.40
5" x 12"	L1@0.40	Ea	362.00	16.40	378.40

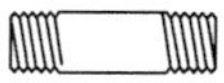

Material	Craft@Hrs	Unit	Material Cost	Labor Cost	Installed Cost
6" x close	L1@0.60	Ea	86.70	24.60	111.30
6" x 5"	L1@0.60	Ea	153.00	24.60	177.60
6" x 6"	L1@0.60	Ea	192.00	24.60	216.60
6" x 8"	L1@0.60	Ea	215.00	24.60	239.60
6" x 10"	L1@0.60	Ea	248.00	24.60	272.60
6" x 12"	L1@0.60	Ea	273.00	24.60	297.60

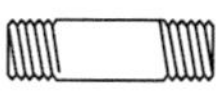

Use these figures to estimate the cost of nipples installed on GRS conduit under the conditions described on pages 5 and 6. Costs listed are for each nipple installed. The crew is one electrician working at a cost of $41.03 per manhour. These costs include removal of the knockout, layout, material handling, and normal waste. Add for terminations, couplings, sales tax, delivery, supervision, mobilization, demobilization, cleanup, overhead and profit. Note: Nipples are factory made, not field made. In many cases a coupling will be needed with a nipple.

Aluminum Rigid Conduit (ARC), Elbows and Nipples

Aluminum rigid conduit

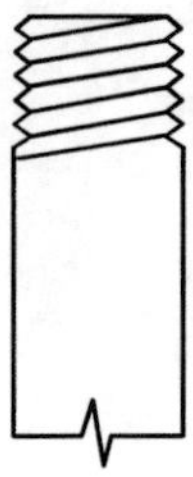

Material	Craft@Hrs	Unit	Material Cost	Labor Cost	Installed Cost
1/2"	L1@3.75	CLF	251.00	154.00	405.00
3/4"	L1@4.00	CLF	336.00	164.00	500.00
1"	L1@4.50	CLF	478.00	185.00	663.00
1-1/4"	L1@6.00	CLF	666.00	246.00	912.00
1-1/2"	L1@7.00	CLF	610.00	287.00	897.00
2"	L1@8.50	CLF	1,040.00	349.00	1,389.00
2-1/2"	L2@10.0	CLF	1,410.00	410.00	1,820.00
3"	L2@12.0	CLF	1,780.00	492.00	2,272.00
3-1/2"	L2@14.0	CLF	2,110.00	574.00	2,684.00
4"	L2@16.0	CLF	2,510.00	656.00	3,166.00
5"	L2@20.0	CLF	3,820.00	821.00	4,641.00
6"	L2@25.0	CLF	5,270.00	1,030.00	6,300.00

90 degree aluminum rigid conduit elbows

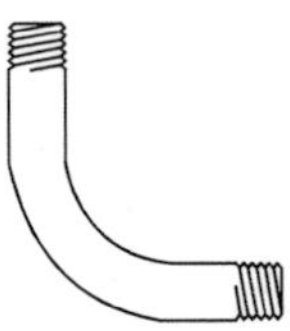

Material	Craft@Hrs	Unit	Material Cost	Labor Cost	Installed Cost
1/2"	L1@0.10	Ea	17.60	4.10	21.70
3/4"	L1@0.10	Ea	22.70	4.10	26.80
1"	L1@0.10	Ea	37.40	4.10	41.50
1-1/4"	L1@0.15	Ea	38.80	6.15	44.95
1-1/2"	L1@0.15	Ea	148.00	6.15	154.15
2"	L1@0.20	Ea	220.00	8.21	228.21
2-1/2"	L2@0.20	Ea	370.00	8.21	378.21
3"	L2@0.25	Ea	571.00	10.30	581.30
3-1/2"	L2@0.25	Ea	892.00	10.30	902.30
4"	L2@0.30	Ea	1,510.00	12.30	1,522.30
5"	L2@0.40	Ea	3,120.00	16.40	3,136.40
6"	L2@0.70	Ea	4,310.00	28.70	4,338.70

Aluminum rigid conduit nipples

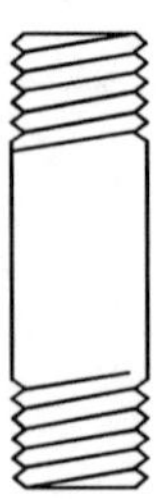

Material	Craft@Hrs	Unit	Material Cost	Labor Cost	Installed Cost
1/2" x close	L1@0.05	Ea	19.40	2.05	21.45
1/2" x 1-1/2"	L1@0.05	Ea	14.70	2.05	16.75
1/2" x 2"	L1@0.05	Ea	15.90	2.05	17.95
1/2" x 2-1/2"	L1@0.05	Ea	18.90	2.05	20.95
1/2" x 3"	L1@0.05	Ea	19.70	2.05	21.75
1/2" x 3-1/2"	L1@0.05	Ea	21.50	2.05	23.55
1/2" x 4"	L1@0.05	Ea	23.10	2.05	25.15
1/2" x 5"	L1@0.05	Ea	26.20	2.05	28.25
1/2" x 6"	L1@0.05	Ea	27.80	2.05	29.85
1/2" x 8"	L1@0.05	Ea	37.30	2.05	39.35
1/2" x 10"	L1@0.05	Ea	45.10	2.05	47.15
1/2" x 12"	L1@0.05	Ea	52.40	2.05	54.45
3/4" x close	L1@0.06	Ea	19.40	2.46	21.86
3/4" x 2"	L1@0.06	Ea	20.80	2.46	23.26
3/4" x 2-1/2"	L1@0.06	Ea	22.30	2.46	24.76

Use these figures to estimate the cost of aluminum rigid conduit, elbows and nipples installed in a building under the conditions described on pages 5 and 6. Costs listed are for each 100 linear feet of conduit or each fitting installed. The crew is one electrician for conduit sizes to 2" and two electricians for conduit over 2". The labor cost is $41.03 per manhour. These costs include conduit bending, one coupling for each length of conduit, layout, material handling, and normal waste. Add for extra couplings, straps, terminations, wire, sales tax, delivery, supervision, mobilization, demobilization, cleanup, overhead and profit. Note: Elbows and nipples are factory made. Do not install ARC in concrete or masonry construction. Conduit runs are assumed to be 50' long. Installation costs per linear foot will be less on longer runs and more on shorter runs.

Aluminum rigid conduit nipples

Material	Craft@Hrs	Unit	Material Cost	Labor Cost	Installed Cost
3/4" x 3"	L1@0.06	Ea	24.10	2.46	26.56
3/4" x 3-1/2"	L1@0.06	Ea	24.80	2.46	27.26
3/4" x 4"	L1@0.06	Ea	26.00	2.46	28.46
3/4" x 5"	L1@0.06	Ea	31.30	2.46	33.76
3/4" x 6"	L1@0.06	Ea	35.60	2.46	38.06
3/4" x 8"	L1@0.06	Ea	47.10	2.46	49.56
3/4" x 10"	L1@0.06	Ea	55.10	2.46	57.56
3/4" x 12"	L1@0.06	Ea	67.40	2.46	69.86
1" x close	L1@0.08	Ea	23.50	3.28	26.78
1" x 2"	L1@0.08	Ea	26.00	3.28	29.28
1" x 2-1/2"	L1@0.08	Ea	28.50	3.28	31.78
1" x 3"	L1@0.08	Ea	30.80	3.28	34.08
1" x 3-1/2"	L1@0.08	Ea	34.50	3.28	37.78
1" x 4"	L1@0.08	Ea	38.10	3.28	41.38
1" x 5"	L1@0.08	Ea	45.00	3.28	48.28
1" x 6"	L1@0.08	Ea	52.90	3.28	56.18
1" x 8"	L1@0.08	Ea	65.90	3.28	69.18
1" x 10"	L1@0.08	Ea	82.80	3.28	86.08
1" x 12"	L1@0.08	Ea	98.00	3.28	101.28
1-1/4" x close	L1@0.10	Ea	31.70	4.10	35.80
1-1/4" x 2"	L1@0.10	Ea	32.50	4.10	36.60
1-1/4" x 2-1/2"	L1@0.10	Ea	35.50	4.10	39.60
1-1/4" x 3"	L1@0.10	Ea	40.10	4.10	44.20
1-1/4" x 3-1/2"	L1@0.10	Ea	45.80	4.10	49.90
1-1/4" x 4"	L1@0.10	Ea	59.00	4.10	63.10
1-1/4" x 5"	L1@0.10	Ea	68.70	4.10	72.80
1-1/4" x 6"	L1@0.10	Ea	68.70	4.10	72.80
1-1/4" x 8"	L1@0.10	Ea	87.40	4.10	91.50
1-1/4" x 10"	L1@0.10	Ea	106.00	4.10	110.10
1-1/4" x 12"	L1@0.10	Ea	125.00	4.10	129.10
1-1/2" x close	L1@0.10	Ea	39.40	4.10	43.50
1-1/2" x 2"	L1@0.10	Ea	40.30	4.10	44.40
1-1/2" x 2-1/2"	L1@0.10	Ea	43.00	4.10	47.10
1-1/2" x 3"	L1@0.10	Ea	48.80	4.10	52.90
1-1/2" x 3-1/2"	L1@0.10	Ea	61.60	4.10	65.70
1-1/2" x 4"	L1@0.10	Ea	61.80	4.10	65.90
1-1/2" x 5"	L1@0.10	Ea	70.80	4.10	74.90
1-1/2" x 6"	L1@0.10	Ea	81.80	4.10	85.90
1-1/2" x 8"	L1@0.10	Ea	105.00	4.10	109.10
1-1/2" x 10"	L1@0.10	Ea	128.00	4.10	132.10
1-1/2" x 12"	L1@0.10	Ea	150.00	4.10	154.10

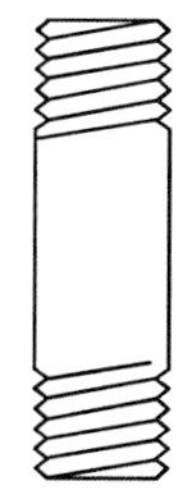

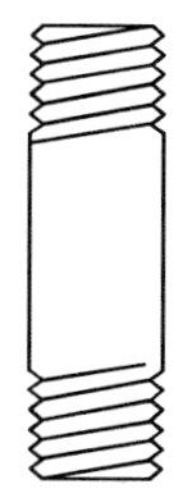

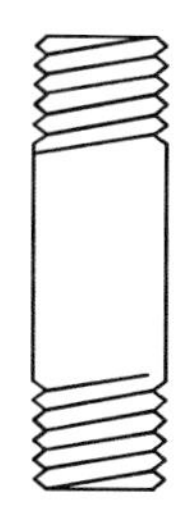

Use these figures to estimate the cost of ARC nipples installed on ARC conduit under the conditions described on pages 5 and 6. Costs listed are for each nipple installed. The crew is one electrician at a labor cost of $41.03 per manhour. These costs include removing the knockout, layout, material handling, and normal waste. Add for extra couplings, straps, boxes, sales tax, delivery, supervision, mobilization, demobilization, cleanup, overhead and profit. Note: Material costs assume the purchase of full packages.

ARC Nipples

Aluminum rigid conduit nipples (continued)

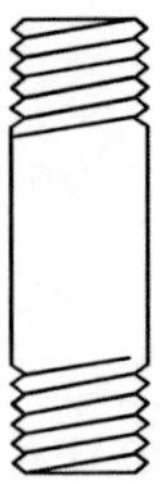

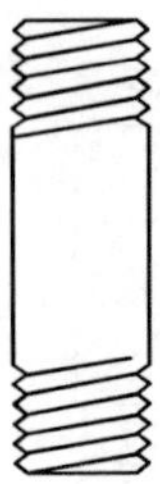

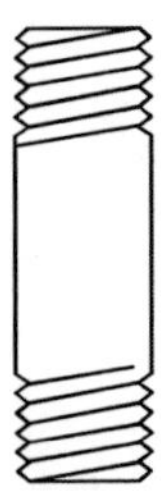

Material	Craft@Hrs	Unit	Material Cost	Labor Cost	Installed Cost
2" x close	L1@0.15	Ea	40.30	6.15	46.45
2" x 2-1/2"	L1@0.15	Ea	56.00	6.15	62.15
2" x 3"	L1@0.15	Ea	62.60	6.15	68.75
2" x 3-1/2"	L1@0.15	Ea	73.80	6.15	79.95
2" x 4"	L1@0.15	Ea	77.10	6.15	83.25
2" x 5"	L1@0.15	Ea	77.10	6.15	83.25
2" x 6"	L1@0.15	Ea	106.00	6.15	112.15
2" x 8"	L1@0.15	Ea	134.00	6.15	140.15
2" x 10"	L1@0.15	Ea	162.00	6.15	168.15
2" x 12"	L1@0.15	Ea	195.00	6.15	201.15
2-1/2" x close	L1@0.15	Ea	111.00	6.15	117.15
2-1/2" x 3"	L1@0.15	Ea	114.00	6.15	120.15
2-1/2" x 3-1/2"	L1@0.15	Ea	126.00	6.15	132.15
2-1/2" x 4"	L1@0.15	Ea	133.00	6.15	139.15
2-1/2" x 5"	L1@0.15	Ea	150.00	6.15	156.15
2-1/2" x 6"	L1@0.15	Ea	162.00	6.15	168.15
2-1/2" x 8"	L1@0.15	Ea	208.00	6.15	214.15
2-1/2" x 10"	L1@0.15	Ea	252.00	6.15	258.15
2-1/2" x 12"	L1@0.15	Ea	283.00	6.15	289.15
3" x close	L1@0.20	Ea	72.00	8.21	80.21
3" x 3-1/2"	L1@0.20	Ea	96.00	8.21	104.21
3" x 4"	L1@0.20	Ea	100.00	8.21	108.21
3" x 5"	L1@0.20	Ea	114.00	8.21	122.21
3" x 6"	L1@0.20	Ea	131.00	8.21	139.21
3" x 8"	L1@0.20	Ea	168.00	8.21	176.21
3" x 10"	L1@0.20	Ea	203.00	8.21	211.21
3" x 12"	L1@0.20	Ea	241.00	8.21	249.21
3-1/2" x close	L1@0.25	Ea	91.70	10.30	102.00
3-1/2" x 4"	L1@0.25	Ea	116.00	10.30	126.30
3-1/2" x 5"	L1@0.25	Ea	138.00	10.30	148.30
3-1/2" x 6"	L1@0.25	Ea	160.00	10.30	170.30
3-1/2" x 8"	L1@0.25	Ea	200.00	10.30	210.30
3-1/2" x 10"	L1@0.25	Ea	248.00	10.30	258.30
3-1/2" x 12"	L1@0.25	Ea	289.00	10.30	299.30

Use these figures to estimate the cost of ARC nipples installed on ARC conduit under the conditions described on pages 5 and 6. Costs listed are for each nipple installed. The crew is one electrician at a labor cost of $41.03 per manhour. These costs include removing the knockout, layout, material handling, and normal waste. Add for extra couplings, straps, boxes, sales tax, delivery, supervision, mobilization, demobilization, cleanup, overhead and profit. Note: Material costs assume the purchase of full packages. Nipples are factory made, not field made. In many cases a coupling will be needed with each nipple. Do not install aluminum fittings in concrete or masonry. The bending, cutting and threading tools for aluminum conduit are the same as used for GRS. Don't mix aluminum fittings with other types of fittings.

Material	Craft@Hrs	Unit	Material Cost	Labor Cost	Installed Cost
Aluminum rigid conduit nipples (continued)					
4" x close	L1@0.25	Ea	96.40	10.30	106.70
4" x 4"	L1@0.25	Ea	120.00	10.30	130.30
4" x 5"	L1@0.25	Ea	138.00	10.30	148.30
4" x 6"	L1@0.25	Ea	160.00	10.30	170.30
4" x 8"	L1@0.25	Ea	203.00	10.30	213.30
4" x 10"	L1@0.25	Ea	248.00	10.30	258.30
4" x 12"	L1@0.25	Ea	292.00	10.30	302.30
5" x close	L1@0.40	Ea	224.00	16.40	240.40
5" x 5"	L1@0.40	Ea	263.00	16.40	279.40
5" x 6"	L1@0.40	Ea	275.00	16.40	291.40
5" x 8"	L1@0.40	Ea	350.00	16.40	366.40
5" x 10"	L1@0.40	Ea	423.00	16.40	439.40
5" x 12"	L1@0.40	Ea	486.00	16.40	502.40
6" x close	L1@0.60	Ea	271.00	24.60	295.60
6" x 5"	L1@0.60	Ea	317.00	24.60	341.60
6" x 6"	L1@0.60	Ea	346.00	24.60	370.60
6" x 8"	L1@0.60	Ea	471.00	24.60	495.60
6" x 10"	L1@0.60	Ea	564.00	24.60	588.60
6" x 12"	L1@0.60	Ea	623.00	24.60	647.60
Aluminum locknuts					
1/2"	L1@0.02	Ea	.57	.82	1.39
3/4"	L1@0.02	Ea	1.04	.82	1.86
1"	L1@0.02	Ea	1.57	.82	2.39
1-1/4"	L1@0.03	Ea	2.08	1.23	3.31
1-1/2"	L1@0.03	Ea	2.94	1.23	4.17
2"	L1@0.05	Ea	4.70	2.05	6.75
2-1/2"	L1@0.05	Ea	8.96	2.05	11.01
3"	L1@0.07	Ea	9.91	2.87	12.78
3-1/2"	L1@0.07	Ea	27.90	2.87	30.77
4"	L1@0.09	Ea	30.60	3.69	34.29
5"	L1@0.10	Ea	77.90	4.10	82.00
6"	L1@0.20	Ea	138.00	8.21	146.21
Aluminum bushings					
1/2"	L1@0.02	Ea	7.30	.82	8.12
3/4"	L1@0.02	Ea	13.40	.82	14.22
1"	L1@0.03	Ea	17.60	1.23	18.83

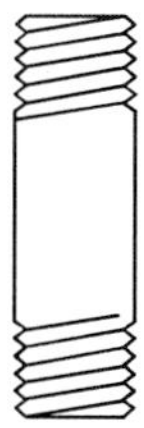

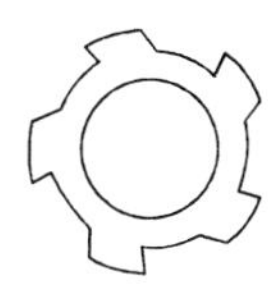

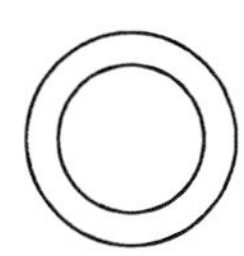

Use these figures to estimate the cost of ARC nipples, locknuts and bushings installed on ARC conduit under the conditions described on pages 5 and 6. Costs listed are for each fitting installed. The crew is one electrician working at a labor cost of $41.03 per manhour. These costs include removing the knockout, layout, material handling, and normal waste. Add for extra couplings, sales tax, delivery, supervision, mobilization, demobilization, cleanup, overhead and profit. Note: Material costs are based on purchase of full packages. Nipples are factory made, not field made. In many cases a coupling will be needed with each nipple. Do not install aluminum fittings in concrete or masonry.

Aluminum Bushings and Terminations

Material	Craft@Hrs	Unit	Material Cost	Labor Cost	Installed Cost

Aluminum bushings (continued)

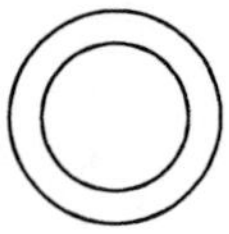

Material	Craft@Hrs	Unit	Material Cost	Labor Cost	Installed Cost
1-1/4"	L1@0.04	Ea	27.60	1.64	29.24
1-1/2"	L1@0.04	Ea	35.00	1.64	36.64
2"	L1@0.05	Ea	43.10	2.05	45.15
2-1/2"	L1@0.05	Ea	58.00	2.05	60.05
3"	L1@0.07	Ea	61.50	2.87	64.37
3-1/2"	L1@0.07	Ea	123.00	2.87	125.87
4"	L1@0.09	Ea	143.00	3.69	146.69
5"	L1@0.10	Ea	237.00	4.10	241.10
6"	L1@0.20	Ea	364.00	8.21	372.21

Insulated aluminum ground bushings

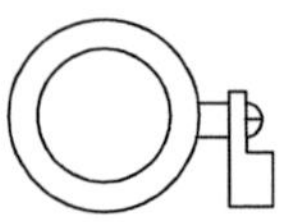

Material	Craft@Hrs	Unit	Material Cost	Labor Cost	Installed Cost
1/2"	L1@0.10	Ea	17.60	4.10	21.70
3/4"	L1@0.10	Ea	19.80	4.10	23.90
1"	L1@0.10	Ea	28.80	4.10	32.90
1-1/4"	L1@0.15	Ea	29.20	6.15	35.35
1-1/2"	L1@0.15	Ea	37.00	6.15	43.15
2"	L1@0.20	Ea	49.70	8.21	57.91
2-1/2"	L1@0.20	Ea	89.40	8.21	97.61
3"	L1@0.25	Ea	136.00	10.30	146.30
3-1/2"	L1@0.25	Ea	163.00	10.30	173.30
4"	L1@0.30	Ea	220.00	12.30	232.30
5"	L1@0.40	Ea	346.00	16.40	362.40
6"	L1@0.50	Ea	534.00	20.50	554.50

Conduit termination, two aluminum locknuts & one plastic bushing

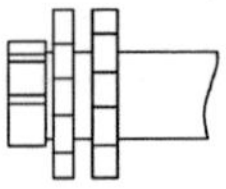

Material	Craft@Hrs	Unit	Material Cost	Labor Cost	Installed Cost
1/2"	L1@0.05	Ea	5.75	2.05	7.80
3/4"	L1@0.06	Ea	8.87	2.46	11.33
1"	L1@0.08	Ea	14.50	3.28	17.78
1-1/4"	L1@0.10	Ea	18.00	4.10	22.10
1-1/2"	L1@0.10	Ea	42.40	4.10	46.50
2"	L1@0.15	Ea	46.60	6.15	52.75
2-1/2"	L1@0.15	Ea	77.00	6.15	83.15
3"	L1@0.20	Ea	89.90	8.21	98.11
3-1/2"	L1@0.20	Ea	143.00	8.21	151.21
4"	L1@0.25	Ea	158.00	10.30	168.30
5"	L1@0.40	Ea	199.00	16.40	215.40
6"	L1@0.60	Ea	351.00	24.60	375.60

Use these figures to estimate the cost of aluminum bushings, ground bushings, and terminations under the conditions described on pages 5 and 6. Costs listed are for each fitting installed. The crew is one electrician working at a labor cost of $41.03 per manhour. These costs include removal of knockouts, layout, material handling, and normal waste. Add for sales tax, delivery, supervision, mobilization, demobilization, cleanup, overhead and profit. Note: Material costs are based on purchase of full boxes. One locknut is used outside the box and inside the box on each conduit termination. A bushing is needed at each conduit end to protect the wire.

Cast Metal Entrance Elbows and Conduit Bodies

Cast metal Type SLB entrance elbows

Material	Craft@Hrs	Unit	Material Cost	Labor Cost	Installed Cost
1/2"	L1@0.10	Ea	8.01	4.10	12.11
3/4"	L1@0.15	Ea	9.85	6.15	16.00
1"	L1@0.15	Ea	17.90	6.15	24.05
1-1/4"	L1@0.20	Ea	27.50	8.21	35.71
1-1/2"	L1@0.20	Ea	49.30	8.21	57.51
2"	L1@0.25	Ea	56.30	10.30	66.60
2-1/2"	L1@0.30	Ea	200.00	12.30	212.30
3"	L1@0.40	Ea	257.00	16.40	273.40

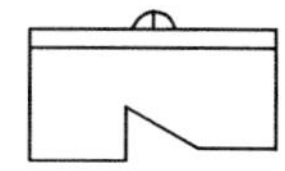

Galvanized cast metal Types LB, LL or LR conduit bodies

1/2"	L1@0.10	Ea	12.90	4.10	17.00
3/4"	L1@0.15	Ea	15.20	6.15	21.35
1"	L1@0.20	Ea	22.70	8.21	30.91
1-1/4"	L1@0.25	Ea	39.10	10.30	49.40
1-1/2"	L1@0.25	Ea	51.20	10.30	61.50
2"	L1@0.30	Ea	85.50	12.30	97.80
2-1/2"	L1@0.40	Ea	171.00	16.40	187.40
3"	L1@0.50	Ea	227.00	20.50	247.50
3-1/2"	L1@0.70	Ea	385.00	28.70	413.70
4"	L1@1.00	Ea	435.00	41.00	476.00

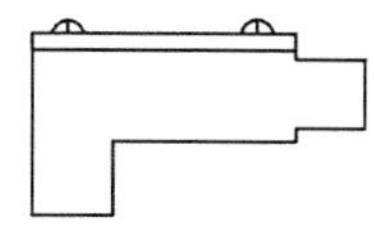

Galvanized cast metal Type T conduit bodies

1/2"	L1@0.15	Ea	11.10	6.15	17.25
3/4"	L1@0.20	Ea	15.20	8.21	23.41
1"	L1@0.25	Ea	22.40	10.30	32.70
1-1/4"	L1@0.30	Ea	33.40	12.30	45.70
1-1/2"	L1@0.30	Ea	50.20	12.30	62.50
2"	L1@0.40	Ea	77.50	16.40	93.90
2-1/2"	L1@0.50	Ea	154.00	20.50	174.50
3"	L1@0.70	Ea	204.00	28.70	232.70
3-1/2"	L1@0.90	Ea	529.00	36.90	565.90
4"	L1@1.25	Ea	679.00	51.30	730.30

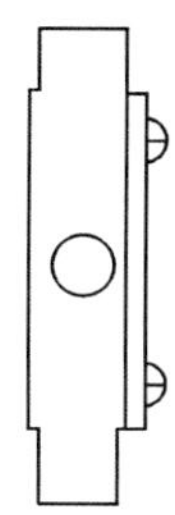

Galvanized cast metal Type X conduit bodies

1/2"	L1@0.20	Ea	37.10	8.21	45.31
3/4"	L1@0.25	Ea	43.50	10.30	53.80
1"	L1@0.30	Ea	60.10	12.30	72.40
1-1/4"	L1@0.40	Ea	84.20	16.40	100.60
1-1/2"	L1@0.40	Ea	106.00	16.40	122.40
2"	L1@0.50	Ea	188.00	20.50	208.50

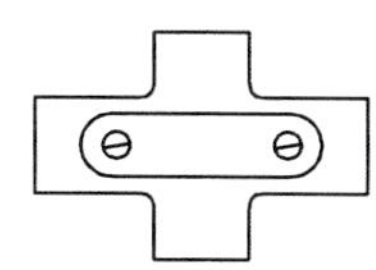

Use these figures to estimate the cost of conduit bodies installed on EMT or GRS conduit under the conditions described on pages 5 and 6. Costs listed are for each body installed. The crew is one electrician working at a labor cost of $41.03 per manhour. These costs include layout, material handling, and normal waste. Add for conduit, nipples, boxes, covers, gaskets, sales tax, delivery, supervision, mobilization, demobilization, cleanup, overhead and profit. Note: Using a larger conduit body or a mogul size can reduce the installation time when wire sizes are larger.

Blank Conduit Body Covers

Steel blank conduit body covers

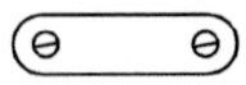

Material	Craft@Hrs	Unit	Material Cost	Labor Cost	Installed Cost
1/2"	L1@0.05	Ea	1.86	2.05	3.91
3/4"	L1@0.05	Ea	4.14	2.05	6.19
1"	L1@0.05	Ea	3.39	2.05	5.44
1-1/4"	L1@0.10	Ea	4.86	4.10	8.96
1-1/2"	L1@0.10	Ea	5.94	4.10	10.04
2"	L1@0.10	Ea	8.93	4.10	13.03
2-1/2" - 3"	L1@0.15	Ea	12.70	6.15	18.85
2-1/2" - 4"	L1@0.20	Ea	23.00	8.21	31.21

Malleable blank conduit body covers

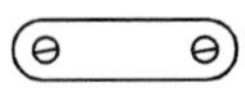

Material	Craft@Hrs	Unit	Material Cost	Labor Cost	Installed Cost
1/2"	L1@0.05	Ea	6.94	2.05	8.99
3/4"	L1@0.05	Ea	5.78	2.05	7.83
1"	L1@0.10	Ea	9.27	4.10	13.37
1-1/4"	L1@0.10	Ea	11.20	4.10	15.30
1-1/2"	L1@0.10	Ea	13.10	4.10	17.20
2"	L1@0.15	Ea	25.90	6.15	32.05
2-1/2" - 3"	L1@0.20	Ea	41.50	8.21	49.71
2-1/2" - 4"	L1@0.25	Ea	59.50	10.30	69.80

Aluminum blank conduit body covers

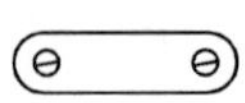

Material	Craft@Hrs	Unit	Material Cost	Labor Cost	Installed Cost
1/2"	L1@0.05	Ea	3.02	2.05	5.07
3/4"	L1@0.05	Ea	4.14	2.05	6.19
1"	L1@0.05	Ea	4.99	2.05	7.04
1-1/4"	L1@0.10	Ea	6.70	4.10	10.80
1-1/2"	L1@0.10	Ea	9.92	4.10	14.02
2"	L1@0.10	Ea	13.10	4.10	17.20
2-1/2" - 3"	L1@0.15	Ea	20.10	6.15	26.25
2-1/2" - 4"	L1@0.20	Ea	24.40	8.21	32.61

Use these figures to estimate the cost of blank conduit body covers installed on conduit bodies under the conditions described on pages 5 and 6. Costs listed are for each cover installed. The crew is one electrician working at a labor cost of $41.03 per manhour. These costs include layout, material handling, and normal waste. Add for conduit bodies, other fittings, sales tax, delivery, supervision, mobilization, demobilization, cleanup, overhead and profit. Note: These figures assume that the conduit body is readily accessible.

Conduit Body Gaskets, Conduit Bodies and Capped Elbows

Material	Craft@Hrs	Unit	Material Cost	Labor Cost	Installed Cost
Conduit body gaskets					
1/2"	L1@0.02	Ea	3.34	.82	4.16
3/4"	L1@0.02	Ea	3.76	.82	4.58
1"	L1@0.03	Ea	4.14	1.23	5.37
1-1/4"	L1@0.05	Ea	4.55	2.05	6.60
1-1/2"	L1@0.05	Ea	5.29	2.05	7.34
2"	L1@0.07	Ea	5.57	2.87	8.44
2-1/2" - 3"	L1@0.10	Ea	10.40	4.10	14.50
2-1/2" - 4"	L1@0.15	Ea	12.40	6.15	18.55
Type LB, LL or LR aluminum conduit bodies with covers					
1/2"	L1@0.10	Ea	17.90	4.10	22.00
3/4"	L1@0.15	Ea	21.30	6.15	27.45
1"	L1@0.15	Ea	31.50	6.15	37.65
1-1/4"	L1@0.20	Ea	49.90	8.21	58.11
1-1/2"	L1@0.20	Ea	64.80	8.21	73.01
2"	L1@0.25	Ea	107.00	10.30	117.30
2-1/2"	L1@0.30	Ea	223.00	12.30	235.30
3"	L1@0.40	Ea	299.00	16.40	315.40
Type LB, LL or LR mogul aluminum conduit bodies with covers & gaskets					
1"	L1@0.25	Ea	135.00	10.30	145.30
1-1/4"	L1@0.30	Ea	142.00	12.30	154.30
1-1/2"	L1@0.30	Ea	252.00	12.30	264.30
2"	L1@0.50	Ea	387.00	20.50	407.50
2-1/2"	L1@0.70	Ea	592.00	28.70	620.70
3"	L1@0.75	Ea	901.00	30.80	931.80
3-1/2"	L1@1.00	Ea	1,040.00	41.00	1,081.00
4"	L1@1.00	Ea	1,160.00	41.00	1,201.00
Galvanized capped elbows					
1/2"	L1@0.10	Ea	17.70	4.10	21.80
3/4"	L1@0.15	Ea	26.60	6.15	32.75
1"	L1@0.20	Ea	33.00	8.21	41.21
1-1/4"	L1@0.25	Ea	40.20	10.30	50.50
1-1/2"	L1@0.25	Ea	52.40	10.30	62.70

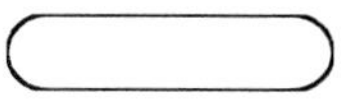

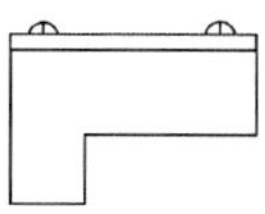

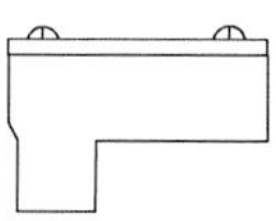

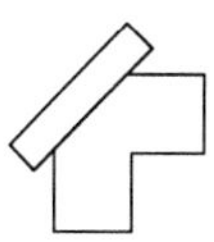

Use these figures to estimate the cost of conduit body gaskets, aluminum conduit bodies and capped elbows installed with covers and aluminum conduit under the conditions described on pages 5 and 6. Costs listed are for each fitting installed. The crew is one electrician working at a labor cost of $41.03 per manhour. These costs include layout, material handling, and normal waste. Add for covers, conduit, nipples, sales tax, delivery, supervision, mobilization, demobilization, cleanup, overhead and profit. Note: Standard conduit bodies do not include covers and gaskets. Cost of mogul bodies includes covers and gaskets.

Galvanized Cast Boxes

Galvanized cast boxes with threaded hubs

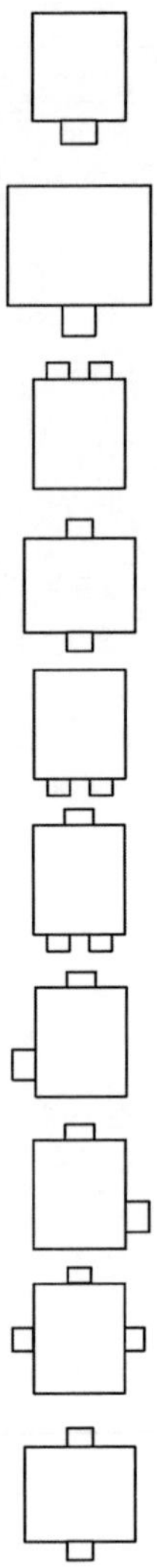

Material		Craft@Hrs	Unit	Material Cost	Labor Cost	Installed Cost
FS-1	1/2" one gang	L1@0.20	Ea	17.90	8.21	26.11
FS-2	3/4" one gang	L1@0.25	Ea	17.70	10.30	28.00
FS-3	1" one gang	L1@0.30	Ea	20.30	12.30	32.60
FS-12	1/2" two gang	L1@0.25	Ea	30.40	10.30	40.70
FS-22	3/4" two gang	L1@0.30	Ea	32.60	12.30	44.90
FS-32	1" two gang	L1@0.35	Ea	34.30	14.40	48.70
FSC-1	1/2" one gang	L1@0.25	Ea	30.70	10.30	41.00
FSC-2	3/4" one gang	L1@0.30	Ea	33.60	12.30	45.90
FSC-3	1" one gang	L1@0.35	Ea	41.60	14.40	56.00
FSC-12	1/2" two gang	L1@0.30	Ea	37.50	12.30	49.80
FSC-22	3/4" two gang	L1@0.35	Ea	33.60	14.40	48.00
FSC-32	1" two gang	L1@0.40	Ea	44.60	16.40	61.00
FSCC-1	1/2" one gang	L1@0.35	Ea	24.30	14.40	38.70
FSCC-2	3/4" one gang	L1@0.40	Ea	41.30	16.40	57.70
FSCT-1	1/2" one gang	L1@0.35	Ea	26.60	14.40	41.00
FSCT-2	3/4" one gang	L1@0.40	Ea	33.30	16.40	49.70
FSL-1	1/2" one gang	L1@0.30	Ea	20.50	12.30	32.80
FSL-2	3/4" one gang	L1@0.35	Ea	22.80	14.40	37.20
FSR-1	1/2" one gang	L1@0.30	Ea	23.20	12.30	35.50
FSR-2	3/4" one gang	L1@0.35	Ea	25.00	14.40	39.40
FSS-1	1/2" one gang	L1@0.35	Ea	21.80	14.40	36.20
FSS-2	3/4" one gang	L1@0.40	Ea	23.60	16.40	40.00
FST-1	1/2" one gang	L1@0.35	Ea	21.80	14.40	36.20
FST-2	3/4" one gang	L1@0.40	Ea	23.60	16.40	40.00
FSX-1	1/2" one gang	L1@0.40	Ea	20.50	16.40	36.90
FSX-2	3/4" one gang	L1@0.45	Ea	22.80	18.50	41.30
FD-1	1/2" one gang	L1@0.25	Ea	28.90	10.30	39.20
FD-2	3/4" one gang	L1@0.30	Ea	22.40	12.30	34.70
FD-3	1" one gang	L1@0.35	Ea	24.00	14.40	38.40
FDC-1	1/2" one gang	L1@0.30	Ea	26.90	12.30	39.20
FDC-2	3/4" one gang	L1@0.35	Ea	29.20	14.40	43.60
FDC-3	1" one gang	L1@0.40	Ea	34.40	16.40	50.80

Use these figures to estimate the cost of galvanized cast boxes installed on conduit under the conditions described on pages 5 and 6. Costs listed are for each box installed. The crew is one electrician working at a labor cost of $41.03 per manhour. These costs include box mounting, layout, material handling, and normal waste. Add for covers, gaskets, sales tax, delivery, supervision, mobilization, demobilization, cleanup, overhead and profit. Note: Boxes are raintight or weatherproof when fitted with the proper cover. These figures assume that the boxes are surface mounted in accessible locations.

Covers for Galvanized Cast Boxes

Single gang stamped metal covers

Material	Craft@Hrs	Unit	Material Cost	Labor Cost	Installed Cost
DS21 single receptacle	L1@0.05	Ea	4.93	2.05	6.98
DS23 duplex receptacle	L1@0.05	Ea	4.93	2.05	6.98
DS32 switch	L1@0.05	Ea	4.93	2.05	6.98
DS100 blank	L1@0.05	Ea	3.93	2.05	5.98

Two gang stamped metal covers

Material	Craft@Hrs	Unit	Material Cost	Labor Cost	Installed Cost
S322 2 switches	L1@0.06	Ea	8.96	2.46	11.42
S1002 blank	L1@0.06	Ea	8.96	2.46	11.42
S32212 duplex	L1@0.06	Ea	8.96	2.46	11.42
S32232 Sw & duplex	L1@0.06	Ea	8.96	2.46	11.42

Single gang cast metal covers

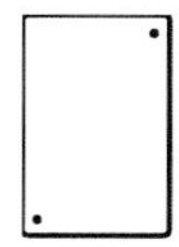

Material	Craft@Hrs	Unit	Material Cost	Labor Cost	Installed Cost
DS100G switch	L1@0.05	Ea	10.70	2.05	12.75
DS100G blank	L1@0.05	Ea	12.00	2.05	14.05

Two gang cast metal covers

Material	Craft@Hrs	Unit	Material Cost	Labor Cost	Installed Cost
S322G 2 switches	L1@0.06	Ea	35.40	2.46	37.86
S1002G blank	L1@0.06	Ea	32.10	2.46	34.56

Single gang cast weatherproof covers

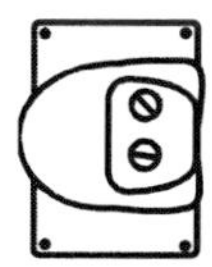

Material	Craft@Hrs	Unit	Material Cost	Labor Cost	Installed Cost
DS128 Sw rod type	L1@0.10	Ea	41.10	4.10	45.20
DS181 Sw rocker type	L1@0.10	Ea	44.40	4.10	48.50

Two gang cast weatherproof covers

Material	Craft@Hrs	Unit	Material Cost	Labor Cost	Installed Cost
DS1282 2 Sw rod type	L1@0.15	Ea	75.10	6.15	81.25

Single gang cast with hinged cover weatherproof

Material	Craft@Hrs	Unit	Material Cost	Labor Cost	Installed Cost
WLRS-1 single recept	L1@0.10	Ea	42.60	4.10	46.70
WLRD-1 duplex recept	L1@0.10	Ea	46.80	4.10	50.90

Use these figures to estimate the cost of covers installed on galvanized boxes under the conditions described on pages 5 and 6. Costs listed are for each cover installed. The crew is one electrician working at a labor cost of $41.03 per manhour. These costs include the cover, mounting, layout, material handling, and normal waste. Add for sales tax, delivery, supervision, mobilization, demobilization, cleanup, overhead and profit. These figures assume that the boxes for the covers are surface mounted in accessible locations.

Galvanized Cast Expansion Fittings and Jumpers

Galvanized 4" cast expansion fitting

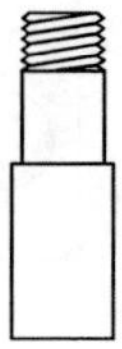

Material	Craft@Hrs	Unit	Material Cost	Labor Cost	Installed Cost
1/2"	L1@0.25	Ea	75.80	10.30	86.10
3/4"	L1@0.30	Ea	77.80	12.30	90.10
1"	L1@0.40	Ea	95.00	16.40	111.40
1-1/4"	L1@0.50	Ea	126.00	20.50	146.50
1-1/2"	L1@0.50	Ea	140.00	20.50	160.50
2"	L1@0.60	Ea	208.00	24.60	232.60
2-1/2"	L1@0.70	Ea	412.00	28.70	440.70
3"	L1@0.70	Ea	411.00	28.70	439.70
3-1/2"	L1@0.80	Ea	646.00	32.80	678.80
4"	L1@1.00	Ea	879.00	41.00	920.00

Galvanized 8" cast expansion fitting

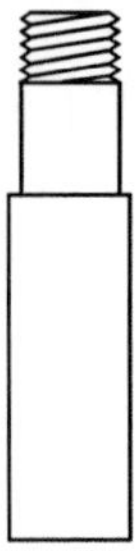

Material	Craft@Hrs	Unit	Material Cost	Labor Cost	Installed Cost
1/2"	L1@0.30	Ea	154.00	12.30	166.30
3/4"	L1@0.40	Ea	169.00	16.40	185.40
1"	L1@0.50	Ea	225.00	20.50	245.50
1-1/4"	L1@0.60	Ea	274.00	24.60	298.60
1-1/2"	L1@0.60	Ea	417.00	24.60	441.60
2"	L1@0.70	Ea	582.00	28.70	610.70
2-1/2"	L1@0.80	Ea	985.00	32.80	1,017.80
3"	L1@1.00	Ea	1,210.00	41.00	1,251.00
3-1/2"	L1@1.25	Ea	1,640.00	51.30	1,691.30
4"	L1@1.30	Ea	1,810.00	53.30	1,863.30

4" bonding jumpers for galvanized cast expansion fitting

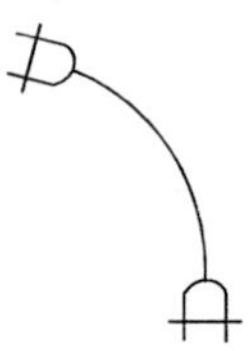

Material	Craft@Hrs	Unit	Material Cost	Labor Cost	Installed Cost
1/2" - 3/4"	L1@0.15	Ea	64.20	6.15	70.35
1" - 1-1/4"	L1@0.20	Ea	65.20	8.21	73.41
1-1/2" - 2"	L1@0.30	Ea	80.50	12.30	92.80
2-1/2" - 3"	L1@0.40	Ea	85.40	16.40	101.80
3-1/2" - 4"	L1@0.50	Ea	176.00	20.50	196.50

8" bonding jumpers for galvanized cast expansion fitting

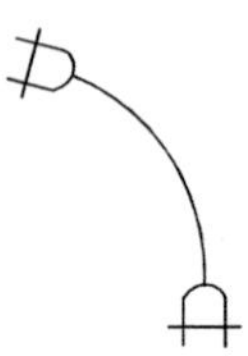

Material	Craft@Hrs	Unit	Material Cost	Labor Cost	Installed Cost
1/2" - 3/4"	L1@0.15	Ea	67.10	6.15	73.25
1" - 1-1/4"	L1@0.25	Ea	77.90	10.30	88.20
1-1/2" - 2"	L1@0.35	Ea	93.30	14.40	107.70
2-1/2" - 3"	L1@0.45	Ea	128.00	18.50	146.50
3-1/2" - 4"	L1@0.60	Ea	131.00	24.60	155.60
5"	L1@0.80	Ea	187.00	32.80	219.80

Use these figures to estimate the cost of expansion fittings and bonding jumpers installed on conduit under the conditions described on pages 5 and 6. Costs listed are for each fitting installed. The crew is one electrician working at a labor cost of $41.03 per manhour. These costs include layout, material handling, and normal waste. Add for conduit, supports, sales tax, delivery, supervision, mobilization, demobilization, cleanup, overhead and profit. Note: These fittings are installed at construction expansion joints and are suitable for installation in concrete. The bonding jumper provides grounding continuity.

Steel or malleable reducing bushings

Material	Craft@Hrs	Unit	Material Cost	Labor Cost	Installed Cost
3/4" - 1/2"	L1@0.05	Ea	1.60	2.05	3.65
1" - 1/2"	L1@0.05	Ea	2.38	2.05	4.43
1" - 3/4"	L1@0.05	Ea	2.38	2.05	4.43
1-1/4" - 1/2"	L1@0.06	Ea	4.19	2.46	6.65
1-1/4" - 3/4"	L1@0.06	Ea	4.19	2.46	6.65
1-1/4" - 1"	L1@0.06	Ea	4.19	2.46	6.65
1-1/2" - 1/2"	L1@0.08	Ea	5.41	3.28	8.69
1-1/2" - 3/4"	L1@0.08	Ea	5.41	3.28	8.69
1-1/2" - 1"	L1@0.08	Ea	5.41	3.28	8.69
1-1/2" - 1-1/4"	L1@0.08	Ea	5.41	3.28	8.69
2" - 1/2"	L1@0.10	Ea	11.60	4.10	15.70
2" - 3/4"	L1@0.10	Ea	11.60	4.10	15.70
2" - 1"	L1@0.10	Ea	14.50	4.10	18.60
2" - 1-1/4"	L1@0.10	Ea	10.50	4.10	14.60
2" - 1-1/2"	L1@0.10	Ea	11.60	4.10	15.70
2-1/2" - 1-1/2"	L1@0.15	Ea	18.10	6.15	24.25
2-1/2" - 2"	L1@0.15	Ea	18.10	6.15	24.25
3" - 2-1/2"	L1@0.20	Ea	49.20	8.21	57.41
3-1/2" - 2"	L1@0.25	Ea	49.20	10.30	59.50
3-1/2" - 2-1/2"	L1@0.25	Ea	49.20	10.30	59.50
3-1/2" - 3"	L1@0.25	Ea	54.10	10.30	64.40
4" - 2-1/2"	L1@0.30	Ea	43.50	12.30	55.80
4" - 3"	L1@0.30	Ea	43.50	12.30	55.80
4" - 3-1/2"	L1@0.30	Ea	40.10	12.30	52.40

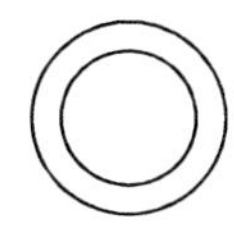

Aluminum reducing bushings

Material	Craft@Hrs	Unit	Material Cost	Labor Cost	Installed Cost
3/4" - 1/2"	L1@0.05	Ea	5.23	2.05	7.28
1" - 1/2"	L1@0.05	Ea	5.26	2.05	7.31
1" - 3/4"	L1@0.05	Ea	5.26	2.05	7.31
1-1/4" - 1/2"	L1@0.06	Ea	10.30	2.46	12.76
1-1/4" - 3/4"	L1@0.06	Ea	10.20	2.46	12.66
1-1/4" - 1"	L1@0.06	Ea	11.40	2.46	13.86
1-1/2" - 1/2"	L1@0.08	Ea	17.10	3.28	20.38
1-1/2" - 3/4"	L1@0.08	Ea	17.10	3.28	20.38
1-1/2" - 1"	L1@0.08	Ea	17.10	3.28	20.38
1-1/2" - 1-1/4"	L1@0.08	Ea	17.10	3.28	20.38
2" - 1/2"	L1@0.10	Ea	23.00	4.10	27.10
2" - 3/4"	L1@0.10	Ea	23.00	4.10	27.10
2" - 1"	L1@0.10	Ea	23.00	4.10	27.10
2" - 1-1/4"	L1@0.10	Ea	23.00	4.10	27.10
2" - 1-1/2"	L1@0.10	Ea	23.00	4.10	27.10

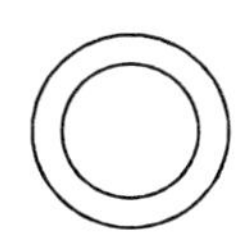

Use these figures to estimate the cost of reducing bushings installed on conduit under the conditions described on pages 5 and 6. Costs listed are for each bushing installed. The crew is one electrician working at a labor cost of $41.03 per manhour. These costs include layout, material handling, and normal waste. Add for sales tax, delivery, supervision, mobilization, demobilization, cleanup, overhead and profit. Note: Material cost is based on purchase of full boxes. These bushings are used to reduce the threaded hub size in cast boxes when smaller conduit is used.

Reducing Bushings and Reducing Washers

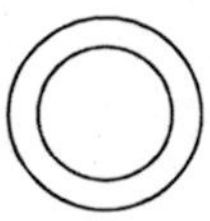

Aluminum reducing bushings

Material	Craft@Hrs	Unit	Material Cost	Labor Cost	Installed Cost
2-1/2" - 1"	L1@0.15	Ea	22.00	6.15	28.15
2-1/2" - 1-1/4"	L1@0.15	Ea	22.00	6.15	28.15
2-1/2" - 1-1/2"	L1@0.15	Ea	22.00	6.15	28.15
2-1/2" - 2"	L1@0.15	Ea	22.00	6.15	28.15
3" - 1-1/4"	L1@0.20	Ea	45.50	8.21	53.71
3" - 1-1/2"	L1@0.20	Ea	45.50	8.21	53.71
3" - 2"	L1@0.20	Ea	45.50	8.21	53.71
3" - 2-1/2"	L1@0.20	Ea	45.50	8.21	53.71
3-1/2" - 2"	L1@0.25	Ea	50.30	10.30	60.60
3-1/2" - 2-1/2"	L1@0.25	Ea	50.30	10.30	60.60
3-1/2" - 3"	L1@0.25	Ea	50.30	10.30	60.60
4" - 2"	L1@0.30	Ea	77.80	12.30	90.10
4" - 2-1/2"	L1@0.30	Ea	77.80	12.30	90.10
4" - 3"	L1@0.30	Ea	77.80	12.30	90.10
4" - 3-1/2"	L1@0.30	Ea	77.80	12.30	90.10

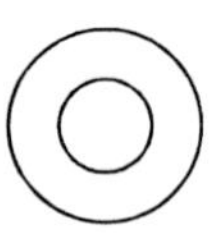

Steel reducing washers, set of 2

Material	Craft@Hrs	Unit	Material Cost	Labor Cost	Installed Cost
3/4" - 1/2"	L1@0.05	Pr	.45	2.05	2.50
1" - 1/2"	L1@0.06	Pr	.69	2.46	3.15
1" - 3/4"	L1@0.06	Pr	.64	2.46	3.10
1-1/4" - 1/2"	L1@0.08	Pr	1.23	3.28	4.51
1-1/4" - 3/4"	L1@0.08	Pr	1.12	3.28	4.40
1-1/4" - 1"	L1@0.08	Pr	1.15	3.28	4.43
1-1/2" - 1/2"	L1@0.10	Pr	1.43	4.10	5.53
1-1/2" - 3/4"	L1@0.10	Pr	1.60	4.10	5.70
1-1/2" - 1"	L1@0.10	Pr	1.35	4.10	5.45
1-1/2" - 1-1/4"	L1@0.10	Pr	1.37	4.10	5.47
2" - 1/2"	L1@0.15	Pr	2.34	6.15	8.49
2" - 3/4"	L1@0.15	Pr	2.07	6.15	8.22
2" - 1"	L1@0.15	Pr	1.94	6.15	8.09
2" - 1-1/4"	L1@0.15	Pr	1.94	6.15	8.09
2" - 1-1/2"	L1@0.15	Pr	1.94	6.15	8.09
2-1/2" - 1"	L1@0.20	Pr	2.55	8.21	10.76
2-1/2" - 1-1/4"	L1@0.20	Pr	2.55	8.21	10.76
2-1/2" - 1-1/2"	L1@0.20	Pr	2.55	8.21	10.76
2-1/2" - 2"	L1@0.20	Pr	2.55	8.21	10.76
3" - 1-1/4"	L1@0.25	Pr	3.22	10.30	13.52
3" - 1-1/2"	L1@0.25	Pr	3.22	10.30	13.52
3" - 2"	L1@0.25	Pr	3.22	10.30	13.52
3" - 2-1/2"	L1@0.25	Pr	3.22	10.30	13.52
3-1/2" - 2"	L1@0.30	Pr	9.25	12.30	21.55
3-1/2" - 2-1/2"	L1@0.30	Pr	9.25	12.30	21.55
3-1/2" - 3"	L1@0.30	Pr	9.25	12.30	21.55
4" - 2"	L1@0.35	Pr	26.20	14.40	40.60
4" - 2-1/2"	L1@0.35	Pr	26.20	14.40	40.60
4" - 3"	L1@0.35	Pr	26.20	14.40	40.60
4" - 3-1/2"	L1@0.35	Pr	26.20	14.40	40.60

Use these figures to estimate the cost of reducing bushings and reducing washers installed on conduit under the conditions described on pages 5 and 6. Costs for bushings are for each bushing installed. Costs for reducing washers are per pair of washers installed. The crew is one electrician working at a labor cost of $41.03 per manhour. These costs include layout, material handling, and normal waste. Add for sales tax, delivery, supervision, mobilization, demobilization, cleanup, overhead and profit. Note: Material cost is based on purchase of full boxes. These bushings are used to reduce the threaded hub size in cast boxes when smaller conduit is used.

Die cast bushed nipples

Material	Craft@Hrs	Unit	Material Cost	Labor Cost	Installed Cost
1/2"	L1@0.05	Ea	.41	2.05	2.46
3/4"	L1@0.06	Ea	.72	2.46	3.18
1"	L1@0.08	Ea	1.39	3.28	4.67
1-1/4"	L1@0.10	Ea	2.11	4.10	6.21
1-1/2"	L1@0.10	Ea	2.98	4.10	7.08
2"	L1@0.15	Ea	4.62	6.15	10.77
2-1/2"	L1@0.20	Ea	7.51	8.21	15.72
3"	L1@0.20	Ea	12.20	8.21	20.41
3-1/2"	L1@0.25	Ea	21.00	10.30	31.30
4"	L1@0.25	Ea	22.00	10.30	32.30

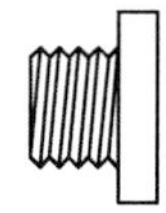

Malleable bushed nipples

1/2"	L1@0.05	Ea	1.02	2.05	3.07
3/4"	L1@0.06	Ea	1.94	2.46	4.40
1"	L1@0.08	Ea	3.51	3.28	6.79
1-1/4"	L1@0.10	Ea	3.42	4.10	7.52
1-1/2"	L1@0.10	Ea	3.66	4.10	7.76
2"	L1@0.15	Ea	4.83	6.15	10.98
2-1/2"	L1@0.20	Ea	8.86	8.21	17.07
3"	L1@0.20	Ea	18.10	8.21	26.31
3-1/2"	L1@0.25	Ea	28.20	10.30	38.50
4"	L1@0.25	Ea	45.80	10.30	56.10

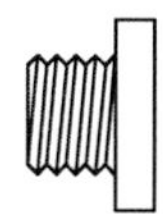

Insulated die cast bushed nipples

1/2"	L1@0.05	Ea	.45	2.05	2.50
3/4"	L1@0.06	Ea	.85	2.46	3.31
1"	L1@0.08	Ea	1.56	3.28	4.84
1-1/4"	L1@0.10	Ea	2.35	4.10	6.45
1-1/2"	L1@0.10	Ea	3.30	4.10	7.40
2"	L1@0.15	Ea	5.16	6.15	11.31
2-1/2"	L1@0.20	Ea	8.33	8.21	16.54
3"	L1@0.20	Ea	13.60	8.21	21.81
3-1/2"	L1@0.25	Ea	23.40	10.30	33.70
4"	L1@0.25	Ea	26.20	10.30	36.50

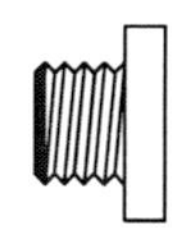

Insulated malleable bushed nipples

1/2"	L1@0.05	Ea	.94	2.05	2.99
3/4"	L1@0.06	Ea	1.73	2.46	4.19
1"	L1@0.08	Ea	3.22	3.28	6.50
1-1/4"	L1@0.10	Ea	4.83	4.10	8.93
1-1/2"	L1@0.10	Ea	6.43	4.10	10.53
2"	L1@0.15	Ea	8.53	6.15	14.68

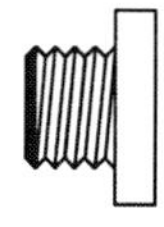

Use these figures to estimate the cost of bushed nipples installed on conduit under the conditions described on pages 5 and 6. Costs listed are for each nipple installed. The crew is one electrician working at a labor cost of $41.03 per manhour. These costs include layout, material handling, and normal waste. Add for locknut, bushing, sales tax, delivery, supervision, mobilization, demobilization, cleanup, overhead and profit. Note: Material cost is based on purchase of full boxes. Bushed nipples are often used in threaded hubs.

Bushed Nipples, Couplings and Offset Nipples

Insulated malleable bushed nipples

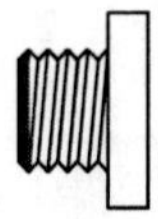

Material	Craft@Hrs	Unit	Material Cost	Labor Cost	Installed Cost
2-1/2"	L1@0.20	Ea	11.10	8.21	19.31
3"	L1@0.20	Ea	35.10	8.21	43.31
3-1/2"	L1@0.25	Ea	49.50	10.30	59.80
4"	L1@0.25	Ea	77.80	10.30	88.10
5"	L1@0.30	Ea	235.00	12.30	247.30
6"	L1@0.40	Ea	357.00	16.40	373.40

Malleable three-piece couplings or unions

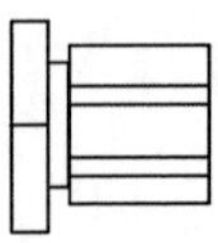

Material	Craft@Hrs	Unit	Material Cost	Labor Cost	Installed Cost
1/2"	L1@0.10	Ea	5.27	4.10	9.37
3/4"	L1@0.10	Ea	8.59	4.10	12.69
1"	L1@0.15	Ea	13.10	6.15	19.25
1-1/4"	L1@0.20	Ea	23.50	8.21	31.71
1-1/2"	L1@0.20	Ea	29.20	8.21	37.41
2"	L1@0.25	Ea	57.50	10.30	67.80
2-1/2"	L1@0.30	Ea	139.00	12.30	151.30
3"	L1@0.30	Ea	191.00	12.30	203.30
3-1/2"	L1@0.50	Ea	322.00	20.50	342.50
4"	L1@0.50	Ea	393.00	20.50	413.50
5"	L1@1.00	Ea	575.00	41.00	616.00
6"	L1@1.25	Ea	876.00	51.30	927.30

Malleable offset nipples

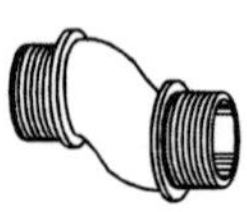

Material	Craft@Hrs	Unit	Material Cost	Labor Cost	Installed Cost
1/2"	L1@0.10	Ea	10.70	4.10	14.80
3/4"	L1@0.10	Ea	11.30	4.10	15.40
1"	L1@0.15	Ea	14.10	6.15	20.25
1-1/4"	L1@0.20	Ea	31.60	8.21	39.81
1-1/2"	L1@0.20	Ea	38.70	8.21	46.91
2"	L1@0.25	Ea	61.50	10.30	71.80

Die cast offset nipples

Material	Craft@Hrs	Unit	Material Cost	Labor Cost	Installed Cost
1/2"	L1@0.10	Ea	3.91	4.10	8.01
3/4"	L1@0.15	Ea	5.48	6.15	11.63
1"	L1@0.20	Ea	7.02	8.21	15.23
1-1/4"	L1@0.25	Ea	10.10	10.30	20.40
1-1/2"	L1@0.25	Ea	12.70	10.30	23.00
2"	L1@0.30	Ea	26.90	12.30	39.20

Use these figures to estimate the cost of bushed nipples, unions, and offset nipples installed on conduit under the conditions described on pages 5 and 6. Costs listed are for each fitting installed. The crew is one electrician working at a labor cost of $41.03 per manhour. These costs include layout, material handling, and normal waste. Add for locknut, bushing, sales tax, delivery, supervision, mobilization, demobilization, cleanup, overhead and profit. Note: Material cost is based on purchase of full boxes. Three-piece couplings are made to fit the flat thread used on electrical fittings. Unions made for plumbing pipe should not be used in electrical systems.

Die cast offset nipples (continued)

Material	Craft@Hrs	Unit	Material Cost	Labor Cost	Installed Cost
2-1/2"	L1@0.35	Ea	99.30	14.40	113.70
3"	L1@0.35	Ea	124.00	14.40	138.40
3-1/2"	L1@0.40	Ea	181.00	16.40	197.40
4"	L1@0.45	Ea	224.00	18.50	242.50

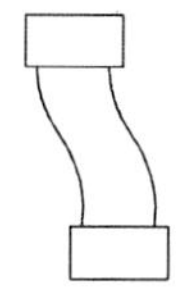

Malleable threadless connectors

Material	Craft@Hrs	Unit	Material Cost	Labor Cost	Installed Cost
1/2"	L1@0.05	Ea	2.91	2.05	4.96
3/4"	L1@0.06	Ea	4.80	2.46	7.26
1"	L1@0.08	Ea	6.80	3.28	10.08
1-1/4"	L1@0.10	Ea	13.50	4.10	17.60
1-1/2"	L1@0.10	Ea	18.70	4.10	22.80
2"	L1@0.15	Ea	36.70	6.15	42.85
2-1/2"	L1@0.25	Ea	173.00	10.30	183.30
3"	L1@0.30	Ea	230.00	12.30	242.30
3-1/2"	L1@0.35	Ea	303.00	14.40	317.40
4"	L1@0.40	Ea	370.00	16.40	386.40

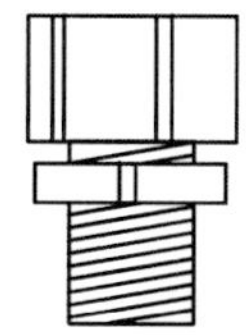

Insulated malleable threadless connectors

Material	Craft@Hrs	Unit	Material Cost	Labor Cost	Installed Cost
1/2"	L1@0.05	Ea	6.76	2.05	8.81
3/4"	L1@0.06	Ea	10.80	2.46	13.26
1"	L1@0.08	Ea	16.00	3.28	19.28
1-1/4"	L1@0.10	Ea	30.60	4.10	34.70
1-1/2"	L1@0.10	Ea	42.50	4.10	46.60
2"	L1@0.15	Ea	99.30	6.15	105.45
2-1/2"	L1@0.25	Ea	244.00	10.30	254.30
3"	L1@0.30	Ea	322.00	12.30	334.30
3-1/2"	L1@0.35	Ea	417.00	14.40	431.40
4"	L1@0.40	Ea	490.00	16.40	506.40

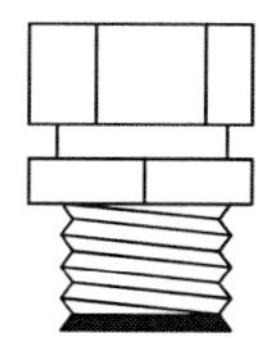

Malleable threadless couplings

Material	Craft@Hrs	Unit	Material Cost	Labor Cost	Installed Cost
1/2"	L1@0.10	Ea	4.62	4.10	8.72
3/4"	L1@0.10	Ea	7.09	4.10	11.19
1"	L1@0.15	Ea	11.90	6.15	18.05
1-1/4"	L1@0.20	Ea	19.80	8.21	28.01
1-1/2"	L1@0.20	Ea	24.80	8.21	33.01
2"	L1@0.25	Ea	55.30	10.30	65.60

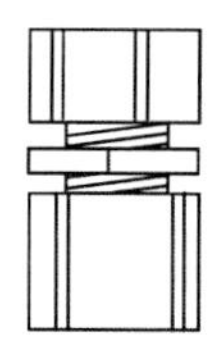

Use these figures to estimate the cost of offset nipples, connectors, and couplings installed in conduit systems under the conditions described on pages 5 and 6. Costs listed are for each fitting installed. The crew is one electrician working at a labor cost of $41.03 per manhour. These costs include removing the knockout, layout, material handling, and normal waste. Add for locknuts, bushings, sales tax, delivery, supervision, mobilization, demobilization, cleanup, overhead and profit. Note: Material costs are based on purchase of full boxes. Threadless fittings are made for rigid conduit only and do not fit EMT conduit. They're rated for raintight or weatherproof applications.

Couplings and Connectors

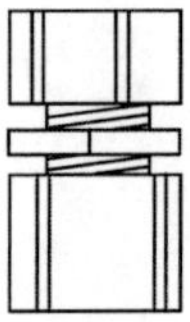

Material	Craft@Hrs	Unit	Material Cost	Labor Cost	Installed Cost
Malleable threadless couplings (continued)					
2-1/2"	L1@0.30	Ea	242.00	12.30	254.30
3"	L1@0.40	Ea	332.00	16.40	348.40
3-1/2"	L1@0.50	Ea	427.00	20.50	447.50
4"	L1@0.60	Ea	558.00	24.60	582.60

Malleable set screw couplings

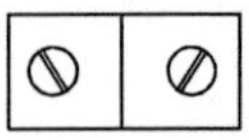

Material	Craft@Hrs	Unit	Material Cost	Labor Cost	Installed Cost
1/2"	L1@0.10	Ea	5.33	4.10	9.43
3/4"	L1@0.10	Ea	7.23	4.10	11.33
1"	L1@0.15	Ea	12.20	6.15	18.35
1-1/4"	L1@0.20	Ea	17.60	8.21	25.81
1-1/2"	L1@0.20	Ea	22.70	8.21	30.91
2"	L1@0.25	Ea	51.00	10.30	61.30
2-1/2"	L1@0.30	Ea	106.00	12.30	118.30
3"	L1@0.40	Ea	128.00	16.40	144.40
3-1/2"	L1@0.50	Ea	167.00	20.50	187.50
4"	L1@0.60	Ea	213.00	24.60	237.60

Steel set screw connectors

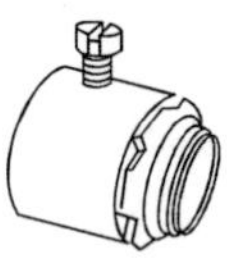

Material	Craft@Hrs	Unit	Material Cost	Labor Cost	Installed Cost
1/2"	L1@0.05	Ea	5.24	2.05	7.29
3/4"	L1@0.06	Ea	6.31	2.46	8.77
1"	L1@0.08	Ea	8.39	3.28	11.67
1-1/4"	L1@0.10	Ea	15.20	4.10	19.30
1-1/2"	L1@0.10	Ea	22.10	4.10	26.20
2"	L1@0.15	Ea	33.40	6.15	39.55
2-1/2"	L1@0.25	Ea	121.00	10.30	131.30
3"	L1@0.30	Ea	154.00	12.30	166.30
3-1/2"	L1@0.35	Ea	230.00	14.40	244.40
4"	L1@0.40	Ea	286.00	16.40	302.40

Insulated steel set screw connectors

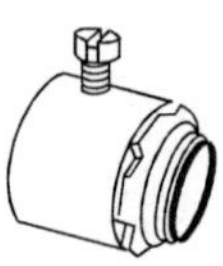

Material	Craft@Hrs	Unit	Material Cost	Labor Cost	Installed Cost
1/2"	L1@0.05	Ea	5.50	2.05	7.55
3/4"	L1@0.06	Ea	6.65	2.46	9.11
1"	L1@0.08	Ea	8.63	3.28	11.91
1-1/4"	L1@0.10	Ea	15.40	4.10	19.50
1-1/2"	L1@0.10	Ea	22.40	4.10	26.50
2"	L1@0.15	Ea	36.70	6.15	42.85
2-1/2"	L1@0.25	Ea	124.00	10.30	134.30
3"	L1@0.30	Ea	160.00	12.30	172.30
3-1/2"	L1@0.35	Ea	247.00	14.40	261.40
4"	L1@0.40	Ea	323.00	16.40	339.40

Use these figures to estimate the cost of couplings and connectors installed on conduit under the conditions described on pages 5 and 6. Costs listed are for each fitting installed. The crew is one electrician working at a labor cost of $41.03 per manhour. These costs include removing the knockout, the locknut, layout, material handling, and normal waste. Add for bushings, sales tax, delivery, supervision, mobilization, demobilization, cleanup, overhead and profit. Note: Material cost is based on purchase of full boxes.

Malleable set screw connectors

Material	Craft@Hrs	Unit	Material Cost	Labor Cost	Installed Cost
1/2"	L1@0.05	Ea	3.84	2.05	5.89
3/4"	L1@0.06	Ea	5.33	2.46	7.79
1"	L1@0.08	Ea	8.57	3.28	11.85
1-1/4"	L1@0.10	Ea	14.90	4.10	19.00
1-1/2"	L1@0.10	Ea	21.50	4.10	25.60
2"	L1@0.15	Ea	42.80	6.15	48.95
2-1/2"	L1@0.25	Ea	125.00	10.30	135.30
3"	L1@0.30	Ea	166.00	12.30	178.30
3-1/2"	L1@0.35	Ea	228.00	14.40	242.40
4"	L1@0.40	Ea	279.00	16.40	295.40

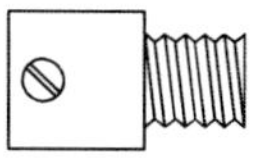

Insulated malleable set screw connectors

Material	Craft@Hrs	Unit	Material Cost	Labor Cost	Installed Cost
1/2"	L1@0.05	Ea	6.29	2.05	8.34
3/4"	L1@0.06	Ea	7.60	2.46	10.06
1"	L1@0.08	Ea	12.30	3.28	15.58
1-1/4"	L1@0.10	Ea	17.30	4.10	21.40
1-1/2"	L1@0.10	Ea	27.20	4.10	31.30
2"	L1@0.15	Ea	46.20	6.15	52.35
2-1/2"	L1@0.25	Ea	138.00	10.30	148.30
3"	L1@0.30	Ea	194.00	12.30	206.30
3-1/2"	L1@0.35	Ea	244.00	14.40	258.40
4"	L1@0.40	Ea	305.00	16.40	321.40

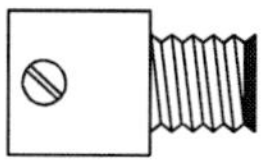

Steel one hole straps

Material	Craft@Hrs	Unit	Material Cost	Labor Cost	Installed Cost
1/2"	L1@0.05	Ea	.18	2.05	2.23
3/4"	L1@0.06	Ea	.26	2.46	2.72
1"	L1@0.08	Ea	.39	3.28	3.67
1-1/4"	L1@0.10	Ea	.58	4.10	4.68
1-1/2"	L1@0.10	Ea	.87	4.10	4.97
2"	L1@0.10	Ea	1.09	4.10	5.19
2-1/2"	L1@0.15	Ea	1.95	6.15	8.10
3"	L1@0.20	Ea	2.36	8.21	10.57
3-1/2"	L1@0.20	Ea	3.09	8.21	11.30
4"	L1@0.20	Ea	3.95	8.21	12.16

Malleable one hole straps

Material	Craft@Hrs	Unit	Material Cost	Labor Cost	Installed Cost
1/2"	L1@0.05	Ea	.35	2.05	2.40
3/4"	L1@0.06	Ea	.48	2.46	2.94
1"	L1@0.08	Ea	.79	3.28	4.07
1-1/4"	L1@0.10	Ea	1.13	4.10	5.23
1-1/2"	L1@0.10	Ea	1.63	4.10	5.73
2"	L1@0.10	Ea	2.60	4.10	6.70
2-1/2"	L1@0.15	Ea	5.06	6.15	11.21
3"	L1@0.20	Ea	7.61	8.21	15.82
3-1/2"	L1@0.20	Ea	11.50	8.21	19.71
4"	L1@0.25	Ea	20.70	10.30	31.00

Use these figures to estimate the cost of connectors and straps installed on conduit under the conditions described on pages 5 and 6. Costs listed are for each fitting installed. The crew is one electrician working at a labor cost of $41.03 per manhour. These costs include removing the knockout, the locknut, layout, material handling, and normal waste. Add for bushings, sales tax, delivery, supervision, mobilization, demobilization, cleanup, overhead and profit. Note: Material cost is based on purchase of full boxes.

Conduit Clamps and Entrance Caps

Material	Craft@Hrs	Unit	Material Cost	Labor Cost	Installed Cost
Steel two hole straps					
1/2"	L1@0.05	Ea	.16	2.05	2.21
3/4"	L1@0.06	Ea	.22	2.46	2.68
1"	L1@0.08	Ea	.36	3.28	3.64
1-1/4"	L1@0.10	Ea	.49	4.10	4.59
1-1/2"	L1@0.10	Ea	.65	4.10	4.75
2"	L1@0.10	Ea	.92	4.10	5.02
2-1/2"	L1@0.15	Ea	2.10	6.15	8.25
3"	L1@0.20	Ea	2.97	8.21	11.18
3-1/2"	L1@0.25	Ea	4.08	10.30	14.38
4"	L1@0.25	Ea	4.18	10.30	14.48
Aluminum one hole straps					
1/2"	L1@0.05	Ea	1.02	2.05	3.07
3/4"	L1@0.06	Ea	1.49	2.46	3.95
1"	L1@0.08	Ea	2.22	3.28	5.50
1-1/4"	L1@0.10	Ea	4.11	4.10	8.21
1-1/2"	L1@0.10	Ea	4.29	4.10	8.39
2"	L1@0.10	Ea	8.82	4.10	12.92
2-1/2"	L1@0.15	Ea	17.50	6.15	23.65
3"	L1@0.20	Ea	24.80	8.21	33.01
3-1/2"	L1@0.20	Ea	32.50	8.21	40.71
4"	L1@0.25	Ea	38.50	10.30	48.80
Malleable clamp backs					
1/2"	L1@0.05	Ea	.48	2.05	2.53
3/4"	L1@0.05	Ea	.54	2.05	2.59
1"	L1@0.05	Ea	.81	2.05	2.86
1-1/4"	L1@0.10	Ea	1.22	4.10	5.32
1-1/2"	L1@0.10	Ea	1.57	4.10	5.67
2"	L1@0.10	Ea	2.56	4.10	6.66
2-1/2"	L1@0.15	Ea	6.14	6.15	12.29
3"	L1@0.15	Ea	9.65	6.15	15.80
3-1/2"	L1@0.20	Ea	31.90	8.21	40.11
4"	L1@0.20	Ea	43.80	8.21	52.01
5"	L1@0.25	Ea	109.00	10.30	119.30
6"	L1@0.25	Ea	112.00	10.30	122.30

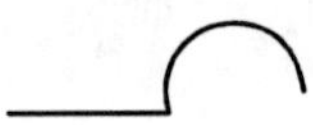

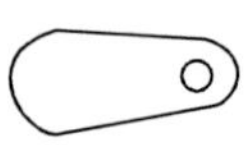

Use these figures to estimate the cost of straps and spacers installed on conduit under the conditions described on pages 5 and 6. Costs listed are for each fitting installed. The crew is one electrician working at a labor cost of $41.03 per manhour. These costs include layout, material handling, and normal waste. Add for screws, bolts, anchors, sales tax, delivery, supervision, mobilization, demobilization, cleanup, overhead and profit. Note: Material cost is based on purchase of full boxes.

Conduit Clamps and Entrance Caps

Rigid steel conduit clamps without bolts

Material	Craft@Hrs	Unit	Material Cost	Labor Cost	Installed Cost
1/2"	L1@0.05	Ea	.49	2.05	2.54
3/4"	L1@0.06	Ea	.55	2.46	3.01
1"	L1@0.08	Ea	.86	3.28	4.14
1-1/4"	L1@0.10	Ea	1.05	4.10	5.15
1-1/2"	L1@0.10	Ea	1.08	4.10	5.18
2"	L1@0.10	Ea	1.41	4.10	5.51
2-1/2"	L1@0.15	Ea	2.09	6.15	8.24
3"	L1@0.15	Ea	2.38	6.15	8.53
3-1/2"	L1@0.20	Ea	3.02	8.21	11.23
4"	L1@0.20	Ea	3.58	8.21	11.79

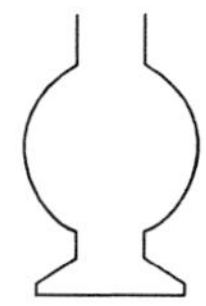

Rigid steel conduit clamps with bolts

Material	Craft@Hrs	Unit	Material Cost	Labor Cost	Installed Cost
1/2"	L1@0.05	Ea	.65	2.05	2.70
3/4"	L1@0.06	Ea	.69	2.46	3.15
1"	L1@0.08	Ea	1.00	3.28	4.28
1-1/4"	L1@0.10	Ea	1.34	4.10	5.44
1-1/2"	L1@0.10	Ea	1.61	4.10	5.71
2"	L1@0.10	Ea	1.82	4.10	5.92
2-1/2"	L1@0.15	Ea	2.63	6.15	8.78
3"	L1@0.15	Ea	2.67	6.15	8.82
3-1/2"	L1@0.20	Ea	3.14	8.21	11.35
4"	L1@0.20	Ea	3.85	8.21	12.06

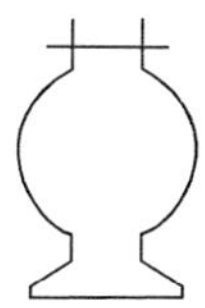

Clamp-type entrance caps

Material	Craft@Hrs	Unit	Material Cost	Labor Cost	Installed Cost
1/2"	L1@0.15	Ea	14.70	6.15	20.85
3/4"	L1@0.20	Ea	17.10	8.21	25.31
1"	L1@0.25	Ea	20.30	10.30	30.60
1-1/4"	L1@0.30	Ea	22.80	12.30	35.10
1-1/2"	L1@0.30	Ea	38.70	12.30	51.00
2"	L1@0.50	Ea	53.00	20.50	73.50
2-1/2"	L1@0.60	Ea	186.00	24.60	210.60
3"	L1@0.75	Ea	299.00	30.80	329.80
3-1/2"	L1@1.00	Ea	367.00	41.00	408.00
4"	L1@1.25	Ea	384.00	51.30	435.30

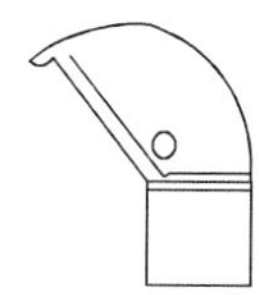

Slip fitter entrance caps

Material	Craft@Hrs	Unit	Material Cost	Labor Cost	Installed Cost
1/2"	L1@0.15	Ea	12.50	6.15	18.65
3/4"	L1@0.20	Ea	15.60	8.21	23.81
1"	L1@0.25	Ea	18.30	10.30	28.60
1-1/4"	L1@0.30	Ea	22.40	12.30	34.70
1-1/2"	L1@0.30	Ea	40.10	12.30	52.40
2"	L1@0.50	Ea	73.60	20.50	94.10

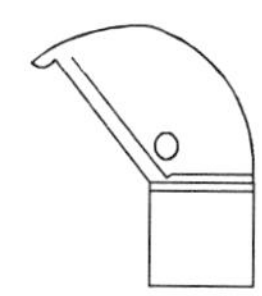

Use these figures to estimate the cost of clamps and entrance caps installed on conduit under the conditions described on pages 5 and 6. Costs listed are for each fitting installed. The crew is one electrician working at a labor cost of $41.03 per manhour. These costs include layout, material handling, and normal waste. Add for screws, bolts, anchors, sales tax, delivery, supervision, mobilization, demobilization, cleanup, overhead and profit. Note: Material cost is based on purchase of full boxes. Many other types of fittings are available. Those listed here are the most common.

PVC Coated Conduit, Elbows and Couplings

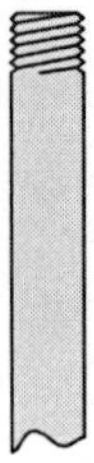

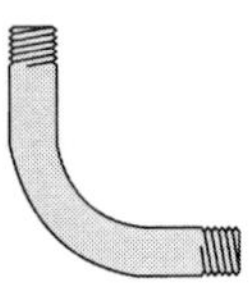

Material	Craft@Hrs	Unit	Material Cost	Labor Cost	Installed Cost
PVC coated steel conduit, 40 mil coating					
1/2"	L1@4.50	CLF	656.00	185.00	841.00
3/4"	L1@5.50	CLF	761.00	226.00	987.00
1"	L1@7.00	CLF	989.00	287.00	1,276.00
1-1/4"	L1@9.00	CLF	1,640.00	369.00	2,009.00
1-1/2"	L1@11.0	CLF	1,530.00	451.00	1,981.00
2"	L1@13.0	CLF	1,980.00	533.00	2,513.00
2-1/2"	L2@15.0	CLF	3,930.00	615.00	4,545.00
3"	L2@17.0	CLF	3,800.00	698.00	4,498.00
3-1/2"	L2@19.0	CLF	6,090.00	780.00	6,870.00
4"	L2@21.0	CLF	5,470.00	862.00	6,332.00
5"	L2@25.0	CLF	12,700.00	1,030.00	13,730.00
PVC coated steel 90 degree elbows					
1/2"	L1@0.10	Ea	31.90	4.10	36.00
3/4"	L1@0.10	Ea	33.20	4.10	37.30
1"	L1@0.15	Ea	23.20	6.15	29.35
1-1/4"	L1@0.20	Ea	46.80	8.21	55.01
1-1/2"	L1@0.20	Ea	57.30	8.21	65.51
2"	L1@0.25	Ea	49.50	10.30	59.80
2-1/2"	L2@0.30	Ea	99.10	12.30	111.40
3"	L2@0.35	Ea	159.00	14.40	173.40
3-1/2"	L2@0.40	Ea	221.00	16.40	237.40
4"	L2@0.50	Ea	415.00	20.50	435.50
5"	L2@0.75	Ea	639.00	30.80	669.80
PVC coated steel couplings					
1/2"	L1@0.05	Ea	7.68	2.05	9.73
3/4"	L1@0.06	Ea	8.06	2.46	10.52
1"	L1@0.08	Ea	10.50	3.28	13.78
1-1/4"	L1@0.10	Ea	12.20	4.10	16.30
1-1/2"	L1@0.10	Ea	14.50	4.10	18.60
2"	L1@0.15	Ea	21.30	6.15	27.45
2-1/2"	L2@0.20	Ea	52.60	8.21	60.81
3"	L2@0.20	Ea	63.90	8.21	72.11
3-1/2"	L2@0.25	Ea	82.10	10.30	92.40
4"	L2@0.25	Ea	95.80	10.30	106.10
5"	L2@0.30	Ea	310.00	12.30	322.30

Use these figures to estimate the cost of PVC coated conduit, elbows and couplings installed in corrosive areas under the conditions described on pages 5 and 6. Costs listed are for each 100 linear feet of conduit or for each fitting installed. The crew is one electrician for sizes up to 2" and two electricians for sizes over 2". The labor cost is $41.03 per manhour. These costs include cutting and threading, one coupling for each length of conduit, layout, material handling, and normal waste. Add for straps, locknuts, bushings, sales tax, delivery, supervision, mobilization, demobilization, cleanup, overhead and profit. Note: PVC patching material is available in spray cans for repairing any damaged PVC coating. Bending tools must be ground out when used on PVC conduit. Threading equipment must be modified for use on PVC conduit.

PVC coated steel one hole straps

Material	Craft@Hrs	Unit	Material Cost	Labor Cost	Installed Cost
1/2"	L1@0.05	Ea	14.60	2.05	16.65
3/4"	L1@0.06	Ea	14.60	2.46	17.06
1"	L1@0.08	Ea	14.90	3.28	18.18
1-1/4"	L1@0.10	Ea	21.80	4.10	25.90
1-1/2"	L1@0.10	Ea	23.30	4.10	27.40
2"	L1@0.15	Ea	33.60	6.15	39.75
2-1/2"	L1@0.20	Ea	31.80	8.21	40.01
3"	L1@0.25	Ea	42.70	10.30	53.00
3-1/2"	L1@0.30	Ea	76.60	12.30	88.90
4"	L1@0.40	Ea	81.60	16.40	98.00

PVC coated malleable one hole straps

Material	Craft@Hrs	Unit	Material Cost	Labor Cost	Installed Cost
1/2"	L1@0.05	Ea	11.80	2.05	13.85
3/4"	L1@0.06	Ea	11.90	2.46	14.36
1"	L1@0.10	Ea	12.20	4.10	16.30
1-1/4"	L1@0.10	Ea	17.70	4.10	21.80
1-1/2"	L1@0.15	Ea	27.40	6.15	33.55
2"	L1@0.20	Ea	28.60	8.21	36.81
2-1/2"	L1@0.20	Ea	51.40	8.21	59.61
3"	L1@0.25	Ea	87.80	10.30	98.10
3-1/2"	L1@0.30	Ea	124.00	12.30	136.30
4"	L1@0.30	Ea	132.00	12.30	144.30

PVC coated right angle beam clamps

Material	Craft@Hrs	Unit	Material Cost	Labor Cost	Installed Cost
1/2"	L1@0.10	Ea	22.00	4.10	26.10
3/4"	L1@0.15	Ea	22.20	6.15	28.35
1"	L1@0.20	Ea	29.80	8.21	38.01
1-1/4"	L1@0.25	Ea	30.10	10.30	40.40
1-1/2"	L1@0.25	Ea	37.10	10.30	47.40
2"	L1@0.30	Ea	48.10	12.30	60.40
2-1/2"	L1@0.40	Ea	52.90	16.40	69.30
3"	L1@0.50	Ea	59.30	20.50	79.80
3-1/2"	L1@0.60	Ea	61.20	24.60	85.80
4"	L1@0.60	Ea	66.90	24.60	91.50

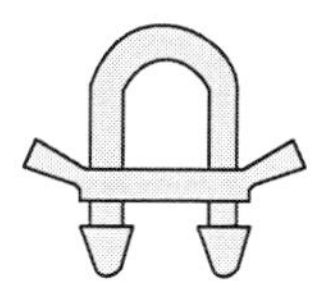

PVC coated parallel beam clamps

Material	Craft@Hrs	Unit	Material Cost	Labor Cost	Installed Cost
1/2"	L1@0.10	Ea	24.10	4.10	28.20
3/4"	L1@0.15	Ea	24.50	6.15	30.65
1"	L1@0.20	Ea	30.10	8.21	38.31
1-1/4"	L1@0.25	Ea	33.20	10.30	43.50
1-1/2"	L1@0.25	Ea	37.30	10.30	47.60
2"	L1@0.30	Ea	46.30	12.30	58.60
2-1/2"	L1@0.30	Ea	56.40	12.30	68.70
3"	L1@0.40	Ea	64.00	16.40	80.40
3-1/2"	L1@0.60	Ea	67.20	24.60	91.80
4"	L1@0.60	Ea	67.70	24.60	92.30

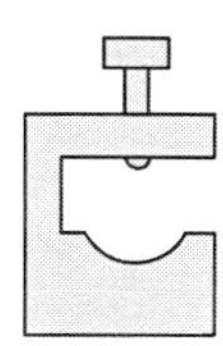

Use these figures to estimate the cost of PVC coated straps and clamps installed on PVC coated conduit under the conditions described on pages 5 and 6. Costs listed are for each fitting installed. The crew is one electrician working at a labor cost of $41.03 per manhour. These costs include screws, anchors, layout, material handling, and normal waste. Add for sales tax, delivery, supervision, mobilization, demobilization, cleanup, overhead and profit. Note: PVC patching material is available in spray cans for repairing any damaged PVC coating.

PVC Coated Clamps, U-bolts and Unions

PVC coated edge-type beam clamps

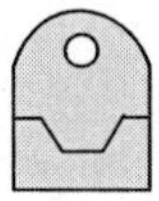

Material	Craft@Hrs	Unit	Material Cost	Labor Cost	Installed Cost
1/2"	L1@0.10	Ea	20.60	4.10	24.70
3/4"	L1@0.15	Ea	22.00	6.15	28.15
1"	L1@0.20	Ea	36.50	8.21	72.56
1-1/4"	L1@0.25	Ea	61.70	10.30	72.00
1-1/2"	L1@0.25	Ea	79.40	10.30	89.70
2"	L1@0.30	Ea	86.60	12.30	98.90

PVC coated U-bolts

Material	Craft@Hrs	Unit	Material Cost	Labor Cost	Installed Cost
1/2"	L1@0.10	Ea	9.34	4.10	13.44
3/4"	L1@0.15	Ea	9.34	6.15	15.49
1"	L1@0.20	Ea	9.34	8.21	17.55
1-1/4"	L1@0.25	Ea	9.69	10.30	19.99
1-1/2"	L1@0.25	Ea	9.87	10.30	20.17
2"	L1@0.30	Ea	12.10	12.30	24.40
2-1/2"	L1@0.35	Ea	21.00	14.40	35.40
3"	L1@0.35	Ea	22.20	14.40	36.60
3-1/2"	L1@0.40	Ea	23.60	16.40	40.00
4"	L1@0.50	Ea	31.20	20.50	51.70
5"	L1@0.60	Ea	43.80	24.60	68.40

PVC coated female conduit unions

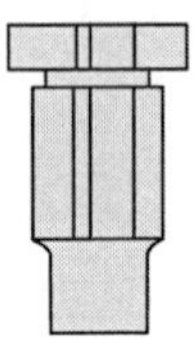

Material	Craft@Hrs	Unit	Material Cost	Labor Cost	Installed Cost
1/2"	L1@0.10	Ea	49.70	4.10	53.80
3/4"	L1@0.10	Ea	50.60	4.10	54.70
1"	L1@0.15	Ea	96.80	6.15	102.95
1-1/4"	L1@0.20	Ea	102.00	8.21	110.21
1-1/2"	L1@0.20	Ea	187.00	8.21	195.21
2"	L1@0.25	Ea	250.00	10.30	260.30
2-1/2"	L1@0.30	Ea	265.00	12.30	277.30
3"	L1@0.30	Ea	367.00	12.30	379.30
3-1/2"	L1@0.35	Ea	455.00	14.40	469.40
4"	L1@0.40	Ea	483.00	16.40	499.40

PVC coated male conduit unions

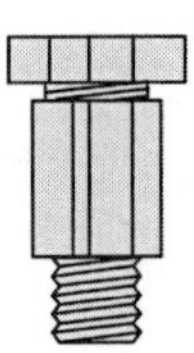

Material	Craft@Hrs	Unit	Material Cost	Labor Cost	Installed Cost
1/2"	L1@0.10	Ea	51.50	4.10	55.60
3/4"	L1@0.10	Ea	57.00	4.10	61.10
1"	L1@0.15	Ea	71.50	6.15	77.65
1-1/4"	L1@0.20	Ea	116.00	8.21	124.21
1-1/2"	L1@0.20	Ea	140.00	8.21	148.21
2"	L1@0.25	Ea	175.00	10.30	185.30
2-1/2"	L1@0.30	Ea	310.00	12.30	322.30
3"	L1@0.30	Ea	411.00	12.30	423.30
3-1/2"	L1@0.35	Ea	536.00	14.40	550.40
4"	L1@0.40	Ea	664.00	16.40	680.40

Use these figures to estimate the cost of PVC coated clamps, U-bolts and unions installed on PVC coated conduit under the conditions described on pages 5 and 6. Costs listed are for each fitting installed. The crew is one electrician working at a labor cost of $41.03 per manhour. These costs include screws, bolts, nuts, layout, material handling, and normal waste. Add for sales tax, delivery, supervision, mobilization, demobilization, cleanup, overhead and profit. Note: PVC patching material is available in spray cans for repairing any damaged PVC coating. PVC conduit fittings are rigid conduit fittings that have a PVC bonded coating for corrosion protection.

PVC Coated Couplings and Conduit Bodies

PVC coated reducing couplings

Material	Craft@Hrs	Unit	Material Cost	Labor Cost	Installed Cost
3/4"- 1/2"	L1@0.05	Ea	33.50	2.05	35.55
1"- 1/2"	L1@0.06	Ea	36.20	2.46	38.66
1"- 3/4"	L1@0.06	Ea	42.10	2.46	44.56
1-1/4"- 3/4"	L1@0.08	Ea	53.20	3.28	56.48
1-1/4"- 1"	L1@0.08	Ea	56.90	3.28	60.18
1-1/2"- 3/4"	L1@0.10	Ea	45.40	4.10	49.50
1-1/2"- 1"	L1@0.10	Ea	53.40	4.10	57.50
1-1/2"- 1-1/4"	L1@0.10	Ea	90.00	4.10	94.10
2"- 3/4"	L1@0.15	Ea	99.40	6.15	105.55
2"- 1"	L1@0.15	Ea	96.10	6.15	102.25
2"- 1-1/4"	L1@0.15	Ea	102.00	6.15	108.15
2"- 1-1/2"	L1@0.20	Ea	118.00	8.21	126.21
3"- 2"	L1@0.20	Ea	190.00	8.21	198.21
3-1/2"- 2-1/2"	L1@0.25	Ea	239.00	10.30	249.30
4"- 3"	L1@0.30	Ea	471.00	12.30	486.30
5"- 4"	L1@0.40	Ea	599.00	16.40	615.40

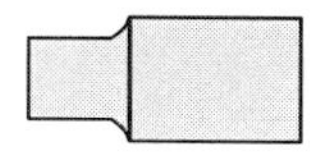

PVC coated Type C conduit bodies

Material	Craft@Hrs	Unit	Material Cost	Labor Cost	Installed Cost
C-17 1/2"	L1@0.20	Ea	48.30	8.21	56.51
C-27 3/4"	L1@0.25	Ea	50.70	10.30	61.00
C-37 1"	L1@0.30	Ea	70.70	12.30	83.00
C-47 1-1/4"	L1@0.35	Ea	106.00	14.40	120.40
C-57 1-1/2"	L1@0.40	Ea	129.00	16.40	145.40
C-67 2"	L1@0.40	Ea	183.00	16.40	199.40
C-77 2-1/2"	L1@0.50	Ea	352.00	20.50	372.50
CLF-87 3"	L1@0.60	Ea	445.00	24.60	469.60
CLF-97 3-1/2"	L1@0.70	Ea	664.00	28.70	692.70
CLF-107 4"	L1@1.00	Ea	749.00	41.00	790.00

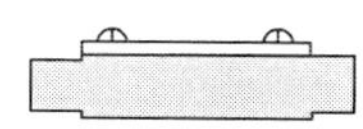

PVC coated Type LB conduit bodies

Material	Craft@Hrs	Unit	Material Cost	Labor Cost	Installed Cost
LB-17 1/2"	L1@0.20	Ea	47.80	8.21	56.01
LB-27 3/4"	L1@0.25	Ea	48.90	10.30	59.20
LB-37 1"	L1@0.30	Ea	64.20	12.30	76.50
LB-47 1-1/4"	L1@0.35	Ea	105.00	14.40	119.40
LB-57 1-1/2"	L1@0.35	Ea	128.00	14.40	142.40
LB-67 2"	L1@0.40	Ea	181.00	16.40	197.40
LB-77 2-1/2"	L1@0.50	Ea	349.00	20.50	369.50
LB-87 3"	L1@0.60	Ea	437.00	24.60	461.60
LB-97 3-1/2"	L1@0.70	Ea	644.00	28.70	672.70
LB-107 4"	L1@1.00	Ea	721.00	41.00	762.00

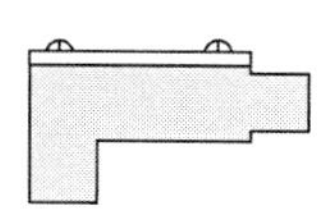

Use these figures to estimate the cost of PVC coated couplings and conduit bodies installed on PVC coated conduit under the conditions described on pages 5 and 6. Costs listed are for each fitting installed. The crew is one electrician working at a labor cost of $41.03 per manhour. These costs include covers, layout, material handling, and normal waste. Add for sales tax, delivery, supervision, mobilization, demobilization, cleanup, overhead and profit. Note: PVC patching material is available in spray cans for repairing any damaged PVC coating.

PVC Coated Conduit Bodies

Material	Craft@Hrs	Unit	Material Cost	Labor Cost	Installed Cost

PVC coated Type T conduit bodies

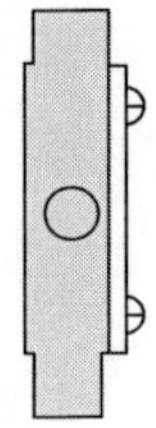

T-17 1/2"	L1@0.25	Ea	63.60	10.30	73.90
T-27 3/4"	L1@0.30	Ea	72.10	12.30	84.40
T-37 1"	L1@0.35	Ea	104.00	14.40	118.40
T-47 1-1/4"	L1@0.40	Ea	174.00	16.40	190.40
T-57 1-1/2"	L1@0.40	Ea	184.00	16.40	200.40
T-67 2"	L1@0.45	Ea	334.00	18.50	352.50
T-77 2-1/2"	L1@0.60	Ea	565.00	24.60	589.60
T-87 3"	L1@0.70	Ea	756.00	28.70	784.70
T-97 3-1/2"	L1@1.00	Ea	1,090.00	41.00	1,131.00
T-107 4"	L1@1.50	Ea	1,190.00	61.50	1,251.50

PVC coated Type TB conduit bodies

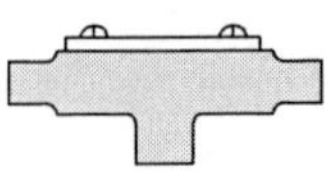

TB-17 1/2"	L1@0.25	Ea	72.10	10.30	82.40
TB-27 3/4"	L1@0.30	Ea	87.80	12.30	100.10
TB-37 1"	L1@0.35	Ea	94.90	14.40	109.30
TB-47 1-1/4"	L1@0.40	Ea	174.00	16.40	190.40
TB-57 1-1/2"	L1@0.40	Ea	184.00	16.40	200.40
TB-67 2"	L1@0.45	Ea	334.00	18.50	325.50

PVC coated Type X conduit bodies

X-17 1/2"	L1@0.30	Ea	75.00	12.30	87.30
X-27 3/4"	L1@0.35	Ea	85.30	14.40	99.70
X-37 1"	L1@0.40	Ea	96.50	16.40	112.90
X-47 1-1/4"	L1@0.45	Ea	231.00	18.50	249.50
X-57 1-1/2"	L1@0.45	Ea	301.00	18.50	319.50
X-67 2"	L1@0.50	Ea	433.00	20.50	453.50

PVC coated steel conduit body covers

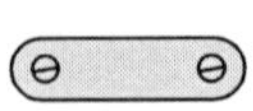

1/2"	L1@0.05	Ea	19.90	2.05	21.95
3/4"	L1@0.06	Ea	21.80	2.46	24.26
1"	L1@0.08	Ea	28.60	3.28	31.88
1-1/4"	L1@0.10	Ea	37.70	4.10	41.80
1-1/2"	L1@0.10	Ea	41.30	4.10	45.40
2"	L1@0.10	Ea	52.40	4.10	56.50
2-1/2" - 3"	L1@0.15	Ea	70.80	6.15	76.95
2-1/2" - 4"	L1@0.15	Ea	139.00	6.15	145.15

Use these figures to estimate the cost of PVC coated conduit bodies installed on PVC coated conduit and PVC coated body covers installed on conduit bodies under the conditions described on pages 5 and 6. Costs listed are for each fitting installed. The crew is one electrician working at a labor cost of $41.03 per manhour. These costs include layout, material handling, and normal waste. Add for sales tax, delivery, supervision, mobilization, demobilization, cleanup, overhead and profit. Note: PVC patching material is available in spray cans for repairing any damaged PVC coating.

PVC coated Type GUAB junction boxes with covers

Material	Craft@Hrs	Unit	Material Cost	Labor Cost	Installed Cost
1/2" - 2" dia.	L1@0.35	Ea	139.00	14.40	153.40
1/2" - 3" dia.	L1@0.40	Ea	157.00	16.40	173.40
3/4" - 2" dia.	L1@0.40	Ea	171.00	16.40	187.40
3/4" - 3" dia.	L1@0.45	Ea	174.00	18.50	192.50
1" - 3" dia.	L1@0.50	Ea	199.00	20.50	219.50
1-1/4" - 3-5/8" dia.	L1@0.60	Ea	320.00	24.60	344.60
1-1/2" - 5" dia.	L1@0.75	Ea	518.00	30.80	548.80
2" - 5" dia.	L1@1.00	Ea	583.00	41.00	624.00

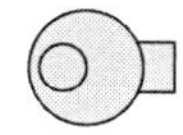

PVC coated Type GUAC junction boxes with covers

Material	Craft@Hrs	Unit	Material Cost	Labor Cost	Installed Cost
1/2" - 2" dia.	L1@0.35	Ea	139.00	14.40	153.40
1/2" - 3" dia.	L1@0.40	Ea	157.00	16.40	173.40
3/4" - 2" dia.	L1@0.40	Ea	171.00	16.40	187.40
3/4" - 3" dia.	L1@0.45	Ea	174.00	18.50	192.50
1" - 3" dia.	L1@0.50	Ea	199.00	20.50	219.50
1-1/4" - 3-5/8" dia.	L1@0.60	Ea	320.00	24.60	344.60
1-1/2" - 5" dia.	L1@0.75	Ea	518.00	30.80	548.80
2" - 5" dia.	L1@1.00	Ea	583.00	41.00	624.00

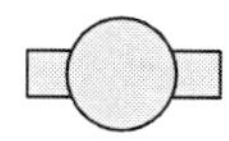

PVC coated Type GUAL junction boxes with covers

Material	Craft@Hrs	Unit	Material Cost	Labor Cost	Installed Cost
1/2" - 2" dia.	L1@0.35	Ea	139.00	14.40	153.40
1/2" - 3" dia.	L1@0.40	Ea	157.00	16.40	173.40
3/4" - 2" dia.	L1@0.40	Ea	171.00	16.40	187.40
3/4" - 3" dia.	L1@0.45	Ea	174.00	18.50	192.50
1" - 3" dia.	L1@0.50	Ea	199.00	20.50	219.50
1-1/4" - 3-5/8" dia.	L1@0.60	Ea	320.00	24.60	344.60
1-1/2" - 5" dia.	L1@0.75	Ea	518.00	30.80	548.80
2" - 5" dia.	L1@1.00	Ea	583.00	41.00	624.00

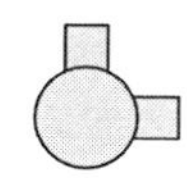

PVC coated Type GUAN junction boxes with covers

Material	Craft@Hrs	Unit	Material Cost	Labor Cost	Installed Cost
1/2" - 2" dia.	L1@0.35	Ea	139.00	14.40	153.40
1/2" - 3" dia.	L1@0.40	Ea	157.00	16.40	173.40
3/4" - 2" dia.	L1@0.40	Ea	171.00	16.40	187.40
3/4" - 3" dia.	L1@0.45	Ea	174.00	18.50	192.50
1" - 3" dia.	L1@0.50	Ea	199.00	20.50	219.50
1-1/4" - 3-5/8" dia.	L1@0.60	Ea	320.00	24.60	344.60
1-1/2" - 5" dia.	L1@0.75	Ea	518.00	30.80	548.80
2" - 5" dia.	L1@1.00	Ea	583.00	41.00	624.00

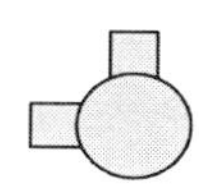

PVC coated Type GUAW junction boxes with covers

Material	Craft@Hrs	Unit	Material Cost	Labor Cost	Installed Cost
1/2" - 2" dia.	L1@0.40	Ea	186.00	16.40	202.40
1/2" - 3" dia.	L1@0.45	Ea	221.00	18.50	239.50
3/4" - 2" dia.	L1@0.45	Ea	195.00	18.50	213.50
3/4" - 3" dia.	L1@0.50	Ea	225.00	20.50	245.50

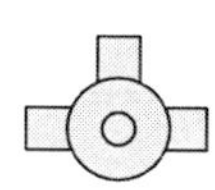

Use these figures to estimate the cost of PVC coated junction boxes installed on PVC coated conduit under the conditions described on pages 5 and 6. Costs listed are for each fitting installed. The crew is one electrician working at a labor cost of $41.03 per manhour. These costs include the box cover, layout, material handling, and normal waste. Add for sales tax, delivery, supervision, mobilization, demobilization, cleanup, overhead and profit. Note: PVC patching material is available in spray cans for repairing damaged PVC coating.

PVC Coated Junction Boxes and Sealing Fittings

PVC coated Type GUAT junction boxes with covers

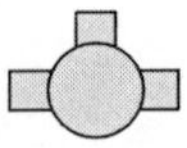

Material	Craft@Hrs	Unit	Material Cost	Labor Cost	Installed Cost
1/2" - 2" dia.	L1@0.40	Ea	195.00	16.40	211.40
1/2" - 3" dia.	L1@0.45	Ea	220.00	18.50	238.50
3/4" - 2" dia.	L1@0.50	Ea	210.00	20.50	230.50
3/4" - 3" dia.	L1@0.55	Ea	236.00	22.60	258.60
1" - 3" dia.	L1@0.60	Ea	369.00	24.60	393.60
1-1/4" - 3-5/8" dia.	L1@0.65	Ea	717.00	26.70	743.70
1-1/2" - 5" dia.	L1@0.80	Ea	800.00	32.80	832.80
2" - 5" dia.	L1@1.10	Ea	841.00	45.10	886.10

PVC coated Type GUAX junction boxes with covers

Material	Craft@Hrs	Unit	Material Cost	Labor Cost	Installed Cost
1/2" - 2" dia.	L1@0.45	Ea	192.00	18.50	210.50
1/2" - 3" dia.	L1@0.50	Ea	216.00	20.50	236.50
3/4" - 2" dia.	L1@0.55	Ea	203.00	22.60	225.60
3/4" - 3" dia.	L1@0.60	Ea	225.00	24.60	249.60
1" - 3" dia.	L1@0.65	Ea	234.00	26.70	260.70
1-1/4" - 3-5/8" dia.	L1@0.70	Ea	252.00	28.70	280.70
1-1/2" - 5" dia.	L1@0.90	Ea	728.00	36.90	764.90
2" - 5" dia.	L1@1.25	Ea	801.00	51.30	852.30

PVC coated Type EYD female sealing fittings

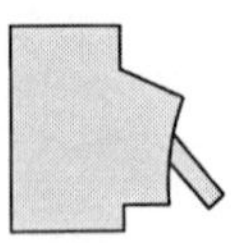

Material	Craft@Hrs	Unit	Material Cost	Labor Cost	Installed Cost
1/2"	L1@0.35	Ea	139.00	14.40	153.40
3/4"	L1@0.40	Ea	140.00	16.40	156.40
1"	L1@0.45	Ea	162.00	18.50	180.50
1-1/4"	L1@0.50	Ea	177.00	20.50	197.50
1-1/2"	L1@0.60	Ea	223.00	24.60	247.60
2"	L1@0.75	Ea	243.00	30.80	273.80

PVC coated Type EYD male-female sealing fittings

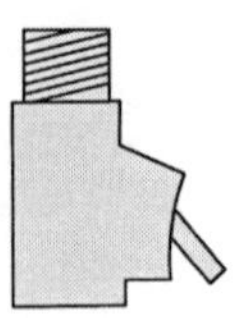

Material	Craft@Hrs	Unit	Material Cost	Labor Cost	Installed Cost
1/2"	L1@0.35	Ea	139.00	14.40	153.40
3/4"	L1@0.40	Ea	140.00	16.40	156.40
1"	L1@0.45	Ea	162.00	18.50	180.50
1-1/4"	L1@0.50	Ea	183.00	20.50	203.50
1-1/2"	L1@0.60	Ea	232.00	24.60	256.60
2"	L1@0.75	Ea	255.00	30.80	285.80
2-1/2"	L1@0.90	Ea	372.00	36.90	408.90
3"	L1@1.00	Ea	507.00	41.00	548.00
3-1/2"	L1@1.25	Ea	507.00	51.30	558.30
4"	L1@1.40	Ea	1,200.00	57.40	1,257.40

Use these figures to estimate the cost of PVC coated junction boxes and sealing fittings installed on PVC coated conduit under the conditions described on pages 5 and 6. Costs listed are for each fitting installed. The crew is one electrician working at a labor cost of $41.03 per manhour. These costs include the cover, layout, material handling, and normal waste. Add for sales tax, delivery, supervision, mobilization, demobilization, cleanup, overhead and profit. Note: PVC patching material is available in spray cans for repairing damaged PVC coating.

PVC coated Type EYS female sealing fittings

Material	Craft@Hrs	Unit	Material Cost	Labor Cost	Installed Cost
1/2"	L1@0.30	Ea	77.60	12.30	89.90
3/4"	L1@0.35	Ea	71.90	14.40	86.30
1"	L1@0.40	Ea	86.20	16.40	102.60
1-1/4"	L1@0.50	Ea	123.00	20.50	143.50
1-1/2"	L1@0.50	Ea	158.00	20.50	178.50
2"	L1@0.60	Ea	175.00	24.60	199.60

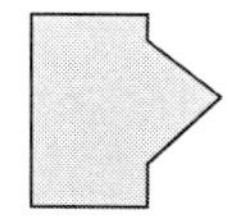

PVC coated Type EYS male-female sealing fittings

Material	Craft@Hrs	Unit	Material Cost	Labor Cost	Installed Cost
1/2"	L1@0.30	Ea	78.40	12.30	90.70
3/4"	L1@0.35	Ea	72.60	14.40	87.00
1"	L1@0.40	Ea	86.80	16.40	103.20
1-1/4"	L1@0.50	Ea	125.00	20.50	145.50
1-1/2"	L1@0.50	Ea	163.00	20.50	183.50
2"	L1@0.60	Ea	182.00	24.60	206.60
2-1/2"	L1@0.75	Ea	282.00	30.80	312.80
3"	L1@0.90	Ea	378.00	36.90	414.90
3-1/2"	L1@1.00	Ea	1,120.00	41.00	1,161.00
4"	L1@1.25	Ea	1,460.00	51.30	1,511.30

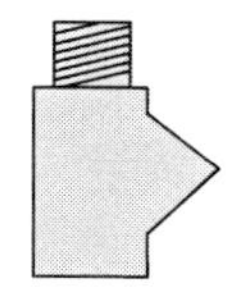

PVC coated Type EZS female sealing fittings

Material	Craft@Hrs	Unit	Material Cost	Labor Cost	Installed Cost
1/2"	L1@0.30	Ea	91.20	12.30	103.50
3/4"	L1@0.35	Ea	119.00	14.40	133.40
1"	L1@0.40	Ea	142.00	16.40	158.40
1-1/4"	L1@0.50	Ea	162.00	20.50	182.50
1-1/2"	L1@0.50	Ea	311.00	20.50	331.50
2"	L1@0.60	Ea	648.00	24.60	672.60

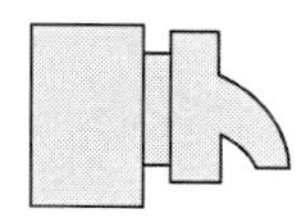

PVC coated Type EZS male-female sealing fittings

Material	Craft@Hrs	Unit	Material Cost	Labor Cost	Installed Cost
1/2"	L1@0.30	Ea	91.90	12.30	104.20
3/4"	L1@0.35	Ea	120.00	14.40	134.40
1"	L1@0.40	Ea	142.00	16.40	158.40
1-1/4"	L1@0.50	Ea	169.00	20.50	189.50
1-1/2"	L1@0.50	Ea	243.00	20.50	263.50
2"	L1@0.60	Ea	218.00	24.60	242.60
2-1/2"	L1@0.70	Ea	411.00	28.70	439.70
3"	L1@0.90	Ea	680.00	36.90	716.90

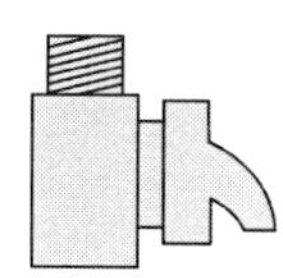

Use these figures to estimate the cost of PVC coated sealing fittings installed on PVC coated conduit under the conditions described on pages 5 and 6. Costs listed are for each fitting installed. The crew is one electrician working at a labor cost of $41.03 per manhour. These costs include layout, material handling, and normal waste. Add for sales tax, delivery, supervision, mobilization, demobilization, cleanup, overhead and profit. Note: PVC patching material is available in spray cans for repairing damaged PVC coating.

Hanger Fittings

Plated threaded rod

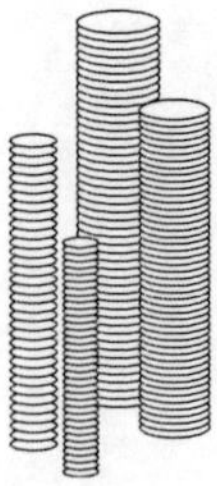

Material	Craft@Hrs	Unit	Material Cost	Labor Cost	Installed Cost
1/4-20 x 6'	L1@1.25	CLF	215.00	51.30	266.30
1/4-20 x 10'	L1@1.25	CLF	345.00	51.30	396.30
1/4-20 x 12'	L1@1.15	CLF	416.00	47.20	463.20
3/8-16 x 6'	L1@1.30	CLF	236.00	53.30	289.30
3/8-16 x 10'	L1@1.30	CLF	580.00	53.30	633.30
3/8-16 x 12'	L1@1.30	CLF	726.00	53.30	779.30
1/2-13 x 6'	L1@1.50	CLF	401.00	61.50	462.50
1/2-13 x 10'	L1@1.50	CLF	1,040.00	61.50	1,101.50
1/2-13 x 12'	L1@1.50	CLF	1,250.00	61.50	1,311.50
5/8-11 x 6'	L1@1.75	CLF	1,090.00	71.80	1,161.80
5/8-11 x 10'	L1@1.75	CLF	1,550.00	71.80	1,621.80
5/8-11 x 12'	L1@1.75	CLF	1,900.00	71.80	1,971.80

Rod couplings

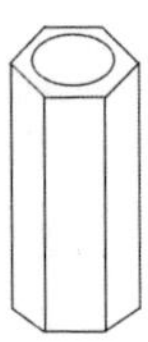

Material	Craft@Hrs	Unit	Material Cost	Labor Cost	Installed Cost
1/4-20	L1@0.05	Ea	1.89	2.05	3.94
3/8-16	L1@0.05	Ea	3.83	2.05	5.88
1/2-13	L1@0.08	Ea	3.93	3.28	7.21
5/8-11	L1@0.10	Ea	9.16	4.10	13.26

Toggle bolts, wing nuts

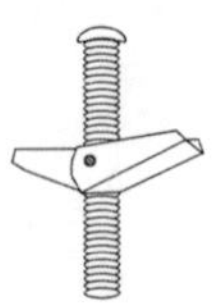

Material	Craft@Hrs	Unit	Material Cost	Labor Cost	Installed Cost
1/8 x 3"	L1@0.10	Ea	.17	4.10	4.27
3/16 x 3"	L1@0.10	Ea	.27	4.10	4.37
1/4 x 4"	L1@0.15	Ea	.41	6.15	6.56
3/8 x 4"	L1@0.20	Ea	.49	8.21	8.70

Expansion anchors, flush type

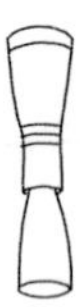

Material	Craft@Hrs	Unit	Material Cost	Labor Cost	Installed Cost
1/4-20	L1@0.15	Ea	.42	6.15	6.57
3/8-16	L1@0.15	Ea	.68	6.15	6.83
1/2-13	L1@0.25	Ea	1.83	10.30	12.13
5/8-11	L1@0.30	Ea	2.62	12.30	14.92

Steel hex nuts

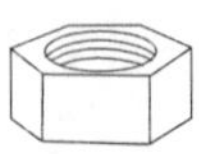

Material	Craft@Hrs	Unit	Material Cost	Labor Cost	Installed Cost
1/4-20	L1@0.02	Ea	.27	.82	1.09
3/8-16	L1@0.03	Ea	.28	1.23	1.51
1/2-13	L1@0.05	Ea	.35	2.05	2.40
5/8-11	L1@0.10	Ea	.41	4.10	4.51

Fender washers, 1-1/2" diameter

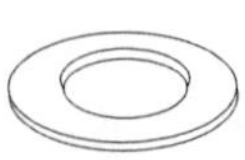

Material	Craft@Hrs	Unit	Material Cost	Labor Cost	Installed Cost
1/4"	L1@0.02	Ea	.03	.82	.85
3/8"	L1@0.03	Ea	.06	1.23	1.29
1/2"	L1@0.04	Ea	.09	1.64	1.73

Use these figures to estimate the cost of installing steel hanger fittings for hanging or mounting conduit or electrical equipment under the conditions described on pages 5 and 6. Costs listed are for each 100 linear feet or steel channel strut, or each fitting installed. The crew is one electrical working at a labor cost of $41.03 per manhour. These costs include layout, material handling, and normal waste. Add for sales tax, delivery, supervision, mobilization, demobilization, cleanup, overhead and profit.

Material	Craft@Hrs	Unit	Material Cost	Labor Cost	Installed Cost
14 gauge steel channel					
13/16" x 1-5/8" plated	L1@4.00	CLF	391.00	164.00	555.00
13/16" x 1-5/8" galvanized	L1@4.00	CLF	486.00	164.00	650.00
1-5/8" x 1-5/8" plated	L1@6.00	CLF	559.00	246.00	805.00
1-5/8" x 1-5/8" galvanized	L1@6.00	CLF	597.00	246.00	843.00
14 gauge steel channel with 9/16" holes, 1-7/8" oc					
13/16" x 1-5/8" plated	L1@4.00	CLF	392.00	164.00	556.00
13/16" x 1-5/8" galvanized	L1@4.00	CLF	519.00	164.00	683.00
1-5/8" x 1-5/8" plated	L1@6.00	CLF	529.00	246.00	775.00
1-5/8" x 1-5/8" galvanized	L1@6.00	CLF	677.00	246.00	923.00
12 gauge steel channel					
13/16" x 1-5/8" plated	L1@4.00	CLF	391.00	164.00	555.00
13/16" x 1-5/8" galvanized	L1@4.00	CLF	466.00	164.00	630.00
1-5/8" x 1-5/8" plated	L1@6.00	CLF	559.00	246.00	805.00
1-5/8" x 1-5/8" galvanized	L1@6.00	CLF	674.00	246.00	920.00
12 gauge steel channel with 9/16" holes, 1-7/8" oc					
13/16" x 1-5/8" plated	L1@4.00	CLF	463.00	164.00	627.00
13/16" x 1-5/8" galvanized	L1@4.00	CLF	483.00	164.00	647.00
1-5/8" x 1-5/8" plated	L1@6.00	CLF	509.00	246.00	755.00
1-5/8" x 1-5/8" galvanized	L1@6.00	CLF	509.00	246.00	755.00
Channel nuts					
1/4-20 13/16" strut	L1@0.05	Ea	1.28	2.05	3.33
3/8-16 13/16" strut	L1@0.05	Ea	1.31	2.05	3.36
1/2-13 13/16" strut	L1@0.05	Ea	1.32	2.05	3.37
1/4-20 1-5/8" strut	L1@0.05	Ea	1.46	2.05	3.51
3/8-16 1-5/8" strut	L1@0.05	Ea	1.70	2.05	3.75
1/2-13 1-5/8" strut	L1@0.05	Ea	1.86	2.05	3.91
Channel spring nuts					
1/4-20 13/16" strut	L1@0.06	Ea	1.57	2.46	4.03
3/8-16 13/16" strut	L1@0.06	Ea	1.57	2.46	4.03
1/2-13 13/16" strut	L1@0.06	Ea	1.69	2.46	4.15
1/4-20 1-5/8" strut	L1@0.06	Ea	2.15	2.46	4.61
3/8-16 1-5/8" strut	L1@0.06	Ea	2.28	2.46	4.74
1/2-13 1-5/8" strut	L1@0.06	Ea	2.47	2.46	4.93

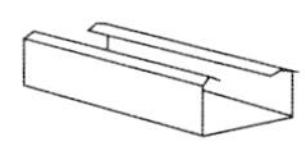

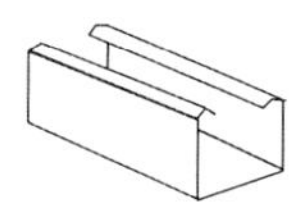

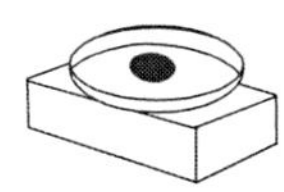

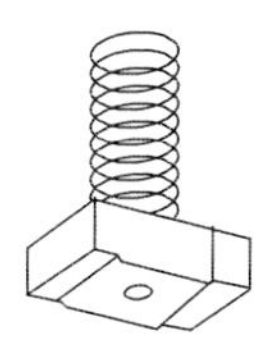

Use these figures to estimate the cost of installing steel channel strut and fittings for hanging or mounting conduit or electrical equipment under the conditions described on pages 5 and 6. Costs listed are for each 100 linear feet or steel channel strut, or each fitting installed. The crew is one electrical working at a labor cost of $41.03 per manhour. These costs include layout, material handling, and normal waste. Add for sales tax, delivery, supervision, mobilization, demobilization, cleanup, overhead and profit.

Section 2: Wire and Cable

Wire and cable come in many types and sizes. Fortunately, only a few of these are of major concern to electrical estimators.

Article 310 of the *National Electrical Code* lists minimum conductor requirements for electrical systems. Tables in the *NEC* show installation standards that will apply in most communities where the *NEC* is followed. But be sure to use the version of the code that the inspectors are using. Even though the *NEC* is revised every three years, some cities and counties don't get around to adopting the current version until months or years after each revision.

In this book we're concerned with building construction exclusively. The scope of the *NEC* is much broader. We'll ignore *NEC* requirements that relate to ship building, aircraft manufacturing and the automobile industry.

In the construction industry, the most common wire types are identified as THW, THHN, THWN, XHHW, MTW, TF, TFF, TFFN, AF, and USE-RHH-RHW. These code letters identify the type of insulation on the wire.

Wire sizes used in building electrical systems range from #18 (called ***18 gauge*** under the American Wire Gauge system) to 1000 kcmil. The abbreviation kcmil stands for 1000 circular mils.

The smaller the gauge number, the larger the wire. #14 gauge is smaller than #12 gauge, for example. Gauges larger than 1 follow the "ought" scale. One ought (usually written as ***1/0***) is larger than #1 gauge. Two ought (2/0) is larger still. Above four ought, wire size is identified in thousands of circular mils. A circular mil is the area of a circle that's one mil (1/1000th of an inch) in diameter. Wire that has a cross section area of 250,000 circular mils (250 kcmil), for example, weighs about 12 ounces per linear foot. 1000 kcmil wire weighs more than 3 pounds per linear foot.

Conductors in wire are either solid (one piece surrounded by insulation) or stranded. Stranded copper wire can be tinned as part of the manufacturing process and might be so specified.

Type THW

Type THW wire is Moisture- and Heat-Resistant Thermoplastic. It has a maximum operating temperature of 75 degrees C or 167 degrees F for dry or wet locations and 90 degrees C or 194 degrees F when used in electric discharge lighting equipment of 1000 open-circuit volts or less. The insulation is flame-retardant, moisture- and heat-resistant thermoplastic. It has no outer jacket or covering and is sized from #14 gauge with either solid or stranded conductors.

Type THHN

Type THHN wire is Heat-Resistant Thermoplastic and has a maximum operating temperature of 90 degrees C or 194 degrees F for dry locations. The insulation is flame-retardant, heat-resistant thermoplastic. It has a nylon jacket or the equivalent and is sized from #14 gauge with either solid or stranded conductors.

Type XHHW

Type XHHW wire is Moisture- and Heat-Resistant Thermoplastic and has a maximum operating temperature of 90 degrees C or 194 degrees F for dry locations, and 75 degrees C or 167 degrees F for wet locations. The insulation is flame-retardant cross-linked synthetic polymer. It has no outer jacket or covering and is sized from #14 gauge in either solid or stranded conductors.

Type USE

Type USE is Underground Service-Entrance Cable, Single Conductor, and has a maximum operating temperature of 75 degrees C or 167 degrees F for applications as listed in Article 338, Service-Entrance Cable in the *NEC*. The insulation is heat- and moisture-resistant. It has a jacket or outer covering rated as moisture-resistant nonmetallic and is sized from #12 gauge in either solid or stranded conductors.

Type MTW

Type MTW is Moisture-, Heat-, and Oil-Resistant Thermoplastic and has a maximum operating temperature of 60 degrees C or 140 degrees F. It's used for machine tool wiring in wet locations and is covered by *NEC* Article 670, Metal-Working Machine Tools. The insulation is flame-retardant, moisture-, heat- and oil-resistant thermoplastic. It's sized from #14 gauge with stranded conductors. There are two thicknesses of insulation. One is a little thinner than the other. The thicker insulation, Type A, has no outer jacket or covering. The thinner insulation, Type B, has a nylon jacket or equivalent covering.

Type TF

Type TF is Thermoplastic-Covered Fixture Wire and has a maximum operating temperature of 60 degrees C or 140 degrees F. It's used for fixture wiring as permitted by the *NEC*. The insulation is thermoplastic. Sizes are #18 or #16 gauge with either solid or stranded conductors. It has no outer jacket or covering.

Type TFF

Type TFF is Thermoplastic-Covered Fixture Wire, Flexible Stranded. It has a maximum operating temperature of 60 degrees C or 140 degrees F and is used for fixture wiring, with some *NEC* restrictions. The insulation is thermoplastic. Sizes are either #18 or #16 gauge with stranded conductors. It has no outer jacket or covering.

Type TFFN

Type TFFN is Heat-Resistant Thermoplastic-Covered Fixture Wire, Flexible Stranded. It has a maximum operating temperature of 90 degrees C or 194 degrees F and is used for fixture wiring as permitted by the *NEC*. The insulation is thermoplastic and is sized at #18 or #16 gauge with solid or stranded conductors. It has a nylon or equivalent outer jacket.

Type AF

Type AF wire is Asbestos Covered Heat-Resistant Fixture Wire. It has a maximum operating temperature of 150 degrees C or 302 degrees F for fixture wiring. It's limited to 300 volts and indoor dry locations. The insulation is impregnated asbestos. Gauges run from #18 to #14 with stranded conductors. Moisture-resistant and impregnated asbestos insulation is sized from #12 to #10 gauge with stranded conductors.

Many other types of insulation are listed in the tables of the *NEC*. The types just listed are the ones most commonly used in building construction. Many other types of insulated wire were used in the past and have been discontinued. The *NEC* still lists some types that are seldom used and are no longer stocked by most dealers. For example, the smaller sizes of THW are no longer popular. Type THHN or THHN-THWN has become more popular as the industry standard changed to newer types of insulation. The insulation jacket on THHN-THWN is much thinner. Yet the wire has a higher ampacity than older THW. The *NEC* tables "Maximum Number of Conductors in Trade Sizes of Conduit or Tubing" permit more current-carrying capacity in conduit of a given size when using THHN wire than when using THW wire.

Flexible Cords

Many types of flexible power cords are listed in the *NEC*. Some of these apply to the construction industry and are important to the electrical estimator.

Type SJ

The trade name for Type SJ is Junior Hard Service Cord. It's made in sizes #18 or #10 with two, three or four conductors. The insulation is thermoset plastic with an outer jacket or covering of thermoset plastic. It's used for portable or pendant fixtures in damp locations where hard usage is expected. The insulation is rated at 300 volts for 60 degrees C.

Type S

The trade name for Type S is Hard Service Cord. It's made in gauges from #18 to #2 with two or more conductors. The insulation is thermoset plastic with an outer jacket or covering of thermoset plastic. It's used for portable or pendant fixtures in damp locations where extra hard use is expected. The insulation is rated at 600 volts for 60 degrees C.

Type SO

The trade name for Type SO is also Hard Service Cord. It's made in gauges from #18 to #2 with two or more conductors. The insulation is thermoset plastic with an outer jacket or covering of oil-resistant thermoset plastic. It's used for portable or pendant fixtures in damp locations where extra hard usage is expected. The insulation is rated at 600 volts for 60 degrees C.

Type STO

The trade name for Type STO is also Hard Service Cord. It's made in gauges from #18 to #2 with two or more conductors. The insulation is thermoplastic or thermoset with an outer jacket or covering of oil-resistant thermoplastic. It's used for portable or pendant fixtures in damp locations where extra hard usage is expected. The insulation is rated at 600 volts for 60 degrees C.

Type NM

The trade name for Type NM is Non-Metallic-Sheathed Cable. It's made in gauges from #14 to #2 with from two to four conductors and either with or without a ground wire. The ground wire may be either insulated or bare. The insulation is as listed in *NEC* Table 310-13. The cable has an outer jacket or covering of a flame-retardant and moisture-resistant plastic. Type NM is commonly used in house wiring. The insulation is rated at 600 volts for 60 degrees C.

Type UF

The trade name for Type UF Cable is Underground Feeder and Branch Circuit Cable. It's made in gauges from #14 to #4/0 with one, two, three or four conductors. Gauges from #14 to #10 have 60 mil insulation. Gauges from #8 to #2 have 80 mil insulation. Gauges from #1 to #4/0 have 95 mil insulation. It comes with or without a ground wire. Single conductor is rated for 600 volt, 60 degrees C or 140 degrees F. The insulation is moisture-resistant and is integral with the jacket or outer covering. Multiple conductor cable is rated at 600 volts and 75 degrees C or 167 degrees F. The insulation is moisture- and heat-resistant.

Type SEU

The trade name for Type SEU is Service-Entrance Cable. It's made in gauges from #12 to #4/0 with one or more conductors. The cable is rated 600 volts. The temperature rating of the wire is the same as the rating of the conductor itself. The cable has a flame-retardant, moisture-resistant jacket or covering.

Bare Copper Wire

Bare copper wire is made in gauges from #14 to #4/0 in either soft drawn, medium hard drawn or hard drawn temper. It's usually sold by the hundredweight, which is 100 pounds. It comes in solid strands in sizes up to #4 and stranded up to #4/0 gauge. The most common use for bare copper wire is in grounding electrical systems.

The *NEC* has a table that gives weights for 1000 linear feet of soft drawn bare copper wire. This simplifies the conversion of lengths to weights so you can convert the hundredweight price to a cost for the length you need.

Metal-Clad Cable

Type MC cable is a factory assembled cable of one or more insulated conductors enclosed in a flexible interlocking metallic sheath, or a smooth or corrugated tube. The conductors have a moisture-resistant and flame-resistant fibrous cover. An internal bonding strip of copper or aluminum is in contact with the metallic covering for the entire length of the cable. It's rated for 600 volts and comes in gauges from #14 to #4 in single or multiple conductors. An approved insulating bushing or the equivalent must be installed at terminations in the cable between the conductor insulation and the outer metallic covering.

There are some restrictions on the use of armored cable. It's prohibited in theaters, places of assembly, motion picture studios, hazardous locations, where exposure to corrosive fumes or vapors is expected, on cranes or hoists, in storage battery rooms, in hoistways or elevators and in commercial garages. Check for exceptions in the *NEC* when working with MC armored cable.

Other types of insulated conductors and cables are available and used occasionally in construction. But be sure you refer to the code and discuss the materials with the inspection authority before including them in your bid. Some cities and counties prohibit certain conductors and cables.

High Voltage Wire and Cable

Medium and high voltage conductors and cables require special consideration. In most projects they're specified for a particular use or service. Many types of high voltage conductors are produced by some manufacturers. Some are stocked only by the factory. Others are not stocked at all and have to be made to order. These will be very expensive unless the factory run is quite large. In any case, be prepared to wait many weeks for delivery.

Terminations for high voltage conductors are also a specialty item. Installation should be done only by an experienced craftsman, usually called a cable splicer. Shielded conductors will require a stress cone for termination at the point of connection to switching equipment or some other device. Stress cones can be made as a kit that's adapted to the end of the cable to relieve stress from the shield.

Splicing cable is also a specialty that should be done by a qualified cable splicer. The splicer prepares the conductor and either uses a splice kit or tapes a splice. The process may take many hours to complete. Once a stress cone or a splice has been started, it should be completed as quickly as possible to keep moisture out of the splice.

Most professional electrical estimators have to figure the cost of extending an existing underground electrical system occasionally. For example, you may have to figure the cost of splicing new XLP cable to three conductor lead covered cable. An experienced cable splicer will be needed to make this splice properly. In an underground system that's subject to moisture or even continuous submersion in brackish water, this splice will have to be made correctly.

High voltage splicing requires experience and precision. If the cable is damaged or if the splice isn't made correctly, replacing the cable and the splice can be very expensive. If you can't locate an experienced cable splicer, consider subcontracting the work to a firm that specializes in this work. Having it done by a qualified expert can preserve many nights of sound sleep.

Most specs require a high-potential test on the cable after it's been installed. The test is done with a special portable test unit. An experienced operator is needed to run the test. Follow the test procedure recommended by the cable manufacturer.

There are good reasons for running this test. Of course, every reel of electrical cable gets a high potential test at the factory. But cable can be damaged between the factory and the job site. Your cable may have been cut from the original reel and re-coiled on a smaller reel for shipment to the job site. It probably sat in storage for at least several months and may have been exposed to adverse conditions for the entire time. When your electricians finally install the cable, it may be defective, even though it looks fine. But if moisture has entered from a cut end or if the cable was damaged in transit, it's much better to find out before the cable is energized. A fault in a high voltage cable can cause damage to other cable in the conduit or to other parts of the electrical system.

The high potential test is not a destructive test. If the cable is sound before the test begins, it will be unchanged by the test. The test proves that the insulation is in good condition and that the conductor will safely carry the rated voltage.

Here's the usual test procedure. First, identify the conductors by circuit and by phase (A, B, C, etc.). Voltage is applied for a few minutes and is increased to a certain value over a set amount of time. The operator records the voltage, time and amount of leakage up to a predetermined voltage. The maximum test voltage will be more than the rated voltage of the cable. The technician performing the test enters results on a special test form and submits it to the designing engineer for review. The form should then be filed with permanent project records.

If a fault shows up during the high potential test, the operator should have equipment available to locate the fault. Usually the problem will be at a stress cone or in a splice. One method of locating a fault in the cable is to install a ***thumper***, a large capacitor-type unit, in the line. The thumper gives off an electrical discharge every few seconds. The charge will create a pop or snap when it hits the fault, making location easy.

Repair the problem by resplicing. If the fault is in the cable, pull out the cable and examine it carefully. It was probably damaged during installation. If the cable jacket is good and doesn't have signs of damage, the cable itself may be at fault. The manufacturer may want to test the cable to determine the cause of failure. Most manufacturers will replace at no charge cable that fails a high potential test ***before*** being put into service. Once the cable is put into use, it's hard to get a free replacement, even if there is no apparent cable damage.

Aluminum Wire

The types of insulation listed for copper wire are the same as used on aluminum wire. There are two major differences between copper and aluminum wire: First, aluminum usually costs less for a similar gauge. Second, the ampacity of aluminum wire is less than that of the copper wire. That means that aluminum wire has to be heavier gauge to carry the same current. Tables in the *NEC* show the allowable current rating for aluminum wire.

Aluminum wire is much lighter than copper wire and can be installed in less time. Compare the difference in the labor units in this book for copper and aluminum wire. Many electricians prefer to install aluminum wire on some types of applications. But aluminum wire comes with a major problem: it oxidizes when not covered with insulation. When the wire is stripped and left bare, it develops a thin white oxide coating that resists the flow current. Eventually the wire begins to heat up at the point of connection. Heat expands the conductor. Cooling causes it to contract. Constant expansion and contraction makes the connection looser and accelerates oxidation. Eventually the conductor will fail. Sometimes overheating will cause a fire.

There are ways to reduce the problem of oxidation. An anti-oxidation material can be applied immediately after the cable is stripped and before it's terminated in the connector. The material can also be applied to the connector. Then the aluminum wire is inserted into the connector and tightened. A good practice is to retighten the connector after a few days and again after the circuit has been in operation for about a year. Some wire and cable connectors are made with an oxidation inhibitor material inside the area where the wire will be inserted. Only connectors rated for aluminum wire should be used on aluminum circuits.

Today, aluminum wire is seldom used on branch circuits to convenience receptacles or lighting switches. But it's common on feeder circuits such as runs from distribution switchboards to lighting or power panels.

When you decide to change a feeder from copper wire to aluminum wire, be sure to compare the carrying capacity of the two materials. You'll find that it takes larger aluminum wire to carry the same amperage as a copper wire. This may mean that you'll have to use larger conduit size to accommodate the larger wire diameter. But using aluminum on feeder circuits will probably still reduce the labor and material cost.

Wire Connectors

Many types of wire connectors and lugs are available. Some are insulated, made for multiple conductors, or intended to be watertight. Be sure the connections you price will comply with the code and job specs. Not all types of connections are listed in this manual. But you'll find a representative selection here that gives you a good idea of how to price most of the common wire connectors. It would take a 200-page book just to list labor and material costs for all the connectors and lugs that are available.

Pulling Wire in Conduit

The best way to thread wire through conduit depends on the type of wire and the type of conduit. Some conduit runs are several hundred feet long. On a job like that, power equipment is recommended. If the circuit length is closer to 50 feet and the wire gauge is #6 or less, a fish tape will be the only tool you need.

Pulling wire is the essence of simplicity: Thread a line through the full length of conduit from box to box, bend wire around the end of that line and pull it back the other way.

Several kinds of fish tape are made for pulling building wire. Tape used on EMT, PVC and rigid steel conduit is usually 1/8" spring steel and will be 100 feet long. Lengths of 200 feet are also available. For larger conduit, use 1/4" spring steel fish tape in 200 feet lengths.

Form a pulling eye in spring steel tape by bending back the last several inches of tape. Heat the tape before bending to keep the steel from breaking as you make the eye.

For flexible conduit, it's easier to use a special fish tape made specifically for flex. The end of the tape comes with an eye to receive the building wire.

Before attaching building wire to the pulling eye, always strip insulation off the last 3 or 4 inches of wire. Feed the bare conductors through the eye, bend the conductors back tightly and wrap the ends with plastic tape. Then start the pulling.

Any time you pull wire, be sure wire is feeding smoothly into the box or conduit at the far end. If there's too much pulling resistance, stop pulling. Something's wrong. You don't want to strip insulation off the wire. Coat the wire with special pulling compound if necessary.

Other Pulling Methods

Some electricians prefer to use a small fishing vacuum unit that draws string through the conduit. A small sponge (called a ***mouse***) sized for the conduit is attached to a fine plastic string. The mouse has a steel eye fitted at one or both ends. String is tied to the mouse eye and then the mouse is sucked through the conduit with the vacuum.

The mouse can also be blown through conduit with air pressure rather than a vacuum. This is the common way of pulling wire in long feeder conduit and underground ducts between manholes or handholes. Some electricians prefer using a parachute in place of the mouse. The parachute is blown through the duct with a plastic line attached.

When pulling heavy wire or pulling wire on longer runs, the thin line first pulled through the conduit may not be strong enough to pull the wire. In that case, attach the first line pulled to some stronger string. Pull that string through the entire length. Then pull the wire with the stronger string.

On underground runs it's good practice to pull a mandrel and wire brush through the duct before pulling any wire. This guarantees that the conduit is free of debris. The mandrel should be at least 4" long and 1/4" smaller than the duct. Tie a heavy line to both ends of the mandrel and brush when it's pulled. If the mandrel gets hung up, you want to be able to pull it back.

Don't pull wire in conduit that you know is damaged. Obstructions or breaks in the duct will probably damage any wire you pull. Make repairs first. Then pull the wire. That's the easiest way.

Getting Set Up

Start each pull by running fish tape through the conduit. Set the rolls of wire where they can be handled into place with minimum effort. When pulling starts, it should continue without interruption until complete.

Pulling is easier if you buy smaller gauge building wire (sizes from AWG 18 to 6) on small disposable reels. Each has from 500 to 2500 feet of continuous wire. Mount these reels on a reel frame or dolly that will hold as many reels as there are conductors in the pull. Set the reel dolly at the feed end of the conduit.

Larger reels have to be set on reel jacks. Run conduit or pipe through the center axle of the reel and rest the pipe on the jacks. Then lift the reel so it's off the floor. The reel should be mounted so wire feeds from the top of the coil. That makes it easier to handle the wire when necessary. More bending or stooping is needed if wire feeds from the bottom of the reel. As wire is fed into the conduit or duct, someone should apply pulling compound to ease the pulling effort.

Power Pulling Equipment

A pulling machine is usually required when pulling heavy wire for feeder circuits. Power pulling rigs are available in several sizes. A truck-mounted winch can also be used for outdoor work. Whatever power equipment is used, be sure to have good communication between both ends of the conduit. If the pull has to be stopped quickly, the operator should know right away.

The best pull rate when using power equipment is between 10 and 15 feet per minute. It's easy to pull faster, until something goes wrong. Then, the faster the pull, the more likely damage will result before pulling can be stopped.

On feeder pulls, a nylon pulling rope may be the best choice if the pulling rig has a capstan. Wrap two to four turns of pulling rope around the capstan. Have the electrician at the feed end hold the rope taut as the slack is being taken out. Begin the pull. If the load increases during the pull, have the operator pull a little slack and let the rope slip on the capstan. That slows the pull to avoid damaging the wire.

There are other ways to pull wire. But those listed here are the most common methods for conduit and duct. Use whatever method helps you get the pulling done most efficiently on the job at hand. No single pulling method is ideal for all situations. Most electrical contractors have several types of pulling equipment and use the method that's most appropriate for job conditions.

Adjusting the Cost Tables

The tables in this section show labor and material costs and labor hours per 1000 linear feet of wire when three conductors are pulled at the same time. That's the most common case on most jobs. The labor cost will be higher per 1000 linear feet of wire when only two conductors are pulled at once. It will usually be lower when more than three conductors are pulled at once.

When pulling three conductors simultaneously, use the tables in this section without modification. If there are two, three and four wire pulls on a job, and if the average pull is three wires, use the tables without modification. Only if two or four wire pulls predominate will it be necessary to adjust the cost tables.

If you find that most pulls on a job will be only two wires, add about 10 percent to the labor cost per 1000 linear feet of wire. If most pulls will be more than three wires, reduce the labor cost by about 10 percent for each conductor over three.

Increasing Productivity

A good supervisor can improve pulling efficiency by planning the job before work begins. Think about the best place to set up the reels. Setting up at the panels is usually a good idea because you can serve many pulls from there. But it may be easier to fish from other locations back to the panel. Avoid pulling uphill. Do whatever is necessary to make the pull easier. Be sure there's enough workspace at the pull end to make pulling possible.

String used with mouse fishing can be reused many times if it's handled properly. Some electrical contractors feel that string is cheaper than the labor needed to coil it for reuse. For most contractors it's better to save the string.

When the wire is pulled, leave plenty of extra wire at both ends to make the connection. Leave 10 feet at full size panels, 10 feet to 15 feet at motor control centers, and 18 inches to 24 inches at outlets and fixtures. Your take-off should reflect these allowances, of course.

Flexible Cords

The tables that follow include prices for flexible cords. These cords aren't installed in conduit. They're used to connect portable equipment, to extend power temporarily, or with certain types of lighting fixtures. Power cords are available on reels with lengths from 25 feet to 100 feet. Larger reels can be ordered also. Flexible cords must be protected from physical damage. Check the *NEC* for restrictions on each type of cord.

Taking Off Wire

Wire take-off should be based on the conduit take-off. Add the length of all conduit runs that have the same number of conductors and use the same wire gauge. Then multiply the computed length by the number of conductors and add for the extra wire needed at each outlet box and panel. Don't worry about the colors at this point. Get the total quantity so you can price the wire and labor.

If you get the job, the first step is to check the panel schedules. Find out which circuits are single phase and which are three phase. If the panels are three phase, wire colors will be black, red, blue and white for 120/208V and brown, orange, yellow and white for 277/480V. If the panels are 120/240V, wire colors will be black, red and white. You may want to set aside another color for switch legs.

When figuring the quantity of each color, figure about 80 percent of the wire will be the primary colors noted above. Divide the 80 percent equally among the colors needed. The other 20 percent will be white. Wire colors for switch legs can be figured separately.

Type THW 600 volt solid copper building wire

Material	Craft@Hrs	Unit	Material Cost	Labor Cost	Installed Cost
# 14	L2@6.00	KLF	168.00	246.00	414.00
# 12	L2@7.00	KLF	241.00	287.00	528.00
# 10	L2@8.00	KLF	339.00	328.00	667.00

Type THW 600 volt stranded copper building wire

Material	Craft@Hrs	Unit	Material Cost	Labor Cost	Installed Cost
# 14	L2@6.00	KLF	197.00	246.00	443.00
# 12	L2@7.00	KLF	279.00	287.00	566.00
# 10	L2@8.00	KLF	390.00	328.00	718.00
# 8	L2@9.00	KLF	770.00	369.00	1,139.00
# 6	L2@10.0	KLF	907.00	410.00	1,317.00
# 4	L2@12.0	KLF	1,410.00	492.00	1,902.00
# 2	L3@13.0	KLF	2,200.00	533.00	2,733.00
# 1	L3@14.0	KLF	6,690.00	574.00	7,264.00
# 1/0	L3@15.0	KLF	3,480.00	615.00	4,095.00
# 2/0	L3@16.0	KLF	4,310.00	656.00	4,966.00
# 3/0	L3@17.0	KLF	5,420.00	698.00	6,118.00
# 4/0	L3@18.0	KLF	6,790.00	739.00	7,529.00
# 250 KCMIL	L4@20.0	KLF	7,510.00	821.00	8,331.00
# 300 KCMIL	L4@23.0	KLF	8,240.00	944.00	9,184.00
# 350 KCMIL	L4@24.0	KLF	9,630.00	985.00	10,615.00
# 400 KCMIL	L4@25.0	KLF	10,900.00	1,030.00	11,930.00
# 500 KCMIL	L4@26.0	KLF	13,400.00	1,070.00	14,470.00
# 600 KCMIL	L4@30.0	KLF	15,800.00	1,230.00	17,030.00
# 750 KCMIL	L4@32.0	KLF	27,900.00	1,310.00	29,210.00
#1000 KCMIL	L4@36.0	KLF	36,800.00	1,480.00	38,280.00

Type THHN 600 volt solid copper building wire

Material	Craft@Hrs	Unit	Material Cost	Labor Cost	Installed Cost
# 14	L2@6.00	KLF	116.00	246.00	362.00
# 12	L2@7.00	KLF	166.00	287.00	453.00
# 10	L2@8.00	KLF	254.00	328.00	582.00

Use these figures to estimate the cost of copper THW solid and stranded, and THHN solid wire installed in conduit under the conditions described on pages 5 and 6. Costs listed are for each 1,000 linear feet installed. The crew is two electricians for sizes up to #4, three electricians for sizes over #4 to #4/0 and four electricians for sizes over #4/0. The labor cost per manhour is $41.03. These costs include fishing string, circuit make-up, splices on wire up to #6, reel set-up, pulling compound, phase identification, circuit testing, layout, material handling, and normal waste. Add for a bonding wire run in non-metallic or flexible conduit, sales tax, delivery, supervision, mobilization, demobilization, cleanup, overhead and profit. Note: These costs are for conduit runs of less than 100 feet and assume that three wires of the same size are pulled at the same time. A deposit is often required on larger-wire reels. These costs assume that larger reels are returned for credit. Order large-size wire in longest lengths available to reduce waste.

See Wire Conversion Table On Page 537

Copper to Aluminum Wire Ampacities

Copper Building Wire

Type THHN 600 volt stranded copper building wire

Material	Craft@Hrs	Unit	Material Cost	Labor Cost	Installed Cost
# 14	L2@6.00	KLF	93.90	246.00	339.90
# 12	L2@7.00	KLF	134.00	287.00	421.00
# 10	L2@8.00	KLF	202.00	328.00	530.00
# 8	L2@9.00	KLF	392.00	369.00	761.00
# 6	L2@10.0	KLF	584.00	410.00	994.00
# 4	L2@12.0	KLF	1,090.00	492.00	1,582.00
# 2	L3@13.0	KLF	1,710.00	533.00	2,243.00
# 1	L3@14.0	KLF	2,170.00	574.00	2,744.00
# 1/0	L3@15.0	KLF	2,520.00	615.00	3,135.00
# 2/0	L3@16.0	KLF	3,190.00	656.00	3,846.00
# 3/0	L3@17.0	KLF	4,010.00	698.00	4,708.00
# 4/0	L3@18.0	KLF	5,030.00	739.00	5,769.00
#250 KCMIL	L4@20.0	KLF	5,990.00	821.00	6,811.00
#300 KCMIL	L4@23.0	KLF	7,130.00	944.00	8,074.00
#350 KCMIL	L4@24.0	KLF	8,340.00	985.00	9,325.00
#400 KCMIL	L4@25.0	KLF	9,500.00	1,030.00	10,530.00
#500 KCMIL	L4@26.0	KLF	11,800.00	1,070.00	12,870.00

Type XHHW 600 volt solid copper building wire

Material	Craft@Hrs	Unit	Material Cost	Labor Cost	Installed Cost
# 14	L2@6.00	KLF	397.00	246.00	643.00
# 12	L2@7.00	KLF	609.00	287.00	896.00
# 10	L2@8.00	KLF	763.00	328.00	1,091.00

Type XHHW 600 volt stranded copper building wire

Material	Craft@Hrs	Unit	Material Cost	Labor Cost	Installed Cost
# 14	L2@6.00	KLF	314.00	246.00	560.00
# 12	L2@7.00	KLF	462.00	287.00	749.00
# 10	L2@8.00	KLF	690.00	328.00	1,018.00
# 8	L2@9.00	KLF	950.00	369.00	1,319.00
# 6	L2@10.0	KLF	1,010.00	410.00	1,420.00
# 4	L2@12.0	KLF	1,610.00	492.00	2,102.00
# 2	L3@13.0	KLF	2,490.00	533.00	3,023.00
# 1	L3@14.0	KLF	3,240.00	574.00	3,814.00
# 1/0	L3@15.0	KLF	3,920.00	615.00	4,535.00
# 2/0	L3@16.0	KLF	4,900.00	656.00	5,556.00
# 3/0	L3@17.0	KLF	6,190.00	698.00	6,888.00
# 4/0	L3@18.0	KLF	7,780.00	739.00	8,519.00
# 250 KCMIL	L4@20.0	KLF	9,220.00	821.00	10,041.00
# 300 KCMIL	L4@23.0	KLF	11,000.00	944.00	11,944.00
# 350 KCMIL	L4@24.0	KLF	13,000.00	985.00	13,985.00
# 400 KCMIL	L4@25.0	KLF	14,600.00	1,030.00	15,630.00
# 500 KCMIL	L4@26.0	KLF	16,900.00	1,070.00	17,970.00
# 600 KCMIL	L4@30.0	KLF	21,700.00	1,230.00	22,930.00
# 750 KCMIL	L4@32.0	KLF	39,200.00	1,310.00	40,510.00
#1000 KCMIL	L4@36.0	KLF	52,200.00	1,480.00	53,680.00

Use these figures to estimate the cost of copper THHN stranded, XHHW solid and stranded wire installed in conduit under the conditions described on pages 5 and 6. Costs listed are for each 1,000 linear feet installed. The crew is two electricians for sizes up to #4, three electricians for sizes over #4 to #4/0 and four electricians for sizes over #4/0. The labor cost per manhour is $41.03. These costs include fishing string, circuit make-up, splices on wire up to #6, reel set-up, pulling compound, phase identification, circuit testing, layout, material handling, and normal waste. Add for a bonding wire run in non-metallic or flexible conduit, sales tax, delivery, supervision, mobilization, demobilization, cleanup, overhead and profit. Note: These costs are for conduit runs of less than 100 feet and assume that three wires of the same size are pulled at the same time. A deposit is often required on larger-wire reels. These costs assume that larger reels are returned for credit. Order large-size wire in longest lengths available to reduce waste.

Type USE, RHH-RHW 600 volt solid copper building wire

Material	Craft@Hrs	Unit	Material Cost	Labor Cost	Installed Cost
# 12	L2@7.00	KLF	324.00	287.00	611.00
# 10	L2@8.00	KLF	409.00	328.00	737.00
# 8	L2@9.00	KLF	596.00	369.00	965.00

Type USE, RHH-RHW 600 volt stranded copper building wire

Material	Craft@Hrs	Unit	Material Cost	Labor Cost	Installed Cost
# 14	L2@6.00	KLF	324.00	246.00	570.00
# 12	L2@7.00	KLF	409.00	287.00	696.00
# 10	L2@8.00	KLF	596.00	328.00	924.00
# 8	L2@9.00	KLF	767.00	369.00	1,136.00
# 6	L2@10.0	KLF	1,290.00	410.00	1,700.00
# 4	L2@12.0	KLF	2,020.00	492.00	2,512.00
# 2	L3@13.0	KLF	3,190.00	533.00	3,723.00
# 1	L3@14.0	KLF	4,050.00	574.00	4,624.00
# 1/0	L3@15.0	KLF	5,060.00	615.00	5,675.00
# 2/0	L3@16.0	KLF	5,520.00	656.00	6,176.00
# 3/0	L3@17.0	KLF	6,930.00	698.00	7,628.00
# 4/0	L3@18.0	KLF	7,520.00	739.00	8,259.00
# 250 KCMIL	L4@20.0	KLF	9,560.00	821.00	10,381.00
# 300 KCMIL	L4@23.0	KLF	11,400.00	944.00	12,344.00
# 350 KCMIL	L4@24.0	KLF	13,300.00	985.00	14,285.00
# 400 KCMIL	L4@25.0	KLF	15,000.00	1,030.00	16,030.00
# 500 KCMIL	L4@26.0	KLF	18,900.00	1,070.00	19,970.00
# 600 KCMIL	L4@30.0	KLF	22,700.00	1,230.00	23,930.00
# 750 KCMIL	L4@32.0	KLF	40,500.00	1,310.00	41,810.00
#1000 KCMIL	L4@36.0	KLF	53,600.00	1,480.00	55,080.00

Type MTW 600 volt 90 degree stranded copper building wire

Material	Craft@Hrs	Unit	Material Cost	Labor Cost	Installed Cost
# 18	L2@5.00	KLF	102.00	205.00	307.00
# 16	L2@5.50	KLF	220.00	226.00	446.00
# 14	L2@6.00	KLF	295.00	246.00	541.00
# 12	L2@7.00	KLF	421.00	287.00	708.00
# 10	L2@8.00	KLF	667.00	328.00	995.00
# 8	L2@9.00	KLF	1,420.00	369.00	1,789.00

Use these figures to estimate the cost of copper USE RHH-RHW solid and stranded, MTW stranded 90-degree and MTW heavy insulation stranded 90-degree wire installed in conduit under the conditions described on pages 5 and 6. Costs listed are for each 1,000 linear feet installed. The crew is two electricians for sizes up to #4, three electricians for sizes over #4 to #4/0 and four electricians for sizes over #4/0. The labor cost per manhour is $41.03. These costs include fishing string, circuit make-up, splices on wire up to #6, reel set-up, pulling compound, phase identification, circuit testing, layout, material handling, and normal waste. Add for a bonding wire run in non-metallic or flexible conduit, sales tax, delivery, supervision, mobilization, demobilization, cleanup, overhead and profit. Note: These costs are for conduit runs of less than 100 feet and assume that three wires of the same size are pulled at the same time. Type MTW is machine tool wire and can be used for control wiring. A deposit is often required on larger-wire reels. These costs assume that larger reels are returned for credit. Order large-size wire in longest lengths available to reduce waste.

Copper Flexible Wire and Flexible Cords

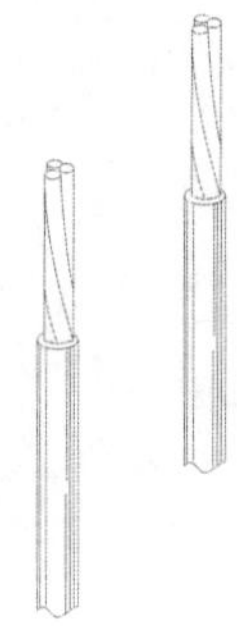

Material	Craft@Hrs	Unit	Material Cost	Labor Cost	Installed Cost

Type TFFN 90 degree stranded copper wire, 600 volt

Material	Craft@Hrs	Unit	Material Cost	Labor Cost	Installed Cost
#18, 16 strand	L2@5.00	KLF	74.00	205.00	279.00
#16, 26 strand	L2@5.25	KLF	87.80	215.00	302.80

Type AWM, 90 degree copper appliance wire, 1,000 volt

Material	Craft@Hrs	Unit	Material Cost	Labor Cost	Installed Cost
#14, 19 strand	L2@5.00	KLF	265.00	205.00	470.00

Type AWM, 105 degree copper appliance wire, 600 volt

Material	Craft@Hrs	Unit	Material Cost	Labor Cost	Installed Cost
#18 stranded	L2@5.00	KLF	172.00	205.00	377.00
#16 stranded	L2@5.25	KLF	199.00	215.00	414.00
#14 stranded	L2@8.00	KLF	265.00	328.00	593.00
#12 stranded	L2@10.0	KLF	378.00	410.00	788.00
#10 stranded	L2@12.0	KLF	597.00	492.00	1,089.00
# 8 stranded	L2@14.0	KLF	695.00	574.00	1,269.00

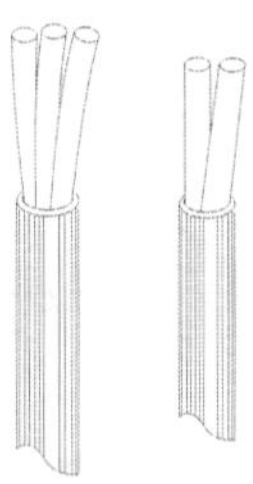

Type SJ 300 volt 60 degree flexible cord

Material	Craft@Hrs	Unit	Material Cost	Labor Cost	Installed Cost
# 18-2	L2@8.00	KLF	599.00	328.00	927.00
# 16-2	L2@9.00	KLF	681.00	369.00	1,050.00
# 14-2	L2@10.0	KLF	944.00	410.00	1,354.00
# 18-3	L2@8.25	KLF	760.00	338.00	1,098.00
# 16-3	L2@9.25	KLF	941.00	380.00	1,321.00
# 14-3	L2@10.3	KLF	1,410.00	423.00	1,833.00
# 18-4	L1@8.50	KLF	1,250.00	349.00	1,599.00
# 16-4	L1@9.50	KLF	1,460.00	390.00	1,850.00
# 14-4	L1@10.5	KLF	2,430.00	431.00	2,861.00

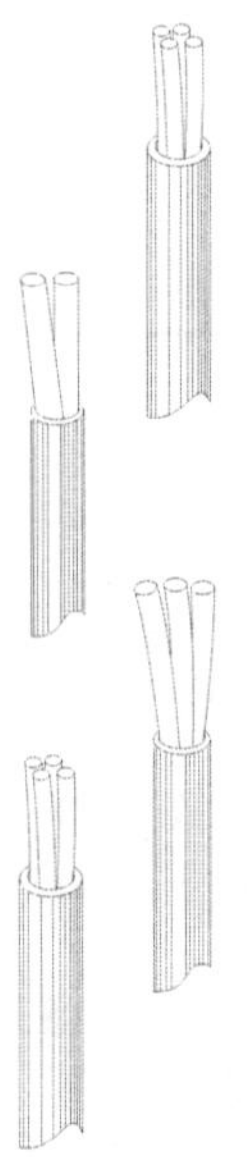

Type S 600 volt 60 degree flexible cord

Material	Craft@Hrs	Unit	Material Cost	Labor Cost	Installed Cost
# 18-2	L1@8.00	KLF	822.00	328.00	1,150.00
# 16-2	L1@9.00	KLF	961.00	369.00	1,330.00
# 14-2	L1@10.0	KLF	1,370.00	410.00	1,780.00
# 12-2	L1@12.0	KLF	1,740.00	492.00	2,232.00
# 10-2	L1@14.0	KLF	2,240.00	574.00	2,814.00
# 18-3	L1@8.25	KLF	865.00	338.00	1,203.00
# 16-3	L1@9.25	KLF	1,020.00	380.00	1,400.00
# 14-3	L1@10.3	KLF	1,630.00	423.00	2,053.00
# 12-3	L1@12.3	KLF	1,950.00	505.00	2,455.00
# 10-3	L1@14.3	KLF	2,390.00	587.00	2,977.00
# 18-4	L1@8.50	KLF	1,260.00	349.00	1,609.00
# 16-4	L1@9.50	KLF	2,040.00	390.00	2,430.00
# 14-4	L1@10.5	KLF	2,100.00	431.00	2,531.00
# 12-4	L1@12.5	KLF	2,280.00	513.00	2,793.00
# 10-4	L1@14.5	KLF	2,790.00	595.00	3,385.00

Use these figures to estimate the cost of copper building wire installed in conduit systems and for flexible cords under the conditions described on pages 5 and 6. Costs listed are for each 1,000 linear feet installed. The crew is two electricians working at a labor cost of $41.03 per manhour. These costs include fishing string, circuit make-up, stripping, phase identification, circuit testing, layout, material handling, and normal waste. Add for a cord connectors, supports, sales tax, delivery, supervision, mobilization, demobilization, cleanup, overhead and profit. Note: Costs for wire pulled in conduit assume conduit runs of less than 100 feet and assume that three wires of the same size are pulled at the same time. Small-gauge wire is usually packed on spools in 500-foot lengths. Some manufacturers offer lengths of 1,500 and 2,500 feet. Flexible cord is usually sold in 250-foot lengths. The *NEC* places limits on how flexible cords can be used. Never pull flexible cord in conduit.

Material	Craft@Hrs	Unit	Material Cost	Labor Cost	Installed Cost
Type SJTO 300 volt 60 degree flexible yellow cord					
# 18-2	L1@8.00	KLF	873.00	328.00	1,201.00
# 16-2	L1@9.00	KLF	1,040.00	369.00	1,409.00
# 14-2	L1@10.0	KLF	1,400.00	410.00	1,810.00
# 18-3	L1@8.25	KLF	804.00	338.00	1,142.00
# 16-3	L1@9.25	KLF	1,060.00	380.00	1,440.00
# 14-3	L1@10.3	KLF	1,480.00	423.00	1,903.00
# 18-4	L1@8.50	KLF	1,940.00	349.00	2,289.00
# 16-4	L1@9.50	KLF	1,710.00	390.00	2,100.00
# 14-4	L1@10.5	KLF	2,170.00	431.00	2,601.00
Type STO 600 volt 60 degree flexible yellow cord					
# 18-2	L1@8.00	KLF	1,470.00	328.00	1,798.00
# 16-2	L1@9.00	KLF	1,680.00	369.00	2,049.00
# 14-2	L1@10.0	KLF	2,100.00	410.00	2,510.00
# 12-2	L1@12.0	KLF	2,660.00	492.00	3,152.00
# 10-2	L1@14.0	KLF	3,290.00	574.00	3,864.00
# 18-3	L1@8.25	KLF	1,370.00	338.00	1,708.00
# 16-3	L1@9.25	KLF	1,410.00	380.00	1,790.00
# 14-3	L1@10.3	KLF	2,390.00	423.00	2,813.00
# 12-3	L1@12.3	KLF	3,330.00	505.00	3,835.00
# 10-3	L1@14.3	KLF	3,990.00	587.00	4,577.00
# 18-4	L1@8.50	KLF	1,780.00	349.00	2,129.00
# 16-4	L1@9.50	KLF	1,990.00	390.00	2,380.00
# 14-4	L1@10.5	KLF	2,970.00	431.00	3,401.00
# 12-4	L1@12.5	KLF	3,650.00	513.00	4,163.00
# 10-4	L1@14.5	KLF	4,900.00	595.00	5,495.00

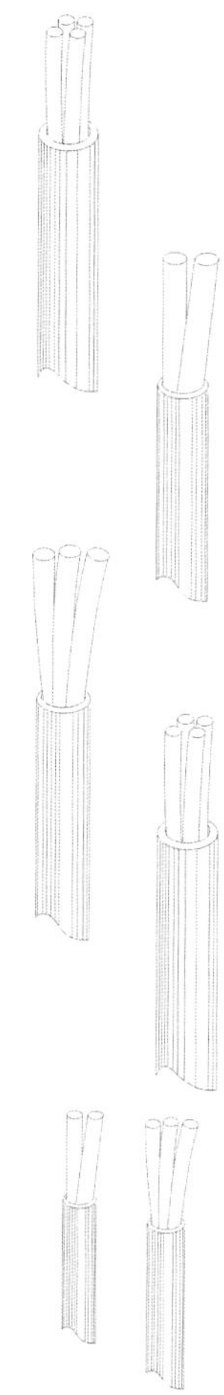

Use these figures to estimate the cost of copper flexible cords installed exposed in buildings under the conditions described on pages 5 and 6. Costs listed are for each 1,000 linear feet installed. The crew is one electrician working at a labor cost of $41.03 per manhour. These costs include circuit make-up, reel set-up, phase identification, circuit testing, stripping, layout, material handling, and normal waste. Add for cord connectors, supports, sales tax, delivery, supervision, mobilization, demobilization, cleanup, overhead and profit. Note: The *NEC* places limits on how flexible cords can be used. Never install flexible cord in conduit.

Flexible Cords and Romex

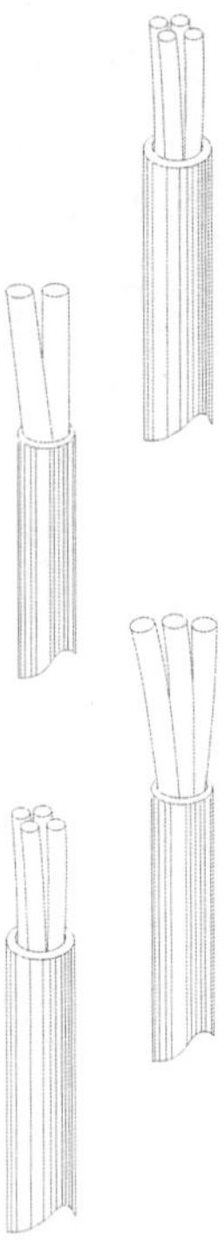

Type W 600 volt portable power cable with neoprene jacket, without ground

Material	Craft@Hrs	Unit	Material Cost	Labor Cost	Installed Cost
# 8-2	L1@9.00	KLF	2,520.00	369.00	2,889.00
# 6-2	L1@10.0	KLF	3,200.00	410.00	3,610.00
# 4-2	L1@11.0	KLF	4,540.00	451.00	4,991.00
# 2-2	L1@12.0	KLF	6,220.00	492.00	6,712.00
# 1-2	L1@13.0	KLF	8,070.00	533.00	8,603.00
# 8-3	L1@9.25	KLF	3,360.00	380.00	3,740.00
# 6-3	L1@10.3	KLF	4,360.00	423.00	4,783.00
# 4-3	L1@11.3	KLF	5,890.00	464.00	6,354.00
# 2-3	L1@12.3	KLF	8,250.00	505.00	8,755.00
# 1-3	L1@13.3	KLF	10,800.00	546.00	11,346.00
# 8-4	L1@9.50	KLF	4,040.00	390.00	4,430.00
# 6-4	L1@10.5	KLF	5,050.00	431.00	5,481.00
# 4-4	L1@11.5	KLF	7,240.00	472.00	7,712.00
# 2-4	L1@12.5	KLF	10,400.00	513.00	10,913.00
# 1-4	L1@13.5	KLF	13,000.00	554.00	13,554.00

Type NM 600 volt non-metallic sheathed cable (Romex) without ground

Material	Craft@Hrs	Unit	Material Cost	Labor Cost	Installed Cost
# 14-2	L1@6.00	KLF	379.00	246.00	625.00
# 12-2	L1@6.50	KLF	583.00	267.00	850.00
# 10-2	L1@7.00	KLF	919.00	287.00	1,206.00
# 14-3	L1@6.50	KLF	529.00	267.00	796.00
# 12-3	L1@7.00	KLF	821.00	287.00	1,108.00
# 10-3	L1@7.50	KLF	1,300.00	308.00	1,608.00
# 8-3	L1@10.0	KLF	2,250.00	410.00	2,660.00
# 6-3	L1@14.0	KLF	3,610.00	574.00	4,184.00
# 4-3	L1@16.0	KLF	6,380.00	656.00	7,036.00

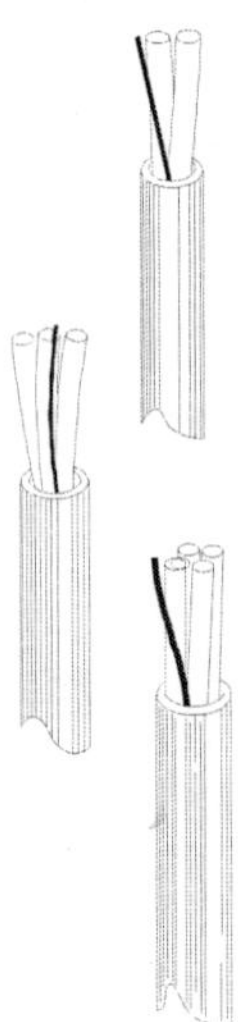

Type NM 600 volt non-metallic sheathed cable (Romex) with ground

Material	Craft@Hrs	Unit	Material Cost	Labor Cost	Installed Cost
#14-2	L1@6.00	KLF	334.00	246.00	580.00
#12-2	L1@6.50	KLF	507.00	267.00	774.00
#10-2	L1@7.00	KLF	850.00	287.00	1,137.00
# 8-2	L1@9.00	KLF	1,340.00	369.00	1,709.00
# 6-2	L1@12.0	KLF	2,130.00	492.00	2,622.00
#14-3	L1@6.50	KLF	467.00	267.00	734.00
#12-3	L1@7.00	KLF	767.00	287.00	1,054.00
#10-3	L1@7.50	KLF	1,220.00	308.00	1,528.00
# 8-3	L1@10.0	KLF	1,970.00	410.00	2,380.00
# 6-3	L1@14.0	KLF	3,160.00	574.00	3,734.00
# 4-3	L1@16.0	KLF	5,730.00	656.00	6,386.00

Use these figures to estimate the cost of copper flexible cords and Type NM cable installed in buildings under the conditions described on pages 5 and 6. Costs listed are for each 1,000 linear feet installed. The crew is one electrician working at a labor cost of $41.03 per manhour. These costs include circuit make-up, reel set-up, phase identification, circuit testing, stripping, boring wood studs, layout, material handling, and normal waste. Add for supports, staples, connectors, sales tax, delivery, supervision, mobilization, demobilization, cleanup, overhead and profit. Note: The *NEC* places limits on how flexible cords and NM cable can be used. Non-metallic (NM) cable is limited to residences. Never install flexible cord or non-metallic cable in conduit.

Strain relief grips

Material	Craft@Hrs	Unit	Material Cost	Labor Cost	Installed Cost
1/2" x 4-3/4"	L1@0.10	Ea	33.70	4.10	37.80
3/4" x 5-3/4"	L1@0.15	Ea	36.30	6.15	42.45
1" x 7"	L1@0.20	Ea	44.40	8.21	52.61

Insulated strain relief grips

Material	Craft@Hrs	Unit	Material Cost	Labor Cost	Installed Cost
1/2" x 4-3/4"	L1@0.10	Ea	41.10	4.10	45.20
3/4" x 5-3/4"	L1@0.15	Ea	46.00	6.15	52.15
1" x 7"	L1@0.20	Ea	49.40	8.21	57.61
1-1/4" x 9"	L1@0.25	Ea	61.50	10.30	71.80
1-1/2" x 11-3/4"	L1@0.25	Ea	103.00	10.30	113.30
2" x 13-1/4"	L1@0.30	Ea	129.00	12.30	141.30
2-1/2" x 13-3/4"	L1@0.35	Ea	251.00	14.40	265.40

HD grips, straight connector

Material	Craft@Hrs	Unit	Material Cost	Labor Cost	Installed Cost
3/8"	L1@0.10	Ea	93.80	4.10	97.90
1/2"	L1@0.10	Ea	91.50	4.10	95.60
3/4"	L1@0.15	Ea	112.00	6.15	118.15
1"	L1@0.20	Ea	133.00	8.21	141.21
1-1/4"	L1@0.25	Ea	206.00	10.30	216.30
1-1/2"	L1@0.25	Ea	242.00	10.30	252.30
2"	L1@0.30	Ea	323.00	12.30	335.30

HD grips, 90 degree

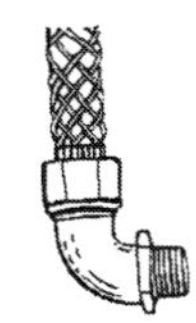

Material	Craft@Hrs	Unit	Material Cost	Labor Cost	Installed Cost
1/2"	L1@0.15	Ea	135.00	6.15	141.15
3/4"	L1@0.20	Ea	173.00	8.21	181.21
1"	L1@0.25	Ea	184.00	10.30	194.30

Service drop grips

Material	Craft@Hrs	Unit	Material Cost	Labor Cost	Installed Cost
3-3/4" long	L1@0.15	Ea	22.10	6.15	28.25
4-1/4" long	L1@0.20	Ea	22.40	8.21	30.61
4-3/4" long	L1@0.25	Ea	25.40	10.30	35.70
5" long	L1@0.25	Ea	25.40	10.30	35.70
5-1/4" long	L1@0.25	Ea	31.90	10.30	42.20
6-1/4" long	L1@0.25	Ea	35.20	10.30	45.50
7-1/4" long	L1@0.25	Ea	37.90	10.30	48.20
8-1/4" long	L1@0.25	Ea	41.70	10.30	52.00
8-3/4" long	L1@0.25	Ea	41.70	10.30	52.00
9" long	L1@0.25	Ea	41.70	10.30	52.00
9-1/2" long	L1@0.25	Ea	41.70	10.30	52.00

Use these figures to estimate the cost of strain relief grip connectors installed under the conditions described on pages 5 and 6. Costs listed are for each grip installed. The crew is one electrician working at a labor cost of $41.03 per manhour. These costs include the grip connector only (and no other material), wire cutting, layout, material handling and normal waste. Add for sales tax, delivery, supervision, mobilization, demobilization, cleanup, overhead and profit.

Copper 600 Volt Non-metallic Sheathed Cable

Type USE-RHH-RHW, XLP 600 volt solid copper direct burial cable

Material	Craft@Hrs	Unit	Material Cost	Labor Cost	Installed Cost
#14	L1@4.50	KLF	491.00	185.00	676.00
#12	L1@4.75	KLF	625.00	195.00	820.00
#10	L1@5.00	KLF	910.00	205.00	1,115.00

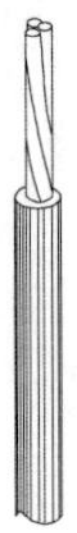

Type USE-RHH-RHW, XLP 600 volt stranded copper direct burial cable

Material	Craft@Hrs	Unit	Material Cost	Labor Cost	Installed Cost
# 14	L1@4.50	KLF	612.00	185.00	797.00
# 12	L1@4.75	KLF	734.00	195.00	929.00
# 10	L1@5.00	KLF	1,090.00	205.00	1,295.00
# 8	L1@6.00	KLF	1,500.00	246.00	1,746.00
# 6	L1@7.00	KLF	2,540.00	287.00	2,827.00
# 4	L1@8.00	KLF	3,980.00	328.00	4,308.00
# 2	L1@10.0	KLF	6,210.00	410.00	6,620.00
# 1	L1@12.0	KLF	7,940.00	492.00	8,432.00
# 1/0	L1@14.0	KLF	9,950.00	574.00	10,524.00
# 2/0	L1@15.0	KLF	12,500.00	615.00	13,115.00
# 3/0	L1@16.0	KLF	15,500.00	656.00	16,156.00
# 4/0	L1@18.0	KLF	17,800.00	739.00	18,539.00
#250 KCMIL	L1@19.0	KLF	18,700.00	780.00	19,480.00
#300 KCMIL	L1@20.0	KLF	22,500.00	821.00	23,321.00
#350 KCMIL	L1@21.0	KLF	26,200.00	862.00	27,062.00
#400 KCMIL	L1@21.0	KLF	30,000.00	862.00	30,862.00
#500 LCMIL	L1@22.0	KLF	37,100.00	903.00	38,003.00

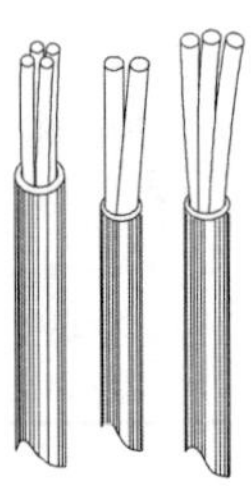

Type UF 600 volt copper solid direct burial cable

Material	Craft@Hrs	Unit	Material Cost	Labor Cost	Installed Cost
#14	L1@4.50	KLF	747.00	185.00	932.00
#12	L1@4.75	KLF	1,140.00	195.00	1,335.00
#10	L1@5.00	KLF	1,820.00	205.00	2,025.00
# 8	L1@6.50	KLF	2,810.00	267.00	3,077.00
# 6	L1@7.00	KLF	4,410.00	287.00	4,697.00

Type UF 600 volt copper stranded direct burial cable

Material	Craft@Hrs	Unit	Material Cost	Labor Cost	Installed Cost
#8	L1@6.00	KLF	4,140.00	246.00	4,386.00

Use these figures to estimate the cost of copper Type NM cable installed in residences and UF cable installed underground under the conditions described on pages 5 and 6. Costs listed are for each 1,000 linear feet installed. The crew is one electrician except for UF cable over #1 when two electricians are needed. The labor cost per manhour is $41.03. These costs include circuit make-up, reel set-up, phase identification, circuit testing, stripping, layout, material handling, and normal waste. Add for connectors, staples, trenching, sales tax, delivery, supervision, mobilization, demobilization, cleanup, overhead and profit. Note: Always bury a warning tape about 12 inches above UF cable.

Copper 600 Volt Non-metallic Sheathed Cable

Type UF 600 volt copper stranded direct burial cable with ground

Material	Craft@Hrs	Unit	Material Cost	Labor Cost	Installed Cost
#14-2	L1@5.00	KLF	747.00	205.00	952.00
#12-2	L1@5.25	KLF	1,140.00	215.00	1,355.00
#10-2	L1@6.00	KLF	1,150.00	246.00	1,396.00
# 8-2	L1@7.00	KLF	1,820.00	287.00	2,107.00
# 6-2	L1@8.00	KLF	4,410.00	328.00	4,738.00
#14-3	L1@5.25	KLF	1,020.00	215.00	1,235.00
#12-3	L1@5.50	KLF	1,560.00	226.00	1,786.00
#10-3	L1@6.25	KLF	2,520.00	256.00	2,776.00
# 8-3	L1@7.25	KLF	3,760.00	297.00	4,057.00
# 6-3	L1@8.25	KLF	6,090.00	338.00	6,428.00

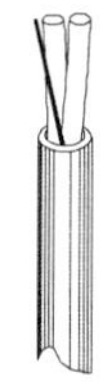

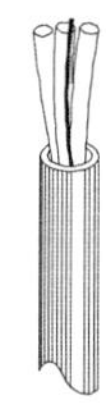

Type SEU 600 volt copper stranded service entrance cable

Material	Craft@Hrs	Unit	Material Cost	Labor Cost	Installed Cost
#10-3	L1@8.50	KLF	950.00	349.00	1,299.00
# 8-3	L1@11.0	KLF	1,500.00	451.00	1,951.00
# 6-3	L1@13.0	KLF	2,570.00	533.00	3,103.00
# 4-3	L1@14.0	KLF	4,250.00	574.00	4,824.00
# 2-3	L1@16.0	KLF	6,310.00	656.00	6,966.00
# 1-3	L1@18.0	KLF	8,100.00	739.00	8,839.00
#1/0-3	L1@20.0	KLF	10,100.00	821.00	10,921.00
#2/0-3	L1@22.0	KLF	12,600.00	903.00	13,503.00
#3/0-3	L1@24.0	KLF	15,800.00	985.00	16,785.00
#4/0-3	L1@26.0	KLF	17,600.00	1,070.00	18,670.00
# 6-2 & 8-1	L1@13.0	KLF	2,270.00	533.00	2,803.00
# 4-2 & 6-1	L1@14.5	KLF	3,560.00	595.00	4,155.00
# 2-2 & 4-1	L1@16.0	KLF	5,530.00	656.00	6,186.00

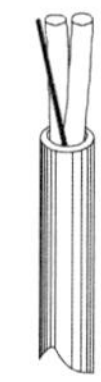

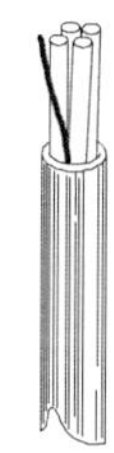

Use these figures to estimate the cost of copper UF and SEU cable installed underground, and SEU in buildings, under the conditions described on pages 5 and 6. Costs listed are for each 1,000 linear feet installed. The crew is one electrician working at a labor cost of $41.03 per manhour. These costs include circuit make-up, reel set-up, phase identification, circuit testing, stripping, layout, material handling, and normal waste. Add for trenching supports, connectors, the SEU service entrance cap, sales tax, delivery, supervision, mobilization, demobilization, cleanup, overhead and profit. Note: Always bury a warning tape about 12 inches above UF cable.

Copper 600 Volt Non-metallic Sheathed Cable

Material	Craft@Hrs	Unit	Material Cost	Labor Cost	Installed Cost

Type SER 600 copper cable, style U

Material	Craft@Hrs	Unit	Material Cost	Labor Cost	Installed Cost
# 6-4	L1@14.5	KLF	4,190.00	595.00	4,785.00
# 4-3 & 6-1	L1@16.0	KLF	6,020.00	656.00	6,676.00
# 3-3 & 5-1	L1@17.0	KLF	7,820.00	698.00	8,518.00
# 2-3 & 4-1	L1@18.0	KLF	9,500.00	739.00	10,239.00
# 1-3 & 3-1	L1@21.0	KLF	11,900.00	862.00	12,762.00
# 1/0-3 & 2-1	L1@22.0	KLF	14,900.00	903.00	15,803.00
# 2/0-3 & 1-1	L1@24.0	KLF	18,700.00	985.00	19,685.00
# 3/0-1/0-1	L1@26.0	KLF	23,400.00	1,070.00	24,470.00
# 4/0-2/0-1	L1@28.0	KLF	29,500.00	1,150.00	30,650.00

Soft drawn solid bare copper wire

Material	Craft@Hrs	Unit	Material Cost	Labor Cost	Installed Cost
#14	L1@4.00	KLF	188.00	164.00	352.00
#12	L1@4.25	KLF	295.00	174.00	469.00
#10	L1@4.50	KLF	467.00	185.00	652.00
# 8	L1@4.75	KLF	753.00	195.00	948.00
# 6	L1@5.00	KLF	1,320.00	205.00	1,525.00
# 4	L1@6.00	KLF	2,240.00	246.00	2,486.00

Soft drawn stranded bare copper wire

Material	Craft@Hrs	Unit	Material Cost	Labor Cost	Installed Cost
# 8	L1@4.75	KLF	823.00	195.00	1,018.00
# 6	L1@5.00	KLF	1,430.00	205.00	1,635.00
# 4	L1@6.00	KLF	2,280.00	246.00	2,526.00
# 2	L1@8.00	KLF	3,640.00	328.00	3,968.00
# 1	L1@10.0	KLF	5,210.00	410.00	5,620.00
# 1/0	L1@11.0	KLF	6,010.00	451.00	6,461.00
# 2/0	L1@12.0	KLF	6,580.00	492.00	7,072.00
# 3/0	L1@13.0	KLF	8,270.00	533.00	8,803.00
# 4/0	L1@14.0	KLF	10,400.00	574.00	10,974.00

Use these figures to estimate the cost of copper SEU service entrance cable and bare copper wire installed in buildings, under the conditions described on pages 5 and 6. Costs listed are for each 1,000 linear feet installed. The crew is one electrician working at a labor cost of $41.03 per manhour. These costs include circuit make-up, reel set-up, phase identification, circuit testing, stripping, layout, material handling, and normal waste. Add for connectors, supports, sales tax, delivery, supervision, mobilization, demobilization, cleanup, overhead and profit. Note: The *NEC* permits installation of bare copper wire in conduit with conductors. But bare copper wire installed outside conduit must be protected against damage. Some dealers sell bare copper wire by weight rather than length. The table at the bottom of the page shows weights per 1,000 linear feet.

Approximate weight per 1000' for soft drawn bare copper wire (pounds)

Size		Weight	Size		Weight
#14	solid	12.40	#2	stranded	204.90
#12	solid	19.80	#1	stranded	258.40
#10	solid	31.43	#1/0	stranded	325.80
#8	stranded	50.97	#2/0	stranded	410.90
#6	stranded	81.05	#3/0	stranded	518.10
#4	stranded	128.90	#4/0	stranded	653.30

Copper Armored Cable (MC)

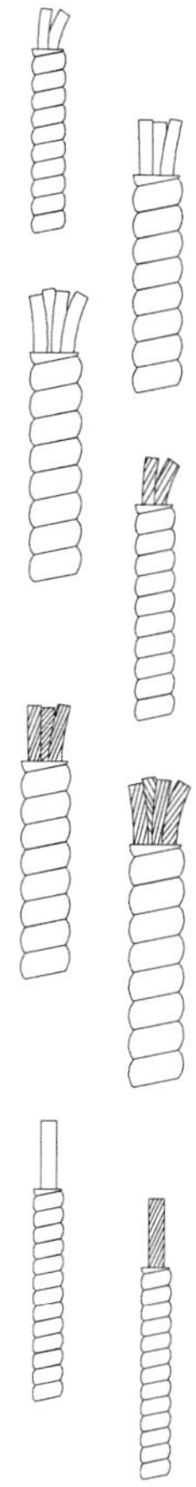

Copper solid armored (MC) cable

Material	Craft@Hrs	Unit	Material Cost	Labor Cost	Installed Cost
#14-2	L1@8.50	KLF	1,510.00	349.00	1,859.00
#12-2	L1@9.50	KLF	1,560.00	390.00	1,950.00
#10-2	L1@11.0	KLF	2,730.00	451.00	3,181.00
#14-3	L1@9.50	KLF	2,360.00	390.00	2,750.00
#12-3	L1@10.5	KLF	2,590.00	431.00	3,021.00
#10-3	L1@14.0	KLF	5,260.00	574.00	5,834.00
#14-4	L1@11.0	KLF	3,190.00	451.00	3,641.00
#12-4	L1@13.0	KLF	3,510.00	533.00	4,043.00
#10-4	L1@15.0	KLF	7,060.00	615.00	7,675.00

Copper stranded armored (MC) cable

Material	Craft@Hrs	Unit	Material Cost	Labor Cost	Installed Cost
# 8-2	L1@12.0	KLF	6,820.00	492.00	7,312.00
# 6-2	L1@15.0	KLF	10,600.00	615.00	11,215.00
# 8-3	L1@15.0	KLF	8,730.00	615.00	9,345.00
# 6-3	L1@17.0	KLF	14,000.00	698.00	14,698.00
# 4-3	L1@20.0	KLF	19,700.00	821.00	20,521.00
# 2-3	L1@24.0	KLF	28,300.00	985.00	29,285.00
# 8-4	L1@17.0	KLF	14,100.00	698.00	14,798.00
# 6-4	L1@20.0	KLF	19,400.00	821.00	20,221.00
# 4-4	L1@24.0	KLF	26,900.00	985.00	27,885.00

Copper solid armored (MC) cable with bare copper ground

Material	Craft@Hrs	Unit	Material Cost	Labor Cost	Installed Cost
# 8	L1@10.0	KLF	2,170.00	410.00	2,580.00
# 6	L1@12.0	KLF	2,310.00	492.00	2,802.00

Copper stranded armored (MC) cable with bare copper ground

Material	Craft@Hrs	Unit	Material Cost	Labor Cost	Installed Cost
# 4	L1@14.0	KLF	4,090.00	574.00	4,664.00

Anti-short bushings

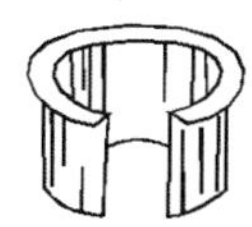

Material	Craft@Hrs	Unit	Material Cost	Labor Cost	Installed Cost
3/8"	L1@0.02	Ea	.09	.82	.91
1/2"	L1@0.02	Ea	.11	.82	.93
3/4"	L1@0.02	Ea	.25	.82	1.07

MC connectors, 1 screw squeeze type

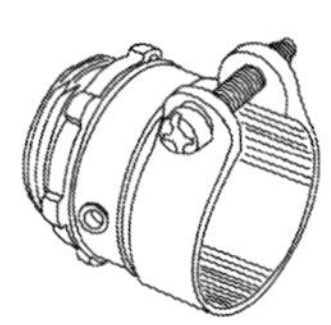

Material	Craft@Hrs	Unit	Material Cost	Labor Cost	Installed Cost
3/8" standard cable	L1@0.05	Ea	1.04	2.05	3.09
3/8" oversize cable	L1@0.05	Ea	1.39	2.05	3.44
1/2" standard cable	L1@0.05	Ea	1.51	2.05	3.56
3/4" standard cable	L1@0.06	Ea	2.22	2.46	4.68

MC connectors, 2 screw

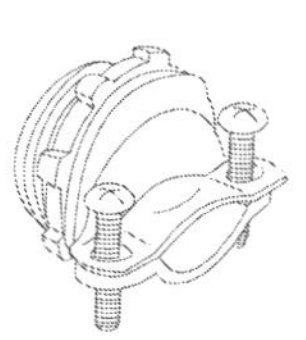

Material	Craft@Hrs	Unit	Material Cost	Labor Cost	Installed Cost
3/8" standard cable	L1@0.05	Ea	.92	2.05	2.97
1/2" standard cable	L1@0.05	Ea	1.86	2.05	3.91
3/4" standard cable	L1@0.05	Ea	3.31	2.05	5.36

Use these figures to estimate the cost of copper MC cable installed in buildings under the conditions described on pages 5 and 6. Costs listed are for each 1000 linear feet of cable, bushing or connector installed. The crew is one electrician working at a labor cost of $41.03 per manhour. These costs include boring, notching, stripping, circuit makeup, layout, material handling, and normal waste. Add for supports, sales tax, delivery, supervision, mobilization, demobilization, cleanup, overhead and profit. Note: the NEC and your local code restrict the ways MC can be used.

Single Conductor Copper Power Cable

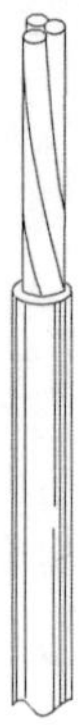

5000 volt cross-linked polyethylene non-shielded copper (XLP) power cable

Material	Craft@Hrs	Unit	Material Cost	Labor Cost	Installed Cost
# 6	L2@14.0	KLF	4,290.00	574.00	4,864.00
# 4	L2@16.0	KLF	5,430.00	656.00	6,086.00
# 2	L2@18.0	KLF	6,820.00	739.00	7,559.00
# 1/0	L3@22.0	KLF	12,600.00	903.00	13,503.00
# 2/0	L3@24.0	KLF	15,600.00	985.00	16,585.00
# 4/0	L3@28.0	KLF	19,200.00	1,150.00	20,350.00
#350 KCMIL	L3@34.0	KLF	32,700.00	1,400.00	34,100.00
#500 KCMIL	L3@36.0	KLF	43,300.00	1,480.00	44,780.00

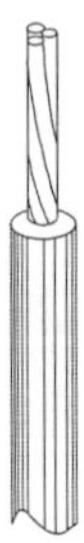

5000 volt cross-linked polyethylene tape shielded copper (XLP) power cable

Material	Craft@Hrs	Unit	Material Cost	Labor Cost	Installed Cost
# 6	L2@14.0	KLF	8,770.00	574.00	9,344.00
# 4	L2@16.0	KLF	9,090.00	656.00	9,746.00
# 2	L2@18.0	KLF	9,510.00	739.00	10,249.00
# 1/0	L3@22.0	KLF	16,500.00	903.00	17,403.00
# 2/0	L3@24.0	KLF	23,600.00	985.00	24,585.00
# 4/0	L3@28.0	KLF	27,800.00	1,150.00	28,950.00
# 350 KCMIL	L3@34.0	KLF	41,700.00	1,400.00	43,100.00
# 500 KCMIL	L3@36.0	KLF	55,100.00	1,480.00	56,580.00

15000 volt ethylene-propylene-rubber tape shielded copper (EPR) power cable

Material	Craft@Hrs	Unit	Material Cost	Labor Cost	Installed Cost
# 2	L2@20.0	KLF	19,400.00	821.00	20,221.00
# 1/0	L3@26.0	KLF	24,200.00	1,070.00	25,270.00
# 2/0	L3@28.0	KLF	28,900.00	1,150.00	30,050.00
# 4/0	L3@32.0	KLF	35,200.00	1,310.00	36,510.00
# 350 KCMIL	L3@38.0	KLF	49,800.00	1,560.00	51,360.00
# 500 KCMIL	L3@42.0	KLF	57,400.00	1,720.00	59,120.00

Use these figures to estimate the cost of medium-voltage copper power cable used for the primary service under the conditions described on pages 5 and 6. Costs listed are for each 1,000 linear feet installed. The crew is two electricians for sizes to #1 and three or four electricians for sizes over #1. Cost per manhour is $41.03. These costs include fishing string, reel set-up, pulling gear set-up, phase identification, pulling compound, layout, material handling, and normal waste. Add for terminations, splices, fire-proofing, high-potential testing, sales tax, delivery, supervision, mobilization, demobilization, cleanup, overhead and profit. Note: These figures assume that cable is pulled in conduit runs of 100 feet or less and that three conductors are pulled at the same time. Keep medium-voltage cable sealed against moisture at all times.

Material	Craft@Hrs	Unit	Material Cost	Labor Cost	Installed Cost
Type THW 600 volt 75 degree stranded aluminum wire					
# 6	L2@9.00	KLF	497.00	369.00	866.00
# 4	L2@10.0	KLF	620.00	410.00	1,030.00
# 2	L2@11.0	KLF	838.00	451.00	1,289.00
# 1	L2@12.0	KLF	1,210.00	492.00	1,702.00
# 1/0	L3@13.0	KLF	1,460.00	533.00	1,993.00
# 2/0	L3@14.0	KLF	1,720.00	574.00	2,294.00
# 3/0	L3@15.0	KLF	2,160.00	615.00	2,775.00
# 4/0	L3@16.0	KLF	2,390.00	656.00	3,046.00
# 250 KCMIL	L3@18.0	KLF	2,910.00	739.00	3,649.00
# 300 KCMIL	L3@19.0	KLF	4,040.00	780.00	4,820.00
# 350 KCMIL	L3@20.0	KLF	4,090.00	821.00	4,911.00
# 400 KCMIL	L3@21.0	KLF	4,830.00	862.00	5,692.00
# 500 KCMIL	L3@22.0	KLF	5,300.00	903.00	6,203.00
# 600 KCMIL	L3@23.0	KLF	6,770.00	944.00	7,714.00
# 700 KCMIL	L3@24.0	KLF	7,760.00	985.00	8,745.00
# 750 KCMIL	L3@25.0	KLF	7,910.00	1,030.00	8,940.00
#1000 KCMIL	L3@30.0	KLF	11,600.00	1,230.00	12,830.00
Type THHN-THWN 600 volt 90 degree stranded aluminum wire					
# 6	L2@9.00	KLF	497.00	369.00	866.00
# 4	L2@10.0	KLF	620.00	410.00	1,030.00
# 2	L2@11.0	KLF	838.00	451.00	1,289.00
# 1	L2@12.0	KLF	1,190.00	492.00	1,682.00
# 1/0	L3@13.0	KLF	1,460.00	533.00	1,993.00
# 2/0	L3@14.0	KLF	1,720.00	574.00	2,294.00
# 3/0	L3@15.0	KLF	2,160.00	615.00	2,775.00
# 4/0	L3@16.0	KLF	2,390.00	656.00	3,046.00
# 250 KCMIL	L3@18.0	KLF	2,910.00	739.00	3,649.00
# 300 KCMIL	L3@19.0	KLF	4,040.00	780.00	4,820.00
# 350 KCMIL	L3@20.0	KLF	4,090.00	821.00	4,911.00
# 400 KCMIL	L3@21.0	KLF	4,830.00	862.00	5,692.00
# 500 KCMIL	L3@22.0	KLF	5,300.00	903.00	6,203.00
# 600 KCMIL	L3@23.0	KLF	6,770.00	944.00	7,714.00
# 700 KCMIL	L3@24.0	KLF	7,760.00	985.00	8,745.00
# 750 KCMIL	L3@25.0	KLF	7,910.00	1,030.00	8,940.00
#1000 KCMIL	L3@30.0	KLF	11,600.00	1,230.00	12,830.00

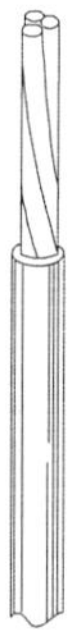

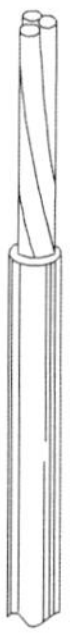

Use these figures to estimate the cost of aluminum THW and THHN-THWN wire installed in conduit under the conditions described on pages 5 and 6. Costs listed are for each 1,000 linear feet installed. The crew is two electricians for sizes to #1 and three electricians for sizes over #1. The labor cost per manhour is $41.03. These costs include fishing string, reel set-up, pulling gear set-up, phase identification, pulling compound, layout, material handling, and normal waste. Add for splicing, anti-oxidation compound, sales tax, delivery, supervision, mobilization, demobilization, cleanup, overhead and profit. Note: These figures assume that wire is pulled in conduit runs of 100 feet or less and that three conductors are pulled at the same time. Use anti-oxidation compound on all aluminum connections. The ampacity of copper wire is greater than the ampacity of aluminum wire of the same size. Check the *NEC* before substituting aluminum wire for copper wire. Terminations for aluminum wire must be made with approved fittings.

See Wire Conversion Table on Page 537
Copper To Aluminum Wire Ampacities

600 Volt Aluminum Wire

Type XHHW cross-linked polyethylene 600 volt 90 degree (XLP) aluminum wire

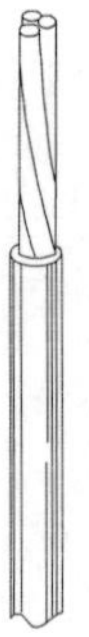

Material	Craft@Hrs	Unit	Material Cost	Labor Cost	Installed Cost
# 6	L2@9.00	KLF	497.00	369.00	866.00
# 4	L2@10.0	KLF	620.00	410.00	1,030.00
# 2	L2@11.0	KLF	838.00	451.00	1,289.00
# 1	L2@12.0	KLF	1,210.00	492.00	1,702.00
# 1/0	L3@13.0	KLF	1,460.00	533.00	1,993.00
# 2/0	L3@14.0	KLF	1,720.00	574.00	2,294.00
# 3/0	L3@15.0	KLF	2,160.00	615.00	2,775.00
# 4/0	L3@16.0	KLF	2,390.00	656.00	3,046.00
# 250 KCMIL	L3@18.0	KLF	2,910.00	739.00	3,649.00
# 300 KCMIL	L3@19.0	KLF	4,040.00	780.00	4,820.00
# 350 KCMIL	L3@20.0	KLF	4,090.00	821.00	4,911.00
# 400 KCMIL	L3@21.0	KLF	4,830.00	862.00	5,692.00
# 500 KCMIL	L3@22.0	KLF	5,300.00	903.00	6,203.00
# 600 KCMIL	L3@23.0	KLF	6,770.00	944.00	7,714.00
# 700 KCMIL	L3@24.0	KLF	7,760.00	985.00	8,745.00
# 750 KCMIL	L3@25.0	KLF	7,910.00	1,030.00	8,940.00
#1000 KCMIL	L3@30.0	KLF	11,600.00	1,230.00	12,830.00

Type USE, RHH-RHW 600 volt cross-linked polyethylene (XLP) aluminum wire

Material	Craft@Hrs	Unit	Material Cost	Labor Cost	Installed Cost
# 6	L2@9.00	KLF	431.00	369.00	800.00
# 4	L2@10.0	KLF	497.00	410.00	907.00
# 2	L2@11.0	KLF	686.00	451.00	1,137.00
# 1	L2@12.0	KLF	534.00	492.00	1,026.00
# 1/0	L3@13.0	KLF	1,180.00	533.00	1,713.00
# 2/0	L3@14.0	KLF	1,380.00	574.00	1,954.00
# 3/0	L3@15.0	KLF	1,630.00	615.00	2,245.00
# 4/0	L3@16.0	KLF	1,830.00	656.00	2,486.00
# 250 KCMIL	L3@18.0	KLF	2,450.00	739.00	3,189.00
# 300 KCMIL	L3@19.0	KLF	2,860.00	780.00	3,640.00
# 350 KCMIL	L3@20.0	KLF	3,270.00	821.00	4,091.00
# 400 KCMIL	L3@21.0	KLF	3,980.00	862.00	4,842.00
# 500 KCMIL	L3@22.0	KLF	4,390.00	903.00	5,293.00
# 600 KCMIL	L3@23.0	KLF	5,110.00	944.00	6,054.00
# 700 KCMIL	L3@24.0	KLF	5,820.00	985.00	6,805.00
# 750 KCMIL	L3@25.0	KLF	6,530.00	1,030.00	7,560.00
#1000 KCMIL	L3@30.0	KLF	7,610.00	1,230.00	8,840.00

Use these figures to estimate the cost of aluminum XHHW wire installed in conduit or USE RHH-RHW wire installed under the conditions described on pages 5 and 6. Costs listed are for each 1,000 linear feet installed. The crew is two electricians for sizes to #1 and three electricians for sizes over #1. The labor cost per manhour is $41.03. These costs include fishing string, reel set-up, pulling gear set-up, phase identification, pulling compound, layout, material handling, and normal waste. Add for splicing, anti-oxidation compound, sales tax, delivery, supervision, mobilization, demobilization, cleanup, overhead and profit. Note: The figures for XHHW wire assume that wire is pulled in conduit runs of 100 feet or less and that three conductors are pulled at the same time. Use anti-oxidation compound on all aluminum connections. Type USE, RHH-RHW wire is approved for use overhead, direct burial or in underground duct systems. The ampacity of copper wire is greater than the ampacity of aluminum wire of the same size. Check the *NEC* before substituting aluminum wire for copper wire. Terminations for aluminum wire must be made with approved fittings.

Type SEU 600 volt plastic jacket service entrance aluminum cable

Material	Craft@Hrs	Unit	Material Cost	Labor Cost	Installed Cost
# 8-3	L2@10.0	KLF	2,040.00	410.00	2,450.00
# 6-3	L2@11.0	KLF	2,060.00	451.00	2,511.00
# 4-2 & 6	L2@12.0	KLF	2,470.00	492.00	2,962.00
# 4-3	L2@13.0	KLF	2,640.00	533.00	3,173.00
# 2-2 & 4	L2@14.0	KLF	3,220.00	574.00	3,794.00
# 1-2 & 3	L2@16.0	KLF	3,530.00	656.00	4,186.00
# 1-3	L2@17.0	KLF	4,780.00	698.00	5,478.00
# 1/0-3	L2@19.0	KLF	5,340.00	780.00	6,120.00
# 2/0-3	L2@21.0	KLF	6,200.00	862.00	7,062.00
# 3/0-3	L2@23.0	KLF	8,240.00	944.00	9,184.00
# 4/0-3	L2@25.0	KLF	8,660.00	1,030.00	9,690.00

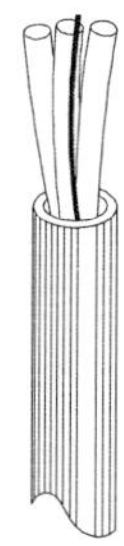

Type SE-SER 600 volt plastic jacket aluminum cable

Material		Craft@Hrs	Unit	Material Cost	Labor Cost	Installed Cost
# 8-3	#2 Gr	L2@11.0	KLF	1,550.00	451.00	2,001.00
# 6-3	#6 Gr	L2@12.0	KLF	1,750.00	492.00	2,242.00
# 4-3	#6 Gr	L2@14.0	KLF	1,970.00	574.00	2,544.00
# 2-3	#4 Gr	L2@16.0	KLF	2,890.00	656.00	3,546.00
# 1-3	#3 Gr	L2@18.0	KLF	3,760.00	739.00	4,499.00
# 1/0-3	#2 Gr	L2@20.0	KLF	4,370.00	821.00	5,191.00
# 2/0-3	#2 Gr	L2@22.0	KLF	5,150.00	903.00	6,053.00
# 3/0-3	#1/0 Gr	L2@24.0	KLF	6,340.00	985.00	7,325.00
# 4/0-3	#2/0 Gr	L2@26.0	KLF	7,330.00	1,070.00	8,400.00

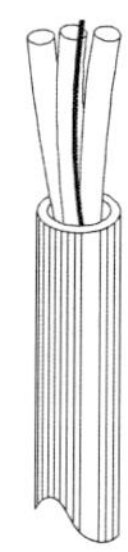

Weatherproof polyethylene 600 volt solid aluminum wire

Material	Code Name	Craft@Hrs	Unit	Material Cost	Labor Cost	Installed Cost
#6	Apple	L2@8.00	KLF	511.00	328.00	839.00
#4	Pear	L2@9.00	KLF	582.00	369.00	951.00
#2	Cherry	L2@10.0	KLF	766.00	410.00	1,176.00

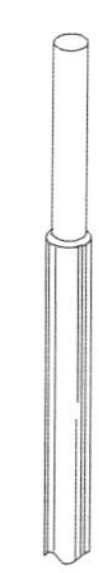

Use these figures to estimate the cost of aluminum service entrance cable installed under the conditions described on pages 5 and 6. Costs listed are for each 1,000 linear feet installed. The crew is two electricians working at the cost of $41.03 per manhour. These costs include stripping, phase identification, layout, material handling, and normal waste. Add for service entrance cap, anti-oxidation compound, supports, sales tax, delivery, supervision, mobilization, demobilization, cleanup, overhead and profit. Note: Use anti-oxidation compound on all aluminum connections. The ampacity of copper wire is greater than the ampacity of aluminum wire of the same size. Check the *NEC* before substituting aluminum wire for copper wire. Terminations for aluminum wire must be made with approved fittings.

See Wire Conversion Table On Page 537
Copper to Aluminum Wire Ampacities

600 Volt Aluminum Wire

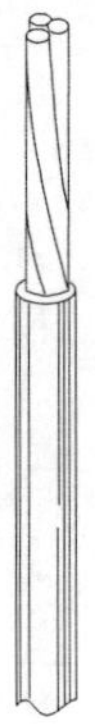

Weatherproof polyethylene 600 volt stranded aluminum wire

Material	Code Name	Strands	Craft@Hrs	Unit	Material Cost	Labor Cost	Installed Cost
#6	Plum	7	L2@8.00	KLF	155.00	328.00	483.00
#4	Apricot	7	L2@9.00	KLF	201.00	369.00	570.00
#2	Peach	7	L2@9.25	KLF	264.00	380.00	644.00
#1/0	Quince	7	L4@10.0	KLF	374.00	410.00	784.00
#2/0	Orange	7	L4@10.3	KLF	492.00	423.00	915.00
#3/0	Fig	7	L4@12.0	KLF	610.00	492.00	1,102.00
#4/0	Olive	7	L4@13.0	KLF	842.00	533.00	1,375.00

600 volt aluminum conductor steel reinforced (ACSR) aluminum wire

Material	Code Name	Strands	Craft@Hrs	Unit	Material Cost	Labor Cost	Installed Cost
#6	Walnut	6	L4@8.00	KLF	470.00	328.00	798.00
#4	Butternut	6	L4@9.00	KLF	634.00	369.00	1,003.00
#4	Hickory	7	L4@9.00	KLF	610.00	369.00	979.00
#2	Pignut	6	L4@10.0	KLF	923.00	410.00	1,333.00
#2	Beech	7	L4@10.0	KLF	989.00	410.00	1,399.00
#1/0	Almond	6	L5@12.0	KLF	1,140.00	492.00	1,632.00
#2/0	Pecan	6	L5@13.0	KLF	1,350.00	533.00	1,883.00
#3/0	Filbert	6	L5@14.0	KLF	1,620.00	574.00	2,194.00
#4/0	Buckeye	6	L5@15.0	KLF	1,910.00	615.00	2,525.00

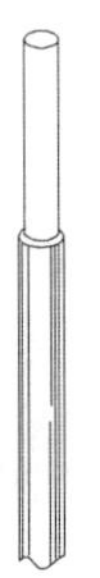

600 volt cross-linked polyethylene (XLP) solid aluminum wire

Material	Code Name	Craft@Hrs	Unit	Material Cost	Labor Cost	Installed Cost
#6	Apple-XLP	L2@8.00	KLF	234.00	328.00	562.00
#4	Pear-XLP	L2@9.00	KLF	348.00	369.00	717.00
#2	Cherry-XLP	L2@10.0	KLF	520.00	410.00	930.00

Use these figures to estimate the cost of aluminum wire installed on overhead supports under the conditions described on pages 5 and 6. Costs listed are for each 1,000 linear feet installed. For wire other than ACSR, the crew is two electricians for wire up to #2 and four electricians for wire over #2. For ACSR wire on overhead poles, use four electricians on wire to #2 and five electricians on wire over #2. The labor cost is $41.03 per manhour. These costs include reel set-up, pulling gear set-up, tensioning, layout, material handling, and normal waste. Add for insulators, terminations, line hardware, anti-oxidation compound, pre-formed ties, sales tax, delivery, supervision, mobilization, demobilization, cleanup, overhead and profit. Note: Use anti-oxidation compound on all aluminum connections, fittings and lugs. The ampacity of copper wire is greater than the ampacity of aluminum wire of the same size. Check the *NEC* before substituting aluminum wire for copper wire. Terminations for aluminum wire must be made with approved fittings.

See Wire Conversion Table On Page 537
Copper to Aluminum Wire Ampacities

600 volt cross-linked polyethylene (XLP) stranded aluminum wire

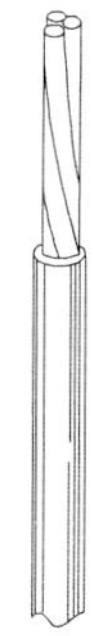

Material	Code Name	Strands	Craft@Hrs	Unit	Material Cost	Labor Cost	Installed Cost
#6	Plum-XLP	7	L2@8.00	KLF	178.00	328.00	506.00
#4	Apricot-XLP	7	L2@9.00	KLF	201.00	369.00	570.00
#2	Peach-XLP	7	L2@10.0	KLF	296.00	410.00	706.00
#1/0	Quince-XLP	7	L4@12.0	KLF	436.00	492.00	928.00
#2/0	Orange-XLP	7	L4@13.0	KLF	620.00	533.00	1,153.00
#3/0	Fig-XLP	7	L4@14.0	KLF	776.00	574.00	1,350.00
#4/0	Olive-XLP	7	L4@15.0	KLF	842.00	615.00	1,457.00

600 volt cross-linked polyethylene ACSR aluminum wire

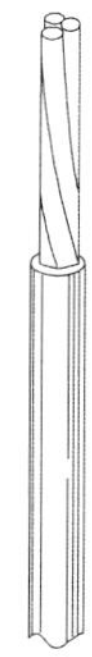

Material	Code Name	Strands	Craft@Hrs	Unit	Material Cost	Labor Cost	Installed Cost
#6	Walnut-XLP	6	L4@8.00	KLF	501.00	328.00	829.00
#4	Butternut-XLP	6	L4@9.00	KLF	271.00	369.00	640.00
#4	Hickory-XLP	7	L4@9.00	KLF	582.00	369.00	951.00
#2	Pignut-XLP	6	L4@10.0	KLF	366.00	410.00	776.00
#2	Beech-XLP	7	L4@10.0	KLF	443.00	410.00	853.00
#1/0	Almond-XLP	6	L5@12.0	KLF	520.00	492.00	1,012.00
#2/0	Pecan-XLP	6	L5@13.0	KLF	582.00	533.00	1,115.00
#3/0	Filbert-XLP	6	L5@14.0	KLF	738.00	574.00	1,312.00
#4/0	Buckeye-XLP	6	L5@15.0	KLF	866.00	615.00	1,481.00

600 volt polyethylene duplex aluminum service drop wire

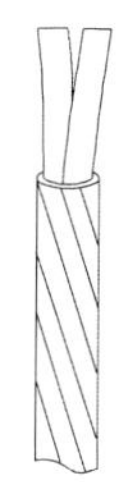

Material	Code Name	Strands	Craft@Hrs	Unit	Material Cost	Labor Cost	Installed Cost
#6	Pekingese	1	L2@12.0	KLF	492.00	492.00	984.00
#6	Collie	7	L2@11.0	KLF	530.00	451.00	981.00
#4	Spaniel	7	L2@14.0	KLF	719.00	574.00	1,293.00
#2	Doberman	7	L2@16.0	KLF	1,160.00	656.00	1,816.00
#1/0	Malamute	19	L4@18.0	KLF	1,840.00	739.00	2,579.00

600 volt triplex aluminum service drop wire, ACSR

Material	Code Name	Strands	Craft@Hrs	Unit	Material Cost	Labor Cost	Installed Cost
#4	Oyster	7	L4@14.0	KLF	1,180.00	574.00	1,754.00
#2	Clam	7	L4@16.0	KLF	1,510.00	656.00	2,166.00
#1/0	Murex	7	L5@18.0	KLF	2,330.00	739.00	3,069.00
#2/0	Nassa	7	L5@20.0	KLF	2,710.00	821.00	3,531.00
#4/0	Portunas	19	L5@24.0	KLF	4,220.00	985.00	5,205.00

Use these figures to estimate the cost of aluminum wire installed on overhead supports under the conditions described on pages 5 and 6. Costs listed are for each 1,000 linear feet installed. For wire other than ACSR, the crew is two electricians for wire up to #2 and four electricians for wire over #2. For ACSR wire on overhead poles, use four electricians on wire to #2 and five electricians on wire over #2. The labor cost is $41.03 per manhour. These costs include reel set-up, pulling gear set-up, tensioning, layout, material handling, and normal waste. Add for insulators, terminations, line hardware, anti-oxidation compound, pre-formed ties, sales tax, delivery, supervision, mobilization, demobilization, cleanup, overhead and profit. Note: Use anti-oxidation compound on all aluminum connections, fittings and lugs. The ampacity of copper wire is greater than the ampacity of aluminum wire of the same size. Check the *NEC* before substituting aluminum wire for copper wire. Terminations for aluminum wire must be made with approved fittings.

See Wire Conversion Table On Page 537
Copper to Aluminum Wire Ampacities

600 Volt Aluminum Service Drop Wire

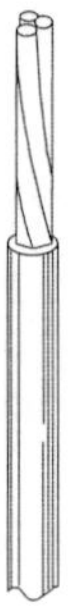

600 volt triplex aluminum service drop wire, ACSR

Material	Code Name	Strands	Craft@Hrs	Unit	Material Cost	Labor Cost	Installed Cost
#6	Paludina	1	L3@14.0	KLF	866.00	574.00	1,440.00
#6	Voluta	7	L3@14.0	KLF	1,020.00	574.00	1,594.00
#4	Periwinkle	7	L3@16.0	KLF	1,380.00	656.00	2,036.00
#2	Conch	7	L3@18.0	KLF	1,990.00	739.00	2,729.00
#1/0	Neritina	7	L3@20.0	KLF	2,680.00	821.00	3,501.00
#1/0	Cenia	19	L3@20.0	KLF	2,920.00	821.00	3,741.00
#2/0	Runcina	7	L3@24.0	KLF	3,060.00	985.00	4,045.00
#4/0	Zurara	19	L3@28.0	KLF	4,780.00	1,150.00	5,930.00

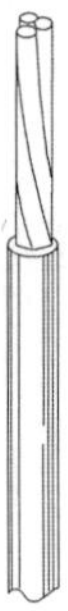

600 volt triplex aluminum service drop wire with reduced neutral

Material	Code Name	Strands	Craft@Hrs	Unit	Material Cost	Labor Cost	Installed Cost
#4	Scallop	1	L3@16.0	KLF	1,180.00	656.00	1,836.00
#4	Strombus	7	L3@16.0	KLF	1,380.00	656.00	2,036.00
#2	Cockle	7	L3@18.0	KLF	1,770.00	739.00	2,509.00
#1/0	Janthina	7	L3@20.0	KLF	2,650.00	821.00	3,471.00
#2/0	Clio	19	L3@24.0	KLF	3,370.00	985.00	4,355.00
#4/0	Cerapus	19	L3@28.0	KLF	4,830.00	1,150.00	5,980.00

600 volt quadruplex aluminum service drop wire

Material	Code Name	Strands	Craft@Hrs	Unit	Material Cost	Labor Cost	Installed Cost
#4	Pinto	7	L3@18.0	KLF	1,700.00	739.00	2,439.00
#2	Mustang	7	L3@20.0	KLF	2,340.00	821.00	3,161.00
#1/0	Criollo	7	L3@24.0	KLF	4,040.00	985.00	5,025.00
#2/0	Percheron	19	L3@26.0	KLF	4,870.00	1,070.00	5,940.00
#4/0	Oldenberg	19	L3@30.0	KLF	6,820.00	1,230.00	8,050.00

Use these figures to estimate the cost of aluminum service drop wire installed on overhead supports under the conditions described on pages 5 and 6. Costs listed are for each 1,000 linear feet installed. The crew is three electricians working at a labor rate of $41.03 per manhour. These costs include reel set-up, pulling gear set-up, tensioning, layout, material handling, and normal waste. Add for insulators, terminations, line hardware, anti-oxidation compound, pre-formed ties, sales tax, delivery, supervision, mobilization, demobilization, cleanup, overhead and profit. Note: Use anti-oxidation compound on all aluminum connections, fittings and lugs. The ampacity of copper wire is greater than the ampacity of aluminum wire of the same size. Check the *NEC* before substituting aluminum wire for copper wire. Terminations for aluminum wire must be made with approved fittings. Multi-conductor service drop is factory-twisted and sold on spools of 500 and 1,000 feet. Many suppliers will sell service drop wire in special lengths at a modest additional charge.

600 volt quadruplex aluminum service drop wire, ACSR neutral

Material	Code Name	Strands	Craft@Hrs	Unit	Material Cost	Labor Cost	Installed Cost
#6	Chola	7	L3@16.0	KLF	1,300.00	656.00	1,956.00
#4	Hackney	7	L3@18.0	KLF	1,340.00	739.00	2,079.00
#2	Palomino	7	L3@20.0	KLF	2,250.00	821.00	3,071.00
#1/0	Costena	19	L3@24.0	KLF	3,950.00	985.00	4,935.00
#2/0	Grullo	19	L3@26.0	KLF	4,640.00	1,070.00	5,710.00
#3/0	Suffolk	19	L3@28.0	KLF	5,490.00	1,150.00	6,640.00
#4/0	Appaloosa	19	L3@30.0	KLF	6,530.00	1,230.00	7,760.00

600 volt cross-linked polyethylene duplex aluminum service drop wire with aluminum neutral

Material	Code Name	Strands	Craft@Hrs	Unit	Material Cost	Labor Cost	Installed Cost
#6	Pekingese	1	L3@13.0	KLF	610.00	533.00	1,143.00
#6	Collie	7	L3@14.0	KLF	610.00	574.00	1,184.00
#4	Spaniel	7	L3@15.0	KLF	776.00	615.00	1,391.00
#2	Doberman	7	L3@17.0	KLF	1,190.00	698.00	1,888.00
#1/0	Malamute	19	L3@19.0	KLF	1,980.00	780.00	2,760.00

600 volt cross-linked polyethylene duplex aluminum service drop wire with ACSR neutral

Material	Code Name	Strands	Craft@Hrs	Unit	Material Cost	Labor Cost	Installed Cost
#6	Setter	1	L3@13.0	KLF	668.00	533.00	1,201.00
#6	Shepherd	7	L3@14.0	KLF	733.00	574.00	1,307.00
#4	Terrier	7	L3@15.0	KLF	947.00	615.00	1,562.00
#2	Chow	7	L3@17.0	KLF	1,210.00	698.00	1,908.00

600 volt cross-linked polyethylene triplex aluminum service drop wire with aluminum neutral

Material	Code Name	Strands	Craft@Hrs	Unit	Material Cost	Labor Cost	Installed Cost
#4	Oyster	7	L3@16.0	KLF	610.00	656.00	1,266.00
#2	Clam	7	L3@18.0	KLF	719.00	739.00	1,458.00
#1/0	Murex	7	L3@20.0	KLF	1,060.00	821.00	1,881.00
#2/0	Nassa	7	L3@24.0	KLF	1,340.00	985.00	2,325.00
#4/0	Portunas	19	L3@28.0	KLF	1,600.00	1,150.00	2,750.00

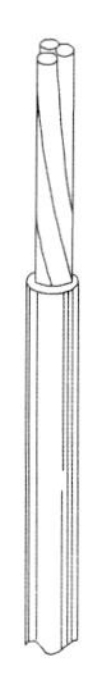

Use these figures to estimate the cost of aluminum service drop wire installed on overhead supports under the conditions described on pages 5 and 6. Costs listed are for each 1,000 linear feet installed. The crew is three electricians working at a labor rate of $41.03 per manhour. These costs include reel set-up, pulling gear set-up, tensioning, layout, material handling, and normal waste. Add for insulators, terminations, line hardware, anti-oxidation compound, pre-formed ties, sales tax, delivery, supervision, mobilization, demobilization, cleanup, overhead and profit. Note: Use anti-oxidation compound on all aluminum connections, fittings and lugs. The ampacity of copper wire is greater than the ampacity of aluminum wire of the same size. Check the *NEC* before substituting aluminum wire for copper wire. Terminations for aluminum wire must be made with approved fittings. Multi-conductor service drop is factory-twisted and sold on spools of 500 and 1,000 feet. Many suppliers will sell service drop wire in special lengths at a modest additional charge.

600 Volt Aluminum Service Drop Wire

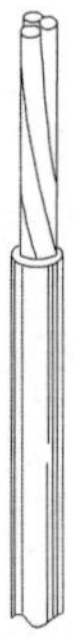

Cross-linked polyethylene (XLP) triplex aluminum service drop wire with ACSR neutral

Material	Code Name	Strands	Craft@Hrs	Unit	Material Cost	Labor Cost	Installed Cost
#6	Paludina	1	L3@14.0	KLF	437.00	574.00	1,011.00
#6	Voluta	7	L3@14.0	KLF	497.00	574.00	1,071.00
#4	Periwinkle	7	L3@16.0	KLF	691.00	656.00	1,347.00
#2	Conch	7	L3@18.0	KLF	866.00	739.00	1,605.00
#1/0	Neritina	7	L3@20.0	KLF	1,010.00	821.00	1,831.00
#1/0	Cenia	19	L3@20.0	KLF	1,140.00	821.00	1,961.00
#2/0	Runcina	7	L3@24.0	KLF	1,350.00	985.00	2,335.00
#4/0	Zurara	19	L3@28.0	KLF	2,060.00	1,150.00	3,210.00

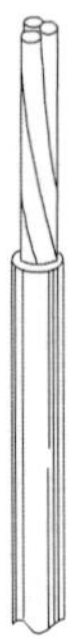

Cross-linked polyethylene (XLP) triplex aluminum service drop wire with reduced neutral

Material	Code Name	Strands	Craft@Hrs	Unit	Material Cost	Labor Cost	Installed Cost
#4	Scallop	6	L3@16.0	KLF	634.00	656.00	1,290.00
#4	Strombus	6	L3@16.0	KLF	1,040.00	656.00	1,696.00
#2	Cockle	4	L3@18.0	KLF	1,490.00	739.00	2,229.00
#1/0	Janthina	2	L3@20.0	KLF	1,770.00	821.00	2,591.00
#2/0	Clio	1	L3@24.0	KLF	2,130.00	985.00	3,115.00
#4/0	Cerapus	2/0	L3@28.0	KLF	3,210.00	1,150.00	4,360.00

Cross-linked polyethylene (XLP) quadruplex aluminum service drop wire with aluminum neutral

Material	Code Name	Strands	Craft@Hrs	Unit	Material Cost	Labor Cost	Installed Cost
#4	Pinto	4	L3@18.0	KLF	923.00	739.00	1,662.00
#2	Mustang	4	L3@20.0	KLF	1,840.00	821.00	2,661.00
#1/0	Criollo	1/0	L3@24.0	KLF	2,400.00	985.00	3,385.00
#2/0	Percheron	2/0	L3@26.0	KLF	2,650.00	1,070.00	3,720.00
#4/0	Oldenberg	4/0	L3@30.0	KLF	3,390.00	1,230.00	4,620.00

Cross-linked polyethylene (XLP) quadruplex aluminum service drop wire with ACSR neutral

Material	Code Name	Strands	Craft@Hrs	Unit	Material Cost	Labor Cost	Installed Cost
#6	Chola	6	L3@16.0	KLF	1,010.00	656.00	1,666.00
#4	Hackney	4	L3@18.0	KLF	1,550.00	739.00	2,289.00
#2	Palomino	2	L3@20.0	KLF	1,760.00	821.00	2,581.00
#1/0	Costena	1/0	L3@24.0	KLF	2,850.00	985.00	3,835.00
#2/0	Grullo	2/0	L3@26.0	KLF	3,430.00	1,070.00	4,500.00
#3/0	Suffolk	3/0	L3@28.0	KLF	3,770.00	1,150.00	4,920.00
#4/0	Appaloosa	4/0	L3@30.0	KLF	4,970.00	1,230.00	6,200.00

Use these figures to estimate the cost of aluminum service drop wire installed on overhead supports under the conditions described on pages 5 and 6. Costs listed are for each 1,000 linear feet installed. The crew is three electricians working at a labor rate of $41.03 per manhour. These costs include reel set-up, pulling gear set-up, tensioning, layout, material handling, and normal waste. Add for insulators, terminations, line hardware, anti-oxidation compound, pre-formed ties, sales tax, delivery, supervision, mobilization, demobilization, cleanup, overhead and profit. Note: Use anti-oxidation compound on all aluminum connections, fittings and lugs. The ampacity of copper wire is greater than the ampacity of aluminum wire of the same size. Check the *NEC* before substituting aluminum wire for copper wire. Terminations for aluminum wire must be made with approved fittings. Multi-conductor service drop is factory-twisted and sold on spools of 500 and 1,000 feet. Many suppliers will sell service drop wire in special lengths at a modest additional charge.

Aluminum Type URD 600 Volt Underground Distribution Cable

Type URD 600 volt aluminum underground cable, two phase conductors

Material	Code Name	Strands	Craft@Hrs	Unit	Material Cost	Labor Cost	Installed Cost
#4	Taft	4	L4@10.0	KLF	1,240.00	410.00	1,650.00
#2	Wells	4	L4@11.0	KLF	1,360.00	451.00	1,811.00
#2	Juilliard	2	L4@11.0	KLF	1,990.00	451.00	2,441.00
#1/0	Marion	2	L4@12.0	KLF	2,870.00	492.00	3,362.00
#1/0	Montclair	1/0	L4@13.0	KLF	3,220.00	533.00	3,753.00
#2/0	Bliss	1/0	L4@14.0	KLF	3,440.00	574.00	4,014.00
#2/0	Bloomfield	2/0	L4@15.0	KLF	3,600.00	615.00	4,215.00
#3/0	Whittier	1/0	L4@16.0	KLF	3,890.00	656.00	4,546.00
#3/0	Pace	3/0	L4@17.0	KLF	4,290.00	698.00	4,988.00
#4/0	Regis	2/0	L4@18.0	KLF	4,430.00	739.00	5,169.00
#4/0	Manhattan	4/0	L4@19.0	KLF	5,010.00	780.00	5,790.00
#250	Adelphi	3/0	L4@20.0	KLF	6,150.00	821.00	6,971.00
#350	Concordia	4/0	L4@24.0	KLF	8,140.00	985.00	9,125.00

Type URD aluminum underground cable, triplex, black ground with yellow stripe

Material	Code Name	Strands	Craft@Hrs	Unit	Material Cost	Labor Cost	Installed Cost
#4	Vassar	4	L4@12.0	KLF	1,930.00	492.00	2,422.00
#2	Stephens	4	L4@13.0	KLF	2,030.00	533.00	2,563.00
#2	Ramapo	2	L4@14.0	KLF	2,120.00	574.00	2,694.00
#1/0	Brenau	2	L4@15.0	KLF	2,500.00	615.00	3,115.00
#1/0	Bergen	1/0	L4@16.0	KLF	2,730.00	656.00	3,386.00
#2/0	Converse	1	L4@17.0	KLF	3,080.00	698.00	3,778.00
#2/0	Hunter	2/0	L4@18.0	KLF	3,660.00	739.00	4,399.00
#3/0	Hollins	1/0	L4@19.0	KLF	3,950.00	780.00	4,730.00
#3/0	Rockland	3/0	L4@20.0	KLF	4,380.00	821.00	5,201.00
#4/0	Sweetbriar	4/0	L4@21.0	KLF	4,920.00	862.00	5,782.00
#4/0	Monmouth	4/0	L4@22.0	KLF	5,250.00	903.00	6,153.00
#250	Pratt	3/0	L4@24.0	KLF	6,820.00	985.00	7,805.00
#250	Wesleyan	4/0	L4@26.0	KLF	7,950.00	1,070.00	9,020.00

Use these figures to estimate the cost of aluminum underground distribution cable installed in an open trench under the conditions described on pages 5 and 6. Costs listed are for each 1,000 linear feet installed. The crew is four electricians working at a labor rate of $41.03 per manhour. These costs include reel set-up in a pickup truck, layout, material handling, and normal waste. Add for terminations, splicing, warning tape, hardware, anti-oxidation compound, trenching, sales tax, delivery, supervision, mobilization, demobilization, cleanup, overhead and profit. Note: Use anti-oxidation compound on all aluminum connections. The ampacity of copper wire is greater than the ampacity of aluminum wire of the same size. Check the *NEC* before substituting aluminum wire for copper wire. Terminations for aluminum wire must be made in approved fittings.

Aluminum Wire and Steel Messenger Strand

Bare aluminum conductor

Material	Code Name	Strands	Craft@Hrs	Unit	Material Cost	Labor Cost	Installed Cost
#4	Swan	6	L2@5.50	KLF	158.00	226.00	384.00
#4	Swanate	7	L2@5.50	KLF	204.00	226.00	430.00
#2	Sparrow	6	L2@6.50	KLF	287.00	267.00	554.00
#2	Sparate	7	L2@6.50	KLF	213.00	267.00	480.00
#1/0	Raven	6	L4@9.00	KLF	483.00	369.00	852.00
#2/0	Quail	6	L4@10.0	KLF	573.00	410.00	983.00
#3/0	Pigeon	6	L4@11.0	KLF	691.00	451.00	1,142.00
#4/0	Penguin	6	L4@12.7	KLF	800.00	521.00	1,321.00
266, #800	Partridge	6	L4@13.0	KLF	1,930.00	533.00	2,463.00
336, #400	Merlin	18	L4@15.0	KLF	2,390.00	615.00	3,005.00
336, #400	Linnet	26	L4@16.0	KLF	2,640.00	656.00	3,296.00

Galvanized steel messenger strand

Material		Craft@Hrs	Unit	Material Cost	Labor Cost	Installed Cost
Common Grade	Tensile Strength					
1/4"	3150	L2@6.00	KLF	183.00	246.00	429.00
5/16"	5350	L2@7.00	KLF	265.00	287.00	552.00
3/8"	6950	L2@8.00	KLF	403.00	328.00	731.00
7/16"	9350	L2@9.00	KLF	643.00	369.00	1,012.00
1/2"	12100	L2@10.0	KLF	793.00	410.00	1,203.00
High Strength	Tensile Strength					
1/4"	4750	L2@6.25	KLF	307.00	256.00	563.00
5/16"	8000	L2@7.25	KLF	378.00	297.00	675.00
3/8"	10800	L2@8.25	KLF	433.00	338.00	771.00
7/16"	14500	L2@9.25	KLF	554.00	380.00	934.00
1/2"	18800	L2@10.3	KLF	858.00	423.00	1,281.00
Extra High Strength	Tensile Strength					
1/4"	6650	L2@6.50	KLF	285.00	267.00	552.00
5/16"	11200	L2@7.50	KLF	368.00	308.00	676.00
3/8"	15400	L2@8.50	KLF	421.00	349.00	770.00
7/16"	20800	L2@9.50	KLF	669.00	390.00	1,059.00
1/2"	26900	L2@10.5	KLF	874.00	431.00	1,305.00
Utility Grade	Tensile Strength					
5/16"	6000	L2@8.00	KLF	410.00	328.00	738.00
3/8"	11500	L2@9.00	KLF	441.00	369.00	810.00
7/16"	18000	L2@10.0	KLF	669.00	410.00	1,079.00
1/2"	25000	L2@12.0	KLF	849.00	492.00	1,341.00

Use these figures to estimate the cost of bare aluminum wire and galvanized steel messenger strand installed in overhead line construction under the conditions described on pages 5 and 6. Costs listed are for each 1,000 linear feet installed. The crew is two electricians for wire up to #2 and four electricians for wire over #2. The labor cost per manhour is $41.03. These costs include reel set-up, pulling gear set-up, tensioning, layout, material handling, and normal waste. Add for insulators, line hardware, anti-oxidation compound, pre-formed ties, sales tax, delivery, supervision, mobilization, demobilization, cleanup, overhead and profit. Note: Use anti-oxidation compound on all aluminum connections. The ampacity of copper wire is greater than the ampacity of aluminum wire of the same size. Check the *NEC* before substituting aluminum wire for copper wire. Terminations for aluminum wire must be made in approved fittings.

Weights of galvanized steel messenger strand per 1000 feet for common grade, high strength, extra high strength and utility grades are: 1/4" - 121 pounds; 5/16" - 205 pounds; 3/8" - 273 pounds; 7/16" - 399 pounds; and 1/2" - 517 pounds.

Insulated screw-on connectors

Material	Craft@Hrs	Unit	Material Cost	Labor Cost	Installed Cost
Wire Size					
#22 - #12	L1@3.00	100	25.30	123.00	148.30
#22 - #12	L1@15.0	500	101.00	615.00	716.00
#18 - #10	L1@3.00	100	40.30	123.00	163.30
#18 - #10	L1@15.0	500	132.00	615.00	747.00
#14 - # 8	L1@3.00	100	69.30	123.00	192.30
#14 - # 8	L1@15.0	500	286.00	615.00	901.00
#12 - # 6	L1@3.00	100	114.00	123.00	237.00

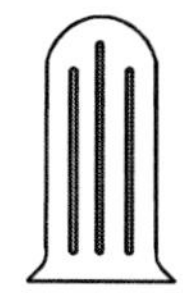

Insulated screw-on self stripping connectors

Material	Craft@Hrs	Unit	Material Cost	Labor Cost	Installed Cost
Wire Size					
#22 - #18	L1@3.00	100	41.00	123.00	164.00
#22 - #18	L1@15.0	500	207.00	615.00	822.00
#18 - #14	L1@3.00	100	24.10	123.00	147.10
#18 - #14	L1@15.0	500	120.00	615.00	735.00
#18 - #14	L1@30.0	1000	241.00	1,230.00	1,471.00
#12 - #10	L1@3.00	100	44.90	123.00	167.90

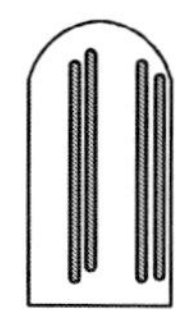

Wire nuts

Material	Craft@Hrs	Unit	Material Cost	Labor Cost	Installed Cost
Gray, very small	L1@3.00	100	8.59	123.00	131.59
Blue, small	L1@3.00	100	10.90	123.00	133.90
Orange, medium	L1@3.00	100	11.90	123.00	134.90
Yellow, large	L1@3.00	100	14.40	123.00	137.40
Red, extra large	L1@3.00	100	28.90	123.00	151.90

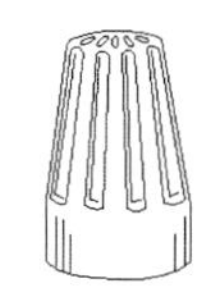

Insulated crimp sleeves

Material	Craft@Hrs	Unit	Material Cost	Labor Cost	Installed Cost
Wire Size					
#22 - #14	L1@3.00	100	54.00	123.00	177.00
#22 - #14	L1@30.0	1000	286.00	1,230.00	1,516.00
#18 - #10	L1@1.50	50	31.20	61.50	92.70
#18 - #10	L1@15.0	500	142.00	615.00	757.00

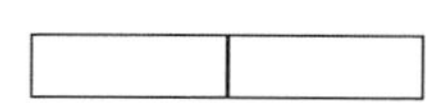

Uninsulated crimp sleeves

Material	Craft@Hrs	Unit	Material Cost	Labor Cost	Installed Cost
Wire Size					
#18 - #10	L1@3.00	100	23.20	123.00	146.20
#18 - #10	L1@30.0	1000	234.00	1,230.00	1,464.00

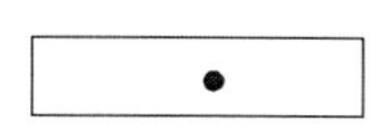

Use these figures to estimate the cost of wire connectors installed under the conditions described on pages 5 and 6. Costs listed are for 50, 100, 500 or 1,000 units. The crew is one electrician working at a labor cost of $41.03 per manhour. These costs include wire stripping, layout, material handling, and normal waste. Add for sales tax, delivery, supervision, mobilization, demobilization, cleanup, overhead and profit.

Wire Connectors

Material	Craft@Hrs	Unit	Material Cost	Labor Cost	Installed Cost

Two way wire connectors

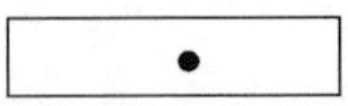

Wire Size	Craft@Hrs	Unit	Material Cost	Labor Cost	Installed Cost
# 8	L1@0.03	Ea	1.72	1.23	2.95
# 6	L1@0.04	Ea	1.97	1.64	3.61
# 4	L1@0.05	Ea	2.61	2.05	4.66
# 2	L1@0.08	Ea	8.41	3.28	11.69
# 1	L1@0.10	Ea	10.10	4.10	14.20
# 1/0	L1@0.10	Ea	11.00	4.10	15.10
# 2/0	L1@0.15	Ea	11.80	6.15	17.95
# 3/0	L1@0.15	Ea	13.90	6.15	20.05
# 4/0	L1@0.15	Ea	14.90	6.15	21.05
# 250 KCMIL	L1@0.20	Ea	16.70	8.21	24.91
# 300 KCMIL	L1@0.20	Ea	18.90	8.21	27.11
# 350 KCMIL	L1@0.20	Ea	19.70	8.21	27.91
# 400 KCMIL	L1@0.25	Ea	23.70	10.30	34.00
# 500 KCMIL	L1@0.25	Ea	29.00	10.30	39.30
# 600 KCMIL	L1@0.30	Ea	44.00	12.30	56.30
# 700 KCMIL	L1@0.30	Ea	46.90	12.30	59.20
# 750 KCMIL	L1@0.35	Ea	52.00	14.40	66.40
#1000 KCMIL	L1@0.40	Ea	70.20	16.40	86.60

Split bolt connectors for copper wire only

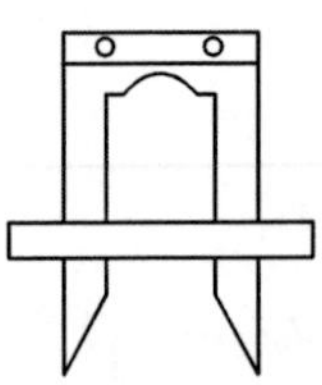

Wire Size	Craft@Hrs	Unit	Material Cost	Labor Cost	Installed Cost
# 10	L1@0.20	Ea	6.48	8.21	14.69
# 8	L1@0.20	Ea	6.58	8.21	14.79
# 6	L1@0.25	Ea	7.01	10.30	17.31
# 4	L1@0.30	Ea	8.68	12.30	20.98
# 2	L1@0.30	Ea	12.80	12.30	25.10
# 1	L1@0.40	Ea	13.30	16.40	29.70
# 1/0	L1@0.50	Ea	13.70	20.50	34.20
# 2/0	L1@0.50	Ea	10.70	20.50	31.20
# 3/0	L1@0.60	Ea	15.70	24.60	40.30
# 4/0	L1@0.60	Ea	18.20	24.60	42.80
# 250 KCMIL	L1@0.70	Ea	49.10	28.70	77.80
# 350 KCMIL	L1@0.70	Ea	89.50	28.70	118.20
# 500 KCMIL	L1@0.75	Ea	117.00	30.80	147.80
# 750 KCMIL	L1@1.00	Ea	138.00	41.00	179.00
#1000 KCMIL	L1@1.25	Ea	293.00	51.30	344.30

Use these figures to estimate the cost of wire connectors installed under the conditions described on pages 5 and 6. Costs listed are for each connector installed. The crew is one electrician working at a labor cost of $41.03 per manhour. These costs include cutting the wire, insulation stripping, taping for 600-volt capacity, layout, material handling, and normal waste. Add for insulating the connector, special encapsulating kits, sales tax, delivery, supervision, mobilization, demobilization, cleanup, overhead and profit. Note: Crimping and high pressure tools are needed when installing connectors.

Material	Craft@Hrs	Unit	Material Cost	Labor Cost	Installed Cost

Split bolt connectors with spacer for copper or aluminum wire

Wire Size					
# 8	L1@0.20	Ea	8.98	8.21	17.19
# 6	L1@0.20	Ea	4.98	8.21	13.19
# 4	L1@0.25	Ea	5.84	10.30	16.14
# 2	L1@0.30	Ea	7.06	12.30	19.36
# 1	L1@0.40	Ea	9.47	16.40	25.87
# 1/0	L1@0.50	Ea	12.20	20.50	32.70
# 4/0	L1@0.50	Ea	33.30	20.50	53.80
# 350 KCMIL	L1@0.70	Ea	56.30	28.70	85.00
# 500 KCMIL	L1@0.75	Ea	77.20	30.80	108.00
# 750 KCMIL	L1@1.00	Ea	214.00	41.00	255.00
#1000 KCMIL	L1@1.25	Ea	290.00	51.30	341.30

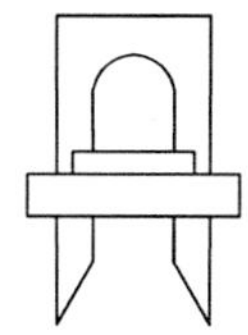

Two bolt connectors for copper wire only

# 4/0	L1@0.70	Ea	60.50	28.70	89.20
# 350 KCMIL	L1@1.00	Ea	119.00	41.00	160.00
# 500 KCMIL	L1@1.25	Ea	138.00	51.30	189.30
# 800 KCMIL	L1@1.40	Ea	189.00	57.40	246.40
#1000 KCMIL	L1@1.50	Ea	266.00	61.50	327.50

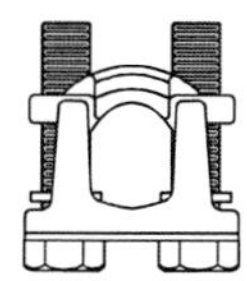

Two bolt connectors for copper or aluminum wire

# 2/0	L1@0.50	Ea	70.20	20.50	90.70
# 250 KCMIL	L1@0.70	Ea	117.00	28.70	145.70
# 350 KCMIL	L1@0.70	Ea	126.00	28.70	154.70
# 500 KCMIL	L1@0.75	Ea	166.00	30.80	196.80
# 800 KCMIL	L1@1.50	Ea	236.00	61.50	297.50

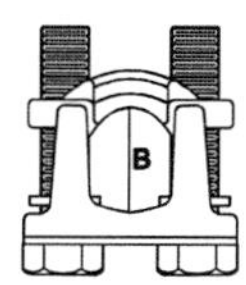

Two bolt connectors with spacer for copper or aluminum wire

# 2/0	L1@0.55	Ea	32.60	22.60	55.20
# 250 KCMIL	L1@0.75	Ea	51.50	30.80	82.30
# 350 KCMIL	L1@0.75	Ea	118.00	30.80	148.80
# 500 KCMIL	L1@0.80	Ea	161.00	32.80	193.80
# 800 KCMIL	L1@1.10	Ea	207.00	45.10	252.10
#1000 KCMIL	L1@1.60	Ea	226.00	65.60	291.60

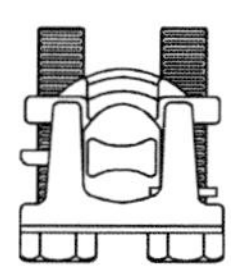

Use these figures to estimate the cost of split bolt and two bolt connectors installed under the conditions described on pages 5 and 6. Costs listed are for each connector installed. The crew is one electrician working at a labor cost of $41.03 per manhour. These costs include cutting the wire, insulation, stripping, taping for 600-volt capacity, forming the wire in a pull box or panel, layout, material handling, and normal waste. Add for insulating the connector, sales tax, delivery, supervision, mobilization, demobilization, cleanup, overhead and profit. Note: Split bolt connectors are installed with pump pliers. Tap the connector with a hammer after tightening and then retighten.

Copper Wire Connector Lugs

Material	Craft@Hrs	Unit	Material Cost	Labor Cost	Installed Cost

One hole solder type connector lugs for copper wire

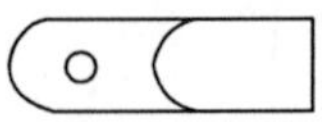

Material	Craft@Hrs	Unit	Material Cost	Labor Cost	Installed Cost
# 10	L1@0.10	Ea	1.71	4.10	5.81
# 8	L1@0.15	Ea	1.80	6.15	7.95
# 6	L1@0.15	Ea	1.81	6.15	7.96
# 4	L1@0.20	Ea	2.46	8.21	10.67
# 2	L1@0.20	Ea	2.90	8.21	11.11
# 1/0	L1@0.25	Ea	4.16	10.30	14.46
# 2/0	L1@0.25	Ea	5.09	10.30	15.39
# 3/0	L1@0.30	Ea	6.93	12.30	19.23
# 4/0	L1@0.30	Ea	8.89	12.30	21.19
# 250 KCMIL	L1@0.40	Ea	17.10	16.40	33.50
# 400 KCMIL	L1@0.50	Ea	25.30	20.50	45.80
# 500 KCMIL	L1@0.50	Ea	39.20	20.50	59.70
# 600 KCMIL	L1@0.50	Ea	41.60	20.50	62.10
# 800 KCMIL	L1@0.60	Ea	69.80	24.60	94.40
#1000 KCMIL	L1@0.70	Ea	76.40	28.70	105.10
#1500 KCMIL	L1@0.75	Ea	139.00	30.80	169.80

One hole solderless type connector lugs for copper wire

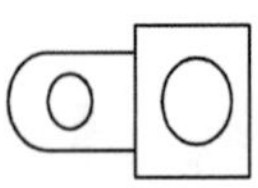

Material	Craft@Hrs	Unit	Material Cost	Labor Cost	Installed Cost
# 10	L1@0.05	Ea	1.81	2.05	3.86
# 6	L1@0.08	Ea	2.07	3.28	5.35
# 4	L1@0.10	Ea	2.90	4.10	7.00
# 2	L1@0.10	Ea	2.98	4.10	7.08
# 1/0	L1@0.15	Ea	5.71	6.15	11.86
# 3/0	L1@0.20	Ea	12.50	8.21	20.71
# 4/0	L1@0.20	Ea	15.20	8.21	23.41
# 350 KCMIL	L1@0.30	Ea	22.00	12.30	34.30
# 500 KCMIL	L1@0.40	Ea	35.00	16.40	51.40
#1000 KCMIL	L1@0.60	Ea	83.60	24.60	108.20

One hole double conductor solderless type connector lugs for copper wire

Material	Craft@Hrs	Unit	Material Cost	Labor Cost	Installed Cost
# 2	L1@0.20	Ea	53.00	8.21	61.21
# 1	L1@0.25	Ea	62.20	10.30	72.50
# 2/0	L1@0.30	Ea	89.00	12.30	101.30
# 4/0	L1@0.40	Ea	120.00	16.40	136.40

Two hole three conductor solderless type connector lugs for copper wire

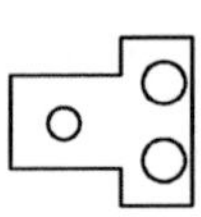

Material	Craft@Hrs	Unit	Material Cost	Labor Cost	Installed Cost
# 4	L1@0.30	Ea	61.30	12.30	73.60
# 1	L1@0.50	Ea	74.70	20.50	95.20
# 2/0	L1@0.75	Ea	107.00	30.80	137.80
# 4/0	L1@1.00	Ea	143.00	41.00	184.00
# 500 KCMIL	L1@1.50	Ea	290.00	61.50	351.50

Use these figures to estimate the cost of conductor lugs installed under the conditions described on pages 5 and 6. Costs listed are for each lug installed. The crew is one electrician working at a labor cost of $41.03 per manhour. These costs include the lug only (and no other material) wire cutting, insulation, stripping for the termination, layout, material handling, and normal waste. Add for insulating the lug (if needed), sales tax, delivery, supervision, mobilization, demobilization, cleanup, overhead and profit. Note: Always recheck solderless lugs for tightness.

Section 3: Outlet Boxes

In construction wiring, switches and electrical receptacles (known as wiring devices) are mounted in boxes which are secured to the wall, ceiling or floor. The box protects the device from damage and gives access to the wiring even after construction is complete. Once the box is installed and wired, the device is mounted and connected to the conductors. Finally, a box cover is installed to prevent accidental contact with conductors in the box.

Manufacturers of electrical equipment make boxes for every conceivable application. Some are used only in residences. Others are intended for commercial and industrial jobs. The *National Electrical Code* and your local code dictate the type of boxes that may be used in your community.

The size of the outlet box or junction box is determined by the size and number of conductors to be installed, and the size of the wiring device needed. When a plaster ring or switch ring is to be set against the box, the box size, the type, and the number of wiring devices to be used determine the size of the ring. The depth of the plaster ring or the switch ring will vary with the thickness of the final wall finish.

Boxes with plaster rings can be installed in ceilings and walls in wood construction, unit masonry, tile or concrete. Special outlet boxes and special plaster and switch rings are made for unit masonry and tile work. The ring depth is greater so the ring lip is flush with the surface of the finished wall. Check the architectural details of the plan to be sure you're pricing the right box and plaster ring. Some types cost considerably more than the standard grade used on most jobs.

This section does not deal with outlet boxes used in hazardous installations. Hazardous conditions require special outlet boxes and fittings that are approved for special environments.

Handy Boxes

Handy boxes are generally intended for surface mounting, though some are made with brackets attached that permit flush mounting. Handy boxes are 2" wide by 4" high. They come in 1 1/2", 1 7/8", 2 1/8" and 2 13/16" depths. The box is one-piece sheet metal with knockouts on all sides and the back. The tapped ears for adapting devices are turned inward into the box.

The boxes and covers for handy boxes are plated steel. The covers are stamped. Cover types include blank, single switch, single receptacle, duplex receptacle, four-wire twist-lock receptacle, and despard device with three knockout squares. Price labor and material cost for the wiring device separately.

Handy boxes have punched holes for mounting with round-head wood screws, round-head machine screws, dome-head stove bolts, tie wire or masonry anchors.

Sectional Switch Boxes

Sectional switch boxes are generally installed flush with the wall in wood frame walls. The finish surface can be either plaster, drywall, insulating board or paneling. The boxes can be ganged together by removing side panels. They're made of plated steel with punched holes for mounting. Boxes can have knockouts in all sides and back or may have Romex or MC clamps for holding the cable in place.

Sectional boxes are made 2", 2 1/2" and 2 3/4" deep. The tapped mounting ears are turned outward away from the box opening and the ears are spaced for standard wiring devices. Sectional switch boxes can be mounted in the wall or ceiling with round head wood screws, round head machine screws, dome head stove bolts, tie wire or masonry anchors. Be sure your electricians set the box flush with the level of the finished surface.

Welded Switch Boxes

Welded switch boxes look like sectional boxes. The difference is that they can't be ganged by removing the side panels. They're made with Romex or MC clamps for holding the cable in place.

Welded switch boxes are available with mounting brackets attached to the sides for rough-in mounting to the construction framing. The boxes are stamped, welded, punched, plated steel. The mounting ears are turned outward away from the box opening and are spaced for mounting standard wiring devices. The box can be mounted with nails, round-head wood screws, round-head machine screws, dome-head stove bolts, tie wire or masonry anchors.

Welded switch boxes come in single gang or two gang. They're 2 1/2", 2 13/16" or 3 13/16" deep.

Octagon Boxes

Octagon boxes are one-piece pressed plated steel boxes. They're punched for knockouts and mounting and are available with mounting brackets that are attached to the box for flush mounting in a wall or ceiling.

The box's diameter is either three (called ***3-0***) or four (called 4-0) inches. The mounting ears are turned inward into the box for mounting covers or fixtures. The boxes can be mounted on fixed or adjustable bar hangers which are nailed to the framing. The box is attached to the bar hanger with dome-head stove bolts or round-head machine screws.

Some bar hangers have a 3/8" fixture stud for securing the lighting fixture directly to the bar hanger for support. The center knockout is removed from the box, the fixture stud is inserted into the box through the removed knockout, and a 3/8" locknut is used to secure the box to the stud.

Octagon boxes are 1 1/2" or 2 1/8" deep. A very shallow box called a ***pancake box*** is also made in 3-0 and 4-0 sizes. The pancake box is used when space is restricted and only one circuit is needed.

Another octagon box is used in concrete construction. Octagon concrete rings are stamped plated steel and are generally used in overhead concrete construction. The rings have 1/2" and 3/4" knockouts and mounting ears that turn into the box for adapting a top cover and finish cover or lighting fixture. The box has tabs for nailing to concrete forms.

The octagon rings are 2 1/2", 3", 3 1/2", 4", 5" or 6" deep. The top cover can be blank with 1/2" or 3/4" knockouts or the cover can be purchased with a fixture stud attached for mounting a lighting fixture directly to the stud. Octagon concrete rings are 4-0.

Square Boxes

The smallest square box is 4" x 4" x 1½" deep. It can be either a one-piece unit or welded, stamped plated steel. The box has ½" or ¾" knockouts or comes with a combination of ½" and ¾" knockouts. The 4-S box can purchased with mounting brackets and with Romex or MC clamps.

4-S boxes are also made 2 1/8" deep with the same knockouts as a 4-S by 1½" box.

The tapped ears are in diagonal corners and have 8-32 tapped holes for attaching a cover or an extension ring. There are punched holes for mounting the box to the framing. Machine screws come with the box.

The extension ring is 1½" deep with ½" or ¾" or a combination of ½" and ¾" knockouts. The bottom diagonal corners have cutouts for mounting to the 4-S box. The top has tapped ears in diagonal corners with 8-32 tapped holes for attaching a cover. Again, machine screws come with the box.

4-S boxes are also made with two sets of mounting ears and 6-32 tapped holes for mounting two wiring devices. These two-ganged 4-S boxes can be purchased with mounting brackets for flush construction. 4-S boxes make good junction boxes in conduit systems. They're used for interior, dry, accessible locations and can be used in concrete walls or ceilings if the mounting holes are covered to keep liquid concrete from filling the box. When used in concrete construction, the point where conduit enters the box should be sealed to keep concrete out of the conduit.

Another square box is the 4 11/16" x 4 11/16" box. It's made the same as the 4-S box except that knockouts in the 2 1/8" deep box can be ½", ¾", 1" or 1¼". Combinations of knockouts are ½" and ¾", 1" and 1¼". It has four tapped tabs for mounting a cover or an extension ring. The tapped holes are for 10-32 machine screws, which come with the box.

4 11/16" boxes can be purchased with flush mounting brackets. They're used for interior, dry, accessible locations.

Bar Hangers

Bar hangers support an outlet box that has to be placed between two framing members. They're offset to allow for the depth of the box. Two lengths are available: one for nailing to studs 16" on center; the other for nailing to studs 24" on center.

There are two types of bar hangers. One comes with stove bolts for attaching the box to the bar hanger. The bar hanger has one long slot for adjustment or centering the box. The stove bolts are passed through the box and through the slot in the bar hanger. A square nut is used for tightening each bolt, thus securing the box to the bar hanger in the correct position.

The other type bar hanger has a 3/8" fixture stud that can be moved on the bar hanger. This permits adjustment of the box position. The center knockout is removed from the back of the outlet box, a 3/8" fixture stud is inserted into the box and a 3/8" locknut is used to secure the box to the bar hanger in the position desired.

Outlet Box Covers

Covers for outlet boxes mentioned so far in this section are made of plated steel. They're stamped with mounting holes lined up to fit the outlet box. Some covers are raised plaster rings or raised switch rings that extend the box rough-in to the wall or ceiling surface. The switch rings are made in single- or two-gang for one or two wiring devices, either lighting switches or convenience receptacles.

The covers for 3" octagon boxes are:

- 3-0 flat blank covers
- 3-0 duplex receptacle covers
- 3-0 flat with ½" knockout covers
- 3-0 single receptacle covers

The covers for 4" octagon boxes are:

- 4-0 flat blank covers
- 4-0 duplex receptacle covers
- 4-0 flat with ½" knockout covers
- 4-0 flat with ¾" knockout covers
- 4-0 single receptacle covers
- 4-0 twist-lock receptacle covers
- 4-0 raised sign receptacle covers
- 4-0 ¼" raised plaster ring
- 4-0 ½" raised plaster ring
- 4-0 ⅝" raised plaster ring
- 4-0 ¾" raised plaster ring
- 4-0 1" raised plaster ring
- 4-0 1¼" raised plaster ring
- 4-0 ½" raised blank cover
- 4-0 ½" raised one device

Covers for 4-S outlet boxes are similar to 4-0 covers except that they're square and many more types are available.

Square covers for $4^{11}/_{16}$" boxes also come in types that will meet almost any application.

Special Outlet Boxes

Special outlet boxes are available to cover applications not mentioned so far in this section. Ganged boxes are used when more than one wiring device is to be installed.

Multi-ganged boxes can be one-piece or welded plated steel. They have ½" and ¾" knockouts and the mounting holes are punched. Most have two 8-32 machine screws at one end and two holding tabs at the other end to hold the switch ring. The switch ring is usually raised ¾".

The multi-ganged box ranges from two-gang to ten-gang. They're generally used for multiple switches. But occasionally you'll see them used for multiple receptacles.

Another special outlet box is the masonry box. It's made of plated steel, is stamped with knockouts and has mounting holes punched. The boxes are usually single- or two-gang and are deep enough so that they can be flush mounted and still intercept conduit that's run through the cavity of a masonry unit. The top and bottom edge of the box are turned inward and tapped for 6-32 machine screws that hold the wiring device in place.

There are through boxes for masonry work that permit mounting devices back-to-back on both sides of a wall. They measure 3½" high and 7½" deep. Both the front and back of the box are made to receive wiring devices.

Non-Metallic Outlet Boxes

Non-metallic outlet boxes are made of either fiberglass or PVC in shapes and sizes like sheet metal boxes.

Fiberglass outlet boxes are generally used in residential concealed wiring. Many come with nails in place in a bracket for attaching to studs or joists. They have threaded metal inserts placed for standard wiring devices. Both 3" and 4" diameter boxes are available. These boxes use fiberglass covers that are similar to steel covers made for steel boxes.

Fiberglass switch boxes are available in single-gang, two-gang, three-gang and four-gang. They all can be purchased with or without mounting brackets and Romex clamps.

Fiberglass 4-S boxes are either $1^{5}/_{8}$", $2^{1}/_{4}$" or $2^{1}/_{2}$" deep. Many covers and plaster rings are available. These are also made of fiberglass.

Another series of non-metallic outlet boxes is made of PVC plastic. These are generally used for surface wiring but can be flush mounted. The switch boxes resemble FS condulets except that they don't have threaded hubs. PVC hubs are threadless because PVC conduit is glued or cemented in place. Other PVC boxes are made as junction boxes. They can be installed in concrete and are approved for corrosive conditions. But they're not acceptable for hazardous environments.

Special covers are available for PVC boxes. Most are weatherproof. There are blank covers, single receptacle covers, duplex receptacle covers, single switch covers and ground fault interrupter (GFI) covers.

I've described many of the common types of outlet boxes. But other types are available. What's listed here is representative of what's stocked by most dealers. Unless you handle a lot of exotic applications, this section has covered nearly all the boxes you use regularly.

Taking Off Outlet Boxes

Many estimators don't count the outlet boxes when making their take-off. Instead, they figure the number of wiring devices that will need a single-gang box, a two-gang box, etc. They use this device count to indicate the number of boxes needed. If you follow this method, be sure to add one outlet box for each single fixture and one for each row of fixtures.

I can't recommend this shortcut. Each box type and size has its own labor and material cost. Counting boxes on the plans is the only way to find the right number of each type of box and the right number of plaster rings.

When making your count, be sure to figure the right size box at each location. *NEC* Article 370, ***Outlet, Switch and Junction Boxes and Fittings*** limits the number of wires allowed in each box. To be accurate, your take-off must consider the number of conductors in each box. Even if the specs allow a 4" square by 1½" box, you may have to use a larger box to comply with the code. Either an extension ring or a 4$^{11}/_{16}$" square box may be required.

It's also worth your time to make an accurate count of plaster and switch rings. The architectural plans should show the wall treatment at each box. On masonry walls, use 1$^{1}/_{2}$" to 2" deep rings. If the wall has ½" thick drywall over $^{1}/_{4}$" plywood, use a $^{3}/_{4}$" deep ring.

Take the time to make your box estimate as accurate as possible. Little omissions compounded many times can make a big difference. The material cost of a box isn't very much. But add in the labor cost and multiply by a hundred boxes or more. The result will usually be more than your profit on most jobs. You can see why counting boxes is worth your time.

Labor for Outlet Box Installation

The labor costs in this section are based on standardized labor units. Using these units should both speed and simplify your box estimates. There are too many different types of boxes and installation conditions to develop an estimate for each case. Instead, use the labor standard that applies to the situation that's closest to the box being estimated.

The labor standards in this section include all the time needed for layout, handling and installation of a box intended for a duplex receptacle and a wall switch next to a door or fixture outlet. Labor for the ring and cover aren't included, of course. See the tables for rings and covers.

I realize that the labor units in this section are higher than many electrical contractors use. But I find that they're accurate when the time required to receive, store, move and handle each box is considered. And most jobs have at least one box that's a real problem. You may set the first fifty boxes faster than the times listed in this section. But the fifty-first unit is going to take 20 minutes if your electrician has no room to work and has to chip out space for the box.

Watch for These Opportunities

Some specs permit a 2" x 4" outlet box as the end device on a circuit. At the end of the circuit only one conduit will enter the box. Usually this will be noted on the plan. Check the symbol list also. Sometimes the symbol list will specify outlet box minimum sizes.

Any time an outlet box has nearly as many conductors as permitted by the code, it may save labor to install the next larger size box. That makes it easier to tuck wires back into the box and install the device.

Material	Craft@Hrs	Unit	Material Cost	Labor Cost	Installed Cost
Handy boxes					
1-1/2" deep 1/2" KO	L1@0.15	Ea	4.03	6.15	10.18
1-7/8" deep 1/2" KO	L1@0.17	Ea	4.23	6.98	11.21
1-7/8" deep 3/4" KO	L1@0.17	Ea	6.20	6.98	13.18
1-7/8" flat bracket	L1@0.17	Ea	7.71	6.98	14.69
1-7/8" extension	L1@0.17	Ea	5.08	6.98	12.06
2-1/2" deep 1/2" KO	L1@0.20	Ea	4.69	8.21	12.90
2-1/2" deep 3/4" KO	L1@0.20	Ea	7.45	8.21	15.66
2-1/8" angle bracket	L1@0.20	Ea	8.34	8.21	16.55
2-1/8" flat bracket	L1@0.20	Ea	10.50	8.21	18.71
Handy box covers					
Blank	L1@0.03	Ea	1.48	1.23	2.71
Switch	L1@0.03	Ea	1.75	1.23	2.98
Single receptacle	L1@0.03	Ea	2.39	1.23	3.62
Duplex receptacle	L1@0.03	Ea	2.42	1.23	3.65
4 wire twistlock	L1@0.03	Ea	3.42	1.23	4.65
Handy boxes, large size					
1-5/8" deep 1/2" KO	L1@0.20	Ea	7.59	8.21	15.80
2-3/16" deep 1/2" KO	L1@0.25	Ea	7.69	10.30	17.99
Handy box covers, large size					
Blank	L1@0.03	Ea	3.60	1.23	4.83
Blank with 1/2" KO	L1@0.03	Ea	4.69	1.23	5.92
Switch	L1@0.03	Ea	3.60	1.23	4.83
Single receptacle	L1@0.03	Ea	4.33	1.23	5.56
Duplex receptacle	L1@0.03	Ea	3.60	1.23	4.83
4 wire twistlock	L1@0.03	Ea	4.33	1.23	5.56
Sectional switch boxes, gangable with ears					
1-1/2" deep 1/2" KO	L1@0.15	Ea	6.15	6.15	12.30
2" deep 1/2" KO	L1@0.15	Ea	6.86	6.15	13.01
2-1/2" deep 1/2" KO	L1@0.17	Ea	6.63	6.98	13.61
2-3/4" deep 1/2" KO	L1@0.20	Ea	8.32	8.21	16.53
2-3/4" deep 3/4" KO	L1@0.20	Ea	8.39	8.21	16.60
3-1/2" deep 1/2" KO	L1@0.25	Ea	8.47	10.30	18.77
3-1/2" deep 3/4" KO	L1@0.25	Ea	8.51	10.30	18.81

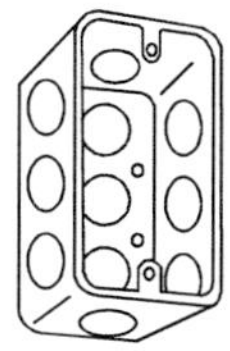

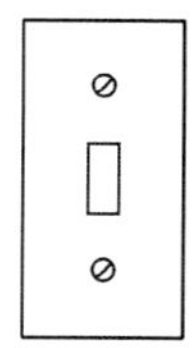

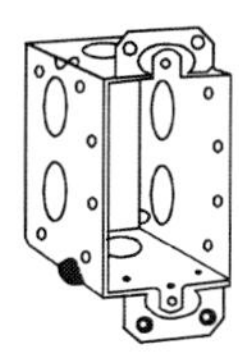

Use these figures to estimate the cost of handy boxes and sectional switch boxes connected to conduit under the conditions described on pages 5 and 6. Costs listed are for each box installed. The crew is one electrician working at a labor cost of $41.03 per manhour. These costs include installing the box flush with the wall or on the wall surface, layout, material handling, and normal waste. Add for box support, extension boxes, plaster rings, switch rings, sales tax, delivery, supervision, mobilization, demobilization, cleanup, overhead and profit. Note: Be sure the box size you select meets *NEC* requirements for the number of conductors to be connected. Sectional boxes can be ganged together for multiple devices.

Switch Boxes

Material	Craft@Hrs	Unit	Material Cost	Labor Cost	Installed Cost
Sectional switch boxes, gangable, no ears					
2" deep 1/2" KO	L1@0.15	Ea	5.75	6.15	11.90
2-1/2" deep 1/2" KO	L1@0.17	Ea	4.33	6.98	11.31
2-3/4" deep 1/2" KO	L1@0.20	Ea	5.56	8.21	13.77
3-1/2" deep 1/2" KO	L1@0.25	Ea	5.94	10.30	16.24
3-1/2" deep 3/4" KO	L1@0.25	Ea	6.85	10.30	17.15
Old work box with ears and side cleats					
1-1/2" KO	L1@0.30	Ea	7.11	12.30	19.41
Switch boxes for conduit					
2-1/2" with long bracket	L1@0.17	Ea	5.94	6.98	12.92
2-1/2" with flat bracket	L1@0.17	Ea	5.99	6.98	12.97
2-3/4" with flat bracket	L1@0.20	Ea	7.27	8.21	15.48
Thru boxes, plaster walls, with brackets					
2-1/8" x 5" x 4"	L1@0.25	Ea	20.20	10.30	30.50
Thru boxes, drywall, with brackets					
2-1/8" x 4-5/16" x 4"	L1@0.25	Ea	19.80	10.30	30.10
Sectional boxes, beveled corners, for Romex					
2-1/4" deep no ears	L1@0.15	Ea	4.75	6.15	10.90
2-1/4" deep with ears	L1@0.15	Ea	4.60	6.15	10.75
2-1/4" with long bracket	L1@0.15	Ea	9.18	6.15	15.33
2-1/4" with flat bracket	L1@0.15	Ea	9.18	6.15	15.33
2-1/4" 2-gang, flat bracket	L1@0.20	Ea	13.70	8.21	21.91
Switch boxes, beveled corners, non-gangable, for Romex, with ears					
2-1/4" deep	L1@0.15	Ea	5.10	6.15	11.25
2-1/4" deep with nails	L1@0.15	Ea	5.67	6.15	11.82
Switch boxes, square corners, non-gangable, for Romex					
1-1/2" deep with ears	L1@0.15	Ea	4.03	6.15	10.18
3-1/8" deep with ears	L1@0.20	Ea	11.10	8.21	19.31
2-1/2" deep no ears	L1@0.15	Ea	4.69	6.15	10.84
2-7/8" deep no ears	L1@0.17	Ea	6.38	6.98	13.36
2-1/2" deep with nails	L1@0.17	Ea	6.75	6.98	13.73
2-3/4" deep with nails	L1@0.20	Ea	8.01	8.21	16.22
3-1/8" deep with nails	L1@0.20	Ea	12.40	8.21	20.61
3-13/32" deep with nails	L1@0.25	Ea	13.90	10.30	24.20

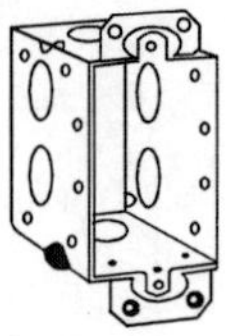
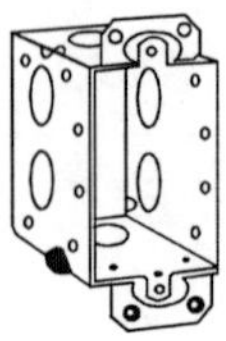
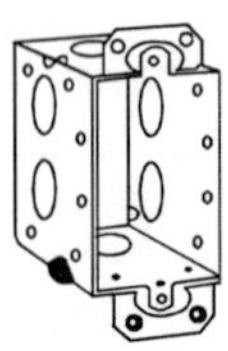
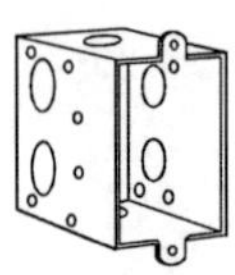

Use these figures to estimate the cost of sectional switch boxes and welded switch boxes connected to conduit under the conditions described on pages 5 and 6. Costs listed are for each box installed. The crew is one electrician working at a labor cost of $41.03 per manhour. These costs include installing the box flush with the wall or on the wall surface, layout, material handling, and normal waste. Add for box support, extension boxes, plaster rings, switch rings, sales tax, delivery, supervision, mobilization, demobilization, cleanup, overhead and profit. Note: Be sure the box size you select meets *NEC* requirements for the number of conductors to be connected. Sectional boxes can be ganged together for multiple devices. Romex and BX boxes have special clamps to hold the cable in place.

Material	Craft@Hrs	Unit	Material Cost	Labor Cost	Installed Cost

Switch boxes, square corners, non-gangable, for Romex

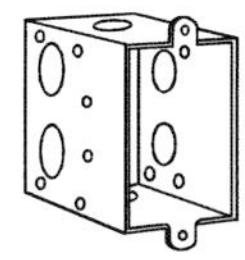

Material	Craft@Hrs	Unit	Material Cost	Labor Cost	Installed Cost
2-13/16" d with flat bracket	L1@0.17	Ea	10.30	6.98	17.28
3-1/8" d with flat bracket	L1@0.20	Ea	11.10	8.21	19.31

Switch boxes, square corners, non-gangable, for MC

Material	Craft@Hrs	Unit	Material Cost	Labor Cost	Installed Cost
2" deep with ears	L1@0.15	Ea	6.16	6.15	12.31
2-1/2" deep with ears	L1@0.17	Ea	5.14	6.98	12.12
2-3/4" deep with ears	L1@0.17	Ea	5.71	6.98	12.69
3-1/2" deep with ears	L1@0.20	Ea	6.77	8.21	14.98
2-1/2" deep no ears	L1@0.17	Ea	3.75	6.98	10.73
2-1/2" with long bracket	L1@0.17	Ea	7.55	6.98	14.53
2-1/2" with flat bracket	L1@0.17	Ea	6.11	6.98	13.09
2-3/4" with flat bracket	L1@0.17	Ea	7.11	6.98	14.09
3-1/2" with flat bracket	L1@0.20	Ea	6.65	8.21	14.86

3" octagon boxes, 1-1/2" deep except pancake box

Material	Craft@Hrs	Unit	Material Cost	Labor Cost	Installed Cost
3-0 pancake 1/2" deep	L1@0.15	Ea	4.20	6.15	10.35
3-0 for Romex	L1@0.15	Ea	6.20	6.15	12.35
3-0 1/2" KO	L1@0.15	Ea	4.97	6.15	11.12
3-0 extension 1/2" KO	L1@0.15	Ea	7.90	6.15	14.05

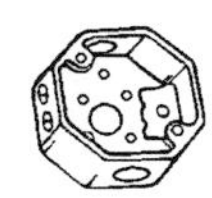

4" octagon boxes, 1-1/2" deep except pancake box

Material	Craft@Hrs	Unit	Material Cost	Labor Cost	Installed Cost
4-0 pancake 1/2" deep	L1@0.17	Ea	4.83	6.98	11.81
4-0 1/2" KO	L1@0.20	Ea	3.81	8.21	12.02
4-0 3/4" KO	L1@0.20	Ea	4.45	8.21	12.66
4-0 1/2" & 3/4" KO	L1@0.20	Ea	4.40	8.21	12.61
4-0 BX flat bracket 1/2" KO	L1@0.20	Ea	8.01	8.21	16.22
4-0 for Romex	L1@0.20	Ea	7.70	8.21	15.91
4-0 Romex with ears	L1@0.20	Ea	8.01	8.21	16.22
4-0 Romex J-bracket	L1@0.20	Ea	9.00	8.21	17.21
4-0 Romex flat bracket	L1@0.20	Ea	8.01	8.21	16.22
4-0 extension ring 1/2" KO	L1@0.20	Ea	5.63	8.21	13.84
4-0 ext. 1/2" & 3/4" KO	L1@0.20	Ea	6.76	8.21	14.97

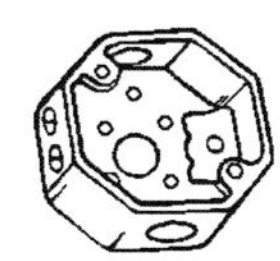

Use these figures to estimate the cost of welded switch boxes and octagon boxes connected to conduit under the conditions described on pages 5 and 6. Costs listed are for each box installed. The crew is one electrician working at a labor cost of $41.03 per manhour. These costs include installing the box flush with the wall or on the wall surface, layout, material handling, and normal waste. Add for box support, extension boxes, plaster rings, switch rings, sales tax, delivery, supervision, mobilization, demobilization, cleanup, overhead and profit. Note: Be sure the box size you select meets *NEC* requirements for the number of conductors to be connected. Sectional boxes can be ganged together for multiple devices.

Octagon Boxes, Concrete Rings and Box Hangers

Material	Craft@Hrs	Unit	Material Cost	Labor Cost	Installed Cost
4" octagon boxes, 2-1/8" deep					
4-0 1/2" KO	L1@0.20	Ea	6.51	8.21	14.72
4-0 3/4" KO	L1@0.20	Ea	6.59	8.21	14.80
4-0 1/2" & 3/4" KO	L1@0.20	Ea	6.54	8.21	14.75
4-0 1" KO	L1@0.20	Ea	9.37	8.21	17.58
4-0 with clamps	L1@0.20	Ea	6.69	8.21	14.90
4-0 with clamps & J-bkt	L1@0.20	Ea	8.02	8.21	16.23
4-0 clamps & flat bkt	L1@0.20	Ea	8.55	8.21	16.76

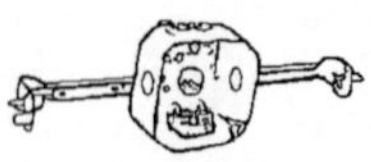

4" octagon boxes, 1-1/2" deep with adjustable hanger and stud					
4-0 1/2" KO	L1@0.25	Ea	9.38	10.30	19.68
4" octagon boxes, 1-1/2" deep with adjustable hanger and clip					
4-0 1/2" KO	L1@0.25	Ea	9.55	10.30	19.85

4-3/8" octagon rings, 1/2" and 3/4" KO					
4-3/8" 2" deep	L1@0.25	Ea	8.87	10.30	19.17
4-3/8" 2-1/2" deep	L1@0.25	Ea	8.54	10.30	18.84
4-3/8" 3" deep	L1@0.25	Ea	8.64	10.30	18.94
4-3/8" 3-1/2" deep	L1@0.30	Ea	9.31	12.30	21.61
4-3/8" 4" deep	L1@0.30	Ea	13.20	12.30	25.50
4-3/8" 5" deep	L1@0.35	Ea	17.90	14.40	32.30
4-3/8" 6" deep	L1@0.35	Ea	20.60	14.40	35.00

Concrete ring plates					
4-3/8" with 1/2" & 3/4" KO	L1@0.05	Ea	2.56	2.05	4.61
4-3/8" with 3/8" stud	L1@0.05	Ea	4.99	2.05	7.04

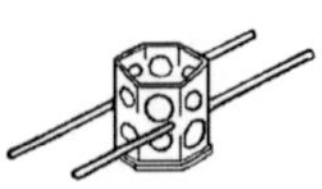

Drop ceiling boxes with 1/2" and 3/4" KO					
4-0 with 18" hanger bars	L1@0.25	Ea	5.56	10.30	15.86
4-0 with 24" hanger bars	L1@0.25	Ea	5.76	10.30	16.06
4-0 with 30" hanger bars	L1@0.25	Ea	5.65	10.30	15.95

Use these figures to estimate the cost of octagon boxes connected to conduit under the conditions described on pages 5 and 6. Costs listed are for each box installed. The crew is one electrician working at a labor cost of $41.03 per manhour. These costs include installing the box flush with the wall or on the wall surface, layout, material handling, and normal waste. Add for box support, extension boxes, plaster rings, switch rings, sales tax, delivery, supervision, mobilization, demobilization, cleanup, overhead and profit. Note: Be sure the box size you select meets *NEC* requirements for the number of conductors to be connected.

Material	Craft@Hrs	Unit	Material Cost	Labor Cost	Installed Cost
4" x 4" x 1-1/4" deep square boxes					
4-S 1/2 KO	L1@0.25	Ea	5.09	10.30	15.39
4" x 4" x 1-1/2" deep square boxes					
4-S 1/2 KO	L1@0.25	Ea	4.63	10.30	14.93
4-S 3/4 KO	L1@0.25	Ea	4.65	10.30	14.95
4-S 1/2 & 3/4 KO	L1@0.25	Ea	4.67	10.30	14.97
4-S bracket	L1@0.25	Ea	7.11	10.30	17.41
4-S Romex	L1@0.25	Ea	5.89	10.30	16.19
4-S Romex, bracket	L1@0.25	Ea	6.17	10.30	16.47
4-S BX, bracket	L1@0.25	Ea	8.27	10.30	18.57
4-S ext. 1/2 KO	L1@0.15	Ea	4.67	6.15	10.82
4-S ext. 1/2 & 3/4 KO	L1@0.15	Ea	4.74	6.15	10.89
4" x 4" x 2-1/8" deep square boxes					
4-S 1/2 KO	L1@0.27	Ea	6.99	11.10	18.09
4-S 3/4 KO	L1@0.27	Ea	7.90	11.10	19.00
4-S 1 KO	L1@0.27	Ea	8.55	11.10	19.65
4-S 1/2 & 3/4 KO	L1@0.27	Ea	7.46	11.10	18.56
4-S 1/2 bracket	L1@0.27	Ea	9.80	11.10	20.90
4-S 3/4 bracket	L1@0.27	Ea	9.80	11.10	20.90
4-S 1/2 & 3/4 bracket	L1@0.27	Ea	9.80	11.10	20.90
4-S Romex	L1@0.27	Ea	11.10	11.10	22.20
4-S Romex, bracket	L1@0.27	Ea	11.10	11.10	22.20
4-11/16" x 4-11/16" x 1-1/2" deep square boxes					
4-11/16 1/2 KO	L1@0.30	Ea	11.50	12.30	23.80
4-11/16 3/4 KO	L1@0.30	Ea	10.40	12.30	22.70
4-11/16 1/2 & 3/4 KO	L1@0.30	Ea	10.10	12.30	22.40
4-11/16 ext 1/2 KO	L1@0.25	Ea	10.00	10.30	20.30
4-11/16 ext 1/2 & 3/4 KO	L1@0.25	Ea	10.00	10.30	20.30

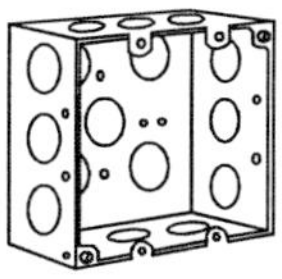

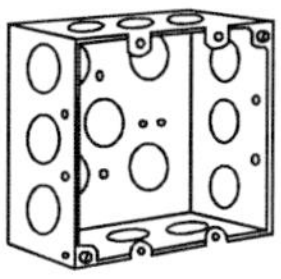

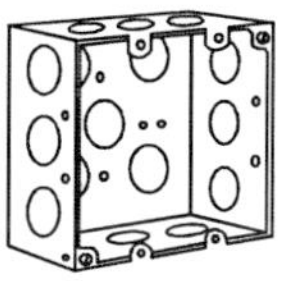

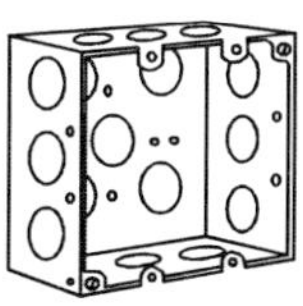

Use these figures to estimate the cost of square boxes connected to conduit under the conditions described on pages 5 and 6. Costs listed are for each box installed. The crew is one electrician working at a labor cost of $41.03 per manhour. These costs include installing the box flush with the wall or on the wall surface, layout, material handling, and normal waste. Add for box support, extension boxes, plaster rings, switch rings, sales tax, delivery, supervision, mobilization, demobilization, cleanup, overhead and profit. Note: Be sure the box size you select meets *NEC* requirements for the number of conductors to be connected.

Square Boxes and Covers

Material	Craft@Hrs	Unit	Material Cost	Labor Cost	Installed Cost
4-11/16" x 4-11/16" x 2-1/8" deep square boxes					
4-11/16 1/2 KO	L1@0.33	Ea	10.70	13.50	24.20
4-11/16 3/4 KO	L1@0.33	Ea	10.70	13.50	24.20
4-11/16 1/2&3/4 KO	L1@0.33	Ea	9.00	13.50	22.50
4-11/16 3/4 & 1 KO	L1@0.33	Ea	11.30	13.50	24.80
4-11/16 1 KO	L1@0.33	Ea	10.70	13.50	24.20
4-11/16 1 1/4 KO	L1@0.33	Ea	15.00	13.50	28.50
4-11/16 ext 1/2 KO	L1@0.25	Ea	11.90	10.30	22.20
4-11/16 ext 3/4 KO	L1@0.25	Ea	12.40	10.30	22.70
4-11/16 ext 1/2 & 3/4 KO	L1@0.25	Ea	15.80	10.30	26.10
Outlet box covers					
3-O flat blank	L1@0.03	Ea	2.74	1.23	3.97
3-O 1/2 KO	L1@0.03	Ea	1.97	1.23	3.20
4-O flat blank	L1@0.05	Ea	2.39	2.05	4.44
4-O 1/2 KO	L1@0.05	Ea	2.39	2.05	4.44
4-O duplex	L1@0.05	Ea	5.38	2.05	7.43
4-O single outlet	L1@0.05	Ea	5.55	2.05	7.60
4-O T/L recept.	L1@0.05	Ea	5.38	2.05	7.43
4-O plaster ring 1/2	L1@0.05	Ea	3.03	2.05	5.08
4-O plaster ring 5/8	L1@0.05	Ea	3.48	2.05	5.53
4-O plaster ring 3/4	L1@0.05	Ea	3.93	2.05	5.98
4-O plaster ring 1	L1@0.05	Ea	8.07	2.05	10.12
4-O plas. ring 1-1/4	L1@0.05	Ea	12.20	2.05	14.25
4-S flat blank	L1@0.05	Ea	1.74	2.05	3.79
4-S flat blank 1/2 KO	L1@0.05	Ea	2.06	2.05	4.11
4-S plaster ring 1/2	L1@0.05	Ea	2.60	2.05	4.65
4-S plaster ring 5/8	L1@0.05	Ea	2.74	2.05	4.79
4-S plaster ring 3/4	L1@0.05	Ea	2.93	2.05	4.98
4-S plaster ring 1	L1@0.05	Ea	11.70	2.05	13.75
4-S plaster ring 1-1/4	L1@0.05	Ea	11.50	2.05	13.55

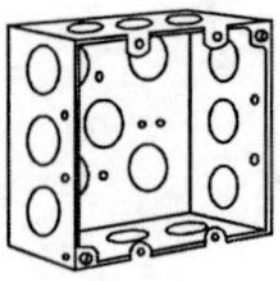

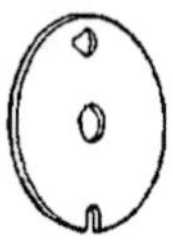

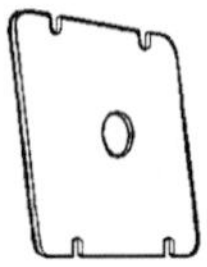

Use these figures to estimate the cost of square boxes and covers installed on conduit under the conditions described on pages 5 and 6. Costs listed are for each unit installed. The crew is one electrician working at a labor cost of $41.03 per manhour. These costs include installing the unit, layout, material handling, and normal waste. Add for box support, extension boxes, plaster rings, switch rings, sales tax, delivery, supervision, mobilization, demobilization, cleanup, overhead and profit. Note: Be sure the box size you select meets *NEC* requirements for the number of conductors to be connected.

Single gang 4" square switch rings

Material	Craft@Hrs	Unit	Material Cost	Labor Cost	Installed Cost
4-S sw ring flat	L1@0.05	Ea	3.79	2.05	5.84
4-S sw ring 1/4	L1@0.05	Ea	4.19	2.05	6.24
4-S sw ring 1/2	L1@0.05	Ea	2.88	2.05	4.93
4-S sw ring 5/8	L1@0.05	Ea	1.69	2.05	3.74
4-S sw ring 3/4	L1@0.05	Ea	1.99	2.05	4.04
4-S sw ring 1	L1@0.05	Ea	2.49	2.05	4.54
4-S sw ring 1-1/4	L1@0.05	Ea	3.85	2.05	5.90
4-S tile 1/2	L1@0.05	Ea	3.21	2.05	5.26
4-S tile 3/4	L1@0.05	Ea	3.29	2.05	5.34
4-S tile 1	L1@0.05	Ea	3.47	2.05	5.52
4-S tile 1-1/4	L1@0.06	Ea	3.81	2.46	6.27
4-S tile 1-1/2	L1@0.06	Ea	6.29	2.46	8.75
4-S tile 2	L1@0.06	Ea	6.57	2.46	9.03

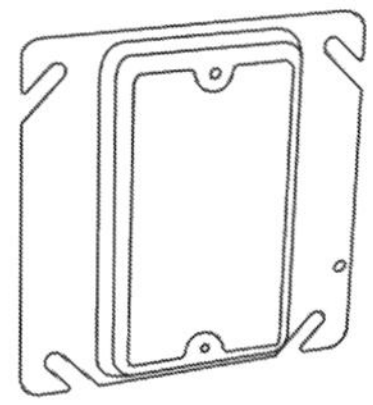

Two gang 4" square switch rings

Material	Craft@Hrs	Unit	Material Cost	Labor Cost	Installed Cost
4-S sw ring flat	L1@0.06	Ea	5.24	2.46	7.70
4-S sw ring 1/4	L1@0.06	Ea	6.27	2.46	8.73
4-S sw ring 1/2	L1@0.06	Ea	2.87	2.46	5.33
4-S sw ring 5/8	L1@0.06	Ea	3.27	2.46	5.73
4-S sw ring 3/4	L1@0.06	Ea	3.61	2.46	6.07
4-S sw ring 1	L1@0.06	Ea	5.62	2.46	8.08
4-S sw ring 1-1/4	L1@0.06	Ea	6.21	2.46	8.67
4-S tile 1/2	L1@0.06	Ea	3.14	2.46	5.60
4-S tile 3/4	L1@0.06	Ea	4.16	2.46	6.62
4-S tile 1	L1@0.06	Ea	5.63	2.46	8.09
4-S tile 1-1/4	L1@0.08	Ea	6.15	3.28	9.43
4-S tile 1-1/2	L1@0.08	Ea	7.87	3.28	11.15
4-S tile 2	L1@0.08	Ea	9.83	3.28	13.11

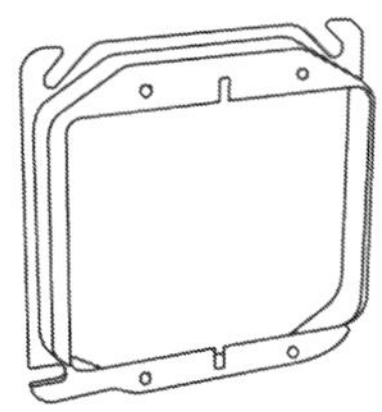

Adjustable bar hangers with 3/8" stud

Material	Craft@Hrs	Unit	Material Cost	Labor Cost	Installed Cost
11-1/2" to 18-1/2"	L1@0.05	Ea	4.80	2.05	6.85
19-1/2" to 26-1/2"	L1@0.05	Ea	6.82	2.05	8.87

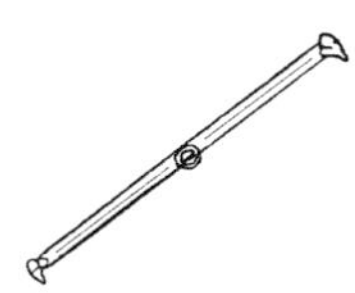

Adjustable bar hanger with clip

Material	Craft@Hrs	Unit	Material Cost	Labor Cost	Installed Cost
11-1/2" to 18-1/2"	L1@0.05	Ea	4.90	2.05	6.95
19-1/2" to 26-1/2"	L1@0.05	Ea	5.42	2.05	7.47

Bar hangers with stove bolts

Material	Craft@Hrs	Unit	Material Cost	Labor Cost	Installed Cost
20" straight	L1@0.05	Ea	7.74	2.05	9.79
24" straight	L1@0.05	Ea	8.39	2.05	10.44
21" 1" offset	L1@0.05	Ea	8.72	2.05	10.77
22" 1-1/2" offset	L1@0.05	Ea	10.60	2.05	12.65

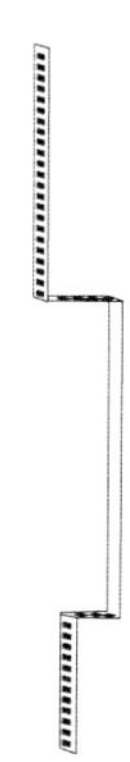

Use these figures to estimate the cost of square switch rings installed on outlet boxes under the conditions described on pages 5 and 6. Costs listed are for each switch ring installed. The crew is one electrician working at a labor cost of $41.03 per manhour. These costs include adapting the cover to the box, layout, material handling, and normal waste. Add for the outlet box, wiring devices, sales tax, delivery, supervision, mobilization, demobilization, cleanup, overhead and profit. Note: Be sure to select plaster and switch rings as specified and that are appropriate for the wall depth. Many types are available.

Square Switch Rings

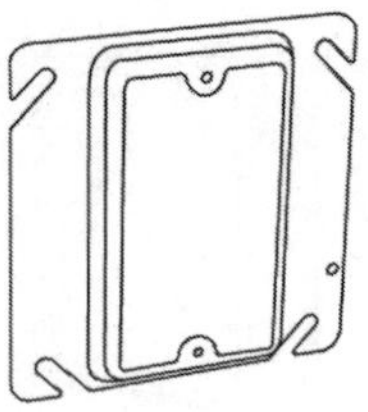

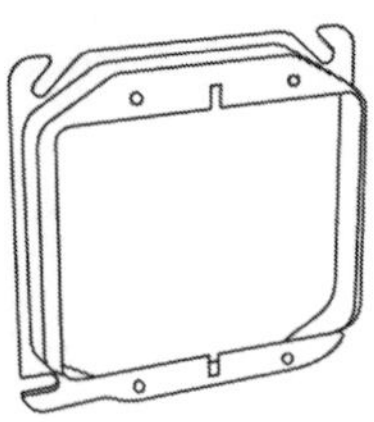

Material	Craft@Hrs	Unit	Material Cost	Labor Cost	Installed Cost
Single gang 4-11/16" square switch rings					
4-11/16 sw ring 1/4	L1@0.06	Ea	8.19	2.46	10.65
4-11/16 sw ring 1/2	L1@0.06	Ea	8.19	2.46	10.65
4-11/16 sw ring 5/8	L1@0.06	Ea	8.19	2.46	10.65
4-11/16 sw ring 3/4	L1@0.06	Ea	8.32	2.46	10.78
4-11/16 sw ring 1	L1@0.06	Ea	8.39	2.46	10.85
4-11/16 sw ring 1-1/4	L1@0.06	Ea	8.80	2.46	11.26
4-11/16 tile 1/2	L1@0.06	Ea	8.32	2.46	10.78
4-11/16 tile 3/4	L1@0.06	Ea	8.32	2.46	10.78
4-11/16 tile 1	L1@0.06	Ea	8.66	2.46	11.12
4-11/16 tile 1-1/4	L1@0.08	Ea	9.09	3.28	12.37
4-11/16 tile 1-1/2	L1@0.08	Ea	15.00	3.28	18.28
4-11/16 tile 2	L1@0.08	Ea	15.00	3.28	18.28
Two gang 4-11/16" square switch rings					
4-11/16 sw ring 1/2	L1@0.08	Ea	10.10	3.28	13.38
4-11/16 sw ring 5/8	L1@0.08	Ea	10.60	3.28	13.88
4-11/16 sw ring 3/4	L1@0.08	Ea	10.70	3.28	13.98
4-11/16 sw ring 1	L1@0.08	Ea	11.10	3.28	14.38
4-11/16 sw ring 1-1/4	L1@0.08	Ea	11.20	3.28	14.48
4-11/16 tile 1/2	L1@0.10	Ea	10.90	4.10	15.00
4-11/16 tile 3/4	L1@0.10	Ea	10.90	4.10	15.00
4-11/16 tile 1	L1@0.10	Ea	11.30	4.10	15.40
4-11/16 tile 1-1/4	L1@0.10	Ea	11.50	4.10	15.60
4-11/16 tile 1-1/2	L1@0.10	Ea	16.50	4.10	20.60
4-11/16 tile 2	L1@0.10	Ea	20.50	4.10	24.60

Use these figures to estimate the cost of square switch rings and surface covers installed on outlet boxes under the conditions described on pages 5 and 6. Costs listed are for each ring or cover installed. The crew is one electrician working at a labor cost of $41.03 per manhour. These costs include adapting the cover to the box, layout, material handling, and normal waste. Add for outlet box, wiring devices, sales tax, delivery, supervision, mobilization, demobilization, cleanup, overhead and profit. Note: Be sure to select plaster and switch rings as specified and that are appropriate to the wall depth. Many types are available.

4" square surface box covers, raised 1/2"

Material	Craft@Hrs	Unit	Material Cost	Labor Cost	Installed Cost
4-S blank	L1@0.05	Ea	1.74	2.05	3.79
4-S one switch	L1@0.05	Ea	3.02	2.05	5.07
4-S two switch	L1@0.05	Ea	3.29	2.05	5.34
4-S one single recept	L1@0.05	Ea	2.97	2.05	5.02
4-S two single recept	L1@0.05	Ea	3.74	2.05	5.79
4-S one duplex recept	L1@0.05	Ea	2.97	2.05	5.02
4-S two duplex recept	L1@0.05	Ea	3.74	2.05	5.79
4-S sw & single recept	L1@0.05	Ea	4.06	2.05	6.11
4-S sw & duplex recept	L1@0.05	Ea	3.74	2.05	5.79
4-S single & duplex recept	L1@0.05	Ea	4.06	2.05	6.11
4-S 3 w T/L recept	L1@0.05	Ea	4.14	2.05	6.19
4-S 4 w T/L recept	L1@0.05	Ea	5.16	2.05	7.21
4-S 3w recept	L1@0.05	Ea	5.24	2.05	7.29
4-S 4w recept	L1@0.05	Ea	5.19	2.05	7.24

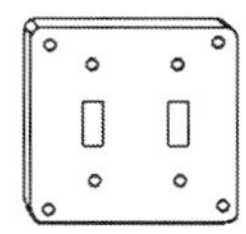

4-11/16" square box covers

Material	Craft@Hrs	Unit	Material Cost	Labor Cost	Installed Cost
4-11/16 flat blank	L1@0.06	Ea	3.25	2.46	5.71
4-11/16 blank 1/2 KO	L1@0.06	Ea	5.65	2.46	8.11
4-11/16 pl ring 1/2	L1@0.06	Ea	8.19	2.46	10.65
4-11/16 pl ring 5/8	L1@0.06	Ea	8.19	2.46	10.65
4-11/16 pl ring 3/4	L1@0.06	Ea	8.19	2.46	10.65
4-11/16 pl ring 1	L1@0.06	Ea	15.90	2.46	18.36
4-11/16 pl ring 1-1/4	L1@0.06	Ea	16.50	2.46	18.96

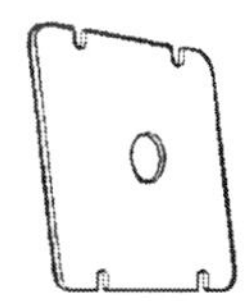

4-11/16" square surface box covers, raised 1/2"

Material	Craft@Hrs	Unit	Material Cost	Labor Cost	Installed Cost
Blank	L1@0.06	Ea	12.20	2.46	14.66
One switch	L1@0.06	Ea	12.20	2.46	14.66
Two switches	L1@0.06	Ea	12.60	2.46	15.06
One duplex	L1@0.06	Ea	13.90	2.46	16.36
Two duplex	L1@0.06	Ea	15.90	2.46	18.36
One switch & duplex	L1@0.06	Ea	15.90	2.46	18.36
One single receptacle	L1@0.06	Ea	18.30	2.46	20.76
Two single receptacle	L1@0.06	Ea	18.30	2.46	20.76
One 30A T/L receptacle	L1@0.06	Ea	18.30	2.46	20.76
One 60A T/L receptacle	L1@0.06	Ea	18.30	2.46	20.76

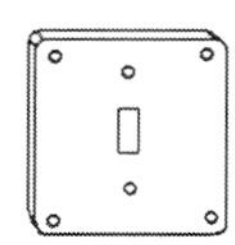

Use these figures to estimate the cost of square and octagon box covers installed on outlet boxes under the conditions described on pages 5 and 6. Costs listed are for each cover installed. The crew is one electrician working at a labor cost of $41.03 per manhour. These costs include adapting the cover to the box, layout, material handling, and normal waste. Add for outlet box, wiring devices, sales tax, delivery, supervision, mobilization, demobilization, cleanup, overhead and profit. Note: Be sure to select plaster and switch rings as specified and that are appropriate to the wall depth. Many types are available.

Gang Switch Boxes and Masonry Boxes

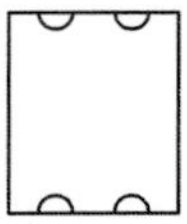

Gang boxes 1-5/8" deep with 1/2" and 3/4" KO

Material	Craft@Hrs	Unit	Material Cost	Labor Cost	Installed Cost
2 gang	L1@0.25	Ea	35.20	10.30	45.50
3 gang	L1@0.30	Ea	39.60	12.30	51.90
4 gang	L1@0.40	Ea	52.70	16.40	69.10
5 gang	L1@0.50	Ea	82.40	20.50	102.90
6 gang	L1@0.60	Ea	150.00	24.60	174.60

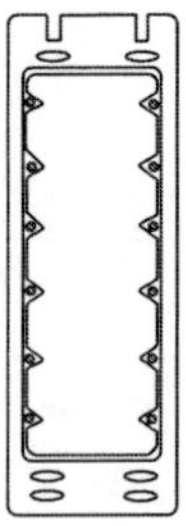

Gang switch rings, 1/2" raised

Material	Craft@Hrs	Unit	Material Cost	Labor Cost	Installed Cost
2 gang	L1@0.05	Ea	11.80	2.05	13.85
3 gang	L1@0.10	Ea	18.00	4.10	22.10
4 gang	L1@0.10	Ea	26.00	4.10	30.10
5 gang	L1@0.10	Ea	36.00	4.10	40.10
6 gang	L1@0.10	Ea	61.50	4.10	65.60

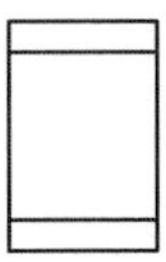

One gang masonry boxes, 1/2" and 3/4" knockouts

Material	Craft@Hrs	Unit	Material Cost	Labor Cost	Installed Cost
2-1/2" deep, 1 gang	L1@0.20	Ea	11.70	8.21	19.91
3-1/2" deep, 1 gang	L1@0.25	Ea	11.40	10.30	21.70
3-1/2" deep, thru box	L1@0.30	Ea	14.60	12.30	26.90
5-1/2" deep, thru box	L1@0.40	Ea	17.20	16.40	33.60
7-1/2" deep, thru box	L1@0.50	Ea	19.90	20.50	40.40

Multi-gang masonry boxes, 1/2" and 3/4" knockouts

Material	Craft@Hrs	Unit	Material Cost	Labor Cost	Installed Cost
2-1/2" deep, 2 gang	L1@0.30	Ea	21.90	12.30	34.20
3-1/2" deep, 2 gang	L1@0.35	Ea	17.90	14.40	32.30
2-1/2" deep, 3 gang	L1@0.40	Ea	26.40	16.40	42.80
3-1/2" deep, 3 gang	L1@0.40	Ea	25.40	16.40	41.80
2-1/2" deep, 4 gang	L1@0.50	Ea	30.40	20.50	50.90
3-1/2" deep, 4 gang	L1@0.50	Ea	32.30	20.50	52.80

Tomic bolt hangers

Material	Craft@Hrs	Unit	Material Cost	Labor Cost	Installed Cost
5/16" x 4"	L1@0.06	Ea	6.02	2.46	8.48
5/16" x 5"	L1@0.06	Ea	6.79	2.46	9.25

Use these figures to estimate the cost of knockout boxes and masonry boxes installed in buildings under the conditions described on pages 5 and 6. Costs listed are for each box installed. The crew is one electrician working at a labor cost of $41.03 per manhour. These costs include switch rings, layout, material handling, and normal waste. Add for box support, wiring devices, sales tax, delivery, supervision, mobilization, demobilization, cleanup, overhead and profit. Note: Ganged switch boxes are made in one piece.

Boxes made for masonry construction come as a single piece and have provisions for attaching devices to the box. But there's no separate switch ring. The knockout for conduit is near the back of the box so conduit can be run from the masonry cells to the box with little or no bending. Usually a short piece of rebar is used to secure the box in place. All other holes in the box should be sealed with masking tape to prevent mortar from entering the box. Be sure that conduit entering the bottom of the box is also sealed.

Material	Craft@Hrs	Unit	Material Cost	Labor Cost	Installed Cost
Round pan box					
3-3/8" x 5/8"	L1@0.15	Ea	4.95	6.15	11.10
Round box					
3-1/2" x 1-3/4"	L1@0.15	Ea	1.81	6.15	7.96
3-1/2" x 2-3/8"	L1@0.15	Ea	2.65	6.15	8.80
4" x 2-3/8"	L1@0.15	Ea	4.95	6.15	11.10
Round box with clamps w/o mounting flange					
3-1/2" x 1-3/4"	L1@0.15	Ea	9.25	6.15	15.40
Round box with ground strap w/o mounting flange					
3-3/8" x 5/8"	L1@0.15	Ea	5.96	6.15	12.11
Round box with mounting flange					
3-1/2" x 1-3/4"	L1@0.15	Ea	3.16	6.15	9.31
3-1/2" x 2-3/4"	L1@0.15	Ea	4.05	6.15	10.20
Round box with mounting flange and grounding strap					
3-1/2" x 1-3/4"	L1@0.15	Ea	3.44	6.15	9.59
3-1/2" x 2-3/4"	L1@0.15	Ea	4.62	6.15	10.77
Round box with mounting flange, grounding strap and clamps					
3-1/2" x 2-3/4"	L1@0.15	Ea	11.10	6.15	17.25
Round box with snap-in bracket					
3-1/2" x 1-3/4"	L1@0.15	Ea	5.13	6.15	11.28
3-1/2" x 2"	L1@0.15	Ea	5.60	6.15	11.75
Round box with snap-in bracket and clamps					
3-1/2" x 1-3/4"	L1@0.15	Ea	6.40	6.15	12.55
3-1/2" x 2"	L1@0.15	Ea	6.84	6.15	12.99

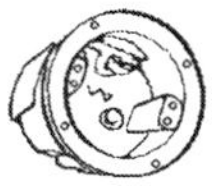

Use these figures to estimate the cost of fiberglass outlet boxes installed in residential buildings under the conditions described on pages 5 and 6. Costs listed are for each box installed. The crew is one electrician working at a labor cost of $41.03 per manhour. These costs include layout, material handling, and normal waste. Add for wiring devices, covers, sales tax, delivery, supervision, mobilization, demobilization, cleanup, overhead, and profit. Note: These boxes are used for Romex (type NM) cable in residences and are generally used for lighting fixtures and junction boxes. They're usually installed flush with the wall or ceiling surface. A lighting fixture can be attached directly to the box with 8-32 machine screws. Be sure the outlet box is secured to the building frame well enough to support the weight of the lighting fixture. The outlet box can either be nailed to a stud or fastened with wood screws.

Fiberglass Outlet Boxes

Material	Craft@Hrs	Unit	Material Cost	Labor Cost	Installed Cost
Round box with snap-in bracket and grounding strap					
3-1/2" x 1-3/4"	L1@0.15	Ea	5.93	6.15	12.08
3-1/2" x 2"	L1@0.15	Ea	6.01	6.15	12.16
3-1/2" x 2-3/4"	L1@0.15	Ea	9.32	6.15	15.47
Round box with snap-in bracket, grounding strap and clamps					
3-1/2" x 2-5/16"	L1@0.15	Ea	11.60	6.15	17.75
Round box with flush mounting hanger					
4" x 2-5/16"	L1@0.15	Ea	9.58	6.15	15.73
Round box with nails					
3-1/2" x 2-7/8"	L1@0.15	Ea	1.95	6.15	8.10
4" x 1-5/8"	L1@0.15	Ea	2.68	6.15	8.83
4" x 2-3/8"	L1@0.15	Ea	2.14	6.15	8.29
Round box with nails and clamps					
3-1/2" x 2-7/8"	L1@0.15	Ea	2.42	6.15	8.57
4" x 2-3/8"	L1@0.15	Ea	2.54	6.15	8.69
Round box with nails and ground strap					
3-1/2" x 2-7/8"	L1@0.15	Ea	2.42	6.15	8.57
4" x 2-3/8"	L1@0.15	Ea	2.58	6.15	8.73
Round box with nails, ground strap and clamps					
3-1/2" x 2-7/8"	L1@0.15	Ea	2.95	6.15	9.10
4" x 2-5/16"	L1@0.15	Ea	3.03	6.15	9.18
Round box with L hanger					
4" x 2-5/16"	L1@0.15	Ea	2.43	6.15	8.58
Round box with L hanger and clamps					
4" x 2-5/16"	L1@0.15	Ea	2.96	6.15	9.11

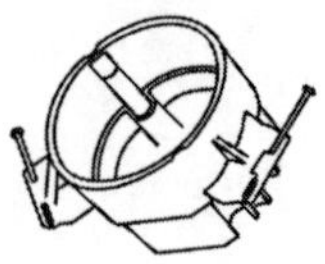

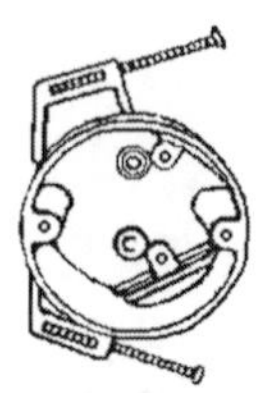

Use these figures to estimate the cost of fiberglass outlet boxes installed in residential buildings under the conditions described on pages 5 and 6. Costs listed are for each box installed. The crew is one electrician working at a labor cost of $41.03 per manhour. These costs include layout, material handling, and normal waste. Add for wiring devices, covers, sales tax, delivery, supervision, mobilization, demobilization, cleanup, overhead, and profit. Note: These boxes are used for Romex (type NM) cable in residences and are generally used for lighting fixtures and junction boxes. They're usually installed flush with the wall or ceiling surface. A lighting fixture can be attached directly to the box with 8-32 machine screws. Be sure the outlet box is secured to the building frame well enough to support the weight of the lighting fixture. The outlet box can either be nailed to a stud or fastened with wood screws.

Material	Craft@Hrs	Unit	Material Cost	Labor Cost	Installed Cost
Round box with adjustable bar hanger					
3-1/2" x 2-7/8"	L1@0.20	Ea	4.21	8.21	12.42
4" x 1-5/8"	L1@0.20	Ea	3.63	8.21	11.84
4" x 2-3/8"	L1@0.20	Ea	3.85	8.21	12.06
Round box with adjustable bar hanger and clamps					
4" x 2-3/8"	L1@0.20	Ea	4.39	8.21	12.60
Round box with adjustable bar hanger and ground strap					
3-1/2" x 2-7/8"	L1@0.20	Ea	4.62	8.21	12.83
4" x 2-3/8"	L1@0.20	Ea	4.25	8.21	12.46
Round box with adjustable bar hanger, ground strap and clamps					
3-1/2" x 2-7/8"	L1@0.20	Ea	4.25	8.21	12.46
Round box with 24" adjustable bar hanger					
3-1/2" x 2-7/8"	L1@0.20	Ea	4.25	8.21	12.46
4" x 1-5/8"	L1@0.20	Ea	3.68	8.21	11.89
4" x 2-5/16"	L1@0.20	Ea	3.93	8.21	12.14
Round box with 24" adjustable bar hanger and clamps					
3-1/2" x 2-7/8"	L1@0.20	Ea	4.25	8.21	12.46
4" x 2-5/16"	L1@0.20	Ea	4.44	8.21	12.65
Round box with 24" adjustable bar hanger and ground strap					
3-1/2" x 2-7/8"	L1@0.20	Ea	4.76	8.21	12.97
4" x 2-5/16"	L1@0.20	Ea	4.38	8.21	12.59
Round box with 24" adjustable bar hanger, ground strap and clamp					
3-1/2" x 2-7/8"	L1@0.20	Ea	4.76	8.21	12.97
4" x 2-5/16"	L1@0.20	Ea	5.19	8.21	13.40

Use these figures to estimate the cost of fiberglass outlet boxes installed in residential buildings under the conditions described on pages 5 and 6. Costs listed are for each box installed. The crew is one electrician working at a labor cost of $41.03 per manhour. These costs include layout, material handling, and normal waste. Add for wiring devices, covers, sales tax, delivery, supervision, mobilization, demobilization, cleanup, overhead, and profit. Note: These boxes are used for Romex (type NM) cable in residences and are generally used for lighting fixtures and junction boxes. They're usually installed flush with the wall or ceiling surface. A lighting fixture can be attached directly to the box with 8-32 machine screws. Be sure the outlet box is secured to the building frame well enough to support the weight of the lighting fixture. The outlet box can either be nailed to a stud or fastened with wood screws.

Fiberglass Covers and Outlet Boxes

Material	Craft@Hrs	Unit	Material Cost	Labor Cost	Installed Cost
Round box covers					
4" plaster ring	L1@0.05	Ea	3.16	2.05	5.21
4" blank cover	L1@0.05	Ea	1.64	2.05	3.69
Square outlet boxes with clamps					
4" x 2-1/4"	L1@0.20	Ea	5.78	8.21	13.99
Square outlet boxes with flush mount hanger					
4" x 1-5/8"	L1@0.20	Ea	6.08	8.21	14.29
4" x 2-1/4"	L1@0.20	Ea	4.60	8.21	12.81
Square outlet boxes with nails					
4" x 2-1/2"	L1@0.20	Ea	4.56	8.21	12.77
Square outlet boxes with nails and clamps					
4" x 2-1/2"	L1@0.20	Ea	5.41	8.21	13.62
Square outlet boxes with nail-on hanger					
4" x 1-5/8"	L1@0.20	Ea	4.26	8.21	12.47
4" x 2-1/4"	L1@0.20	Ea	4.49	8.21	12.70
Square outlet boxes with Z-bracket					
4" x 1-1/4"	L1@0.20	Ea	3.72	8.21	11.93
Square outlet boxes with Z-bracket flush with face of plaster ring					
4" x 1-1/4"	L1@0.20	Ea	4.42	8.21	12.63
4" square outlet box covers					
Blank	L1@0.05	Ea	1.80	2.05	3.85
1/2" 1-gang sw ring	L1@0.05	Ea	2.66	2.05	4.71
1/2" 2-gang sw ring	L1@0.05	Ea	3.81	2.05	5.86
1/2" plaster ring	L1@0.05	Ea	2.18	2.05	4.23

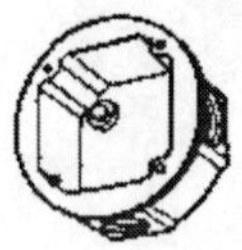

Use these figures to estimate the cost of fiberglass outlet boxes and covers installed in residential buildings under the conditions described on pages 5 and 6. Costs listed are for each box or cover installed. The crew is one electrician working at a labor cost of $41.03 per manhour. These costs include layout, material handling, and normal waste. Add for wiring devices, sales tax, delivery, supervision, mobilization, demobilization, cleanup, overhead, and profit. Note: These boxes are used with Romex (type NM) cable in residences. Attach grounding wire to the device grounding terminal. These boxes are installed flush with the wall surface. The box can either be nailed to a stud or fastened with wood screws.

Material	Craft@Hrs	Unit	Material Cost	Labor Cost	Installed Cost

One gang boxes with ears and speed clamps

1-1/8" x 2-1/4"	L1@0.15	Ea	2.60	6.15	8.75
1-1/2" x 2-3/8"	L1@0.15	Ea	1.90	6.15	8.05
1-3/4" x 2-3/8"	L1@0.15	Ea	2.14	6.15	8.29
2-1/2" x 2-3/8"	L1@0.15	Ea	2.60	6.15	8.75
2-7/8" x 2-3/8"	L1@0.15	Ea	2.33	6.15	8.48

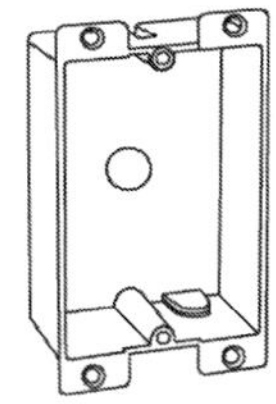

One gang boxes with side bracket

1-1/8" x 2-1/4"	L1@0.15	Ea	2.25	6.15	8.40

One gang boxes with ears and wire clamps

1-3/4" x 2-3/8"	L1@0.15	Ea	3.02	6.15	9.17
2-1/2" x 2-3/8"	L1@0.15	Ea	3.04	6.15	9.19
2-3/4" x 2-3/8"	L1@0.15	Ea	3.01	6.15	9.16
2-7/8" x 2-3/8"	L1@0.15	Ea	3.01	6.15	9.16

One gang boxes with ears, snap-in bracket and speed clamps

2-1/2" x 2-3/8"	L1@0.15	Ea	3.04	6.15	9.19
2-3/4" x 2-3/8"	L1@0.15	Ea	2.97	6.15	9.12
2-7/8" x 2-3/8"	L1@0.15	Ea	2.97	6.15	9.12

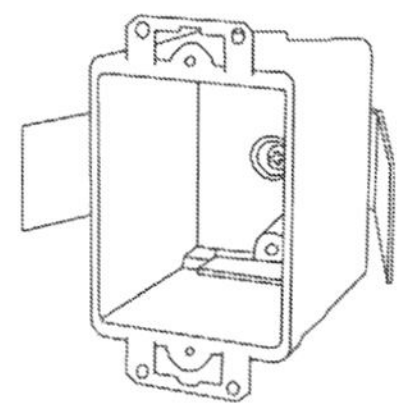

One gang boxes with ears, snap-in bracket and wire clamps

2-1/2" x 2-3/8"	L1@0.15	Ea	3.73	6.15	9.88
2-3/4" x 2-3/8"	L1@0.15	Ea	3.68	6.15	9.83
2-7/8" x 2-3/8"	L1@0.15	Ea	3.68	6.15	9.83

One gang boxes with Z-bracket, 1/4" offset

3" x 2-1/4"	L1@0.15	Ea	2.93	6.15	9.08
3-1/4" x 2-1/4"	L1@0.15	Ea	2.68	6.15	8.83

One gang boxes with Z-bracket, 1/2" offset

3" x 2-1/4"	L1@0.15	Ea	2.68	6.15	8.83
3-1/4" x 2-1/4"	L1@0.15	Ea	2.68	6.15	8.83

Use these figures to estimate the cost of fiberglass outlet boxes installed in residential buildings under the conditions described on pages 5 and 6. Costs listed are for each box installed. The crew is one electrician working at a labor cost of $41.03 per manhour. These costs include layout, material handling, and normal waste. Add for wiring devices, covers, sales tax, delivery, supervision, mobilization, demobilization, cleanup, overhead, and profit. Note: These boxes are used with Romex (type NM) cable in residences. Attach grounding wire to the device grounding terminal. These boxes are installed flush with the wall surface. The box can either be nailed to a stud or fastened with wood screws.

Fiberglass Outlet Switch Boxes

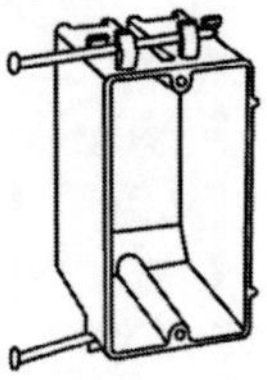

Material	Craft@Hrs	Unit	Material Cost	Labor Cost	Installed Cost
One gang boxes with nails					
2-1/4" x 2-1/4"	L1@0.15	Ea	1.24	6.15	7.39
3" x 2-1/4"	L1@0.15	Ea	.71	6.15	6.86
3-1/4" x 2-1/4"	L1@0.15	Ea	.73	6.15	6.88
3-1/4" x 2-3/8"	L1@0.15	Ea	1.39	6.15	7.54
One gang boxes with ears, wing bracket and speed clamps					
1-11/16" x 4-15/16"	L1@0.15	Ea	3.02	6.15	9.17
2-1/2" x 2-3/8"	L1@0.15	Ea	1.52	6.15	7.67
2-7/8" x 2-3/8"	L1@0.15	Ea	1.38	6.15	7.53
3-3/16" x 2-3/8"	L1@0.15	Ea	1.82	6.15	7.97
One gang boxes with ears, wing bracket and wire clamps					
2-7/8" x 2-3/8"	L1@0.15	Ea	2.17	6.15	8.32
Two gang boxes with ears and speed clamps					
2-1/2" x 4" square	L1@0.20	Ea	2.39	8.21	10.60
Two gang boxes with ears, wing bracket and speed clamps					
2-1/2" x 4" square	L1@0.20	Ea	2.96	8.21	11.17
Two gang boxes with ears, wing bracket and wire clamps					
2-1/2" x 4" square	L1@0.20	Ea	3.44	8.21	11.65
 **Two gang boxes with nails and speed clamps**					
3" x 4" square	L1@0.20	Ea	2.25	8.21	10.46
3-7/16" x 4" square	L1@0.20	Ea	2.53	8.21	10.74
Two gang boxes with nails and wire clamps					
3" x 4" square	L1@0.20	Ea	2.87	8.21	11.08
Two gang boxes with Z-bracket, 1/4" offset and speed clamps					
3" x 4" square	L1@0.20	Ea	3.09	8.21	11.30

Use these figures to estimate the cost of fiberglass outlet boxes installed in residential buildings under the conditions described on pages 5 and 6. Costs listed are for each box installed. The crew is one electrician working at a labor cost of $41.03 per manhour. These costs include layout, material handling, and normal waste. Add for wiring devices, covers, sales tax, delivery, supervision, mobilization, demobilization, cleanup, overhead, and profit. Note: These boxes are used with Romex (type NM) cable in residences. Attach grounding wire to the device grounding terminal. These boxes are installed flush with the wall surface. The box can either be nailed to a stud or fastened with wood screws.

Fiberglass Outlet Switch Boxes

Material	Craft@Hrs	Unit	Material Cost	Labor Cost	Installed Cost

Three gang boxes with nails and wire clamps

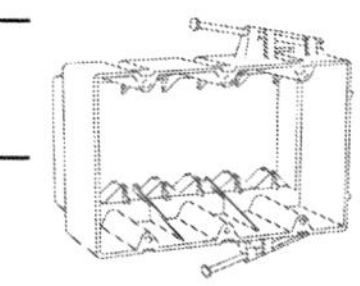

3" x 5-11/16" wide	L1@0.25	Ea	3.90	10.30	14.20
3-9/16" x 5-11/16 wide	L1@0.25	Ea	3.52	10.30	13.82

Three gang boxes with nails and speed clamps

3" x 5-11/16" wide	L1@0.25	Ea	3.44	10.30	13.74
3-9/16" x 5-11/16" wide	L1@0.25	Ea	3.63	10.30	13.93

Three gang boxes with nails, adjustable stabilizing bar and wire clamps

3" x 5-11/16" wide	L1@0.25	Ea	5.01	10.30	15.31

Three gang boxes with nails, adjustable stabilizing bar and speed clamps

3" x 5 11/16" wide	L1@0.25	Ea	4.39	10.30	14.69
3-9/16" x 5-11/16" wide	L1@0.25	Ea	4.61	10.30	14.91

Three gang boxes with Z-bracket, 1/4" offset and speed clamps

3" x 5-11/16" wide	L1@0.25	Ea	3.93	10.30	14.23

Three gang boxes with Z-bracket, 1/2" offset and speed clamps

3" x 5-11/16" wide	L1@0.25	Ea	4.91	10.30	15.21

Four gang boxes with nails and wire clamps

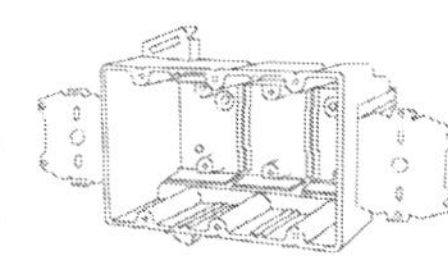

3" x 7-1/2" wide	L1@0.30	Ea	5.99	12.30	18.29

Four gang boxes with nails and speed clamps

3" x 7-1/2" wide	L1@0.30	Ea	4.86	12.30	17.16

Four gang boxes with nails, adjustable stabilizing bar and wire clamps

3" x 7-1/2" wide	L1@0.30	Ea	6.84	12.30	19.14

Four gang boxes with nails, adjustable stabilizing bar and speed clamps

3" x 7-1/2" wide	L1@0.30	Ea	5.62	12.30	17.92

Four gang boxes with Z-bracket, 1/4" offset, adjustable stabilizing bar and speed clamps

3" x 7-1/5" wide	L1@0.30	Ea	5.40	12.30	17.70

Use these figures to estimate the cost of fiberglass outlet boxes installed in residential buildings under the conditions described on pages 5 and 6. Costs listed are for each box installed. The crew is one electrician working at a labor cost of $41.03 per manhour. These costs include layout, material handling, and normal waste. Add for wiring devices, covers, sales tax, delivery, supervision, mobilization, demobilization, cleanup, overhead, and profit. Note: These boxes are used with Romex (type NM) cable in residences. Attach grounding wire to the device grounding terminal. These boxes are installed flush with the wall surface. The box can either be nailed to a stud or fastened with wood screws.

Plastic Boxes

Material	Craft@Hrs	Unit	Material Cost	Labor Cost	Installed Cost
Round boxes with nails					
3" x 2-3/4" deep	L1@0.20	Ea	2.45	8.21	10.66
4" x 2-3/8"	L1@0.20	Ea	2.45	8.21	10.66
Round boxes with nails, and ground lug					
3" x 2-3/4" deep	L1@0.20	Ea	3.05	8.21	11.26
4" x 2-3/8" deep	L1@0.20	Ea	3.18	8.21	11.39
Round boxes with L bracket					
4" x 2-3/8" deep	L1@0.20	Ea	5.38	8.21	13.59
Round boxes with 16" adjustable bar hanger					
4" x 2-3/8" deep	L1@0.25	Ea	5.48	10.30	15.78
Round boxes with 16" adjustable bar hanger and ground lug					
4" x 2-3/8" deep	L1@0.25	Ea	5.96	10.30	16.26
Round boxes with 24" adjustable bar hanger					
4" x 2-3/8" deep	L1@0.25	Ea	5.48	10.30	15.78
Round boxes with 24" adjustable bar hanger and ground lug					
4" x 2-3/8" deep	L1@0.25	Ea	5.77	10.30	16.07
Round boxes with zip-mount retainers for old work					
3" x 2-3/4" deep	L1@0.30	Ea	4.28	12.30	16.58
Square outlet boxes with nails					
4" x 1-5/8" deep	L1@0.20	Ea	3.65	8.21	11.86
Square outlet boxes with nails and ground lug					
4" x 2-5/8" deep	L1@0.20	Ea	4.57	8.21	12.78
Square boxes with stud mounting bracket					
4" x 1-1/2" deep	L1@0.20	Ea	3.20	8.21	11.41

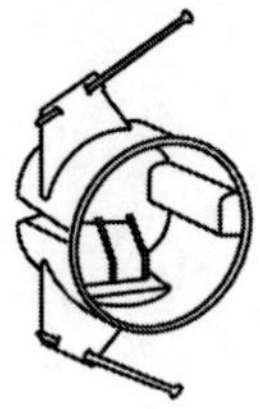

Use these figures to estimate the cost of plastic boxes installed on plastic conduit under the conditions described on pages 5 and 6. Costs listed are for each box installed. The crew is one electrician working at a labor cost of $41.03 per manhour. These costs include layout, material, and normal waste. Add for sales tax, delivery, supervision, mobilization, demobilization, cleanup, overhead, and profit.

Material	Craft@Hrs	Unit	Material Cost	Labor Cost	Installed Cost
Square boxes with stud mounting bracket					
4" x 2-3/8" deep	L1@0.20	Ea	4.69	8.21	12.90
Square boxes with nails					
4" x 1-1/2" deep	L1@0.20	Ea	2.49	8.21	10.70
4" x 2-3/8" deep	L1@0.20	Ea	3.88	8.21	12.09
One gang switch boxes 2-1/4" wide with nails					
3-3/4" x 2-7/8 deep	L1@0.15	Ea	.81	6.15	6.96
One gang switch boxes 2-1/4" wide with nails, OK for masonry walls					
3-3/4" x 2-7/8" deep	L1@0.15	Ea	.81	6.15	6.96
One gang switch boxes 2-1/4" wide with nails and bracket					
3-3/4" x 2-7/8" deep	L1@0.15	Ea	3.43	6.15	9.58
Two gang switch boxes 4" wide with nails					
2-7/8" deep	L1@0.20	Ea	2.50	8.21	10.71
Two gang switch boxes 4" wide with nails, OK for masonry walls					
2-7/8" deep	L1@0.20	Ea	2.50	8.21	10.71
Two gang switch boxes 4" wide with nails and clips					
3-3/4" x 3" deep	L1@0.20	Ea	7.78	8.21	15.99
Two gang switch boxes 4" wide with lateral bracket, 3/8" offset					
3-3/4" x 3-3/32" deep	L1@0.20	Ea	8.15	8.21	16.36
Two gang switch boxes 4" wide with nails and clamps					
2-7/8" deep	L1@0.20	Ea	2.49	8.21	10.70

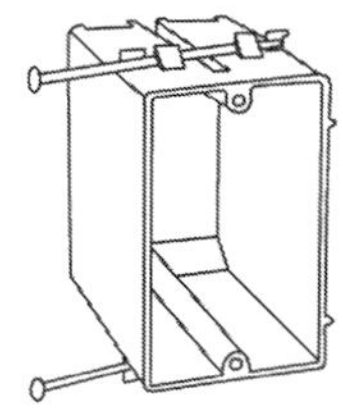

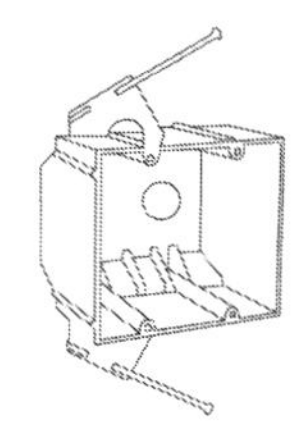

Use these figures to estimate the cost of plastic boxes installed on plastic conduit under the conditions described on pages 5 and 6. Costs listed are for each box installed. The crew is one electrician working at a labor cost of $41.03 per manhour. These costs include layout, material, and normal waste. Add for sales tax, delivery, supervision, mobilization, demobilization, cleanup, overhead, and profit.

Plastic Boxes

Material	Craft@Hrs	Unit	Material Cost	Labor Cost	Installed Cost
Three gang switch boxes 5-13/16" wide with nails and clamps					
3-3/4" x 2-3/4" deep	L1@0.25	Ea	7.47	10.30	17.77
Three gang switch boxes 5-13/16" wide with bracket and clamps					
3-3/4" x 2-3/4" deep	L1@0.25	Ea	12.00	10.30	22.30
Four gang switch boxes 7-9/16" wide with nails and bracket					
3-3/4" x 2-3/4" deep	L1@0.30	Ea	18.30	12.30	30.60
Four gang switch boxes 7-9/16" wide with nails and lateral offset bracket					
3-3/4" x 2-3/4" deep	L1@0.30	Ea	24.00	12.30	36.30
Old work switch boxes					
1-gang, 2 clamps	L1@0.25	Ea	2.76	10.30	13.06
2-gang, 4 clamps	L1@0.30	Ea	5.14	12.30	17.44
One gang switch rings 4" square					
1/2" raised	L1@0.05	Ea	2.18	2.05	4.23
5/8" raised	L1@0.05	Ea	2.45	2.05	4.50
3/4" raised	L1@0.05	Ea	2.76	2.05	4.81
1" raised	L1@0.05	Ea	3.05	2.05	5.10
1-1/4" raised	L1@0.05	Ea	3.60	2.05	5.65
Two gang switch rings 4" square					
1/2" raised	L1@0.05	Ea	3.12	2.05	5.17
5/8" raised	L1@0.05	Ea	3.38	2.05	5.43
3/4" raised	L1@0.05	Ea	3.63	2.05	5.68
Blank covers					
4" round	L1@0.05	Ea	1.46	2.05	3.51
4" square	L1@0.05	Ea	1.34	2.05	3.39

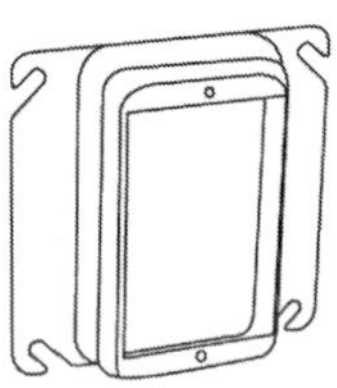

Use these figures to estimate the cost of plastic boxes and covers installed on plastic conduit under the conditions described on pages 5 and 6. Costs listed are for each box or cover installed. The crew is one electrician working at a labor cost of $41.03 per manhour. These costs include layout, material, and normal waste. Add for sales tax, delivery, supervision, mobilization, demobilization, cleanup, overhead, and profit.

Material	Craft@Hrs	Unit	Material Cost	Labor Cost	Installed Cost
3" round cast weatherproof boxes with four 1/2" hubs and mounting feet					
Less cover	L1@0.25	Ea	13.70	10.30	24.00
With blank cover	L1@0.30	Ea	22.30	12.30	34.60
One hub in cover	L1@0.30	Ea	23.50	12.30	35.80
Three holes in cover	L1@0.30	Ea	27.00	12.30	39.30
3" round cast aluminum cover					
Cover	L1@0.05	Ea	6.43	2.05	8.48
4" round cast weatherproof boxes less cover, with mounting feet					
Five 1/2" hubs	L1@0.25	Ea	14.00	10.30	24.30
Six 1/2" hubs	L1@0.30	Ea	37.30	12.30	49.60
4" round cast weatherproof boxes with blank cover and mounting feet					
Five 1/2" hubs	L1@0.30	Ea	20.60	12.30	32.90
Six 1/2" hubs	L1@0.30	Ea	44.40	12.30	56.70
4" round cast weatherproof boxes with one 1/2" hub in cover and mounting feet					
Five 1/2" hubs	L1@0.30	Ea	22.70	12.30	35.00
4" round cast weatherproof boxes with three 1/2" hubs in cover and mounting feet					
Five 1/2" hubs	L1@0.30	Ea	24.30	12.30	36.60
4" round cast weatherproof boxes with mounting feet, less cover					
Five 3/4" hubs	L1@0.25	Ea	23.40	10.30	33.70
4" round cast weatherproof boxes with mounting feet with blank cover					
Five 3/4" hubs	L1@0.30	Ea	29.50	12.30	41.80
4" round cast weatherproof boxes with mounting feet and 3/4" hub in cover					
Five 3/4" hubs	L1@0.30	Ea	30.70	12.30	43.00

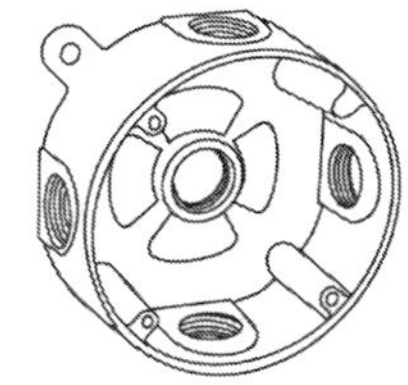

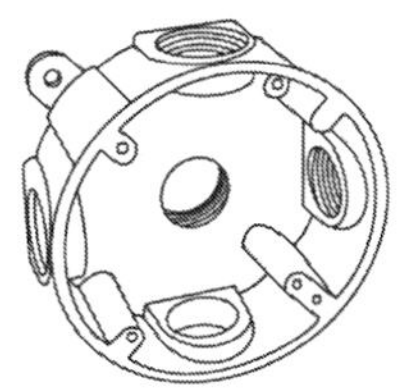

Use these figures to estimate the cost of cast aluminum boxes and covers installed under the conditions described on pages 5 and 6. Costs listed are for each box or cover installed. The crew is one electrician working at a labor cost of $41.03 per manhour. These costs include layout, material, and normal waste. Add for sales tax, delivery, supervision, mobilization, demobilization, cleanup, overhead, and profit.

Cast Aluminum Boxes

4" round cast weatherproof covers

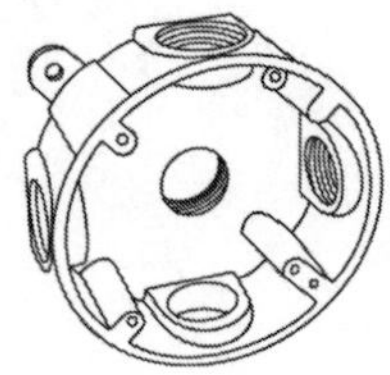

Material	Craft@Hrs	Unit	Material Cost	Labor Cost	Installed Cost
Blank	L1@0.05	Ea	5.94	2.05	7.99
1/2" hub	L1@0.05	Ea	6.75	2.05	8.80
3/4" hub	L1@0.05	Ea	10.40	2.05	12.45
Two 1/2" hubs	L1@0.05	Ea	8.34	2.05	10.39
Three 1/2" hubs	L1@0.05	Ea	9.27	2.05	11.32

One gang cast weatherproof boxes

Material	Craft@Hrs	Unit	Material Cost	Labor Cost	Installed Cost
Three 1/2" hubs	L1@0.20	Ea	9.29	8.21	17.50
Four 1/2" hubs	L1@0.20	Ea	11.70	8.21	19.91
Five 1/2" hubs	L1@0.20	Ea	14.90	8.21	23.11
Three 3/4" hubs	L1@0.25	Ea	11.20	10.30	21.50
Four 3/4" hubs	L1@0.25	Ea	12.10	10.30	22.40
Five 3/4" hubs	L1@0.25	Ea	15.90	10.30	26.20

One gang cast weatherproof boxes with mounting feet

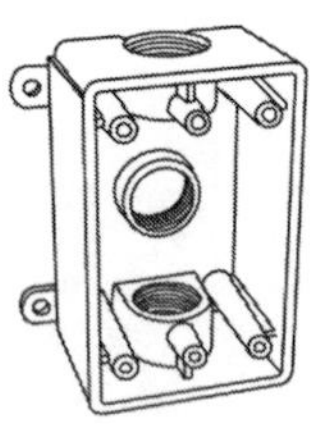

Material	Craft@Hrs	Unit	Material Cost	Labor Cost	Installed Cost
Three 1/2" hubs	L1@0.20	Ea	9.90	8.21	18.11
Four 1/2" hubs	L1@0.20	Ea	12.00	8.21	20.21
Five 1/2" hubs	L1@0.20	Ea	15.30	8.21	23.51
Three 3/4" hubs	L1@0.25	Ea	11.50	10.30	21.80
Four 3/4" hubs	L1@0.25	Ea	13.00	10.30	23.30
Five 3/4" hubs	L1@0.25	Ea	16.70	10.30	27.00

One gang extension rings, no hubs

Material	Craft@Hrs	Unit	Material Cost	Labor Cost	Installed Cost
3/4" deep	L1@0.10	Ea	15.30	4.10	19.40
1" deep	L1@0.10	Ea	17.50	4.10	21.60

One gang extension rings, 1-1/2" deep

Material	Craft@Hrs	Unit	Material Cost	Labor Cost	Installed Cost
Four 1/2" hubs	L1@0.15	Ea	18.00	6.15	24.15
Four 3/4" hubs	L1@0.15	Ea	26.40	6.15	32.55

Use these figures to estimate the cost of cast aluminum boxes, covers and rings installed under the conditions described on pages 5 and 6. Costs listed are for each box, cover, or ring installed. The crew is one electrician working at a labor cost of $41.03 per manhour. These costs include layout, material, and normal waste. Add for sales tax, delivery, supervision, mobilization, demobilization, cleanup, overhead, and profit.

Material	Craft@Hrs	Unit	Material Cost	Labor Cost	Installed Cost
Two gang cast weatherproof boxes					
Three 1/2" hubs	L1@0.25	Ea	23.00	10.30	33.30
Four 1/2" hubs	L1@0.25	Ea	23.80	10.30	34.10
Five 1/2" hubs	L1@0.25	Ea	23.70	10.30	34.00
Six 1/2" hubs	L1@0.25	Ea	28.70	10.30	39.00
Seven 1/2" hubs	L1@0.25	Ea	33.90	10.30	44.20
Three 3/4" hubs	L1@0.30	Ea	23.70	12.30	36.00
Four 3/4" hubs	L1@0.30	Ea	24.90	12.30	37.20
Five 3/4" hubs	L1@0.30	Ea	26.60	12.30	38.90
Six 3/4" hubs	L1@0.30	Ea	33.70	12.30	46.00
Seven 3/4" hubs	L1@0.30	Ea	59.30	12.30	71.60
Two gang extension rings					
No hubs	L1@0.10	Ea	26.40	4.10	30.50

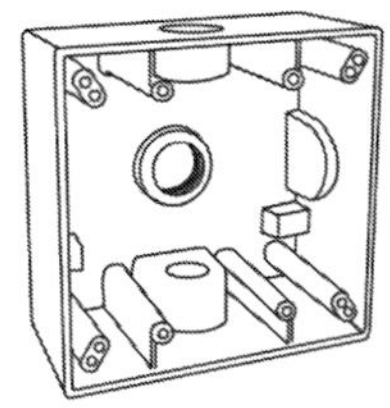

Use these figures to estimate the cost of aluminum boxes and rings installed under the conditions described on pages 5 and 6. Costs listed are for each box or ring installed. The crew is one electrician working at a labor cost of $41.03 per manhour. These costs include layout, material, and normal waste. Add for sales tax, delivery, supervision, mobilization, demobilization, cleanup, overhead, and profit.

Sheet Metal Pull Boxes

Material	Craft@Hrs	Unit	Material Cost	Labor Cost	Installed Cost
NEMA 1 surface or flush mounted screw cover pull boxes					
4 x 4 x 4	L1@0.25	Ea	20.50	10.30	30.80
6 x 6 x 3	L1@0.30	Ea	35.00	12.30	47.30
6 x 6 x 4	L1@0.35	Ea	27.50	14.40	41.90
8 x 6 x 4	L1@0.40	Ea	33.30	16.40	49.70
8 x 8 x 4	L1@0.45	Ea	37.40	18.50	55.90
10 x 8 x 4	L1@0.50	Ea	43.30	20.50	63.80
10 x 10 x 4	L1@0.55	Ea	52.00	22.60	74.60
12 x 6 x 4	L1@0.45	Ea	75.40	18.50	93.90
12 x 8 x 4	L1@0.50	Ea	52.00	20.50	72.50
12 x 10 x 4	L1@0.55	Ea	57.80	22.60	80.40
12 x 12 x 4	L1@0.55	Ea	63.60	22.60	86.20
16 x 12 x 4	L1@0.60	Ea	79.50	24.60	104.10
18 x 12 x 4	L1@0.60	Ea	97.60	24.60	122.20
18 x 18 x 4	L1@0.70	Ea	133.00	28.70	161.70
24 x 12 x 4	L1@0.70	Ea	136.00	28.70	164.70
24 x 18 x 4	L1@0.75	Ea	209.00	30.80	239.80
24 x 24 x 4	L1@0.80	Ea	268.00	32.80	300.80
6 x 6 x 6	L1@0.40	Ea	34.60	16.40	51.00
8 x 6 x 6	L1@0.45	Ea	37.40	18.50	55.90
8 x 8 x 6	L1@0.50	Ea	46.10	20.50	66.60
10 x 8 x 6	L1@0.50	Ea	53.50	20.50	74.00
10 x 10 x 6	L1@0.55	Ea	59.30	22.60	81.90
12 x 6 x 6	L1@0.60	Ea	58.10	24.60	82.70
12 x 8 x 6	L1@0.55	Ea	72.70	22.60	95.30
12 x 10 x 6	L1@0.60	Ea	73.70	24.60	98.30
12 x 12 x 6	L1@0.60	Ea	74.70	24.60	99.30

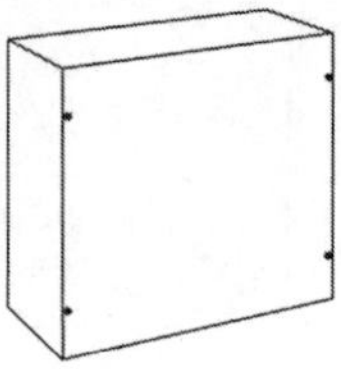

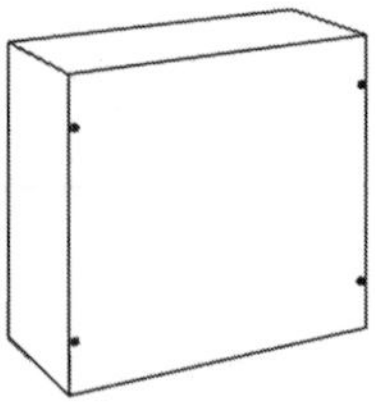

Use these figures to estimate the cost of sheet metal pull boxes installed on conduit under the conditions described on pages 5 and 6. Costs listed are for each pull box installed. The crew is one electrician working at a labor cost of $41.03 per manhour. These costs include the cover, layout, material handling, and normal waste. Add for supports, sales tax, delivery, supervision, mobilization, demobilization, cleanup, overhead, and profit. Note: Pull boxes are used as junction boxes when smaller boxes would be too crowded with wire and when a long run must be divided into two or more pulls. Pull boxes are also used when conduit runs exceed 360 degrees of bends.

Material	Craft@Hrs	Unit	Material Cost	Labor Cost	Installed Cost
NEMA 1 surface or flush mounted screw cover pull boxes					
16 x 12 x 6	L1@0.65	Ea	104.00	26.70	130.70
16 x 16 x 6	L1@0.75	Ea	126.00	30.80	156.80
18 x 12 x 6	L1@0.70	Ea	117.00	28.70	145.70
18 x 18 x 6	L1@0.75	Ea	123.00	30.80	153.80
24 x 12 x 6	L1@0.75	Ea	117.00	30.80	147.80
24 x 18 x 6	L1@0.80	Ea	185.00	32.80	217.80
24 x 24 x 6	L1@1.00	Ea	258.00	41.00	299.00
36 x 24 x 6	L1@1.25	Ea	535.00	51.30	586.30
8 x 8 x 8	L1@0.55	Ea	66.70	22.60	89.30
12 x 12 x 8	L1@0.65	Ea	92.40	26.70	119.10
18 x 12 x 8	L1@0.75	Ea	97.60	30.80	128.40
18 x 18 x 8	L1@0.80	Ea	219.00	32.80	251.80
24 x 12 x 8	L1@0.80	Ea	207.00	32.80	239.80
24 x 18 x 8	L1@1.00	Ea	153.00	41.00	194.00
24 x 24 x 8	L1@1.25	Ea	286.00	51.30	337.30
18 x 18 x 10	L1@1.00	Ea	215.00	41.00	256.00
24 x 24 x 10	L1@1.50	Ea	399.00	61.50	460.50
18 x 18 x 12	L1@1.00	Ea	222.00	41.00	263.00
24 x 18 x 12	L1@1.50	Ea	526.00	61.50	587.50
24 x 24 x 12	L1@1.75	Ea	603.00	71.80	674.80
NEMA 1 hinge cover pull boxes					
6 x 6 x 4	L1@0.35	Ea	26.70	14.40	41.10
8 x 8 x 4	L1@0.45	Ea	44.10	18.50	62.60
10 x 10 x 4	L1@0.50	Ea	58.10	20.50	78.60
12 x 10 x 4	L1@0.55	Ea	65.70	22.60	88.30
12 x 12 x 4	L1@0.55	Ea	74.70	22.60	97.30

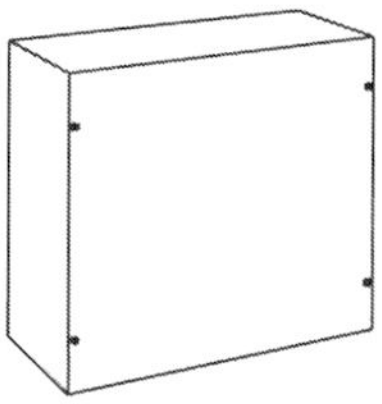

Use these figures to estimate the cost of sheet metal pull boxes installed on conduit under the conditions described on pages 5 and 6. Costs listed are for each pull box installed. The crew is one electrician working at a labor cost of $41.03 per manhour. These costs include the cover, layout, material handling, and normal waste. Add for supports, sales tax, delivery, supervision, mobilization, demobilization, cleanup, overhead, and profit. Note: Pull boxes are used as junction boxes when smaller boxes would be too crowded with wire and when a long run must be divided into two or more pulls. Pull boxes are also used when conduit runs exceed 360 degrees of bends.

Sheet Metal Pull Boxes

Material	Craft@Hrs	Unit	Material Cost	Labor Cost	Installed Cost
NEMA 1 hinge cover pull boxes					
18 x 12 x 4	L1@0.65	Ea	100.00	26.70	126.70
6 x 6 x 6	L1@0.40	Ea	39.10	16.40	55.50
8 x 8 x 6	L1@0.50	Ea	52.90	20.50	73.40
12 x 12 x 6	L1@0.60	Ea	87.30	24.60	111.90
18 x 12 x 6	L1@0.70	Ea	117.00	28.70	145.70
NEMA 3R raintight screw cover pull boxes					
4 x 4 x 4	L1@0.30	Ea	42.10	12.30	54.40
6 x 4 x 4	L1@0.35	Ea	49.80	14.40	64.20
6 x 6 x 4	L1@0.35	Ea	52.10	14.40	66.50
8 x 6 x 4	L1@0.40	Ea	73.70	16.40	90.10
8 x 8 x 4	L1@0.45	Ea	78.00	18.50	96.50
10 x 8 x 4	L1@0.50	Ea	80.90	20.50	101.40
10 x 10 x 4	L1@0.50	Ea	47.10	20.50	67.60
12 x 8 x 4	L1@0.50	Ea	90.90	20.50	111.40
12 x 10 x 4	L1@0.55	Ea	94.10	22.60	116.70
12 x 12 x 4	L1@0.55	Ea	106.00	22.60	128.60
16 x 12 x 4	L1@0.50	Ea	172.00	20.50	192.50
18 x 12 x 4	L1@0.60	Ea	182.00	24.60	206.60
18 x 18 x 4	L1@0.70	Ea	210.00	28.70	238.70
24 x 12 x 4	L1@0.75	Ea	256.00	30.80	286.80
6 x 6 x 6	L1@0.40	Ea	55.90	16.40	72.30
8 x 6 x 6	L1@0.45	Ea	72.20	18.50	90.70
8 x 8 x 6	L1@0.50	Ea	81.90	20.50	102.40
10 x 8 x 6	L1@0.50	Ea	92.80	20.50	113.30
10 x 10 x 6	L1@0.55	Ea	105.00	22.60	127.60

Use these figures to estimate the cost of sheet metal pull boxes installed on conduit under the conditions described on pages 5 and 6. Costs listed are for each pull box installed. The crew is one electrician working at a labor cost of $41.03 per manhour. These costs include the cover, layout, material handling, and normal waste. Add for supports, sales tax, delivery, supervision, mobilization, demobilization, cleanup, overhead, and profit. Note: Pull boxes are used as junction boxes when smaller boxes would be too crowded with wire and when a long run must be divided into two or more pulls. Pull boxes are also used when conduit runs exceed 360 degrees of bends.

Material	Craft@Hrs	Unit	Material Cost	Labor Cost	Installed Cost

NEMA 3R raintight screw cover pull boxes

Material	Craft@Hrs	Unit	Material Cost	Labor Cost	Installed Cost
12 x 10 x 6	L1@0.60	Ea	114.00	24.60	138.60
12 x 12 x 6	L1@0.60	Ea	126.00	24.60	150.60
16 x 12 x 6	L1@0.65	Ea	167.00	26.70	193.70
18 x 12 x 6	L1@0.70	Ea	181.00	28.70	209.70
18 x 18 x 6	L1@0.75	Ea	257.00	30.80	287.80
24 x 12 x 6	L1@0.75	Ea	201.00	30.80	231.80
24 x 18 x 6	L1@0.80	Ea	208.00	32.80	240.80
24 x 24 x 6	L1@1.00	Ea	209.00	41.00	250.00
12 x 12 x 8	L1@0.65	Ea	146.00	26.70	172.70
18 x 12 x 8	L1@0.75	Ea	188.00	30.80	218.80
24 x 16 x 8	L1@0.90	Ea	241.00	36.90	277.90
24 x 18 x 8	L1@1.00	Ea	307.00	41.00	348.00
24 x 24 x 8	L1@1.25	Ea	430.00	51.30	481.30
18 x 18 x 10	L1@1.00	Ea	143.00	41.00	184.00
24 x 18 x 10	L1@1.25	Ea	232.00	51.30	283.30
24 x 24 x 10	L1@1.40	Ea	477.00	57.40	534.40
18 x 18 x 12	L1@1.25	Ea	338.00	51.30	389.30
24 x 18 x 12	L1@1.75	Ea	491.00	71.80	562.80
24 x 24 x 12	L1@2.00	Ea	508.00	82.10	590.10

NEMA 3R raintight hinge cover pull boxes

Material	Craft@Hrs	Unit	Material Cost	Labor Cost	Installed Cost
6 x 6 x 4	L1@0.35	Ea	45.30	14.40	59.70
8 x 8 x 4	L1@0.45	Ea	57.80	18.50	76.30
10 x 10 x 4	L1@0.50	Ea	72.20	20.50	92.70
12 x 12 x 4	L1@0.55	Ea	147.00	22.60	169.60
12 x 12 x 6	L1@0.65	Ea	184.00	26.70	210.70
18 x 12 x 6	L1@0.70	Ea	241.00	28.70	269.70

Use these figures to estimate the cost of sheet metal pull boxes installed on conduit under the conditions described on pages 5 and 6. Costs listed are for each pull box installed. The crew is one electrician working at a labor cost of $41.03 per manhour. These costs include the cover, layout, material handling, and normal waste. Add for supports, sales tax, delivery, supervision, mobilization, demobilization, cleanup, overhead, and profit. Note: Pull boxes are used as junction boxes when smaller boxes would be too crowded with wire and when a long run must be divided into two or more pulls. Pull boxes are also used when conduit runs exceed 360 degrees of bends.

Sheet Metal Pull Boxes

Material	Craft@Hrs	Unit	Material Cost	Labor Cost	Installed Cost
NEMA 3R raintight hinge cover pull boxes					
24 x 24 x 6	L1@1.00	Ea	555.00	41.00	596.00
18 x 18 x 8	L1@0.80	Ea	301.00	32.80	333.80
24 x 24 x 10	L1@1.50	Ea	579.00	61.50	640.50
NEMA 4 hinged clamping cover JIC wiring boxes, no panels					
4 x 4 x 3	L1@0.30	Ea	92.50	12.30	104.80
6 x 6 x 3	L1@0.40	Ea	105.00	16.40	121.40
6 x 4 x 4	L1@0.30	Ea	109.00	12.30	121.30
6 x 6 x 4	L1@0.35	Ea	116.00	14.40	130.40
8 x 6 x 3.5	L1@0.40	Ea	119.00	16.40	135.40
10 x 8 x 4	L1@0.60	Ea	157.00	24.60	181.60
12 x 10 x 5	L1@0.70	Ea	187.00	28.70	215.70
14 x 12 x 6	L1@0.80	Ea	237.00	32.80	269.80
16 x 14 x 6	L1@0.90	Ea	294.00	36.90	330.90
NEMA 12 lift-off cover JIC wiring boxes					
4 x 4 x 3	L1@0.40	Ea	72.40	16.40	88.80
4 x 4 x 4	L1@0.45	Ea	74.80	18.50	93.30
6 x 4 x 3	L1@0.50	Ea	82.80	20.50	103.30
6 x 4 x 4	L1@0.50	Ea	86.50	20.50	107.00
6 x 6 x 4	L1@0.50	Ea	90.30	20.50	110.80
8 x 6 x 3.5	L1@0.55	Ea	106.00	22.60	128.60
10 x 8 x 4	L1@0.70	Ea	116.00	28.70	144.70
12 x 10 x 5	L1@0.80	Ea	155.00	32.80	187.80
14 x 12 x 6	L1@0.90	Ea	192.00	36.90	228.90
16 x 14 x 6	L1@1.00	Ea	431.00	41.00	472.00

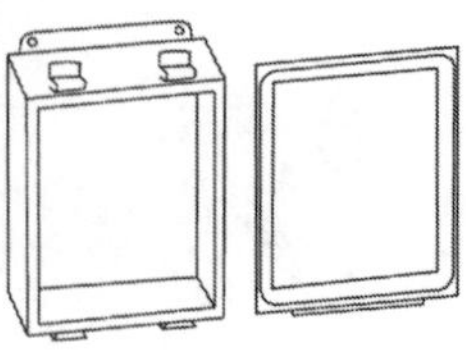

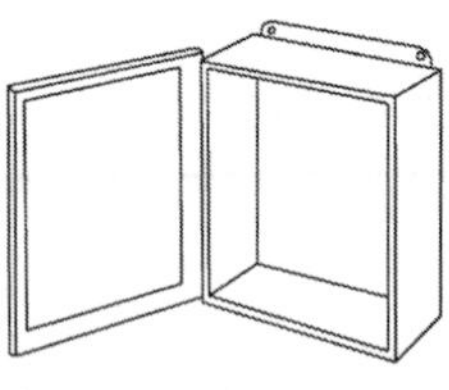

Use these figures to estimate the cost of sheet metal pull boxes and JIC wiring boxes installed on conduit under the conditions described on pages 5 and 6. Costs listed are for each pull box installed. The crew is one electrician working at a labor cost of $41.03 per manhour. These costs include the cover, layout, material handling, and normal waste. Add for supports, sales tax, delivery, supervision, mobilization, demobilization, cleanup, overhead, and profit. Note: Pull boxes are used as junction boxes when smaller boxes would be too crowded with wire and when a long run must be divided into two or more pulls. Pull boxes are also used when conduit runs exceed 360 degrees of bend.

NEMA 12 hinge cover JIC wiring boxes, no panels

Material	Craft@Hrs	Unit	Material Cost	Labor Cost	Installed Cost
4 x 4 x 3	L1@0.40	Ea	95.20	16.40	111.60
4 x 4 x 4	L1@0.40	Ea	103.00	16.40	119.40
6 x 4 x 3	L1@0.50	Ea	111.00	20.50	131.50
6 x 4 x 4	L1@0.50	Ea	116.00	20.50	136.50
6 x 6 x 4	L1@0.45	Ea	123.00	18.50	141.50
8 x 6 x 3.5	L1@0.50	Ea	137.00	20.50	157.50
8 x 6 x 6	L1@0.50	Ea	159.00	20.50	179.50
8 x 8 x 4	L1@0.50	Ea	155.00	20.50	175.50
10 x 8 x 4	L1@0.70	Ea	174.00	28.70	202.70
10 x 8 x 6	L1@0.70	Ea	184.00	28.70	212.70
10 x 10 x 6	L1@0.70	Ea	221.00	28.70	249.70
12 x 6 x 4	L1@0.80	Ea	174.00	32.80	206.80
12 x 10 x 5	L1@0.80	Ea	223.00	32.80	255.80
12 x 12 x 6	L1@0.80	Ea	249.00	32.80	281.80
14 x 12 x 6	L1@0.90	Ea	279.00	36.90	315.90
16 x 14 x 6	L1@1.00	Ea	324.00	41.00	365.00
16 x 14 x 8	L1@1.00	Ea	359.00	41.00	400.00

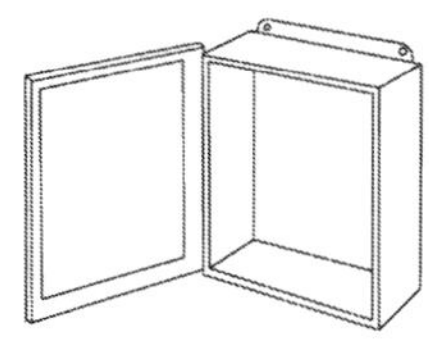

NEMA 12 hinged quick-release cover JIC wiring boxes, no panels

Material	Craft@Hrs	Unit	Material Cost	Labor Cost	Installed Cost
6 x 4 x 3	L1@0.40	Ea	130.00	16.40	146.40
6 x 6 x 4	L1@0.40	Ea	149.00	16.40	165.40
8 x 6 x 3.5	L1@0.45	Ea	163.00	18.50	181.50
10 x 8 x 4	L1@0.60	Ea	202.00	24.60	226.60
14 x 12 x 6	L1@0.80	Ea	320.00	32.80	352.80
16 x 14 x 6	L1@0.90	Ea	368.00	36.90	404.90

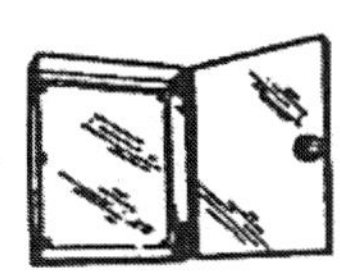

Use these figures to estimate the cost of sheet metal JIC wiring boxes installed on conduit under the conditions described on pages 5 and 6. Costs listed are for each JIC wiring box installed. The crew is one electrician working at a labor cost of $41.03 per manhour. These costs include the cover, layout, material handling, and normal waste. Add for supports, sales tax, delivery, supervision, mobilization, demobilization, cleanup, overhead, and profit.

Material	Craft@Hrs	Unit	Material Cost	Labor Cost	Installed Cost
Panels only for JIC enclosures					
4.87 x 2.87	L1@0.05	Ea	6.31	2.05	8.36
4.87 x 4.87	L1@0.05	Ea	6.64	2.05	8.69
6.87 x 4.87	L1@0.05	Ea	7.40	2.05	9.45
6.87 x 6.87	L1@0.05	Ea	8.64	2.05	10.69
8.87 x 6.87	L1@0.05	Ea	9.35	2.05	11.40
8.87 x 8.87	L1@0.05	Ea	11.60	2.05	13.65
10.87 x 8.87	L1@0.06	Ea	13.50	2.46	15.96
10.87 x 10.87	L1@0.06	Ea	16.20	2.46	18.66
12.87 x 10.87	L1@0.08	Ea	18.50	3.28	21.78
14.87 x 12.87	L1@0.08	Ea	24.80	3.28	28.08

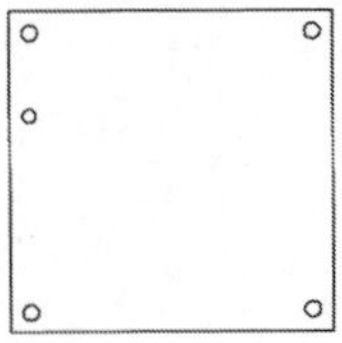

Use these figures to estimate the cost of panels installed in JIC wiring boxes under the conditions described on pages 5 and 6. Costs listed are for each panel installed. The crew is one electrician working at a cost of $41.03 per manhour. These costs include layout, material handling, and normal waste. Add for sales tax, delivery, supervision, mobilization, demobilization, cleanup, overhead, and profit.

Adjustable sheet metal round floor outlet boxes

Opening	Depth	Craft@Hrs	Unit	Material Cost	Labor Cost	Installed Cost
4-3/16"	3"	L1@1.00	Ea	121.00	41.00	162.00
4-3/16"	4"	L1@1.00	Ea	150.00	41.00	191.00

Semi-adjustable sheet metal round floor outlet boxes

Opening	Depth	Craft@Hrs	Unit	Material Cost	Labor Cost	Installed Cost
3-5/16"	2-1/2"	L1@1.00	Ea	126.00	41.00	167.00
3-5/16"	3"	L1@1.00	Ea	159.00	41.00	200.00

Adjustable cast iron round floor outlet boxes

Opening	Depth	Craft@Hrs	Unit	Material Cost	Labor Cost	Installed Cost
4-3/16"	2-5/8"	L1@1.00	Ea	134.00	41.00	175.00
4-3/16"	2-13/16"	L1@1.00	Ea	94.80	41.00	135.80
4-3/16"	3"	L1@1.00	Ea	172.00	41.00	213.00
4-3/16"	3-1/16"	L1@1.00	Ea	108.00	41.00	149.00
4-3/16	3-3/4"	L1@1.00	Ea	187.00	41.00	228.00
4-3/16"	4-3/4"	L1@1.25	Ea	306.00	51.30	357.30

Semi-adjustable cast iron round floor outlet boxes

Opening	Depth	Craft@Hrs	Unit	Material Cost	Labor Cost	Installed Cost
3-3/4"	2"	L1@1.00	Ea	159.00	41.00	200.00

Non-adjustable cast iron round floor outlet boxes

Opening	Depth	Craft@Hrs	Unit	Material Cost	Labor Cost	Installed Cost
4-3/16"	3-3/4"	L1@1.00	Ea	106.00	41.00	147.00

Use these figures to estimate the cost of floor boxes installed in building floors under the conditions described on pages 5 and 6. Costs listed are for each floor box installed. The crew is one electrician working at a labor cost of $41.03 per manhour. These costs include layout, material handling, and normal waste. Add for supports, wiring devices, special adapters, covers, sales tax, delivery, supervision, mobilization, demobilization, cleanup, overhead and profit. Note: The labor cost assumes that floor boxes are being installed in a concrete floor slab on grade. Deduct 10% for formed concrete floor slabs installed above grade. Many other sizes of floor boxes are available. Cast iron boxes have threaded hubs of various sizes. Some boxes can be installed in wood floor framing.

Adjustable Floor Boxes and Covers

Adjustable round floor box covers

Diameter	Use	Craft@Hrs	Unit	Material Cost	Labor Cost	Installed Cost
3-5/16"	3/4" hole	L1@0.10	Ea	82.40	4.10	86.50
3-5/16"	1" hole	L1@0.10	Ea	88.10	4.10	92.20
3-5/16"	SR	L1@0.10	Ea	84.40	4.10	88.50
3-3/4"	3/4" hole	L1@0.10	Ea	150.00	4.10	154.10
3-3/4"	1" hole	L1@0.10	Ea	104.00	4.10	108.10
3-3/4"	SR	L1@0.10	Ea	79.50	4.10	83.60
4-3/16"	3/4" hole	L1@0.10	Ea	76.90	4.10	81.00
4-3/16"	1" hole	L1@0.10	Ea	98.20	4.10	102.30
4-3/16"	2-1/8 S R	L1@0.10	Ea	72.20	4.10	76.30
4-3/16"	2-3/8 S R	L1@0.10	Ea	146.00	4.10	150.10
4-3/16"	DR	L1@0.10	Ea	111.00	4.10	115.10
4-3/16"	duplex flap	L1@0.10	Ea	128.00	4.10	132.10

Adjustable rectangle sheet metal floor boxes

Opening	Depth	Craft@Hrs	Unit	Material Cost	Labor Cost	Installed Cost
3-3/16 x 4-3/8	3"	L1@1.25	Ea	187.00	51.30	238.30
3-3/16 x 4-3/8	4"	L1@1.25	Ea	152.00	51.30	203.30

Adjustable rectangle one gang cast iron floor boxes

Opening	Depth	Craft@Hrs	Unit	Material Cost	Labor Cost	Installed Cost
3-3/16 x 4-3/8	2-7/8"	L1@1.25	Ea	157.00	51.30	208.30
3-3/16 x 4-3/8	3-1/8"	L1@1.25	Ea	172.00	51.30	223.30

Adjustable rectangle two gang cast iron floor boxes

Opening	Depth	Craft@Hrs	Unit	Material Cost	Labor Cost	Installed Cost
6-3/8 x 4-3/8	3-3/8"	L1@1.50	Ea	345.00	61.50	406.50

Adjustable rectangle three gang cast iron floor boxes

Opening	Depth	Craft@Hrs	Unit	Material Cost	Labor Cost	Installed Cost
9-9/16 x 4-3/8	3-3/8"	L1@1.75	Ea	482.00	71.80	553.80

Use these figures to estimate the cost of floor boxes installed in building floors under the conditions described on pages 5 and 6. Costs listed are for each floor box installed. The crew is one electrician working at a labor cost of $41.03 per manhour. These costs include layout, material handling, and normal waste. Add for supports, wiring devices, special adapters, covers, sales tax, delivery, supervision, mobilization, demobilization, cleanup, overhead and profit. Note: The labor cost assumes that floor boxes are being installed in a concrete floor slab on grade. Deduct 10% for formed concrete floor slabs installed above grade. Many other sizes of floor boxes are available. Cast iron boxes have threaded hubs of various sizes. Some boxes can be installed in wood floor framing. SR is single receptacle, DR is duplex receptacle.

Semi-adjustable Floor Boxes and Covers

Semi-adjustable rectangle one gang cast iron floor boxes

Material		Craft@Hrs	Unit	Material Cost	Labor Cost	Installed Cost
Opening	Depth					
3-3/16 x 4-3/8	2"	L1@1.25	Ea	164.00	51.30	215.30

Semi-adjustable rectangle two gang cast iron floor boxes

Material		Craft@Hrs	Unit	Material Cost	Labor Cost	Installed Cost
Opening	Depth					
6-3/8 x 4-3/8	2"	L1@1.50	Ea	242.00	61.50	303.50

Semi-adjustable rectangle three gang cast iron floor boxes

Material		Craft@Hrs	Unit	Material Cost	Labor Cost	Installed Cost
Opening	Depth					
9-9/16 x 4-3/8	2"	L1@1.75	Ea	398.00	71.80	469.80

Semi-adjustable rectangle floor box covers

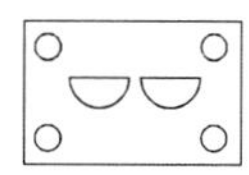

Material	Craft@Hrs	Unit	Material Cost	Labor Cost	Installed Cost
Opening					
2-1/8 - 3/4" combination	L1@0.15	Ea	88.10	6.15	94.25
2-1/8 - 1" combination	L1@0.15	Ea	98.80	6.15	104.95
2-1/8 - single	L1@0.15	Ea	86.50	6.15	92.65
2-3/8 - single	L1@0.15	Ea	202.00	6.15	208.15
Duplex	L1@0.15	Ea	101.00	6.15	107.15
Duplex, flip	L1@0.15	Ea	181.00	6.15	187.15
Pedestal 3/4 x 3"	L1@0.10	Ea	72.20	4.10	76.30
Pedestal 3/4 x 6"	L1@0.10	Ea	120.00	4.10	124.10
Pedestal duplex 3/4"	L1@0.25	Ea	404.00	10.30	414.30
Pedestal duplex 1"	L1@0.25	Ea	404.00	10.30	414.30
Pedestal duplex 2G 3/4"	L1@0.35	Ea	396.00	14.40	410.40
Pedestal duplex 2G 1"	L1@0.25	Ea	396.00	10.30	406.30
Pedestal telephone	L1@0.25	Ea	246.00	10.30	256.30
Pedestal single receptacle	L1@0.25	Ea	260.00	10.30	270.30
Pedestal flange, 3/4"	L1@0.15	Ea	120.00	6.15	126.15
Pedestal flange, 1"	L1@0.15	Ea	148.00	6.15	154.15
Mound, telephone, 1 side	L1@0.25	Ea	184.00	10.30	194.30
Mound, telephone, 2 sides	L1@0.25	Ea	198.00	10.30	208.30
Mound, duplex, 1 side	L1@0.25	Ea	189.00	10.30	199.30
Mound, duplex 2 sides	L1@0.25	Ea	208.00	10.30	218.30

Use these figures to estimate the cost of floor boxes and accessories installed in building floors under the conditions described on pages 5 and 6. Costs listed are for each unit installed. The crew is one electrician working at a labor cost of $41.03 per manhour. These costs include layout, material handling, and normal waste. Add for supports, wiring devices, special adapters, sales tax, delivery, supervision, mobilization, demobilization, cleanup, overhead and profit.

Floor Box Accessories

Round brass carpet flanges for floor boxes

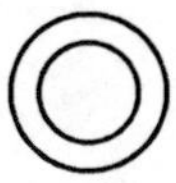

Material	Craft@Hrs	Unit	Material Cost	Labor Cost	Installed Cost
Carpet thickness					
3/8"	L1@0.10	Ea	130.00	4.10	134.10
1/2"	L1@0.10	Ea	136.00	4.10	140.10
5/8"	L1@0.10	Ea	145.00	4.10	149.10
3/4"	L1@0.10	Ea	150.00	4.10	154.10

Clear polycarbonate carpet flanges for floor boxes

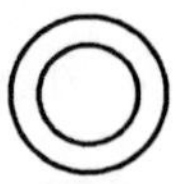

Material	Craft@Hrs	Unit	Material Cost	Labor Cost	Installed Cost
Round	L1@0.10	Ea	33.30	4.10	37.40
Rectangle, one gang	L1@0.15	Ea	36.10	6.15	42.25
Rectangle, two gang	L1@0.17	Ea	44.80	6.98	51.78
Rectangle, three gang	L1@0.20	Ea	56.80	8.21	65.01

Brass crown plugs for floor boxes

Material	Craft@Hrs	Unit	Material Cost	Labor Cost	Installed Cost
3/4"	L1@0.05	Ea	16.40	2.05	18.45
1"	L1@0.06	Ea	26.30	2.46	28.76

Brass flush plugs for floor boxes

Material	Craft@Hrs	Unit	Material Cost	Labor Cost	Installed Cost
3/4"	L1@0.05	Ea	31.30	2.05	33.35
1"	L1@0.06	Ea	50.90	2.46	53.36
1-1/2"	L1@0.10	Ea	41.90	4.10	46.00
2-1/8"	L1@0.10	Ea	36.10	4.10	40.20
2-3/8"	L1@0.10	Ea	60.20	4.10	64.30

Use these figures to estimate the cost of floor box accessories installed in building floors under the conditions described on pages 5 and 6. Costs listed are for each unit installed. The crew is one electrician working at a labor cost of $41.03 per manhour. These costs include layout, material handling, and normal waste. Add for supports, wiring devices, special adapters, sales tax, delivery, supervision, mobilization, demobilization, cleanup, overhead and profit. Note: Some manufacturers offer aluminum finish floor box parts that fit their floor boxes. The labor for these accessories will be the same as for the brass accessories listed above. Adjustable floor boxes are usually installed in concrete floors. Non-adjustable boxes are usually installed in wood floors. The boxes and accessories listed in this section are only a few of the many items available.

Section 4: Lighting Fixtures

If you handle many commercial jobs, you'll spend a lot of time estimating lighting fixtures. Light fixtures are a major cost item on many commercial jobs. Designers scour manufacturer's catalogs to find exactly the right fixture that shows off their client's merchandise to best advantage. When you see a particular fixture specified for an upscale store, you can bet that thought and consideration went into its selection.

There's a vogue in lighting fixtures, just as there is in hair styles and Paris gowns. And, from time to time, a new edition of the National Electrical Code changes the way some lighting fixtures are made and installed. That's why lighting fixtures become obsolete and are replaced by more modern, efficient or attractive units.

In the last 10 years *NEC* restrictions and state law have set limits on lighting in residential and commercial buildings. The purpose of these regulations is to promote the conservation of electricity. Standards have been set for energy use in most work environments. Lighting levels have been decreased and better controlled in many areas. In response, manufacturers have designed energy-efficient lighting fixtures for both new construction and retrofit jobs. Don't underestimate the care needed when estimating electrical lighting fixtures. Start with a complete take-off of the lighting system as it's designed. Then review the installation details for each fixture. Count all the special components that will be required: stem and canopy hangers, ceiling spacers, end-of-row caps, special support material, the type and quantity of lamps, mounting height, safety clips, safety wires or cables, guards, finish color, etc. There are so many choices and components that it's easy to overlook something.

Stem and Canopies

Many lighting fixtures are suspended on a stem device. The stem is usually attached to the outlet box and both supports the fixture and serves as the wireway from the outlet box to the fixture. Some engineers require that the stem include a swivel so the fixture can swing free in an earthquake or if hit by something. Otherwise, the stem might break, dropping the fixture on whatever is directly below. The swivel is usually a ball aligner fitting that's attached to an outlet box fitted with a threaded hub to receive the stem. Ball aligner fittings come in several sizes that fit most of the popular stem sizes.

Ball aligners are made for a certain degree of movement. The maximum movement permitted varies with the manufacturer and with the style of swivel. Be sure the aligner you install meets job specs.

The stem canopy is an escutcheon that serves as trim between the connecting part of the stem and the outlet box. Some canopy assemblies have a swivel built in. Some are referred to as **earthquake stem and canopies**. Some stem and canopy sets also have a swivel fitting at the point of connection to the lighting fixture. This dual swivel system is common on fluorescent fixtures where more than one stem is used on the same fixture, or when fluorescent fixtures are suspended in rows of two or more.

Four-foot-long fluorescent fixtures mounted in rows may need only one stem and canopy per fixture, plus an additional stem and canopy for the end fixture. For example, if you find a row of three fixtures on the plans, figure on using four stem and canopy sets.

Your supplier will quote a separate price for stem and canopy sets. The price is probably for a standard stem length of two feet, with an additional cost per foot for stems over two feet. If you use long stems, it might be less expensive to buy the stem material yourself and cut individual stems to order in your shop.

Most inspection authorities don't permit couplings in a stem. A coupling becomes the weak spot that would be the point of rupture under stress. Stem and canopy sets are usually painted with white enamel at the factory. Any other color will probably cost extra. Enamel white is the standard finish because it matches the standard color for fluorescent fixtures.

Ceiling Spacers

Incorrect mounting of some types of lighting fixtures can be a fire hazard. For example, some fixtures have a ballast that's in direct contact with the top of the fixture housing. The housing is going to get pretty hot. You can see that mounting that fixture flush against a combustible ceiling material is a bad idea, even if the inspector doesn't catch the problem.

The code requires spacers that provide a gap between the fixture and the ceiling. The spacer is usually $1^1/_2$" deep. It's painted white to match the fixture. It attaches to the outlet box and joins it to the fixture. Some spacers can be attached directly to the ceiling material. Your supplier will quote prices on several common types of fixture spacers.

Ideally, support for the fixture is an integral part of the ceiling, such as in a suspended grid ceiling. The grid is designed to include ceiling fixtures and may provide all the support needed. But in many cases you'll have to provide additional support for the fixtures you install.

Usually wood screws are used to attach the fixture or a suspension hanger to the ceiling frame or blocking. Be sure the screws extend through the ceiling surface and into the framing. In an emergency, your support should hold up both the fixture and the weight of a man.

Ganging Fluorescent Fixtures

You'll need some additional parts when fluorescent fixtures are installed in rows of two or more. An end-of-row cap is needed at the end of each row and joiners are needed between fixtures. Usually these parts have to be ordered separately and will be an additional cost item.

When you've finished counting all the fluorescent lighting fixtures during the take-off, go back and count the joiners and end caps. And don't forget the additional labor required to install these fittings.

In some cases your supplier will offer fixtures that are factory-built into rows of fixtures. The fixture cost will be higher, but the installed cost will probably be lower than assemblies of individual fixtures built-up on the job. Long rows of fixtures can be made with fixture channels that reduce the number of ballasts needed and eliminate separate joiners and caps.

Special Support Material

Recessed lighting fixtures in plaster ceilings require plaster frames. These frames are part of the rough-in because they have to be installed before the lath is hung. The plaster frame provides a finished opening for the lighting fixture. When you order recessed fixtures, order the plaster frames at the same time. Plaster frames cost extra, so don't forget to list them separately on your take-off sheet.

Some recessed fixtures intended for recessed mounting in drywall ceilings also require a factory-made frame that has to be set before the drywall is hung. The drywall installer cuts an opening in the drywall to fit around the frame.

Lighting fixtures installed in some types of drop ceilings may need more support than the drop ceiling offers. Usually furring channels are used to support the ceiling adjacent to the fixture. The support spreads the weight of the fixture over a wider area and helps hold the fixture in position. When a fixture is installed in a concealed spline suspended ceiling system, the fixture will be attached directly to the additional support.

Some fluorescent fixtures have **bat wings** attached to the housing sides. When the fixture is installed, the bat wings are extended to provide support. Adjusting screws in the bat wings are used to level the fixture flush with the ceiling tile. When this type of fixture is needed, the designer probably specified a fixture with bat wings. But check the exact mounting method to be sure that you're pricing what's needed.

Type and Quantity of Lamps

The lighting fixtures aren't finished until lamps are installed. Check the lighting fixture schedule on the plans or in the specs to see what lamps are needed. The lamps are usually ordered separately — though some fixtures come with lamps. When lamps are included in the fixture price, the supplier's catalog should note that fact.

Sometimes it will save money to have the supplier ship the fixtures with lamps already installed. Consider asking for a quote on this service if you're installing a large quantity of fluorescent fixtures. Some suppliers can provide fixtures with both lamps and a flex conduit tail already installed. This speeds field hookup and may reduce your installed cost on a larger job.

Your take-off for fixtures should leave space for listing the lamps required. Note the quantity of each type of fixture and find the desired lamp. Later, you'll add material prices and installation labor for both fixtures and lamps.

Don't omit the labor for lamp installation. Every lamp has to be ordered, received, stored, handled, distributed on the job and installed. And there will be occasional breakage and waste. If you don't include an estimate of the time required for installing lamps, and include this cost in your bid, you're running a charity, not a business. True, you could include lamping labor in the labor time for fixture installation. But this manual doesn't. Some fixtures arrive on the job with lamps already installed. And occasionally you may have a job where the lamps are not included in the electrical work. That's why it's best to estimate the lamping labor separately.

Mounting Height

Somewhere in the plans, the fixture schedule, or the specs, you'll find the mounting height for suspended and wall-mounted lighting fixtures. Usually the installation height is noted on the plans in the area where the fixture is to be mounted. Note that the mounting height has an effect on installation cost. The longer the suspension system, the more material required and the higher your cost. Higher ceilings have the same effect. Working from the floor is fastest. Using a 6 foot step ladder takes more time. The greater the height, the more equipment needed and the more up and down time required, even if you use a modern power lift.

Safety Clips

Safety clips may be required on fluorescent fixtures installed in suspended ceilings. The clips probably won't show up in the plans or specs. But they're a cost item that should be in your estimate. Safety clips are usually required in exposed "T" bar grid ceilings. The ceiling tile rests in exposed runners. Where a lighting fixture is required, the fixture replaces one of the ceiling tiles. The fixture should have a safety clip attached to either the fixture or the exposed runners. The clip keeps the fixture from falling if the grid is displaced or damaged in some way. In seismic zones these clips are called earthquake clips.

Safety Wire and Cable

On many jobs, safety wires or cable supports will be required in addition to the regular fixture support. Safety wires are often used on fluorescent fixtures mounted in an exposed "T" bar grid ceiling. The wires run from the fixture to an anchor point above the suspended ceiling. Small cable is sometimes used for this purpose. Usually only two wires are needed per fixture. The wires run from diagonal corners of the fixture and should have a little slack so the fixture doesn't lift the "T" bar runners. Safety wires are usually required even though safety clips are also installed. That may seem like wearing both a belt and suspenders to keep your pants up. But I guess you can't be too careful.

Safety cables may also be required on high bay lighting fixtures. If the fixture is mounted on a hook arrangement, the safety cable is attached to the fixture housing and runs either to the framing above or to the conduit system.

Guards

Guards protect either a fixture or the lamp from damage. Some guards are like a simple wire basket. Others are more substantial. Some outdoor wall bracket fixtures have guards to protect the glass globe from damage. Storage room lights may have guards to prevent accidental damage to the lamps. Auditorium fixtures usually have guards to protect the public if glass in the fixture should shatter. Watch for notes in the plans or specs on what guards are required. Most fixtures are available both with and without guards. Omitting the guard from your estimate can be an expensive oversight.

Finish Color

Many designers specify a certain finish or color for the exposed parts of fixtures. Finishes include anodizing, colored enamel, simulated wood finishes or real wood frames. Note that the finish specified may not be the standard finish or may come at additional cost. In some cases, you may have to arrange for custom finishing or do the finishing yourself. Special finishes are expensive. Be sure you understand what finish is required before pricing each fixture.

Estimating Lighting Fixtures

Many electrical estimators start each labor and material take-off by counting the lighting fixtures. Scanning the plans for fixtures is a quick and easy introduction to the project. Count and list each type of lighting fixture on your work sheet. On larger jobs, there may be several sheets with fixtures. As you count the fixtures, study the type of ceiling they will be mounted on or in. Look for the mounting height of suspended and wall fixtures. Look for the fluorescents that are in rows and can be tandem mounted. Watch the voltage and switching controls and look for dimmers. If fluorescent fixtures are controlled from a dimmer, you'll need to price fixtures with special dimming ballasts.

Some specs require lighting fixtures with radio suppressors. The manufacturer will install suppressors, but you have to order the right material and include the extra cost in the estimate.

Some fixtures have remote ballasts that are mounted away from the fixture. Be sure to list these units separately because the labor cost is higher and more materials are needed.

Watch out for lighting fixtures that have to be installed in coving built by the contractor. Also note carefully any special diffusers required on lighting fixtures. They're sure to be an additional cost item. Sometimes the diffuser is supplied by the contractor who builds the coving. That may or may not be the electrical contractor. Be sure you know what your work includes.

When you've taken off the lighting fixtures throughout the project, you know a lot about the building. That makes fixture take-off a good introduction to the project.

Using a Pricing Service

Some electrical suppliers have the fixture take-off done by a fixture estimating service. The supplier may quote a competitive package price for all fixtures on the job. The service lists all fixtures — and may even remember to include most of the additional parts and pieces required for installation.

Some of these fixture services are reasonably accurate. And the package price may be attractive because they substitute the lowest-cost fixtures whenever permitted by the plans and specs. But it's hard for you to check their work and verify fixture prices. And, since the supplier won't be doing the installation, they may omit important items like supports and clips. Your own fixture estimate is more likely to include everything that's needed to comply with the code and do the job.

Doing Your Take-off

Use a separate take-off form to list all lighting fixtures and the accessories. Across the top of the page list the names of each supplier you plan to get fixture quotes from. When you've assembled all quotes and extended figures to a total, select the most competitive supplier. But watch out for those alternates some suppliers like to slip in. It's their duty to advise you when they offer an alternate to what's specified. It's always a gamble to accept an alternate bid without first checking with the designing engineer. If the engineer rejects the alternate, you may have to supply the item specified at your own cost. The supplier won't make good on the difference if the alternate was rejected.

Work carefully when taking off lighting fixtures. Counting the fixtures is easy — deceptively so. Remembering to include everything needed for the lighting system is much harder.

Figuring the Manhours

The labor units listed in this section are based on good installation conditions as defined in the section "How to Use This Book" at the front of this manual. If you expect unusual job conditions to delay production, increase the labor time accordingly.

Here are some conditions that will increase your labor cost:

- A work area littered with construction debris
- Fixtures stored in unmarked boxes
- Fixtures stored at a considerable distance from the points of installation
- Extra high ceilings
- Restricted work area
- Smaller room size
- Working around furniture and equipment in an occupied room. Remodeling work almost always takes more labor per unit than new construction.

It takes judgment to spot conditions that will slow the work. Every job isn't "standard." Don't expect that the manhours in this manual will cover every situation without modification. Be ready to increase (or reduce) the figures offered here when appropriate. It's better to be cautious when in doubt. There are old electrical estimators and there are bold electrical estimators. But there are no **old and bold** electrical estimators.

Surface mounted porcelain keyless incandescent light fixtures

Material	Craft@Hrs	Unit	Material Cost	Labor Cost	Installed Cost
250V 3-3/4" dia.	L1@0.20	Ea	3.78	8.21	11.99
250V 4-1/2" dia.	L1@0.20	Ea	3.50	8.21	11.71
250V 5-1/2" dia.	L1@0.20	Ea	5.53	8.21	13.74
600V 4-1/2" dia.	L1@0.20	Ea	4.55	8.21	12.76

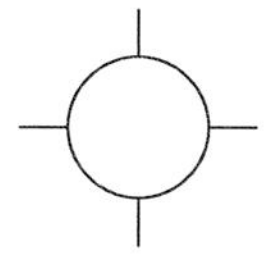

Surface mounted porcelain chain pull incandescent light fixtures

Material	Craft@Hrs	Unit	Material Cost	Labor Cost	Installed Cost
250V 4-1/2" dia.	L1@0.20	Ea	7.76	8.21	15.97
250V 5-1/4" dia.	L1@0.20	Ea	7.85	8.21	16.06

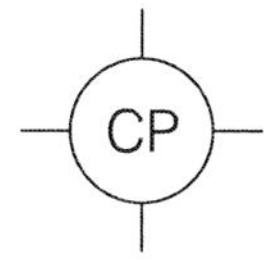

Cover mounted keyless incandescent light fixtures

Material	Craft@Hrs	Unit	Material Cost	Labor Cost	Installed Cost
600V 3-1/2" dia.	L1@0.25	Ea	5.08	10.30	15.38
600V 4-1/8" dia.	L1@0.25	Ea	6.27	10.30	16.57

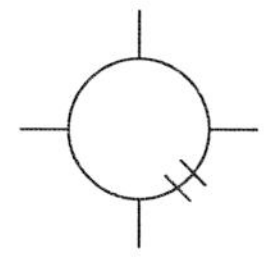

Ceiling square glass incandescent light fixtures, two 60-watt lamps

Material	Craft@Hrs	Unit	Material Cost	Labor Cost	Installed Cost
White 12"	L1@0.30	Ea	12.90	12.30	25.20
White 14"	L1@0.30	Ea	21.10	12.30	33.40
Satin white 12"	L1@0.30	Ea	18.20	12.30	30.50
Satin white 14"	L1@0.30	Ea	26.30	12.30	38.60
Gold applique 12"	L1@0.30	Ea	19.90	12.30	32.20
Gold applique 14"	L1@0.30	Ea	23.40	12.30	35.70
White textured 12"	L1@0.30	Ea	19.40	12.30	31.70
White patterned 12"	L1@0.30	Ea	13.90	12.30	26.20
White patterned 14"	L1@0.30	Ea	19.80	12.30	32.10

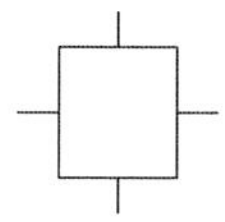

Ceiling square glass incandescent light fixtures, three 60-watt lamps

Material	Craft@Hrs	Unit	Material Cost	Labor Cost	Installed Cost
White, scroll 13"	L1@0.35	Ea	36.20	14.40	50.60
White, patterned 16"	L1@0.35	Ea	36.00	14.40	50.40

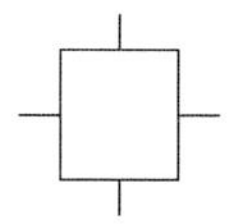

Use these figures to estimate the cost of lighting fixtures installed in buildings under the conditions described on pages 5 and 6. Costs listed are for each fixture installed. The crew is one electrician working at a labor cost of $41.03 per manhour. These costs include layout, material handling, and normal waste. Add for lamp, accessories, sales tax, delivery, supervision, mobilization, demobilization, cleanup, overhead and profit. Note: These costs are based on internally pre-wired fixtures installed in new construction not over 10 feet above the floor. Many types of glass diffusers are available. Many have textured glass with gold designs that can be matched with larger or smaller diffusers.

Ceiling Mounted Incandescent Light Fixtures

Ceiling round glass incandescent light fixtures, two 60-watt lamps

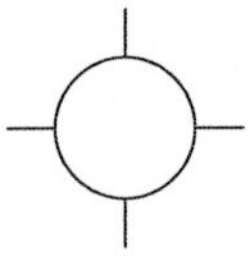

Material		Craft@Hrs	Unit	Material Cost	Labor Cost	Installed Cost
Floral textured	12"	L1@0.30	Ea	20.10	12.30	32.40
Satin white	12"	L1@0.30	Ea	22.40	12.30	34.70
Silver applique	13"	L1@0.30	Ea	27.40	12.30	39.70
White floral	13"	L1@0.30	Ea	27.40	12.30	39.70

Ceiling round glass incandescent light fixtures, three 60-watt lamps

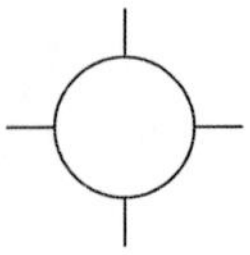

Material		Craft@Hrs	Unit	Material Cost	Labor Cost	Installed Cost
Satin white	15"	L1@0.35	Ea	38.30	14.40	52.70
White scroll	15"	L1@0.35	Ea	38.30	14.40	52.70
Satin floral	15"	L1@0.35	Ea	40.50	14.40	54.90
White floral	15"	L1@0.35	Ea	41.60	14.40	56.00
Silver applique	15"	L1@0.35	Ea	43.50	14.40	57.90
Colorful clown	18"	L1@0.35	Ea	61.40	14.40	75.80
Needlepoint	15"	L1@0.35	Ea	62.80	14.40	77.20
Scalloped	16"	L1@0.35	Ea	67.30	14.40	81.70
Leaf & floral	14"	L1@0.35	Ea	104.00	14.40	118.40
Etched floral	14"	L1@0.35	Ea	116.00	14.40	130.40

Ceiling round glass incandescent light fixtures, four 60-watt lamps

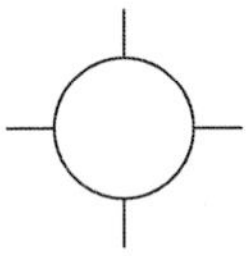

Material		Craft@Hrs	Unit	Material Cost	Labor Cost	Installed Cost
White prismatic	18"	L1@0.40	Ea	62.80	16.40	79.20
White textured	18"	L1@0.40	Ea	76.50	16.40	92.90

Ceiling round glass incandescent light fixtures, two 60-watt lamps, with canopy

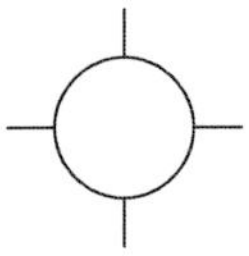

Material		Craft@Hrs	Unit	Material Cost	Labor Cost	Installed Cost
White canopy	12"	L1@0.40	Ea	32.60	16.40	49.00
White canopy	14"	L1@0.40	Ea	36.20	16.40	52.60
White canopy	11"	L1@0.40	Ea	45.50	16.40	61.90
Brass canopy	11"	L1@0.40	Ea	48.90	16.40	65.30
Brass canopy	12"	L1@0.40	Ea	50.10	16.40	66.50

Use these figures to estimate the cost for lighting fixtures installed in buildings under the conditions described on pages 5 and 6. Costs listed are for each fixture installed. The crew is one electrician working at labor cost of $41.03 per manhour. These costs include layout, material handling, and normal waste. Add for the lamp, accessories, sales tax, delivery, supervision, mobilization, demobilization, cleanup, overhead, and profit. Note: These costs are based on internally pre-wired fixtures installed in new construction not over 10 feet above the floor. Many types of glass diffusers are available. Many have textured glass with gold designs that can be matched with larger or smaller diffusers.

Ceiling round incandescent light fixtures, three 60-watt lamps, with canopy

Material		Craft@Hrs	Unit	Material Cost	Labor Cost	Installed Cost
Brass canopy	13"	L1@0.40	Ea	47.90	16.40	64.30
Brass canopy	14"	L1@0.40	Ea	65.20	16.40	81.60
Brass canopy	15"	L1@0.40	Ea	65.20	16.40	81.60
White canopy	13"	L1@0.40	Ea	61.30	16.40	77.70
Brass/jeweled	15"	L1@0.40	Ea	74.90	16.40	91.30
Wagon wheel	16"	L1@0.40	Ea	83.70	16.40	100.10
Patterned brass	18"	L1@0.40	Ea	83.70	16.40	100.10
Almond canopy	14"	L1@0.40	Ea	80.30	16.40	96.70
Peach floral	17"	L1@0.40	Ea	97.60	16.40	114.00
Cathedral	15"	L1@0.40	Ea	106.00	16.40	122.40

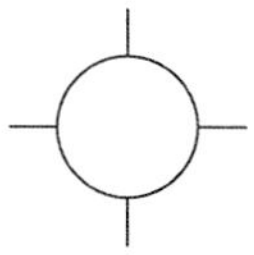

Ceiling white glass incandescent light fixtures, 60- or 100-watt lamp, exposed canopy

Material		Craft@Hrs	Unit	Material Cost	Labor Cost	Installed Cost
Polished brass	5"	L1@0.35	Ea	14.60	14.40	29.00
Antique brass	5"	L1@0.35	Ea	17.50	14.40	31.90
Polished brass	6"	L1@0.35	Ea	20.50	14.40	34.90
Antique brass	6"	L1@0.35	Ea	20.50	14.40	34.90
Polished brass	7"	L1@0.35	Ea	21.20	14.40	35.60
Antique brass	7"	L1@0.35	Ea	21.20	14.40	35.60
White globe	7"	L1@0.40	Ea	23.70	16.40	40.10
Chrome globe	7"	L1@0.40	Ea	25.00	16.40	41.40
Polished brass	9"	L1@0.40	Ea	30.00	16.40	46.40
Antique brass	9"	L1@0.40	Ea	30.00	16.40	46.40
White globe	10"	L1@0.45	Ea	38.30	18.50	56.80
Polished brass	10"	L1@0.45	Ea	39.50	18.50	58.00
Antique brass	10"	L1@0.45	Ea	39.50	18.50	58.00

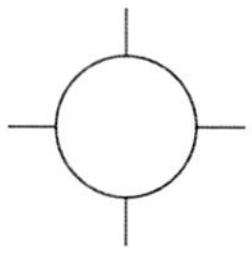

Ceiling opal glass utility incandescent light fixtures, 60-watt lamp

Material		Craft@Hrs	Unit	Material Cost	Labor Cost	Installed Cost
Globe	5"	L1@0.40	Ea	31.60	16.40	48.00
Globe	6"	L1@0.40	Ea	41.90	16.40	58.30

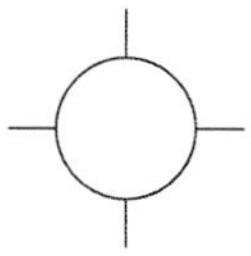

Use these figures to estimate the cost of lighting fixtures installed in buildings under the conditions described on pages 5 and 6. Costs listed are for each fixture installed. The crew is one electrician working at a labor cost of $41.03 per manhour. These costs include layout, material handling, and normal waste. Add for lamp, accessories, sales tax, delivery, supervision, mobilization, demobilization, cleanup, overhead and profit. Note: These costs are based on internally pre-wired fixtures installed in new construction not over 10 feet above the floor. The fixture body for utility lights is satin finish cast aluminum. The opal globe screws into the fixture body.

Incandescent Light Fixtures

Opal glass utility incandescent light fixtures, 100-watt lamp

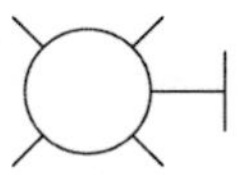

Material		Craft@Hrs	Unit	Material Cost	Labor Cost	Installed Cost
Wall mount sphere	5"	L1@0.40	Ea	36.90	16.40	53.30
Wall mount sphere	6"	L1@0.40	Ea	40.40	16.40	56.80
Double unit	5"	L1@0.50	Ea	80.40	20.50	100.90

Polycarbonate diffuser utility incandescent light fixtures, 60-watt lamp

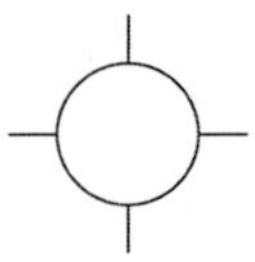

Ceiling	5"	L1@0.40	Ea	46.50	16.40	62.90
Wall	5"	L1@0.40	Ea	48.90	16.40	65.30
Double unit	5"	L1@0.50	Ea	91.90	20.50	112.40

Wall mounted adjustable flood incandescent light fixtures, one lamp

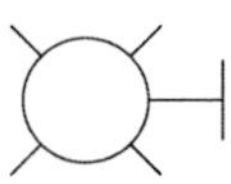

Satin aluminum	100W	L1@0.40	Ea	37.20	16.40	53.60
Satin brass	100W	L1@0.40	Ea	37.20	16.40	53.60
White	50W	L1@0.40	Ea	39.80	16.40	56.20
White sphere	50W	L1@0.40	Ea	66.70	16.40	83.10
White sphere	75W	L1@0.40	Ea	74.60	16.40	91.00
White cylinder	75W	L1@0.40	Ea	74.60	16.40	91.00
White cylinder	150W	L1@0.40	Ea	80.10	16.40	96.50

Wall mounted adjustable flood incandescent light fixtures, two lamps

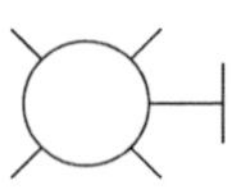

Satin aluminum	100W	L1@0.50	Ea	67.10	20.50	87.60
Satin brass	100W	L1@0.50	Ea	67.10	20.50	87.60
White	100W	L1@0.50	Ea	69.90	20.50	90.40
Antique brass	100W	L1@0.50	Ea	69.90	20.50	90.40

Wraparound glass bathroom incandescent fixtures, two or four 75-watt lamps

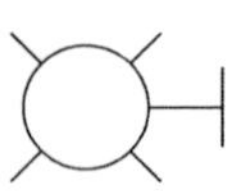

White glass	13"	L1@0.40	Ea	21.00	16.40	37.40
White with outlet	13"	L1@0.40	Ea	23.20	16.40	39.60
White glass	24"	L1@0.45	Ea	35.60	18.50	54.10
White with outlet	24"	L1@0.45	Ea	38.90	18.50	57.40
Ribbed glass	13"	L1@0.40	Ea	25.80	16.40	42.20
Ribbed glass	24"	L1@0.45	Ea	44.20	18.50	62.70

Use these figures to estimate the cost of lighting fixtures installed in buildings under the conditions described on pages 5 and 6. Costs listed are for each fixture installed. The crew is one electrician working at a labor cost of $41.03 per manhour. These costs include layout, material handling, and normal waste. Add for lamp, accessories, sales tax, delivery, supervision, mobilization, demobilization, cleanup, overhead and profit. Note: These costs are based on internally pre-wired fixtures installed in new construction not over 10 feet above the floor.

Incandescent Drum Light Fixtures and Entrance Lights

White glass utility incandescent drum light fixtures with canopy

Material	Craft@Hrs	Unit	Material Cost	Labor Cost	Installed Cost
White 7" 60W	L1@0.40	Ea	19.90	16.40	36.30
Chrome 7" 60W	L1@0.40	Ea	21.70	16.40	38.10
Antique brass 7" 60W	L1@0.40	Ea	21.70	16.40	38.10
White 9" 100W	L1@0.40	Ea	29.10	16.40	45.50
Chrome 9" 100W	L1@0.40	Ea	30.90	16.40	47.30
Antique brass 9" 100W	L1@0.40	Ea	30.90	16.40	47.30
White 11" 2-60W	L1@0.45	Ea	38.50	18.50	57.00
Chrome 11" 2-60W	L1@0.45	Ea	40.40	18.50	58.90
Antique brass 11" 2-60W	L1@0.45	Ea	40.40	18.50	58.90

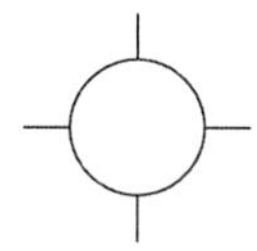

Opal glass utility incandescent drum light fixtures with canopy

Material	Craft@Hrs	Unit	Material Cost	Labor Cost	Installed Cost
Chrome 9" 100W	L1@0.40	Ea	47.70	16.40	64.10
Chrome 11" 2-75W	L1@0.45	Ea	64.10	18.50	82.60
Chrome 13" 3-60W	L1@0.45	Ea	85.90	18.50	104.40
Chrome 14" 3-60W	L1@0.45	Ea	103.00	18.50	121.50

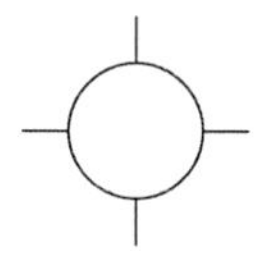

Round opal glass utility incandescent drum light fixtures with canopy

Material	Craft@Hrs	Unit	Material Cost	Labor Cost	Installed Cost
Concealed can 8" 60W	L1@0.40	Ea	45.70	16.40	62.10
White 11" 100W	L1@0.40	Ea	60.70	16.40	77.10
White 12" 2-60W	L1@0.45	Ea	81.50	18.50	100.00

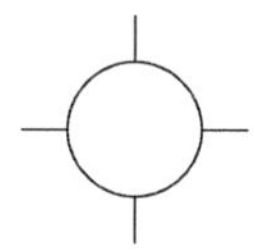

Square opal glass utility incandescent drum light fixtures with canopy

Material	Craft@Hrs	Unit	Material Cost	Labor Cost	Installed Cost
Concealed can 8" 60W	L1@0.40	Ea	51.60	16.40	68.00
White 11" 100W	L1@0.40	Ea	61.40	16.40	77.80
White 12" 2-60W	L1@0.45	Ea	81.10	18.50	99.60

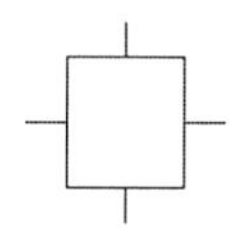

Satin black exterior entrance fixtures, 100-watt lamp

Material	Craft@Hrs	Unit	Material Cost	Labor Cost	Installed Cost
Clear glass 7" H	L1@0.40	Ea	11.10	16.40	27.50
Clear glass 8" H	L1@0.40	Ea	13.50	16.40	29.90
Textured glass 9" H	L1@0.40	Ea	18.60	16.40	35.00
Satin white 7" H	L1@0.40	Ea	22.70	16.40	39.10
Stippled glass 8" H	L1@0.40	Ea	18.00	16.40	34.40
White glass 8" H	L1@0.40	Ea	25.40	16.40	41.80

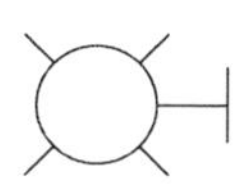

Use these figures to estimate the cost of lighting fixtures installed in buildings under the conditions described on pages 5 and 6. Costs listed are for each fixture installed. The crew is one electrician working at a labor cost of $41.03 per manhour. These costs include layout, material handling, and normal waste. Add for lamp, accessories, sales tax, delivery, supervision, mobilization, demobilization, cleanup, overhead and profit. Note: These costs are based on internally pre-wired fixtures installed in new construction not over 10 feet above the floor.

Recessed Lighting Fixtures with Pre-wired Housings

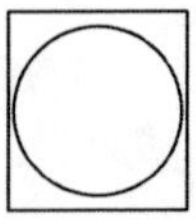
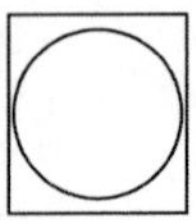
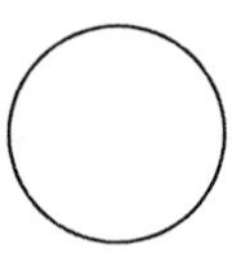

Material		Craft@Hrs	Unit	Material Cost	Labor Cost	Installed Cost
Square white glass recessed light fixtures with metal trim						
Chrome	8" 60W	L1@0.60	Ea	46.50	24.60	71.10
White	9" 100W	L1@0.60	Ea	48.40	24.60	73.00
Chrome	9" 100W	L1@0.60	Ea	49.40	24.60	74.00
White	11" 150W	L1@0.60	Ea	55.30	24.60	79.90
Chrome	11" 150W	L1@0.60	Ea	55.60	24.60	80.20
Square recessed light fixtures with white trim						
Flat glass	9" 100W	L1@0.60	Ea	47.40	24.60	72.00
Drop opal	9" 100W	L1@0.60	Ea	61.30	24.60	85.90
Flat glass	11" 150W	L1@0.60	Ea	61.30	24.60	85.90
Drop opal	11" 150W	L1@0.60	Ea	75.40	24.60	100.00
Flat glass	13" 200W	L1@0.60	Ea	108.00	24.60	132.60
Drop glass	13" 200W	L1@0.60	Ea	115.00	24.60	139.60
Round recessed light fixtures with white trim						
Downlight	7" 150R40	L1@0.60	Ea	42.30	24.60	66.90
Shower	7" 75W	L1@0.60	Ea	54.40	24.60	79.00
Louver	6" 75R30	L1@0.60	Ea	60.30	24.60	84.90
Cone	6" 75R30	L1@0.60	Ea	64.50	24.60	89.10
Baffle	6" 75R30	L1@0.60	Ea	69.10	24.60	93.70
Scoop	8" 100W	L1@0.60	Ea	69.10	24.60	93.70
Baffle	6" 50R20	L1@0.60	Ea	73.30	24.60	97.90
Cone	8" 150R40	L1@0.60	Ea	73.30	24.60	97.90
Louver	8" 150R40	L1@0.60	Ea	75.30	24.60	99.90
Baffle	8" 150R40	L1@0.60	Ea	77.90	24.60	102.50
Downlight	6" 50R20	L1@0.60	Ea	78.80	24.60	103.40
Fresnel	7" 100W	L1@0.60	Ea	83.90	24.60	108.50
Eyeball	6" 50R20	L1@0.60	Ea	89.80	24.60	114.40
Opal	8" 100W	L1@0.60	Ea	101.00	24.60	125.60
Scoop	8" 75R30	L1@0.60	Ea	106.00	24.60	130.60
Eyeball	8" 50R20	L1@0.60	Ea	113.00	24.60	137.60
Adj accent	8" 75R30	L1@0.60	Ea	117.00	24.60	141.60
Downlight	8" 75R30	L1@0.60	Ea	117.00	24.60	141.60
Cone	8" 150W	L1@0.60	Ea	123.00	24.60	147.60
Eyeball	9" 150R40	L1@0.60	Ea	123.00	24.60	147.60
Baffle	7" 150R40	L1@0.60	Ea	123.00	24.60	147.60
Downlight	9" 150R40	L1@0.60	Ea	148.00	24.60	172.60

Use these figures to estimate the cost of lighting fixtures installed in buildings under the conditions described on pages 5 and 6. Costs listed are for each fixture installed. The crew is one electrician working at a labor cost of $41.03 per manhour. These costs include layout, material handling, and normal waste. Add for lamp, accessories, sales tax, delivery, supervision, mobilization, demobilization, cleanup, overhead and profit. Note: These costs are based on internally pre-wired fixtures installed in new construction not over 10 feet above the floor.

Recessed Lighting Fixtures with Pre-wired Housings

Round recessed light fixtures with white trim

Material			Craft@Hrs	Unit	Material Cost	Labor Cost	Installed Cost
Scoop	7"	150R40	L1@0.60	Ea	154.00	24.60	178.60
Reflector	9"	300W	L1@0.60	Ea	157.00	24.60	181.60
Wall washer	7"	150W	L1@0.60	Ea	184.00	24.60	208.60
Wall washer	9"	300W	L1@0.60	Ea	195.00	24.60	219.60

Round recessed light fixtures with white trim and heat guard

Material			Craft@Hrs	Unit	Material Cost	Labor Cost	Installed Cost
Downlight	7"	150R40	L1@0.60	Ea	47.40	24.60	72.00
Shower	7"	75W	L1@0.60	Ea	47.70	24.60	72.30
Louver	6"	75R30	L1@0.60	Ea	65.10	24.60	89.70
Cone	6"	75R30	L1@0.60	Ea	69.10	24.60	93.70
Baffle	6"	75R30	L1@0.60	Ea	74.60	24.60	99.20
Wall washer	8"	150R40	L1@0.60	Ea	74.60	24.60	99.20
Baffle	8"	50R20	L1@0.60	Ea	78.50	24.60	103.10
Cone	8"	150R40	L1@0.60	Ea	78.50	24.60	103.10
Louver	8"	150R40	L1@0.60	Ea	79.90	24.60	104.50
Baffle	8"	150R40	L1@0.60	Ea	82.80	24.60	107.40
Downlight	6"	50R20	L1@0.60	Ea	83.50	24.60	108.10
Fresnel	7"	100W	L1@0.60	Ea	88.90	24.60	113.50
Eyeball	6"	50R20	L1@0.60	Ea	94.80	24.60	119.40
Baffle	9"	300R40	L1@0.60	Ea	99.40	24.60	124.00
Eyeball	8"	75R30	L1@0.60	Ea	106.00	24.60	130.60
Opal	8"	100W	L1@0.60	Ea	106.00	24.60	130.60
Wall washer	8"	75R30	L1@0.60	Ea	109.00	24.60	133.60
Eyeball	8"	50R20	L1@0.60	Ea	116.00	24.60	140.60
Adj accent	8"	75R30	L1@0.60	Ea	120.00	24.60	144.60
Downlight	8"	75R30	L1@0.60	Ea	120.00	24.60	144.60
Cone	8"	150W	L1@0.60	Ea	130.00	24.60	154.60
Eyeball	9"	150R40	L1@0.60	Ea	130.00	24.60	154.60
Baffle	7"	150R40	L1@0.60	Ea	130.00	24.60	154.60
Downlight	9"	150R40	L1@0.60	Ea	149.00	24.60	173.60
Baffle	9"	300R40	L1@0.60	Ea	151.00	24.60	175.60
Wall washer	7"	150R40	L1@0.60	Ea	160.00	24.60	184.60
Reflector	9"	300W	L1@0.60	Ea	161.00	24.60	185.60
Wall washer	7"	150W	L1@0.60	Ea	188.00	24.60	212.60
Wall washer	9"	300W	L1@0.60	Ea	199.00	24.60	223.60

Use these figures to estimate the cost of lighting fixtures installed in buildings under the conditions described on pages 5 and 6. Costs listed are for each fixture installed. The crew is one electrician working at a labor cost of $41.03 per manhour. These costs include layout, material handling, and normal waste. Add for lamp, accessories, sales tax, delivery, supervision, mobilization, demobilization, cleanup, overhead and profit. Note: These costs are based on internally pre-wired fixtures installed in new construction not over 10 feet above the floor.

Track Light Fixtures

Material		Craft@Hrs	Unit	Material Cost	Labor Cost	Installed Cost
Surface mounted single circuit track light, 120 volt						
2' starter, bronze		L1@0.30	Ea	41.10	12.30	53.40
2' starter, white		L1@0.30	Ea	41.10	12.30	53.40
2' starter, black		L1@0.30	Ea	41.10	12.30	53.40
4' starter, bronze		L1@0.40	Ea	61.30	16.40	77.70
4' starter, white		L1@0.40	Ea	61.30	16.40	77.70
4' starter, black		L1@0.40	Ea	61.30	16.40	77.70
8' starter, bronze		L1@0.55	Ea	96.40	22.60	119.00
8' starter, white		L1@0.55	Ea	96.40	22.60	119.00
8' starter, black		L1@0.55	Ea	96.40	22.60	119.00
8' joiner, bronze		L1@0.45	Ea	82.50	18.50	101.00
8' joiner, white		L1@0.45	Ea	82.50	18.50	101.00
8' joiner, black		L1@0.45	Ea	82.50	18.50	101.00
Track light universal lampholder for medium base lamp						
Bronze	5"	L1@0.20	Ea	45.20	8.21	53.41
White	5"	L1@0.20	Ea	45.20	8.21	53.41
Black	5"	L1@0.20	Ea	45.20	8.21	53.41
Chrome	5"	L1@0.20	Ea	45.20	8.21	53.41
Track light shielded universal lampholder for R30 lamp						
Bronze	10"	L1@0.20	Ea	20.70	8.21	28.91
White	10"	L1@0.20	Ea	20.70	8.21	28.91
Black	10"	L1@0.20	Ea	20.70	8.21	28.91
Chrome	10"	L1@0.20	Ea	20.70	8.21	28.91
Track light shielded universal lampholder for R40 or Par38 lamp						
Bronze	11"	L1@0.20	Ea	20.70	8.21	28.91
White	11"	L1@0.20	Ea	20.70	8.21	28.91
Black	11"	L1@0.20	Ea	20.70	8.21	28.91
Chrome	11"	L1@0.20	Ea	20.70	8.21	28.91
Track light continental lampholder for R20 lamp						
Bronze	6"	L1@0.20	Ea	47.30	8.21	55.51
White	6"	L1@0.20	Ea	47.30	8.21	55.51
Black	6"	L1@0.20	Ea	47.30	8.21	55.51
Chrome	6"	L1@0.20	Ea	47.30	8.21	55.51

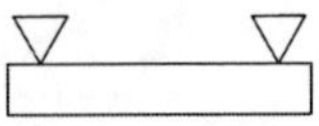
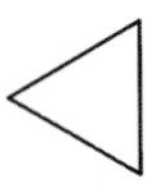
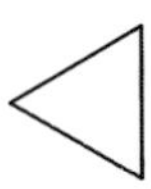
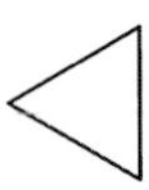

Use these figures to estimate the cost of lighting fixtures installed in buildings under the conditions described on pages 5 and 6. Costs listed are for each fixture installed. The crew is one electrician working at a labor cost of $41.03 per manhour. These costs include layout, material handling, and normal waste. Add for lamp, accessories, sales tax, delivery, supervision, mobilization, demobilization, cleanup, overhead and profit. Note: These costs are based on internally pre-wired fixtures installed in new construction not over 10 feet above the floor. Many types of fittings are available for track lighting.

Track light continental lampholder for R20 lamp

Material		Craft@Hrs	Unit	Material Cost	Labor Cost	Installed Cost
Polished brass	7"	L1@0.20	Ea	72.90	8.21	81.11
Bronze	7"	L1@0.20	Ea	68.20	8.21	76.41
White	7"	L1@0.20	Ea	68.20	8.21	76.41
Black	7"	L1@0.20	Ea	68.20	8.21	76.41

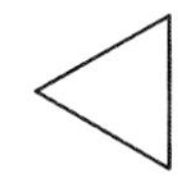

Track light continental lampholder for R30 lamp

Material		Craft@Hrs	Unit	Material Cost	Labor Cost	Installed Cost
Bronze	10"	L1@0.20	Ea	69.60	8.21	77.81
White	10"	L1@0.20	Ea	69.60	8.21	77.81
Black	10"	L1@0.20	Ea	69.60	8.21	77.81

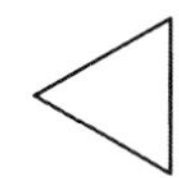

Track light continental lampholder for R40 lamp

Material		Craft@Hrs	Unit	Material Cost	Labor Cost	Installed Cost
Bronze	12"	L1@0.20	Ea	75.80	8.21	84.01
White	12"	L1@0.20	Ea	75.80	8.21	84.01
Black	12"	L1@0.20	Ea	75.80	8.21	84.01
Bronze	9"	L1@0.20	Ea	66.70	8.21	74.91
White	9"	L1@0.20	Ea	66.70	8.21	74.91
Black	9"	L1@0.20	Ea	66.70	8.21	74.91

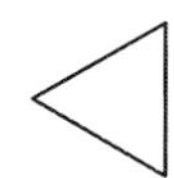

Track light petite cylinder lampholder for R14 lamp

Material		Craft@Hrs	Unit	Material Cost	Labor Cost	Installed Cost
Bronze	5"	L1@0.20	Ea	49.90	8.21	58.11
White	5"	L1@0.20	Ea	49.90	8.21	58.11
Black	5"	L1@0.20	Ea	49.90	8.21	58.11

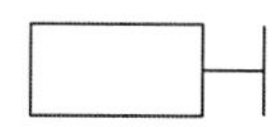

Track light petite round cylinder lampholder for R20 lamp

Material		Craft@Hrs	Unit	Material Cost	Labor Cost	Installed Cost
Bronze	5"	L1@0.20	Ea	54.80	8.21	63.01
White	5"	L1@0.20	Ea	54.80	8.21	63.01
Black	5"	L1@0.20	Ea	54.80	8.21	63.01

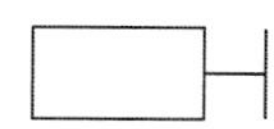

Use these figures to estimate the cost of lighting fixtures installed in buildings under the conditions described on pages 5 and 6. Costs listed are for each fixture installed. The crew is one electrician working at a labor cost of $41.03 per manhour. These costs include layout, material handling, and normal waste. Add for lamp, accessories, sales tax, delivery, supervision, mobilization, demobilization, cleanup, overhead and profit. Note: These costs are based on internally pre-wired fixtures installed in new construction not over 10 feet above the floor.

Track Light Fixtures

Decorator track light fixture for R20 or 100-watt lamp

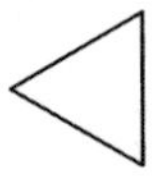

Material		Craft@Hrs	Unit	Material Cost	Labor Cost	Installed Cost
Square	6"	L1@0.20	Ea	69.40	8.21	77.61
Cone	16"	L1@0.20	Ea	119.00	8.21	127.21
Scoop	5"	L1@0.20	Ea	129.00	8.21	137.21

Track light cylinder lampholder for various lamp sizes

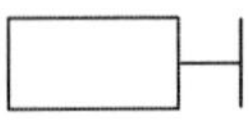

Material		Craft@Hrs	Unit	Material Cost	Labor Cost	Installed Cost
Bronze	5"	L1@0.20	Ea	54.40	8.21	62.61
White	5"	L1@0.20	Ea	54.40	8.21	62.61
Black	5"	L1@0.20	Ea	54.40	8.21	62.61
Bronze	7"	L1@0.20	Ea	62.00	8.21	70.21
White	7"	L1@0.20	Ea	62.00	8.21	70.21
Black	7"	L1@0.20	Ea	62.00	8.21	70.21
Antique brass	7"	L1@0.20	Ea	62.50	8.21	70.71
Bronze	8"	L1@0.20	Ea	71.40	8.21	79.61
White	8"	L1@0.20	Ea	71.40	8.21	79.61
Black	8"	L1@0.20	Ea	71.40	8.21	79.61
Antique brass	8"	L1@0.20	Ea	71.40	8.21	79.61
Bronze	9"	L1@0.20	Ea	96.40	8.21	104.61
White	9"	L1@0.20	Ea	96.40	8.21	104.61
Black	9"	L1@0.20	Ea	96.40	8.21	104.61
Bronze	10"	L1@0.20	Ea	104.00	8.21	112.21
White	10"	L1@0.20	Ea	104.00	8.21	112.21
Black	10"	L1@0.20	Ea	104.00	8.21	112.21

Track light spherical lampholder for various lamp sizes

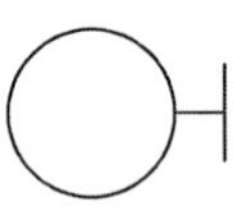

Material		Craft@Hrs	Unit	Material Cost	Labor Cost	Installed Cost
Polished alum.	5"	L1@0.20	Ea	79.90	8.21	88.11
Bronze	5"	L1@0.20	Ea	77.70	8.21	85.91
White	5"	L1@0.20	Ea	77.70	8.21	85.91
Black	5"	L1@0.20	Ea	77.70	8.21	85.91
Polished alum.	6"	L1@0.20	Ea	94.60	8.21	102.81
Bronze	6"	L1@0.20	Ea	84.50	8.21	92.71
White	6"	L1@0.20	Ea	84.50	8.21	92.71
Black	6"	L1@0.20	Ea	84.50	8.21	92.71
Polished alum.	7"	L1@0.20	Ea	116.00	8.21	124.21
Bronze	7"	L1@0.20	Ea	100.00	8.21	108.21
White	7"	L1@0.20	Ea	100.00	8.21	108.21
Black	7"	L1@0.20	Ea	100.00	8.21	108.21

Use these figures to estimate the cost of lighting fixtures installed in buildings under the conditions described on pages 5 and 6. Costs listed are for each fixture installed. The crew is one electrician working at a labor cost of $41.03 per manhour. These costs include layout, material handling, and normal waste. Add for lamp, accessories, sales tax, delivery, supervision, mobilization, demobilization, cleanup, overhead and profit. Note: These costs are based on internally pre-wired fixtures installed in new construction not over 10 feet above the floor.

Track light cylinder lampholders with rounded base, track lighting

Material		Craft@Hrs	Unit	Material Cost	Labor Cost	Installed Cost
Bronze	5"	L1@0.20	Ea	69.40	8.21	77.61
White	5"	L1@0.20	Ea	69.40	8.21	77.61
Black	5"	L1@0.20	Ea	69.40	8.21	77.61
Antique brass	5"	L1@0.20	Ea	69.40	8.21	77.61
Bronze	6"	L1@0.20	Ea	80.40	8.21	88.61
White	6"	L1@0.20	Ea	80.40	8.21	88.61
Black	6"	L1@0.20	Ea	80.40	8.21	88.61
Antique brass	6"	L1@0.20	Ea	80.40	8.21	88.61
Bronze	7"	L1@0.20	Ea	115.00	8.21	123.21
White	7"	L1@0.20	Ea	115.00	8.21	123.21
Black	7"	L1@0.20	Ea	115.00	8.21	123.21
Antique brass	7"	L1@0.20	Ea	115.00	8.21	123.21
Bronze	8"	L1@0.20	Ea	139.00	8.21	147.21
White	8"	L1@0.20	Ea	139.00	8.21	147.21
Black	8"	L1@0.20	Ea	139.00	8.21	147.21
Antique brass	8"	L1@0.20	Ea	139.00	8.21	147.21

Track light stepped base lampholders for R30 or R40 lamps

Material		Craft@Hrs	Unit	Material Cost	Labor Cost	Installed Cost
Bronze	8"	L1@0.20	Ea	129.00	8.21	137.21
White	8"	L1@0.20	Ea	129.00	8.21	137.21
Black	8"	L1@0.20	Ea	129.00	8.21	137.21
Polished brass	8"	L1@0.20	Ea	141.00	8.21	149.21

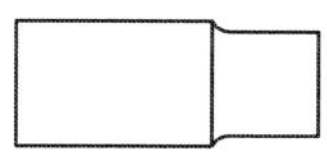

Single face exit fixtures with universal arrows, 20-watt T6 1/2 lamps

Material	Craft@Hrs	Unit	Material Cost	Labor Cost	Installed Cost
Ceiling mounted, red	L1@0.50	Ea	39.10	20.50	59.60
End mounted, red	L1@0.50	Ea	39.10	20.50	59.60
Wall mounted, red	L1@0.50	Ea	39.10	20.50	59.60
Ceiling mounted, green	L1@0.50	Ea	39.10	20.50	59.60
End mounted, green	L1@0.50	Ea	39.10	20.50	59.60
Wall mounted, green	L1@0.50	Ea	39.10	20.50	59.60

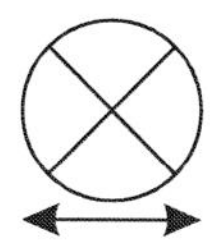

Use these figures to estimate the cost of lighting fixtures installed in buildings under the conditions described on pages 5 and 6. Costs listed are for each fixture installed. The crew is one electrician working at a labor cost of $41.03 per manhour. These costs include layout, material handling, and normal waste. Add for lamp, accessories, sales tax, delivery, supervision, mobilization, demobilization, cleanup, overhead and profit. Note: These costs are based on internally pre-wired fixtures installed in new construction not over 10 feet above the floor.

Exit Lighting Fixtures with Universal Arrows

Double face exit fixtures, 20-watt T6 1/2 lamps

Material	Craft@Hrs	Unit	Material Cost	Labor Cost	Installed Cost
Ceiling mounted, red	L1@0.55	Ea	41.30	22.60	63.90
End mounted, red	L1@0.55	Ea	41.30	22.60	63.90
Ceiling mounted, green	L1@0.55	Ea	41.30	22.60	63.90
End mounted, green	L1@0.55	Ea	41.30	22.60	63.90

Aluminum housing single face exit fixtures, 20-watt T6 1/2 lamps

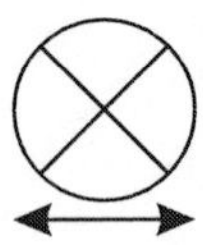

Material	Craft@Hrs	Unit	Material Cost	Labor Cost	Installed Cost
Ceiling mounted, red	L1@0.50	Ea	86.20	20.50	106.70
End mounted, red	L1@0.50	Ea	86.20	20.50	106.70
Wall mounted, red	L1@0.50	Ea	86.20	20.50	106.70
Ceiling mounted, green	L1@0.50	Ea	86.20	20.50	106.70
End mounted, green	L1@0.50	Ea	86.20	20.50	106.70
Wall mounted, green	L1@0.50	Ea	86.20	20.50	106.70

Aluminum housing double face exit fixtures, 20-watt T6 1/2 lamps

Material	Craft@Hrs	Unit	Material Cost	Labor Cost	Installed Cost
Ceiling mounted, red	L1@0.55	Ea	349.00	22.60	371.60
End mounted, red	L1@0.55	Ea	349.00	22.60	371.60
Ceiling mounted, green	L1@0.55	Ea	349.00	22.60	371.60
End mounted, green	L1@0.55	Ea	349.00	22.60	371.60

1 or 2 face exit fixtures with rechargeable batteries

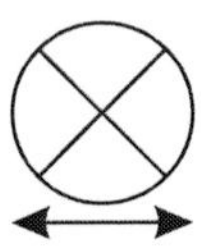

Material	Craft@Hrs	Unit	Material Cost	Labor Cost	Installed Cost
Ceiling mounted, red	L1@0.75	Ea	247.00	30.80	277.80
End mounted, red	L1@0.75	Ea	247.00	30.80	277.80
Wall mounted, red	L1@0.75	Ea	247.00	30.80	277.80
Ceiling mounted, green	L1@0.75	Ea	247.00	30.80	277.80
End mounted, green	L1@0.75	Ea	247.00	30.80	277.80
Wall mounted, green	L1@0.75	Ea	247.00	30.80	277.80

Use these figures to estimate the cost of self illuminating exit fixtures installed in buildings under the conditions described on pages 5 and 6. Costs listed are for each fixture installed. The crew is one electrician working at a labor cost of $41.03 per manhour. These costs include layout, and material handling. Add for sales tax, delivery, supervision, mobilization, demobilization, cleanup, overhead and profit. Note: These costs are based on fixtures installed in new construction not over 10 feet above the floor. Check the job specifications carefully when selecting emergency fixtures. Those listed here are standard grade and may not meet rigid specifications. Get advice from your supplier on the fixtures needed.

Material	Craft@Hrs	Unit	Material Cost	Labor Cost	Installed Cost
Single face exit fixtures, red, green, or black, no wiring					
10 Year Life	L1@0.30	Ea	350.00	12.30	362.30
12 Year Life	L1@0.30	Ea	409.00	12.30	421.30
20 Year Life	L1@0.30	Ea	465.00	12.30	477.30
Double face exit fixtures, red, green, or black, no wiring					
10 Year Life	L1@0.50	Ea	641.00	20.50	661.50
12 Year Life	L1@0.50	Ea	699.00	20.50	719.50
20 Year Life	L1@0.50	Ea	739.00	20.50	759.50

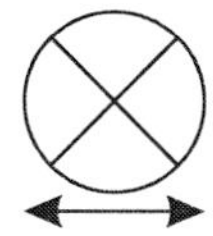

Use these figures to estimate the cost of self illuminating exit fixtures installed in buildings under the conditions described on pages 5 and 6. Costs listed are for each fixture installed. The crew is one electrician working at a labor cost of $41.03 per manhour. These costs include layout, and material handling. Add for sales tax, delivery, supervision, mobilization, demobilization, cleanup, overhead and profit. Note: These costs are based on fixtures installed in new construction not over 10 feet above the floor. Check the job specifications carefully when selecting emergency fixtures. Get advice from your supplier on the fixtures needed.

Surface Mounted 120 Volt Fluorescent Lighting Fixtures

Surface mounted 120 volt fluorescent single-lamp strip lighting fixtures

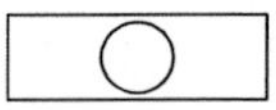

Material	Craft@Hrs	Unit	Material Cost	Labor Cost	Installed Cost
18"	L1@0.25	Ea	19.90	10.30	30.20
24"	L1@0.30	Ea	26.70	12.30	39.00
36"	L1@0.35	Ea	29.10	14.40	43.50
48"	L1@0.40	Ea	30.30	16.40	46.70
48" energy saving	L1@0.40	Ea	30.30	16.40	46.70
72" slimline	L1@0.50	Ea	52.50	20.50	73.00
96" slimline	L1@0.60	Ea	59.50	24.60	84.10
96" 2-48" tandem	L1@0.70	Ea	94.40	28.70	123.10

Surface mounted 120 volt fluorescent two-lamp strip lighting fixtures

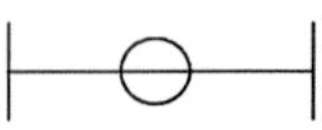

Material	Craft@Hrs	Unit	Material Cost	Labor Cost	Installed Cost
24"	L1@0.35	Ea	27.90	14.40	42.30
36"	L1@0.40	Ea	44.20	16.40	60.60
48"	L1@0.45	Ea	64.10	18.50	82.60
48" energy saving	L1@0.45	Ea	64.10	18.50	82.60
72" slimline	L1@0.55	Ea	52.50	22.60	75.10
96" slimline	L1@0.65	Ea	61.90	26.70	88.60
96" 2-48" tandem	L1@0.75	Ea	105.00	30.80	135.80

Side-mounted 120 volt fluorescent single-lamp lighting fixtures

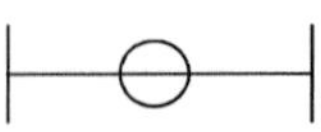

Material	Craft@Hrs	Unit	Material Cost	Labor Cost	Installed Cost
18"	L1@0.25	Ea	35.00	10.30	45.30
24"	L1@0.30	Ea	35.00	12.30	47.30
36"	L1@0.35	Ea	40.90	14.40	55.30
48"	L1@0.40	Ea	40.90	16.40	57.30

Side-mounted 120 volt fluorescent single-lamp on both sides fixtures

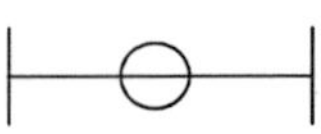

Material	Craft@Hrs	Unit	Material Cost	Labor Cost	Installed Cost
18"	L1@0.30	Ea	52.50	12.30	64.80
24"	L1@0.35	Ea	56.00	14.40	70.40
36"	L1@0.40	Ea	61.90	16.40	78.30
48"	L1@0.45	Ea	65.20	18.50	83.70

Use these figures to estimate the cost of lighting fixtures installed in buildings under the conditions described on pages 5 and 6. Costs listed are for each fixture installed. The crew is one electrician working at a labor cost of $41.03 per manhour. These costs include layout, material handling, and normal waste. Add for lamp, accessories, sales tax, delivery, supervision, mobilization, demobilization, cleanup, overhead and profit. Note: These costs are based on internally pre-wired fixtures installed in new construction not over 10 feet above the floor.

Two-lamp industrial fluorescent lighting fixtures with enameled reflector

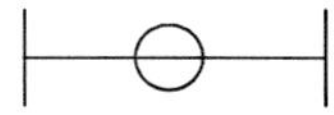

Material	Craft@Hrs	Unit	Material Cost	Labor Cost	Installed Cost
24"	L1@0.35	Ea	35.00	14.40	49.40
48"	L1@0.45	Ea	40.20	18.50	58.70
48" energy saving	L1@0.45	Ea	38.50	18.50	57.00
72" slimline	L1@0.55	Ea	93.40	22.60	116.00
96" slimline	L1@0.65	Ea	100.00	26.70	126.70
96" slimline, energy saving	L1@0.65	Ea	95.40	26.70	122.10

Two-lamp industrial fluorescent lighting fixtures with 15% uplight

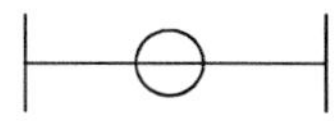

Material	Craft@Hrs	Unit	Material Cost	Labor Cost	Installed Cost
48"	L1@0.45	Ea	81.30	18.50	99.80
48" high output	L1@0.45	Ea	129.00	18.50	147.50
48" very high output	L1@0.45	Ea	199.00	18.50	217.50
48" powergroove	L1@0.45	Ea	163.00	18.50	181.50
72" slimline	L1@0.55	Ea	132.00	22.60	154.60
96" slimline	L1@0.65	Ea	136.00	26.70	162.70
96" high output	L1@0.65	Ea	153.00	26.70	179.70
96" 2-48" tandem	L1@0.65	Ea	153.00	26.70	179.70
96" powergroove	L1@0.65	Ea	201.00	26.70	227.70

Three-lamp industrial fluorescent lighting fixtures with 15% uplight

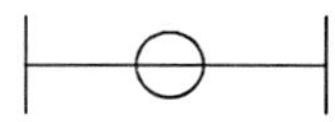

Material	Craft@Hrs	Unit	Material Cost	Labor Cost	Installed Cost
48"	L1@0.50	Ea	104.00	20.50	124.50
96" slimline	L1@0.70	Ea	175.00	28.70	203.70
96" high output	L1@0.70	Ea	220.00	28.70	248.70
96" powergroove	L1@0.70	Ea	227.00	28.70	255.70

Four-lamp industrial fluorescent lighting fixtures with 15% uplight

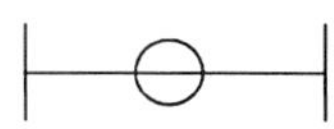

Material	Craft@Hrs	Unit	Material Cost	Labor Cost	Installed Cost
48"	L1@0.55	Ea	90.90	22.60	113.50
96" slimline	L1@0.75	Ea	139.00	30.80	169.80

Two-lamp industrial fluorescent lighting, enameled reflector with 25% uplight

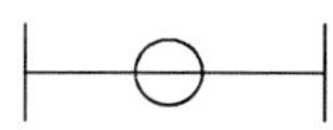

Material	Craft@Hrs	Unit	Material Cost	Labor Cost	Installed Cost
48"	L1@0.40	Ea	115.00	16.40	131.40
96" slimline	L1@0.65	Ea	189.00	26.70	215.70
96" high output	L1@0.65	Ea	221.00	26.70	247.70
96" very high output	L1@0.65	Ea	247.00	26.70	273.70
96" 2-48" tandem	L1@0.65	Ea	207.00	26.70	233.70

Use these figures to estimate the cost of lighting fixtures installed in buildings under the conditions described on pages 5 and 6. Costs listed are for each fixture installed. The crew is one electrician working at a labor cost of $41.03 per manhour. These costs include layout, material handling, and normal waste. Add for lamp, accessories, sales tax, delivery, supervision, mobilization, demobilization, cleanup, overhead and profit. Note: These costs are based on internally pre-wired fixtures installed in new construction not over 10 feet above the floor. Industrial fluorescent fixtures are made with many uplight hues and paint finishes. On some jobs additional accessories will be needed: joiners, hanging fittings, stems and canopies.

Fluorescent Lighting Fixtures

Two-lamp enclosed-gasketed fluorescent fixtures for damp locations

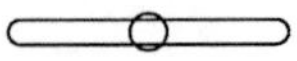

Material	Craft@Hrs	Unit	Material Cost	Labor Cost	Installed Cost
48"	L1@0.50	Ea	117.00	20.50	137.50
48" high output	L1@0.50	Ea	201.00	20.50	221.50
96" slimline	L1@0.70	Ea	222.00	28.70	250.70
96" high output	L1@0.70	Ea	286.00	28.70	314.70
96" 2-48" tandem	L1@0.70	Ea	251.00	28.70	279.70

Two-lamp enclosed-gasketed fluorescent lighting fixtures for wet locations

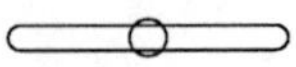

Material	Craft@Hrs	Unit	Material Cost	Labor Cost	Installed Cost
48"	L1@0.50	Ea	126.00	20.50	146.50
48" high output	L1@0.50	Ea	210.00	20.50	230.50
96" slimline	L1@0.70	Ea	241.00	28.70	269.70
96" high output	L1@0.70	Ea	278.00	28.70	306.70
96" 2-48" tandem	L1@0.70	Ea	257.00	28.70	285.70

Wall mounted single-lamp fluorescent lighting fixtures with acrylic lens

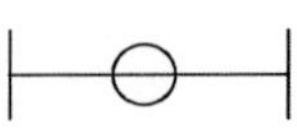

Material	Craft@Hrs	Unit	Material Cost	Labor Cost	Installed Cost
24"	L1@0.40	Ea	46.50	16.40	62.90
36"	L1@0.50	Ea	81.50	20.50	102.00
48"	L1@0.55	Ea	88.50	22.60	111.10

Wall mounted two-lamp fluorescent lighting fixtures with acrylic lens

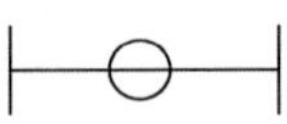

Material	Craft@Hrs	Unit	Material Cost	Labor Cost	Installed Cost
24"	L1@0.45	Ea	69.90	18.50	88.40
36"	L1@0.55	Ea	93.40	22.60	116.00
48"	L1@0.60	Ea	96.70	24.60	121.30

Two-lamp wraparound injection-molded corridor fluorescent fixtures

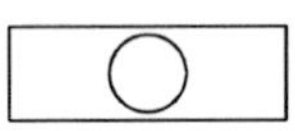

Material	Craft@Hrs	Unit	Material Cost	Labor Cost	Installed Cost
48"	L1@0.60	Ea	81.50	24.60	106.10
96" 2-48" tandem	L1@0.75	Ea	154.00	30.80	184.80

Four-lamp wraparound injection-molded corridor fluorescent fixtures

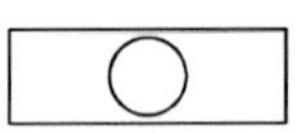

Material	Craft@Hrs	Unit	Material Cost	Labor Cost	Installed Cost
48"	L1@0.70	Ea	139.00	28.70	167.70
96" 2-48" tandem	L1@0.90	Ea	279.00	36.90	315.90

Two-lamp wraparound ceiling fluorescent lighting fixtures

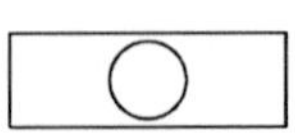

Material	Craft@Hrs	Unit	Material Cost	Labor Cost	Installed Cost
48"	L1@0.60	Ea	75.80	24.60	100.40
48" energy saving	L1@0.60	Ea	103.00	24.60	127.60
96" 2-48" tandem	L1@0.75	Ea	146.00	30.80	176.80
96" 2-48" energy saving	L1@0.75	Ea	186.00	30.80	216.80

Four-lamp wraparound ceiling fluorescent lighting fixtures

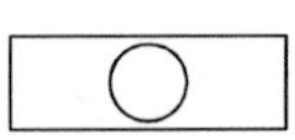

Material	Craft@Hrs	Unit	Material Cost	Labor Cost	Installed Cost
48"	L1@0.70	Ea	137.00	28.70	165.70
48" energy savings	L1@0.70	Ea	188.00	28.70	216.70
96" 2-48" tandem	L1@0.90	Ea	200.00	36.90	236.90
96" 2-48" energy saving	L1@0.90	Ea	232.00	36.90	268.90

Use these figures to estimate the cost of lighting fixtures installed in buildings under the conditions described on pages 5 and 6. Costs listed are for each fixture installed. The crew is one electrician working at a labor cost of $41.03 per manhour. These costs include layout, material handling, and normal waste. Add for lamp, accessories, sales tax, delivery, supervision, mobilization, demobilization, cleanup, overhead and profit. Note: These costs are based on internally pre-wired fixtures installed in new construction not over 10 feet above the floor. Many styles of surface mounted fluorescent fixtures are available. Prices range from economy to architectural quality.

Framed Troffer Lay-in T-Bar Fluorescent Fixtures

24"-wide lay-in T-bar fluorescent fixtures, steel frame

Material	Craft@Hrs	Unit	Material Cost	Labor Cost	Installed Cost
48" 2 lamp	L1@0.70	Ea	145.00	28.70	173.70
48" 3 lamp	L1@0.75	Ea	163.00	30.80	193.80
48" 4 lamp	L1@0.80	Ea	160.00	32.80	192.80
48" 2 lamp, energy saver	L1@0.70	Ea	160.00	28.70	188.70
48" 3 lamp, energy saver	L1@0.75	Ea	201.00	30.80	231.80
48" 4 lamp, energy saver	L1@0.80	Ea	198.00	32.80	230.80

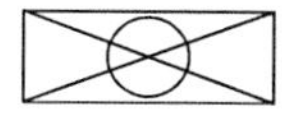

Recessed 24"-wide lay-in T-bar fluorescent fixtures, steel frame

Material	Craft@Hrs	Unit	Material Cost	Labor Cost	Installed Cost
48" 2 lamp	L1@0.70	Ea	154.00	28.70	182.70
48" 3 lamp	L1@0.75	Ea	176.00	30.80	206.80
48" 4 lamp	L1@0.80	Ea	172.00	32.80	204.80
48" 2 lamp, energy saver	L1@0.70	Ea	174.00	28.70	202.70
48" 3 lamp, energy saver	L1@0.75	Ea	220.00	30.80	250.80
48" 4 lamp, energy saver	L1@0.80	Ea	210.00	32.80	242.80

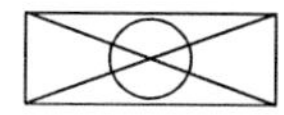

24"-wide lay-in T-bar fluorescent fixtures, aluminum frame

Material	Craft@Hrs	Unit	Material Cost	Labor Cost	Installed Cost
48" 2 lamp	L1@0.70	Ea	153.00	28.70	181.70
48" 3 lamp	L1@0.75	Ea	171.00	30.80	201.80
48" 4 lamp	L1@0.80	Ea	179.00	32.80	211.80
48" 2 lamp, energy saver	L1@0.70	Ea	184.00	28.70	212.70
48" 3 lamp, energy saver	L1@0.75	Ea	222.00	30.80	252.80
48" 4 lamp, energy saver	L1@0.80	Ea	228.00	32.80	260.80

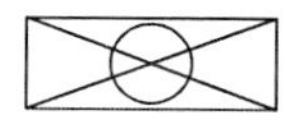

Recessed 24"-wide lay-in T-bar fluorescent fixtures, aluminum frame

Material	Craft@Hrs	Unit	Material Cost	Labor Cost	Installed Cost
48" 2 lamp	L1@0.70	Ea	162.00	28.70	190.70
48" 3 lamp	L1@0.75	Ea	183.00	30.80	213.80
48" 4 lamp	L1@0.80	Ea	184.00	32.80	216.80
48" 2 lamp, energy saver	L1@0.70	Ea	188.00	28.70	216.70
48" 3 lamp, energy saver	L1@0.75	Ea	234.00	30.80	264.80
48" 4 lamp, energy saver	L1@0.80	Ea	233.00	32.80	265.80

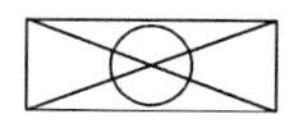

24"-wide air-handling lay-in T-bar fluorescent fixtures

Material	Craft@Hrs	Unit	Material Cost	Labor Cost	Installed Cost
48" 2 lamp	L1@0.70	Ea	188.00	28.70	216.70
48" 3 lamp	L1@0.75	Ea	233.00	30.80	263.80
48" 4 lamp	L1@0.80	Ea	234.00	32.80	266.80
48" 2 lamp, energy saver	L1@0.70	Ea	221.00	28.70	249.70
48" 3 lamp, energy saver	L1@0.75	Ea	245.00	30.80	275.80
48" 4 lamp, energy saver	L1@0.80	Ea	247.00	32.80	279.80

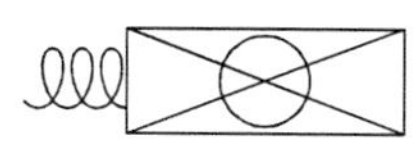

Use these figures to estimate the cost of fluorescent fixtures installed in buildings under the conditions described on pages 5 and 6. Costs listed are for each fixture installed. The crew is one electrician working at a labor cost of $41.03 per manhour. These costs include layout, material handling, and normal waste. Add for lamp, accessories, sales tax, delivery, supervision, mobilization, demobilization, cleanup, overhead and profit. Note: These costs are based on internally pre-wired fixtures installed in new construction not over 10 feet above the floor. These fixtures can be ordered with the lamps installed. Be sure to specify the lamp color needed. Lay-in fixtures are available both with flex tails ready for make-up in J-boxes and for soft wiring systems. Soft wiring is a special plug-in system used when wiring fixtures in a suspended ceiling system. The inter-connecting cables are factory assembled in various lengths with circuiting to meet most job requirements. Soft wiring systems carry the UL label and are accepted in most code jurisdictions. Energy saving (ES) ballasts are available and should be installed at the factory to save field labor hours. When ES fluorescent lighting fixtures are installed, be sure to use ES lamps. Special surface mounted and semi-flush fluorescent fixtures with wood siding are available for designer applications. You can also order parabolic louvers with colored anodized finishes.

Framed Troffer Lay-in T-Bar Fluorescent Fixtures

Material	Craft@Hrs	Unit	Material Cost	Labor Cost	Installed Cost
48"-wide air-handling lay-in T-bar fluorescent fixtures					
48" 4 lamp	L1@1.25	Ea	665.00	51.30	716.30
48" 6 lamp	L1@1.30	Ea	733.00	53.30	786.30
48" 8 lamp	L1@1.35	Ea	770.00	55.40	825.40
48" 4 lamp, energy saver	L1@1.25	Ea	757.00	51.30	808.30
48" 6 lamp, energy saver	L1@1.30	Ea	813.00	53.30	866.30
48" 8 lamp, energy saver	L1@1.35	Ea	873.00	55.40	928.40
12"-wide heat-recovery lay-in T-bar fluorescent fixtures					
48" 1 lamp	L1@0.55	Ea	151.00	22.60	173.60
48" 2 lamp	L1@0.60	Ea	154.00	24.60	178.60
48" 3 lamp	L1@0.65	Ea	192.00	26.70	218.70
48" 1 lamp, energy saver	L1@0.55	Ea	182.00	22.60	204.60
48" 2 lamp, energy saver	L1@0.60	Ea	184.00	24.60	208.60
48" 3 lamp, energy saver	L1@0.65	Ea	207.00	26.70	233.70
20"-wide heat-recovery lay-in T-bar fluorescent fixtures					
48" 2 lamp	L1@0.65	Ea	159.00	26.70	185.70
48" 3 lamp	L1@0.70	Ea	198.00	28.70	226.70
48" 4 lamp	L1@0.75	Ea	233.00	30.80	263.80
48" 2 lamp, energy saver	L1@0.65	Ea	188.00	26.70	214.70
48" 3 lamp, energy saver	L1@0.70	Ea	263.00	28.70	291.70
48" 4 lamp, energy saver	L1@0.75	Ea	258.00	30.80	288.80
24"-square heat-recovery lay-in T-bar fluorescent fixtures					
F40 U bent, 1 lamp	L1@0.50	Ea	151.00	20.50	171.50
F40 U bent, 2 lamp	L1@0.55	Ea	157.00	22.60	179.60
F40 U bent, 3 lamp	L1@0.60	Ea	198.00	24.60	222.60
F20 4 lamp	L1@0.60	Ea	215.00	24.60	239.60
24"-wide heat-recovery lay-in T-bar fluorescent fixtures					
48" 2 lamp	L1@0.70	Ea	172.00	28.70	200.70
48" 3 lamp	L1@0.75	Ea	209.00	30.80	239.80
48" 4 lamp	L1@0.80	Ea	211.00	32.80	243.80
48" 2 lamp, energy saver	L1@0.70	Ea	202.00	28.70	230.70
48" 3 lamp, energy saver	L1@0.75	Ea	258.00	30.80	288.80
48" 4 lamp, energy saver	L1@0.80	Ea	263.00	32.80	295.80

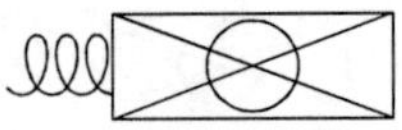
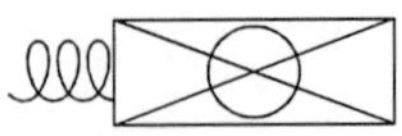
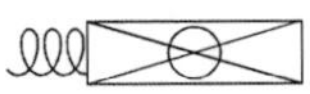
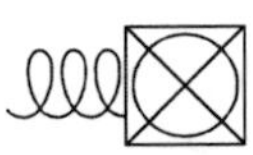
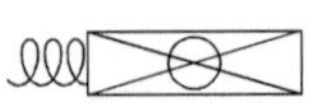

Use these figures to estimate the cost of fluorescent fixtures installed in buildings under the conditions described on pages 5 and 6. Costs listed are for each fixture installed. The crew is one electrician working at a labor cost of $41.03 per manhour. These costs include layout, material handling, and normal waste. Add for lamp, accessories, sales tax, delivery, supervision, mobilization, demobilization, cleanup, overhead and profit. Note: These costs are based on internally pre-wired fixtures installed in new construction not over 10 feet above the floor. These fixtures can be ordered with the lamps installed. Be sure to specify the lamp color needed. Lay-in fixtures are available both with flex tails ready for make-up in J-boxes and for soft wiring systems. Soft wiring is a special plug-in system used when wiring fixtures in a suspended ceiling system. The inter-connecting cables are factory assembled in various lengths with circuiting to meet most job requirements. Soft wiring systems carry the UL label and are accepted in most code jurisdictions. Energy saving (ES) ballasts are available and should be installed at the factory to save field labor hours. When ES fluorescent lighting fixtures are installed, be sure to use ES lamps. Special surface mounted and semi-flush fluorescent fixtures with wood siding are available for designer applications. You can also order parabolic louvers with colored anodized finishes.

Framed Troffer Lay-in T-Bar Fluorescent Fixtures

24"-wide return-air-handling lay-in T-bar fluorescent fixtures

Material	Craft@Hrs	Unit	Material Cost	Labor Cost	Installed Cost
48" 2 lamp	L1@0.70	Ea	160.00	28.70	188.70
48" 3 lamp	L1@0.75	Ea	186.00	30.80	216.80
48" 4 lamp	L1@0.80	Ea	189.00	32.80	221.80
48" 2 lamp, energy saver	L1@0.70	Ea	186.00	28.70	214.70
48" 3 lamp, energy saver	L1@0.75	Ea	241.00	30.80	271.80
48" 4 lamp, energy saver	L1@0.80	Ea	247.00	32.80	279.80

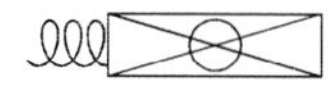

12"-wide lay-in T-bar fluorescent fixture with vandal resistant lens

Material	Craft@Hrs	Unit	Material Cost	Labor Cost	Installed Cost
48" 1 lamp	L1@0.55	Ea	209.00	22.60	231.60
48" 2 lamp	L1@0.60	Ea	211.00	24.60	235.60
48" 3 lamp	L1@0.65	Ea	247.00	26.70	273.70
48" 1 lamp, energy saver	L1@0.55	Ea	233.00	22.60	255.60
48" 2 lamp, energy saver	L1@0.60	Ea	236.00	24.60	260.60
48" 3 lamp, energy saver	L1@0.65	Ea	263.00	26.70	289.70

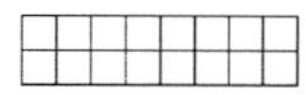

24"-square lay-in T-bar fluorescent fixture with vandal resistant lens

Material	Craft@Hrs	Unit	Material Cost	Labor Cost	Installed Cost
F40 U bent, 1 lamp	L1@0.50	Ea	209.00	20.50	229.50
F40 U bent, 2 lamp	L1@0.55	Ea	212.00	22.60	234.60
F40 U bent, 3 lamp	L1@0.60	Ea	247.00	24.60	271.60
F20 4 lamp	L1@0.60	Ea	274.00	24.60	298.60

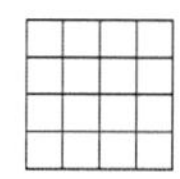

24"-wide lay-in T-bar fluorescent fixture with vandal resistant lens

Material	Craft@Hrs	Unit	Material Cost	Labor Cost	Installed Cost
48" 2 lamp	L1@0.70	Ea	234.00	28.70	262.70
48" 3 lamp	L1@0.75	Ea	267.00	30.80	297.80
48" 4 lamp	L1@0.80	Ea	276.00	32.80	308.80
48" 2 lamp, energy saver	L1@0.70	Ea	259.00	28.70	287.70
48" 3 lamp, energy saver	L1@0.75	Ea	314.00	30.80	344.80
48" 4 lamp, energy saver	L1@0.80	Ea	324.00	32.80	356.80

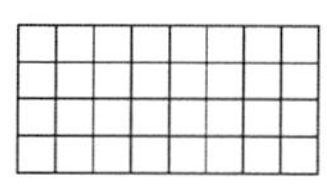

Use these figures to estimate the cost of fluorescent fixtures installed in buildings under the conditions described on pages 5 and 6. Costs listed are for each fixture installed. The crew is one electrician working at a labor cost of $41.03 per manhour. These costs include layout, material handling, and normal waste. Add for lamp, accessories, sales tax, delivery, supervision, mobilization, demobilization, cleanup, overhead and profit. Note: These costs are based on internally pre-wired fixtures installed in new construction not over 10 feet above the floor. These fixtures can be ordered with the lamps installed. Be sure to specify the lamp color needed. Lay-in fixtures are available both with flex tails ready for make-up in J-boxes and for soft wiring systems. Soft wiring is a special plug-in system used when wiring fixtures in a suspended ceiling system. The inter-connecting cables are factory assembled in various lengths with circuiting to meet most job requirements. Soft wiring systems carry the UL label and are accepted in most code jurisdictions. Energy saving (ES) ballasts are available and should be installed at the factory to save field labor hours. When ES fluorescent lighting fixtures are installed, be sure to use ES lamps. Special surface-mounted and semi-flush fluorescent fixtures with wood siding are available for designer applications. You can also order parabolic louvers with colored anodized finishes.

Recessed High Intensity Discharge Fixtures

14" high bay recessed HID open reflector fixtures

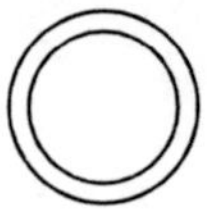

Material	Craft@Hrs	Unit	Material Cost	Labor Cost	Installed Cost
400W MV, 480V	L1@0.65	Ea	333.00	26.70	359.70
400W MV, Taps	L1@0.65	Ea	289.00	26.70	315.70
400W MV, Taps	L1@0.65	Ea	312.00	26.70	338.70
400W MH, Taps	L1@0.65	Ea	326.00	26.70	352.70
250W HPS, 480V	L1@0.65	Ea	331.00	26.70	357.70
250W HPS, Taps	L1@0.65	Ea	347.00	26.70	373.70
400W HPS, 480V	L1@0.65	Ea	371.00	26.70	397.70
400W HPS, Taps	L1@0.65	Ea	382.00	26.70	408.70

18" high bay recessed HID open reflector fixtures

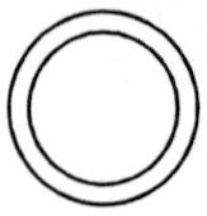

Material	Craft@Hrs	Unit	Material Cost	Labor Cost	Installed Cost
250W MV, Taps	L1@0.70	Ea	372.00	28.70	400.70
400W MV, Taps	L1@0.70	Ea	378.00	28.70	406.70
400W MH, Taps	L1@0.70	Ea	571.00	28.70	599.70
200W HPS, Taps	L1@0.70	Ea	486.00	28.70	514.70
250W HPS, Taps	L1@0.70	Ea	582.00	28.70	610.70
310W HPS, Taps	L1@0.70	Ea	536.00	28.70	564.70
400W HPS, Taps	L1@0.70	Ea	549.00	28.70	577.70

23" high bay recessed HID open reflector fixtures

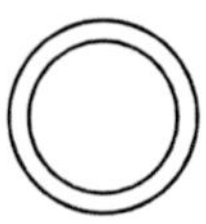

Material	Craft@Hrs	Unit	Material Cost	Labor Cost	Installed Cost
1000W MV, Taps	L1@0.75	Ea	746.00	30.80	776.80
1000W MH, Taps	L1@0.75	Ea	873.00	30.80	903.80
1000W HPS, Taps	L1@0.75	Ea	1,020.00	30.80	1,050.80

11" low bay recessed HID open reflector retrofit fixtures

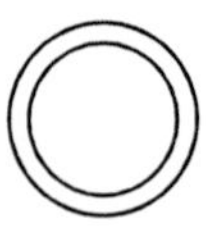

Material	Craft@Hrs	Unit	Material Cost	Labor Cost	Installed Cost
35W HPS, MSB	L1@0.15	Ea	189.00	6.15	195.15
50W HPS, MSB	L1@0.15	Ea	198.00	6.15	204.15
70W HPS, MSB	L1@0.15	Ea	199.00	6.15	205.15
35W HPS, CM	L1@0.35	Ea	189.00	14.40	203.40
50W HPS, CM	L1@0.35	Ea	198.00	14.40	212.40
70W HPS, CM	L1@0.35	Ea	200.00	14.40	214.40
35W HPS, OBM	L1@0.25	Ea	189.00	10.30	199.30
50W HPS, OBM	L1@0.25	Ea	198.00	10.30	208.30
70W HPS, OBM	L1@0.25	Ea	201.00	10.30	211.30

Use these figures to estimate the cost of lighting fixtures installed in buildings under the conditions described on pages 5 and 6. Costs listed are for each fixture installed. The crew is one electrician working at a labor cost of $41.03 per manhour. These costs include layout, material handling, and normal waste. Add for lamp, accessories, sales tax, delivery, supervision, mobilization, demobilization, cleanup, overhead and profit. Note: These costs are based on internally pre-wired fixtures installed in new construction not over 16 feet above the floor. Abbreviations: MV indicates mercury vapor, MH is metal halide, HPS is high pressure sodium.

14" low bay recessed HID open reflector fixtures

Material	Craft@Hrs	Unit	Material Cost	Labor Cost	Installed Cost
100W MV	L1@0.65	Ea	264.00	26.70	290.70
175W MV	L1@0.65	Ea	267.00	26.70	293.70
250W MV	L1@0.65	Ea	269.00	26.70	295.70
175W MH	L1@0.65	Ea	316.00	26.70	342.70
50W HPS	L1@0.65	Ea	346.00	26.70	372.70
70W HPS	L1@0.65	Ea	350.00	26.70	376.70
100W HPS	L1@0.65	Ea	350.00	26.70	376.70
150W HPS	L1@0.65	Ea	359.00	26.70	385.70

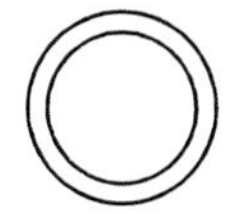

10"-diameter recessed high intensity work area enclosed fixtures

Material	Craft@Hrs	Unit	Material Cost	Labor Cost	Installed Cost
50W HPS	L1@0.50	Ea	274.00	20.50	294.50
70W HPS	L1@0.50	Ea	278.00	20.50	298.50

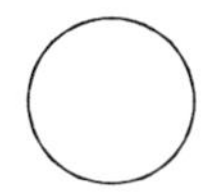

10"-diameter recessed high intensity work fixture for retrofit

Material	Craft@Hrs	Unit	Material Cost	Labor Cost	Installed Cost
50W HPS	L1@0.15	Ea	306.00	6.15	312.15
70W HPS	L1@0.15	Ea	308.00	6.15	314.15

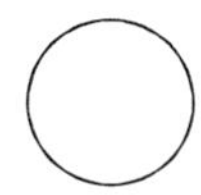

18"-diameter recessed HID industrial enclosed fixtures

Material	Craft@Hrs	Unit	Material Cost	Labor Cost	Installed Cost
100W MV	L1@0.70	Ea	302.00	28.70	330.70
150W MV	L1@0.70	Ea	308.00	28.70	336.70
250W MV	L1@0.70	Ea	310.00	28.70	338.70
175W MH	L1@0.70	Ea	313.00	28.70	341.70
100W HPS	L1@0.70	Ea	391.00	28.70	419.70
150W HPS	L1@0.70	Ea	397.00	28.70	425.70

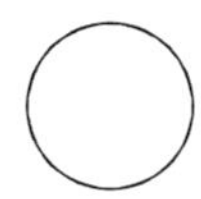

Use these figures to estimate the cost of lighting fixtures installed in buildings under the conditions described on pages 5 and 6. Costs listed are for each fixture installed. The crew is one electrician working at a labor cost of $41.03 per manhour. These costs include layout, material handling, and normal waste. Add for lamp, accessories, sales tax, delivery, supervision, mobilization, demobilization, cleanup, overhead and profit. Note: These costs are based on internally pre-wired fixtures installed in new construction not over 16 feet above the floor. Abbreviations: Mercury vapor is MV, metal halide is MH, high pressure sodium is HPS. Some high bay high intensity fixtures are available with a variety of mounting accessories: power hook assemblies for quick servicing, soft wiring connectors, and flexible tails for connections to J-boxes. Some manufacturers offer a computerized lighting design plan to help select the right fixture and the best location. In some cases it may be necessary to have high intensity light fixtures made to order to meet specific requirements.

Interior Industrial High Intensity Discharge Luminaires

Material	Craft@Hrs	Unit	Material Cost	Labor Cost	Installed Cost
16"-diameter enclosed 120 volt interior industrial HID fixtures					
250W MH	L1@0.70	Ea	451.00	28.70	479.70
200W HPS	L1@0.70	Ea	526.00	28.70	554.70
250W HPS	L1@0.70	Ea	548.00	28.70	576.70
29"-diameter enclosed 120, 208, 240 or 277 volt interior industrial HID fixtures					
400W MV	L1@0.75	Ea	415.00	30.80	445.80
250W MH	L1@0.75	Ea	417.00	30.80	447.80
400W MH	L1@0.75	Ea	539.00	30.80	569.80
200W HPS	L1@0.75	Ea	583.00	30.80	613.80
250W HPS	L1@0.75	Ea	592.00	30.80	622.80
310W HPS	L1@0.75	Ea	596.00	30.80	626.80
400W HPS	L1@0.75	Ea	606.00	30.80	636.80
16"-diameter enclosed 120, 208, 240 or 277 volt interior industrial HID fixtures for fast installation					
400W MV	L1@0.60	Ea	515.00	24.60	539.60
250W MH	L1@0.60	Ea	566.00	24.60	590.60
400W MH	L1@0.60	Ea	570.00	24.60	594.60
200W HPS	L1@0.60	Ea	641.00	24.60	665.60
250W HPS	L1@0.60	Ea	650.00	24.60	674.60
310W HPS	L1@0.60	Ea	665.00	24.60	689.60
400W HPS	L1@0.60	Ea	674.00	24.60	698.60

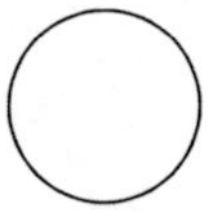

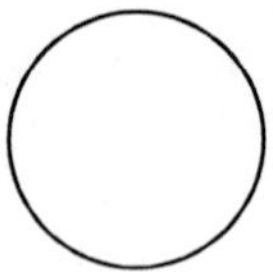

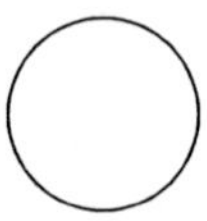

Use these figures to estimate the cost of lighting fixtures installed in buildings under the conditions described on pages 5 and 6. Costs listed are for each fixture installed. The crew is one electrician working at a labor cost of $41.03 per manhour. These costs include layout, material handling, and normal waste. Add for lamp, accessories, sales tax, delivery, supervision, mobilization, demobilization, cleanup, overhead and profit. Note: These costs are based on internally pre-wired fixtures installed in new construction not over 16 feet above the floor. Abbreviations: Mercury vapor is MV, metal halide is MH, high pressure sodium is HPS. Some high bay high intensity fixtures are available with a variety of mounting accessories: power hook assemblies for quick servicing, soft wiring connectors, and flexible tails for connections to J-boxes. Some manufacturers offer a computerized lighting design plan to help select the right fixture and the best location. In some cases it may be necessary to have high intensity light fixtures made to order to meet specific requirements.

HID bracket-mounted exterior floodlights with lamp base down

Material	Craft@Hrs	Unit	Material Cost	Labor Cost	Installed Cost
250W Mercury vapor	L1@1.25	Ea	513.00	51.30	564.30
400W Mercury vapor	L1@1.25	Ea	539.00	51.30	590.30
1000W Mercury vapor	L1@1.25	Ea	737.00	51.30	788.30
400W Metal halide	L1@1.25	Ea	592.00	51.30	643.30
1000W Metal halide	L1@1.25	Ea	802.00	51.30	853.30
200W Hi pres sodium	L1@1.25	Ea	619.00	51.30	670.30
250W Hi pres sodium	L1@1.25	Ea	644.00	51.30	695.30
400W Hi pres sodium	L1@1.25	Ea	675.00	51.30	726.30
1000W Hi pres sodium	L1@1.25	Ea	994.00	51.30	1,045.30

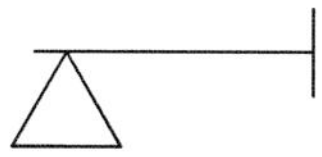

HID bracket-mounted exterior floodlights with lamp base horizontal

Material	Craft@Hrs	Unit	Material Cost	Labor Cost	Installed Cost
250W Mercury vapor	L1@1.25	Ea	516.00	51.30	567.30
400W Mercury vapor	L1@1.25	Ea	562.00	51.30	613.30
1000W Mercury vapor	L1@1.25	Ea	739.00	51.30	790.30
400W Metal halide	L1@1.25	Ea	583.00	51.30	634.30
1000W Metal halide	L1@1.25	Ea	802.00	51.30	853.30
200W Hi pres sodium	L1@1.25	Ea	619.00	51.30	670.30
250W Hi pres sodium	L1@1.25	Ea	641.00	51.30	692.30
400W Hi pres sodium	L1@1.25	Ea	675.00	51.30	726.30
1000W Hi pres sodium	L1@1.25	Ea	1,100.00	51.30	1,151.30

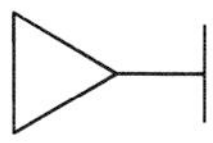

Small HID bracket-mounted exterior floodlights

Material	Craft@Hrs	Unit	Material Cost	Labor Cost	Installed Cost
35W Hi pres sodium	L1@1.00	Ea	239.00	41.00	280.00
50W Hi pres sodium	L1@1.00	Ea	241.00	41.00	282.00
70W Hi pres sodium	L1@1.00	Ea	246.00	41.00	287.00
100W Hi pres sodium	L1@1.00	Ea	250.00	41.00	291.00

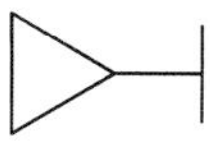

Heavy duty HID bracket-mounted floodlights with lamp base horizontal

Material	Craft@Hrs	Unit	Material Cost	Labor Cost	Installed Cost
400W Mercury vapor	L1@1.30	Ea	1,130.00	53.30	1,183.30
1000W Mercury vapor	L1@1.30	Ea	1,400.00	53.30	1,453.30
400W Metal halide	L1@1.30	Ea	1,130.00	53.30	1,183.30
1000W Metal halide	L1@1.30	Ea	1,400.00	53.30	1,453.30

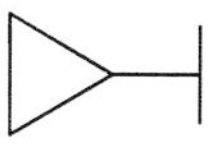

Use these figures to estimate the cost of bracket-mounted area lighting installed under the conditions described on pages 5 and 6. Costs listed are for each fixture installed. The crew is one electrician working at a labor cost of $41.03 per manhour. These costs include layout, material handling, and normal waste. Add for lamp, accessories, sales tax, delivery, supervision, mobilization, demobilization, cleanup, overhead and profit. Note: Many of these lighting fixtures can be mounted in clusters to reduce the number of poles needed. Some manufacturers offer a computerized layout service that will find the most efficient combination of lamps and locations that meets job requirements.

HID Lamp Base Horizontal Pole-Mounted Floodlights

HID pole-mounted floodlights with lamp base horizontal without poles

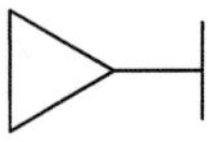

Material		Craft@Hrs	Unit	Material Cost	Labor Cost	Installed Cost
100W	Mercury vapor	L1@0.70	Ea	436.00	28.70	464.70
175W	Mercury vapor	L1@0.70	Ea	561.00	28.70	589.70
250W	Mercury vapor	L1@0.70	Ea	601.00	28.70	629.70
400W	Mercury vapor	L1@0.70	Ea	640.00	28.70	668.70
1000W	Mercury vapor	L1@0.70	Ea	922.00	28.70	950.70
250W	Metal halide	L1@0.70	Ea	686.00	28.70	714.70
400W	Metal halide	L1@0.70	Ea	687.00	28.70	715.70
1000W	Metal halide	L1@0.70	Ea	976.00	28.70	1,004.70
175W	Super metalarc	L1@0.70	Ea	659.00	28.70	687.70
250W	Super metalarc	L1@0.70	Ea	699.00	28.70	727.70
400W	Super metalarc	L1@0.70	Ea	703.00	28.70	731.70
100W	Hi pres sodium	L1@0.70	Ea	640.00	28.70	668.70
150W	Hi pres sodium	L1@0.70	Ea	665.00	28.70	693.70
200W	Hi pres sodium	L1@0.70	Ea	703.00	28.70	731.70
250W	Hi pres sodium	L1@0.70	Ea	734.00	28.70	762.70
400W	Hi pres sodium	L1@0.70	Ea	814.00	28.70	842.70
1000W	Hi pres sodium	L1@0.70	Ea	1,160.00	28.70	1,188.70

HID floodlights with lamp base horizontal, including typical poles

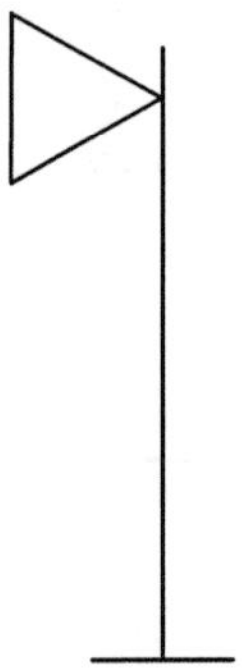

175W	Mercury vapor	L1@0.70	Ea	733.00	28.70	761.70
250W	Mercury vapor	L1@0.70	Ea	851.00	28.70	879.70
400W	Mercury vapor	L1@0.70	Ea	1,170.00	28.70	1,198.70
1000W	Mercury vapor	L1@0.80	Ea	2,020.00	32.80	2,052.80
250W	Metal halide	L1@0.70	Ea	1,170.00	28.70	1,198.70
400W	Metal halide	L1@0.70	Ea	1,020.00	28.70	1,048.70
1000W	Metal halide	L1@0.80	Ea	1,470.00	32.80	1,502.80
175W	Super metalarc	L1@0.70	Ea	873.00	28.70	901.70
250W	Super metalarc	L1@0.70	Ea	1,190.00	28.70	1,218.70
400W	Super metalarc	L1@0.70	Ea	1,290.00	28.70	1,318.70
18W	Lo pres sodium	L1@0.70	Ea	698.00	28.70	726.70
35W	Lo pres sodium	L1@0.70	Ea	814.00	28.70	842.70
100W	Hi pres sodium	L1@0.70	Ea	763.00	28.70	791.70
150W	Hi pres sodium	L1@0.70	Ea	929.00	28.70	957.70
250W	Hi pres sodium	L1@0.70	Ea	1,250.00	28.70	1,278.70
400W	Hi pres sodium	L1@0.70	Ea	1,200.00	28.70	1,228.70
1000W	Hi pres sodium	L1@0.80	Ea	1,660.00	32.80	1,692.80

Use these figures to estimate the cost of pole-mounted area lighting installed under the conditions described on pages 5 and 6. Costs listed are for each fixture installed. The crew is one electrician working at a labor cost of $41.03 per manhour. These costs include layout, material handling, and normal waste. Add for lamp, pole, accessories, sales tax, delivery, supervision, mobilization, demobilization, cleanup, overhead and profit. Note: Many of these lighting fixtures can be mounted in clusters to reduce the number of poles needed. Some manufacturers offer a computerized layout service that will find the most efficient combination of lamps and locations that meets job requirements.

Pole-mounted sphere HID exterior floodlights, lamp base down

Material	Craft@Hrs	Unit	Material Cost	Labor Cost	Installed Cost
75W Mercury vapor	L1@0.70	Ea	361.00	28.70	389.70
100W Mercury vapor	L1@0.70	Ea	364.00	28.70	392.70
175W Mercury vapor	L1@0.70	Ea	371.00	28.70	399.70
175W Metal halide	L1@0.70	Ea	375.00	28.70	403.70
35W Hi pres sodium	L1@0.70	Ea	391.00	28.70	419.70
50W Hi pres sodium	L1@0.70	Ea	396.00	28.70	424.70
70W Hi pres sodium	L1@0.70	Ea	429.00	28.70	457.70
100W Hi pres sodium	L1@0.70	Ea	413.00	28.70	441.70

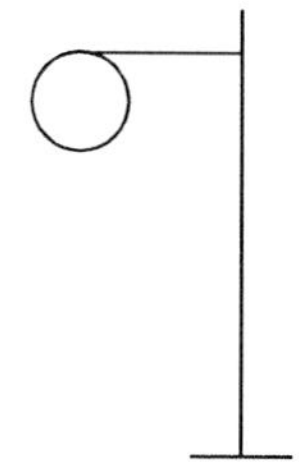

Wall-mounted cylinder HID exterior floodlights, lamp base down

Material	Craft@Hrs	Unit	Material Cost	Labor Cost	Installed Cost
75W Mercury vapor	L1@0.80	Ea	420.00	32.80	452.80
100W Mercury vapor	L1@0.80	Ea	427.00	32.80	459.80
175W Mercury vapor	L1@0.80	Ea	430.00	32.80	462.80
175W Metal halide	L1@0.80	Ea	436.00	32.80	468.80
35W Hi pres sodium	L1@0.80	Ea	463.00	32.80	495.80
50W Hi pres sodium	L1@0.80	Ea	469.00	32.80	501.80
70W Hi pres sodium	L1@0.80	Ea	487.00	32.80	519.80
100W Hi pres sodium	L1@0.80	Ea	496.00	32.80	528.80
150W Hi pres sodium	L1@0.80	Ea	501.00	32.80	533.80

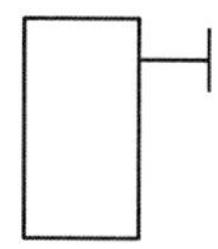

Wall-mounted round HID exterior floodlights, lamp base horizontal

Material	Craft@Hrs	Unit	Material Cost	Labor Cost	Installed Cost
175W Mercury vapor	L1@0.75	Ea	873.00	30.80	903.80
250W Mercury vapor	L1@0.75	Ea	909.00	30.80	939.80
400W Mercury vapor	L1@0.75	Ea	1,010.00	30.80	1,040.80
250W Metal halide	L1@0.75	Ea	1,030.00	30.80	1,060.80
400W Metal halide	L1@0.75	Ea	1,030.00	30.80	1,060.80
175W Super metalarc	L1@0.75	Ea	941.00	30.80	971.80
250W Super metalarc	L1@0.75	Ea	1,030.00	30.80	1,060.80
400W Super metalarc	L1@0.75	Ea	1,030.00	30.80	1,060.80
70W Hi pres sodium	L1@0.75	Ea	922.00	30.80	952.80
100W Hi pres sodium	L1@0.75	Ea	924.00	30.80	954.80
150W Hi pres sodium	L1@0.75	Ea	971.00	30.80	1,001.80
250W Hi pres sodium	L1@0.75	Ea	1,200.00	30.80	1,230.80
400W Hi pres sodium	L1@0.75	Ea	1,200.00	30.80	1,230.80

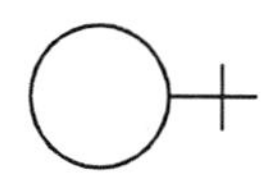

Use these figures to estimate the cost of area lighting installed under the conditions described on pages 5 and 6. Costs listed are for each fixture installed. The crew is one electrician working at a labor cost of $41.03 per manhour. These costs include layout, material handling, and normal waste. Add for lamp, pole, accessories, sales tax, delivery, supervision, mobilization, demobilization, cleanup, overhead and profit. Note: Many of these lighting fixtures can be mounted in clusters to reduce the number of poles needed. Some manufacturers offer a computerized layout service that will find the most efficient combination of lamps and locations that meets job requirements.

HID Off-Street Area Lighting, Lamp Base Up

HID off-street 120 volt area lighting with photocell control, lamp base up

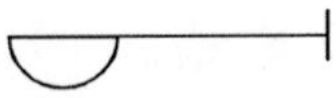

Material	Craft@Hrs	Unit	Material Cost	Labor Cost	Installed Cost
175W Mercury vapor	L1@0.60	Ea	148.00	24.60	172.60
50W Hi pres sodium	L1@0.60	Ea	325.00	24.60	349.60
70W Hi pres sodium	L1@0.60	Ea	325.00	24.60	349.60
100W Hi pres sodium	L1@0.60	Ea	333.00	24.60	357.60
150W Hi pres sodium	L1@0.60	Ea	338.00	24.60	362.60

HID off-street area lighting 120, 208, 240, 277 or 480 volt open luminaires

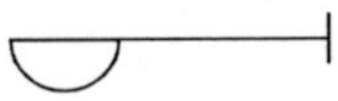

Material	Craft@Hrs	Unit	Material Cost	Labor Cost	Installed Cost
1000W Mercury vapor	L1@1.25	Ea	1,040.00	51.30	1,091.30
1000W Metal halide	L1@1.25	Ea	1,040.00	51.30	1,091.30
400W Hi pres sodium	L1@1.25	Ea	853.00	51.30	904.30
1000W Hi pres sodium	L1@1.25	Ea	1,190.00	51.30	1,241.30

HID off-street area lighting 120, 208, 240, 277 or 480 volt flat lens

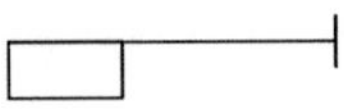

Material	Craft@Hrs	Unit	Material Cost	Labor Cost	Installed Cost
1000W Mercury vapor	L1@1.30	Ea	1,080.00	53.30	1,133.30
1000W Metal halide	L1@1.30	Ea	1,080.00	53.30	1,133.30
400W Hi pres sodium	L1@1.30	Ea	954.00	53.30	1,007.30
1000W Hi pres sodium	L1@1.30	Ea	1,400.00	53.30	1,453.30

HID off-street area lighting 120, 208, 240, 277 or 480 volt closed asymmetric lens

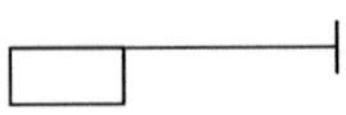

Material	Craft@Hrs	Unit	Material Cost	Labor Cost	Installed Cost
1000W Mercury vapor	L1@1.30	Ea	1,220.00	53.30	1,273.30
1000W Metal halide	L1@1.30	Ea	1,220.00	53.30	1,273.30
400W Hi pres sodium	L1@1.30	Ea	1,420.00	53.30	1,473.30
1000W Hi pres sodium	L1@1.30	Ea	1,510.00	53.30	1,563.30

Use these figures to estimate the cost of bracket-mounted area lighting installed under the conditions described on pages 5 and 6. Costs listed are for each fixture installed. The crew is one electrician working at a labor cost of $41.03 per manhour. These costs include layout, material handling, and normal waste. Add for lamp, accessories, sales tax, delivery, supervision, mobilization, demobilization, cleanup, overhead and profit. Note: Many of these lighting fixtures can be mounted in clusters to reduce the number of poles needed. Some manufacturers offer a computerized layout service that will find the most efficient combination of lamps and locations that meets job requirements. The luminaires on this page are intended for mounting at 60' above ground level. All high intensity discharge luminaires must go through a warm-up period when first energized and a cool-down period when de-energized. Always burn-in the lamp for at least an hour on the first turn-on. Some HID lamps (especially Metal halide) will not restart quickly after being disconnected from their power source. They must go through cool-down and warm-up all over again.

HID exterior pole-mounted rectangular architectural horizontal floodlights

Material	Craft@Hrs	Unit	Material Cost	Labor Cost	Installed Cost
100W Mercury vapor	L1@0.85	Ea	828.00	34.90	862.90
175W Mercury vapor	L1@0.85	Ea	837.00	34.90	871.90
250W Mercury vapor	L1@0.85	Ea	859.00	34.90	893.90
400W Mercury vapor	L1@0.85	Ea	885.00	34.90	919.90
400W Metal halide	L1@0.85	Ea	928.00	34.90	962.90
175W Super metalarc	L1@0.85	Ea	903.00	34.90	937.90
250W Super metalarc	L1@0.85	Ea	939.00	34.90	973.90
400W Super metalarc	L1@0.85	Ea	954.00	34.90	988.90
70W Hi pressure sodium	L1@0.85	Ea	941.00	34.90	975.90
100W Hi pressure sodium	L1@0.85	Ea	959.00	34.90	993.90
150W Hi pressure sodium	L1@0.85	Ea	965.00	34.90	999.90
200W Hi pressure sodium	L1@0.85	Ea	1,020.00	34.90	1,054.90
250W Hi pressure sodium	L1@0.85	Ea	1,030.00	34.90	1,064.90
310W Hi pressure sodium	L1@0.85	Ea	1,040.00	34.90	1,074.90
400W Hi pressure sodium	L1@0.85	Ea	1,050.00	34.90	1,084.90

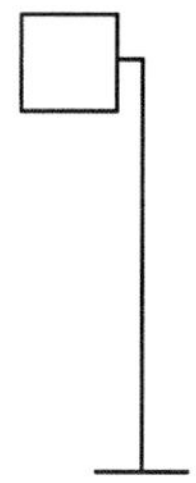

HID exterior pole-mounted architectural horizontal floodlights, 45 degree cutoff

Material	Craft@Hrs	Unit	Material Cost	Labor Cost	Installed Cost
400W Metal halide	L1@1.00	Ea	1,230.00	41.00	1,271.00
1000W Metal halide	L1@1.00	Ea	1,630.00	41.00	1,671.00
400W Hi pressure sodium	L1@1.00	Ea	1,410.00	41.00	1,451.00
1000W Hi pressure sodium	L1@1.00	Ea	1,830.00	41.00	1,871.00

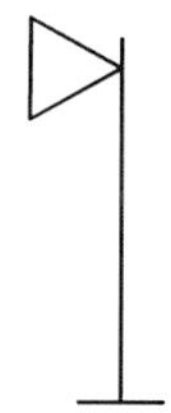

HID exterior pole-mounted architectural square floodlights, lamp base down

Material	Craft@Hrs	Unit	Material Cost	Labor Cost	Installed Cost
1000W Mercury vapor	L1@1.00	Ea	1,290.00	41.00	1,331.00
400W Metal halide	L1@1.00	Ea	1,200.00	41.00	1,241.00
1000W Metal halide	L1@1.00	Ea	1,290.00	41.00	1,331.00
250W Hi pressure sodium	L1@1.00	Ea	1,260.00	41.00	1,301.00
400W Hi pressure sodium	L1@1.00	Ea	1,300.00	41.00	1,341.00

Use these figures to estimate the cost of pole-mounted area lighting installed under the conditions described on pages 5 and 6. Costs listed are for each fixture installed. The crew is one electrician working at a labor cost of $41.03 per manhour. These costs include layout, material handling, and normal waste. Add for the pole, lamp, excavation, accessories, sales tax, delivery, supervision, mobilization, demobilization, cleanup, overhead and profit. Note: Many of these lighting fixtures can be mounted in clusters to reduce the number of poles needed. Some manufacturers offer a computerized layout service that will find the most efficient combination of lamps and locations that meets job requirements.

Exterior HID Walkway and Wall Fixtures

HID exterior square walkway 120 volt bollards, lamp base down

Material	Craft@Hrs	Unit	Material Cost	Labor Cost	Installed Cost
75W Mercury vapor	L1@1.75	Ea	986.00	71.80	1,057.80
100W Mercury vapor	L1@1.75	Ea	998.00	71.80	1,069.80
35W Hi pres sodium	L1@1.75	Ea	851.00	71.80	922.80
50W Hi pres sodium	L1@1.75	Ea	865.00	71.80	936.80
70W Hi pres sodium	L1@1.75	Ea	1,060.00	71.80	1,131.80
100W Hi pres sodium	L1@1.75	Ea	1,080.00	71.80	1,151.80

HID exterior round walkway 120 volt bollards, lamp base down

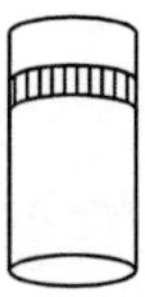

Material	Craft@Hrs	Unit	Material Cost	Labor Cost	Installed Cost
70W Mercury vapor	L1@1.75	Ea	977.00	71.80	1,048.80
100W Mercury vapor	L1@1.75	Ea	990.00	71.80	1,061.80
35W Hi pres sodium	L1@1.75	Ea	845.00	71.80	916.80
50W Hi pres sodium	L1@1.75	Ea	859.00	71.80	930.80
70W Hi pres sodium	L1@1.75	Ea	1,070.00	71.80	1,141.80
100W Hi pres sodium	L1@1.75	Ea	1,090.00	71.80	1,161.80

HID exterior prismatic lens 120 to 480 volt wall luminaires, lamp base horizontal

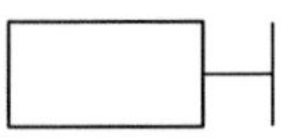

Material	Craft@Hrs	Unit	Material Cost	Labor Cost	Installed Cost
100W Mercury vapor	L1@1.00	Ea	347.00	41.00	388.00
175W Mercury vapor	L1@1.00	Ea	350.00	41.00	391.00
175W Metal halide	L1@1.00	Ea	364.00	41.00	405.00
18W Lo pres sodium	L1@1.00	Ea	408.00	41.00	449.00
35W Hi pres sodium	L1@1.00	Ea	373.00	41.00	414.00
50W Hi pres sodium	L1@1.00	Ea	384.00	41.00	425.00
70W Hi pres sodium	L1@1.00	Ea	427.00	41.00	468.00
100W Hi pres sodium	L1@1.00	Ea	441.00	41.00	482.00

HID exterior clear lens 120 to 480 volt wall luminaires, lamp base horizontal

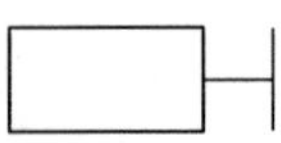

Material	Craft@Hrs	Unit	Material Cost	Labor Cost	Installed Cost
100W Mercury vapor	L1@1.00	Ea	411.00	41.00	452.00
175W Mercury vapor	L1@1.00	Ea	411.00	41.00	452.00
175W Metal halide	L1@1.00	Ea	423.00	41.00	464.00
18W Lo pres sodium	L1@1.00	Ea	467.00	41.00	508.00
35W Hi pres sodium	L1@1.00	Ea	427.00	41.00	468.00
50W Hi pres sodium	L1@1.00	Ea	441.00	41.00	482.00
70W Hi pres sodium	L1@1.00	Ea	489.00	41.00	530.00
100W Hi pres sodium	L1@1.00	Ea	500.00	41.00	541.00

Use these figures to estimate the cost of bracket mounted area lighting installed under the conditions described on pages 5 and 6. Costs listed are for each fixture installed. The crew is one electrician working at a labor cost of $41.03 per manhour. These costs include layout, material handling, and normal waste. Add for lamp, accessories, sales tax, delivery, supervision, mobilization, demobilization, cleanup, overhead and profit. Note: Many of these lighting fixtures can be mounted in clusters to reduce the number of poles needed. Some manufacturers offer a computerized layout service that will find the most efficient combination of lamps and locations that meets job requirements.

HID exterior pole-mounted flat lens square arm 120 volt luminaires

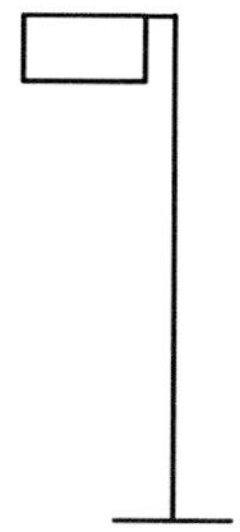

Material	Craft@Hrs	Unit	Material Cost	Labor Cost	Installed Cost
1000W Mercury vapor	L1@1.50	Ea	1,510.00	61.50	1,571.50
1000W Metal halide	L1@1.50	Ea	1,590.00	61.50	1,651.50
1000W Hi pres sodium	L1@1.50	Ea	2,000.00	61.50	2,061.50

HID pole-mounted square exterior walkway 120 volt luminaires, base down

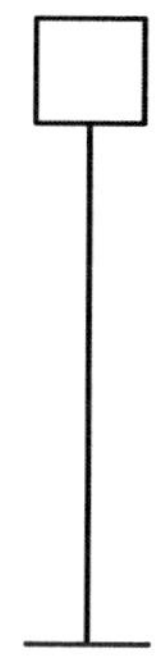

Material	Craft@Hrs	Unit	Material Cost	Labor Cost	Installed Cost
100W Mercury vapor	L1@0.70	Ea	1,690.00	28.70	1,718.70
175W Mercury vapor	L1@0.70	Ea	1,720.00	28.70	1,748.70
250W Mercury vapor	L1@0.70	Ea	1,740.00	28.70	1,768.70
400W Mercury vapor	L1@0.70	Ea	1,880.00	28.70	1,908.70
175W Metal halide	L1@0.70	Ea	1,860.00	28.70	1,888.70
250W Metal halide	L1@0.70	Ea	1,890.00	28.70	1,918.70
400W Metal halide	L1@0.70	Ea	1,890.00	28.70	1,918.70
70W Hi pres sodium	L1@0.70	Ea	1,720.00	28.70	1,748.70
100W Hi pres sodium	L1@0.70	Ea	1,740.00	28.70	1,768.70
150W Hi pres sodium	L1@0.70	Ea	1,880.00	28.70	1,908.70
250W Hi pres sodium	L1@0.70	Ea	2,010.00	28.70	2,038.70
400W Hi pres sodium	L1@0.70	Ea	2,070.00	28.70	2,098.70

HID pole-mounted round exterior walkway 120 volt luminaires, base down

Material	Craft@Hrs	Unit	Material Cost	Labor Cost	Installed Cost
100W Mercury vapor	L1@0.70	Ea	1,720.00	28.70	1,748.70
175W Mercury vapor	L1@0.70	Ea	1,760.00	28.70	1,788.70
400W Mercury vapor	L1@0.70	Ea	1,950.00	28.70	1,978.70
175W Metal halide	L1@0.70	Ea	1,880.00	28.70	1,908.70
250W Metal halide	L1@0.70	Ea	1,940.00	28.70	1,968.70
400W Metal halide	L1@0.70	Ea	2,010.00	28.70	2,038.70
70W Hi pres sodium	L1@0.70	Ea	1,750.00	28.70	1,778.70
100W Hi pres sodium	L1@0.70	Ea	1,760.00	28.70	1,788.70
150W Hi pres sodium	L1@0.70	Ea	1,950.00	28.70	1,978.70
250W Hi pres sodium	L1@0.70	Ea	2,080.00	28.70	2,108.70
400W Hi pres sodium	L1@0.70	Ea	2,120.00	28.70	2,148.70

Use these figures to estimate the cost of pole-mounted area lighting installed under the conditions described on pages 5 and 6. Costs listed are for each fixture installed. The crew is one electrician working at a labor cost of $41.03 per manhour. These costs include layout, material handling, and normal waste. Add for lamppole, accessories, sales tax, delivery, supervision, mobilization, demobilization, cleanup, overhead and profit. Note: Many of these lighting fixtures can be mounted in clusters to reduce the number of poles needed. Some manufacturers offer a computerized layout service that will find the most efficient combination of lamps and locations that meets job requirements.

Pole-Mounted HID Square Luminaires, 480 Volts

HID pole-mounted 480 volt square luminaires with flat glass lens

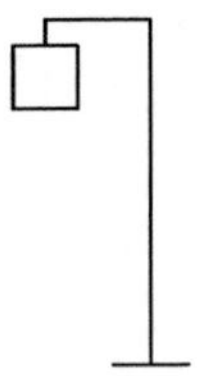

Material	Craft@Hrs	Unit	Material Cost	Labor Cost	Installed Cost
1000W Mercury vapor	L1@1.50	Ea	1,290.00	61.50	1,351.50
400W Metal halide	L1@1.50	Ea	1,290.00	61.50	1,351.50
1000W Metal halide	L1@1.50	Ea	1,300.00	61.50	1,361.50
250W Hi pres sodium	L1@1.50	Ea	1,330.00	61.50	1,391.50
400W Hi pres sodium	L1@1.50	Ea	1,370.00	61.50	1,431.50
1000W Hi pres sodium	L1@1.50	Ea	1,530.00	61.50	1,591.50

HID pole-mounted 480 volt square luminaires with convex glass lens

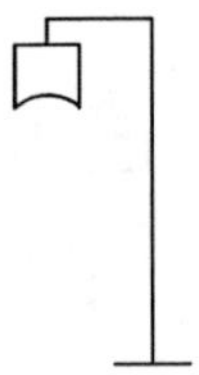

Material	Craft@Hrs	Unit	Material Cost	Labor Cost	Installed Cost
1000W Mercury vapor	L1@1.50	Ea	1,400.00	61.50	1,461.50
400W Metal halide	L1@1.50	Ea	1,410.00	61.50	1,471.50
1000W Metal halide	L1@1.50	Ea	1,370.00	61.50	1,431.50
250W Hi pres sodium	L1@1.50	Ea	1,400.00	61.50	1,461.50
400W Hi pres sodium	L1@1.50	Ea	1,410.00	61.50	1,471.50
1000W Hi pres sodium	L1@1.50	Ea	1,610.00	61.50	1,671.50

HID pole-mounted 480 volt round luminaires with flat glass lens

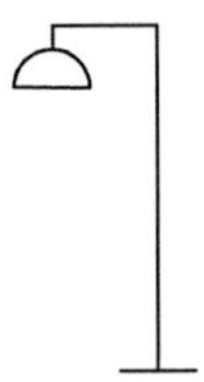

Material	Craft@Hrs	Unit	Material Cost	Labor Cost	Installed Cost
1000W Mercury vapor	L1@1.50	Ea	1,300.00	61.50	1,361.50
400W Metal halide	L1@1.50	Ea	1,250.00	61.50	1,311.50
1000W Metal halide	L1@1.50	Ea	1,300.00	61.50	1,361.50
250W Hi pres sodium	L1@1.50	Ea	1,300.00	61.50	1,361.50
400W Hi pres sodium	L1@1.50	Ea	1,330.00	61.50	1,391.50
1000W Hi pres sodium	L1@1.50	Ea	1,530.00	61.50	1,591.50

HID pole-mounted 480 volt round luminaires with convex glass lens

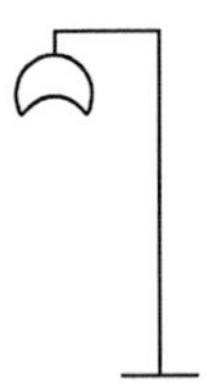

Material	Craft@Hrs	Unit	Material Cost	Labor Cost	Installed Cost
1000W Mercury vapor	L1@1.50	Ea	1,370.00	61.50	1,431.50
400W Metal halide	L1@1.50	Ea	1,290.00	61.50	1,351.50
1000W Metal halide	L1@1.50	Ea	1,370.00	61.50	1,431.50
250W Hi pres sodium	L1@1.50	Ea	1,370.00	61.50	1,431.50
400W Hi pres sodium	L1@1.50	Ea	1,370.00	61.50	1,431.50
1000W Hi pres sodium	L1@1.50	Ea	1,590.00	61.50	1,651.50

Use these figures to estimate the cost of pole-mounted area lighting installed under the conditions described on pages 5 and 6. Costs listed are for each fixture installed. The crew is one electrician working at a labor cost of $41.03 per manhour. These costs include layout, material handling, and normal waste. Add for lamp pole, accessories, sales tax, delivery, supervision, mobilization, demobilization, cleanup, overhead and profit. Note: These luminaires are intended to be mounted on 40' to 60' poles.

HID street and roadway luminaires, 28"-long housing

Material	Craft@Hrs	Unit	Material Cost	Labor Cost	Installed Cost
100W Mercury vapor	L1@0.80	Ea	267.00	32.80	299.80
175W Mercury vapor	L1@0.80	Ea	316.00	32.80	348.80
250W Mercury vapor	L1@0.80	Ea	369.00	32.80	401.80
50W Hi pres sodium	L1@0.80	Ea	304.00	32.80	336.80
70W Hi pres sodium	L1@0.80	Ea	378.00	32.80	410.80
100W Hi pres sodium	L1@0.80	Ea	317.00	32.80	349.80
150W Hi pres sodium	L1@0.80	Ea	346.00	32.80	378.80
200W Hi pres sodium	L1@0.80	Ea	474.00	32.80	506.80
250W Hi pres sodium	L1@0.80	Ea	499.00	32.80	531.80

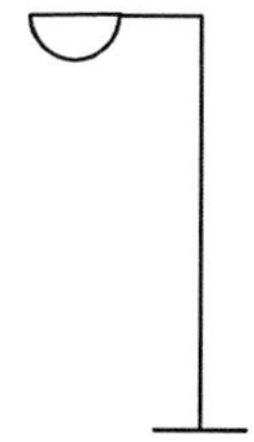

HID street and roadway luminaires, 29"-long housing

Material	Craft@Hrs	Unit	Material Cost	Labor Cost	Installed Cost
100W Mercury vapor	L1@0.85	Ea	325.00	34.90	359.90
175W Mercury vapor	L1@0.85	Ea	328.00	34.90	362.90
250W Mercury vapor	L1@0.85	Ea	372.00	34.90	406.90
70W Hi pres sodium	L1@0.85	Ea	337.00	34.90	371.90
100W Hi pres sodium	L1@0.85	Ea	339.00	34.90	373.90
150W Hi pres sodium	L1@0.85	Ea	350.00	34.90	384.90
200W Hi pres sodium	L1@0.85	Ea	452.00	34.90	486.90
250W Hi pres sodium	L1@0.85	Ea	478.00	34.90	512.90

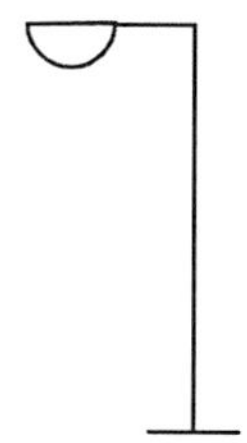

HID street and roadway luminaires, 33"-long housing

Material	Craft@Hrs	Unit	Material Cost	Labor Cost	Installed Cost
400W Mercury vapor	L1@1.00	Ea	350.00	41.00	391.00
400W Metal halide	L1@1.00	Ea	469.00	41.00	510.00
200W Hi pres sodium	L1@1.00	Ea	498.00	41.00	539.00
250W Hi pres sodium	L1@1.00	Ea	509.00	41.00	550.00
310W Hi pres sodium	L1@1.00	Ea	545.00	41.00	586.00
400W Hi pres sodium	L1@1.00	Ea	553.00	41.00	594.00

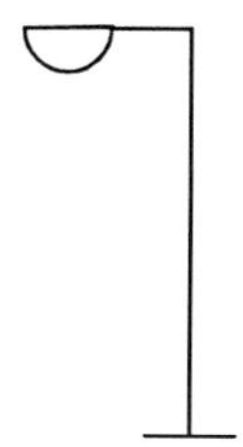

Use these figures to estimate the cost of street and roadway luminaires installed under the conditions described on pages 5 and 6. Costs listed are for each unit installed. The crew is one electrician working at a labor cost of $41.03 per manhour. These costs include layout, material handling, and normal waste. Add for pole, arm, lamp, accessories, sales tax, delivery, supervision, mobilization, demobilization, cleanup, overhead and profit. Note: These luminaires mount on an arm that supports the fixture housing in a horizontal position. Usually the arm is attached to an 18' to 30' pole or to a wall mounting bracket. The housing has provision for installing a photo cell control switch.

HID Street Lighting and Ballasts

HID street luminaires, 33"-long housing, lamp base horizontal

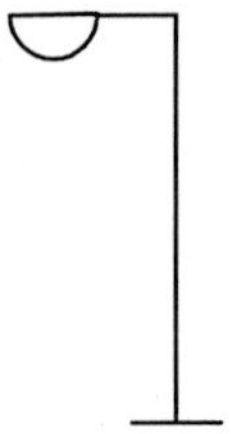

Material		Craft@Hrs	Unit	Material Cost	Labor Cost	Installed Cost
400W	Mercury vapor	L1@1.25	Ea	778.00	51.30	829.30
200W	Hi pres sodium	L1@1.25	Ea	944.00	51.30	995.30
250W	Hi pres sodium	L1@1.25	Ea	990.00	51.30	1,041.30
310W	Hi pres sodium	L1@1.25	Ea	1,050.00	51.30	1,101.30
400W	Hi pres sodium	L1@1.25	Ea	1,080.00	51.30	1,131.30

HID street luminaires, 40"-long housing, lamp base horizontal

Material		Craft@Hrs	Unit	Material Cost	Labor Cost	Installed Cost
1000W	Mercury vapor	L1@1.50	Ea	828.00	61.50	889.50
1000W	Metal halide	L1@1.50	Ea	1,710.00	61.50	1,771.50
1000W	Hi pres sodium	L1@1.50	Ea	1,170.00	61.50	1,231.50

Enclosed indoor mercury vapor ballasts

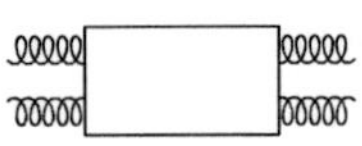

Material		Craft@Hrs	Unit	Material Cost	Labor Cost	Installed Cost
400W	MV 1 lamp	L1@0.40	Ea	338.00	16.40	354.40
400W	MV 2 lamp	L1@0.40	Ea	489.00	16.40	505.40
1000W	MV 1 lamp	L1@0.45	Ea	579.00	18.50	597.50
1000W	MV 2 lamp	L1@0.45	Ea	1,030.00	18.50	1,048.50

Exterior weatherproof potted high intensity ballasts

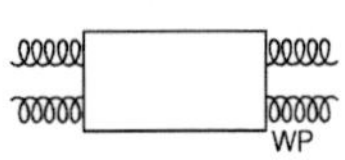

Material		Craft@Hrs	Unit	Material Cost	Labor Cost	Installed Cost
175W	MV 1 lamp	L1@0.50	Ea	364.00	20.50	384.50
250W	MV 1 lamp	L1@0.50	Ea	411.00	20.50	431.50
400W	MV 1 lamp	L1@0.50	Ea	410.00	20.50	430.50
400W	MV 2 lamp	L1@0.60	Ea	936.00	24.60	960.60
1000W	MV 1 lamp	L1@0.55	Ea	760.00	22.60	782.60
1000W	MV 2 lamp	L1@0.60	Ea	1,400.00	24.60	1,424.60
400W	MH 1 lamp	L1@0.50	Ea	539.00	20.50	559.50
400W	MH 2 lamp	L1@0.60	Ea	1,010.00	24.60	1,034.60
1000W	MH 1 lamp	L1@0.55	Ea	1,100.00	22.60	1,122.60
250W	HPS 1 lamp	L1@0.50	Ea	724.00	20.50	744.50
400W	HPS 1 lamp	L1@0.50	Ea	869.00	20.50	889.50
1000W	HPS 1 lamp	L1@0.60	Ea	1,610.00	24.60	1,634.60

Use these figures to estimate the cost of street and roadway luminaires and ballasts installed under the conditions described on pages 5 and 6. Costs listed are for each unit installed. The crew is one electrician working at a labor cost of $41.03 per manhour. These costs include layout, material handling, and normal waste. Add for pole, arm, lamp, special accessories, sales tax, delivery, supervision, mobilization, demobilization, cleanup, overhead and profit. Note: These luminaires mount on an arm that supports the fixture housing in a horizontal position. Usually the arm is attached to an 18' to 30' pole or to a wall mounting bracket. The housing has provision for installing a photo cell control switch.

Steel Light Poles with Square Mounting Base

Round steel street light poles with square mounting base

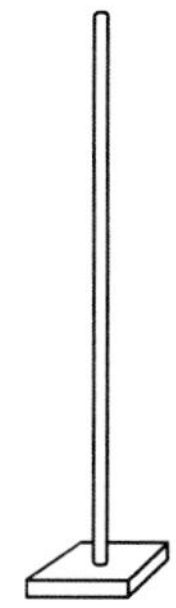

Material	Craft@Hrs	Unit	Material Cost	Labor Cost	Installed Cost
8' high, 4" diameter	L2@0.70	Ea	767.00	28.70	795.70
10' high, 4" diameter	L2@0.75	Ea	880.00	30.80	910.80
12' high, 4" diameter	L2@0.90	Ea	861.00	36.90	897.90
14' high, 4" diameter	L2@1.00	Ea	994.00	41.00	1,035.00
16' high, 4" diameter	L2@1.20	Ea	1,070.00	49.20	1,119.20
18' high, 4" diameter	L2@1.50	Ea	1,110.00	61.50	1,171.50
Add for 6" diameter	—	—	60.0%	10.0%	—

Round tapered steel street light poles with square mounting base

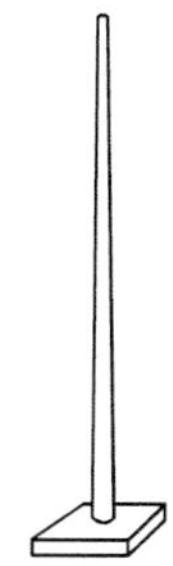

Material	Craft@Hrs	Unit	Material Cost	Labor Cost	Installed Cost
20' high	L2@1.25	Ea	2,150.00	51.30	2,201.30
25' high	L2@1.50	Ea	2,430.00	61.50	2,491.50
30' high	L2@1.75	Ea	2,770.00	71.80	2,841.80
35' high	L2@2.00	Ea	3,430.00	82.10	3,512.10
39' high	L2@2.50	Ea	3,870.00	103.00	3,973.00
50' high	L2@5.00	Ea	7,580.00	205.00	7,785.00
Add for heavy duty	—	—	25.0%	23.0%	—

Square steel street light poles with square mounting base

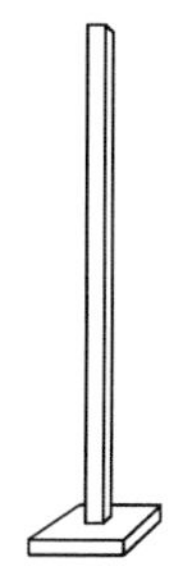

Material	Craft@Hrs	Unit	Material Cost	Labor Cost	Installed Cost
10' high	L2@0.75	Ea	967.00	30.80	997.80
12' high	L2@0.90	Ea	1,090.00	36.90	1,126.90
14' high	L2@1.00	Ea	1,200.00	41.00	1,241.00
16' high	L2@1.20	Ea	1,270.00	49.20	1,319.20
20' high	L2@1.60	Ea	1,860.00	65.60	1,925.60
24' high	L2@1.75	Ea	2,530.00	71.80	2,601.80
30' high	L2@2.00	Ea	2,770.00	82.10	2,852.10
35' high	L2@2.50	Ea	3,280.00	103.00	3,383.00

Use these figures to estimate the cost of street and yard light poles installed under the conditions described on pages 5 and 6. Costs listed are for each pole installed. The crew is two electricians working at a labor cost of $41.03 per manhour. These costs include four anchor bolts, layout, material handling, and normal waste. Add for the fixture, lamps, pole foundation, brackets, arms, excavation, sales tax, delivery, supervision, mobilization, demobilization, cleanup, overhead and profit. Note: Hoisting equipment may be necessary on the larger poles. Read the specs carefully when pricing poles. Some jobs will require items that are not included in these costs: decorative bases, anchor bolt covers, ballast bases, weatherproof receptacles, pole vibration dampers, or lowering devices.

Light Poles with Square Mounting Base

Material	Craft@Hrs	Unit	Material Cost	Labor Cost	Installed Cost

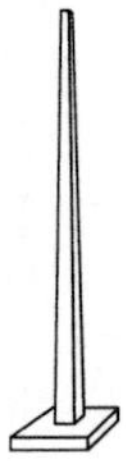

Square tapered steel light poles with square mounting base

Material	Craft@Hrs	Unit	Material Cost	Labor Cost	Installed Cost
30' high	L2@2.00	Ea	2,530.00	82.10	2,612.10
35' high	L2@2.50	Ea	3,400.00	103.00	3,503.00
39' high	L2@3.00	Ea	3,510.00	123.00	3,633.00
50' high	L2@6.00	Ea	6,150.00	246.00	6,396.00
60' high	L2@7.00	Ea	7,820.00	287.00	8,107.00
Add for heavy duty	—	—	25.0%	22.0%	—

Hinged square steel light poles with square mounting base

Material	Craft@Hrs	Unit	Material Cost	Labor Cost	Installed Cost
20' high	L2@1.60	Ea	2,530.00	65.60	2,595.60
24' high	L2@1.80	Ea	3,040.00	73.90	3,113.90
30' high	L2@2.20	Ea	3,590.00	90.30	3,680.30
35' high	L2@3.00	Ea	4,410.00	123.00	4,533.00
39' high	L2@3.50	Ea	5,520.00	144.00	5,664.00

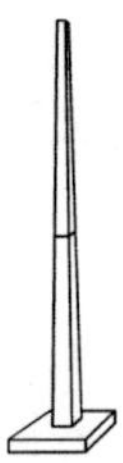

Hinged square tapered steel light poles with square mounting base

Material	Craft@Hrs	Unit	Material Cost	Labor Cost	Installed Cost
30' high	L2@2.25	Ea	4,560.00	92.30	4,652.30
35' high	L2@2.55	Ea	5,030.00	105.00	5,135.00
39' high	L2@3.10	Ea	6,460.00	127.00	6,587.00

Round tapered aluminum light poles with square mounting base

Material	Craft@Hrs	Unit	Material Cost	Labor Cost	Installed Cost
6' high	L2@0.50	Ea	499.00	20.50	519.50
8' high	L2@0.65	Ea	573.00	26.70	599.70
10' high	L2@0.70	Ea	644.00	28.70	672.70
12' high	L2@0.80	Ea	954.00	32.80	986.80
15' high	L2@0.85	Ea	821.00	34.90	855.90
18' high	L2@1.00	Ea	1,100.00	41.00	1,141.00
20' high	L2@1.15	Ea	1,340.00	47.20	1,387.20
25' high	L2@1.30	Ea	2,130.00	53.30	2,183.30
Add for heavy duty	—	—	33.0%	10.0%	—

Round tapered aluminum light poles without arms, square base

Material	Craft@Hrs	Unit	Material Cost	Labor Cost	Installed Cost
30' high	L2@1.40	Ea	1,660.00	57.40	1,717.40
35' high	L2@1.75	Ea	3,610.00	71.80	3,681.80
40' high	L2@2.25	Ea	4,840.00	92.30	4,932.30
45' high	L2@2.75	Ea	5,780.00	113.00	5,893.00
50' high	L2@3.50	Ea	6,670.00	144.00	6,814.00
Add for heavy duty	—	—	33.0%	10.0%	—

Use these figures to estimate the cost of street and yard light poles installed under the conditions described on pages 5 and 6. Costs listed are for each pole installed. The crew is two electricians working at a labor cost of $41.03 per manhour. These costs include four anchor bolts, layout, material handling, and normal waste. Add for the fixture, lamps, pole foundation, brackets, arms, excavation, sales tax, delivery, supervision, mobilization, demobilization, cleanup, overhead and profit. Note: Hoisting equipment may be necessary on the larger poles. Read the specs carefully when pricing poles. Some jobs will require items that are not included in these costs: decorative bases, anchor bolt covers, ballast bases, weatherproof receptacles, pole vibration dampers, or lowering devices.

Round tapered aluminum light poles with one arm, square base

Material	Craft@Hrs	Unit	Material Cost	Labor Cost	Installed Cost
20' pole, 4' arm	L2@1.30	Ea	1,510.00	53.30	1,563.30
25' pole, 4' arm	L2@1.40	Ea	2,100.00	57.40	2,157.40
30' pole, 4' arm	L2@1.50	Ea	2,760.00	61.50	2,821.50
35' pole, 4' arm	L2@1.80	Ea	3,330.00	73.90	3,403.90
Add for 6' arm	—	—	5.0%	4.0%	—
Add for 8' arm	—	—	15.0%	7.0%	—

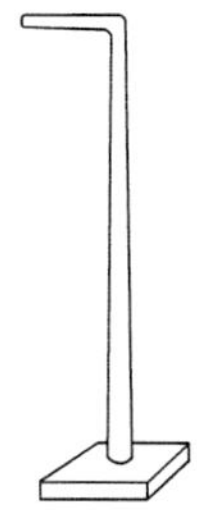

Round tapered aluminum light poles with two arms, square base

20' pole, 4' arms	L2@1.40	Ea	1,830.00	57.40	1,887.40
25' pole, 4' arms	L2@1.60	Ea	2,340.00	65.60	2,405.60
30' pole, 4' arms	L2@1.80	Ea	2,980.00	73.90	3,053.90
35' pole, 4' arms	L2@2.00	Ea	3,540.00	82.10	3,622.10
Add for 6' arms	—	—	7.0%	10.0%	—
Add for 8' arms	—	—	10.0%	10.0%	—

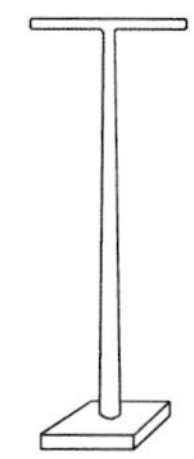

Round tapered aluminum light poles with three arms, square base

30' pole, 4' arms	L2@1.90	Ea	3,490.00	78.00	3,568.00
35' pole, 4' arms	L2@2.30	Ea	3,910.00	94.40	4,004.40
Add for 6' arms	—	—	8.0%	10.0%	—
Add for 8' arms	—	—	10.0%	15.0%	—

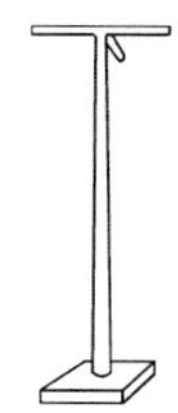

Round tapered aluminum light poles with four arms, square base

30' pole, 4' arms	L2@2.00	Ea	3,770.00	82.10	3,852.10
35' pole, 4' arms	L2@2.60	Ea	4,320.00	107.00	4,427.00
Add for 6' arms	—	—	9.0%	10.0%	—
Add for 8' arms	—	—	13.0%	20.0%	—

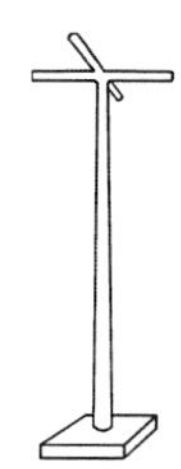

Square aluminum light poles without arms

8' high	L2@0.65	Ea	1,210.00	26.70	1,236.70
10' high	L2@0.70	Ea	1,480.00	28.70	1,508.70
12' high	L2@0.80	Ea	1,420.00	32.80	1,452.80
15' high	L2@0.85	Ea	1,570.00	34.90	1,604.90
18' high	L2@1.00	Ea	2,520.00	41.00	2,561.00
20' high	L2@1.15	Ea	2,280.00	47.20	2,327.20
25' high	L2@1.30	Ea	3,490.00	53.30	3,543.30
30' high	L2@1.40	Ea	5,750.00	57.40	5,807.40
Add for heavy duty	—	—	45.0%	8.0%	—

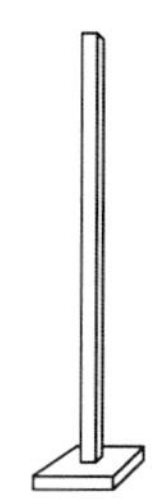

Use these figures to estimate the cost of street and yard light poles installed under the conditions described on pages 5 and 6. Costs listed are for each pole installed. The crew is two electricians working at a labor cost of $41.03 per manhour. These costs include four anchor bolts, layout, material handling, and normal waste. Add for the fixture, lamps, pole foundation, brackets, arms (except as noted above), excavation, sales tax, delivery, supervision, mobilization, demobilization, cleanup, overhead and profit. Note: Hoisting equipment may be necessary on the larger poles. Read the specs carefully when pricing poles. Some jobs will require items that are not included in these costs: decorative bases, anchor bolt covers, ballast bases, weatherproof receptacles, pole vibration dampers, or lowering devices.

LED Light Fixtures

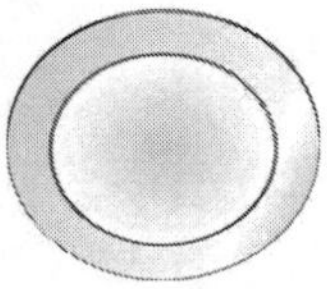

Material	Craft@Hrs	Unit	Material Cost	Labor Cost	Installed Cost
Recessed-LED 12 volt accent lights, aluminum reflector, 2-1/2" glass lens and pigtail connection, dimmable					
1 watt	L1@1.15	Ea	17.90	47.20	65.10
5 watt	L1@1.15	Ea	27.20	47.20	74.40
6 watt, waterproof	L1@1.15	Ea	54.40	47.20	101.60
Recessed LED downlight, 100 to 240 volt, frosted dome lens, white baffle trim, dimmable					
5" dia. 9W, 60W equiv.	L1@1.21	Ea	24.10	49.60	73.70
6" dia. 15W, 7 W equiv.	L1@1.21	Ea	27.20	49.60	76.80
8" dia. 24W, 190W equiv.	L1@1.21	Ea	21.60	49.60	71.20
6" square, 9W, 60W equiv.	L1@1.21	Ea	16.30	49.60	65.90
Flush mount-LED ceiling lights, frosted dome lens, white aluminum trim, dimmable					
5-1/2" dia. 10W, 60W equiv.	L1@1.21	Ea	17.90	49.60	67.50
7" dia. 12W, 75W equiv.	L1@1.21	Ea	17.90	49.60	67.50
9" dia. 18W, 100W equiv.	L1@1.21	Ea	21.60	49.60	71.20
13" dia. 25W, 100W equiv.	L1@1.21	Ea	21.60	49.60	71.20
16" dia. 23W, 100W, bronze trim	L1@1.21	Ea	53.90	49.60	103.50
14" square, 10W, 60W equiv.	L1@1.21	Ea	21.60	49.60	71.20
LED panel lights, frosted dome lens, white aluminum trim, dimmable					
5-1/2" dia. 10W, 60W equiv.	L1@1.21	Ea	17.90	49.60	67.50
7" dia. 12W, 75W equiv.	L1@1.21	Ea	17.90	49.60	67.50
9" dia. 18W, 100W equiv.	L1@1.21	Ea	21.60	49.60	71.20
13" dia. 25W, 100W equiv.	L1@1.21	Ea	21.60	49.60	71.20
16" dia. 23W, 100W, bronze trim	L1@1.21	Ea	53.90	49.60	103.50
14" square, 10W, 60W equiv.	L1@1.21	Ea	21.60	49.60	71.20
Industrial LED wall packs, aluminum housing, bronze finish, polycarbonate lens, weatherproof, with lamps					
20 watt, 70W halide equiv.	L1@9.55	Ea	35.20	392.00	427.20
28 watt, 70W halide equiv.	L1@9.55	Ea	46.70	392.00	438.70
40 watt, 250W halide equiv.	L1@9.55	Ea	90.50	392.00	482.50
80 watt, 400W halide equiv.	L1@9.55	Ea	101.00	392.00	493.00
80 watt, 400W halide equiv., rotatable	L1@9.55	Ea	141.00	392.00	533.00
120 watt, 400W halide equiv.	L1@9.55	Ea	131.00	392.00	523.00
Add for photo cell control, factory-installed	—	Ea	15.50	—	15.50

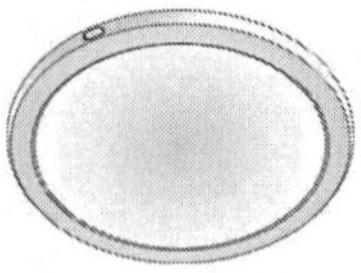

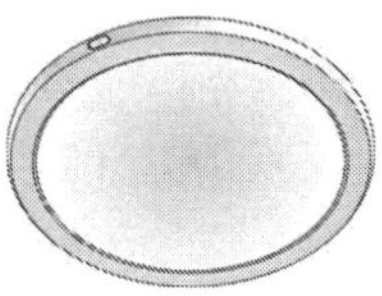

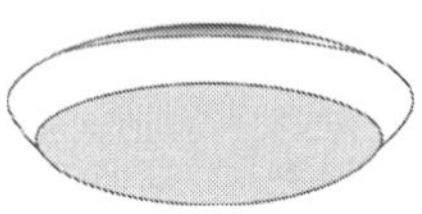

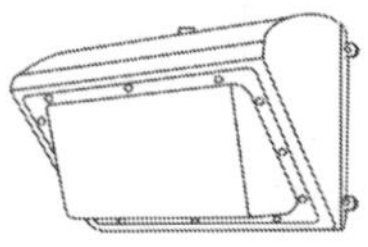

Use these figures to estimate the cost of lighting fixtures installed in buildings under the conditions described on pages 5 and 6. Costs listed are for each fixture installed. The crew is one electrician working at a labor cost of $41.03 per manhour. These costs include layout, material handling, and normal waste. Add for lamp, accessories, sales tax, delivery, supervision, mobilization, demobilization, cleanup, overhead and profit. Note: These costs are based on internally pre-wired fixtures installed in new construction not over 10 feet above the floor.

Material	Craft@Hrs	Unit	Material Cost	Labor Cost	Installed Cost

High-bay LED industrial fixture, cast aluminum housing, 10" diameter tempered glass lens, suspended or surface mounted, with lamp, U-bracket, eye hook, and safety cable

Material	Craft@Hrs	Unit	Material Cost	Labor Cost	Installed Cost
100 watt, 250W halide equiv.	L1@1.30	Ea	151.00	53.30	204.30
150 watt, 400W halide equiv.	L1@1.30	Ea	181.00	53.30	234.30
200 watt, 750W halide equiv.	L1@1.30	Ea	211.00	53.30	264.30
300 watt, 1,000W halide equiv.	L1@1.30	Ea	311.00	53.30	364.30
400 watt, 1,500W halide equiv.	L1@1.30	Ea	512.00	53.30	565.30
500 watt, 1,500W halide equiv.	L1@1.30	Ea	613.00	53.30	666.30
Add for wire guard on lens	L1@.250	Ea	76.00	10.30	86.30

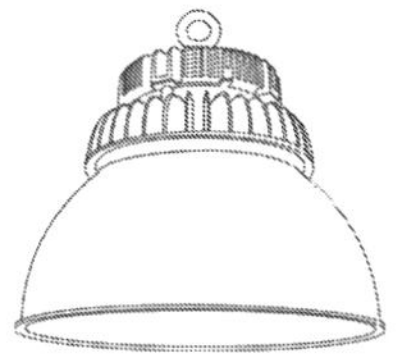

High-bay LED linear fixtures, white aluminum frame, frosted poly lens, suspension mount, by metal halide lamp equivalent

Material	Craft@Hrs	Unit	Material Cost	Labor Cost	Installed Cost
80 watt, 3 lamps, 24"x12", 250W	L1@1.39	Ea	90.50	57.00	147.50
100 watt, 4 lamps, 48"x12", 250W	L1@1.39	Ea	80.30	57.00	137.30
110 watt, 3 lamps, 24"x12", 320W	L1@1.39	Ea	101.00	57.00	158.00
150 watt, 4 lamps, 48"x12", 250W	L1@1.39	Ea	101.00	57.00	158.00
165 watt, 4 lamps, 24"x12", 400W	L1@1.39	Ea	156.00	57.00	213.00
225 watt, 6 lamps, 48"x12", 400W	L1@1.39	Ea	171.00	57.00	228.00
250 watt, 8 lamps, 48"x20", 400W	L1@1.39	Ea	238.00	57.00	295.00
300 watt, 10 lamps, 48"x20", 1,000W	L1@1.39	Ea	277.00	57.00	334.00
Add for conduit mounting	L1@.250	Ea	6.92	10.30	17.22
Add for emergency driver, 25W backup power for 90 minutes, with test switch	L1@1.71	Ea	144.00	70.20	214.20

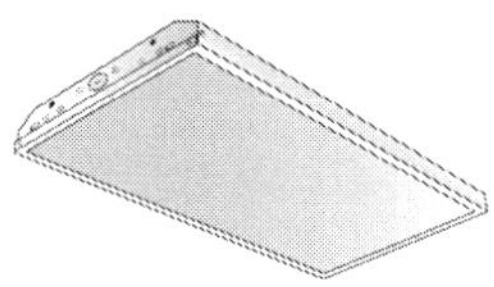

LED area flood light fixtures, U-bracket mount, aluminum frame, poly lens, weatherproof

Material	Craft@Hrs	Unit	Material Cost	Labor Cost	Installed Cost
30 watt, 150W halide equiv., 4" x 6"	L1@1.07	Ea	35.20	43.90	79.10
50 watt, 175W halide equiv.	L1@1.07	Ea	50.30	43.90	94.20
100 watt, 250W halide equiv.	L1@1.07	Ea	120.00	43.90	163.90
150 watt, 400W halide equiv.	L1@1.07	Ea	170.00	43.90	213.90
200 watt, 600W halide equiv., 18" x 14"	L1@1.07	Ea	201.00	43.90	244.90

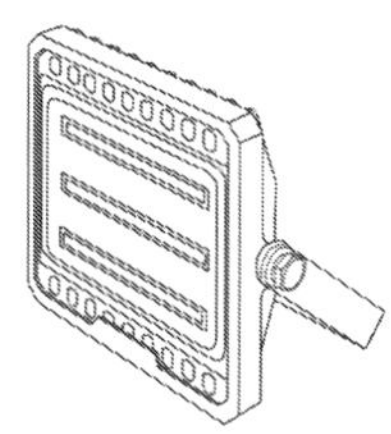

Vapor tight LED light fixtures, weatherproof aluminum housing and impact-resistant frosted polycarbonate lens. For use in harsh environments: gas stations, car washes, industrial kitchens, loading docks

Material	Craft@Hrs	Unit	Material Cost	Labor Cost	Installed Cost
30 watt, 2' long	L1@1.07	Ea	70.40	43.90	114.30
50 watt, 4' long	L1@1.21	Ea	110.00	49.60	159.60
60 watt, 4' long	L1@1.21	Ea	131.00	49.60	180.60
20 watt caged jelly jar fixture	L1@1.07	Ea	80.40	43.90	124.30

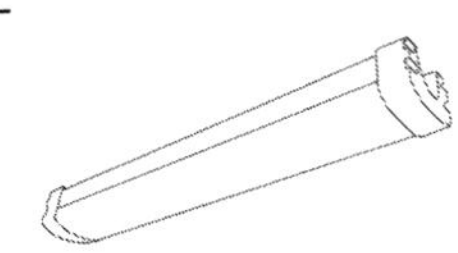

Use these figures to estimate the cost of lighting fixtures installed in buildings under the conditions described on pages 5 and 6. Costs listed are for each fixture installed. The crew is one electrician working at a labor cost of $41.03 per manhour. These costs include layout, material handling, and normal waste. Add for lamp, accessories, sales tax, delivery, supervision, mobilization, demobilization, cleanup, overhead and profit. Note: These costs are based on internally pre-wired fixtures installed in new construction not over 10 feet above the floor.

LED Light Fixtures

Material	Craft@Hrs	Unit	Material Cost	Labor Cost	Installed Cost

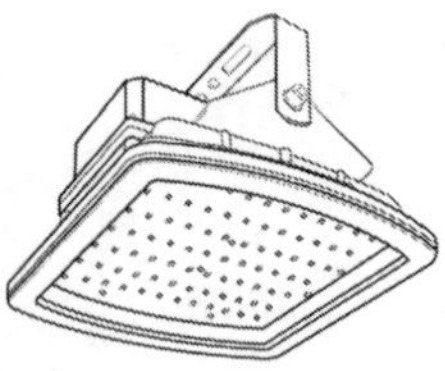

Explosion proof LED light fixtures, rated for Class 1 Division 2 hazards (granaries, mills, mines, gas stations, airplane hangars, auto shops) cast aluminum housing, U-bracket mount, tempered glass lens

Material	Craft@Hrs	Unit	Material Cost	Labor Cost	Installed Cost
100 watt, 175W halide equiv.	L1@2.57	Ea	603.00	105.00	708.00
200 watt, 400W halide equiv.	L1@2.57	Ea	905.00	105.00	1,010.00

Explosion proof LED light fixtures, rated for Class 1 Division 1 hazards (vapors and gases are constantly present, i.e. paint spray booth) cast aluminum housing, U-bracket mount, tempered glass lens

Material	Craft@Hrs	Unit	Material Cost	Labor Cost	Installed Cost
150 watt, 175 halide equiv.	L1@2.57	Ea	1,890.00	105.00	1,995.00

Yard and Street Lighting Includes setting and connecting prewired fixtures with ballast but no pole, concrete work, excavation, wire, lamps or conduit. See pole costs below.

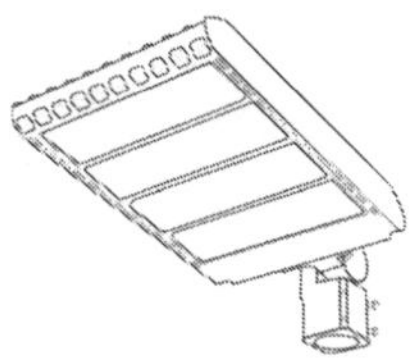

Rectangular LED floodlights, die-cast aluminum housing, tempered glass lens, photoelectric cell, add the cost of mounting pole or bracket

Material	Craft@Hrs	Unit	Material Cost	Labor Cost	Installed Cost
100 watt, 250W halide equiv.	L1@1.55	Ea	176.00	63.60	239.60
150 watt, 400W halide equiv.	L1@1.55	Ea	191.00	63.60	254.60
200 watt, 750W halide equiv.	L1@1.55	Ea	237.00	63.60	300.60
300 watt, 750W halide equiv.	L1@1.55	Ea	327.00	63.60	390.60
Add for pole mount	L1@.547	Ea	15.10	22.40	37.50
Add for wall mount bracket	L1@1.07	Ea	30.20	43.90	74.10

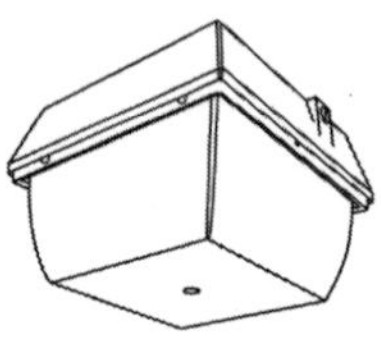

Canopy-mounted LED floodlights, die-cast anodized aluminum housing with shatterproof polycarbonate lens, flush or surface mount. Security lighting for parking garages, gas stations, drive-thrus and covered walkways

Material	Craft@Hrs	Unit	Material Cost	Labor Cost	Installed Cost
60 watt, 100W halide equiv.	L1@2.71	Ea	70.40	111.00	181.40
100 watt, 250W halide equiv.	L1@2.71	Ea	85.40	111.00	196.40
150 watt, 400W halide equiv.	L1@2.71	Ea	183.00	111.00	294.00

Use these figures to estimate the cost of lighting fixtures installed in buildings under the conditions described on pages 5 and 6. Costs listed are for each fixture installed. The crew is one electrician working at a labor cost of $41.03 per manhour. These costs include layout, material handling, and normal waste. Add for lamp, accessories, sales tax, delivery, supervision, mobilization, demobilization, cleanup, overhead and profit. Note: These costs are based on internally pre-wired fixtures installed in new construction not over 10 feet above the floor.

Material	Craft@Hrs	Unit	Material Cost	Labor Cost	Installed Cost
LED lamps, 120 volt, by watts, bulb type, watt equivalent and Kelvin					
3/8/18A21, 3-Way 2700K	L1@0.05	Ea	25.70	2.05	27.75
19A21, 100W 2700K	L1@0.05	Ea	29.90	2.05	31.95
19A21, 100W 5000K	L1@0.05	Ea	29.90	2.05	31.95
18A21, 100W 2700K	L1@0.05	Ea	23.70	2.05	25.75
18A21, 100W 5000K	L1@0.05	Ea	24.60	2.05	26.65
10.2A19, 100W 2700K	L1@0.05	Ea	48.30	2.05	50.35
15A19, 75W 5000K	L1@0.05	Ea	17.70	2.05	19.75
12A19, 75W 3000K	L1@0.05	Ea	24.80	2.05	26.85
14A19, Omni, 75W 5000K	L1@0.05	Ea	18.20	2.05	20.25
13.5A19, 75W 5500K	L1@0.05	Ea	24.60	2.05	26.65
10A19, 60W 5000K	L1@0.05	Ea	6.98	2.05	9.03
9.5A19, 60W 2700K	L1@0.05	Ea	10.20	2.05	12.25
9.5A19, Omni,60W 2700K	L1@0.05	Ea	11.80	2.05	13.85
13A19, 60W 2700K	L1@0.05	Ea	18.90	2.05	20.95
6A19, Omni, 60W 5500K	L1@0.05	Ea	25.70	25.70	
60W 5500K	L1@0.05	Ea	31.60	2.05	33.65
8A19, 50W 2700K	L1@0.05	Ea	6.44	2.05	8.49
8A19, 40W 2700K	L1@0.05	Ea	10.10	2.05	12.15
8A19, 40W 5000K	L1@0.05	Ea	9.66	2.05	11.71
3.5A17, 25W 2700K	L1@0.05	Ea	13.50	2.05	15.55
4A19, 25W 5000K	L1@0.05	Ea	8.91	2.05	10.96
0.5G40, E7, 7W 2700K	L1@0.05	Ea	3.43	2.05	5.48
4B11, E12, 25W 5000K	L1@0.05	Ea	6.23	2.05	8.28
14PAR38, 75W 2700K	L1@0.05	Ea	27.00	2.05	29.05
16PAR38, 90W Wet	L1@0.05	Ea	30.10	2.05	32.15
15PAR38, 100W 2700K	L1@0.05	Ea	26.60	2.05	28.65
15PAR38, 120W 4000K	L1@0.05	Ea	34.30	2.05	36.35
23PAR38, 150W 2700K	L1@0.05	Ea	50.00	2.05	52.05
9.5PAR30, 75W 3000K	L1@0.05	Ea	18.20	2.05	20.25
14PAR30, 75W Wet	L1@0.05	Ea	29.00	2.05	31.05
11PAR30, 90W 3000K	L1@0.05	Ea	22.60	2.05	24.65
6.5PAR20, 30W 3000K	L1@0.05	Ea	8.80	2.05	10.85
8PAR20, 45W 3000K	L1@0.05	Ea	18.10	2.05	20.15
8PAR20, 50W Wet	L1@0.05	Ea	19.30	2.05	21.35
9PAR20, 50W 4100K	L1@0.05	Ea	25.10	2.05	27.15
8PAR20, 60W 2700K	L1@0.05	Ea	23.10	2.05	25.15
10PAR20, 60W 25 Deg	L1@0.05	Ea	49.40	2.05	51.45
7PAR16, 35W 4100K	L1@0.05	Ea	13.30	2.05	15.35

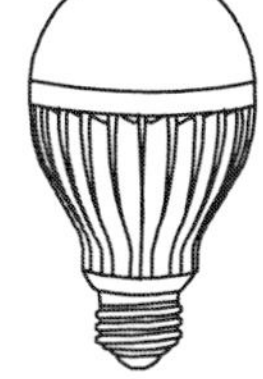

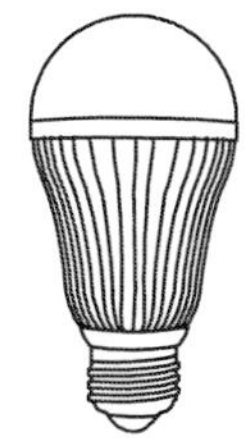

Use these figures to estimate the cost of screw base LED lamps installed in lighting fixtures under the conditions described on pages 5 and 6. Costs listed are per lamp showing watts, bulb type and color temperature ("K"). The crew is one electrician working at a labor cost of $41.03 per manhour. These costs include layout, material handling, and normal waste. Add for the lighting fixture, sales tax, delivery, supervision, mobilization, demobilization, cleanup, overhead and profit. Note: Some special voltage fixtures come with the right lamp packaged in the fixture box. If the lamp isn't supplied with the fixture, be sure to select the lamp with the right voltage and wattage. If the specifications require a lamp in a fixture, only a certain lamp will meet job requirements.

LED Lamps

Material		Craft@Hrs	Unit	Material Cost	Labor Cost	Installed Cost
LED lamps, 120 volt, by watts, bulb type, watt equivalent and Kelvin						
7.5R20, 50W	2400K	L1@0.05	Ea	19.60	2.05	21.65
9R20, 50W	2400K	L1@0.05	Ea	28.30	2.05	30.35
10BR30, 65W	2400K	L1@0.05	Ea	22.60	2.05	24.65
13BR30, 65W	2700K	L1@0.05	Ea	11.80	2.05	13.85
14BR30, 85W	2400K	L1@0.05	Ea	36.50	2.05	38.55
10BR40, 65W	2400K	L1@0.05	Ea	29.50	2.05	31.55
12BR40, 85W	2400K	L1@0.05	Ea	33.80	2.05	35.85
Standard voltage incandescent lamps, ES = Energy-Saving						
150A/21	120V	L1@0.05	Ea	1.67	2.05	3.72
150A/21	130V	L1@0.05	Ea	1.67	2.05	3.72
150A21/CL	120V	L1@0.05	Ea	2.90	2.05	4.95
150A21/CL	130V	L1@0.05	Ea	3.55	2.05	5.60
150A21/RS	120V	L1@0.05	Ea	3.78	2.05	5.83
150A21/RS	130V	L1@0.05	Ea	5.66	2.05	7.71
150A21/F	120V	L1@0.05	Ea	.94	2.05	2.99
200A21/CL	120V	L1@0.05	Ea	5.26	2.05	7.31
200A21/CL	130V	L1@0.05	Ea	5.80	2.05	7.85
200A21/F	120V	L1@0.05	Ea	4.63	2.05	6.68
200A21/F	130V	L1@0.05	Ea	4.66	2.05	6.71
200A21/IF	120V	L1@0.05	Ea	6.15	2.05	8.20
200A21/IF	130V	L1@0.05	Ea	6.93	2.05	8.98
200IF	120V	L1@0.05	Ea	5.37	2.05	7.42
200IF	130V	L1@0.05	Ea	7.22	2.05	9.27
200PAR46/3MFL	120V	L1@0.08	Ea	50.30	3.28	53.58
200PAR46/3MFL	130V	L1@0.08	Ea	51.70	3.28	54.98
200PAR56/NSP	120V	L1@0.08	Ea	30.40	3.28	33.68
200PAR56/NSP	130V	L1@0.08	Ea	31.30	3.28	34.58
200PAR64	120V	L1@0.05	Ea	106.00	2.05	108.05
200PAR64	130V	L1@0.05	Ea	155.00	2.05	157.05
250PAR58/NSP	120V	L1@0.08	Ea	30.40	3.28	33.68
300M/PS25/IF/99	120V	L1@0.06	Ea	11.90	2.46	14.36
300M/PS25/IF/99	130V	L1@0.06	Ea	17.80	2.46	20.26
300M/PS25/IF	120V	L1@0.06	Ea	4.92	2.46	7.38
300M/PS25/IF	130V	L1@0.06	Ea	7.33	2.46	9.79

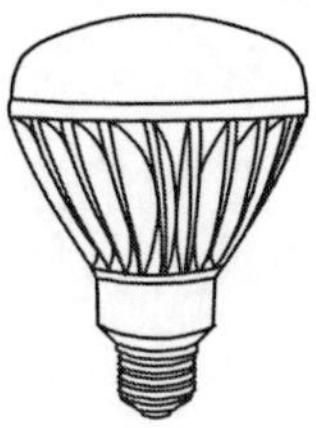

Use these figures to estimate the cost of lamps installed in lighting fixtures under the conditions described on pages 5 and 6. Costs listed are for each lamp installed. The crew is one electrician working at a labor cost of $41.03 per manhour. These costs include layout, material handling, and normal waste. Add for the lighting fixture, sales tax, delivery, supervision, mobilization, demobilization, cleanup, overhead and profit. Note: Some special voltage fixtures come with the right lamp packaged in the fixture box. If the lamp isn't supplied with the fixture, be sure to select the lamp with the right voltage and wattage. If the specifications require a lamp in a fixture, only a certain lamp will meet job requirements.

Material		Craft@Hrs	Unit	Material Cost	Labor Cost	Installed Cost
Compact fluorescent lamps						
CFL9/40/W	120V	L1@0.06	Ea	6.15	2.46	8.61
CFL9/40/BW	120V	L1@0.06	Ea	7.37	2.46	9.83
CFL14/60/W	120V	L1@0.06	Ea	5.22	2.46	7.68
CFL14/60/BW	120V	L1@0.06	Ea	7.12	2.46	9.58
CF5DS 5W	120V	L1@0.10	Ea	2.02	4.10	6.12
CF7DS 7W	120V	L1@0.10	Ea	1.61	4.10	5.71
CF9DS 9W	120V	L1@0.10	Ea	1.61	4.10	5.71
CF13DS 13W	120V	L1@0.10	Ea	1.78	4.10	5.88
CF13DD/E 13W	120V	L1@0.10	Ea	5.23	4.10	9.33
CF18DD/E 18W	120V	L1@0.06	Ea	5.23	2.46	7.69
CF26DD/E 26W	120V	L1@0.06	Ea	5.23	2.46	7.69
PAR38/23W/27K	120V	L1@0.06	Ea	6.84	2.46	9.30
PAR38/23W/30K	120V	L1@0.06	Ea	6.84	2.46	9.30
PAR38/23W/41K	120V	L1@0.08	Ea	14.00	3.28	17.28
PAR38/23W/50K	120V	L1@0.10	Ea	14.00	4.10	18.10
FLE15/A2/R30	120V	L1@0.10	Ea	13.40	4.10	17.50
FLE15/2/R30/S	120V	L1@0.10	Ea	5.56	4.10	9.66
FLE/15/2/DV/R30	120V	L1@0.10	Ea	44.20	4.10	48.30
BC-EL/A-R30	120V	L1@0.10	Ea	39.50	4.10	43.60
BC-EL/A-R30	120V	L1@0.10	Ea	30.90	4.10	35.00
BC-EL/A-R30	120V	L1@0.10	Ea	61.80	4.10	65.90
R30/E26/27K	120V	L1@0.10	Ea	3.31	4.10	7.41
R30/E26/41K	120V	L1@0.10	Ea	3.31	4.10	7.41
R30/E26/50K	120V	L1@0.10	Ea	3.31	4.10	7.41
R40/E26/27K	120V	L1@0.10	Ea	4.40	4.10	8.50
R40/E26/41K	120V	L1@0.15	Ea	4.40	6.15	10.55
R40/E26/50K	120V	L1@0.15	Ea	4.40	6.15	10.55
R40/A4/FLE23	120V	L1@0.15	Ea	11.90	6.15	18.05
R40/2/DV/FLE26	120V	L1@0.15	Ea	26.00	6.15	32.15
Special voltage incandescent lamps						
25A/17	12V	L1@0.05	Ea	1.06	2.05	3.11
25A/17	34V	L1@0.05	Ea	1.06	2.05	3.11
25A17/RS	75V	L1@0.05	Ea	4.69	2.05	6.74
25A/17	230V	L1@0.05	Ea	4.64	2.05	6.69
25PAR36	4.7V	L1@0.08	Ea	14.10	3.28	17.38
25PAR36	6V	L1@0.08	Ea	10.50	3.28	13.78
25PAR36	12.8V	L1@0.08	Ea	18.20	3.28	21.48
25PAR36	13V	L1@0.08	Ea	30.40	3.28	33.68
25PAR36	28V	L1@0.08	Ea	30.10	3.28	33.38

Use these figures to estimate the cost of lamps installed in lighting fixtures under the conditions described on pages 5 and 6. Costs listed are for each lamp installed. The crew is one electrician working at a labor cost of $41.03 per manhour. These costs include layout, material handling, and normal waste. Add for the lighting fixture, sales tax, delivery, supervision, mobilization, demobilization, cleanup, overhead and profit. Note: Some special voltage fixtures come with the right lamp packaged in the fixture box. If the lamp isn't supplied with the fixture, be sure to select the lamp with the right voltage and wattage. If the specifications require a lamp in a fixture, only a certain lamp will meet job requirements.

Incandescent Lamps

Special voltage incandescent lamps (continued)

Material		Craft@Hrs	Unit	Material Cost	Labor Cost	Installed Cost
50A19/RS	75V	L1@0.05	Ea	5.46	2.05	7.51
50A19	250V	L1@0.05	Ea	4.75	2.05	6.80
50A19/RS	250V	L1@0.05	Ea	6.36	2.05	8.41
50A21	12V	L1@0.05	Ea	4.98	2.05	7.03
50A21	30V	L1@0.05	Ea	5.21	2.05	7.26
50A21	34V	L1@0.05	Ea	3.90	2.05	5.95
50A21	300V	L1@0.05	Ea	6.14	2.05	8.19
60A21	230V	L1@0.05	Ea	5.51	2.05	7.56
60PAR46	38V	L1@0.08	Ea	34.00	3.28	37.28
75A21	12V	L1@0.05	Ea	5.66	2.05	7.71
100A21	230V	L1@0.05	Ea	4.87	2.05	6.92
100A21	250V	L1@0.05	Ea	4.87	2.05	6.92
100A21/RS	250V	L1@0.05	Ea	5.37	2.05	7.42
100A21	277V	L1@0.05	Ea	7.27	2.05	9.32
100PAR38/FL	12V	L1@0.08	Ea	38.60	3.28	41.88
100PAR46	60V	L1@0.10	Ea	42.30	4.10	46.40
100R30/CL	12V	L1@0.08	Ea	23.30	3.28	26.58
200PS30	250V	L1@0.05	Ea	8.58	2.05	10.63
240PAR56/MFL	12V	L1@0.08	Ea	38.60	3.28	41.88
240PAR56/WFL	12V	L1@0.08	Ea	38.60	3.28	41.88
240PAR56/NSP	12V	L1@0.08	Ea	38.60	3.28	41.88
300R40/FL	250V	L1@0.10	Ea	37.70	4.10	41.80
500PAR64/MFL	230V	L1@0.10	Ea	107.00	4.10	111.10
500PAR64/WFL	230V	L1@0.10	Ea	107.00	4.10	111.10
500R40/FL	250V	L1@0.10	Ea	46.80	4.10	50.90
500R52	250V	L1@0.10	Ea	55.30	4.10	59.40
1000PS52	250V	L1@0.15	Ea	37.10	6.15	43.25
1000PS52	277V	L1@0.15	Ea	58.60	6.15	64.75

Standard voltage incandescent halogen lamps

Material		Craft@Hrs	Unit	Material Cost	Labor Cost	Installed Cost
45PAR38/FL	120V	L1@0.08	Ea	16.60	3.28	19.88
45PAR38/FL	130V	L1@0.08	Ea	17.60	3.28	20.88
45PAR38/SP	120V	L1@0.08	Ea	16.60	3.28	19.88
45PAR38/SP	130V	L1@0.08	Ea	19.80	3.28	23.08
50PAR38/FL	120V	L1@0.08	Ea	12.40	3.28	15.68
50PAR38/FL	130V	L1@0.08	Ea	15.70	3.28	18.98
50PAR38/SP	120V	L1@0.08	Ea	13.70	3.28	16.98
50PAR38/SP	130V	L1@0.08	Ea	17.30	3.28	20.58

Use these figures to estimate the cost of special purpose incandescent lamps installed in lighting fixtures under the conditions described on pages 5 and 6. Costs listed are for each lamp installed. The crew is one electrician working at a labor cost of $41.03 per manhour. These costs include layout, material handling, and normal waste. Add for the lighting fixture, sales tax, delivery, supervision, mobilization, demobilization, cleanup, overhead and profit. Note: Some special voltage fixtures come with the right lamp packaged in the fixture box. If the lamp isn't supplied with the fixture, be sure to select the lamp with the right voltage and wattage. If the specifications require a lamp in a fixture, only a certain lamp will meet job requirements.

Standard voltage incandescent halogen lamps

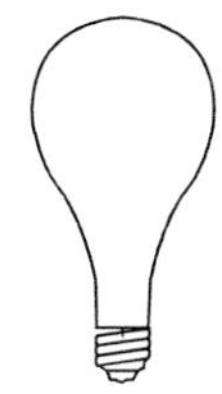

Material	Craft@Hrs	Unit	Material Cost	Labor Cost	Installed Cost
90PAR38/FL 120V	L1@0.08	Ea	14.00	3.28	17.28
90PAR38/FL 130V	L1@0.08	Ea	16.60	3.28	19.88
90PAR38/SP 120V	L1@0.08	Ea	14.00	3.28	17.28
90PAR38/SP 130V	L1@0.08	Ea	16.60	3.28	19.88
100PAR38/FL 120V	L1@0.08	Ea	10.40	3.28	13.68
100PAR38/FL 130V	L1@0.08	Ea	14.70	3.28	17.98
100PAR38/SP 120V	L1@0.08	Ea	12.70	3.28	15.98
100PAR38/SP 130V	L1@0.08	Ea	16.20	3.28	19.48

Tubular quartz lamps

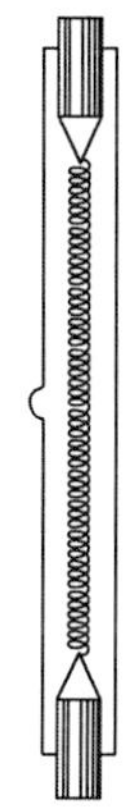

Material	Craft@Hrs	Unit	Material Cost	Labor Cost	Installed Cost
100T3/CL 120V	L1@0.06	Ea	18.80	2.46	21.26
150T3/CL 120V	L1@0.06	Ea	19.30	2.46	21.76
200T3/CL 120V	L1@0.06	Ea	57.00	2.46	59.46
250T3/CL 120V	L1@0.06	Ea	17.70	2.46	20.16
300T3/CL 120V	L1@0.06	Ea	18.80	2.46	21.26
425T3/CL 120V	L1@0.06	Ea	42.10	2.46	44.56
500T3/CL 120V	L1@0.06	Ea	15.70	2.46	18.16
500T3/CL 130V	L1@0.06	Ea	17.30	2.46	19.76
1000T3/CL 220V	L1@0.08	Ea	75.10	3.28	78.38
1000T3/CL 240V	L1@0.08	Ea	75.10	3.28	78.38
1500T3/CL 208V	L1@0.10	Ea	59.50	4.10	63.60
1500T3/CL 220V	L1@0.10	Ea	59.50	4.10	63.60
1500T3/CL 240V	L1@0.10	Ea	23.00	4.10	27.10
1500T3/CL 277V	L1@0.10	Ea	24.40	4.10	28.50
6000T3/CL 480V	L1@0.15	Ea	87.90	6.15	94.05

LED lamps, special purpose, by watts, bulb type, watt equivalent, base type and volts

Material	Craft@Hrs	Unit	Material Cost	Labor Cost	Installed Cost
.7T4, 5W, bi-pin base, 12V	L1@0.05	Ea	4.95	2.05	7.00
2T3, 10W, bi-pin base, 12V	L1@0.05	Ea	5.72	2.05	7.77
6T3, 35W, bi-pin base, 12V	L1@0.05	Ea	20.10	2.05	22.15
7T3, 50W, bi-pin base, 12V	L1@0.05	Ea	19.10	2.05	21.15
2T3, 20W, wedge base, 12V	L1@0.05	Ea	19.60	2.05	21.65
3.3MR16, 20W, GU10, 120V	L1@0.05	Ea	7.12	2.05	9.17
7MR16, 35W, GU10, 120V	L1@0.05	Ea	12.30	2.05	14.35
5.2MR16, 20W, GU5.3, 12V	L1@0.05	Ea	9.95	2.05	12.00
4MR16, 20W, GU5.3, 12V	L1@0.05	Ea	21.20	2.05	23.25
35 Post, 150W, E26, 120V	L1@0.05	Ea	152.00	2.05	154.05
45 Post, 175W, E26, 120V	L1@0.05	Ea	173.00	2.05	175.05
45 Wall, 175W, E39, 120V	L1@0.05	Ea	182.00	2.05	184.05

Use these figures to estimate the cost of special purpose incandescent and LED lamps installed in lighting fixtures under the conditions described on pages 5 and 6. Costs listed are for each lamp installed. The crew is one electrician working at a labor cost of $41.03 per manhour. These costs include layout, material handling, and normal waste. Add for the lighting fixture, sales tax, delivery, supervision, mobilization, demobilization, cleanup, overhead and profit. Note: Some special voltage fixtures come with the right lamp packaged in the fixture box. If the lamp isn't supplied with the fixture, be sure to select the lamp with the right voltage and wattage. If the specifications require a lamp in a fixture, only a certain lamp will meet job requirements. Counsel your installers that they have to keep their fingers off the socket metal of quartz lamps. Oil from their skin increases electrical resistance which makes the point of contact overheat.

High Intensity Discharge Lamps

Material	Craft@Hrs	Unit	Material Cost	Labor Cost	Installed Cost
Mercury vapor lamps					
HR40/50/B17/DX	L1@0.10	Ea	89.90	4.10	94.00
HR75/B17/DX	L1@0.10	Ea	89.90	4.10	94.00
HR100/A23/CL	L1@0.10	Ea	82.20	4.10	86.30
HR100/B17/DX	L1@0.10	Ea	86.60	4.10	90.70
HR100/ED23/CL	L1@0.10	Ea	54.60	4.10	58.70
HR100/ED23/DX	L1@0.10	Ea	48.40	4.10	52.50
HR100/ED23/WDX	L1@0.10	Ea	85.90	4.10	90.00
HR100/PAR38/FL	L1@0.06	Ea	150.00	2.46	152.46
HR100/PAR38/SP	L1@0.06	Ea	150.00	2.46	152.46
HR100/R40/FL	L1@0.06	Ea	130.00	2.46	132.46
HR100/R40/DX/WFL	L1@0.06	Ea	133.00	2.46	135.46
HR175/ED28/CL	L1@0.10	Ea	37.00	4.10	41.10
HR175/ED28/DX	L1@0.10	Ea	36.10	4.10	40.20
HR175/ED28/WDX	L1@0.10	Ea	80.90	4.10	85.00
HR175/R40/FL	L1@0.06	Ea	133.00	2.46	135.46
HR175/R40/RFL	L1@0.06	Ea	110.00	2.46	112.46
HR250/ED28/CL	L1@0.10	Ea	71.40	4.10	75.50
HR250/ED28/DX	L1@0.10	Ea	64.10	4.10	68.20
HR250/ED28/WDX	L1@0.10	Ea	124.00	4.10	128.10
HR400/BT37/DX	L1@0.10	Ea	74.50	4.10	78.60
HR400/ED37/CL	L1@0.10	Ea	47.50	4.10	51.60
HR400/ED37/DX	L1@0.10	Ea	51.50	4.10	55.60
HR400/ED37/WDX	L1@0.10	Ea	99.50	4.10	103.60
HR400/R52/DX	L1@0.10	Ea	154.00	4.10	158.10
HR400/R60/DX/WFL	L1@0.10	Ea	215.00	4.10	219.10
HR400/R60/SP	L1@0.10	Ea	194.00	4.10	198.10
HR400/T16/CL	L1@0.10	Ea	183.00	4.10	187.10
HR1000/BT65/CL	L1@0.10	Ea	135.00	4.10	139.10
HR1000/BT56/DX	L1@0.10	Ea	118.00	4.10	122.10
HR175/ED28/DX	L1@0.10	Ea	109.00	4.10	113.10
HR250/ED28/DX	L1@0.10	Ea	133.00	4.10	137.10
HR400/ED28/DX	L1@0.10	Ea	123.00	4.10	127.10

Material	Craft@Hrs	Unit	Material Cost	Labor Cost	Installed Cost
Self-ballasted lamps (retrofit)					
HSB160/ED24/DX, 120V	L1@0.10	Ea	128.00	4.10	132.10
HSB250/ED28/DX, 120V	L1@0.10	Ea	202.00	4.10	206.10
HSB250/ED28/DX, 130V	L1@0.10	Ea	202.00	4.10	206.10
HSB450/BT37/DX, 120V	L1@0.10	Ea	360.00	4.10	364.10
HSB750/R57/DX, 120V	L1@0.10	Ea	528.00	4.10	532.10

Use these figures to estimate the cost of high intensity discharge lamps installed in lighting fixtures under the conditions described on pages 5 and 6. Costs listed are for each lamp installed. The crew is one electrician working at a labor cost of $41.03 per manhour. These costs include layout, material handling, and normal waste. Add for the lighting fixture, sales tax, delivery, supervision, mobilization, demobilization, cleanup, overhead and profit.

Metal Halide High Intensity Discharge Lamps

Metal halide high intensity discharge lamps, ES = Energy-Saving

Material	Craft@Hrs	Unit	Material Cost	Labor Cost	Installed Cost
MXR50/BD17/CL	L1@0.10	Ea	67.90	4.10	72.00
MXR50/BD17/CTD	L1@0.10	Ea	71.90	4.10	76.00
MXR70/BD17/CL	L1@0.10	Ea	64.50	4.10	68.60
MXR70/BD17/CTD	L1@0.10	Ea	68.60	4.10	72.70
MXR70/ED17/CL	L1@0.10	Ea	74.70	4.10	78.80
MXR70/ED17/CTD	L1@0.10	Ea	77.50	4.10	81.60
MXR70/PAR38/FL	L1@0.08	Ea	74.70	3.28	77.98
MXR70/PAR38/WFL	L1@0.08	Ea	74.70	3.28	77.98
MXR70/PAR38/SP	L1@0.08	Ea	74.70	3.28	77.98
MXR100/BD17/CL	L1@0.10	Ea	64.50	4.10	68.60
MXR100/BD17/CTD	L1@0.10	Ea	68.60	4.10	72.70
MXR100/ED17/CL	L1@0.10	Ea	74.70	4.10	78.80
MXR100/ED17/CTD	L1@0.10	Ea	77.50	4.10	81.60
MXR100/PAR38/FL	L1@0.10	Ea	74.70	4.10	78.80
MXR100/PAR38/WFL	L1@0.10	Ea	74.70	4.10	78.80
MXR100/PAR38/SP	L1@0.10	Ea	74.70	4.10	78.80
MXR150/BD17/CL	L1@0.10	Ea	71.20	4.10	75.30
MXR150/BD17/CTD	L1@0.10	Ea	75.30	4.10	79.40
MVR175/BD17/CL	L1@0.10	Ea	63.50	4.10	67.60
MVR175/BD17/CTD	L1@0.10	Ea	66.50	4.10	70.60
MVR175/ED28/CL	L1@0.10	Ea	46.70	4.10	50.80
MVR175/ED28/CTD	L1@0.10	Ea	49.70	4.10	53.80
MVR250/ED28/CL	L1@0.10	Ea	52.60	4.10	56.70
MVR250/ED28/CTD	L1@0.10	Ea	55.60	4.10	59.70
MVR360/ED37/CL/ES	L1@0.10	Ea	62.90	4.10	67.00
MVR360/ED37/CTD/ES	L1@0.10	Ea	66.20	4.10	70.30
MVR400/ED28/CL	L1@0.10	Ea	76.80	4.10	80.90
MVR400/ED28/CTD	L1@0.10	Ea	85.80	4.10	89.90
MVR400/ED37/CL	L1@0.10	Ea	44.90	4.10	49.00
MVR400/ED37/CTD	L1@0.10	Ea	56.70	4.10	60.80
MVR1000/BT56/CL	L1@0.10	Ea	119.00	4.10	123.10
MVR1000/BT56/CTD	L1@0.10	Ea	133.00	4.10	137.10

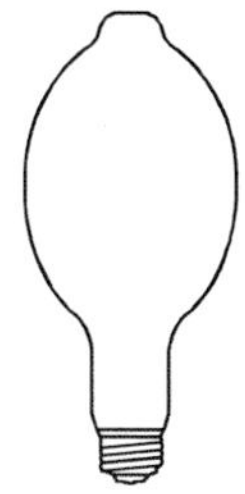

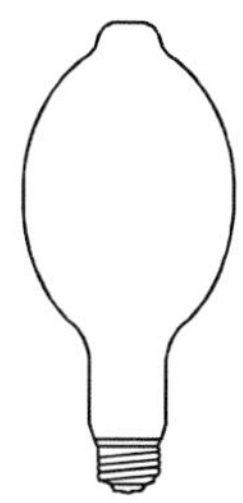

Metal halide high intensity discharge ballasts, encased

Material	Craft@Hrs	Unit	Material Cost	Labor Cost	Installed Cost
175W, 120V to 480V	L1@1.00	Ea	336.00	41.00	377.00
250W, 120V to 480V	L1@1.00	Ea	387.00	41.00	428.00
400W, 120V to 480V	L1@1.00	Ea	382.00	41.00	423.00
1000W, 120V to 480V	L1@1.00	Ea	880.00	41.00	921.00

Use these figures to estimate the cost of high intensity discharge lamps installed in lighting fixtures under the conditions described on pages 5 and 6. Costs listed are for each lamp installed. The crew is one electrician working at a labor cost of $41.03 per manhour. These costs include layout, material handling, and normal waste. Add for the lighting fixture, sales tax, delivery, supervision, mobilization, demobilization, cleanup, overhead and profit. Note: The lamps above are identified by both wattage and by the shape of the bulb.

High Intensity Discharge Ballasts

Mercury vapor single lamp indoor ballasts

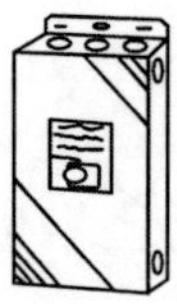

Material	Craft@Hrs	Unit	Material Cost	Labor Cost	Installed Cost
100 W 120/208/240277V	L1@1.00	Ea	185.00	41.00	226.00
100W 480V	L1@1.00	Ea	185.00	41.00	226.00
175W 120/208/240/277V	L1@1.00	Ea	173.00	41.00	214.00
175W 480V	L1@1.00	Ea	173.00	41.00	214.00
250W 120/208/240/277V	L1@1.00	Ea	189.00	41.00	230.00
250W 480V	L1@1.00	Ea	189.00	41.00	230.00
400W 120/208/240/277V	L1@1.00	Ea	226.00	41.00	267.00
400W 480V	L1@1.00	Ea	226.00	41.00	267.00
1000W 120/208/240/277V	L1@1.00	Ea	360.00	41.00	401.00
1000W 480V	L1@1.00	Ea	360.00	41.00	401.00

Weatherproof high intensity lighting ballasts

WP
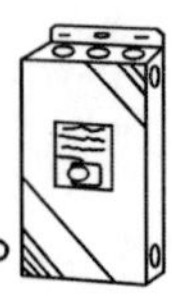

Material	Craft@Hrs	Unit	Material Cost	Labor Cost	Installed Cost
100W 120/208/240277V	L1@1.00	Ea	250.00	41.00	291.00
100W 480V	L1@1.00	Ea	250.00	41.00	291.00
175W 120/208/240/277V	L1@1.00	Ea	262.00	41.00	303.00
175W 480V	L1@1.00	Ea	262.00	41.00	303.00
250W 120/208/240/277V	L1@1.00	Ea	280.00	41.00	321.00
250W 480V	L1@1.00	Ea	280.00	41.00	321.00
400W 120/208/240/277V	L1@1.00	Ea	257.00	41.00	298.00
400W 480V	L1@1.00	Ea	290.00	41.00	331.00
1000W 120/208/240/277V	L1@1.00	Ea	572.00	41.00	613.00
1000W 480V	L1@1.00	Ea	572.00	41.00	613.00

High intensity ballast core and coil with capacitor

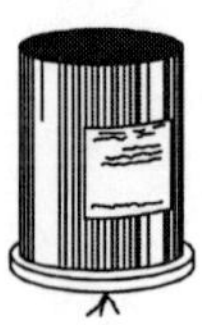

Material	Craft@Hrs	Unit	Material Cost	Labor Cost	Installed Cost
100 W 120/208/240277V	L1@1.25	Ea	115.00	51.30	166.30
100W 480V	L1@1.25	Ea	115.00	51.30	166.30
175W 120/208/240/277V	L1@1.25	Ea	85.00	51.30	136.30
175W 480V	L1@1.25	Ea	90.40	51.30	141.70
250W 120/208/240/277V	L1@1.25	Ea	102.00	51.30	153.30
250W 480V	L1@1.25	Ea	118.00	51.30	169.30
400W 120/208/240/277V	L1@1.25	Ea	116.00	51.30	167.30
400W 480V	L1@1.25	Ea	121.00	51.30	172.30
1000W 120/208/240/277V	L1@1.25	Ea	182.00	51.30	233.30
1000W 480V	L1@1.25	Ea	186.00	51.30	237.30

Use these figures to estimate the cost of installing high intensity discharge ballasts under the conditions described on pages 5 and 6. Costs listed are for each ballast installed. The crew is one electrician working at a labor cost of $41.03 per manhour. These costs include layout, material handling, and normal waste. Add for the wiring connections, enclosures if needed, sales tax, delivery, supervision, mobilization, demobilization, cleanup, overhead and profit.

Metal Halide High Intensity Discharge Lamps and Ballasts

Metal halide high intensity lamps

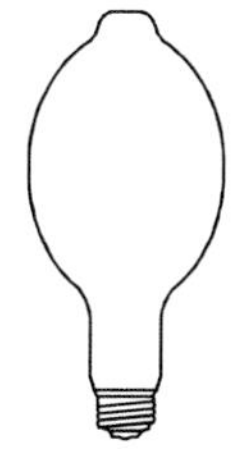

Material		Craft@Hrs	Unit	Material Cost	Labor Cost	Installed Cost
MVR175/C/U	E28	L1@0.10	Ea	98.10	4.10	102.20
MVR175/C	E28	L1@0.10	Ea	92.70	4.10	96.80
MVR250/C/U	E28	L1@0.10	Ea	109.00	4.10	113.10
MVR250U	E28	L1@0.10	Ea	104.00	4.10	108.10
MVR325/C/I/U/ES	E28	L1@0.10	Ea	121.00	4.10	125.10
MVR325/I/U/ES	E37	L1@0.10	Ea	158.00	4.10	162.10
MVT325/I/U/ES	E37	L1@0.10	Ea	223.00	4.10	227.10
MVT325/C/I/U/ES	E37	L1@0.10	Ea	234.00	4.10	238.10
MVR400/VBU	E37	L1@0.10	Ea	121.00	4.10	125.10
MVR400/C/I/U	E37	L1@0.10	Ea	158.00	4.10	162.10
MVR400/C/U	E37	L1@0.10	Ea	112.00	4.10	116.10
MVR400/C/VBD	E37	L1@0.10	Ea	138.00	4.10	142.10
MVR400/C/VBU	E37	L1@0.10	Ea	119.00	4.10	123.10
MVR400/I/U	E37	L1@0.10	Ea	145.00	4.10	149.10
MVR400/U	E37	L1@0.10	Ea	89.80	4.10	93.90
MVR400V BD	E37	L1@0.10	Ea	141.00	4.10	145.10
MVR950/I/VBD	BT56	L1@0.10	Ea	361.00	4.10	365.10
MVR960/I/VBU	BT56	L1@0.10	Ea	361.00	4.10	365.10
MVR1000/C/U	BT56	L1@0.10	Ea	257.00	4.10	261.10
MVR1000/U	BT56	L1@0.10	Ea	232.00	4.10	236.10
MVR1500/HBD/E	BT56	L1@0.10	Ea	295.00	4.10	299.10
MVR1500/HBU/E	BT56	L1@0.10	Ea	264.00	4.10	268.10

Metal halide high intensity indoor ballasts, encased

Material	Craft@Hrs	Unit	Material Cost	Labor Cost	Installed Cost
175W 120/208/240/277V	L1@1.00	Ea	339.00	41.00	380.00
175W 480V	L1@1.00	Ea	339.00	41.00	380.00
250W 120/208/240/277V	L1@1.00	Ea	416.00	41.00	457.00
250W 480V	L1@1.00	Ea	416.00	41.00	457.00
400W 120/208/240/277V	L1@1.00	Ea	409.00	41.00	450.00
400W 480V	L1@1.00	Ea	409.00	41.00	450.00
1000W 120/208/240/277V	L1@1.00	Ea	933.00	41.00	974.00
1000W 480V	L1@1.00	Ea	933.00	41.00	974.00

Use these figures to estimate the cost of metal halide discharge lamps and ballasts installed in or with lighting fixtures under the conditions described on pages 5 and 6. Costs listed are for each lamp or ballast installed. The crew is one electrician working at a labor cost of $41.03 per manhour. These costs include layout, material handling, and normal waste. Add for the lighting fixture, sales tax, delivery, supervision, mobilization, demobilization, cleanup, overhead and profit. Note: Metal halide lamps have a high rated lamp life when installed in some fixtures.

Sodium Lamps

Material		Craft@Hrs	Unit	Material Cost	Labor Cost	Installed Cost
Low pressure sodium lamps						
18W	T16	L1@0.15	Ea	120.00	6.15	126.15
35W	T16	L1@0.15	Ea	133.00	6.15	139.15
55W	T16	L1@0.15	Ea	145.00	6.15	151.15
90W	T21	L1@0.15	Ea	171.00	6.15	177.15
135W	T21	L1@0.15	Ea	219.00	6.15	225.15
High pressure sodium lamps, ES = Energy-Saving						
35W	B17	L1@0.10	Ea	90.10	4.10	94.20
35W/D	B17	L1@0.10	Ea	93.50	4.10	97.60
50W	B17	L1@0.10	Ea	90.10	4.10	94.20
50W/D	B17	L1@0.10	Ea	111.00	4.10	115.10
50W	ED23.5	L1@0.10	Ea	129.00	4.10	133.10
50W/D	ED23.5	L1@0.10	Ea	93.50	4.10	97.60
70W	B17	L1@0.10	Ea	90.10	4.10	94.20
70W/D	B17	L1@0.10	Ea	93.50	4.10	97.60
70W	ED23.5	L1@0.10	Ea	87.20	4.10	91.30
70W/D	ED23.5	L1@0.10	Ea	129.00	4.10	133.10
100W	B17	L1@0.10	Ea	92.10	4.10	96.20
100W/D	B17	L1@0.10	Ea	106.00	4.10	110.10
100W	ED23.5	L1@0.10	Ea	92.10	4.10	96.20
100W/D	ED23.5	L1@0.10	Ea	133.00	4.10	137.10
110W/ES	ED23.5	L1@0.10	Ea	170.00	4.10	174.10
150W	B17	L1@0.10	Ea	97.00	4.10	101.10
150W/D	B17	L1@0.10	Ea	108.00	4.10	112.10
150W	ED23.5	L1@0.10	Ea	93.50	4.10	97.60
150W/D	ED23.5	L1@0.10	Ea	124.00	4.10	128.10
150W/ES	ED28	L1@0.10	Ea	156.00	4.10	160.10
150W	ED28	L1@0.10	Ea	145.00	4.10	149.10
200W	ED18	L1@0.10	Ea	124.00	4.10	128.10
215W/ES	ED28	L1@0.10	Ea	169.00	4.10	173.10
250W	ED18	L1@0.10	Ea	101.00	4.10	105.10
250W/D	ED18	L1@0.10	Ea	144.00	4.10	148.10
250W/S	ED18	L1@0.10	Ea	172.00	4.10	176.10
250W/SBY	ED18	L1@0.10	Ea	242.00	4.10	246.10
310W	ED18	L1@0.10	Ea	191.00	4.10	195.10
400W	ED18	L1@0.10	Ea	104.00	4.10	108.10
400W/SBY	ED18	L1@0.10	Ea	250.00	4.10	254.10
400W/D	ED37	L1@0.10	Ea	154.00	4.10	158.10
400W	T7-R7s	L1@0.10	Ea	245.00	4.10	249.10
1000W	E25	L1@0.10	Ea	319.00	4.10	323.10
1000W	T7-R7s	L1@0.10	Ea	339.00	4.10	343.10

Use these figures to estimate the cost of high intensity discharge lamps installed in lighting fixtures under the conditions described on pages 5 and 6. Costs listed are for each lamp installed. The crew is one electrician working at a labor cost of $41.03 per manhour. These costs include layout, material handling, and normal waste. Add for the lighting fixture, sales tax, delivery, supervision, mobilization, demobilization, cleanup, overhead and profit. Note: Be sure the lamp selected is compatible with the installed ballast.

Material	Craft@Hrs	Unit	Material Cost	Labor Cost	Installed Cost
High intensity discharge core and coil with capacitor ballasts					
175W 120/208/240/277V	L1@1.25	Ea	113.00	51.30	164.30
175W 480V	L1@1.25	Ea	149.00	51.30	200.30
250W 120/208/240/277V	L1@1.25	Ea	149.00	51.30	200.30
250W 480V	L1@1.25	Ea	149.00	51.30	200.30
400W 120/208/240/277V	L1@1.25	Ea	168.00	51.30	219.30
400W 480V	L1@1.25	Ea	168.00	51.30	219.30
1000W 120/208/240/277V	L1@1.25	Ea	230.00	51.30	281.30
1000W 480V	L1@1.25	Ea	230.00	51.30	281.30
High pressure sodium indoor encased ballasts					
250W 120/208/240/277V	L1@1.00	Ea	262.00	41.00	303.00
250W 480V	L1@1.00	Ea	206.00	41.00	247.00
400W 120/208/240/277V	L1@1.00	Ea	326.00	41.00	367.00
400W 480V	L1@1.00	Ea	326.00	41.00	367.00
High pressure sodium weatherproof ballasts					
250W 120/208/240/277V	L1@1.00	Ea	313.00	41.00	354.00
250W 480V	L1@1.00	Ea	237.00	41.00	278.00
400W 120/208/240/277V	L1@1.00	Ea	403.00	41.00	444.00
400W 480V	L1@1.00	Ea	403.00	41.00	444.00

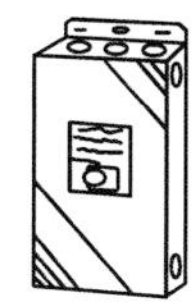

WP
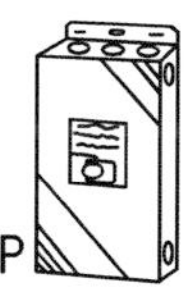

Use these figures to estimate the cost of installing high intensity discharge lamps and ballasts under the conditions described on pages 5 and 6. Costs listed are for each lamp or ballast installed. The crew is one electrician working at a labor cost of $41.03 per manhour. These costs include layout, material handling, and normal waste. Add for the wiring connections, enclosures if needed, sales tax, delivery, supervision, mobilization, demobilization, cleanup, overhead and profit. Note: Be sure the lamp selected is compatible with the installed ballast.

Fluorescent Lamps

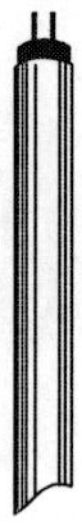

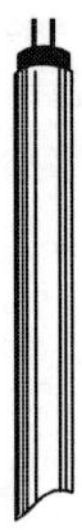

Material	Craft@Hrs	Unit	Material Cost	Labor Cost	Installed Cost
Preheat fluorescent lamps for use with starters					
F4T5/CW	L1@0.02	Ea	10.50	.82	11.32
F6T5/CW	L1@0.02	Ea	9.63	.82	10.45
F6T5/D	L1@0.02	Ea	15.30	.82	16.12
F8T5/CW	L1@0.03	Ea	9.50	1.23	10.73
F8T5/WW	L1@0.03	Ea	13.70	1.23	14.93
F8T5/D	L1@0.03	Ea	13.90	1.23	15.13
F13T5/CW	L1@0.03	Ea	11.80	1.23	13.03
F13T5/WW	L1@0.03	Ea	15.90	1.23	17.13
F13T8/CW	L1@0.04	Ea	15.50	1.64	17.14
F14T8/CW	L1@0.04	Ea	16.70	1.64	18.34
F14T8/D	L1@0.04	Ea	17.70	1.64	19.34
F15T8/CW	L1@0.04	Ea	8.62	1.64	10.26
F15T8/WW	L1@0.04	Ea	11.10	1.64	12.74
F15T8/D	L1@0.04	Ea	11.60	1.64	13.24
F30T8/CW	L1@0.04	Ea	11.30	1.64	12.94
F30T8/WW	L1@0.04	Ea	13.30	1.64	14.94
F30T8/D	L1@0.04	Ea	15.20	1.64	16.84
F14T12/CW	L1@0.04	Ea	10.60	1.64	12.24
F14T12/D	L1@0.04	Ea	12.80	1.64	14.44
F15T12/CW	L1@0.04	Ea	10.30	1.64	11.94
F15T12/WW	L1@0.04	Ea	12.10	1.64	13.74
F15T12/D	L1@0.04	Ea	12.80	1.64	14.44
F20T12/CW	L1@0.04	Ea	9.01	1.64	10.65
F20T12/WW	L1@0.04	Ea	9.63	1.64	11.27
F20T12/D	L1@0.04	Ea	10.70	1.64	12.34

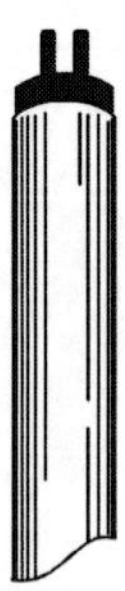

Material	Craft@Hrs	Unit	Material Cost	Labor Cost	Installed Cost
Rapid start fluorescent lamps, ES = Energy-Saving					
F30T12/CW/ES	L1@0.04	Ea	13.30	1.64	14.94
F30T12/WW/ES	L1@0.04	Ea	16.20	1.64	17.84
F30T12/CW	L1@0.04	Ea	11.30	1.64	12.94
F30T12/WW	L1@0.04	Ea	14.50	1.64	16.14
F30T12/D	L1@0.04	Ea	16.80	1.64	18.44
F40T12/CW/ES	L1@0.05	Ea	6.04	2.05	8.09
F40T12/WW/ES	L1@0.05	Ea	7.03	2.05	9.08
F40T12/W/ES	L1@0.05	Ea	6.78	2.05	8.83
F40T12/DX/ES	L1@0.05	Ea	10.50	2.05	12.55
F40T12/CW/U/3/ES	L1@0.05	Ea	27.30	2.05	29.35
F40T12/WW/U/3/ES	L1@0.06	Ea	28.00	2.46	30.46
F40T12/W/U/3/ES	L1@0.06	Ea	28.00	2.46	30.46
F40T12/CW/U/6/ES	L1@0.06	Ea	24.50	2.46	26.96
F40T12/WW/U/6/ES	L1@0.06	Ea	24.50	2.46	26.96
F40T12/W/U/6/ES	L1@0.06	Ea	24.50	2.46	26.96

Use these figures to estimate the cost of fluorescent lamps installed in lighting fixtures under the conditions described on pages 5 and 6. Costs listed are for each lamp installed. The crew is one electrician working at a labor cost of $41.03 per manhour. These costs include layout, material handling, and normal waste. Add for the lighting fixture, sales tax, delivery, supervision, mobilization, demobilization, cleanup, overhead and profit. Note: These figures are typical minimums for most fluorescent fixtures. If the fixture must be dismantled for lamping or if access to the fixture is difficult, the labor cost will be higher.

Material	Craft@Hrs	Unit	Material Cost	Labor Cost	Installed Cost
Slimline fluorescent lamps, ES = Energy-Saving					
F42T6/CW	L1@0.05	Ea	30.50	2.05	32.55
F42T6/WW	L1@0.05	Ea	33.40	2.05	35.45
F64T6/CW	L1@0.08	Ea	31.00	3.28	34.28
F64T6/WW	L1@0.08	Ea	33.90	3.28	37.18
F72T8/CW	L1@0.08	Ea	31.00	3.28	34.28
F72T8/WW	L1@0.08	Ea	33.40	3.28	36.68
F96T8/CW	L1@0.10	Ea	26.60	4.10	30.70
F24T12/CW	L1@0.05	Ea	23.80	2.05	25.85
F36T12/CW	L1@0.05	Ea	23.80	2.05	25.85
F42T12/CW	L1@0.05	Ea	23.80	2.05	25.85
F48T12/CW/ES	L1@0.05	Ea	18.90	2.05	20.95
F48T12/WW/ES	L1@0.05	Ea	20.30	2.05	22.35
F48T12/W/ES	L1@0.05	Ea	20.30	2.05	22.35
F48T12/CW	L1@0.05	Ea	15.50	2.05	17.55
F48T12/D	L1@0.05	Ea	19.40	2.05	21.45
F60T12/CW	L1@0.08	Ea	22.00	3.28	25.28
F60T12/D	L1@0.08	Ea	24.60	3.28	27.88
F64T12/CW	L1@0.08	Ea	22.20	3.28	25.48
F64T12/D	L1@0.08	Ea	24.60	3.28	27.88
F72T12/CW	L1@0.10	Ea	17.30	4.10	21.40
F72T12/CWX	L1@0.10	Ea	23.40	4.10	27.50
F72T12/WW	L1@0.10	Ea	22.20	4.10	26.30
F72T12/D	L1@0.10	Ea	22.20	4.10	26.30
F84T12/CW	L1@0.10	Ea	23.00	4.10	27.10
F96T12/CW/ES	L1@0.10	Ea	13.00	4.10	17.10
F96T12/WW/ES	L1@0.10	Ea	15.70	4.10	19.80
F96T12/W/ES	L1@0.10	Ea	14.50	4.10	18.60
High-Output (HO) fluorescent lamps, ES = Energy-Saving					
F18T12/CW/HO	L1@0.04	Ea	17.80	1.64	19.44
F24T12/CW/HO	L1@0.04	Ea	24.10	1.64	25.74
F24T12/D/HO	L1@0.04	Ea	27.90	1.64	29.54
F30T12/CW/HO	L1@0.04	Ea	21.30	1.64	22.94
F36T12/CW/HO	L1@0.04	Ea	25.50	1.64	27.14
F36T12/D/HO	L1@0.04	Ea	28.20	1.64	29.84
F42T12/CW/HO	L1@0.05	Ea	26.50	2.05	28.55
F42T12/D/HO	L1@0.05	Ea	29.10	2.05	31.15
F48T12/CW/HO	L1@0.05	Ea	17.30	2.05	19.35
F48T12/WW/HO	L1@0.05	Ea	24.10	2.05	26.15
F48T12/D/HO	L1@0.05	Ea	22.60	2.05	24.65
F48T12/W/HO/ES	L1@0.05	Ea	22.60	2.05	24.65
F64T12/CW/HO	L1@0.08	Ea	22.60	3.28	25.88
F64T12/D/HO	L1@0.08	Ea	26.10	3.28	29.38

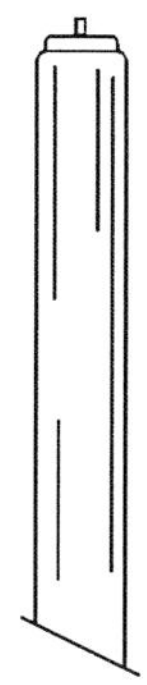

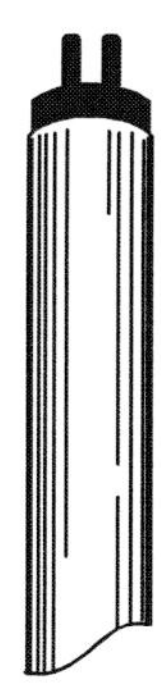

Use these figures to estimate the cost of fluorescent lamps installed in lighting fixtures under the conditions described on pages 5 and 6. Costs listed are for each lamp installed. The crew is one electrician working at a labor cost of $41.03 per manhour. These costs include layout, material handling, and normal waste. Add for the lighting fixture, sales tax, delivery, supervision, mobilization, demobilization, cleanup, overhead and profit. Note: These figures are typical minimums for most fluorescent fixtures. If the fixture must be dismantled for lamping or if access to the fixture is difficult, the labor cost will be higher.

Fluorescent Lamps

Material	Craft@Hrs	Unit	Material Cost	Labor Cost	Installed Cost
High-Output (HO) fluorescent lamps, ES = Energy-Saving (continued)					
F72T12/CW/HO	L1@0.08	Ea	18.10	3.28	21.38
F72T12/WW/HO	L1@0.08	Ea	24.30	3.28	27.58
F72T12/D/HO	L1@0.08	Ea	24.10	3.28	27.38
F84T12/CW/HO	L1@0.08	Ea	19.70	3.28	22.98
F84T12/D/HO	L1@0.08	Ea	24.30	3.28	27.58
F96T12/CW/HO/ES	L1@0.10	Ea	16.20	4.10	20.30
F96T12/WW/HO/ES	L1@0.10	Ea	23.00	4.10	27.10
F96T12/W/HO/ES	L1@0.10	Ea	20.00	4.10	24.10
Very High-Output (VHO) fluorescent lamps, ES = Energy-Saving					
F48T12/CW/VHO	L1@0.05	Ea	35.20	2.05	37.25
F48T12/WW/VHO	L1@0.05	Ea	39.60	2.05	41.65
F72T12/CW/VHO	L1@0.08	Ea	36.40	3.28	39.68
F96T12/CW/VHO/ES	L1@0.10	Ea	34.80	4.10	38.90
F96T12/W/VHO/ES	L1@0.10	Ea	41.10	4.10	45.20
F96T12/CW/VHO	L1@0.10	Ea	33.40	4.10	37.50
F96T12/WW/VHO	L1@0.10	Ea	37.30	4.10	41.40
F96T12/D/VHO	L1@0.10	Ea	43.30	4.10	47.40
Power Groove (PG) fluorescent lamps, ES = Energy-Saving					
F48PG17/CW/ES	L1@0.05	Ea	44.00	2.05	46.05
F48PG17/CW	L1@0.05	Ea	40.10	2.05	42.15
F48PG17/D	L1@0.05	Ea	46.90	2.05	48.95
F72PG17/CW	L1@0.08	Ea	43.40	3.28	46.68
F96PG17/CW/ES	L1@0.10	Ea	41.10	4.10	45.20
F96PG17/W/ES	L1@0.10	Ea	53.90	4.10	58.00
F96PG17/CW	L1@0.10	Ea	38.20	4.10	42.30
F96PG17/D	L1@0.10	Ea	48.00	4.10	52.10
Circular fluorescent lamps					
FC6T9/CW	L1@0.08	Ea	15.50	3.28	18.78
FC6T9/WW	L1@0.08	Ea	15.90	3.28	19.18
FC8T9/CW	L1@0.08	Ea	13.00	3.28	16.28
FC8T9/WW	L1@0.08	Ea	16.20	3.28	19.48
FC12T9/CW	L1@0.10	Ea	16.60	4.10	20.70
FC12T9/WW	L1@0.10	Ea	18.20	4.10	22.30
FC12T9/D	L1@0.10	Ea	20.60	4.10	24.70
FC16T9/CW	L1@0.10	Ea	20.30	4.10	24.40
FC16T9/WW	L1@0.10	Ea	24.40	4.10	28.50
FC16T9/D	L1@0.10	Ea	26.10	4.10	30.20

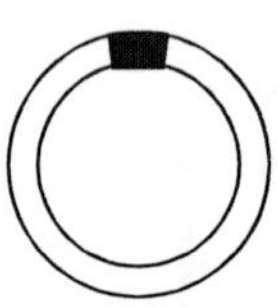

Use these figures to estimate the cost of fluorescent lamps installed in lighting fixtures under the conditions described on pages 5 and 6. Costs listed are for each lamp installed. The crew is one electrician working at a labor cost of $41.03 per manhour. These costs include layout, material handling, and normal waste. Add for the lighting fixture, sales tax, delivery, supervision, mobilization, demobilization, cleanup, overhead and profit. Note: These figures are typical minimums for most fluorescent fixtures. If the fixture must be dismantled for lamping or if access to the fixture is difficult, the labor cost will be higher.

Material	Craft@Hrs	Unit	Material Cost	Labor Cost	Installed Cost
Rapid start energy-saving fluorescent ballasts, 430 M.A.-HPF					
1 F40T12-120V	L1@0.40	Ea	33.30	16.40	49.70
1 F40T12-277V	L1@0.40	Ea	90.20	16.40	106.60
2 F40T12-120V	L1@0.40	Ea	31.40	16.40	47.80
2 F40T12-277V	L1@0.40	Ea	36.70	16.40	53.10
3 F40T12-120V	L1@0.40	Ea	64.30	16.40	80.70
3 F40T12-277V	L1@0.40	Ea	124.00	16.40	140.40
Rapid start energy-saving fluorescent ballasts, 460 M.A.-HPF					
3 F40T12 (34W)-120V	L1@0.40	Ea	95.30	16.40	111.70
3 F40T12 (34W)-277V	L1@0.40	Ea	102.00	16.40	118.40
Rapid start energy-saving fluorescent ballasts, 800 M.A.-HPF					
2 F72T12-120V	L1@0.50	Ea	124.00	20.50	144.50
2 F96T12-120V	L1@0.50	Ea	124.00	20.50	144.50
2 F72T12-277V	L1@0.50	Ea	129.00	20.50	149.50
2 F96T12-277V	L1@0.50	Ea	129.00	20.50	149.50
Electronic energy-saving fluorescent ballasts					
1 F30T12-120V	L1@0.40	Ea	110.00	16.40	126.40
1 F40T12-120V	L1@0.40	Ea	110.00	16.40	126.40
1 F40T10-120V	L1@0.40	Ea	110.00	16.40	126.40
1 F30T12-277V	L1@0.40	Ea	117.00	16.40	133.40
1 F40T12-277V	L1@0.40	Ea	117.00	16.40	133.40
1 F40T10-277V	L1@0.40	Ea	117.00	16.40	133.40
2 F40T12-120V	L1@0.40	Ea	180.00	16.40	196.40
2 F40T10-120V	L1@0.40	Ea	180.00	16.40	196.40
2 F40T12-277V	L1@0.40	Ea	182.00	16.40	198.40
2 F40T10-277V	L1@0.40	Ea	182.00	16.40	198.40
2 F48T12-120V	L1@0.40	Ea	145.00	16.40	161.40
2 F72T12-120V	L1@0.50	Ea	145.00	20.50	165.50
2 F96T12-120V	L1@0.50	Ea	145.00	20.50	165.50
2 F48T10-277V	L1@0.50	Ea	153.00	20.50	173.50
2 F72T12-277V	L1@0.50	Ea	153.00	20.50	173.50
2 F96T12-277V	L1@0.50	Ea	153.00	20.50	173.50
Slimline energy-saving fluorescent ballasts, 425 M.A.-HPF					
2 F72T12-120V	L1@0.50	Ea	47.90	20.50	68.40
2 F96T10-120V	L1@0.50	Ea	47.90	20.50	68.40
2 F72T12-277V	L1@0.50	Ea	112.00	20.50	132.50
2 F96T12-277V	L1@0.50	Ea	132.00	20.50	152.50

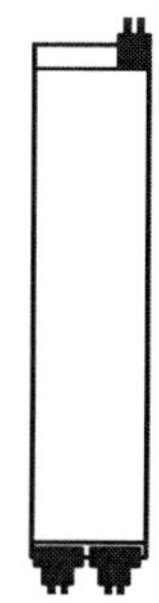

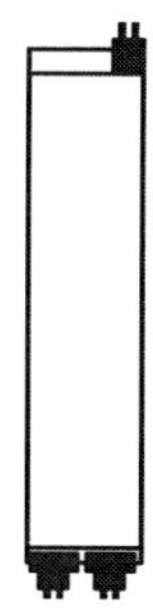

Use these figures to estimate the cost of installing fluorescent ballasts in lighting fixtures under the conditions described on pages 5 and 6. Costs listed are for each ballast installed. The crew is one electrician working at a labor cost of $41.03 per manhour. These costs include layout, material handling, and normal waste. Add for the lighting fixture, wire connections, sales tax, delivery, supervision, mobilization, demobilization, cleanup, overhead and profit. These costs are typical minimums for most fluorescent fixtures. If the ballast compartment requires substantial dismantling and reassembly or if the fixture is difficult to reach, the labor cost will be higher.

Fluorescent Ballasts, Standard

Rapid start standard fluorescent ballasts, 430 M.A.-LPF

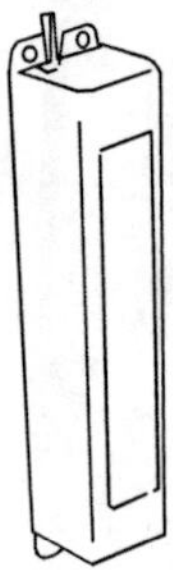

Material	Craft@Hrs	Unit	Material Cost	Labor Cost	Installed Cost
1 F30T12-120V	L1@0.40	Ea	21.40	16.40	37.80
1 F40T12-120V	L1@0.40	Ea	21.40	16.40	37.80

Rapid start standard fluorescent ballasts, 460 M.A.-HPF

Material	Craft@Hrs	Unit	Material Cost	Labor Cost	Installed Cost
1 F30T12-120V	L1@0.40	Ea	66.30	16.40	82.70
1 F30T12-277V	L1@0.40	Ea	70.60	16.40	87.00
1 F40T12-120V	L1@0.40	Ea	32.30	16.40	48.70
1 F40T12-277V	L1@0.40	Ea	37.10	16.40	53.50
2 F30T12-120V	L1@0.40	Ea	58.90	16.40	75.30
2 F30T12-277V	L1@0.40	Ea	63.50	16.40	79.90
2 F40T12-120V	L1@0.40	Ea	31.40	16.40	47.80
2 F40T12-277V	L1@0.40	Ea	37.10	16.40	53.50

Rapid start standard fluorescent ballasts, 800 M.A.-HPF

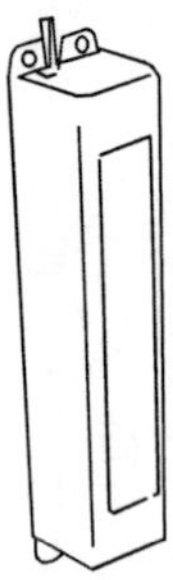

Material	Craft@Hrs	Unit	Material Cost	Labor Cost	Installed Cost
1 F24T12-120V	L1@0.40	Ea	135.00	16.40	151.40
1 F48T12-120V	L1@0.40	Ea	135.00	16.40	151.40
1 F72T12-120V	L1@0.50	Ea	113.00	20.50	133.50
1 F96T12-120V	L1@0.50	Ea	113.00	20.50	133.50
1 F24T12-277V	L1@0.40	Ea	133.00	16.40	149.40
1 F48T12-277V	L1@0.40	Ea	133.00	16.40	149.40
1 F72T12-277V	L1@0.50	Ea	140.00	20.50	160.50
1 F96T12-277V	L1@0.50	Ea	140.00	20.50	160.50
2 F24T12-120V	L1@0.40	Ea	129.00	16.40	145.40
2 F48T12-120V	L1@0.40	Ea	129.00	16.40	145.40
2 F72T12-120V	L1@0.50	Ea	97.90	20.50	118.40
2 F96T12-120V	L1@0.50	Ea	97.90	20.50	118.40
2 F24T12-277V	L1@0.40	Ea	138.00	16.40	154.40
2 F48T12-277V	L1@0.40	Ea	138.00	16.40	154.40
2 F72T12-277V	L1@0.50	Ea	103.00	20.50	123.50
2 F96T12-277V	L1@0.50	Ea	103.00	20.50	123.50

Rapid start standard fluorescent ballasts, 1,500 M.A.-HPF

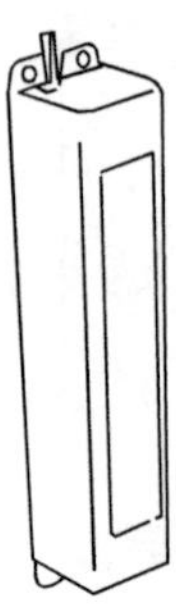

Material	Craft@Hrs	Unit	Material Cost	Labor Cost	Installed Cost
1 F48T12-120V	L1@0.40	Ea	155.00	16.40	171.40
1 F72T12-120V	L1@0.50	Ea	155.00	20.50	175.50
1 F96T12-120V	L1@0.50	Ea	155.00	20.50	175.50
1 F48T12-277V	L1@0.40	Ea	170.00	16.40	186.40
1 F72T12-277V	L1@0.50	Ea	170.00	20.50	190.50
1 F96T12-277V	L1@0.50	Ea	170.00	20.50	190.50
2 F48T12-120V	L1@0.40	Ea	155.00	16.40	171.40
2 F72T12-120V	L1@0.50	Ea	173.00	20.50	193.50
2 F96T12-120V	L1@0.50	Ea	173.00	20.50	193.50
2 F48T12-277V	L1@0.40	Ea	170.00	16.40	186.40
2 F72T12-277V	L1@0.50	Ea	195.00	20.50	215.50
2 F96T12-277V	L1@0.50	Ea	195.00	20.50	215.50

Use these figures to estimate the cost of installing fluorescent ballasts in lighting fixtures under the conditions described on pages 5 and 6. Costs listed are for each ballast installed. The crew is one electrician working at a labor cost of $41.03 per manhour. These costs include layout, material handling, and normal waste. Add for lighting fixture, wire connections, sales tax, delivery, supervision, mobilization, demobilization, cleanup, overhead and profit. These costs are typical minimums for most fluorescent fixtures. If the ballast compartment requires substantial dismantling and reassembly or if the fixture is difficult to reach, the labor cost will be higher.

Instant start standard fluorescent ballasts, slimline, 200 M.A.-LPF

Material	Craft@Hrs	Unit	Material Cost	Labor Cost	Installed Cost
1 F42T6-120V	L1@0.40	Ea	64.00	16.40	80.40

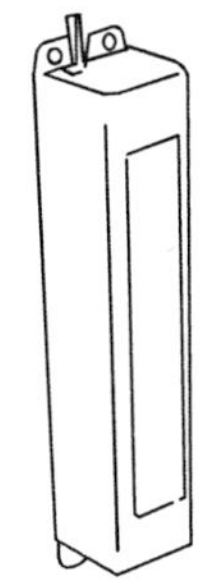

Instant start standard fluorescent ballasts, slimline, 200 M.A.-HPF

Material	Craft@Hrs	Unit	Material Cost	Labor Cost	Installed Cost
1 F42T6-120V	L1@0.40	Ea	62.60	16.40	79.00
1 F64T6-120V	L1@0.40	Ea	89.30	16.40	105.70
1 F72T8-120V	L1@0.50	Ea	89.30	20.50	109.80
1 F96T8-120V	L1@0.50	Ea	89.30	20.50	109.80
2 F42T6-120V	L1@0.40	Ea	148.00	16.40	164.40
2 F64T6-120V	L1@0.40	Ea	97.00	16.40	113.40
2 F72T8-120V	L1@0.50	Ea	103.00	20.50	123.50
2 F96T8-120V	L1@0.50	Ea	103.00	20.50	123.50

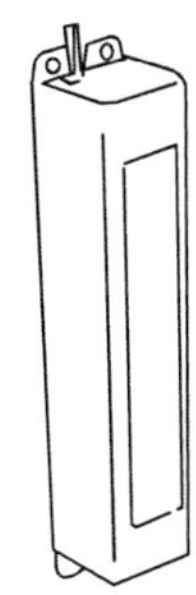

Instant start standard fluorescent ballasts, slimline, 425 M.A.-HPF

Material	Craft@Hrs	Unit	Material Cost	Labor Cost	Installed Cost
1 F24T12-120V	L1@0.40	Ea	69.70	16.40	86.10
1 F40T12-120V	L1@0.40	Ea	69.70	16.40	86.10
1 F48T12-120V	L1@0.40	Ea	69.70	16.40	86.10
1 F60T12-120V	L1@0.40	Ea	60.30	16.40	76.70
1 F64T12-120V	L1@0.40	Ea	60.30	16.40	76.70
1 F72T12-120V	L1@0.50	Ea	60.30	20.50	80.80
1 F96T12-120V	L1@0.50	Ea	60.30	20.50	80.80
1 F24T12-277V	L1@0.40	Ea	84.00	16.40	100.40
1 F40T12-277V	L1@0.40	Ea	84.00	16.40	100.40
1 F48T12-277V	L1@0.40	Ea	84.00	16.40	100.40
1 F60T12-277V	L1@0.40	Ea	60.30	16.40	76.70
1 F64T12-277V	L1@0.40	Ea	60.30	16.40	76.70
1 F72T12-277V	L1@0.50	Ea	64.30	20.50	84.80
1 F96T12-277V	L1@0.50	Ea	64.30	20.50	84.80
2 F24T12-120V	L1@0.40	Ea	58.40	16.40	74.80
2 F40T12-120V	L1@0.40	Ea	58.40	16.40	74.80
2 F48T12-120V	L1@0.40	Ea	58.40	16.40	74.80
2 F60T12-120V	L1@0.40	Ea	66.30	16.40	82.70
2 F64T12-120V	L1@0.40	Ea	66.30	16.40	82.70
2 F72T12-120V	L1@0.50	Ea	51.60	20.50	72.10
2 F96T12-120V	L1@0.50	Ea	51.60	20.50	72.10
2 F40T12-277V	L1@0.40	Ea	66.30	16.40	82.70
2 F48T12-277V	L1@0.40	Ea	66.30	16.40	82.70
2 F60T12-277V	L1@0.40	Ea	72.60	16.40	89.00
2 F64T12-277V	L1@0.40	Ea	72.60	16.40	89.00
2 F72T12-277V	L1@0.50	Ea	141.00	20.50	161.50
2 F96T12-277V	L1@0.50	Ea	141.00	20.50	161.50

Use these figures to estimate the cost of installing fluorescent ballasts in lighting fixtures under the conditions described on pages 5 and 6. Costs listed are for each ballast installed. The crew is one electrician working at a labor cost of $41.03 per manhour. These costs include layout, material handling, and normal waste. Add for lighting fixture, wire connections, sales tax, delivery, supervision, mobilization, demobilization, cleanup, overhead and profit. These cost are typical minimums for most fluorescent fixtures. If the ballast compartment requires substantial dismantling and reassembly or if the fixture is difficult to reach, the labor cost will be higher.

Fluorescent Ballasts

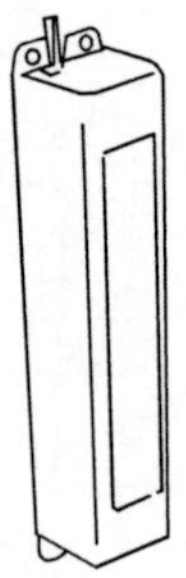

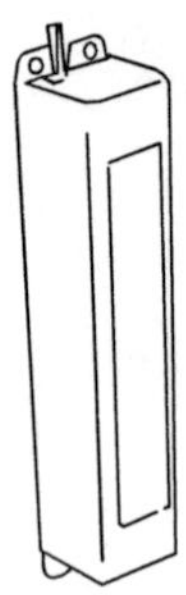

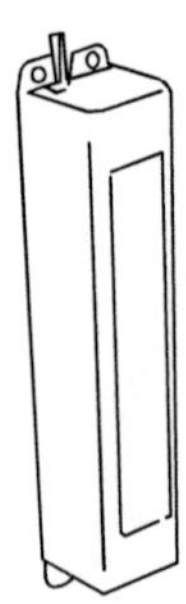

Material	Craft@Hrs	Unit	Material Cost	Labor Cost	Installed Cost
Preheat fluorescent ballasts, LPF					
1 F4T5-120V	L1@0.35	Ea	7.62	14.40	22.02
1 F6T5-120V	L1@0.35	Ea	7.62	14.40	22.02
1 F8T5-120V	L1@0.35	Ea	7.62	14.40	22.02
1 F14T12-120V	L1@0.35	Ea	6.34	14.40	20.74
1 F20T12-120V	L1@0.40	Ea	6.34	16.40	22.74
2 F15T8-120V	L1@0.35	Ea	6.34	14.40	20.74
2 F25T12-120V	L1@0.40	Ea	13.50	16.40	29.90
Preheat fluorescent ballasts, LPF, Class P					
1 F30T8-120V	L1@0.40	Ea	20.10	16.40	36.50
2 F40T12-120V	L1@0.40	Ea	20.10	16.40	36.50
2 F90T17-120V	L1@0.50	Ea	170.00	20.50	190.50
2 F100T17-120V	L1@0.50	Ea	170.00	20.50	190.50
Trigger start fluorescent ballasts, LPF, Class P					
1 F14T12-120V	L1@0.40	Ea	20.10	16.40	36.50
1 F20T12-120V	L1@0.40	Ea	20.10	16.40	36.50
1 F15T8-120V	L1@0.40	Ea	20.10	16.40	36.50
2 F14T12-120V	L1@0.40	Ea	29.40	16.40	45.80
2 F20T12-120V	L1@0.40	Ea	29.40	16.40	45.80
2 F15T8-120V	L1@0.40	Ea	29.40	16.40	45.80
2 F15T12-120V	L1@0.40	Ea	29.40	16.40	45.80
2 F20T12-120V	L1@0.40	Ea	29.40	16.40	45.80
2 F15T8-120V	L1@0.40	Ea	29.40	16.40	45.80
Trigger start fluorescent ballasts, HPF, Class P					
1 F14T12-120V	L1@0.40	Ea	71.10	16.40	87.50
1 F20T12-120V	L1@0.40	Ea	71.10	16.40	87.50
1 F15T8-120V	L1@0.40	Ea	71.10	16.40	87.50
1 F14T12-277V	L1@0.40	Ea	75.70	16.40	92.10
1 F20T12-277V	L1@0.40	Ea	75.70	16.40	92.10
1 F15T8-277V	L1@0.40	Ea	75.70	16.40	92.10
2 F14T12-120V	L1@0.40	Ea	62.70	16.40	79.10
2 F20T12-120V	L1@0.40	Ea	62.70	16.40	79.10
2 F15T8-120V	L1@0.40	Ea	62.70	16.40	79.10
2 F14T12-277V	L1@0.40	Ea	66.30	16.40	82.70
2 F20T12-277V	L1@0.40	Ea	66.30	16.40	82.70
2 F15T8-277V	L1@0.40	Ea	66.30	16.40	82.70

Use these figures to estimate the cost of installing fluorescent ballasts in lighting fixtures under the conditions described on pages 5 and 6. Costs listed are for each ballast installed. The crew is one electrician working at a labor cost of $41.03 per manhour. These costs include layout, material handling, and normal waste. Add for lighting fixture, wire connections, sales tax, delivery, supervision, mobilization, demobilization, cleanup, overhead and profit. These costs are typical minimums for most fluorescent fixtures. If the ballast compartment requires substantial dismantling and reassembly or if the fixture is difficult to reach, the labor cost will be higher.

Material	Craft@Hrs	Unit	Material Cost	Labor Cost	Installed Cost
Ceiling fans, 3-speed reversible motors, four blades, 120 volt, 240 watts, three light					
36" plain, white, oak	L1@1.25	Ea	75.10	51.30	126.40
42" plain, white, oak	L1@1.50	Ea	90.20	61.50	151.70
44" plain, white, oak	L1@1.50	Ea	90.20	61.50	151.70
52" plain, white, oak	L1@1.50	Ea	90.20	61.50	151.70
60" plain, white, oak	L1@1.60	Ea	101.00	65.60	166.60
Ceiling fans, 3-speed reversible motors, five blades, 120 volt, 240 watts, four light					
36" plain, white, oak	L1@1.25	Ea	82.70	51.30	134.00
42" plain, white, oak	L1@1.50	Ea	97.60	61.50	159.10
44" plain, white, oak	L1@1.50	Ea	97.60	61.50	159.10
52" plain, white, oak	L1@1.50	Ea	97.60	61.50	159.10
60" plain, white, oak	L1@1.60	Ea	133.00	65.60	198.60
Ceiling fans, wall or remote control reversible motors, 120 volts, 240 watts, four blades, three or four lights					
36" plain, white, oak	L1@1.35	Ea	189.00	55.40	244.40
42" plain, white, oak	L1@1.60	Ea	187.00	65.60	252.60
44" plain, white, oak	L1@1.60	Ea	187.00	65.60	252.60
52" plain, white, oak	L1@1.60	Ea	187.00	65.60	252.60
60" plain, white, oak	L1@1.70	Ea	201.00	69.80	270.80
Ceiling fans, wall or remote control reversible motors, 120 volts, 500 watts, five blades, three or four lights					
36" plain, white, oak	L1@1.45	Ea	223.00	59.50	282.50
42" plain, white, oak	L1@1.70	Ea	240.00	69.80	309.80
44" plain, white, oak	L1@1.70	Ea	240.00	69.80	309.80
52" plain, white, oak	L1@1.70	Ea	240.00	69.80	309.80
60" plain, white, oak	L1@1.80	Ea	258.00	73.90	331.90

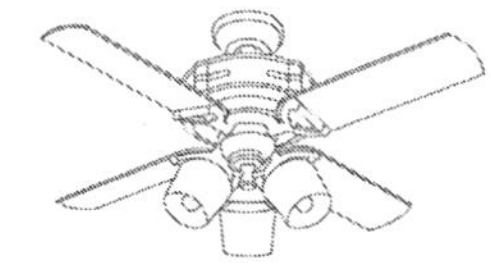

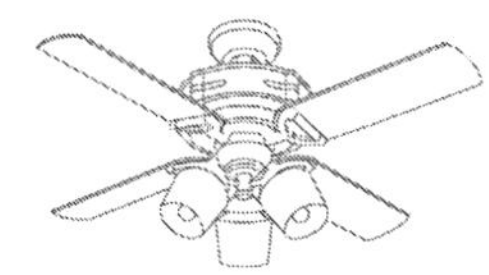

Use these figures to estimate the cost of ceiling fans installed in buildings under the conditions described on pages 5 and 6. Costs listed are for each ceiling fan installed. The crew is one electrician working at a labor cost of $41.03 per manhour. These costs include layout, material handling, and normal waste. Add for lamps, accessories, sales tax, delivery, supervision, mobilization, demobilization, cleanup, overhead and profit. These costs are based on ceiling fans installed not over 10 feet above the floor.

Section 5: Wiring Devices

The term ***wiring devices*** refers to the circuit control switches and convenience outlets in an electrical system. The switches we use to turn lights on and off and the convenience receptacles we use to get power for electrical appliances are wiring devices.

Why the Cost of Devices Varies

Estimating the cost of wiring devices may seem to be a simple matter of accurate counting. On some jobs it is. But pricing the number of devices counted can be more difficult. Many types of wiring devices are available. Some look about the same, serve the same function, but vary widely in price. It's not always easy to be sure you're pricing the right device.

This section describes the more common devices and suggests what to watch out for when pricing the device section of your estimate.

Standard-grade devices used in residences will be the least expensive because they're made in great quantity and carried by every electrical supplier. Less common devices are made to meet special requirements and constructed of special materials. For example, switches and trim plates made of nylon are intended for heavy use. They cost more than standard residential-grade devices and may not be stocked by your local supplier.

Every modern wiring device has a voltage and amperage rating. This rating has an effect on cost. Usually, the greater the ampacity and voltage rating, the higher the cost.

Devices intended to meet exacting specifications or to serve in special applications are said to be ***specification grade***. Usually these devices are made with more expensive contact material, are designed with special features, or are built to be particularly durable. That increases the cost, often to several times the cost of the standard-grade device. But the added cost may be a good investment from the owner's standpoint. For example, one specification-grade convenience outlet has contact material with better shape-retention characteristics. That ensures better electrical contact after years of hard use. You've probably seen duplex outlets that will hardly hold an electrical plug in place after a year or two of service.

Most manufacturers offer wiring devices with smooth surfaces that stay cleaner longer — and are easier to clean when they do get soiled. Older switch covers had decorative grooves that accumulated smudges and were hard to clean.

Standard colors for wiring devices are ivory, white and brown. Other colors are available but will cost more.

Some devices have built-in grounding. There are two common types of grounded devices. One has a screw or terminal for attaching a ground conductor. The other has a special spring on one of the mounting screws. This spring provides a positive ground through the device frame to the mounting screw and on to a metal-backed outlet box. Automatic grounding devices cost a little more but install a little faster than the type with a grounding terminal or grounding screw, because the electrician saves one connection.

Wiring Device Descriptions

Duplex receptacles are the most common wiring device. Many types are available. The traditional duplex outlet has screw terminals around which wire is wrapped to make contact. Then the screw is tightened. Some have holes designed to receive and hold a stripped wire conductor. That eliminates wrapping the wire around a screw contact and tightening the screw. Other receptacles have both screw terminals and insert holes. Some are side wired. Others are wired at the back.

Many duplex receptacles have a removable tab between the screw terminals on each side. Remove the tab to isolate one part of the receptacle from the other. These are commonly used for switched outlets. Some residential rooms are wired so that a switch by the entrance controls one or more outlets where table lamps will be installed. Outlets with tabs can also be wired to two different circuits.

Clock hanger receptacles are usually mounted high on a wall. They have a recessed single outlet for a clock. They also have a tab to hold the clock. This is a decorative item, so many colors and styles are available. Clock hanger outlets are usually connected to an unswitched circuit.

Locking receptacles come in a variety of sizes and configurations. Each service load capacity has a different size and configuration as set by NEMA (National Electrical Manufacturers' Association) standards. They're made that way so that appliances that use the receptacles can use only receptacles rated at the correct capacity.

Locking receptacles can be wall-mounted in a box or may be attached to an extension cord.

Be careful when selecting trim plates for locking receptacles. The size of the receptacle face varies with the service rating. A 20-amp, 4-wire receptacle is larger than a 20-amp, 3-wire receptacle. Locking receptacles are commonly called ***twist-lock*** devices because they must be rotated or twisted when inserted and removed.

Switches are classified by the number of poles. Figure 5-1 shows the most simple type, the one-pole switch. Notice that the switch interrupts only one conductor when open. A one-pole switch can control only one circuit.

Two-pole switches can control two separate circuits at the same time. For example, a large room may need two lighting circuits. If only one switch is needed, use a single two-pole switch. Several two-pole switches can be ganged to control a number of circuits. This reduces the labor, switch, and box cost.

Many offices require two-level lighting. Regulations designed to reduce energy consumption sometimes require two-level lighting in all office buildings. Each four-lamp fluorescent fixture will have two lamps controlled from one switch and two lamps controlled by another switch located immediately beside the first. When there are more than two circuits involved, two-pole switches can be used in the same two-gang switch box.

Three-way switches allow switching a lighting circuit from two separate locations. For example, in a hall you want lighting controlled from switches at either end. On a stairway you want lights controlled from both upstairs and downstairs. Figure 5-2 shows that both switches 1 and 2 can control the circuit no matter what position the other switch is in.

Four-way switches are used where three switches need control of the circuit. The four-way switch is connected into the switching circuits between two three-way switches. In Figure 5-3, switch 3 can open or close the circuit no matter what position switches 1 and 2 are in.

There are two types of switches with illuminated handles. One lights when the switch is in the ***on*** position. The other lights when the switch is in the ***off*** position. Either can be used as a pilot light. But be sure you order the right switch. They're not interchangeable and can't be modified on site.

The switch that lights when on is used where the switch and the fixture or appliance are in different rooms — such as the light for a storage room or closet.

The switch that lights when the switch is off is used in rooms that are normally dark when unlighted. When the switch is on, the handle isn't illuminated.

Quiet (silent) switches are a good choice for better-quality homes and offices. You'll hear these called ***mercury switches***, but they may or may not be made with mercury contacts. Quiet switches come in voltages to 277V.

Heavy duty "T" rated switches are made for heavy loads. The circuit is closed by a contact armature that moves very fast to minimize arcing when the switch changes position.

Decorative wiring devices are offered by some manufacturers. Unlike standard devices, decorative devices are generally rectangular. The trim plate is designed to fit only a decorative style device. And note that trim plates made by some manufacturers for decorative switches won't fit decorative duplex receptacles by the same manufacturer. That creates a problem when placing a decorative switch and a decorative duplex receptacle in the same box.

Horsepower-rated switches are used to control small motor loads, such as exhaust fans, where manual motor starters are not required. The motor has a built-in overload protector.

Key-operated switches are common in public areas where lighting shouldn't be controlled by the public — for example, in a public restroom where the light has to be on while the building is open to the public. Key switches come with either a simple stamped 2-prong key or a conventional lockset.

Duplex switches have two switches mounted in a housing shaped like a duplex outlet. They can be wired to control two separate circuits or to control one circuit with two loads. Duplex switches can be used to control two-level lighting in an office.

Momentary-contact switches provide current only for an instant. When not being activated, the switch is always open. Momentary switches are commonly used to throw another switch like a lighting contactor that controls a large bank of fluorescent lights. If the lighting contactor has a close and open set of coils, a momentary-center-off switch is used. The contactor is closed when the momentary switch handle is lifted to the ***on*** position. When the switch handle is released, it returns to the center position. To de-energize the lighting contactor, the momentary switch is pushed down to the ***off*** position. Any time the momentary switch handle is released, it returns to the center position.

Maintain-contact switches are used when current is to flow in both the ***on*** and ***off*** positions. There is a center ***off*** position. But the switch handle must be moved to that center position. It doesn't return to center by just releasing the handle. One application for this type of switch is a conveyor that can be operated in two directions but also has to be stopped when not in use. There are many other applications.

Dimming switches have to be sized for the anticipated lighting load. Incandescent lighting loads and incandescent dimmers are rated in

Simple Switch Control Schemes

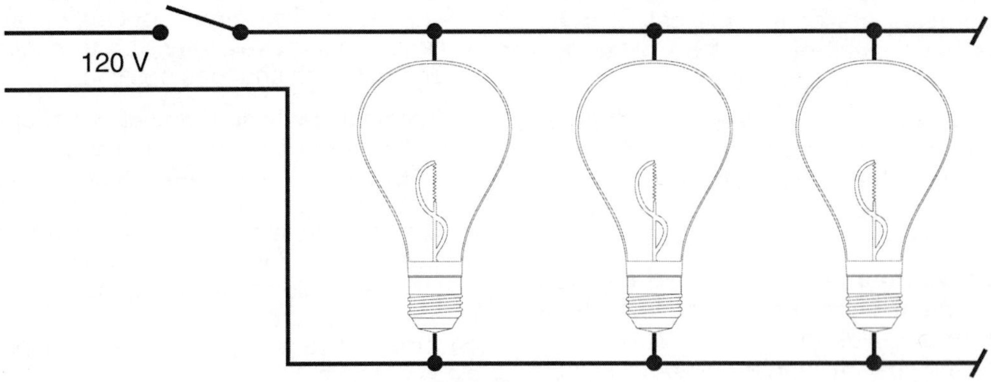

Figure 5-1
Single-Pole Switch

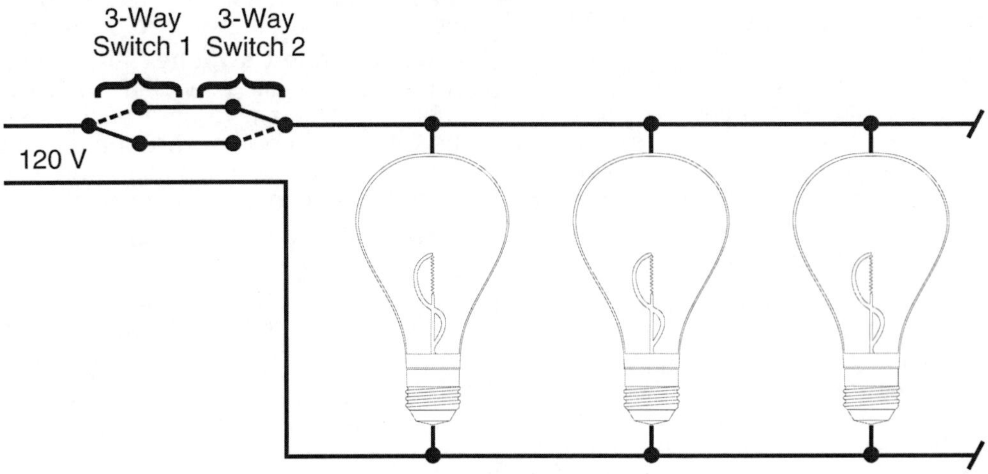

Figure 5-2
Three-Way Switch

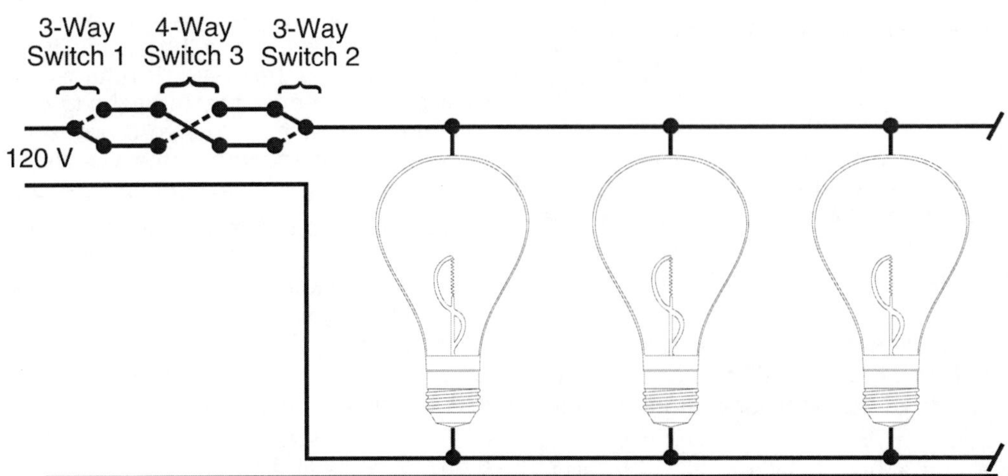

Figure 5-3
Four-Way Switch

watts. Incandescent dimmers can't be used with fluorescent lighting. Fluorescent dimmers are rated by the number of lamps in the dimming circuit. Fluorescent lighting fixtures need special dimming ballasts. Standard fluorescent ballasts won't work on a dimming circuit.

There are many types of dimming systems, from a simple manual wheel to large motorized units. Large dimming systems, such as stage lighting, are custom engineered for the lighting load expected.

Photoelectric switches are made to energize a circuit from dusk until dawn. Exterior lighting is usually controlled with photoelectric switches built into the luminaire. Photoelectric switches can also be used to control latches on entryways or gates. When the light beam is interrupted, the switch operates.

Device plates come in as many sizes, shapes and colors and are made of as many materials as the devices they cover. But if you need a special plate, order it early. Some plates are stocked or made only on special order.

Electrical material catalogs list many plates that are not included in this section. Most are made for specific uses. Some are not interchangeable with any other plate and may require a specific outlet box.

Every month new wiring devices and trim plates are introduced and old devices and plates are discontinued. Keep your catalog current for quick reference.

Many remodeling contractors use semi-jumbo and jumbo device trim plates. Oversize plates hide mistakes and make it unnecessary to repaint just because a device was changed. The standard device trim plate doesn't provide much cover beyond the edge of a recessed outlet box.

Deep-cut trim plates are made with a more concave shape so they fit flush against a wall even if the box protrudes slightly beyond the wall surface. That's a big advantage when a box is being cut into an existing wall. Deep-cut trim plates allow more room for the mounting screws that hold the box to the wall surface.

When a standard box is cut into an existing wall, the mounting yoke or ears at the top and bottom of the box set flush with the wall surface. Adding screws to attach the box to the wall will hold a standard trim plate away from the wall surface. That's unsightly. The easy solution is to use a deep-cut plate. That's much easier than recessing the mounting ears into the finish wall surface.

Galvanized or plated trim plates are used to cover an outlet box that's surface mounted. You'll hear these plates called "industrial" covers or "handy box" covers. They're made for switches, receptacles and combinations of both. Material and labor costs for this item are listed in Section 3 of this manual.

Weatherproof plates also come in a wide variety of sizes and descriptions. There are spring-loaded types with a flip cover that returns to the closed position when released. The locking type can be opened only with a key. Some are made for use in corrosive environments. some are intended for use with vertically-mounted duplex receptacles. Others are intended for horizontal mounting.

Take-off and Pricing Wiring Devices

The remainder of this section points out some important considerations when preparing a take-off of wiring devices.

The cost tables in this section cover all the common wiring devices. But many more devices are available. Your supplier can provide a more complete list of the devices that will meet any conceivable need.

Your device take-off should start with the switches. I recommend using an 8½" x 11" lined tablet to list all devices. Draw a line down the left side of the page. The line should be about 1½ inches from the left edge. At the top of the sheet and just to the right of the line you've drawn, write in the number of the first plan sheet you're using. For example, the first plan sheet might be E2. To the left of the vertical line, enter a device symbol, such as **S** for a single pole switch. Under E2, enter the number of single-pole switches counted on sheet E2. Then go on to the next type of switch.

On the next line, to the left of the vertical line, and on the next line below **S**, write in **SS** for two single-pole switches. Continue listing types of switches down the left column until you've covered all of the lighting switches found on sheet E2. Then go on to sheet E3. Head the next column E3 and list each type of switch found on that sheet.

You'll usually find three-way switches (S3 or S3S3 is the common symbol) in pairs. When you find one, look for the mate across the room, down the hall or up a flight of stairs. When you see S3S3, there will most likely be another S3S3 in the circuit across the room.

Switching combinations may be SS3 or SS3S3 or SSS3 or S3SS3 or S3S3S. Custom-designed lighting may include some exotic switching. But the owner and designer probably intend exactly what the plans require. Be sure you install what's on the plans.

Any uncommon symbols should be defined in a schedule on one of the plan sheets. But don't be surprised if you find switch symbols on the plans that aren't on the schedule or symbols on the schedule that aren't on the plans. Electrical designers usually use standard switch schedules without considering what applies or doesn't apply to the job at hand. The schedule you see is probably no more than a standard guide.

When all the lighting switches on all sheets are counted and recorded, start counting duplex receptacles. Count and record all the receptacles. Watch for notes on the plans about special devices such as locking receptacles, special voltages and amperages. Watch for specifications that require ivory duplex receptacles in some areas and brown duplex receptacles in other areas.

Continuing the Take-off

The specifications should list the type of device plate for the switches and receptacles. For example, you may need stainless steel plates in some rooms and smooth plastic plates in other rooms. The plans and specifications shouldn't leave any doubt about the grade or type of wiring device needed.

The specifications may list a wiring device by manufacturer and catalog number only. Look in that manufacturer's catalog to find the grade and color of the device specified. If the specs indicate "or equal" after the catalog number, be very sure any substitute you use is a true equal. Some owners and architects watch this very carefully.

When designer-grade wiring devices are specified, the specs will list the devices to install at each location. Designer devices include both switches and receptacles.

Count devices accurately. Mistakes in counting make problems for the installer. Have enough devices to complete the job without leaving a surplus that has to go back to the shop. Leftovers tend to become misfits that take up valuable storage space and may never be used. When doing your take-off, don't assume that changes will add more devices or that some devices will be lost on the jobsite. Do a good job of counting in the beginning to get the job started off right.

Improving Labor Productivity

Your electricians can't work efficiently if they don't have the right materials. But don't burden the job with excess materials. For example, don't ship finish materials until it's nearly time to install finish materials.

Tell your installers how many hours are estimated for conduit, pulling wire, installing devices, etc. If they can't meet your expectations, you should know why. If your estimate is too high, you should be informed of that fact.

On medium to large jobs, have a good installer set the devices. A quick electrician can really pick up speed if there are several hundred devices to set. But be sure each device is installed flush with the wall. The plate should fit properly. And don't leave any dirty fingerprints or scrapes behind. A dirty or messy job reflects on your company as well as the installer.

On a large job like a hospital, some contractors have the trim crew use a lockable push cart that can be wheeled from room to room. Load the wiring devices, plates, grounding jumpers, extra mounting screws, a selection of trim screws and a trash box on the cart. Be sure the cart is narrow enough to pass easily through a doorway. The cart should be locked when left unattended.

On a large job with a good trim electrician installing devices, reduce the labor costs in this section by 5 to 10 percent. If there are few devices and if they're spaced farther apart, increase the labor cost by 5 percent.

Material	Craft@Hrs	Unit	Material Cost	Labor Cost	Installed Cost

Residential framed toggle push-in & side wired, 15 amp, 120 volt, AC quiet

Material	Craft@Hrs	Unit	Material Cost	Labor Cost	Installed Cost
Single pole, brown	L1@0.20	Ea	.95	8.21	9.16
Single pole, ivory	L1@0.20	Ea	.95	8.21	9.16
Single pole, white	L1@0.20	Ea	.95	8.21	9.16
Three-way, brown	L1@0.25	Ea	1.58	10.30	11.88
Three-way, ivory	L1@0.25	Ea	1.58	10.30	11.88
Three-way, white	L1@0.25	Ea	1.58	10.30	11.88

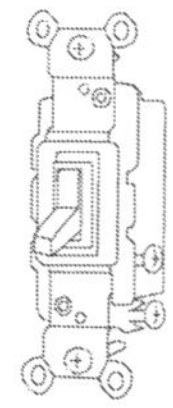

Residential framed toggle push-in & side wired, ground screw, 15 amp, 120 volt, AC quiet

Material	Craft@Hrs	Unit	Material Cost	Labor Cost	Installed Cost
Single pole, brown	L1@0.20	Ea	.98	8.21	9.19
Single pole, ivory	L1@0.20	Ea	.98	8.21	9.19
Single pole, white	L1@0.20	Ea	.98	8.21	9.19
Three-way, brown	L1@0.25	Ea	1.61	10.30	11.91
Three-way, ivory	L1@0.25	Ea	1.61	10.30	11.91
Three-way, white	L1@0.25	Ea	1.61	10.30	11.91

Residential framed toggle push-in & side wired, ground screw, 15 amp, 120 volt, AC quiet

Material	Craft@Hrs	Unit	Material Cost	Labor Cost	Installed Cost
Single pole, brown	L1@0.20	Ea	1.61	8.21	9.82
Single pole, ivory	L1@0.20	Ea	1.61	8.21	9.82
Single pole, white	L1@0.20	Ea	1.61	8.21	9.82
Three-way, brown	L1@0.25	Ea	2.66	10.30	12.96
Three-way, ivory	L1@0.25	Ea	2.66	10.30	12.96
Three-way, white	L1@0.25	Ea	2.66	10.30	12.96

Residential specification-grade, framed toggle side wired, 15 amp, 120 volt, AC quiet

Material	Craft@Hrs	Unit	Material Cost	Labor Cost	Installed Cost
Single pole, brown	L1@0.20	Ea	6.92	8.21	15.13
Single pole, ivory	L1@0.20	Ea	6.92	8.21	15.13
Single pole, white	L1@0.20	Ea	6.92	8.21	15.13
Three-way, brown	L1@0.25	Ea	9.91	10.30	20.21
Three-way, ivory	L1@0.25	Ea	9.91	10.30	20.21
Three-way, white	L1@0.25	Ea	9.91	10.30	20.21

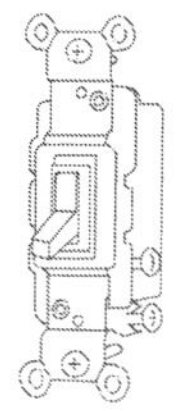

Use these figures to estimate the cost of residential grade lighting switches installed in outlet boxes under the conditions described on pages 5 and 6. Costs listed are for each switch installed. The crew is one electrician working at a labor cost of $41.03 per manhour. These costs include layout, material handling, and normal waste. Add for outlet box and ring, trim plate, sales tax, delivery, supervision, mobilization, demobilization, cleanup, overhead, and profit.

Switches

Material	Craft@Hrs	Unit	Material Cost	Labor Cost	Installed Cost

Commercial specification-grade, side wired with ground screw, 15 amp, 120/277 volt, AC quiet

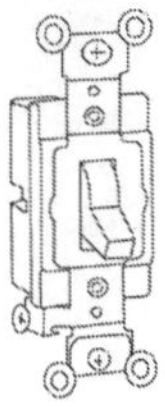

Material	Craft@Hrs	Unit	Material Cost	Labor Cost	Installed Cost
Single pole, brown	L1@0.20	Ea	1.90	8.21	10.11
Single pole, ivory	L1@0.20	Ea	1.90	8.21	10.11
Single pole, white	L1@0.20	Ea	1.90	8.21	10.11
Three-way, brown	L1@0.25	Ea	3.47	10.30	13.77
Three-way, ivory	L1@0.25	Ea	3.47	10.30	13.77
Three-way, white	L1@0.25	Ea	3.47	10.30	13.77
Double pole, brown	L1@0.25	Ea	8.97	10.30	19.27
Double pole, ivory	L1@0.25	Ea	8.97	10.30	19.27
Double pole, white	L1@0.25	Ea	8.97	10.30	19.27
Four-way, brown	L1@0.30	Ea	11.20	12.30	23.50
Four-way, ivory	L1@0.30	Ea	11.20	12.30	23.50
Four-way, white	L1@0.30	Ea	11.20	12.30	23.50

Commercial specification-grade, side wired, 15 amp, 120/277 volt, AC quiet

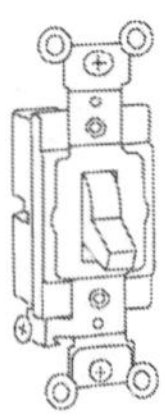

Material	Craft@Hrs	Unit	Material Cost	Labor Cost	Installed Cost
Single pole, brown	L1@0.20	Ea	2.44	8.21	10.65
Single pole ivory	L1@0.20	Ea	2.44	8.21	10.65
Single pole, white	L1@0.20	Ea	2.44	8.21	10.65
Three-way, brown	L1@0.25	Ea	4.36	10.30	14.66
Three-way ivory	L1@0.25	Ea	4.36	10.30	14.66
Three-way, white	L1@0.25	Ea	4.36	10.30	14.66
Double pole, brown	L1@0.25	Ea	7.37	10.30	17.67
Double pole ivory	L1@0.25	Ea	7.37	10.30	17.67
Double pole white	L1@0.25	Ea	7.37	10.30	17.67
Four-way, brown	L1@0.30	Ea	10.80	12.30	23.10
Four-way ivory	L1@0.30	Ea	10.80	12.30	23.10
Four-way, white	L1@0.30	Ea	10.80	12.30	23.10

Use these figures to estimate the cost of lighting switches installed in outlet boxes under conditions described on pages 5 and 6. Costs listed are for each switch installed. The crew is one electrician working at a labor cost of $41.03 per manhour. These costs include layout, material handling, and normal waste. Add for outlet box and ring, trim plate, sales tax, delivery, supervision, mobilization, demobilization, cleanup, overhead, and profit.

Material	Craft@Hrs	Unit	Material Cost	Labor Cost	Installed Cost

Commercial specification-grade, back & side wired, 15 amp, 120/277 volt, AC quiet

Material	Craft@Hrs	Unit	Material Cost	Labor Cost	Installed Cost
Single pole, brown	L1@0.20	Ea	3.56	8.21	11.77
Single pole, ivory	L1@0.20	Ea	3.56	8.21	11.77
Single pole, white	L1@0.20	Ea	3.56	8.21	11.77
Three-way, brown	L1@0.25	Ea	4.79	10.30	15.09
Three-way, ivory	L1@0.25	Ea	4.79	10.30	15.09
Three-way, white	L1@0.25	Ea	4.79	10.30	15.09
Double pole, brown	L1@0.25	Ea	15.20	10.30	25.50
Double pole, ivory	L1@0.25	Ea	15.20	10.30	25.50
Double pole, white	L1@0.25	Ea	15.20	10.30	25.50
Four-way, brown	L1@0.30	Ea	28.00	12.30	40.30
Four-way, ivory	L1@0.30	Ea	28.00	12.30	40.30
Four-way, white	L1@0.30	Ea	28.00	12.30	40.30

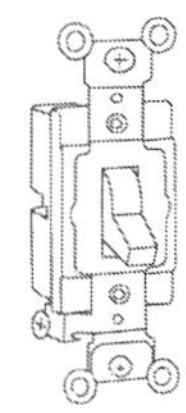

Commercial specification-grade, illuminated frame, 15 amp, 120/277 volt, AC quiet

Material	Craft@Hrs	Unit	Material Cost	Labor Cost	Installed Cost
Single pole, ivory	L1@0.20	Ea	9.54	8.21	17.75
Single pole, white	L1@0.20	Ea	9.54	8.21	17.75
Three-way, ivory	L1@0.25	Ea	15.30	10.30	25.60
Three-way, white	L1@0.25	Ea	15.30	10.30	25.60

Industrial specification-grade, side wired, 15 amp, 120/277 volt, AC quiet

Material	Craft@Hrs	Unit	Material Cost	Labor Cost	Installed Cost
Single pole, brown	L1@0.20	Ea	6.05	8.21	14.26
Single pole, ivory	L1@0.20	Ea	6.05	8.21	14.26
Single pole, white	L1@0.20	Ea	6.05	8.21	14.26
Three-way, brown	L1@0.25	Ea	8.66	10.30	18.96
Three-way, ivory	L1@0.25	Ea	8.66	10.30	18.96
Three-way, white	L1@0.25	Ea	8.66	10.30	18.96
Double pole, brown	L1@0.25	Ea	9.96	10.30	20.26
Double pole, ivory	L1@0.25	Ea	9.96	10.30	20.26
Double pole, white	L1@0.25	Ea	9.96	10.30	20.26
Four-way, brown	L1@0.30	Ea	28.00	12.30	40.30
Four-way, ivory	L1@0.30	Ea	28.00	12.30	40.30
Four-way, white	L1@0.30	Ea	28.00	12.30	40.30

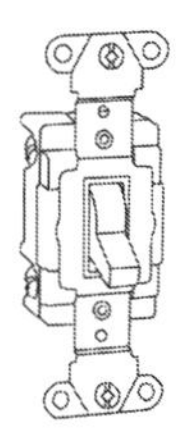

Use these figures to estimate the cost of lighting switches installed in outlet boxes under conditions described on pages 5 and 6. Costs listed are for each switch installed. The crew is one electrician working at a labor cost of $41.03 per manhour. These costs include layout, material handling, and normal waste. Add for outlet box and ring, trim plate, sales tax, delivery, supervision, mobilization, demobilization, cleanup, overhead, and profit.

Switches

Material	Craft@Hrs	Unit	Material Cost	Labor Cost	Installed Cost
Industrial specification-grade, side wired, 15 amp, 120/277 volt, AC quiet					
Single pole, brown	L1@0.20	Ea	6.95	8.21	15.16
Single pole, ivory	L1@0.25	Ea	6.95	10.30	17.25
Single pole, white	L1@0.25	Ea	6.95	10.30	17.25
Three-way, brown	L1@0.25	Ea	10.60	10.30	20.90
Three-way, ivory	L1@0.25	Ea	10.60	10.30	20.90
Three-way, white	L1@0.25	Ea	10.60	10.30	20.90
Double pole, brown	L1@0.25	Ea	12.00	10.30	22.30
Double pole, ivory	L1@0.25	Ea	12.00	10.30	22.30
Double pole, white	L1@0.25	Ea	12.00	10.30	22.30
Four-way, brown	L1@0.40	Ea	28.40	16.40	44.80
Four-way, ivory	L1@0.40	Ea	28.40	16.40	44.80
Four-way, white	L1@0.40	Ea	28.40	16.40	44.80
Commercial specification-grade, side wired with ground screw, 20 amp, 120/277 volt, AC quiet					
Single pole, brown	L1@0.20	Ea	2.49	8.21	10.70
Single pole, ivory	L1@0.20	Ea	2.49	8.21	10.70
Single pole, white	L1@0.20	Ea	2.49	8.21	10.70
Three-way, brown	L1@0.25	Ea	3.83	10.30	14.13
Three-way, ivory	L1@0.25	Ea	3.83	10.30	14.13
Three-way, white	L1@0.25	Ea	3.83	10.30	14.13
Double pole, brown	L1@0.25	Ea	12.00	10.30	22.30
Double pole, ivory	L1@0.25	Ea	12.00	10.30	22.30
Double pole, white	L1@0.25	Ea	12.00	10.30	22.30
Four-way, brown	L1@0.30	Ea	28.40	12.30	40.70
Four-way, ivory	L1@0.30	Ea	28.40	12.30	40.70
Four-way, white	L1@0.30	Ea	28.40	12.30	40.70

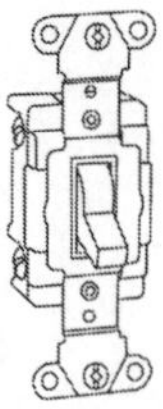

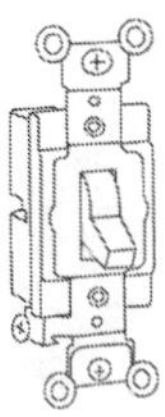

Use these figures to estimate the cost of lighting switches installed in outlet boxes under conditions described on pages 5 and 6. Costs listed are for each switch installed. The crew is one electrician working at a labor cost of $41.03 per manhour. These costs include layout, material handling, and normal waste. Add for outlet box and ring, trim plate, sales tax, delivery, supervision, mobilization, demobilization, cleanup, overhead, and profit.

Commercial specification-grade, shallow back & side wired, 20 amp, 120/277 volt, AC quiet

Material	Craft@Hrs	Unit	Material Cost	Labor Cost	Installed Cost
Single pole, brown	L1@0.20	Ea	3.98	8.21	12.19
Single pole, ivory	L1@0.20	Ea	3.98	8.21	12.19
Single pole, white	L1@0.20	Ea	3.98	8.21	12.19
Three-way, brown	L1@0.25	Ea	4.96	10.30	15.26
Three-way, ivory	L1@0.25	Ea	4.96	10.30	15.26
Three-way, white	L1@0.25	Ea	4.96	10.30	15.26
Double pole, brown	L1@0.25	Ea	14.00	10.30	24.30
Double pole, ivory	L1@0.25	Ea	14.00	10.30	24.30
Double pole, white	L1@0.25	Ea	14.00	10.30	24.30
Four-way, brown	L1@0.30	Ea	25.80	12.30	38.10
Four-way, ivory	L1@0.30	Ea	25.80	12.30	38.10
Four-way, white	L1@0.30	Ea	25.80	12.30	38.10

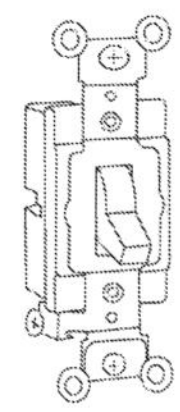

Industrial specification-grade, back & side wired with ground screw, 20 amp, 120/277 volt

Material	Craft@Hrs	Unit	Material Cost	Labor Cost	Installed Cost
Single pole, brown	L1@0.20	Ea	7.03	8.21	15.24
Single pole, ivory	L1@0.20	Ea	7.03	8.21	15.24
Single pole, white	L1@0.20	Ea	7.03	8.21	15.24
Three-way, brown	L1@0.25	Ea	10.60	10.30	20.90
Three-way, ivory	L1@0.25	Ea	10.60	10.30	20.90
Three-way, white	L1@0.25	Ea	10.60	10.30	20.90
Double pole, brown	L1@0.25	Ea	12.00	10.30	22.30
Double pole, ivory	L1@0.25	Ea	12.00	10.30	22.30
Double pole, white	L1@0.25	Ea	12.00	10.30	22.30
Four-way, brown	L1@0.30	Ea	28.40	12.30	40.70
Four-way, ivory	L1@0.30	Ea	28.40	12.30	40.70
Four-way, white	L1@0.30	Ea	28.40	12.30	40.70

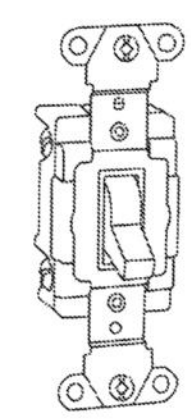

Industrial specification-grade, side wired with ground screw, 20 amp, 120/277 volt

Material	Craft@Hrs	Unit	Material Cost	Labor Cost	Installed Cost
Single pole, brown	L1@0.20	Ea	7.96	8.21	16.17
Single pole, ivory	L1@0.20	Ea	7.96	8.21	16.17
Single pole, white	L1@0.20	Ea	7.96	8.21	16.17

Use these figures to estimate the cost of lighting switches installed in outlet boxes under conditions described on pages 5 and 6. Costs listed are for each switch installed. The crew is one electrician working at a labor cost of $41.03 per manhour. These costs include layout, material handling, and normal waste. Add for outlet box and ring, trim plate, sales tax, delivery, supervision, mobilization, demobilization, cleanup, overhead, and profit.

Industrial specification-grade, side wired with ground screw, 20 amp, 120/277 volt (continued)

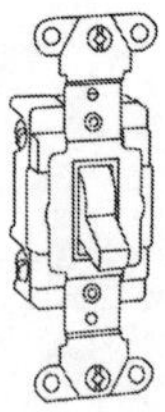

Material	Craft@Hrs	Unit	Material Cost	Labor Cost	Installed Cost
Three-way, brown	L1@0.25	Ea	9.36	10.30	19.66
Three-way, ivory	L1@0.25	Ea	9.36	10.30	19.66
Three-way, white	L1@0.25	Ea	9.36	10.30	19.66
Double pole, brown	L1@0.25	Ea	9.36	10.30	19.66
Double pole, ivory	L1@0.25	Ea	9.36	10.30	19.66
Double pole, white	L1@0.25	Ea	9.36	10.30	19.66
Four-way, brown	L1@0.30	Ea	28.60	12.30	40.90
Four-way, ivory	L1@0.30	Ea	28.60	12.30	40.90
Four-way, white	L1@0.30	Ea	28.60	12.30	40.90

Industrial specification-grade, back & side wired, HP rated, 20 amp, 120/277 volt

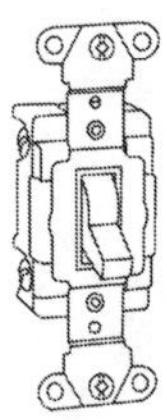

Material	Craft@Hrs	Unit	Material Cost	Labor Cost	Installed Cost
Single pole, brown	L1@0.20	Ea	11.40	8.21	19.61
Single pole, ivory	L1@0.20	Ea	11.40	8.21	19.61
Single pole, white	L1@0.20	Ea	11.40	8.21	19.61
Three-way, brown	L1@0.25	Ea	12.20	10.30	22.50
Three-way, ivory	L1@0.25	Ea	12.20	10.30	22.50
Three-way, white	L1@0.25	Ea	12.20	10.30	22.50
Double pole, brown	L1@0.25	Ea	15.60	10.30	25.90
Double pole, ivory	L1@0.25	Ea	15.60	10.30	25.90
Double pole, white	L1@0.25	Ea	15.60	10.30	25.90
Four-way, brown	L1@0.30	Ea	33.60	12.30	45.90
Four-way, ivory	L1@0.30	Ea	33.60	12.30	45.90
Four-way, white	L1@0.30	Ea	33.60	12.30	45.90

Industrial specification-grade, back & side wired, HP rated, 30 amp, 120 volt

Material	Craft@Hrs	Unit	Material Cost	Labor Cost	Installed Cost
Single pole, brown	L1@0.25	Ea	18.40	10.30	28.70
Single pole, ivory	L1@0.25	Ea	18.40	10.30	28.70
Single pole, white	L1@0.25	Ea	18.40	10.30	28.70
Three-way, brown	L1@0.30	Ea	23.40	12.30	35.70
Three-way, ivory	L1@0.30	Ea	23.40	12.30	35.70
Three-way, white	L1@0.30	Ea	23.40	12.30	35.70

Use these figures to estimate the cost of lighting switches installed in outlet boxes under conditions described on pages 5 and 6. Costs listed are for each switch installed. The crew is one electrician working at a labor cost of $41.03 per manhour. These costs include layout, material handling, and normal waste. Add for outlet box and ring, trim plate, sales tax, delivery, supervision, mobilization, demobilization, cleanup, overhead, and profit.

Industrial specification grade, back & side wired, HP rated, 30 amp, 120 volt

Material	Craft@Hrs	Unit	Material Cost	Labor Cost	Installed Cost
Double pole, brown	L1@0.30	Ea	23.70	12.30	36.00
Double pole, ivory	L1@0.30	Ea	23.70	12.30	36.00
Double pole, white	L1@0.30	Ea	23.70	12.30	36.00

15 amp and 20 amp, 120 volt, lighted handle when OFF

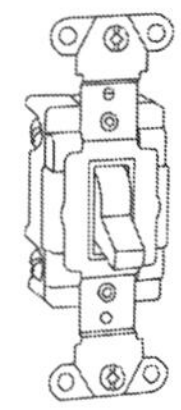

15 amp, single pole, ivory	L1@0.20	Ea	14.70	8.21	22.91
15 amp, three-way, ivory	L1@0.25	Ea	21.50	10.30	31.80
20 amp, single pole, ivory	L1@0.20	Ea	20.00	8.21	28.21
20 amp, three-way, ivory	L1@0.25	Ea	26.10	10.30	36.40

15 amp and 20 amp, 120 volt, lighted handle when ON

15 amp, single pole, ivory	L1@0.20	Ea	13.10	8.21	21.31
20 amp, single pole, ivory	L1@0.25	Ea	20.70	10.30	31.00

Dimmer switches, incandescent with decorative wallplate, 120 volt

600W, slide on/off, brown	L1@0.25	Ea	11.60	10.30	21.90
600W, slide on/off, ivory	L1@0.25	Ea	11.60	10.30	21.90
600W, slide on/off, white	L1@0.25	Ea	11.60	10.30	21.90
600W, touch on/off, brown	L1@0.25	Ea	16.20	10.30	26.50
600W, touch on/off, ivory	L1@0.25	Ea	16.20	10.30	26.50
600W, touch on/off, white	L1@0.25	Ea	16.20	10.30	26.50

Dimmer switches, incandescent with decorative wallplate, 120 volt, three-way

600W, touch on/off, brown	L1@0.25	Ea	32.10	10.30	42.40
600W, touch on/off, ivory	L1@0.25	Ea	32.10	10.30	42.40
600W, touch on/off, white	L1@0.25	Ea	32.10	10.30	42.40
800W, push on/off, ivory	L1@0.25	Ea	28.20	10.30	38.50
800W, push on/off, white	L1@0.25	Ea	28.20	10.30	38.50
1000W, push on/off, ivory	L1@0.25	Ea	39.40	10.30	49.70
1000W, push on/off, white	L1@0.25	Ea	39.40	10.30	49.70

Dimmer switches, fluorescent with decorative wallplate, rotary on/off

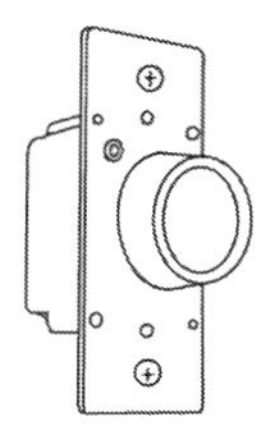

1 to 12 4ORS lamps	L1@0.30	Ea	37.90	12.30	50.20

Use these figures to estimate the cost of lighting switches installed in outlet boxes under conditions described on pages 5 and 6. Costs listed are for each switch installed. The crew is one electrician working at a labor cost of $41.03 per manhour. These costs include layout, material handling, and normal waste. Add for outlet box and ring, trim plate, sales tax, delivery, supervision, mobilization, demobilization, cleanup, overhead, and profit.

Switches

Material	Craft@Hrs	Unit	Material Cost	Labor Cost	Installed Cost
Specification-grade, side wired, keyed type, 15 amp, 120/277 volt, AC quiet					
Single pole	L1@0.20	Ea	9.73	8.21	17.94
Three-way	L1@0.25	Ea	15.80	10.30	26.10
Double pole	L1@0.25	Ea	16.80	10.30	27.10
Four-way	L1@0.30	Ea	32.20	12.30	44.50
Specification-grade, side wired, keyed type, 20 amp, 120/277 volt, AC quiet					
Single pole	L1@0.20	Ea	15.40	8.21	23.61
Three-way	L1@0.25	Ea	16.50	10.30	26.80
Double pole	L1@0.25	Ea	19.40	10.30	29.70
Four-way	L1@0.30	Ea	36.00	12.30	48.30
Specification-grade, back & side wired, keyed type, 15 amp, 120/277 volt, AC quiet					
Single pole	L1@0.20	Ea	14.30	8.21	22.51
Three-way	L1@0.25	Ea	16.80	10.30	27.10
Double pole	L1@0.25	Ea	19.40	10.30	29.70
Four-way	L1@0.30	Ea	36.00	12.30	48.30
Specification-grade, back & side wired, keyed type, 20 amp, 120/277 volt, AC quiet					
Single pole	L1@0.20	Ea	22.80	8.21	31.01
Three-way	L1@0.25	Ea	25.00	10.30	35.30
Double pole	L1@0.25	Ea	25.70	10.30	36.00
Four-way	L1@0.30	Ea	46.70	12.30	59.00
Commercial specification-grade, decorator-style, 15 amp, 120/277 volt, push-in & side wired					
Single pole, brown	L1@0.20	Ea	4.92	8.21	13.13
Single pole, ivory	L1@0.20	Ea	4.92	8.21	13.13
Single pole, white	L1@0.20	Ea	4.92	8.21	13.13
Three pole, brown	L1@0.25	Ea	6.93	10.30	17.23
Three pole, ivory	L1@0.25	Ea	6.93	10.30	17.23
Three pole, white	L1@0.25	Ea	6.93	10.30	17.23

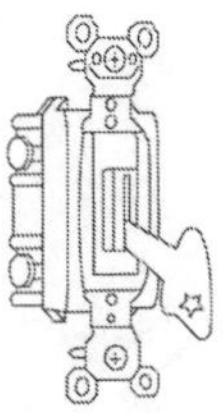

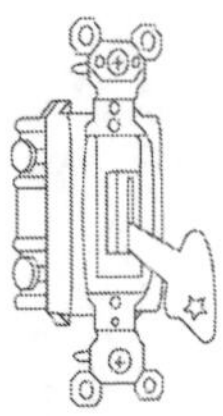

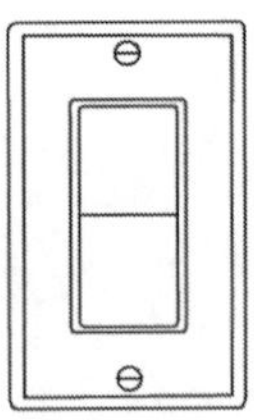

Use these figures to estimate the cost of lighting switches installed in outlet boxes under conditions described on pages 5 and 6. Costs listed are for each switch installed. The crew is one electrician working at a labor cost of $41.03 per manhour. These costs include layout, material handling, and normal waste. Add for outlet box and ring, trim plate, sales tax, delivery, supervision, mobilization, demobilization, cleanup, overhead, and profit.

Decorator style, commercial specification-grade, 15 amp, 120/277 volt, push-in & side wired, ground screw

Material	Craft@Hrs	Unit	Material Cost	Labor Cost	Installed Cost
Single pole, brown	L1@0.20	Ea	4.91	8.21	13.12
Single pole, ivory	L1@0.20	Ea	4.91	8.21	13.12
Single pole, white	L1@0.20	Ea	4.91	8.21	13.12
Three pole, brown	L1@0.25	Ea	6.92	10.30	17.22
Three pole, ivory	L1@0.25	Ea	6.92	10.30	17.22
Three pole, white	L1@0.25	Ea	6.92	10.30	17.22
Double pole, brown	L1@0.25	Ea	19.00	10.30	29.30
Double pole, ivory	L1@0.25	Ea	19.00	10.30	29.30
Double pole, white	L1@0.25	Ea	19.00	10.30	29.30
Four-way, brown	L1@0.30	Ea	19.10	12.30	31.40
Four-way, ivory	L1@0.30	Ea	19.10	12.30	31.40
Four-way, white	L1@0.30	Ea	19.10	12.30	31.40

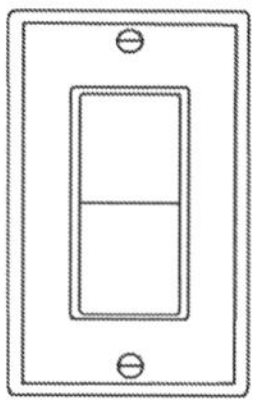

Decorator style, commercial specification-grade 20 amp, 120/277 volt, push-in & side wired, ground screw

Material	Craft@Hrs	Unit	Material Cost	Labor Cost	Installed Cost
Single pole, brown	L1@0.25	Ea	11.30	10.30	21.60
Single pole, ivory	L1@0.25	Ea	11.30	10.30	21.60
Single pole, white	L1@0.25	Ea	11.30	10.30	21.60
Single pole, gray	L1@0.25	Ea	11.30	10.30	21.60
Three-way, brown	L1@0.25	Ea	16.50	10.30	26.80
Three-way, ivory	L1@0.25	Ea	16.50	10.30	26.80
Three-way, white	L1@0.25	Ea	16.50	10.30	26.80
Three-way, gray	L1@0.25	Ea	16.50	10.30	26.80
Double pole, brown	L1@0.30	Ea	26.40	12.30	38.70
Double pole, ivory	L1@0.30	Ea	26.40	12.30	38.70
Double pole, white	L1@0.30	Ea	26.40	12.30	38.70
Double pole, gray	L1@0.30	Ea	26.40	12.30	38.70
Four-way, brown	L1@0.30	Ea	40.70	12.30	53.00
Four-way, ivory	L1@0.30	Ea	40.70	12.30	53.00
Four-way, white	L1@0.30	Ea	40.70	12.30	53.00
Four-way, gray	L1@0.30	Ea	40.70	12.30	53.00

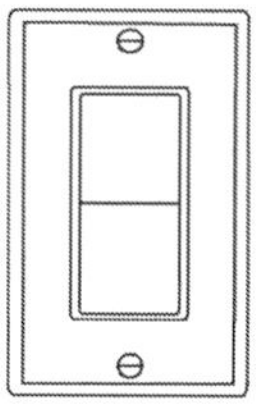

Use these figures to estimate the cost of lighting switches installed in outlet boxes under conditions described on pages 5 and 6. Costs listed are for each switch installed. The crew is one electrician working at a labor cost of $41.03 per manhour. These costs include layout, material handling, and normal waste. Add for outlet box and ring, trim plate, sales tax, delivery, supervision, mobilization, demobilization, cleanup, overhead, and profit.

Switches

Material	Craft@Hrs	Unit	Material Cost	Labor Cost	Installed Cost
Decorator style, commercial specification-grade, 15 amp, 120/277 volt, side wired, ground screw, illuminated					
Single pole, ivory	L1@0.20	Ea	9.13	8.21	17.34
Single pole, white	L1@0.20	Ea	9.13	8.21	17.34
Three-way, ivory	L1@0.25	Ea	12.10	10.30	22.40
Three-way, white	L1@0.25	Ea	12.10	10.30	22.40
Decorator style, commercial specification-grade, 20 amp, 120/277 volt, side wired, ground screw, illuminated					
Single pole, ivory	L1@0.20	Ea	15.10	8.21	23.31
Single pole, white	L1@0.20	Ea	15.10	8.21	23.31
Three-way, ivory	L1@0.25	Ea	20.30	10.30	30.60
Three-way, white	L1@0.25	Ea	20.30	10.30	30.60
Decorator style, commercial specification-grade, 15 amp, 120 volt, combination switch and U-ground receptacle					
Single pole, ivory	L1@0.25	Ea	17.50	10.30	27.80
Single pole, white	L1@0.25	Ea	17.50	10.30	27.80
Three-way, ivory	L1@0.30	Ea	24.70	12.30	37.00
Three-way, white	L1@0.30	Ea	24.70	12.30	37.00
Decorator style, commercial specification-grade, 15 amp, 120 volt, combination switch and neon pilot					
Single pole, brown	L1@0.25	Ea	18.50	10.30	28.80
Single pole, ivory	L1@0.25	Ea	18.50	10.30	28.80
Single pole, white	L1@0.25	Ea	18.50	10.30	28.80
Decorator style, commercial specification-grade, 15 amp, 120 volt, two switches					
Single pole, brown	L1@0.25	Ea	16.00	10.30	26.30
Single pole, ivory	L1@0.25	Ea	16.00	10.30	26.30
Single pole, white	L1@0.25	Ea	16.00	10.30	26.30
Decorator style, commercial specification-grade, 15 amp, 120 volt, combination single pole & three-way switches					
Ivory	L1@0.30	Ea	24.50	12.30	36.80
White	L1@0.30	Ea	24.50	12.30	36.80

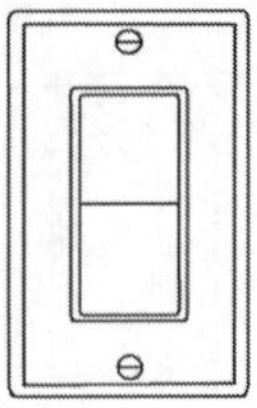

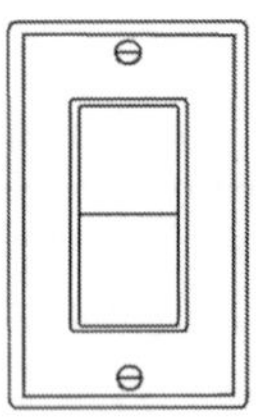

Use these figures to estimate the cost of lighting switches installed in outlet boxes under conditions described on pages 5 and 6. Costs listed are for each switch installed. The crew is one electrician working at a labor cost of $41.03 per manhour. These costs include layout, material handling, and normal waste. Add for outlet box and ring, trim plate, sales tax, delivery, supervision, mobilization, demobilization, cleanup, overhead, and profit.

Material	Craft@Hrs	Unit	Material Cost	Labor Cost	Installed Cost
Decorator style, commercial specification-grade, 15 amp, 120 volt, combination two three-way switches					
Ivory	L1@0.35	Ea	28.10	14.40	42.50
White	L1@0.35	Ea	28.10	14.40	42.50
Combination, commercial specification-grade, 10 amp, 120 volt AC, two single pole switches					
Brown	L1@0.25	Ea	5.43	10.30	15.73
Ivory	L1@0.25	Ea	5.43	10.30	15.73
White	L1@0.25	Ea	5.43	10.30	15.73
Combination, commercial specification-grade, 15 amp, 120 volt AC, one single pole & one three-way switch					
Brown	L1@0.30	Ea	16.10	12.30	28.40
Ivory	L1@0.30	Ea	16.10	12.30	28.40
White	L1@0.30	Ea	16.10	12.30	28.40
Combination, commercial specification-grade, 15 amp, 120 volt AC, two three-way switches					
Brown	L1@0.35	Ea	20.70	14.40	35.10
Ivory	L1@0.35	Ea	20.70	14.40	35.10
White	L1@0.35	Ea	20.70	14.40	35.10
Combination, commercial specification-grade, 15 amp, 120 volt AC, single pole switch & U-ground receptacle					
Brown	L1@0.30	Ea	11.20	12.30	23.50
Ivory	L1@0.30	Ea	11.20	12.30	23.50
White	L1@0.30	Ea	11.20	12.30	23.50
Combination, commercial specification-grade, 15 amp, 120 volt AC, three-way switch & U-ground receptacle					
Brown	L1@0.35	Ea	20.60	14.40	35.00
Ivory	L1@0.35	Ea	20.60	14.40	35.00
White	L1@0.35	Ea	20.60	14.40	35.00
Combination, commercial specification-grade, 15 amp, 120 volt, single pole switch and neon pilot					
Ivory	L1@0.25	Ea	10.60	10.30	20.90
White	L1@0.25	Ea	10.60	10.30	20.90

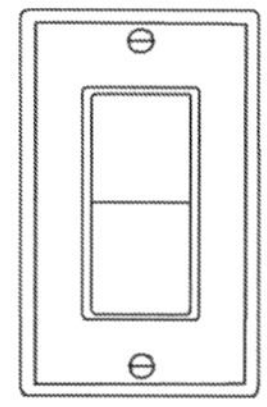

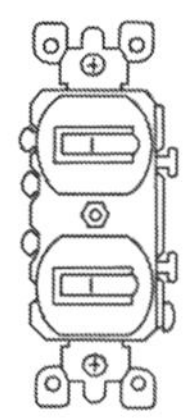

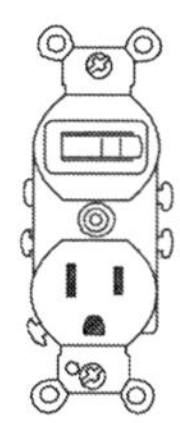

Use these figures to estimate the cost of lighting switches installed in outlet boxes under conditions described on pages 5 and 6. Costs listed are for each switch installed. The crew is one electrician working at a labor cost of $41.03 per manhour. These costs include layout, material handling, and normal waste. Add for outlet box and ring, trim plate, sales tax, delivery, supervision, mobilization, demobilization, cleanup, overhead, and profit.

Switches

Material	Craft@Hrs	Unit	Material Cost	Labor Cost	Installed Cost
Maintain contact, three position, 2 circuit, center-off, 15 amp, 120/277 volt, double throw switches					
Single pole, brown	L1@0.30	Ea	38.50	12.30	50.80
Single pole, ivory	L1@0.30	Ea	38.50	12.30	50.80
Single pole, white	L1@0.30	Ea	38.50	12.30	50.80
Double pole, brown	L1@0.35	Ea	45.70	14.40	60.10
Double pole, ivory	L1@0.35	Ea	45.70	14.40	60.10
Double pole, white	L1@0.35	Ea	45.70	14.40	60.10
Maintain contact, three position, 2 circuit, center-off, 20 amp, 120/277 volt, double throw switches					
Single pole, brown	L1@0.30	Ea	50.50	12.30	62.80
Single pole, ivory	L1@0.30	Ea	50.50	12.30	62.80
Single pole, white	L1@0.30	Ea	50.50	12.30	62.80
Double pole, brown	L1@0.35	Ea	52.30	14.40	66.70
Double pole, ivory	L1@0.35	Ea	52.30	14.40	66.70
Double pole, white	L1@0.35	Ea	52.30	14.40	66.70
Maintain contact, three position, 2 circuit, center-off, 30 amp, 120/277 volt, double throw switches					
Single pole, brown	L1@0.30	Ea	56.30	12.30	68.60
Single pole, ivory	L1@0.30	Ea	56.30	12.30	68.60
Double pole, brown	L1@0.35	Ea	58.80	14.40	73.20
Double pole, ivory	L1@0.35	Ea	58.80	14.40	73.20
Maintain contact, three position, 2 circuit, center-off, keyed type, 15 amp, 120/277 volt, double throw switches					
Single pole, gray	L1@0.30	Ea	49.00	12.30	61.30
Double pole, gray	L1@0.35	Ea	49.00	14.40	63.40
Maintain contact, three position, 2 circuit, center-off, keyed type, 20 amp, 120/277 volt, double throw switches					
Single pole, gray	L1@0.30	Ea	55.20	12.30	67.50
Double pole, gray	L1@0.35	Ea	55.20	14.40	69.60

Use these figures to estimate the cost of maintain contact switches installed in outlet boxes under the conditions described on pages 5 and 6. Costs listed are for each switch installed. The crew is one electrician working at a labor cost of $41.03 per manhour. These costs include layout, material handling, and normal waste. Add for outlet box and switch ring, trim plate, sales tax, delivery, supervision, mobilization, demobilization, cleanup, overhead, and profit.

Material	Craft@Hrs	Unit	Material Cost	Labor Cost	Installed Cost

Momentary contact, three position, center-off, 15 amp, 120/277 volt, double throw switches

Single pole, brown	L1@0.30	Ea	16.90	12.30	29.20
Single pole, ivory	L1@0.30	Ea	16.90	12.30	29.20
Single pole, white	L1@0.30	Ea	16.90	12.30	29.20

Momentary contact, three position, center-off, 20 amp, 120/277 volt, double throw switches

Single pole, brown	L1@0.30	Ea	20.60	12.30	32.90
Single pole, ivory	L1@0.30	Ea	20.60	12.30	32.90

Momentary contact, three position, center-off, 30 amp, 120/277 volt, double throw switches

Single pole, brown	L1@0.30	Ea	50.40	12.30	62.70
Double pole, brown	L1@0.35	Ea	51.00	14.40	65.40

Momentary contact, three position, center-off, keyed type, 15 amp, 120/277 volt, double throw switches

Single pole, gray	L1@0.30	Ea	25.00	12.30	37.30

Momentary contact, three position, 2 circuit, center-off, keyed type, 20 amp, 120/277 volt, double throw switches

Single pole, gray	L1@0.30	Ea	30.30	12.30	42.60

Momentary contact, three position, 2 circuit, center-off, keyed type, 30 amp, 120/277 volt, double throw switches

Single pole, gray	L1@0.30	Ea	53.50	12.30	65.80
Double pole, gray	L1@0.35	Ea	53.50	14.40	67.90

Spare key for keyed type

Key	L1@0.05	Ea	1.39	2.05	3.44

Use these figures to estimate the cost of momentary contact switches installed in outlet boxes under conditions described on pages 5 and 6. Costs listed are for each switch installed. The crew is one electrician working at a labor cost of $41.03 per manhour. These costs include layout, material handling, and normal waste. Add for outlet box and switch ring, trim plate, sales tax, delivery, supervision, mobilization, demobilization, cleanup, overhead, and profit.

Time Switches

Material	Craft@Hrs	Unit	Material Cost	Labor Cost	Installed Cost

24 hour plain dial, 40 amp

Material	Craft@Hrs	Unit	Material Cost	Labor Cost	Installed Cost
120V, SPST	L1@1.00	Ea	57.40	41.00	98.40
208/277V, SPST	L1@1.00	Ea	64.10	41.00	105.10
120V, DPST	L1@1.00	Ea	62.70	41.00	103.70
208/277V, DPST	L1@1.00	Ea	66.60	41.00	107.60

24 hour dial with skip-a-day, 40 amp

Material	Craft@Hrs	Unit	Material Cost	Labor Cost	Installed Cost
120V, SPDT	L1@1.00	Ea	83.70	41.00	124.70
208/277V, SPDT	L1@1.00	Ea	96.40	41.00	137.40
120V, DPST	L1@1.00	Ea	81.00	41.00	122.00
208/277V, DPST	L1@1.00	Ea	93.50	41.00	134.50

24 hour dial with astro-dial and skip-a-day, 40 amp

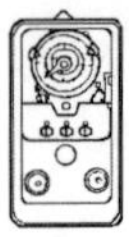

Material	Craft@Hrs	Unit	Material Cost	Labor Cost	Installed Cost
120V, SPDT	L1@1.00	Ea	135.00	41.00	176.00
208/277V, SPDT	L1@1.00	Ea	148.00	41.00	189.00
120V, DPST	L1@1.00	Ea	131.00	41.00	172.00
208/277V, DPST	L1@1.00	Ea	149.00	41.00	190.00

24 hour plain dial, outdoor, heavy-duty, 40 amp

Material	Craft@Hrs	Unit	Material Cost	Labor Cost	Installed Cost
120V, SPST	L1@1.00	Ea	88.00	41.00	129.00
208/277V, SPDT	L1@1.00	Ea	97.30	41.00	138.30
120V, DPST	L1@1.00	Ea	99.60	41.00	140.60
208/277V, DPST	L1@1.00	Ea	105.00	41.00	146.00

24 hour dial with skip-a-day, outdoor, 40 amp

Material	Craft@Hrs	Unit	Material Cost	Labor Cost	Installed Cost
120V, DPST	L1@1.00	Ea	121.00	41.00	162.00
208/277V, DPST	L1@1.00	Ea	127.00	41.00	168.00

24 hour dial with astro-dial and skip-a-day, outdoor, 40 amp

Material	Craft@Hrs	Unit	Material Cost	Labor Cost	Installed Cost
120V, DPST	L1@1.00	Ea	165.00	41.00	206.00

Swimming pool timer, NEMA 3R, 40 amp

Material	Craft@Hrs	Unit	Material Cost	Labor Cost	Installed Cost
120V, SPST	L1@1.25	Ea	91.30	51.30	142.60
208/277V, DPST	L1@1.25	Ea	112.00	51.30	163.30

7 day calendar dial, spring wound, 40 amp

Material	Craft@Hrs	Unit	Material Cost	Labor Cost	Installed Cost
120V, 4PST-4NO	L1@1.50	Ea	364.00	61.50	425.50
120V, 2NO-2NC	L1@1.50	Ea	364.00	61.50	425.50
208/277V, 4PST-4NO	L1@1.50	Ea	392.00	61.50	453.50
208/277V, 2NO-2NC	L1@1.50	Ea	392.00	61.50	453.50

Use these figures to estimate the cost of time switches installed under the conditions described on pages 5 and 6. Costs listed are for each time switch installed. The crew is one electrician working at a labor cost of $41.03 per manhour. These costs include mounting screws, layout, material handling, and normal waste. Add for sales tax, delivery, supervision, mobilization, demobilization, cleanup, overhead and profit.

Material	Craft@Hrs	Unit	Material Cost	Labor Cost	Installed Cost
15 amp, 125 volt, back & side wired, NEMA 5-15R					
Brown	L1@0.20	Ea	2.84	8.21	11.05
Ivory	L1@0.20	Ea	2.37	8.21	10.58
White	L1@0.20	Ea	2.37	8.21	10.58
15 amp, 125 volt, side wired, self grounding, NEMA 5-15R					
Brown	L1@0.20	Ea	6.70	8.21	14.91
Ivory	L1@0.20	Ea	6.70	8.21	14.91
15 amp, 125 volt, back & side wired, self grounding, NEMA 5-15R					
Brown	L1@0.20	Ea	11.70	8.21	19.91
Ivory	L1@0.20	Ea	11.70	8.21	19.91
15 amp, 125 volt, isolated ground, NEMA 5-15R					
Orange	L1@0.20	Ea	23.50	8.21	31.71
15 amp, 125 volt, hospital grade, NEMA 5-15R					
Brown	L1@0.20	Ea	13.90	8.21	22.11
Ivory	L1@0.20	Ea	13.90	8.21	22.11
White	L1@0.20	Ea	13.90	8.21	22.11
Red	L1@0.20	Ea	13.90	8.21	22.11
15 amp, 125 volt, hospital grade, NEMA 5-15R, with isolated ground and surge suppressor					
Ivory	L1@0.25	Ea	20.10	10.30	30.40
White	L1@0.25	Ea	20.10	10.30	30.40
Orange	L1@0.25	Ea	20.10	10.30	30.40
15 amp, 125 volt, clock hanger with cover, NEMA 5-15R					
Brown	L1@0.25	Ea	4.84	10.30	15.14
Ivory	L1@0.25	Ea	4.91	10.30	15.21
Brass	L1@0.25	Ea	5.01	10.30	15.31
Stainless steel	L1@0.25	Ea	13.80	10.30	24.10

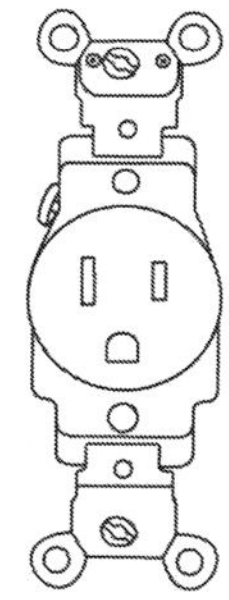

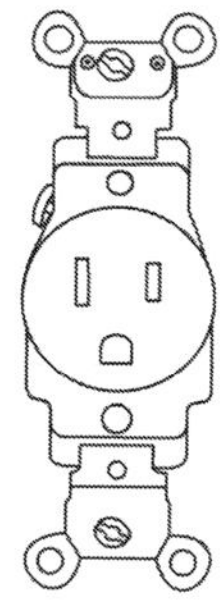

Use these figures to estimate the cost of single receptacles installed in outlet boxes under the conditions described on pages 5 and 6. Costs listed are for each receptacle installed. The crew is one electrician working at a labor cost of $41.03 per manhour. These costs include layout, material handling, and normal waste. Add for the outlet box and ring, sales tax, delivery, supervision, mobilization, demobilization, cleanup, overhead, and profit.

Single Receptacles

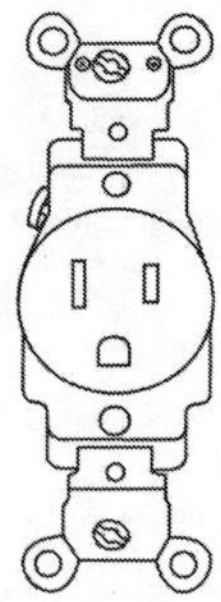

15 amp, 250 volt, side wired, NEMA 6-15R

Material	Craft@Hrs	Unit	Material Cost	Labor Cost	Installed Cost
Ivory	L1@0.20	Ea	3.02	8.21	11.23

15 amp, 250 volt, back & side wired, NEMA 6-15R

Material	Craft@Hrs	Unit	Material Cost	Labor Cost	Installed Cost
Brown	L1@0.20	Ea	7.70	8.21	15.91
Ivory	L1@0.20	Ea	7.70	8.21	15.91

15 amp, 250 volt, side wired, self grounding, NEMA 6-15R

Material	Craft@Hrs	Unit	Material Cost	Labor Cost	Installed Cost
Brown	L1@0.20	Ea	4.49	8.21	12.70
Ivory	L1@0.20	Ea	4.49	8.21	12.70

15 amp, 250 volt, back & side wired, self grounding, NEMA 6-15R

Material	Craft@Hrs	Unit	Material Cost	Labor Cost	Installed Cost
Brown	L1@0.20	Ea	11.00	8.21	19.21
Ivory	L1@0.20	Ea	11.00	8.21	19.21

20 amp, 250 volt, side wired, NEMA 6-20R

Material	Craft@Hrs	Unit	Material Cost	Labor Cost	Installed Cost
Brown	L1@0.20	Ea	3.58	8.21	11.79
Ivory	L1@0.20	Ea	3.58	8.21	11.79
White	L1@0.20	Ea	3.58	8.21	11.79

20 amp, 250 volt, back & side wired, NEMA 6-20R

Material	Craft@Hrs	Unit	Material Cost	Labor Cost	Installed Cost
Brown	L1@0.20	Ea	9.96	8.21	18.17
Ivory	L1@0.20	Ea	9.96	8.21	18.17

20 amp, 250 volt, back & side wired, self grounding, NEMA 6-20R

Material	Craft@Hrs	Unit	Material Cost	Labor Cost	Installed Cost
Brown	L1@0.20	Ea	12.30	8.21	20.51
Ivory	L1@0.20	Ea	12.30	8.21	20.51

Use these figures to estimate the cost of single receptacles installed in outlet boxes under the conditions described on pages 5 and 6. Costs listed are for each receptacle installed. The crew is one electrician working at a labor cost of $41.03 per manhour. These costs include layout, material handling, and normal waste. Add for the outlet box and ring, sales tax, delivery, supervision, mobilization, demobilization, cleanup, overhead, and profit.

Material	Craft@Hrs	Unit	Material Cost	Labor Cost	Installed Cost
15 amp, 125 volt, screw terminals, NEMA 5-15R					
Brown	L1@0.20	Ea	.57	8.21	8.78
Ivory	L1@0.20	Ea	.57	8.21	8.78
White	L1@0.20	Ea	.57	8.21	8.78
15 amp, 125 volt, side wired, NEMA 5-15R					
Brown	L1@0.20	Ea	.58	8.21	8.79
Ivory	L1@0.20	Ea	.58	8.21	8.79
White	L1@0.20	Ea	.58	8.21	8.79
15 amp, 125 volt, push-in, residential grade, NEMA 5-15R					
Brown	L1@0.20	Ea	.61	8.21	8.82
Ivory	L1@0.20	Ea	.61	8.21	8.82
White	L1@0.20	Ea	.61	8.21	8.82
15 amp, 125 volt, back & side wired, corrosion resistant, commercial grade, NEMA 5-15R					
Brown	L1@0.20	Ea	.93	8.21	9.14
Ivory	L1@0.20	Ea	.93	8.21	9.14
White	L1@0.20	Ea	.93	8.21	9.14
Gray	L1@0.20	Ea	.93	8.21	9.14
15 amp, 125 volt, back & side wired, specification-grade, NEMA 5-15R					
Brown	L1@0.20	Ea	2.51	8.21	10.72
Ivory	L1@0.20	Ea	2.51	8.21	10.72
White	L1@0.20	Ea	2.51	8.21	10.72

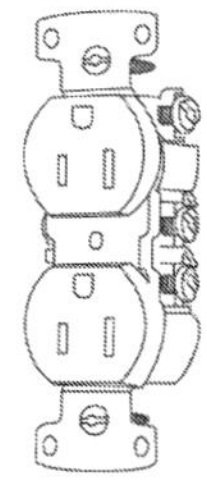

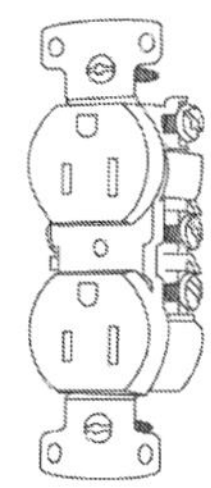

Use these figures to estimate the cost of duplex receptacles installed in outlet boxes under the conditions described on pages 5 and 6. Costs listed are for each receptacle installed. Where tamper-resistant (TR) receptacles are required, add 15% to the material cost. The crew is one electrician working at a labor cost of $41.03 per manhour. These costs include layout, material handling, and normal waste. Add for outlet box and ring, trim plate, sales tax, delivery, supervision, mobilization, demobilization, cleanup, overhead, and profit.

Duplex Receptacles

Material	Craft@Hrs	Unit	Material Cost	Labor Cost	Installed Cost

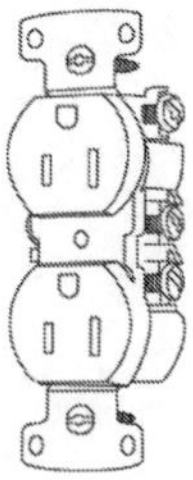

15 amp, 125 volt, side wired, self grounding, specification-grade, NEMA 5-15R

Material	Craft@Hrs	Unit	Material Cost	Labor Cost	Installed Cost
Brown	L1@0.20	Ea	4.56	8.21	12.77
Ivory	L1@0.20	Ea	4.56	8.21	12.77
White	L1@0.20	Ea	4.56	8.21	12.77
Gray	L1@0.20	Ea	4.56	8.21	12.77

15 amp, 125 volt, back & side wired, self grounding, extra hard use specification-grade, NEMA 5-15R

Material	Craft@Hrs	Unit	Material Cost	Labor Cost	Installed Cost
Brown	L1@0.20	Ea	10.50	8.21	18.71
Ivory	L1@0.20	Ea	10.50	8.21	18.71
White	L1@0.20	Ea	12.40	8.21	20.61
Gray	L1@0.20	Ea	12.40	8.21	20.61
Red	L1@0.20	Ea	12.40	8.21	20.61

15 amp, 125 volt, screw terminals, specification-grade, NEMA 5-15R

Material	Craft@Hrs	Unit	Material Cost	Labor Cost	Installed Cost
Brown	L1@0.20	Ea	13.40	8.21	21.61
Ivory	L1@0.20	Ea	13.40	8.21	21.61
White	L1@0.20	Ea	13.40	8.21	21.61
Gray	L1@0.20	Ea	13.40	8.21	21.61
Red	L1@0.20	Ea	15.70	8.21	23.91

15 amp, 125 volt, screw terminals, extra hard use specification-grade, NEMA 5-15R

Material	Craft@Hrs	Unit	Material Cost	Labor Cost	Installed Cost
Brown	L1@0.20	Ea	24.60	8.21	32.81
Ivory	L1@0.20	Ea	24.60	8.21	32.81
White	L1@0.20	Ea	24.60	8.21	32.81
Gray	L1@0.20	Ea	24.60	8.21	32.81

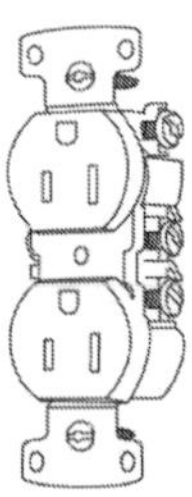

15 amp, 125 volt, screw terminals, surge suppressor, hospital grade, NEMA 5-15R

Material	Craft@Hrs	Unit	Material Cost	Labor Cost	Installed Cost
Brown	L1@0.20	Ea	29.90	8.21	38.11
Ivory	L1@0.20	Ea	29.90	8.21	38.11
White	L1@0.20	Ea	29.90	8.21	38.11
Red	L1@0.20	Ea	29.90	8.21	38.11

15 amp, 125 volt, screw terminals, isolated ground, surge suppressor, hospital grade, NEMA 5-15R

Material	Craft@Hrs	Unit	Material Cost	Labor Cost	Installed Cost
Brown	L1@0.20	Ea	32.40	8.21	40.61
Ivory	L1@0.20	Ea	32.40	8.21	40.61
Orange	L1@0.20	Ea	32.40	8.21	40.61

Use these figures to estimate the cost of duplex receptacles installed in outlet boxes under the conditions described on pages 5 and 6. Costs listed are for each receptacle installed. Where tamper-resistant (TR) receptacles are required, add 15% to the material cost. The crew is one electrician working at a labor cost of $41.03 per manhour. These costs include layout, material handling, and normal waste. Add for outlet box and ring, trim plate, sales tax, delivery, supervision, mobilization, demobilization, cleanup, overhead, and profit.

Material	Craft@Hrs	Unit	Material Cost	Labor Cost	Installed Cost
20 amp, 125 volt, side wired, corrosion resistant, commercial-grade, NEMA 5-20R					
Brown	L1@0.20	Ea	1.17	8.21	9.38
Ivory	L1@0.20	Ea	1.17	8.21	9.38
White	L1@0.20	Ea	1.17	8.21	9.38
Gray	L1@0.20	Ea	1.17	8.21	9.38
20 amp, 125 volt, side wired, self grounding, commercial-grade, NEMA 5-20R					
Brown	L1@0.20	Ea	6.71	8.21	14.92
Ivory	L1@0.20	Ea	6.71	8.21	14.92
White	L1@0.20	Ea	6.71	8.21	14.92
Gray	L1@0.20	Ea	6.71	8.21	14.92
20 amp, 125 volt, back & side wired, self grounding, commercial-grade, NEMA 5-20R					
Brown	L1@0.20	Ea	9.03	8.21	17.24
Ivory	L1@0.20	Ea	9.03	8.21	17.24
White	L1@0.20	Ea	9.03	8.21	17.24
Gray	L1@0.20	Ea	9.03	8.21	17.24
20 amp, 125 volt, back & side wired, self grounding, specification-grade, NEMA 5-20R					
Brown	L1@0.20	Ea	12.40	8.21	20.61
Ivory	L1@0.20	Ea	12.40	8.21	20.61
White	L1@0.20	Ea	12.40	8.21	20.61
Gray	L1@0.20	Ea	12.40	8.21	20.61
Red	L1@0.20	Ea	14.30	8.21	22.51
Orange (isolated group)	L1@0.20	Ea	16.30	8.21	24.51
20 amp, 125 volt, hospital grade, NEMA 5-20R					
Brown	L1@0.20	Ea	17.10	8.21	25.31
Ivory	L1@0.20	Ea	14.60	8.21	22.81
White	L1@0.20	Ea	14.60	8.21	22.81
Gray	L1@0.20	Ea	14.60	8.21	22.81
Red	L1@0.20	Ea	14.60	8.21	22.81

Use these figures to estimate the cost of duplex receptacles installed in outlet boxes under the conditions described on pages 5 and 6. Costs listed are for each receptacle installed. Where tamper-resistant (TR) receptacles are required, add 15% to the material cost. The crew is one electrician working at a labor cost of $41.03 per manhour. These costs include layout, material handling, and normal waste. Add for outlet box and ring, trim plate, sales tax, delivery, supervision, mobilization, demobilization, cleanup, overhead, and profit.

Duplex Receptacles

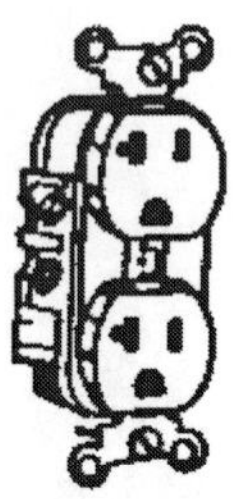

Material	Craft@Hrs	Unit	Material Cost	Labor Cost	Installed Cost
20 amp, 125 volt, side wired, NEMA 5-20R					
Brown	L1@0.20	Ea	2.68	8.21	10.89
Ivory	L1@0.20	Ea	2.68	8.21	10.89
White	L1@0.20	Ea	2.68	8.21	10.89
20 amp, 125 volt, side wired, self grounding, NEMA 5-20R					
Brown	L1@0.20	Ea	6.62	8.21	14.83
Ivory	L1@0.20	Ea	6.62	8.21	14.83
20 amp, 125 volt, back & side wired, isolated ground, NEMA 5-20R					
Brown	L1@0.20	Ea	10.40	8.21	18.61
Ivory	L1@0.20	Ea	10.40	8.21	18.61
White	L1@0.20	Ea	10.40	8.21	18.61
Orange	L1@0.20	Ea	16.10	8.21	24.31
20 amp, 125 volt, back & side wired, hospital grade, NEMA 5-20R					
Brown	L1@0.20	Ea	12.60	8.21	20.81
Ivory	L1@0.20	Ea	12.60	8.21	20.81
White	L1@0.20	Ea	12.60	8.21	20.81
Red	L1@0.20	Ea	12.60	8.21	20.81
20 amp, 125 volt, hospital grade with surge suppressor					
Brown	L1@0.20	Ea	21.90	8.21	30.11
Ivory	L1@0.20	Ea	20.00	8.21	28.21
White	L1@0.20	Ea	20.00	8.21	28.21
Red	L1@0.20	Ea	20.00	8.21	28.21
20 amp, 125 volt, hospital grade with surge suppressor & isolated ground					
White	L1@0.20	Ea	20.00	8.21	28.21
Orange	L1@0.20	Ea	20.00	8.21	28.21

Use these figures to estimate the cost of duplex receptacles installed in outlet boxes under the conditions described on pages 5 and 6. Costs listed are for each receptacle installed. Where tamper-resistant (TR) receptacles are required, add 15% to the material cost. The crew is one electrician working at a labor cost of $41.03 per manhour. These costs include layout, material handling, and normal waste. Add for outlet box and ring, trim plate, sales tax, delivery, supervision, mobilization, demobilization, cleanup, overhead, and profit.

Material	Craft@Hrs	Unit	Material Cost	Labor Cost	Installed Cost
20 amp, 125 volt, screw terminal, with surge suppressor, specification-grade, NEMA 5-20R					
Brown	L1@0.20	Ea	28.40	8.21	36.61
Ivory	L1@0.20	Ea	28.40	8.21	36.61
White	L1@0.20	Ea	28.40	8.21	36.61
20 amp, 125 volt, screw terminal, with surge suppressor, hospital grade, NEMA 5-20R					
Brown	L1@0.20	Ea	32.30	8.21	40.51
Ivory	L1@0.20	Ea	32.30	8.21	40.51
White	L1@0.20	Ea	32.30	8.21	40.51
Red	L1@0.20	Ea	32.30	8.21	40.51
20 amp, 125 volt, screw terminal, with surge suppressor and isolated ground, hospital grade, NEMA 5-20R					
Ivory	L1@0.20	Ea	35.80	8.21	44.01
White	L1@0.20	Ea	35.80	8.21	44.01
Orange	L1@0.20	Ea	35.80	8.21	44.01
20 amp, 250 volt, side wired, specification-grade, NEMA 6-20R					
Brown	L1@0.20	Ea	8.68	8.21	16.89
Ivory	L1@0.20	Ea	8.68	8.21	16.89
White	L1@0.20	Ea	8.68	8.21	16.89
20 amp, 250 volt, back & side wired, specification-grade, NEMA 6-20R					
Brown	L1@0.20	Ea	8.68	8.21	16.89
Ivory	L1@0.20	Ea	8.68	8.21	16.89
White	L1@0.20	Ea	8.68	8.21	16.89
20 amp, 250 volt, back & side wired, self grounding, specification-grade, NEMA 6-20R					
Brown	L1@0.20	Ea	15.70	8.21	23.91
Ivory	L1@0.20	Ea	15.70	8.21	23.91
White	L1@0.20	Ea	15.70	8.21	23.91

Use these figures to estimate the cost of duplex receptacles installed in outlet boxes under the conditions described on pages 5 and 6. Costs listed are for each receptacle installed. The crew is one electrician working at a labor cost of $41.03 per manhour. These costs include layout, material handling, and normal waste. Add for outlet box and ring, trim plate, sales tax, delivery, supervision, mobilization, demobilization, cleanup, overhead, and profit.

Single Decorator Receptacles

Material	Craft@Hrs	Unit	Material Cost	Labor Cost	Installed Cost
15 amp, 125 volt, 2 pole, 3 wire, side wired, grounded, NEMA 5-15R					
Brown	L1@0.20	Ea	4.56	8.21	12.77
Ivory	L1@0.20	Ea	4.56	8.21	12.77
White	L1@0.20	Ea	4.56	8.21	12.77
15 amp, 125 volt, 2 pole, 3 wire, back & side wired, grounded, NEMA 5-15R					
Brown	L1@0.20	Ea	8.22	8.21	16.43
Ivory	L1@0.20	Ea	8.22	8.21	16.43
White	L1@0.20	Ea	8.22	8.21	16.43
20 amp, 125 volt, 2 pole, 3 wire, side wired, grounded, NEMA 5-20R					
Brown	L1@0.20	Ea	6.98	8.21	15.19
Ivory	L1@0.20	Ea	6.98	8.21	15.19
White	L1@0.20	Ea	6.98	8.21	15.19
20 amp, 125 volt, 2 pole, 3 wire, back & side wired, grounded, NEMA 5-20R					
Brown	L1@0.20	Ea	10.50	8.21	18.71
Ivory	L1@0.20	Ea	10.50	8.21	18.71
White	L1@0.20	Ea	10.50	8.21	18.71
20 amp, 250 volt, 2 pole, 3 wire, side wired, grounded, NEMA 6-20R					
Brown	L1@0.20	Ea	7.52	8.21	15.73
Ivory	L1@0.20	Ea	7.52	8.21	15.73
White	L1@0.20	Ea	7.52	8.21	15.73
20 amp, 250 volt, 2 pole, 3 wire, back & side wired, grounded, NEMA 6-20R					
Brown	L1@0.20	Ea	10.50	8.21	18.71
Ivory	L1@0.20	Ea	10.50	8.21	18.71
White	L1@0.20	Ea	10.50	8.21	18.71

Use these figures to estimate the cost of single receptacles installed in outlet boxes under the conditions described on pages 5 and 6. Costs listed are for each receptacle installed. Where tamper-resistant (TR) receptacles are required, add 15% to the material cost. The crew is one electrician working at a labor cost of $41.03 per manhour. These costs include layout, material handling, and normal waste. Add for the outlet box and ring, sales tax, delivery, supervision, mobilization, demobilization, cleanup, overhead, and profit.

Material	Craft@Hrs	Unit	Material Cost	Labor Cost	Installed Cost
15 amp, 125 volt, 2 pole, 3 wire, push-in, grounded, NEMA 5-15R					
Brown	L1@0.20	Ea	2.95	8.21	11.16
Ivory	L1@0.20	Ea	2.95	8.21	11.16
White	L1@0.20	Ea	2.95	8.21	11.16
15 amp, 125 volt, 2 pole, 3 wire, push-in, self grounding, NEMA 5-15R					
Brown	L1@0.20	Ea	4.56	8.21	12.77
Ivory	L1@0.20	Ea	4.56	8.21	12.77
White	L1@0.20	Ea	4.56	8.21	12.77
15 amp, 125 volt, 2 pole, 3 wire, side wired, self grounding, NEMA 5-15R					
Brown	L1@0.20	Ea	4.91	8.21	13.12
Ivory	L1@0.20	Ea	4.91	8.21	13.12
White	L1@0.20	Ea	4.91	8.21	13.12
15 amp, 125 volt, 2 pole, 3 wire, back & side wired, grounding, NEMA 5-15R					
Brown	L1@0.20	Ea	6.35	8.21	14.56
Ivory	L1@0.20	Ea	6.35	8.21	14.56
White	L1@0.20	Ea	6.35	8.21	14.56
20 amp, 125 volt, 2 pole, 3 wire, side wired, grounded, NEMA 5-20R					
Brown	L1@0.20	Ea	8.06	8.21	16.27
Ivory	L1@0.20	Ea	8.06	8.21	16.27
White	L1@0.20	Ea	8.06	8.21	16.27
20 amp, 125 volt, 2 pole, 3 wire, back & side wired, self grounding, NEMA 5-20R					
Brown	L1@0.20	Ea	11.00	8.21	19.21
Ivory	L1@0.20	Ea	11.00	8.21	19.21
White	L1@0.20	Ea	11.00	8.21	19.21
Gray	L1@0.20	Ea	11.00	8.21	19.21

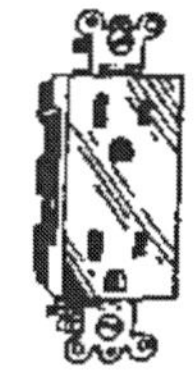

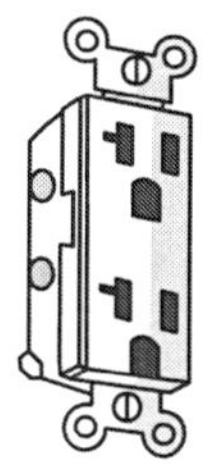

Use these figures to estimate the cost of duplex receptacles installed in outlet boxes under the conditions described on pages 5 and 6. Costs listed are for each receptacle installed. Where tamper-resistant (TR) receptacles are required, add 15% to the material cost. The crew is one electrician working at a labor cost of $41.03 per manhour. These costs include layout, material handling, and normal waste. Add for outlet box and ring, trim plate, sales tax, delivery, supervision, mobilization, demobilization, cleanup, overhead, and profit.

Ground Fault Circuit Interrupter (GFCI) Duplex Receptacles

15 amp, 120 volt AC, commercial specification-grade, duplex less indicating light, with wall plate

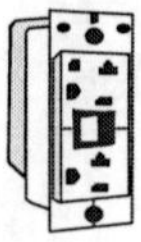

Material	Craft@Hrs	Unit	Material Cost	Labor Cost	Installed Cost
Brown	L1@0.20	Ea	10.40	8.21	18.61
Ivory	L1@0.20	Ea	10.40	8.21	18.61
White	L1@0.20	Ea	10.40	8.21	18.61
Gray	L1@0.20	Ea	10.40	8.21	18.61
Red	L1@0.20	Ea	10.40	8.21	18.61
Black	L1@0.20	Ea	10.40	8.21	18.61
Almond	L1@0.20	Ea	10.40	8.21	18.61

15 amp, 120 volt AC, commercial specification-grade, duplex with indicating light and wall plate

Material	Craft@Hrs	Unit	Material Cost	Labor Cost	Installed Cost
Brown	L1@0.20	Ea	11.60	8.21	19.81
Ivory	L1@0.20	Ea	11.60	8.21	19.81
White	L1@0.20	Ea	11.60	8.21	19.81

20 amp, 120 volt AC, commercial specification-grade, duplex with indicating light and wall plate, feed through

Material	Craft@Hrs	Unit	Material Cost	Labor Cost	Installed Cost
Brown	L1@0.20	Ea	12.60	8.21	20.81
Ivory	L1@0.20	Ea	12.60	8.21	20.81
White	L1@0.20	Ea	12.60	8.21	20.81
Gray	L1@0.20	Ea	12.60	8.21	20.81
Red	L1@0.20	Ea	12.60	8.21	20.81

15 amp, 120 volt AC, hospital grade, duplex with indicating light and wall plate, feed through

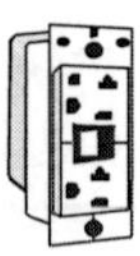

Material	Craft@Hrs	Unit	Material Cost	Labor Cost	Installed Cost
Brown	L1@0.20	Ea	57.60	8.21	65.81
Ivory	L1@0.20	Ea	57.60	8.21	65.81
White	L1@0.20	Ea	57.60	8.21	65.81
Gray	L1@0.20	Ea	57.60	8.21	65.81

20 amp, 120 volt AC, heavy duty, specification-grade, duplex with wall plate, feed through

Material	Craft@Hrs	Unit	Material Cost	Labor Cost	Installed Cost
Brown	L1@0.20	Ea	53.60	8.21	61.81
Ivory	L1@0.20	Ea	53.60	8.21	61.81
White	L1@0.20	Ea	53.60	8.21	61.81
Gray	L1@0.20	Ea	53.60	8.21	61.81

Use these figures to estimate the cost of ground fault circuit interrupter receptacles (GFCI) under the conditions described on pages 5 and 6. Costs listed are for each receptacle installed. Where tamper-resistant (TR) receptacles are required, add 15% to the material cost. The crew is one electrician working at a labor cost of $41.03 per manhour. These costs include layout, material handling, and normal waste. Add for boxes and rings, trip plate, sales tax, supervision, mobilization, demobilization, cleanup, overhead, and profit.

Arc Fault Circuit Interrupter (AFCI) Duplex Receptacles

Material	Craft@Hrs	Unit	Material Cost	Labor Cost	Installed Cost
15 amp, 120 volt AC, commercial specification-grade, duplex with indicating light and wall plate					
Ivory	L1@0.20	Ea	34.50	8.21	42.71
White	L1@0.20	Ea	34.50	8.21	42.71
20 amp, 120 volt AC, commercial specification-grade, duplex with indicating light and wall plate, feed through					
Ivory	L1@0.20	Ea	39.50	8.21	47.71
White	L1@0.20	Ea	39.50	8.21	47.71
Gray	L1@0.20	Ea	39.50	8.21	47.71
15 amp, 120 volt AC, hospital grade, duplex with indicating light and wall plate, feed through					
Ivory	L1@0.20	Ea	43.40	8.21	51.61
White	L1@0.20	Ea	43.40	8.21	51.61
Gray	L1@0.20	Ea	43.40	8.21	51.61
20 amp, 120 volt AC, heavy duty, specification-grade, duplex with wall plate, feed through					
Ivory	L1@0.20	Ea	59.20	8.21	67.41
White	L1@0.20	Ea	59.20	8.21	67.41
Gray	L1@0.20	Ea	59.20	8.21	67.41

Use these figures to estimate the cost of arc fault circuit interrupters in outlet boxes under the conditions described on pages 5 and 6. Costs listed are for each receptacle installed. The crew is one electrician working at a labor cost of $41.03 per manhours. These costs include layout, material handling, and normal waste. Add for the outlet box and ring, trim plate, sales tax, delivery, supervision, mobilization, demobilization, cleanup, overhead, and profit.

Power Cord Receptacles

Material	Craft@Hrs	Unit	Material Cost	Labor Cost	Installed Cost

2 pole, 3 wire single, U-grounding, surface mounted flat blade receptacles

50A, 250V, NEMA 6-50R	L1@0.40	Ea	13.90	16.40	30.30

3 pole, 3 wire single non-grounding polarized surface mounted DRYER flat blade receptacles

30A, 125/250V, NEMA 10-30R	L1@0.35	Ea	4.26	14.40	18.66

3 pole, 3 wire single non-grounding polarized surface mounted RANGE flat blade receptacles

50A, 125/250V, NEMA 10-50R	L1@0.40	Ea	4.26	16.40	20.66

3 pole, 4 wire single U-ground polarized surface mounted DRYER flat blade receptacles

30A, 125/250V, NEMA 14-30R	L1@0.35	Ea	13.10	14.40	27.50

3 pole, 4 wire single U-ground polarized surface mounted RANGE flat blade receptacles

50A, 125/250V, NEMA 14-50R	L1@0.40	Ea	25.10	16.40	41.50

2 pole, 3-wire single, grounding

20A, 277V, NEMA 7-20R	L1@0.25	Ea	9.87	10.30	20.17
30A, 125V, NEMA 5-30R	L1@0.25	Ea	14.10	10.30	24.40
30A, 250V, NEMA 6-30R	L1@0.25	Ea	11.60	10.30	21.90
50A, 250V, NEMA 6-50R	L1@0.25	Ea	11.60	10.30	21.90

2 pole, 3-wire grounding, duplex

15A, 277V, NEMA 7-15R	L1@0.25	Ea	19.00	10.30	29.30

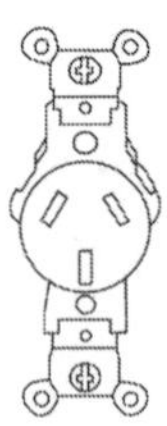

3 pole, 3-wire single, non-grounding, Dryer

20A, 125/250V, NEMA 10-20R	L1@0.25	Ea	17.70	10.30	28.00
30A, 125/250V, NEMA 10-30R	L1@0.25	Ea	2.95	10.30	13.25

3 pole, 3-wire single, non-grounding, Range

50A, 125/250V, NEMA 10-50R	L1@0.30	Ea	6.71	12.30	19.01

3 pole, 4-wire single, grounding, Dryer

20A, 125/250V, NEMA 14-20R	L1@0.25	Ea	18.90	10.30	29.20
30A, 125/250V, NEMA 14-30R	L1@0.25	Ea	19.10	10.30	29.40

3 pole, 4-wire single, grounding, Range

50A, 125/250V, NEMA 14-50R	L1@0.35	Ea	18.60	14.40	33.00

Use these figures to estimate the cost of power cord receptacles in outlet boxes under the conditions described on pages 5 and 6. Costs listed are for each receptacle installed. The crew is one electrician working at a labor cost of $41.03 per manhours. These costs include layout, material handling, and normal waste. Add for the outlet box and ring, trim plate, sales tax, delivery, supervision, mobilization, demobilization, cleanup, overhead, and profit.

Material	Craft@Hrs	Unit	Material Cost	Labor Cost	Installed Cost
3 pole, 4-wire single, 3 phase, grounding					
30A, 250V, NEMA 15-30R	L1@0.30	Ea	25.20	12.30	37.50
50A, 250V, NEMA 15-50R	L1@0.40	Ea	25.20	16.40	41.60
60A, 250V, NEMA 15-60R	L1@0.45	Ea	25.20	18.50	43.70
4 pole, 4-wire single, 3 phase Y, single, non-grounding					
20A, 120/208V, NEMA 18-20R	L1@0.30	Ea	17.00	12.30	29.30
60A, 120/208V, NEMA 18-60R	L1@0.45	Ea	24.20	18.50	42.70
2 pole, 3 wire U-ground parallel slot, connector bodies					
15A, 125V, NEMA 5-15R, Black	L1@0.20	Ea	5.19	8.21	13.40
15A, 250V, NEMA 5-15R, Nylon	L1@0.20	Ea	12.40	8.21	20.61
2 pole, 3 wire U-ground parallel slot, hospital grade connector bodies					
15A, 125V, NEMA 5-15R, Nylon	L1@0.20	Ea	9.96	8.21	18.17
2 pole, 3 wire U-ground tandem slot, connector bodies					
15A, 250V, NEMA 5-20R, Black	L1@0.20	Ea	6.44	8.21	14.65
15A, 125V, NEMA 5-20R, Nylon	L1@0.20	Ea	14.60	8.21	22.81
2 pole, 3 wire U-ground right angle slot, connector bodies					
20A, 125V, NEMA 5-20R, Black	L1@0.25	Ea	6.26	10.30	16.56
20A, 125V, NEMA 5-20R, Nylon, Hospital Grade	L1@0.25	Ea	16.50	10.30	26.80
20A, 125V, NEMA 5-20R, Nylon, with cord grip	L1@0.25	Ea	16.20	10.30	26.50
20A, 250V, NEMA 6-20R, Nylon	L1@0.25	Ea	17.40	10.30	27.70
20A, 250V, NEMA 6-20R, Bk Nylon	L1@0.25	Ea	7.41	10.30	17.71

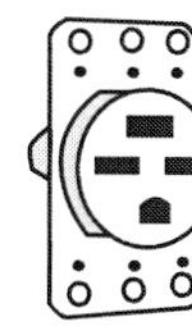

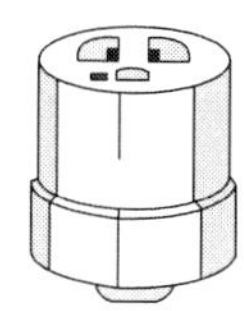

Use these figures to estimate the cost of power cord connectors and plugs on power cord cable under the conditions described on pages 5 and 6. Costs listed are for each connector or plug installed. The crew is one electrician working at a labor cost of $41.03 per manhour. These costs include layout, material handling, and normal waste. Add for sales tax, delivery, supervision, mobilization, demobilization, cleanup, overhead, and profit.

Power Cord Plugs

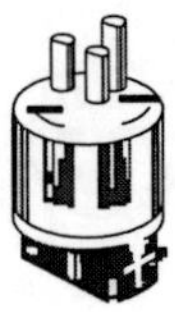

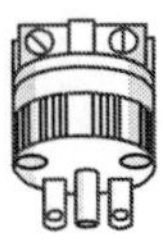

Material	Craft@Hrs	Unit	Material Cost	Labor Cost	Installed Cost
2 wire, 125 volt, parallel blade bakelite plugs					
Brown, flat handle	L1@0.15	Ea	.99	6.15	7.14
Ivory, flat handle	L1@0.15	Ea	.99	6.15	7.14
White, flat handle	L1@0.15	Ea	.99	6.15	7.14
Brown, flat vinyl handle	L1@0.15	Ea	1.53	6.15	7.68
Ivory, flat vinyl handle	L1@0.15	Ea	1.53	6.15	7.68
Black, flat vinyl handle	L1@0.15	Ea	1.53	6.15	7.68
2 wire or 3 wire, 10 amp, 125 volt, parallel blade plugs for 18/2 SPT-1 for 20-20-2 XT cord					
Brown, flat plastic handle	L1@0.15	Ea	1.89	6.15	8.04
Ivory, flat plastic handle	L1@0.15	Ea	1.89	6.15	8.04
2 wire, 15 amp, 125 volt, parallel blade, dead front rubber plugs					
Brown, flat handle	L1@0.15	Ea	1.42	6.15	7.57
Ivory, flat handle	L1@0.15	Ea	1.42	6.15	7.57
White, flat handle	L1@0.15	Ea	1.42	6.15	7.57
2 wire, 15 amp, 125 volt, parallel blade vinyl plugs					
Brown, angle handle	L1@0.15	Ea	1.69	6.15	7.84
White, angle handle	L1@0.15	Ea	1.69	6.15	7.84
Black, angle handle	L1@0.15	Ea	1.69	6.15	7.84
2 wire, 10 amp, 125 volt, parallel blade rubber plugs					
Ivory, flat handle	L1@0.15	Ea	1.69	6.15	7.84
White, flat handle	L1@0.15	Ea	1.69	6.15	7.84
Black, flat handle	L1@0.15	Ea	1.69	6.15	7.84
2 wire, 10 amp, 125 volt, parallel blade side outlet rubber plugs					
Ivory, flat handle	L1@0.15	Ea	1.79	6.15	7.94
White, flat handle	L1@0.15	Ea	1.79	6.15	7.94
Black, flat handle	L1@0.15	Ea	1.79	6.15	7.94
Brown, angle handle	L1@0.15	Ea	1.69	6.15	7.84
White, angle handle	L1@0.15	Ea	1.69	6.15	7.84
Black, angle handle	L1@0.15	Ea	1.69	6.15	7.84
2 wire, 10 amp, 125 volt, parallel blade rubber plugs					
Ivory, flat handle	L1@0.15	Ea	1.69	6.15	7.84
White, flat handle	L1@0.15	Ea	1.69	6.15	7.84
Black, flat handle	L1@0.15	Ea	1.69	6.15	7.84

Use these figures to estimate the cost of power cord plugs installed on power cords under the conditions described on pages 5 and 6. Costs listed are for each plug installed. The crew is one electrician working at a labor cost of $41.03 per manhour. These costs include layout, material handling, and normal waste. Add for sales tax, delivery, supervision, mobilization, demobilization, cleanup, overhead, and profit.

2 wire, 10 amp, 125 volt, parallel blade, side outlet rubber plugs

Material	Craft@Hrs	Unit	Material Cost	Labor Cost	Installed Cost
Ivory, flat handle	L1@0.15	Ea	1.79	6.15	7.94
White, flat handle	L1@0.15	Ea	1.79	6.15	7.94
Black, flat handle	L1@0.15	Ea	1.79	6.15	7.94

2 pole 15 amp, 125 volt, 3-wire, U-ground plugs

Material	Craft@Hrs	Unit	Material Cost	Labor Cost	Installed Cost
Black, NEMA 5-15P	L1@0.15	Ea	2.51	6.15	8.66
Plastic, NEMA 5-15P	L1@0.15	Ea	6.98	6.15	13.13
Hospital type, NEMA 5-15P	L1@0.15	Ea	7.87	6.15	14.02

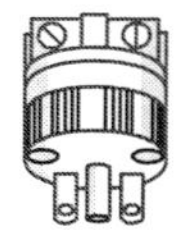

2 pole, 15 amp, 125 volt, 3 wire U-ground plugs

Material	Craft@Hrs	Unit	Material Cost	Labor Cost	Installed Cost
Angle, NEMA 5-15P	L1@0.15	Ea	8.49	6.15	14.64
Hospital type, NEMA 5-15P	L1@0.15	Ea	11.50	6.15	17.65

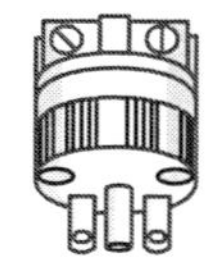

2 pole, 20 amp, 125 volt, 3 wire U-ground plugs

Material	Craft@Hrs	Unit	Material Cost	Labor Cost	Installed Cost
Black, NEMA 5-20P	L1@0.20	Ea	9.54	8.21	17.75
Hospital type, NEMA 5-20P	L1@0.20	Ea	12.60	8.21	20.81
Hospital angle, NEMA 5-15P	L1@0.20	Ea	16.50	8.21	24.71
Yellow, NEMA 5-20P	L1@0.20	Ea	12.80	8.21	21.01

2 pole, 15 amp, 250 volt, 3 wire U-ground plugs

Material	Craft@Hrs	Unit	Material Cost	Labor Cost	Installed Cost
Black, NEMA 6-15P	L1@0.15	Ea	2.87	6.15	9.02
Black, HD, NEMA 6-15P	L1@0.15	Ea	8.41	6.15	14.56

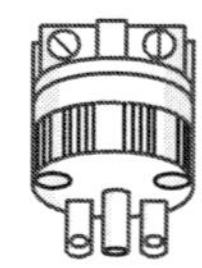

2 pole, 20 amp, 250 volt, 3 wire U-ground plugs

Material	Craft@Hrs	Unit	Material Cost	Labor Cost	Installed Cost
Black, NEMA 6-20P	L1@0.20	Ea	3.85	8.21	12.06
Yellow, NEMA 6-20P	L1@0.20	Ea	11.00	8.21	19.21
Yellow angle, NEMA 6-20P	L1@0.20	Ea	14.60	8.21	22.81

2 pole, 15 amp, 277 volt, 3 wire U-ground plugs

Material	Craft@Hrs	Unit	Material Cost	Labor Cost	Installed Cost
Yellow, NEMA 7-15P	L1@0.15	Ea	12.60	6.15	18.75

2 pole, 20 amp, 277 volt, 3 wire U-ground plugs

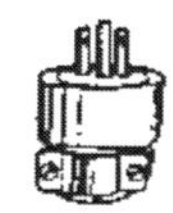

Material	Craft@Hrs	Unit	Material Cost	Labor Cost	Installed Cost
Yellow, NEMA 7-20P	L1@0.20	Ea	9.26	8.21	17.47

Use these figures to estimate the cost of power cord plugs installed on power cords under the conditions described on pages 5 and 6. Costs listed are for each plug installed. The crew is one electrician working at a labor cost of $41.03 per manhour. These costs include layout, material handling, and normal waste. Add for sales tax, delivery, supervision, mobilization, demobilization, cleanup, overhead, and profit.

Power Cord Plugs

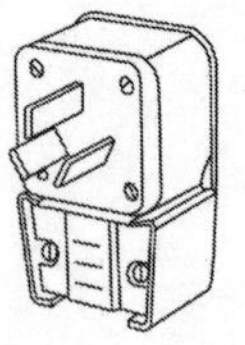

3 pole, 3 wire, non-grounding black polarized angle plugs

Material	Craft@Hrs	Unit	Material Cost	Labor Cost	Installed Cost
20A, 125/250V, NEMA 10-20P	L1@0.20	Ea	14.30	8.21	22.51
30A, 125/250V, NEMA 10-30P	L1@0.25	Ea	22.10	10.30	32.40
50A, 125/250V, NEMA 10-50P	L1@0.30	Ea	41.90	12.30	54.20

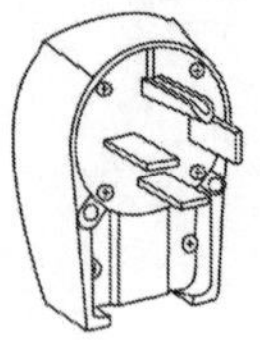

3 pole, 4 wire, U-ground yellow polarized angle plugs

Material	Craft@Hrs	Unit	Material Cost	Labor Cost	Installed Cost
20A, 125/250V, NEMA 14-20P	L1@0.25	Ea	30.50	10.30	40.80
30A, 125/250V, NEMA 10-30P	L1@0.30	Ea	34.00	12.30	46.30
50A, 125/250V, NEMA 10-50P	L1@0.35	Ea	40.10	14.40	54.50
60A, 125/250V, NEMA 10-60P	L1@0.35	Ea	50.50	14.40	64.90

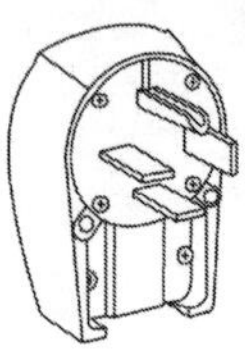

3 pole, 4 wire, 3 phase U-ground yellow polarized angle plugs

Material	Craft@Hrs	Unit	Material Cost	Labor Cost	Installed Cost
20A, 250V, NEMA 15-20P	L1@0.30	Ea	32.30	12.30	44.60
30A, 250V, NEMA 15-30P	L1@0.35	Ea	37.60	14.40	52.00
50A, 250V, NEMA 15-50P	L1@0.40	Ea	51.10	16.40	67.50
60A, 250V, NEMA 15-60P	L1@0.45	Ea	49.20	18.50	67.70

4 pole, 4 wire, 3 phase non-grounding polarized plugs

Material	Craft@Hrs	Unit	Material Cost	Labor Cost	Installed Cost
20A, 120/208V, NEMA 18-20P	L1@0.35	Ea	24.80	14.40	39.20
60A, 120/208V, NEMA 18-60P	L1@0.50	Ea	59.70	20.50	80.20

Use these figures to estimate the cost of power cord plugs installed on power cords under the conditions described on pages 5 and 6. Costs listed are for each plug installed. The crew is one electrician working at a labor cost of $41.03 per manhour. These costs include layout, material handling, and normal waste. Add for sales tax, delivery, supervision, mobilization, demobilization, cleanup, overhead, and profit.

Material	Craft@Hrs	Unit	Material Cost	Labor Cost	Installed Cost
2 pole, 2 wire, single non-grounding locking receptacles					
15A, 125V, NEMA L1-15R	L1@0.20	Ea	8.06	8.21	16.27
20A, 250V, NEMA L2-20R	L1@0.20	Ea	10.70	8.21	18.91
2 pole, 3 wire, single grounding locking receptacles					
15A, 125V, NEMA L5-15R	L1@0.20	Ea	11.70	8.21	19.91
20A, 125V, NEMA L5-20R	L1@0.20	Ea	13.60	8.21	21.81
30A, 125V, NEMA L5-30R	L1@0.20	Ea	19.20	8.21	27.41
15A, 250V, NEMA L6-15R	L1@0.25	Ea	12.00	10.30	22.30
20A, 250V, NEMA L6-20R	L1@0.25	Ea	13.70	10.30	24.00
30A, 250V, NEMA L6-30R	L1@0.25	Ea	20.90	10.30	31.20
15A, 277V, NEMA L7-15R	L1@0.25	Ea	12.20	10.30	22.50
20A, 277V, NEMA L7-20R	L1@0.25	Ea	13.70	10.30	24.00
30A, 277V, NEMA L7-30R	L1@0.30	Ea	20.90	12.30	33.20
20A, 480V, NEMA L8-20R	L1@0.25	Ea	17.10	10.30	27.40
30A, 480V, NEMA L8-30R	L1@0.30	Ea	25.20	12.30	37.50
20A, 600V, NEMA L9-20R	L1@0.25	Ea	11.60	10.30	21.90
30A, 600V, NEMA L9-30R	L1@0.30	Ea	25.40	12.30	37.70
2 pole, 3 wire, single orange isolated ground locking receptacles					
15A, 125V, NEMA L5-15R	L1@0.20	Ea	22.80	8.21	31.01
20A, 125V, NEMA L5-20R	L1@0.25	Ea	16.00	10.30	26.30
30A, 125V, NEMA L5-30R	L1@0.25	Ea	22.00	10.30	32.30
20A, 250V, NEMA L6-20R	L1@0.30	Ea	16.40	12.30	28.70
3 pole, 3 wire, single non-grounding locking receptacles					
20A, 125/250V, NEMA L10-20R	L1@0.25	Ea	18.00	10.30	28.30
30A, 125/250V, NEMA L10-30R	L1@0.30	Ea	23.20	12.30	35.50
3 pole, 3 wire, 3 phase non-grounding locking receptacles					
20A, 250V, NEMA L11-20R	L1@0.25	Ea	20.80	10.30	31.10
30A, 250V, NEMA L11-30R	L1@0.30	Ea	23.20	12.30	35.50
20A, 480V, NEMA L12-20R	L1@0.25	Ea	18.00	10.30	28.30
30A, 600V, NEMA L13-30R	L1@0.30	Ea	23.20	12.30	35.50

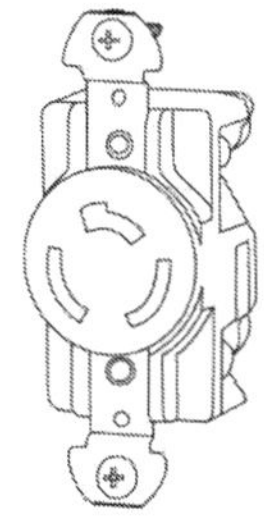

Use these figures to estimate the cost of locking receptacles under the conditions described on pages 5 and 6. Costs listed are for each receptacle installed. The crew is one electrician working at a labor cost of $41.03 per manhour. These costs include layout, material handling, and normal waste. Add for outlet boxes and switch rings, sales tax, delivery, supervision, mobilization, demobilization, cleanup, overhead, and profit. There's a special configuration for each voltage and amperage. Receptacles with special contact materials designed for durability under heavy use can be considerably more expensive than standard grade receptacles.

Locking Receptacles

Material	Craft@Hrs	Unit	Material Cost	Labor Cost	Installed Cost
3 pole, 4 wire, single orange isolated ground locking receptacles					
20A, 125/250V, NEMA L14-20R	L1@0.25	Ea	21.00	10.30	31.30
30A, 125/250V, NEMA L14-30R	L1@0.30	Ea	18.00	12.30	30.30
3 pole, 4 wire, single non-grounding locking receptacles					
20A, 125/250V, NEMA L14-20R	L1@0.25	Ea	21.00	10.30	31.30
30A, 125/250V, NEMA L14-30R	L1@0.30	Ea	18.00	12.30	30.30
3 pole, 4 wire, 3 phase single grounding locking receptacles					
20A, 250V, NEMA L15-20R	L1@0.25	Ea	19.10	10.30	29.40
30A, 250V, NEMA L15-30R	L1@0.30	Ea	30.40	12.30	42.70
20A, 480V, NEMA L16-20R	L1@0.25	Ea	22.80	10.30	33.10
30A, 480V, NEMA L16-30R	L1@0.30	Ea	35.80	12.30	48.10
30A, 600V, NEMA L17-30R	L1@0.30	Ea	34.80	12.30	47.10
4 pole, 4 wire, 3 phase Y single non-grounding locking receptacles					
20A, 120/208V, NEMA L18-20R	L1@0.25	Ea	28.80	10.30	39.10
30A, 120/208V, NEMA L18-30R	L1@0.30	Ea	42.60	12.30	54.90
20A, 277/480V, NEMA L19-20R	L1@0.25	Ea	29.30	10.30	39.60
30A, 277/480V, NEMA L19-30R	L1@0.30	Ea	42.70	12.30	55.00
20A, 600V, NEMA L20-20R	L1@0.25	Ea	29.30	10.30	39.60
30A, 600V, NEMA L20-30R	L1@0.30	Ea	42.80	12.30	55.10
4 pole, 5 wire, 3 phase Y grounding locking receptacles					
20A, 120/208V, NEMA L21-20R	L1@0.25	Ea	23.30	10.30	33.60
30A, 120/208V, NEMA L21-30R	L1@0.30	Ea	32.40	12.30	44.70
20A, 277/480V, NEMA L22-20R	L1@0.25	Ea	27.90	10.30	38.20
30A, 277/480V, NEMA L22-30R	L1@0.30	Ea	38.70	12.30	51.00
20A, 347/600V, NEMA L23-20R	L1@0.25	Ea	27.90	10.30	38.20
30A, 347/600V, NEMA L23-30R	L1@0.30	Ea	48.20	12.30	60.50

Use these figures to estimate the cost of locking receptacles under the conditions described on pages 5 and 6. Costs listed are for each receptacle installed. The crew is one electrician working at a labor cost of $41.03 per manhour. These costs include layout, material handling, and normal waste. Add for outlet boxes and switch rings, sales tax, delivery, supervision, mobilization, demobilization, cleanup, overhead, and profit. There's a special configuration for each voltage and amperage. Receptacles with special contact materials designed for durability under heavy use can be considerably more expensive than standard grade receptacles.

Material	Craft@Hrs	Unit	Material Cost	Labor Cost	Installed Cost
2 pole, 3 wire, grounding locking type connector bodies					
15A, 125V, NEMA L5-15R	L1@0.20	Ea	15.10	8.21	23.31
15A, 250V, NEMA L6-15R	L1@0.20	Ea	15.50	8.21	23.71
20A, 125V, NEMA L5-20R	L1@0.25	Ea	16.60	10.30	26.90
20A, 250V, NEMA L6-20R	L1@0.25	Ea	16.60	10.30	26.90
20A, 277V, NEMA L7-20R	L1@0.25	Ea	16.60	10.30	26.90
20A, 480V, NEMA L8-20R	L1@0.25	Ea	24.90	10.30	35.20
20A, 600V, NEMA L9-20R	L1@0.25	Ea	20.60	10.30	30.90
20A, 125/250V, NEMA L10-20R	L1@0.25	Ea	18.00	10.30	28.30
30A, 125V, NEMA L5-30R	L1@0.30	Ea	33.10	12.30	45.40
30A, 250V, NEMA L6-30R	L1@0.30	Ea	33.40	12.30	45.70
30A, 277V, NEMA L7-30R	L1@0.30	Ea	33.40	12.30	45.70
30A, 480V, NEMA L8-30R	L1@0.30	Ea	39.80	12.30	52.10
30A, 600V, NEMA L9-30R	L1@0.30	Ea	40.00	12.30	52.30
30A, 125/250V, NEMA L10-30R	L1@0.30	Ea	40.10	12.30	52.40
3 pole, 4 wire, grounding locking type connector bodies					
20A, 120/250V, NEMA L14-20R	L1@0.35	Ea	22.80	14.40	37.20
3 pole, 3 phase, grounding locking type connector bodies					
20A, 250V, NEMA L11-20R	L1@0.35	Ea	21.00	14.40	35.40
20A, 480V, NEMA L12-20R	L1@0.35	Ea	18.00	14.40	32.40
30A, 250V, NEMA L11-30R	L1@0.35	Ea	40.10	14.40	54.50
30A, 480V, NEMA L12-30R	L1@0.35	Ea	40.10	14.40	54.50
30A, 600V, NEMA L13-30R	L1@0.35	Ea	40.10	14.40	54.50

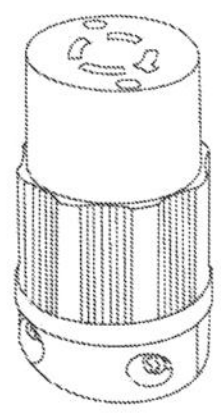

Use these figures to estimate the cost of locking connector bodies under the conditions described on pages 5 and 6. Costs listed are for each connector installed on a power cable. The crew is one electrician working at a labor cost of $41.03 per manhour. These costs include layout, material handling, and normal waste. Add sales tax, supervision, mobilization, demobilization, cleanup, overhead, and profit. Connectors with special contact materials designed for durability under heavy use can be considerably more expensive than standard grade connectors.

Plastic Locking Connectors

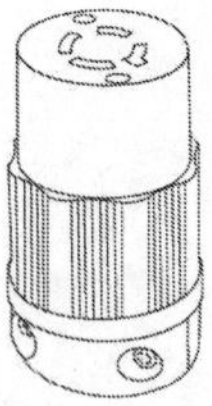

Material	Craft@Hrs	Unit	Material Cost	Labor Cost	Installed Cost
3 pole, 4 wire, 3 phase grounding locking connector bodies					
20A, 250V, NEMA L15-20R	L1@0.40	Ea	22.80	16.40	39.20
20A, 480V, NEMA L16-20R	L1@0.40	Ea	27.40	16.40	43.80
20A, 125/208V, NEMA L18-20R	L1@0.40	Ea	35.40	16.40	51.80
20A, 277/480V, NEMA L19-20R	L1@0.40	Ea	35.40	16.40	51.80
20A, 347/600V, NEMA L20-20R	L1@0.40	Ea	35.40	16.40	51.80
30A, 250V, NEMA L15-30R	L1@0.45	Ea	46.20	18.50	64.70
30A, 480V, NEMA L16-30R	L1@0.45	Ea	53.70	18.50	72.20
30A, 600V, NEMA L17-30R	L1@0.45	Ea	55.70	18.50	74.20
30A, 120/208V, NEMA L18-30R	L1@0.45	Ea	68.00	18.50	86.50
30A, 277/480V, NEMA L19-30R	L1@0.45	Ea	68.00	18.50	86.50
30A, 347/600V, NEMA L20-30R	L1@0.45	Ea	68.00	18.50	86.50
4 pole, 5 wire, 3 phase Y grounding locking connector bodies					
20A, 120/208V, NEMA L21-20R	L1@0.50	Ea	36.00	20.50	56.50
20A, 277/480V, NEMA L22-20R	L1@0.50	Ea	43.20	20.50	63.70
20A, 347/600V, NEMA L23-20R	L1@0.50	Ea	43.30	20.50	63.80
30A, 120/208V, NEMA L21-30R	L1@0.55	Ea	45.60	22.60	68.20
30A, 277/480V, NEMA L22-30R	L1@0.55	Ea	54.70	22.60	77.30
30A, 347/600V, NEMA L23-30R	L1@0.55	Ea	68.10	22.60	90.70

Use these figures to estimate the cost of locking connector bodies under the conditions described on pages 5 and 6. Costs listed are for each connector installed on a power cable. The crew is one electrician working at a labor cost of $41.03 per manhour. These costs include layout, material handling, and normal waste. Add sales tax, supervision, mobilization, demobilization, cleanup, overhead, and profit. Connectors with special contact materials designed for durability under heavy use can be considerably more expensive than standard grade connectors.

Material	Craft@Hrs	Unit	Material Cost	Labor Cost	Installed Cost
2 pole, 2 wire, non-grounding locking plugs					
15A, 125V, NEMA L1-15P	L1@0.15	Ea	9.12	6.15	15.27
20A, 250V, NEMA L2-20P	L1@0.20	Ea	8.06	8.21	16.27
2 pole, 3 wire, grounding locking plugs					
15A, 125V, NEMA L5-15P	L1@0.15	Ea	8.41	6.15	14.56
20A, 125V, NEMA L5-20P	L1@0.20	Ea	10.70	8.21	18.91
30A, 125V, NEMA L5-30P	L1@0.25	Ea	16.60	10.30	26.90
15A, 250V, NEMA L6-15P	L1@0.15	Ea	8.94	6.15	15.09
20A, 250V, NEMA L6-20P	L1@0.20	Ea	10.70	8.21	18.91
30A, 250V, NEMA L6-30P	L1@0.25	Ea	16.60	10.30	26.90
15A, 277V, NEMA L7-15P	L1@0.15	Ea	5.45	6.15	11.60
20A, 277V, NEMA L7-20P	L1@0.20	Ea	10.70	8.21	18.91
30A, 277V, NEMA L7-30P	L1@0.25	Ea	16.40	10.30	26.70
20A, 480V, NEMA L8-20P	L1@0.20	Ea	13.20	8.21	21.41
30A, 480V, NEMA L8-30P	L1@0.25	Ea	19.80	10.30	30.10
20A, 600V, NEMA L9-20P	L1@0.20	Ea	13.20	8.21	21.41
30A, 600V, NEMA L9-30P	L1@0.25	Ea	19.10	10.30	29.40
3 pole, 3 wire, non-grounding locking plugs					
20A, 125/250V, NEMA L10-20P	L1@0.20	Ea	12.60	8.21	20.81
30A, 125/250V, NEMA L10-30P	L1@0.25	Ea	18.30	10.30	28.60
3 pole, 3 wire, 3 phase non-grounding locking plugs					
20A, 250V, NEMA L11-20P	L1@0.20	Ea	14.50	8.21	22.71
30A, 250V, NEMA L11-30P	L1@0.25	Ea	20.50	10.30	30.80
20A, 480V, NEMA L12-20P	L1@0.20	Ea	12.60	8.21	20.81
30A, 480V, NEMA L12-30P	L1@0.25	Ea	20.50	10.30	30.80
30A, 600V, NEMA L13-30P	L1@0.25	Ea	20.50	10.30	30.80

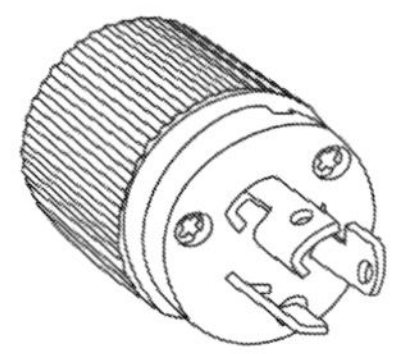

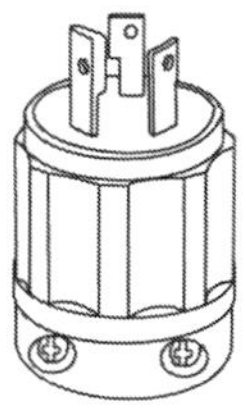

Use these figures to estimate the cost of locking plugs under the conditions described on pages 5 and 6. Costs listed are for each plug installed on a power cord. The crew is one electrician working at a labor cost of $41.03 per manhour. These costs include layout, material handling, and normal waste. Add for sales tax, supervision, mobilization, demobilization, cleanup, overhead, and profit. Plugs with special contact materials designed for durability under heavy use can be considerably more expensive than standard grade plugs.

Plastic Locking Plugs

Material	Craft@Hrs	Unit	Material Cost	Labor Cost	Installed Cost
3 pole, 4 wire, grounding locking plugs					
20A, 125/250V, NEMA L14-20P	L1@0.20	Ea	16.40	8.21	24.61
30A, 125/250V, NEMA L14-30P	L1@0.25	Ea	22.20	10.30	32.50
3 pole, 4 wire, 3 phase grounding locking plugs					
20A, 250V, NEMA L15-20P	L1@0.20	Ea	17.20	8.21	25.41
30A, 250V, NEMA L15-30P	L1@0.25	Ea	22.40	10.30	32.70
20A, 480V, NEMA L16-20P	L1@0.20	Ea	17.60	8.21	25.81
30A, 480V, NEMA L16-30P	L1@0.25	Ea	26.40	10.30	36.70
30A, 600V, NEMAL17-30P	L1@0.25	Ea	26.60	10.30	36.90
4 pole, 4 wire, 3 phase non-grounding locking plugs					
20A, 120/208V, NEMA L18-20P	L1@0.25	Ea	25.50	10.30	35.80
30A, 120/208V, NEMA L18-30P	L1@0.30	Ea	32.20	12.30	44.50
20A, 277/480V, NEMA L19-20P	L1@0.25	Ea	25.50	10.30	35.80
30A, 277/480V, NEMA L19-30P	L1@0.30	Ea	32.20	12.30	44.50
20A, 600V, NEMA L20-20P	L1@0.25	Ea	25.80	10.30	36.10
30A, 600V, NEMA L20-30P	L1@0.30	Ea	32.20	12.30	44.50
4 pole, 5 wire, 3 phase grounding locking plugs					
20A, 120/208V, NEMA L21-20P	L1@0.25	Ea	20.90	10.30	31.20
30A, 120/208V, NEMA L21-30P	L1@0.30	Ea	24.50	12.30	36.80
20A, 277/480V, NEMA L22-20P	L1@0.25	Ea	25.20	10.30	35.50
30A, 277/480V, NEMA L22-30P	L1@0.30	Ea	29.90	12.30	42.20
20A, 600V, NEMA L23-20P	L1@0.25	Ea	25.20	10.30	35.50
30A, 600V, NEMA L23-30P	L1@0.30	Ea	37.00	12.30	49.30

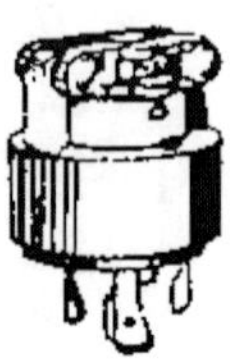

Use these figures to estimate the cost of locking plugs under the conditions described on pages 5 and 6. Costs listed are for each plug installed on a power cord. The crew is one electrician working at a labor cost of $41.03 per manhour. These costs include layout, material handling, and normal waste. Add for sales tax, supervision, mobilization, demobilization, cleanup, overhead, and profit. Plugs with special contact materials designed for durability under heavy use can be considerably more expensive than standard grade plugs.

Material	Craft@Hrs	Unit	Material Cost	Labor Cost	Installed Cost
Flush wire-in type					
1000W, 120V SPST	L1@0.20	Ea	8.76	8.21	16.97
1000W, 208/277V SPST	L1@0.20	Ea	10.60	8.21	18.81
Flush wire-in with single gang wall plate					
1000W, 120V SPST	L1@0.20	Ea	10.60	8.21	18.81
1000W, 208/277V SPST	L1@0.20	Ea	13.80	8.21	22.01
Adjustable light control, wall plate					
1800W, 120V SPST	L1@0.25	Ea	11.50	10.30	21.80
1800W, 208/277V SPST	L1@0.25	Ea	14.10	10.30	24.40
1800W, 480V SPST	L1@0.25	Ea	21.30	10.30	31.60
2000W, 120V SPST	L1@0.25	Ea	11.10	10.30	21.40
2000W, 208/277V SPST	L1@0.25	Ea	13.20	10.30	23.50
2000W, 480V SPST	L1@0.25	Ea	19.80	10.30	30.10
3000W, 120V SPST	L1@0.25	Ea	20.00	10.30	30.30
3000W, 208/277V SPST	L1@0.25	Ea	20.70	10.30	31.00
Plug-in locking type, delayed response, thermal					
1800W, 120V SPST	L1@0.25	Ea	7.16	10.30	17.46
1800W, 208/277V SPST	L1@0.25	Ea	8.76	10.30	19.06
1800W, 480V SPST	L1@0.25	Ea	23.00	10.30	33.30
Plug-in locking type, delayed response, thermal, low maintenance					
1800W, 120V SPST	L1@0.25	Ea	15.80	10.30	26.10
1800W, 208/277V SPST	L1@0.25	Ea	18.40	10.30	28.70
1800W, 480V SPST	L1@0.25	Ea	25.10	10.30	35.40

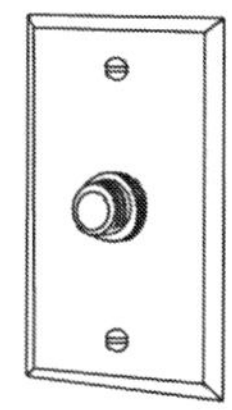

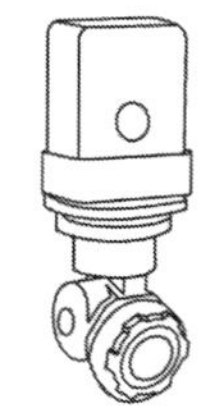

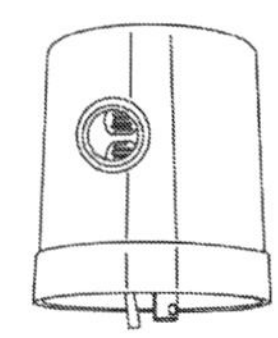

Use these figures to estimate the cost of photo controls under the conditions described on pages 5 and 6. Costs are for each photo control installed on a box or lighting fixture. The crew is one electrician working at a labor cost of $41.03 per manhour. These costs include layout, material handling, and normal waste. Add for boxes, covers, sales tax, supervision, mobilization, demobilization, cleanup, overhead, and profit.

Wiring Device Plates

Switch cover plates

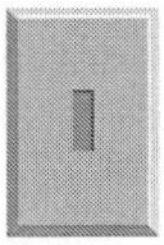

Material	Craft@Hrs	Unit	Material Cost	Labor Cost	Installed Cost
1 gang brown	L1@0.05	Ea	.32	2.05	2.37
1 gang ivory	L1@0.05	Ea	.32	2.05	2.37
1 gang gray	L1@0.05	Ea	.35	2.05	2.40
1 gang white	L1@0.05	Ea	.35	2.05	2.40
1 gang red	L1@0.05	Ea	.35	2.05	2.40
1 gang black	L1@0.05	Ea	.35	2.05	2.40
1 gang brass	L1@0.05	Ea	3.87	2.05	5.92
1 gang aluminum	L1@0.05	Ea	1.93	2.05	3.98
1 gang stainless steel	L1@0.05	Ea	1.53	2.05	3.58

Material	Craft@Hrs	Unit	Material Cost	Labor Cost	Installed Cost
2 gang brown	L1@0.10	Ea	.65	4.10	4.75
2 gang ivory	L1@0.10	Ea	.65	4.10	4.75
2 gang gray	L1@0.10	Ea	.65	4.10	4.75
2 gang white	L1@0.10	Ea	.72	4.10	4.82
2 gang red	L1@0.10	Ea	.72	4.10	4.82
2 gang black	L1@0.10	Ea	.72	4.10	4.82
2 gang brass	L1@0.10	Ea	8.55	4.10	12.65
2 gang aluminum	L1@0.10	Ea	3.87	4.10	7.97
2 gang stainless steel	L1@0.10	Ea	2.94	4.10	7.04

Material	Craft@Hrs	Unit	Material Cost	Labor Cost	Installed Cost
3 gang brown	L1@0.15	Ea	.97	6.15	7.12
3 gang ivory	L1@0.15	Ea	.97	6.15	7.12
3 gang gray	L1@0.15	Ea	.97	6.15	7.12
3 gang white	L1@0.15	Ea	1.06	6.15	7.21
3 gang red	L1@0.15	Ea	1.06	6.15	7.21
3 gang black	L1@0.15	Ea	1.06	6.15	7.21
3 gang brass	L1@0.15	Ea	13.30	6.15	19.45
3 gang aluminum	L1@0.15	Ea	6.61	6.15	12.76
3 gang stainless steel	L1@0.15	Ea	4.38	6.15	10.53

Material	Craft@Hrs	Unit	Material Cost	Labor Cost	Installed Cost
4 gang brown	L1@0.20	Ea	1.42	8.21	9.63
4 gang ivory	L1@0.20	Ea	1.42	8.21	9.63
4 gang gray	L1@0.20	Ea	1.63	8.21	9.84
4 gang white	L1@0.20	Ea	1.63	8.21	9.84
4 gang red	L1@0.20	Ea	1.63	8.21	9.84
4 gang black	L1@0.20	Ea	1.63	8.21	9.84
4 gang brass	L1@0.20	Ea	19.70	8.21	27.91
4 gang aluminum	L1@0.20	Ea	9.98	8.21	18.19
4 gang stainless steel	L1@0.20	Ea	7.54	8.21	15.75

Use these figures to estimate the cost of trim plates installed on outlet boxes under the conditions described on pages 5 and 6. Costs listed are for each plate installed. The crew is one electrician working at a labor cost of $41.03 per manhour. These costs include layout, material handling, and normal waste. Add for the outlet box, switch, plaster ring, sales tax, delivery, supervision, mobilization, demobilization, cleanup, overhead and profit. Note: Be careful to select the right plate color and material. Special plates with non-standard configurations can be ordered for custom applications.

Switch cover plates (continued)

Material	Craft@Hrs	Unit	Material Cost	Labor Cost	Installed Cost
5 gang brown	L1@0.25	Ea	2.66	10.30	12.96
5 gang ivory	L1@0.25	Ea	2.66	10.30	12.96
5 gang gray	L1@0.25	Ea	3.06	10.30	13.36
5 gang white	L1@0.25	Ea	3.06	10.30	13.36
5 gang red	L1@0.25	Ea	3.06	10.30	13.36
5 gang black	L1@0.25	Ea	3.06	10.30	13.36
5 gang brass	L1@0.25	Ea	24.50	10.30	34.80
5 gang aluminum	L1@0.25	Ea	12.40	10.30	22.70
5 gang stainless steel	L1@0.25	Ea	9.36	10.30	19.66
6 gang brown	L1@0.30	Ea	3.14	12.30	15.44
6 gang ivory	L1@0.30	Ea	3.14	12.30	15.44
6 gang gray	L1@0.30	Ea	3.55	12.30	15.85
6 gang white	L1@0.30	Ea	3.55	12.30	15.85
6 gang red	L1@0.30	Ea	3.55	12.30	15.85
6 gang black	L1@0.30	Ea	3.55	12.30	15.85
6 gang brass	L1@0.30	Ea	28.80	12.30	41.10
6 gang aluminum	L1@0.30	Ea	17.30	12.30	29.60
6 gang stainless steel	L1@0.30	Ea	11.20	12.30	23.50

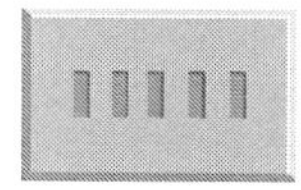

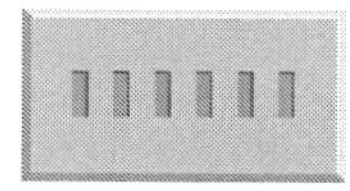

Duplex receptacle cover plates

Material	Craft@Hrs	Unit	Material Cost	Labor Cost	Installed Cost
1 gang brown	L1@0.05	Ea	.32	2.05	2.37
1 gang ivory	L1@0.05	Ea	.32	2.05	2.37
1 gang gray	L1@0.05	Ea	.35	2.05	2.40
1 gang white	L1@0.05	Ea	.35	2.05	2.40
1 gang red	L1@0.05	Ea	.35	2.05	2.40
1 gang black	L1@0.05	Ea	.35	2.05	2.40
1 gang brass	L1@0.05	Ea	3.87	2.05	5.92
1 gang aluminum	L1@0.05	Ea	1.93	2.05	3.98
1 gang stainless steel	L1@0.05	Ea	1.53	2.05	3.58
2 gang brown	L1@0.10	Ea	.72	4.10	4.82
2 gang ivory	L1@0.10	Ea	.72	4.10	4.82
2 gang gray	L1@0.10	Ea	.81	4.10	4.91
2 gang white	L1@0.10	Ea	.81	4.10	4.91
2 gang red	L1@0.10	Ea	.81	4.10	4.91
2 gang black	L1@0.10	Ea	.81	4.10	4.91
2 gang brass	L1@0.10	Ea	9.57	4.10	13.67
2 gang aluminum	L1@0.10	Ea	9.16	4.10	13.26
2 gang stainless steel	L1@0.10	Ea	3.67	4.10	7.77

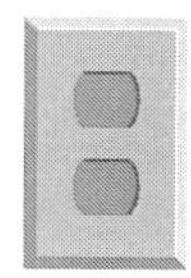

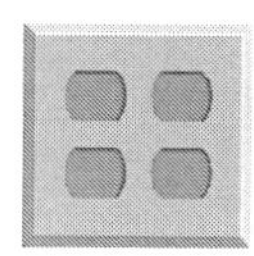

Use these figures to estimate the cost of trim plates installed on outlet boxes under the conditions described on pages 5 and 6. Costs listed are for each plate installed. The crew is one electrician working at a labor cost of $41.03 per manhour. These costs include layout, material handling, and normal waste. Add for the outlet box, switch, plaster ring, sales tax, delivery, supervision, mobilization, demobilization, cleanup, overhead and profit. Note: Be careful to select the right plate color and material. Special plates with non-standard configurations can be ordered for custom applications.

Wiring Device Plates

Combination switch and duplex receptacle cover plates

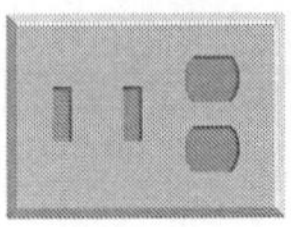

Material	Craft@Hrs	Unit	Material Cost	Labor Cost	Installed Cost
2 gang brown	L1@0.10	Ea	.65	4.10	4.75
2 gang ivory	L1@0.10	Ea	.65	4.10	4.75
2 gang gray	L1@0.10	Ea	.72	4.10	4.82
2 gang white	L1@0.10	Ea	.72	4.10	4.82
2 gang red	L1@0.10	Ea	.72	4.10	4.82
2 gang black	L1@0.10	Ea	.72	4.10	4.82
2 gang brass	L1@0.10	Ea	8.84	4.10	12.94
2 gang aluminum	L1@0.10	Ea	4.38	4.10	8.48
2 gang stainless steel	L1@0.10	Ea	3.25	4.10	7.35
3 gang brown	L1@0.15	Ea	1.14	6.15	7.29
3 gang ivory	L1@0.15	Ea	1.14	6.15	7.29
3 gang gray	L1@0.15	Ea	1.28	6.15	7.43
3 gang white	L1@0.15	Ea	1.28	6.15	7.43
3 gang red	L1@0.15	Ea	1.28	6.15	7.43
3 gang black	L1@0.15	Ea	1.28	6.15	7.43
3 gang brass	L1@0.15	Ea	13.30	6.15	19.45
3 gang aluminum	L1@0.15	Ea	7.42	6.15	13.57
3 gang stainless steel	L1@0.15	Ea	4.88	6.15	11.03

Single receptacle cover plates

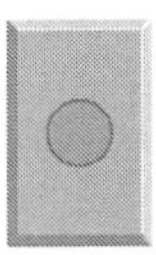

Material	Craft@Hrs	Unit	Material Cost	Labor Cost	Installed Cost
1 gang brown	L1@0.05	Ea	.33	2.05	2.38
1 gang ivory	L1@0.05	Ea	.33	2.05	2.38
1 gang gray	L1@0.05	Ea	.37	2.05	2.42
1 gang white	L1@0.05	Ea	.37	2.05	2.42
1 gang red	L1@0.05	Ea	.37	2.05	2.42
1 gang black	L1@0.05	Ea	.37	2.05	2.42
1 gang brass	L1@0.05	Ea	4.27	2.05	6.32
1 gang aluminum	L1@0.05	Ea	2.03	2.05	4.08
1 gang stainless steel	L1@0.05	Ea	1.63	2.05	3.68

Blank wiring device cover plates

Material	Craft@Hrs	Unit	Material Cost	Labor Cost	Installed Cost
1 gang brown	L1@0.05	Ea	.41	2.05	2.46
1 gang ivory	L1@0.05	Ea	.41	2.05	2.46
1 gang gray	L1@0.05	Ea	.49	2.05	2.54
1 gang white	L1@0.05	Ea	.49	2.05	2.54
1 gang red	L1@0.05	Ea	.49	2.05	2.54
1 gang black	L1@0.05	Ea	.49	2.05	2.54
1 gang brass	L1@0.05	Ea	4.62	2.05	6.67
1 gang aluminum	L1@0.05	Ea	2.03	2.05	4.08
1 gang stainless steel	L1@0.05	Ea	1.63	2.05	3.68
2 gang brown	L1@0.10	Ea	1.44	4.10	5.54
2 gang ivory	L1@0.10	Ea	1.44	4.10	5.54
2 gang gray	L1@0.10	Ea	1.66	4.10	5.76
2 gang white	L1@0.10	Ea	1.66	4.10	5.76
2 gang red	L1@0.10	Ea	1.66	4.10	5.76
2 gang black	L1@0.10	Ea	1.66	4.10	5.76

Use these figures to estimate the cost of trim plates installed on outlet boxes under the conditions described on pages 5 and 6. Costs listed are for each plate installed. The crew is one electrician working at a labor cost of $41.03 per manhour. These costs include layout, material handling, and normal waste. Add for the outlet box, receptacle and switch, plaster ring, sales tax, delivery, supervision, mobilization, demobilization, cleanup, overhead and profit. Note: Be careful to select the right plate color and material. Special plates with non-standard configurations can be ordered for custom applications.

Blank wiring device plates

Material	Craft@Hrs	Unit	Material Cost	Labor Cost	Installed Cost
2 gang brass	L1@0.10	Ea	10.40	4.10	14.50
2 gang aluminum	L1@0.10	Ea	3.32	4.10	7.42
2 gang stainless steel	L1@0.10	Ea	1.55	4.10	5.65

Telephone wiring device plates

Material	Craft@Hrs	Unit	Material Cost	Labor Cost	Installed Cost
1 gang brown	L1@0.05	Ea	.39	2.05	2.44
1 gang ivory	L1@0.05	Ea	.39	2.05	2.44
1 gang gray	L1@0.05	Ea	.47	2.05	2.52
1 gang white	L1@0.05	Ea	.47	2.05	2.52
1 gang red	L1@0.05	Ea	.47	2.05	2.52
1 gang brass	L1@0.05	Ea	4.44	2.05	6.49
1 gang aluminum	L1@0.05	Ea	2.09	2.05	4.14
1 gang stainless steel	L1@0.05	Ea	1.59	2.05	3.64

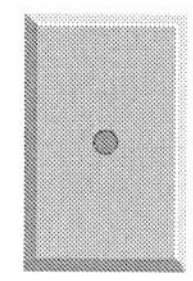

Decorator wiring device plates

Material	Craft@Hrs	Unit	Material Cost	Labor Cost	Installed Cost
1 gang brown	L1@0.05	Ea	.59	2.05	2.64
1 gang ivory	L1@0.05	Ea	.54	2.05	2.59
1 gang gray	L1@0.05	Ea	.72	2.05	2.77
1 gang white	L1@0.05	Ea	.54	2.05	2.59
1 gang red	L1@0.05	Ea	.74	2.05	2.79
1 gang black	L1@0.05	Ea	.72	2.05	2.77
1 gang brass	L1@0.05	Ea	3.78	2.05	5.83
1 gang aluminum	L1@0.05	Ea	2.18	2.05	4.23
1 gang stainless steel	L1@0.05	Ea	1.64	2.05	3.69
2 gang brown	L1@0.10	Ea	1.35	4.10	5.45
2 gang ivory	L1@0.10	Ea	1.20	4.10	5.30
2 gang gray	L1@0.10	Ea	1.44	4.10	5.54
2 gang white	L1@0.10	Ea	1.08	4.10	5.18
2 gang red	L1@0.10	Ea	1.44	4.10	5.54
2 gang black	L1@0.10	Ea	1.44	4.10	5.54
2 gang brass	L1@0.10	Ea	7.63	4.10	11.73
2 gang aluminum	L1@0.10	Ea	4.37	4.10	8.47
2 gang stainless steel	L1@0.10	Ea	4.16	4.10	8.26
3 gang brown	L1@0.15	Ea	3.17	6.15	9.32
3 gang ivory	L1@0.15	Ea	3.17	6.15	9.32
3 gang gray	L1@0.15	Ea	3.17	6.15	9.32
3 gang white	L1@0.15	Ea	3.17	6.15	9.32
3 gang red	L1@0.15	Ea	3.17	6.15	9.32
3 gang black	L1@0.15	Ea	3.17	6.15	9.32
3 gang brass	L1@0.15	Ea	16.30	6.15	22.45
3 gang stainless steel	L1@0.15	Ea	6.37	6.15	12.52

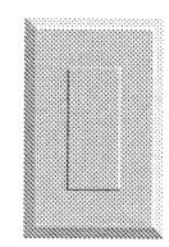

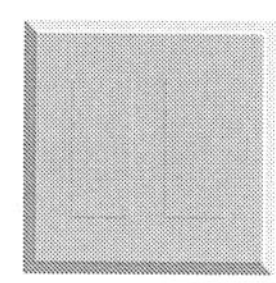

Use these figures to estimate the cost of trim plates installed on outlet boxes under the conditions described on pages 5 and 6. Costs listed are for each plate installed. The crew is one electrician working at a labor cost of $41.03 per manhour. These costs include layout, material handling, and normal waste. Add for the outlet box, plaster ring, sales tax, delivery, supervision, mobilization, demobilization, cleanup, overhead and profit. Note: Be careful to select the right plate color and material. Special plates with non-standard configurations can be ordered for custom applications.

Wiring Device Plates

Decorator wiring device plates

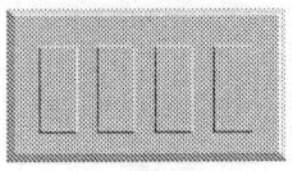

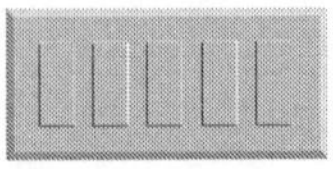

Material	Craft@Hrs	Unit	Material Cost	Labor Cost	Installed Cost
4 gang brown	L1@0.20	Ea	5.68	8.21	13.89
4 gang ivory	L1@0.20	Ea	5.68	8.21	13.89
4 gang gray	L1@0.20	Ea	5.68	8.21	13.89
4 gang white	L1@0.20	Ea	5.68	8.21	13.89
4 gang black	L1@0.20	Ea	5.68	8.21	13.89
4 gang brass	L1@0.20	Ea	20.30	8.21	28.51
4 gang stainless steel	L1@0.20	Ea	8.45	8.21	16.66
5 gang brown	L1@0.25	Ea	7.98	10.30	18.28
5 gang ivory	L1@0.25	Ea	7.98	10.30	18.28
5 gang gray	L1@0.25	Ea	7.98	10.30	18.28
5 gang white	L1@0.25	Ea	7.98	10.30	18.28
5 gang red	L1@0.25	Ea	7.98	10.30	18.28
5 gang black	L1@0.25	Ea	7.98	10.30	18.28
5 gang brass	L1@0.25	Ea	27.40	10.30	37.70
5 gang stainless steel	L1@0.25	Ea	25.40	10.30	35.70

Combination decorator and standard switch plates

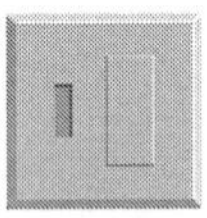

Material	Craft@Hrs	Unit	Material Cost	Labor Cost	Installed Cost
2 gang brown	L1@0.10	Ea	1.39	4.10	5.49
2 gang ivory	L1@0.10	Ea	1.39	4.10	5.49
2 gang gray	L1@0.10	Ea	1.44	4.10	5.54
2 gang white	L1@0.10	Ea	1.44	4.10	5.54
2 gang red	L1@0.10	Ea	1.44	4.10	5.54
2 gang black	L1@0.10	Ea	1.44	4.10	5.54
2 gang brass	L1@0.10	Ea	8.24	4.10	12.34
2 gang aluminum	L1@0.10	Ea	4.37	4.10	8.47
2 gang stainless steel	L1@0.10	Ea	4.16	4.10	8.26

Combination decorator and two standard switch plates

Material	Craft@Hrs	Unit	Material Cost	Labor Cost	Installed Cost
3 gang brown	L1@0.15	Ea	2.80	6.15	8.95
3 gang ivory	L1@0.15	Ea	2.80	6.15	8.95
3 gang gray	L1@0.15	Ea	2.80	6.15	8.95
3 gang white	L1@0.15	Ea	2.80	6.15	8.95
3 gang brass	L1@0.15	Ea	12.40	6.15	18.55
3 gang aluminum	L1@0.15	Ea	6.62	6.15	12.77
3 gang stainless steel	L1@0.15	Ea	6.30	6.15	12.45

Use these figures to estimate the cost of switch trim plates installed on outlet boxes under the conditions described on pages 5 and 6. Costs listed are for each plate installed. The crew is one electrician working at a labor cost of $41.03 per manhour. These costs include layout, material handling, and normal waste. Add for the outlet box, switch, plaster ring, sales tax, delivery, supervision, mobilization, demobilization, cleanup, overhead and profit. Note: Be careful to select the right plate color and material. Special plates with non-standard configurations can be ordered for custom applications.

Combination decorator and three standard switch plates

Material	Craft@Hrs	Unit	Material Cost	Labor Cost	Installed Cost
4 gang brown	L1@0.20	Ea	6.25	8.21	14.46
4 gang ivory	L1@0.20	Ea	6.25	8.21	14.46
4 gang white	L1@0.20	Ea	6.25	8.21	14.46

Semi-jumbo switch plates

Material	Craft@Hrs	Unit	Material Cost	Labor Cost	Installed Cost
1 gang brown	L1@0.05	Ea	.89	2.05	2.94
1 gang ivory	L1@0.05	Ea	.89	2.05	2.94
1 gang white	L1@0.05	Ea	.97	2.05	3.02
1 gang gray	L1@0.05	Ea	.97	2.05	3.02
2 gang brown	L1@0.10	Ea	1.93	4.10	6.03
2 gang ivory	L1@0.10	Ea	1.93	4.10	6.03
2 gang white	L1@0.10	Ea	1.93	4.10	6.03
2 gang gray	L1@0.10	Ea	1.93	4.10	6.03
3 gang brown	L1@0.15	Ea	2.88	6.15	9.03
3 gang ivory	L1@0.15	Ea	2.88	6.15	9.03
3 gang white	L1@0.15	Ea	2.88	6.15	9.03
3 gang gray	L1@0.15	Ea	2.88	6.15	9.03

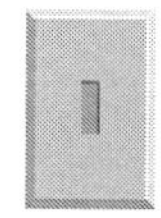

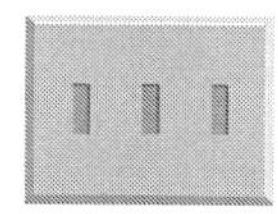

Semi-jumbo duplex wiring device plates

Material	Craft@Hrs	Unit	Material Cost	Labor Cost	Installed Cost
1 gang brown	L1@0.05	Ea	.80	2.05	2.85
1 gang ivory	L1@0.05	Ea	.80	2.05	2.85
1 gang white	L1@0.05	Ea	.97	2.05	3.02
1 gang gray	L1@0.05	Ea	.97	2.05	3.02

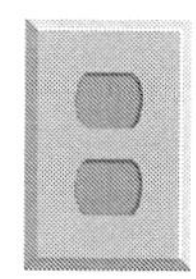

Semi-jumbo single receptacle plates

Material	Craft@Hrs	Unit	Material Cost	Labor Cost	Installed Cost
1 gang brown	L1@0.05	Ea	1.07	2.05	3.12
1 gang ivory	L1@0.05	Ea	1.07	2.05	3.12
1 gang white	L1@0.05	Ea	1.07	2.05	3.12
1 gang gray	L1@0.05	Ea	1.07	2.05	3.12

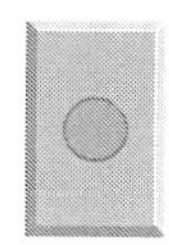

Semi-jumbo double duplex wiring device plates

Material	Craft@Hrs	Unit	Material Cost	Labor Cost	Installed Cost
2 gang brown	L1@0.10	Ea	1.94	4.10	6.04
2 gang ivory	L1@0.10	Ea	1.94	4.10	6.04
2 gang white	L1@0.10	Ea	1.94	4.10	6.04
2 gang gray	L1@0.10	Ea	1.94	4.10	6.04

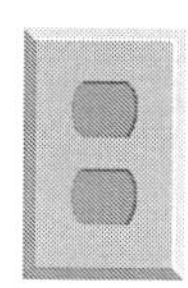

Use these figures to estimate the cost of switch and receptacle trim plates installed on outlet boxes under the conditions described on pages 5 and 6. Costs listed are for each plate installed. The crew is one electrician working at a labor cost of $41.03 per manhour. These costs include layout, material handling, and normal waste. Add for the outlet box, switch and receptacle, plaster ring, sales tax, delivery, supervision, mobilization, demobilization, cleanup, overhead and profit. Note: Be careful to select the right plate color and material. Special plates with non-standard configurations can be ordered for custom applications.

Wiring Device Plates

Semi-jumbo decorator wiring device plates

Material	Craft@Hrs	Unit	Material Cost	Labor Cost	Installed Cost
1 gang brown	L1@0.05	Ea	.97	2.05	3.02
1 gang ivory	L1@0.05	Ea	.97	2.05	3.02
1 gang white	L1@0.05	Ea	.97	2.05	3.02
2 gang brown	L1@0.10	Ea	2.85	4.10	6.95
2 gang ivory	L1@0.10	Ea	2.85	4.10	6.95
2 gang white	L1@0.10	Ea	2.85	4.10	6.95
3 gang brown	L1@0.15	Ea	5.58	6.15	11.73
3 gang ivory	L1@0.15	Ea	5.58	6.15	11.73
3 gang white	L1@0.15	Ea	5.58	6.15	11.73

Jumbo switch plates

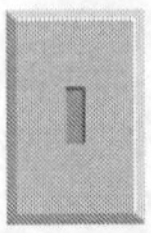
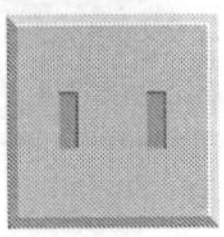

Material	Craft@Hrs	Unit	Material Cost	Labor Cost	Installed Cost
1 gang brown	L1@0.05	Ea	1.29	2.05	3.34
1 gang ivory	L1@0.05	Ea	1.14	2.05	3.19
1 gang white	L1@0.05	Ea	1.29	2.05	3.34
2 gang brown	L1@0.10	Ea	2.84	4.10	6.94
2 gang ivory	L1@0.10	Ea	2.56	4.10	6.66

Jumbo duplex wiring device plates

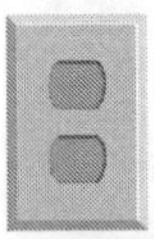
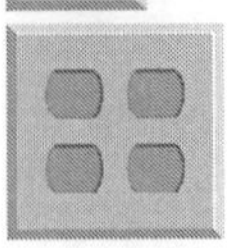

Material	Craft@Hrs	Unit	Material Cost	Labor Cost	Installed Cost
1 gang brown	L1@0.05	Ea	1.29	2.05	3.34
1 gang ivory	L1@0.05	Ea	1.14	2.05	3.19
1 gang white	L1@0.05	Ea	1.29	2.05	3.34
2 gang brown	L1@0.10	Ea	2.84	4.10	6.94
2 gang ivory	L1@0.10	Ea	2.56	4.10	6.66

Jumbo combination switch and duplex receptacle plates

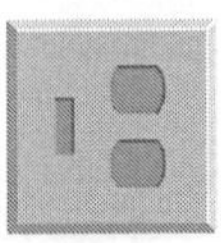

Material	Craft@Hrs	Unit	Material Cost	Labor Cost	Installed Cost
2 gang brown	L1@0.10	Ea	2.84	4.10	6.94
2 gang ivory	L1@0.10	Ea	2.56	4.10	6.66

Jumbo combination switch and blank wiring device plates

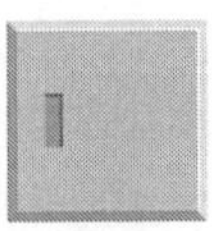

Material	Craft@Hrs	Unit	Material Cost	Labor Cost	Installed Cost
2 gang brown	L1@0.10	Ea	3.38	4.10	7.48
2 gang ivory	L1@0.10	Ea	3.38	4.10	7.48

Jumbo combination duplex and blank wiring device plates

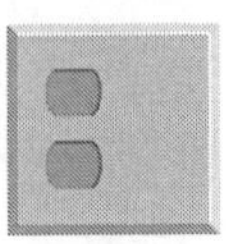

Material	Craft@Hrs	Unit	Material Cost	Labor Cost	Installed Cost
2 gang brown	L1@0.10	Ea	2.96	4.10	7.06
2 gang ivory	L1@0.10	Ea	2.96	4.10	7.06

Jumbo blank wiring device plates

Material	Craft@Hrs	Unit	Material Cost	Labor Cost	Installed Cost
1 gang brown	L1@0.05	Ea	1.72	2.05	3.77
1 gang ivory	L1@0.05	Ea	1.55	2.05	3.60
2 gang brown	L1@0.10	Ea	3.43	4.10	7.53
2 gang ivory	L1@0.10	Ea	3.43	4.10	7.53

Use these figures to estimate the cost of switch and receptacle trim plates installed on outlet boxes under the conditions described on pages 5 and 6. Costs listed are for each plate installed. The crew is one electrician working at a labor cost of $41.03 per manhour. These costs include layout, material handling, and normal waste. Add for the outlet box, switch and receptacle, plaster ring, sales tax, delivery, supervision, mobilization, demobilization, cleanup, overhead and profit. Note: Be careful to select the right plate color and material. Special plates with non-standard configurations can be ordered for custom applications.

Jumbo decorator wiring device plates

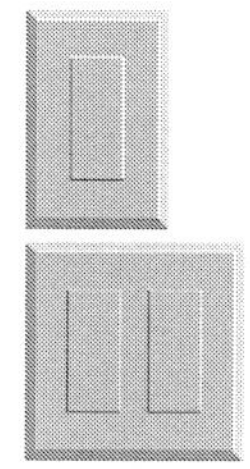

Material	Craft@Hrs	Unit	Material Cost	Labor Cost	Installed Cost
1 gang brown	L1@0.05	Ea	1.40	2.05	3.45
1 gang ivory	L1@0.05	Ea	1.29	2.05	3.34
1 gang white	L1@0.05	Ea	1.40	2.05	3.45
2 gang brown	L1@0.10	Ea	2.85	4.10	6.95
2 gang ivory	L1@0.10	Ea	2.85	4.10	6.95
2 gang white	L1@0.10	Ea	2.85	4.10	6.95

Jumbo combination decorator and blank wiring device plates

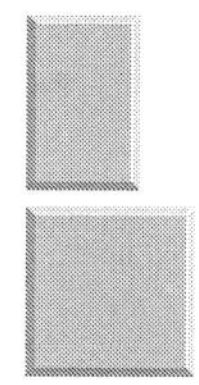

Material	Craft@Hrs	Unit	Material Cost	Labor Cost	Installed Cost
2 gang brown	L1@0.10	Ea	3.38	4.10	7.48
2 gang ivory	L1@0.10	Ea	3.38	4.10	7.48
2 gang white	L1@0.10	Ea	3.38	4.10	7.48

Deep switch plates

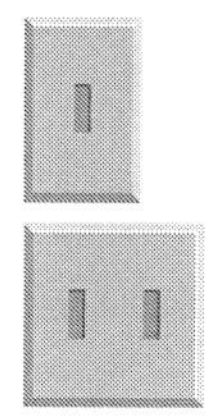

Material	Craft@Hrs	Unit	Material Cost	Labor Cost	Installed Cost
1 gang brown	L1@0.05	Ea	.63	2.05	2.68
1 gang ivory	L1@0.05	Ea	.58	2.05	2.63
2 gang brown	L1@0.10	Ea	1.24	4.10	5.34
2 gang ivory	L1@0.10	Ea	1.13	4.10	5.23

Deep duplex receptacle plates

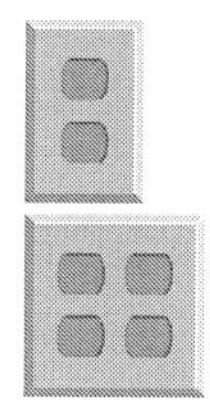

Material	Craft@Hrs	Unit	Material Cost	Labor Cost	Installed Cost
1 gang brown	L1@0.05	Ea	.58	2.05	2.63
1 gang ivory	L1@0.05	Ea	.58	2.05	2.63
2 gang brown	L1@0.10	Ea	1.38	4.10	5.48
2 gang ivory	L1@0.10	Ea	1.13	4.10	5.23

Deep combination switch and duplex receptacle plates

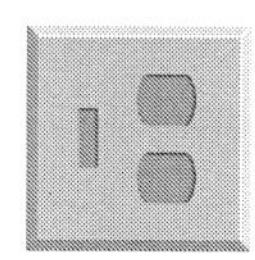

Material	Craft@Hrs	Unit	Material Cost	Labor Cost	Installed Cost
2 gang brown	L1@0.10	Ea	1.50	4.10	5.60
2 gang ivory	L1@0.10	Ea	1.36	4.10	5.46

Deep combination switch and blank wiring device plates

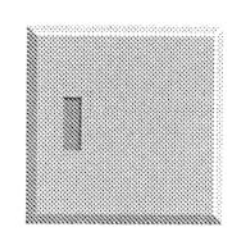

Material	Craft@Hrs	Unit	Material Cost	Labor Cost	Installed Cost
2 gang brown	L1@0.10	Ea	1.50	4.10	5.60
2 gang ivory	L1@0.10	Ea	1.36	4.10	5.46

Deep combination duplex receptacle and blank wiring device plates

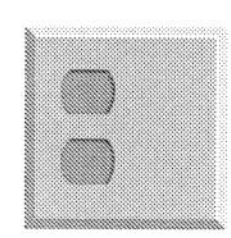

Material	Craft@Hrs	Unit	Material Cost	Labor Cost	Installed Cost
2 gang brown	L1@0.10	Ea	1.50	4.10	5.60
2 gang ivory	L1@0.10	Ea	1.36	4.10	5.46

15 amp size single receptacle plates

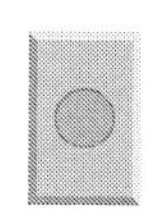

Material	Craft@Hrs	Unit	Material Cost	Labor Cost	Installed Cost
1 gang brass	L1@0.05	Ea	9.59	2.05	11.64
1 gang aluminum	L1@0.05	Ea	2.58	2.05	4.63
1 gang stainless 430	L1@0.05	Ea	2.03	2.05	4.08
1 gang stainless 302	L1@0.05	Ea	3.18	2.05	5.23

Use these figures to estimate the cost of switch and receptacle trim plates installed on outlet boxes under the conditions described on pages 5 and 6. Costs listed are for each plate installed. The crew is one electrician working at a labor cost of $41.03 per manhour. These costs include layout, material handling, and normal waste. Add for the outlet box, switch and receptacle, plaster ring, sales tax, delivery, supervision, mobilization, demobilization, cleanup, overhead and profit. Note: Use deep plates when the wall has been built out away from the outlet box and the screws from a standard plate won't reach the box. This is common in remodeling work when a wall has been covered with new paneling.

Wiring Device Plates

20 amp size single receptacle plates

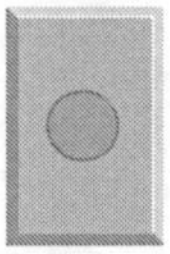

Material	Craft@Hrs	Unit	Material Cost	Labor Cost	Installed Cost
1 gang brass	L1@0.05	Ea	9.60	2.05	11.65
1 gang stainless 430	L1@0.05	Ea	2.33	2.05	4.38
1 gang stainless 302	L1@0.05	Ea	2.33	2.05	4.38

30 amp size single receptacle plates

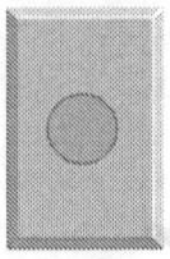

Material	Craft@Hrs	Unit	Material Cost	Labor Cost	Installed Cost
1 gang brass	L1@0.05	Ea	9.59	2.05	11.64
1 gang stainless 430	L1@0.05	Ea	2.62	2.05	4.67
1 gang stainless 302	L1@0.05	Ea	2.69	2.05	4.74

50 amp size single receptacle plates

Material	Craft@Hrs	Unit	Material Cost	Labor Cost	Installed Cost
2 gang brass	L1@0.10	Ea	11.70	4.10	15.80
2 gang stainless 430	L1@0.10	Ea	4.42	4.10	8.52

Weatherproof single receptacle plates

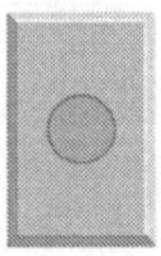

WP

Material	Craft@Hrs	Unit	Material Cost	Labor Cost	Installed Cost
15A cast metal	L1@0.10	Ea	24.00	4.10	28.10
20A cast metal	L1@0.10	Ea	25.40	4.10	29.50

Weatherproof horizontal duplex receptacle plates with twin covers

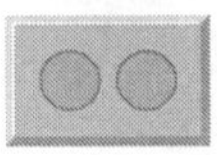

WP

Material	Craft@Hrs	Unit	Material Cost	Labor Cost	Installed Cost
1 gang cast metal	L1@0.10	Ea	21.50	4.10	25.60

Use these figures to estimate the cost of switch and receptacle trim plates installed on outlet boxes under the conditions described on pages 5 and 6. Costs listed are for each plate installed. The crew is one electrician working at a labor cost of $41.03 per manhour. These costs include layout, material handling, and normal waste. Add for the outlet box, switch and receptacle, plaster ring, sales tax, delivery, supervision, mobilization, demobilization, cleanup, overhead and profit. Note: Be careful to select the right plate color and material. Special plates with non-standard configurations can be ordered for custom applications.

Section 6: Service Entrance Equipment

This section deals with the equipment that's located at the point where electrical service enters the building — hence the name service entrance equipment. It includes safety switches, circuit breakers, a meter socket for mounting the utility company's meter, perhaps a transformer, at least one panelboard for distribution to the various loads, and wireway. This gear may be housed in a single steel cabinet or it may consist of several components, each with a separate enclosure.

Service entrance gear is custom-designed for larger commercial and industrial buildings. The type, size and ampacity of the equipment depend on the power that's needed, of course, and on requirements imposed by the local inspection authority and the electrical utility. Both the inspector and the electric company have standards that must be met. Be sure the equipment you're pricing meets those standards.

Description of Service Entrance Equipment

Figure 6-1 shows service entrance equipment for a small home. Figures 6-2 and 6-3 show service entrance gear that might be used in commercial and larger industrial buildings.

Notice in Figures 6-1, 6-2 and 6-3 that all three installations have two main parts: the service section and the distribution section.

The **service section** connects to the utility company power feed and outputs power to the distribution section. It includes either a cable pull section where the feed is pulled into the service entrance cabinet from underground power lines, or conduit to a service entrance cap which receives lines from an overhead distribution system.

The main disconnect or circuit breaker is also located in the service section. It provides a quick and convenient way to cut off power to all circuits when necessary, or when an overload endangers the system. The *NEC* requires that each conductor entering the service entrance section have a readily accessible disconnect. Exceptions are made for very small installations.

Also in the service section will be the meter socket which receives the utility company's meter. The service side may include instrumentation such as a voltmeter or ammeter, a fire alarm breaker, and perhaps a transformer if needed to reduce distribution voltage to the voltage required by loads in the building.

The **distribution section** is usually mounted right beside the service section and is connected by a set of metal bars called **bus bars**. These bars carry power from the service section to the distribution section.

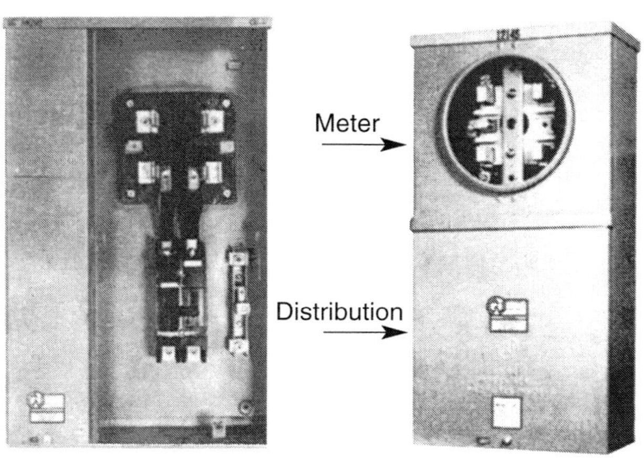

Figure 6-1
Residential Service Entrance Equipment

Figure 6-2
Multi-metering Center

Power Systems Equipment

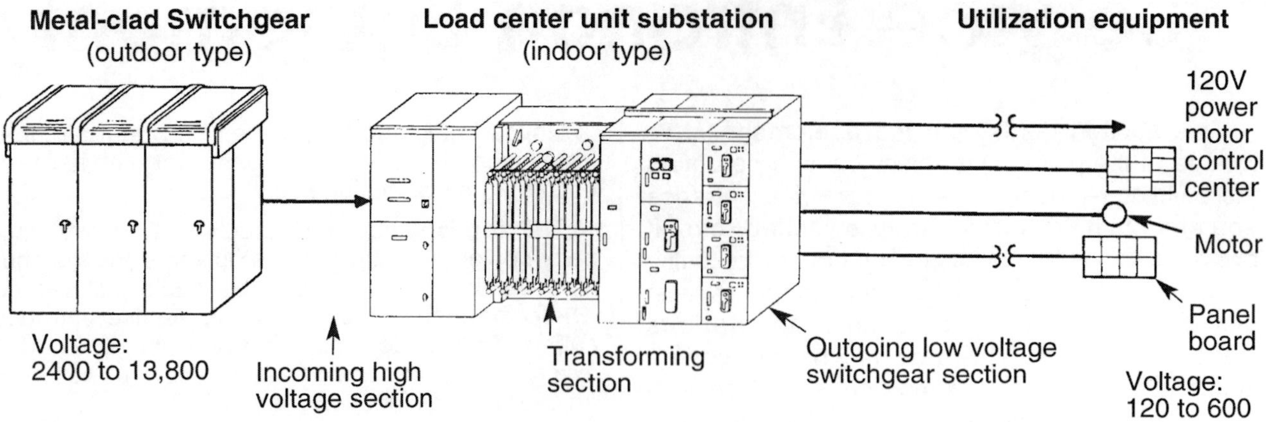

"Express" switchboards

Specially selected offering from the Type FA-1 Switchboard line.
Dependable electrical system distribution for a variety of applications

Figure 6-3
Power Systems Equipment

The purpose of the distribution section is to subdivide electrical power among the individual circuits, each of which is protected by a circuit breaker mounted in the distribution panel. The distribution section in a larger commercial building will include feeder breakers in addition to individual breakers for every circuit in the building's electrical system. If the building covers a large area or has electrical equipment like large motors or air conditioning units, the distribution section may include circuits that feed subpanels located in other parts of the building.

Service entrance equipment listed in this section is only representative of the many types of equipment that are available. But from what's listed here you can see the range of material prices and typical installation costs. Prices quoted for larger pieces of equipment are for the most competitively-priced units. Prices can vary widely. It's good practice to get several quotes.

Much of this equipment is custom-assembled to order. There will be only a few competitive suppliers in your area. But note that multiple units ordered at the same time can reduce costs 20 percent or more.

Safety Switches

Safety switches are also known as **disconnect switches and externally operated (EXO) devices.** A safety switch is a convenient way of shutting off electrical equipment in an emergency. For example, the *NEC* requires that a safety switch be installed at or near heavy machinery. In some cases the switch will be protected with a padlock so the power can't be shut off accidentally.

There are many types of safety switches. Some have specific applications. Others are appropriate for a wide range of uses. Ratings for safety switches are by amperage (30 amp, 60 amp, 100 amp, etc.), and by voltage (120 volt, 240 volt, 480 volt, etc.). Safety switches intended for use with electric motors are rated by motor horsepower. Safety switches are also classified as general duty, heavy duty, and by NEMA (National Electrical Manufacturer's Association) class. The most common classes are 1 (indoor), 3R (weatherproof), 4 and 4X (waterproof and dustproof), and 7 through 9 (for hazardous locations). Explosion-proof enclosures are also available.

Safety switches can be either fusible or non-fusible. The fusible type has a renewable fuse that can be changed to alter the overload protection characteristics of the circuit. A safety switch that isn't fused has no overload protection.

The engineer who designs the electrical system should identify the type and rating of each safety switch on the plans. But note that it's acceptable practice (but an unnecessary expense) to use oversize safety switches. For example, a 100A, 480V, 3-phase NEMA class 3R (weatherproof) safety switch can be installed inside a building to serve a machine with a 20A, 240V load. But it's much less expensive to use a switch that just meets rather than exceeds circuit requirements.

The fusible switches listed in this section are for cartridge fuses. Screw-in fuses can be used in some safety switches, but the cartridge type is much more common. Most suppliers carry cartridge fuses in these ranges. The fuse amperage ratings are as follows for the switch ratings indicated:

30A	fusible switches use fractional amp fuses to 30 amps
60A	fusible switches use 35 amp to 60 amp fuses
100A	fusible switches use 70 amp to 100 amp fuses
200A	fusible switches use 110 amp to 200 amp fuses
400A	fusible switches use 225 amp to 400 amp fuses
600A	fusible switches use 450 amp to 600 amp fuses
800A	fusible switches use 601 amp to 800 amp fuses
1200A	fusible switches use 1000 amp to 1200 amp fuses

The *National Electrical Code* lists several installation rules for safety switches. Some job specifications will include additional requirements for safety switches. For example, some specs require switches with interlocks that prevent opening the switch door or cover when the switch is in the "on" position. The specs might require that the switch have a block for connecting a neutral wire.

Circuit Breakers

Like safety switches, circuit breakers can be used to shut off the power to any circuit. Generally, a circuit breaker can be used anywhere a safety switch is required. But an automatic circuit breaker's most important function is to protect the system by opening when an overload is detected. Unlike a fuse, a breaker doesn't have to be replaced after tripping. Resetting the breaker restores power to the circuit.

Although non-automatic circuit breakers are available, automatic breakers with permanent trip settings are most common. They're rated by amperage capacity and voltage. They're also rated for maximum short-circuit interrupting capacity. The NEC dictates the type of circuit breaker to use for each application.

The most common use for circuit breakers is in loadcenters and panelboards. They can be either the plug-in or bolt-on type. Loadcenters use plug-in breakers. Panelboards use bolt-on breakers that bolt directly to the panelboard bus and to the panel frame for support. The vast majority of our jobs use bolt-on. Most well-engineered jobs require bolt-on breakers. Use bolt-on for high A.I.C. interrupting capacities.

Circuit breakers are available with more than one pole. A single handle opens and closes contact between two or more conductors. But single-pole breakers can be ganged to protect each line. When that's done, the operating handles for all poles on a circuit should be tied together with a bar device. Usually these breakers are made from single-pole breakers riveted together to form a two- or three-pole device.

Circuit breakers can be either thermal, magnetic or a combination of the two. Thermal breakers react to changes in temperature, opening the circuit in response to extra heat from a short. Magnetic breakers react to changes in current flow. A sudden increase in current flow creates enough magnetic force to activate an armature, opening the circuit.

Thermal-magnetic breakers combine the best features of both types of breakers and are the type commonly used in panelboards and loadcenters. Magnetic breakers are generally used in special applications where the breaker must be unaffected by air temperature.

Meter Sockets

Meter sockets are usually installed by the contractor. The power company will usually install the actual kilowatt hour meter.

Be sure to install the right socket. Each type of socket has a certain number of jaws that make contact with the meter. The socket must match the meter the power company intends to use. That depends on the size of the service feed and the service voltage provided.

In some cases the meter will require a current transformer in the service section. The current transformer is connected to the meter base. Test blocks may also be required. Check with your local utility company before buying or installing the meter socket.

Multi-socket assemblies are required in multi-family residences. Usually all the meters will be located in the same panel for convenience. Your supplier will help you select the right meter panel. But it's your responsibility to get approval from the utility company for the panel selected.

Loadcenters and Panelboards

Loadcenters are electrical distribution panels that use ***plug-in*** circuit breakers. They're usually used in residential and light commercial buildings where the electrical loads are not heavy.

Loadcenters are enclosed in a sheet metal box which can be ordered for surface, flush or semi-flush mounting in a wall. The cover has knockouts where breakers are mounted. Most loadcenters come with a door to cover the breaker handles. Both indoor- and outdoor-rated loadcenters are available. Units can have as few as two single-pole spaces or as many as 42 single-pole spaces. Two- and three-pole plug-in breakers can also be inserted. Individual breakers can be sized up to 100 amps.

Panelboards are electrical distribution panels that use ***bolt-on*** circuit breakers. They're usually used in heavy commercial and industrial buildings where higher amperages are needed.

Panelboards are enclosed in a heavy gauge sheet metal box which can be ordered for surface or flush mounting on a wall. Panelboards have interiors like similar loadcenters except that the circuit breakers must be bolted to the interior bus system. An interior cover is placed over the circuit breakers and extends to the panel cover. The panel cover is made of heavy gauge sheet metal and usually has a door which covers the circuit-breaker handles. Two- and three-pole bolt-on breakers can be installed. The breakers can be sized up to 100 amps for lighting panels, and higher for power panels.

Panelboards are rated for total load, such as 100A, 225A, 400A, etc. The circuit breakers installed cannot exceed the panel rating. Single-width panelboards are limited to 42 circuits. If additional circuits are required, a double-width panelboard can be used.

Panelboards are usually custom-assembled at the factory to meet specific job requirements. Loadcenters are usually stock items. Your supplier probably offers a good selection of loadcenter housings and large quantities of the more common plug-in breakers that are inserted into the housing. When you order a loadcenter, they can pull the loadcenter housing from stock and furnish whatever breakers you will insert into the housing.

Wireway

Wireway is an enclosed metal channel with one hinged or removable side so wire can be laid, rather than pulled, into place. Indoor wireway may have either a screw cover or a hinged cover. Both interior and exterior wireways are available in several sizes and lengths.

Wireway is useful in many situations. For example, it's used in service entrance equipment to enclose wire running between the various components. It can be installed in long rows for circuits, control or communications systems. Fittings are available for making turns up, down, right or left. Flange fittings allow connection to pull boxes and cabinets.

Wireway usually has knockouts every few inches on each side wall. These knockouts are spaced every few inches so a junction is possible at nearly all points.

Transformers

The most common transformer for buildings is the dry air-cooled type. Both indoor and outdoor dry transformers are available. Other types of transformers are cooled with oil or silicon compounds. The larger units are generally used at primary service facilities or substations and can be installed on poles, pads or submerged.

Most transformers are available with adjustable taps for making small changes in the output voltage. When the connected load voltage is too high or too low, the adjustable taps are changed to bring the voltage closer to the desired range. Usually the transformers must be de-energized before the taps are changed. The taps are connected inside the transformer. Each tap allows a 2-1/2 percent adjustment. Most transformers have two taps above normal and two taps below the normal setting.

Estimating Service Equipment

Service entrance equipment is sized to meet the needs of the electrical loads served. The design engineer will supply the load calculations. Use these figures to select the right loadcenters or panelboards. The main service equipment must be based on the total building load, plus some capacity for future expansion. Since every building can be different, all service entrance equipment can be different. But you'll notice that most manufacturers offer only certain load capacities, such as 100 amp, 225 amp, 400 amp, etc.

Your estimate for service entrance equipment will always include several components: panels, breakers, and metering. Larger and more specialized installations will include specially-engineered gear.

Making the material estimate is usually a simple matter of getting a supplier to make a take-off and furnish a quote. The supplier will pass the take-off on to the manufacturer's representative for pricing major components. The supplier will prepare a quote on smaller stock items. You'll usually get a single lump sum price. That makes your job easy. But watch out for exceptions in the quote. And be sure the quote you get covers all service equipment and panels.

Labor for Service Equipment

Finding the installation cost for service entrance equipment is much harder than finding the material cost. First, study the plans and specs carefully. Then list all the major components on a pricing sheet.

Suppose the specs show a meter socket and main combination as the first line item. The next line shows a 12-circuit, flush-mounted loadcenter with 12 single-pole plug-in circuit breakers. First, find the labor cost for the meter socket and main combination. Then figure the labor for the loadcenter. The loadcenter is assembled by inserting plug-in circuit breakers into the panel. My usual practice is to have a single labor cost for the panel, housing and cover. The cost of installing breakers is a separate item. Circuit breaker labor in the tables that follow includes connection to the circuit conductor.

Items to Watch For

Most mistakes in estimating service entrance equipment are the result of omissions and oversights. Run through the checklist that follows when your estimate is complete. These questions should help you spot an error before it becomes an expensive mistake.

1) Are the service voltage and size correct?
2) Is it single- or three-phase?
3) Is this an indoor or outdoor installation?
4) Is special grounding needed?
5) Are the interrupt capacities correct?
6) Is it flush or surface mounted?
7) Is this for overhead or underground service?
8) Does it require a special nameplate?
9) What are the ground fault provisions?
10) Is a special corrosion-resisting finish needed?
11) Did you include the fuses?
12) Are there shunt trip breakers?
13) Is the undervoltage protection correct?
14) Are there surge arrestors?
15) What are the load shedding provisions?
16) Are there energy-management provisions?
17) Is any standby power equipment included?
18) Are transfer switches included?
19) Is there any special metering equipment?
20) Do physical clearances comply with the code?
21) Is it accessible for installation?
22) What are the utility company hookup charges?
23) Are there housekeeping pads?
24) Are special equipment pads needed?
25) Will it fit in the space provided?

Installation

Installation times for service equipment in commercial and industrial buildings will vary because the equipment is usually custom designed. Only on housing tracts will you install the same service equipment over and over again.

First, determine where the service equipment and panels are to be installed. Good access to the installation areas is important. If there is a substation in the service equipment, where will it be located? Lifting equipment may be needed to install heavy pieces of service equipment in vaults, basements, and mezzanines. Special rigging may be required. Moving heavy items into awkward places can be expensive.

If you suspect that a crane will be needed, get a quote from a local crane company. There will usually be a minimum charge. They'll probably charge for the entire time the crane is out of the yard. To get a quote, you'll need to describe the weight, maximum height and all obstructions that are likely to affect the lift. The angle of the boom will also be a factor. The longer the horizontal reach, the bigger the crane has to be.

Include an allowance for unloading when installing any heavy service equipment. The manufacturer's quote probably includes delivery. But it's your responsibility to unload it promptly. Always include a note on your purchase order requiring that you be notified at least 24 hours before heavy deliveries are made. You need some time to arrange for unloading equipment.

The labor units in these tables include time required to connect wires or conductors, clean the unit and put it into service. If any special testing is required, add that time separately. The contract documents will outline the testing procedure, if any.

The building inspector will probably have to approve your work before the utility company will install the meter. Usually the inspector is the one who calls for the meter. Before the utility company sets the meter, an application for service may be necessary. They need to know who's going to pay the electric bill.

In some cases the inspection authority will permit installation of a temporary meter while the building is under construction. The meter is later assigned to the owner after final inspection.

Estimates of service entrance equipment require a close study of the plans. It's a good idea to estimate the service gear last. That way you're fully familiar with the project when estimating the installation times.

Put some extra thought into estimates for service equipment placement and hookup. Use the service entrance checklist to spot an error or omission. This is one of the more difficult areas for most electrical estimators.

NEMA 1 general duty non-fused 240 volt safety switches

Material		Craft@Hrs	Unit	Material Cost	Labor Cost	Installed Cost
3P	30A	L1@0.50	Ea	102.00	20.50	122.50
3P	60A	L1@0.70	Ea	128.00	28.70	156.70
3P	100A	L1@1.00	Ea	301.00	41.00	342.00
3P	200A	L1@1.50	Ea	548.00	61.50	609.50
3P	400A	L1@2.00	Ea	1,400.00	82.10	1,482.10
3P	600A	L1@4.00	Ea	2,660.00	164.00	2,824.00

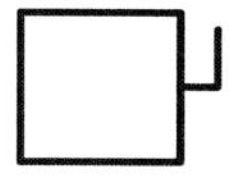

NEMA 3R general duty non-fused 240 volt safety switches

Material		Craft@Hrs	Unit	Material Cost	Labor Cost	Installed Cost
3P	30A	L1@0.50	Ea	184.00	20.50	204.50
3P	60A	L1@0.70	Ea	287.00	28.70	315.70
3P	100A	L1@1.00	Ea	517.00	41.00	558.00
3P	200A	L1@1.50	Ea	922.00	61.50	983.50

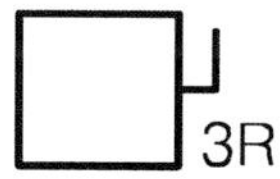

NEMA 1 general duty fusible 240 volt safety switches

Material		Craft@Hrs	Unit	Material Cost	Labor Cost	Installed Cost
2P	30A	L1@0.50	Ea	80.90	20.50	101.40
2P	60A	L1@0.70	Ea	131.00	28.70	159.70
2P	100A	L1@1.00	Ea	269.00	41.00	310.00
2P	200A	L1@1.50	Ea	577.00	61.50	638.50
2P	400A	L1@4.00	Ea	1,800.00	164.00	1,964.00
2P	600A	L1@4.00	Ea	3,390.00	164.00	3,554.00
3P	30A	L1@0.50	Ea	126.00	20.50	146.50
3P	60A	L1@0.70	Ea	207.00	28.70	235.70
3P	100A	L1@1.00	Ea	352.00	41.00	393.00
3P	200A	L1@1.50	Ea	764.00	61.50	825.50
3P	400A	L1@2.00	Ea	2,010.00	82.10	2,092.10
3P	600A	L1@4.00	Ea	3,680.00	164.00	3,844.00

NEMA 3R general duty fusible 240 volt safety switches

Material		Craft@Hrs	Unit	Material Cost	Labor Cost	Installed Cost
3P	30A	L1@0.50	Ea	185.00	20.50	205.50
3P	60A	L1@0.70	Ea	280.00	28.70	308.70
3P	100A	L1@1.00	Ea	517.00	41.00	558.00
3P	200A	L1@1.50	Ea	947.00	61.50	1,008.50
3P	400A	L1@2.00	Ea	2,470.00	82.10	2,552.10
3P	600A	L1@4.00	Ea	5,100.00	164.00	5,264.00

Use these figures to estimate the cost of safety switches installed in buildings under the conditions described on pages 5 and 6. Costs listed are for each switch installed. The crew is one electrician working at a labor cost of $41.03 per manhour. These costs include layout, material handling, and normal waste. Add for the fuses, hubs, supports, sales tax, delivery, supervision, mobilization, demobilization, cleanup, overhead and profit. Note: Safety switches can be purchased with provision for neutral connection and are used for many purposes besides installation as part of the service entrance gear. NEMA class designations are as follows: 1 is for indoor use, 3R is for outdoor use, 4 is dusttight or watertight, 12 is watertight. Class 3R safety switches usually come with blank top hubs. Be sure the conduit size is appropriate for the switch hubs. Many safety switches are available with door interlocks or with eyes for attaching a padlock.

Fuses are on pages 283 to 305.

240 Volt Heavy Duty Safety Switches

NEMA 1 heavy duty non-fused 240 volt safety switches

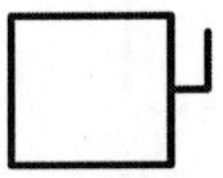

Material		Craft@Hrs	Unit	Material Cost	Labor Cost	Installed Cost
2P	30A	L1@0.50	Ea	232.00	20.50	252.50
2P	60A	L1@0.70	Ea	404.00	28.70	432.70
2P	100A	L1@1.00	Ea	650.00	41.00	691.00
2P	200A	L1@1.50	Ea	1,000.00	61.50	1,061.50
2P	400A	L1@2.00	Ea	2,340.00	82.10	2,422.10
2P	600A	L1@4.00	Ea	4,170.00	164.00	4,334.00
2P	800A	L1@5.00	Ea	8,300.00	205.00	8,505.00
2P	1200A	L1@6.00	Ea	11,100.00	246.00	11,346.00
3P	30A	L1@0.50	Ea	232.00	20.50	252.50
3P	60A	L1@0.70	Ea	404.00	28.70	432.70
3P	100A	L1@1.00	Ea	650.00	41.00	691.00
3P	200A	L1@1.50	Ea	107.00	61.50	168.50
3P	400A	L1@2.00	Ea	2,340.00	82.10	2,422.10
3P	600A	L1@4.00	Ea	4,170.00	164.00	4,334.00
3P	800A	L1@5.00	Ea	8,300.00	205.00	8,505.00
3P	1200A	L1@6.00	Ea	11,100.00	246.00	11,346.00

NEMA 3R heavy duty non-fused 240 volt safety switches

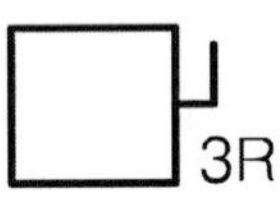

Material		Craft@Hrs	Unit	Material Cost	Labor Cost	Installed Cost
2P	30A	L1@0.50	Ea	404.00	20.50	424.50
2P	60A	L1@0.70	Ea	731.00	28.70	759.70
2P	100A	L1@1.00	Ea	1,020.00	41.00	1,061.00
2P	200A	L1@1.50	Ea	1,240.00	61.50	1,301.50
2P	400A	L1@2.00	Ea	3,170.00	82.10	3,252.10
2P	600A	L1@4.00	Ea	6,350.00	164.00	6,514.00
2P	800A	L1@5.00	Ea	10,800.00	205.00	11,005.00
2P	1200A	L1@6.00	Ea	14,900.00	246.00	15,146.00
3P	30A	L1@0.50	Ea	404.00	20.50	424.50
3P	60A	L1@0.70	Ea	731.00	28.70	759.70
3P	100A	L1@1.00	Ea	1,020.00	41.00	1,061.00
3P	200A	L1@1.50	Ea	1,240.00	61.50	1,301.50
3P	400A	L1@2.00	Ea	3,170.00	82.10	3,252.10
3P	600A	L1@4.00	Ea	6,350.00	164.00	6,514.00
3P	800A	L1@5.00	Ea	10,800.00	205.00	11,005.00
3P	1200A	L1@6.00	Ea	14,900.00	246.00	15,146.00

Use these figures to estimate the cost of safety switches installed in buildings under the conditions described on pages 5 and 6. Costs listed are for each switch installed. The crew is one electrician working at a labor cost of $41.03 per manhour. These costs include layout, material handling, and normal waste. Add for the fuses, hubs, supports, sales tax, delivery, supervision, mobilization, demobilization, cleanup, overhead and profit. Note: Safety switches can be purchased with provision for neutral connection and are used for many purposes besides installation as part of the service entrance gear. NEMA class designations are as follows: 1 is for indoor use, 3R is for outdoor use, 4 is dusttight or watertight, 12 is watertight. Class 3R safety switches usually come with blank top hubs. Be sure the conduit size is appropriate for the switch hubs. Many safety switches are available with door interlocks or with eyes for attaching a padlock.

NEMA 4 heavy duty non-fused 240 volt safety switches

Material		Craft@Hrs	Unit	Material Cost	Labor Cost	Installed Cost
2P	30A	L1@0.60	Ea	1,680.00	24.60	1,704.60
2P	60A	L1@0.80	Ea	1,990.00	32.80	2,022.80
2P	100A	L1@1.25	Ea	4,040.00	51.30	4,091.30
2P	200A	L1@1.70	Ea	5,510.00	69.80	5,579.80
3P	30A	L1@0.60	Ea	1,680.00	24.60	1,704.60
3P	60A	L1@0.80	Ea	1,990.00	32.80	2,022.80
3P	100A	L1@1.25	Ea	4,040.00	51.30	4,091.30
3P	200A	L1@1.70	Ea	5,510.00	69.80	5,579.80

NEMA 12 heavy duty non-fused 240 volt safety switches

Material		Craft@Hrs	Unit	Material Cost	Labor Cost	Installed Cost
2P	30A	L1@0.50	Ea	544.00	20.50	564.50
2P	60A	L1@0.70	Ea	662.00	28.70	690.70
2P	100A	L1@1.00	Ea	945.00	41.00	986.00
2P	200A	L1@1.50	Ea	1,270.00	61.50	1,331.50
2P	400A	L1@2.00	Ea	2,540.00	82.10	2,622.10
2P	600A	L1@4.00	Ea	4,270.00	164.00	4,434.00
3P	30A	L1@0.50	Ea	544.00	20.50	564.50
3P	60A	L1@0.70	Ea	662.00	28.70	690.70
3P	100A	L1@1.00	Ea	945.00	41.00	986.00
3P	200A	L1@1.50	Ea	1,270.00	61.50	1,331.50
3P	400A	L1@2.00	Ea	3,160.00	82.10	3,242.10
3P	600A	L1@4.00	Ea	5,990.00	164.00	6,154.00
4P	30A	L1@0.60	Ea	728.00	24.60	752.60
4P	60A	L1@0.80	Ea	794.00	32.80	826.80
4P	100A	L1@1.25	Ea	1,420.00	51.30	1,471.30
4P	200A	L1@1.70	Ea	2,250.00	69.80	2,319.80
4P	400A	L1@2.25	Ea	4,520.00	92.30	4,612.30

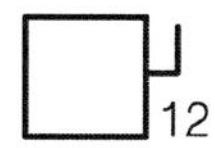

Use these figures to estimate the cost of safety switches installed in buildings under the conditions described on pages 5 and 6. Costs listed are for each switch installed. The crew is one electrician working at a labor cost of $41.03 per manhour. These costs include layout, material handling, and normal waste. Add for the fuses, hubs, supports, sales tax, delivery, supervision, mobilization, demobilization, cleanup, overhead and profit. Note: Safety switches can be purchased with provision for neutral connection and are used for many purposes besides installation as part of the service entrance gear. NEMA class designations are as follows: 1 is for indoor use, 3R is for outdoor use, 4 is dusttight or watertight, 12 is watertight. Class 3R safety switches usually come with blank top hubs. Be sure the conduit size is appropriate for the switch hubs. Many safety switches are available with door interlocks or with eyes for attaching a padlock.

600 Volt Heavy Duty Safety Switches

NEMA 1 heavy duty non-fused 600 volt safety switches

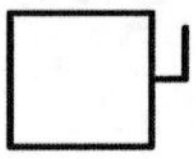

Material		Craft@Hrs	Unit	Material Cost	Labor Cost	Installed Cost
2P	30A	L1@0.50	Ea	232.00	20.50	252.50
2P	60A	L1@0.70	Ea	404.00	28.70	432.70
2P	100A	L1@1.00	Ea	650.00	41.00	691.00
2P	200A	L1@1.50	Ea	1,000.00	61.50	1,061.50
2P	400A	L1@2.00	Ea	2,340.00	82.10	2,422.10
2P	600A	L1@4.00	Ea	4,170.00	164.00	4,334.00
2P	800A	L1@5.00	Ea	8,300.00	205.00	8,505.00
2P	1200A	L1@6.00	Ea	11,100.00	246.00	11,346.00
3P	30A	L1@0.50	Ea	232.00	20.50	252.50
3P	60A	L1@0.70	Ea	404.00	28.70	432.70
3P	100A	L1@1.00	Ea	650.00	41.00	691.00
3P	200A	L1@1.50	Ea	1,000.00	61.50	1,061.50
3P	400A	L1@2.00	Ea	2,340.00	82.10	2,422.10
3P	600A	L1@4.00	Ea	4,170.00	164.00	4,334.00
3P	800A	L1@5.00	Ea	8,300.00	205.00	8,505.00
3P	1200A	L1@6.00	Ea	11,100.00	246.00	11,346.00
4P	30A	L1@0.60	Ea	653.00	24.60	677.60
4P	60A	L1@0.80	Ea	722.00	32.80	754.80
4P	100A	L1@1.25	Ea	1,310.00	51.30	1,361.30
4P	200A	L1@1.70	Ea	1,890.00	69.80	1,959.80
4P	400A	L1@2.25	Ea	4,130.00	92.30	4,222.30
4P	600A	L1@4.50	Ea	7,200.00	185.00	7,385.00

NEMA 3R heavy duty non-fused 600 volt safety switches

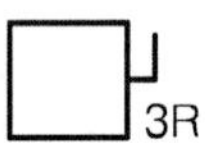

Material		Craft@Hrs	Unit	Material Cost	Labor Cost	Installed Cost
2P	30A	L1@0.50	Ea	404.00	20.50	424.50
2P	60A	L1@0.70	Ea	731.00	28.70	759.70
2P	100A	L1@1.00	Ea	1,020.00	41.00	1,061.00
2P	200A	L1@1.50	Ea	1,240.00	61.50	1,301.50
2P	400A	L1@2.00	Ea	3,170.00	82.10	3,252.10
2P	600A	L1@4.00	Ea	6,350.00	164.00	6,514.00
2P	800A	L1@5.00	Ea	10,800.00	205.00	11,005.00
2P	1200A	L1@6.00	Ea	14,900.00	246.00	15,146.00
3P	30A	L1@0.50	Ea	404.00	20.50	424.50
3P	60A	L1@0.70	Ea	731.00	28.70	759.70
3P	100A	L1@1.00	Ea	1,020.00	41.00	1,061.00
3P	200A	L1@1.50	Ea	1,240.00	61.50	1,301.50
3P	400A	L1@2.00	Ea	3,170.00	82.10	3,252.10
3P	600A	L1@4.00	Ea	6,350.00	164.00	6,514.00
3P	800A	L1@5.00	Ea	10,800.00	205.00	11,005.00
3P	1200A	L1@6.00	Ea	14,900.00	246.00	15,146.00

Use these figures to estimate the cost of safety switches installed in buildings under the conditions described on pages 5 and 6. Costs listed are for each switch installed. The crew is one electrician working at a labor cost of $41.03 per manhour. These costs include layout, material handling, and normal waste. Add for the fuses, hubs, supports, sales tax, delivery, supervision, mobilization, demobilization, cleanup, overhead and profit. Note: Safety switches can be purchased with provision for neutral connection and are used for many purposes besides installation as part of the service entrance gear. NEMA class designations are as follows: 1 is for indoor use, 3R is for outdoor use, 4 is dusttight or watertight, 12 is watertight. Class 3R safety switches usually come with blank top hubs. Be sure the conduit size is appropriate for the switch hubs. Many safety switches are available with door interlocks or with eyes for attaching a padlock.

Material		Craft@Hrs	Unit	Material Cost	Labor Cost	Installed Cost

NEMA 4 heavy duty non-fused 600 volt safety switches

Material		Craft@Hrs	Unit	Material Cost	Labor Cost	Installed Cost
2P	30A	L1@0.60	Ea	1,680.00	24.60	1,704.60
2P	60A	L1@0.80	Ea	1,990.00	32.80	2,022.80
2P	100A	L1@1.25	Ea	4,040.00	51.30	4,091.30
2P	200A	L1@1.70	Ea	5,510.00	69.80	5,579.80
3P	30A	L1@0.60	Ea	1,680.00	24.60	1,704.60
3P	60A	L1@0.80	Ea	1,990.00	32.80	2,022.80
3P	100A	L1@1.25	Ea	4,040.00	51.30	4,091.30
3P	200A	L1@1.70	Ea	5,510.00	69.80	5,579.80

NEMA 12 heavy duty non-fused 600 volt safety switches

Material		Craft@Hrs	Unit	Material Cost	Labor Cost	Installed Cost
2P	30A	L1@0.50	Ea	514.00	20.50	534.50
2P	60A	L1@0.70	Ea	662.00	28.70	690.70
2P	100A	L1@1.00	Ea	945.00	41.00	986.00
2P	200A	L1@1.50	Ea	1,270.00	61.50	1,331.50
2P	400A	L1@2.00	Ea	2,540.00	82.10	2,622.10
2P	600A	L1@4.00	Ea	4,270.00	164.00	4,434.00
3P	30A	L1@0.50	Ea	514.00	20.50	534.50
3P	60A	L1@0.70	Ea	662.00	28.70	690.70
3P	100A	L1@1.00	Ea	945.00	41.00	986.00
3P	200A	L1@1.50	Ea	1,270.00	61.50	1,331.50
3P	400A	L1@2.00	Ea	3,160.00	82.10	3,242.10
3P	600A	L1@4.00	Ea	5,330.00	164.00	5,494.00
4P	30A	L1@0.60	Ea	728.00	24.60	752.60
4P	60A	L1@0.80	Ea	794.00	32.80	826.80
4P	100A	L1@1.25	Ea	1,420.00	51.30	1,471.30
4P	200A	L1@1.70	Ea	2,250.00	69.80	2,319.80
4P	400A	L1@2.25	Ea	4,520.00	92.30	4,612.30

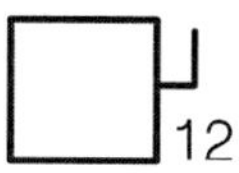

NEMA 1 heavy duty fusible 600 volt safety switches

Material		Craft@Hrs	Unit	Material Cost	Labor Cost	Installed Cost
2P	30A	L1@0.50	Ea	436.00	20.50	456.50
2P	60A	L1@0.70	Ea	533.00	28.70	561.70
2P	100A	L1@1.00	Ea	988.00	41.00	1,029.00
2P	200A	L1@1.50	Ea	1,420.00	61.50	1,481.50
2P	400A	L1@2.00	Ea	3,520.00	82.10	3,602.10

Use these figures to estimate the cost of safety switches installed in buildings under the conditions described on pages 5 and 6. Costs listed are for each switch installed. The crew is one electrician working at a labor cost of $41.03 per manhour. These costs include layout, material handling, and normal waste. Add for the fuses, hubs, supports, sales tax, delivery, supervision, mobilization, demobilization, cleanup, overhead and profit. Note: Safety switches can be purchased with provision for neutral connection and are used for many purposes besides installation as part of the service entrance gear. NEMA class designations are as follows: 1 is for indoor use, 3R is for outdoor use, 4 is dusttight or watertight, 12 is watertight. Class 3R safety switches usually come with blank top hubs. Be sure the conduit size is appropriate for the switch hubs. Many safety switches are available with door interlocks or with eyes for attaching a padlock.

Fuses are on pages 283 to 305.

600 Volt Heavy Duty Safety Switches

NEMA 1 heavy duty fusible 600 volt safety switches

Material		Craft@Hrs	Unit	Material Cost	Labor Cost	Installed Cost
2P	600A	L1@4.00	Ea	5,520.00	164.00	5,684.00
2P	800A	L1@5.00	Ea	8,600.00	205.00	8,805.00
2P	1200A	L1@6.00	Ea	12,200.00	246.00	12,446.00
3P	30A	L1@0.50	Ea	436.00	20.50	456.50
3P	60A	L1@0.70	Ea	533.00	28.70	561.70
3P	100A	L1@1.00	Ea	988.00	41.00	1,029.00
3P	200A	L1@1.50	Ea	1,420.00	61.50	1,481.50
3P	400A	L1@2.00	Ea	3,780.00	82.10	3,862.10
3P	600A	L1@4.00	Ea	6,330.00	164.00	6,494.00
3P	800A	L1@5.00	Ea	11,700.00	205.00	11,905.00
3P	1200A	L1@6.00	Ea	14,400.00	246.00	14,646.00
4P	30A	L1@0.60	Ea	722.00	24.60	746.60
4P	60A	L1@0.80	Ea	843.00	32.80	875.80
4P	100A	L1@1.25	Ea	1,400.00	51.30	1,451.30
4P	200A	L1@1.70	Ea	2,360.00	69.80	2,429.80
4P	400A	L1@2.25	Ea	4,940.00	92.30	5,032.30
4P	600A	L1@4.25	Ea	7,990.00	174.00	8,164.00

NEMA 3R heavy duty fusible 600 volt safety switches

Material		Craft@Hrs	Unit	Material Cost	Labor Cost	Installed Cost
2P	30A	L1@0.50	Ea	741.00	20.50	761.50
2P	60A	L1@0.70	Ea	879.00	28.70	907.70
2P	100A	L1@1.00	Ea	1,370.00	41.00	1,411.00
2P	200A	L1@1.50	Ea	1,870.00	61.50	1,931.50
2P	400A	L1@2.00	Ea	4,570.00	82.10	4,652.10
2P	600A	L1@4.00	Ea	9,070.00	164.00	9,234.00
2P	800A	L1@5.00	Ea	13,600.00	205.00	13,805.00
2P	1200A	L1@6.00	Ea	14,900.00	246.00	15,146.00
3P	30A	L1@0.50	Ea	741.00	20.50	761.50
3P	60A	L1@0.70	Ea	879.00	28.70	907.70
3P	100A	L1@1.00	Ea	1,370.00	41.00	1,411.00
3P	200A	L1@1.50	Ea	1,870.00	61.50	1,931.50
3P	400A	L1@2.00	Ea	4,580.00	82.10	4,662.10
3P	600A	L1@4.00	Ea	9,070.00	164.00	9,234.00
3P	800A	L1@5.00	Ea	13,700.00	205.00	13,905.00
3P	1200A	L1@6.00	Ea	16,500.00	246.00	16,746.00

Use these figures to estimate the cost of safety switches installed in buildings under the conditions described on pages 5 and 6. Costs listed are for each switch installed. The crew is one electrician working at a labor cost of $41.03 per manhour. These costs include layout, material handling, and normal waste. Add for the fuses, hubs, supports, sales tax, delivery, supervision, mobilization, demobilization, cleanup, overhead and profit. Note: Safety switches can be purchased with provision for neutral connection and are used for many purposes besides installation as part of the service entrance gear. NEMA class designations are as follows: 1 is for indoor use, 3R is for outdoor use, 4 is dusttight or watertight, 12 is watertight. Class 3R safety switches usually come with blank top hubs. Be sure the conduit size is appropriate for the switch hubs. Many safety switches are available with door interlocks or with eyes for attaching a padlock.

Fuses are on pages 283 to 305.

Material		Craft@Hrs	Unit	Material Cost	Labor Cost	Installed Cost
NEMA 4 heavy duty fusible 600 volt safety switches						
2P	30A	L1@0.60	Ea	1,990.00	24.60	2,014.60
2P	60A	L1@0.80	Ea	2,190.00	32.80	2,222.80
2P	100A	L1@1.25	Ea	4,320.00	51.30	4,371.30
2P	200A	L1@1.75	Ea	6,090.00	71.80	6,161.80
2P	400A	L1@2.25	Ea	12,200.00	92.30	12,292.30
2P	600A	L1@4.25	Ea	17,200.00	174.00	17,374.00
3P	30A	L1@0.60	Ea	1,990.00	24.60	2,014.60
3P	60A	L1@0.80	Ea	2,190.00	32.80	2,222.80
3P	100A	L1@1.25	Ea	4,320.00	51.30	4,371.30
3P	200A	L1@1.75	Ea	6,090.00	71.80	6,161.80
3P	400A	L1@2.25	Ea	12,200.00	92.30	12,292.30
3P	600A	L1@4.25	Ea	17,700.00	174.00	17,874.00
NEMA 12 heavy duty fusible 600 volt safety switches						
2P	30A	L1@0.50	Ea	780.00	20.50	800.50
2P	60A	L1@0.70	Ea	782.00	28.70	810.70
2P	100A	L1@1.00	Ea	1,220.00	41.00	1,261.00
2P	200A	L1@1.50	Ea	1,890.00	61.50	1,951.50
2P	400A	L1@2.00	Ea	4,040.00	82.10	4,122.10
2P	600A	L1@4.00	Ea	6,040.00	164.00	6,204.00
3P	30A	L1@0.50	Ea	780.00	20.50	800.50
3P	60A	L1@0.70	Ea	782.00	28.70	810.70
3P	100A	L1@1.00	Ea	1,220.00	41.00	1,261.00
3P	200A	L1@1.50	Ea	1,890.00	61.50	1,951.50
3P	400A	L1@2.00	Ea	4,320.00	82.10	4,402.10
3P	600A	L1@4.00	Ea	7,310.00	164.00	7,474.00
4P	30A	L1@0.60	Ea	883.00	24.60	907.60
4P	60A	L1@0.80	Ea	995.00	32.80	1,027.80
4P	100A	L1@1.25	Ea	1,540.00	51.30	1,591.30
4P	200A	L1@1.75	Ea	2,620.00	71.80	2,691.80
4P	400A	L1@2.25	Ea	5,410.00	92.30	5,502.30

Use these figures to estimate the cost of safety switches installed in buildings under the conditions described on pages 5 and 6. Costs listed are for each switch installed. The crew is one electrician working at a labor cost of $41.03 per manhour. These costs include layout, material handling, and normal waste. Add for the fuses, hubs, supports, sales tax, delivery, supervision, mobilization, demobilization, cleanup, overhead and profit. Note: Safety switches can be purchased with provision for neutral connection and are used for many purposes besides installation as part of the service entrance gear. NEMA class designations are as follows: 1 is for indoor use, 3R is for outdoor use, 4 is dusttight or watertight, 12 is watertight. Class 3R safety switches usually come with blank top hubs. Be sure the conduit size is appropriate for the switch hubs. Many safety switches are available with door interlocks or with eyes for attaching a padlock.

Fuses are on pages 283 to 305.

Safety Switches

NEMA 1 heavy duty non-fused 600 volt safety switches

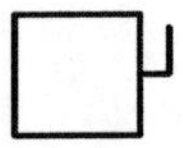

Material		Craft@Hrs	Unit	Material Cost	Labor Cost	Installed Cost
6P	30A	L1@0.70	Ea	2,730.00	28.70	2,758.70
6P	60A	L1@1.00	Ea	3,160.00	41.00	3,201.00
6P	100A	L1@1.50	Ea	3,870.00	61.50	3,931.50
6P	200A	L1@2.00	Ea	8,550.00	82.10	8,632.10

NEMA 7 non-fused 600 volt safety switches

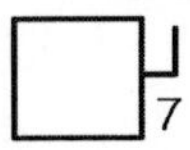

Material		Craft@Hrs	Unit	Material Cost	Labor Cost	Installed Cost
3P	60A	L1@0.90	Ea	2,090.00	36.90	2,126.90
3P	100A	L1@1.25	Ea	2,340.00	51.30	2,391.30
3P	200A	L1@1.50	Ea	5,130.00	61.50	5,191.50

NEMA 1 non-fused 240 volt double throw safety switches

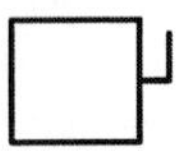

Material		Craft@Hrs	Unit	Material Cost	Labor Cost	Installed Cost
2P	30A	L1@0.75	Ea	461.00	30.80	491.80
2P	60A	L1@1.25	Ea	734.00	51.30	785.30
2P	100A	L1@1.75	Ea	1,310.00	71.80	1,381.80
2P	200A	L1@2.25	Ea	2,060.00	92.30	2,152.30
2P	400A	L1@4.50	Ea	4,240.00	185.00	4,425.00
2P	600A	L1@6.00	Ea	7,880.00	246.00	8,126.00
3P	30A	L1@0.80	Ea	544.00	32.80	576.80
3P	60A	L1@1.30	Ea	792.00	53.30	845.30
3P	100A	L1@1.80	Ea	1,310.00	73.90	1,383.90
3P	200A	L1@2.30	Ea	2,060.00	94.40	2,154.40
3P	400A	L1@4.60	Ea	5,770.00	189.00	5,959.00
3P	600A	L1@6.25	Ea	7,850.00	256.00	8,106.00
4P	30A	L1@1.00	Ea	755.00	41.00	796.00
4P	60A	L1@1.50	Ea	1,180.00	61.50	1,241.50
4P	100A	L1@2.00	Ea	2,390.00	82.10	2,472.10
4P	200A	L1@2.50	Ea	3,340.00	103.00	3,443.00
4P	400A	L1@5.00	Ea	7,610.00	205.00	7,815.00
4P	600A	L1@6.50	Ea	9,170.00	267.00	9,437.00

NEMA 3R non-fused 240 volt double throw safety switches

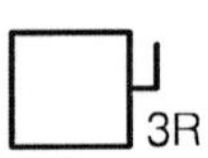

Material		Craft@Hrs	Unit	Material Cost	Labor Cost	Installed Cost
2P	100A	L1@1.75	Ea	653.00	71.80	724.80
2P	200A	L1@2.00	Ea	1,720.00	82.10	1,802.10
2P	400A	L1@2.50	Ea	10,300.00	103.00	10,403.00
3P	100A	L1@2.00	Ea	2,390.00	82.10	2,472.10
3P	200A	L1@2.50	Ea	4,620.00	103.00	4,723.00
3P	400A	L1@2.75	Ea	10,300.00	113.00	10,413.00
3P	600A	L1@5.00	Ea	12,500.00	205.00	12,705.00

Use these figures to estimate the cost of safety switches installed in buildings under the conditions described on pages 5 and 6. Costs listed are for each switch installed. The crew is one electrician working at a labor cost of $41.03 per manhour. These costs include layout, material handling, and normal waste. Add for the fuses, hubs, supports, sales tax, delivery, supervision, mobilization, demobilization, cleanup, overhead and profit. Note: Safety switches can be purchased with provision for neutral connection and are used for many purposes besides installation as part of the service entrance gear. NEMA class designations are as follows: 1 is for indoor use, 3R is for outdoor use, 4 is dusttight or watertight, 12 is watertight. Class 3R safety switches usually come with blank top hubs. Be sure the conduit size is appropriate for the switch hubs. Many safety switches are available with door interlocks or with eyes for attaching a padlock.

Plug Fuses, Fast Acting, 120 Volt, Current Limiting, Class T

Material	Craft@Hrs	Unit	Material Cost	Labor Cost	Installed Cost
Fast acting single element current limiting plug fuses, 120 volt, class T					
1 amp	L1@0.05	Ea	4.08	2.05	6.13
2 amp	L1@0.05	Ea	4.27	2.05	6.32
3 amp	L1@0.05	Ea	4.27	2.05	6.32
5 amp	L1@0.05	Ea	4.27	2.05	6.32
6 amp	L1@0.05	Ea	4.27	2.05	6.32
8 amp	L1@0.05	Ea	4.15	2.05	6.20
10 amp	L1@0.05	Ea	4.27	2.05	6.32
15 amp	L1@0.05	Ea	2.26	2.05	4.31
20 amp	L1@0.05	Ea	1.96	2.05	4.01
25 amp	L1@0.05	Ea	2.54	2.05	4.59
30 amp	L1@0.05	Ea	1.98	2.05	4.03
Fast acting dual element current limiting plug fuses, 120 volt, class T					
3/10 amp	L1@0.05	Ea	15.50	2.05	17.55
4/10 amp	L1@0.05	Ea	15.50	2.05	17.55
1/2 amp	L1@0.05	Ea	15.50	2.05	17.55
6/10 amp	L1@0.05	Ea	15.50	2.05	17.55
8/10 amp	L1@0.05	Ea	15.50	2.05	17.55
1 amp	L1@0.05	Ea	15.70	2.05	17.75
1-1/8 amp	L1@0.05	Ea	15.70	2.05	17.75
1-1/4 amp	L1@0.05	Ea	15.70	2.05	17.75
1-4/10 amp	L1@0.05	Ea	15.70	2.05	17.75
1-6/10 amp	L1@0.05	Ea	15.70	2.05	17.75
1-8/10 amp	L1@0.05	Ea	15.70	2.05	17.75
2 amp	L1@0.05	Ea	15.70	2.05	17.75
2-1/4 amp	L1@0.05	Ea	15.70	2.05	17.75
2-8/10 amp	L1@0.05	Ea	15.70	2.05	17.75
3-2/10 amp	L1@0.05	Ea	15.70	2.05	17.75
3-1/2 amp	L1@0.05	Ea	15.70	2.05	17.75
4 amp	L1@0.05	Ea	14.00	2.05	16.05
4-1/2 amp	L1@0.05	Ea	15.70	2.05	17.75
5 amp	L1@0.05	Ea	14.00	2.05	16.05
5-6/10 amp	L1@0.05	Ea	15.70	2.05	17.75
6-1/4 amp	L1@0.05	Ea	12.00	2.05	14.05
7 amp	L1@0.05	Ea	15.70	2.05	17.75
8 amp	L1@0.05	Ea	12.00	2.05	14.05
9 amp	L1@0.05	Ea	15.80	2.05	17.85
10 amp	L1@0.05	Ea	10.50	2.05	12.55
12 amp	L1@0.05	Ea	14.00	2.05	16.05
14 amp	L1@0.05	Ea	15.70	2.05	17.75
15 amp	L1@0.05	Ea	4.26	2.05	6.31
20 amp	L1@0.05	Ea	4.26	2.05	6.31
25 amp	L1@0.05	Ea	6.87	2.05	8.92
30 amp	L1@0.05	Ea	4.26	2.05	6.31

Use these figures to estimate the cost of fuses installed in safety switches and combination starters under the conditions described on pages 5 and 6. Costs listed are for each fuse installed. The crew size is one electrician working at a labor cost of $41.03 per manhour. These costs include layout, material handling, and normal waste. Add for sales tax, delivery supervision, mobilization, demobilization, cleanup, overhead and profit.

Plug Fuses, 120 Volt

Material	Craft@Hrs	Unit	Material Cost	Labor Cost	Installed Cost
Dual element type "S" 120 volt plug fuses					
3/10 amp	L1@0.05	Ea	15.80	2.05	17.85
4/10 amp	L1@0.05	Ea	15.80	2.05	17.85
1/2 amp	L1@0.05	Ea	15.80	2.05	17.85
6/10 amp	L1@0.05	Ea	15.80	2.05	17.85
8/10 amp	L1@0.05	Ea	15.80	2.05	17.85
1 amp	L1@0.05	Ea	15.80	2.05	17.85
1-1/8 amp	L1@0.05	Ea	15.80	2.05	17.85
1-1/4 amp	L1@0.05	Ea	13.10	2.05	15.15
1-4/10 amp	L1@0.05	Ea	15.80	2.05	17.85
1-6/10 amp	L1@0.05	Ea	15.80	2.05	17.85
1-8/10 amp	L1@0.05	Ea	15.80	2.05	17.85
2 amp	L1@0.05	Ea	15.80	2.05	17.85
2-1/4 amp	L1@0.05	Ea	15.80	2.05	17.85
2-8/10 amp	L1@0.05	Ea	15.80	2.05	17.85
3 amp	L1@0.05	Ea	15.80	2.05	17.85
3-2/10 amp	L1@0.05	Ea	12.00	2.05	14.05
3-1/2 amp	L1@0.05	Ea	15.80	2.05	17.85
4 amp	L1@0.05	Ea	15.10	2.05	17.15
4-1/2 amp	L1@0.05	Ea	15.80	2.05	17.85
5 amp	L1@0.05	Ea	14.20	2.05	16.25
5-6/10 amp	L1@0.05	Ea	15.80	2.05	17.85
6 amp	L1@0.05	Ea	15.80	2.05	17.85
6-1/4 amp	L1@0.05	Ea	12.20	2.05	14.25
7 amp	L1@0.05	Ea	15.80	2.05	17.85
8 amp	L1@0.05	Ea	14.20	2.05	16.25
9 amp	L1@0.05	Ea	15.80	2.05	17.85
10 amp	L1@0.05	Ea	12.20	2.05	14.25
12 amp	L1@0.05	Ea	15.10	2.05	17.15
14 amp	L1@0.05	Ea	15.80	2.05	17.85
15 amp	L1@0.05	Ea	4.32	2.05	6.37
20 amp	L1@0.05	Ea	4.32	2.05	6.37
25 amp	L1@0.05	Ea	6.10	2.05	8.15
30 amp	L1@0.05	Ea	6.10	2.05	8.15
Time delay type "SL" 120 volt plug fuses					
15 amp	L1@0.05	Ea	2.12	2.05	4.17
20 amp	L1@0.05	Ea	2.60	2.05	4.65
25 amp	L1@0.05	Ea	3.87	2.05	5.92
30 amp	L1@0.05	Ea	3.02	2.05	5.07

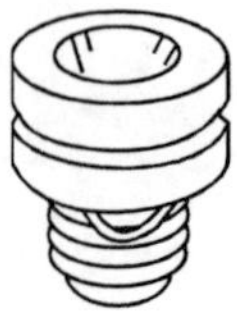

Use these figures to estimate the cost of fuses installed in safety switches and combination starters under the conditions described on pages 5 and 6. Costs listed are for each fuse installed. The crew size is one electrician working at a labor cost of $41.03 per manhour. These costs include layout, material handling, and normal waste. Add for sales tax, delivery supervision, mobilization, demobilization, cleanup, overhead and profit.

Cartridge Fuses, Non-Renewable, 250 Volt, Class H

Material	Craft@Hrs	Unit	Material Cost	Labor Cost	Installed Cost
Ferrule type non-renewable 250 volt class H cartridge fuses					
1 amp	L1@0.06	Ea	3.76	2.46	6.22
2 amp	L1@0.06	Ea	5.61	2.46	8.07
3 amp	L1@0.06	Ea	3.77	2.46	6.23
4 amp	L1@0.06	Ea	5.61	2.46	8.07
5 amp	L1@0.06	Ea	3.77	2.46	6.23
6 amp	L1@0.06	Ea	3.77	2.46	6.23
7 amp	L1@0.06	Ea	5.61	2.46	8.07
8 amp	L1@0.06	Ea	5.61	2.46	8.07
10 amp	L1@0.06	Ea	3.77	2.46	6.23
12 amp	L1@0.06	Ea	5.61	2.46	8.07
15 amp	L1@0.06	Ea	3.26	2.46	5.72
20 amp	L1@0.06	Ea	3.26	2.46	5.72
25 amp	L1@0.06	Ea	4.54	2.46	7.00
30 amp	L1@0.06	Ea	3.26	2.46	5.72
35 amp	L1@0.06	Ea	5.51	2.46	7.97
40 amp	L1@0.06	Ea	6.10	2.46	8.56
45 amp	L1@0.06	Ea	7.22	2.46	9.68
50 amp	L1@0.06	Ea	6.23	2.46	8.69
60 amp	L1@0.06	Ea	5.49	2.46	7.95
Blade type non-renewable 250 volt class H cartridge fuses					
70 amp	L1@0.08	Ea	30.10	3.28	33.38
80 amp	L1@0.08	Ea	32.50	3.28	35.78
90 amp	L1@0.08	Ea	33.80	3.28	37.08
100 amp	L1@0.08	Ea	22.00	3.28	25.28
110 amp	L1@0.08	Ea	82.40	3.28	85.68
125 amp	L1@0.08	Ea	72.90	3.28	76.18
150 amp	L1@0.08	Ea	72.90	3.28	76.18
175 amp	L1@0.08	Ea	82.40	3.28	85.68
200 amp	L1@0.08	Ea	53.50	3.28	56.78
225 amp	L1@0.10	Ea	146.00	4.10	150.10
250 amp	L1@0.10	Ea	137.00	4.10	141.10
300 amp	L1@0.10	Ea	139.00	4.10	143.10
350 amp	L1@0.10	Ea	146.00	4.10	150.10
400 amp	L1@0.15	Ea	109.00	6.15	115.15
450 amp	L1@0.15	Ea	219.00	6.15	225.15
500 amp	L1@0.15	Ea	219.00	6.15	225.15
600 amp	L1@0.15	Ea	212.00	6.15	218.15

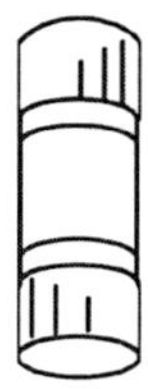

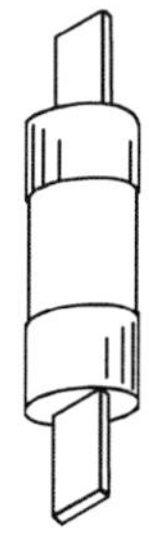

Use these figures to estimate the cost of fuses installed in safety switches and combination starters under the conditions described on pages 5 and 6. Costs listed are for each fuse installed. The crew size is one electrician working at a labor cost of $41.03 per manhour. These costs include layout, material handling, and normal waste. Add for sales tax, delivery supervision, mobilization, demobilization, cleanup, overhead and profit.

Cartridge Fuses, Non-Renewable, 600 Volt, Class H

Material	Craft@Hrs	Unit	Material Cost	Labor Cost	Installed Cost
Ferrule type non-renewable 600 volt class H cartridge fuses					
1 amp	L1@0.06	Ea	24.20	2.46	26.66
2 amp	L1@0.06	Ea	23.10	2.46	25.56
3 amp	L1@0.06	Ea	24.20	2.46	26.66
4 amp	L1@0.06	Ea	23.10	2.46	25.56
5 amp	L1@0.06	Ea	19.20	2.46	21.66
6 amp	L1@0.06	Ea	24.20	2.46	26.66
8 amp	L1@0.06	Ea	24.20	2.46	26.66
10 amp	L1@0.06	Ea	19.20	2.46	21.66
12 amp	L1@0.06	Ea	24.20	2.46	26.66
15 amp	L1@0.06	Ea	15.40	2.46	17.86
20 amp	L1@0.06	Ea	15.40	2.46	17.86
25 amp	L1@0.06	Ea	20.90	2.46	23.36
30 amp	L1@0.06	Ea	19.20	2.46	21.66
35 amp	L1@0.06	Ea	28.70	2.46	31.16
40 amp	L1@0.06	Ea	27.20	2.46	29.66
45 amp	L1@0.06	Ea	33.40	2.46	35.86
50 amp	L1@0.06	Ea	30.70	2.46	33.16
60 amp	L1@0.06	Ea	27.20	2.46	29.66

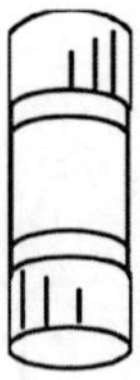

Material	Craft@Hrs	Unit	Material Cost	Labor Cost	Installed Cost
Blade type non-renewable 600 volt class H cartridge fuses					
70 amp	L1@0.08	Ea	70.20	3.28	73.48
80 amp	L1@0.08	Ea	70.20	3.28	73.48
90 amp	L1@0.08	Ea	73.00	3.28	76.28
100 amp	L1@0.08	Ea	59.10	3.28	62.38
110 amp	L1@0.08	Ea	144.00	3.28	147.28
125 amp	L1@0.08	Ea	139.00	3.28	142.28
150 amp	L1@0.08	Ea	139.00	3.28	142.28
175 amp	L1@0.08	Ea	144.00	3.28	147.28
200 amp	L1@0.08	Ea	139.00	3.28	142.28
225 amp	L1@0.10	Ea	289.00	4.10	293.10
250 amp	L1@0.10	Ea	289.00	4.10	293.10
300 amp	L1@0.10	Ea	278.00	4.10	282.10
350 amp	L1@0.10	Ea	289.00	4.10	293.10
400 amp	L1@0.15	Ea	278.00	6.15	284.15
450 amp	L1@0.15	Ea	424.00	6.15	430.15
500 amp	L1@0.15	Ea	424.00	6.15	430.15
600 amp	L1@0.15	Ea	424.00	6.15	430.15

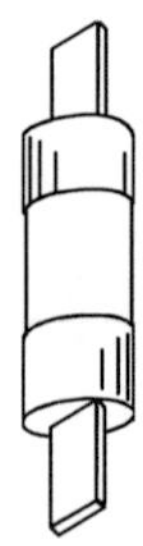

Use these figures to estimate the cost of fuses installed in safety switches and combination starters under the conditions described on pages 5 and 6. Costs listed are for each fuse installed. The crew size is one electrician working at a labor cost of $41.03 per manhour. These costs include layout, material handling, and normal waste. Add for sales tax, delivery supervision, mobilization, demobilization, cleanup, overhead and profit.

Material	Craft@Hrs	Unit	Material Cost	Labor Cost	Installed Cost

Ferrule type renewable 250 volt class H cartridge fuses with links

Material	Craft@Hrs	Unit	Material Cost	Labor Cost	Installed Cost
1 amp	L1@0.06	Ea	24.50	2.46	26.96
2 amp	L1@0.06	Ea	24.50	2.46	26.96
3 amp	L1@0.06	Ea	23.40	2.46	25.86
4 amp	L1@0.06	Ea	24.50	2.46	26.96
5 amp	L1@0.06	Ea	23.40	2.46	25.86
6 amp	L1@0.06	Ea	20.90	2.46	23.36
8 amp	L1@0.06	Ea	24.50	2.46	26.96
10 amp	L1@0.06	Ea	18.90	2.46	21.36
12 amp	L1@0.06	Ea	22.60	2.46	25.06
15 amp	L1@0.06	Ea	20.30	2.46	22.76
20 amp	L1@0.06	Ea	20.30	2.46	22.76
25 amp	L1@0.06	Ea	22.60	2.46	25.06
30 amp	L1@0.06	Ea	15.30	2.46	17.76
35 amp	L1@0.06	Ea	44.40	2.46	46.86
40 amp	L1@0.06	Ea	42.70	2.46	45.16
45 amp	L1@0.06	Ea	44.40	2.46	46.86
50 amp	L1@0.06	Ea	42.70	2.46	45.16
60 amp	L1@0.06	Ea	29.60	2.46	32.06

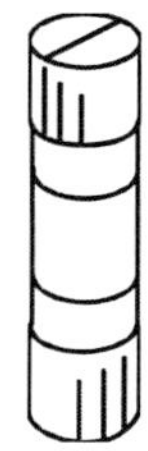

Blade type renewable 250 volt class H cartridge fuses with links

Material	Craft@Hrs	Unit	Material Cost	Labor Cost	Installed Cost
70 amp	L1@0.08	Ea	96.10	3.28	99.38
80 amp	L1@0.08	Ea	99.70	3.28	102.98
90 amp	L1@0.08	Ea	99.70	3.28	102.98
100 amp	L1@0.08	Ea	66.90	3.28	70.18
110 amp	L1@0.08	Ea	226.00	3.28	229.28
125 amp	L1@0.08	Ea	226.00	3.28	229.28
150 amp	L1@0.08	Ea	198.00	3.28	201.28
175 amp	L1@0.08	Ea	226.00	3.28	229.28
200 amp	L1@0.08	Ea	171.00	3.28	174.28
225 amp	L1@0.10	Ea	407.00	4.10	411.10
250 amp	L1@0.10	Ea	407.00	4.10	411.10
300 amp	L1@0.10	Ea	407.00	4.10	411.10
350 amp	L1@0.10	Ea	407.00	4.10	411.10
400 amp	L1@0.15	Ea	273.00	6.15	279.15
450 amp	L1@0.15	Ea	624.00	6.15	630.15
500 amp	L1@0.15	Ea	624.00	6.15	630.15
600 amp	L1@0.15	Ea	624.00	6.15	630.15

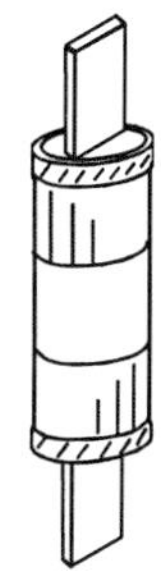

Use these figures to estimate the cost of fuses installed in safety switches and combination starters under the conditions described on pages 5 and 6. Costs listed are for each fuse installed. The crew size is one electrician working at a labor cost of $41.03 per manhour. These costs include layout, material handling, and normal waste. Add for sales tax, delivery supervision, mobilization, demobilization, cleanup, overhead and profit.

Cartridge Fuses, Links, 250 Volt, Class H

Material	Craft@Hrs	Unit	Material Cost	Labor Cost	Installed Cost
Links only for 250 volt class H cartridge fuses					
1 amp	L1@0.06	Ea	5.05	2.46	7.51
2 amp	L1@0.06	Ea	5.05	2.46	7.51
3 amp	L1@0.06	Ea	5.05	2.46	7.51
4 amp	L1@0.06	Ea	5.05	2.46	7.51
5 amp	L1@0.06	Ea	4.85	2.46	7.31
6 amp	L1@0.06	Ea	4.85	2.46	7.31
8 amp	L1@0.06	Ea	5.05	2.46	7.51
10 amp	L1@0.06	Ea	4.47	2.46	6.93
12 amp	L1@0.06	Ea	1.63	2.46	4.09
15 amp	L1@0.06	Ea	1.63	2.46	4.09
20 amp	L1@0.06	Ea	1.22	2.46	3.68
25 amp	L1@0.06	Ea	1.63	2.46	4.09
30 amp	L1@0.06	Ea	1.10	2.46	3.56
35 amp	L1@0.06	Ea	3.07	2.46	5.53
40 amp	L1@0.06	Ea	3.07	2.46	5.53
45 amp	L1@0.06	Ea	3.07	2.46	5.53
50 amp	L1@0.06	Ea	3.07	2.46	5.53
60 amp	L1@0.06	Ea	2.33	2.46	4.79
70 amp	L1@0.08	Ea	5.95	3.28	9.23
80 amp	L1@0.08	Ea	5.95	3.28	9.23
90 amp	L1@0.08	Ea	5.95	3.28	9.23
100 amp	L1@0.08	Ea	4.00	3.28	7.28
110 amp	L1@0.08	Ea	12.60	3.28	15.88
125 amp	L1@0.08	Ea	12.60	3.28	15.88
150 amp	L1@0.08	Ea	12.60	3.28	15.88
175 amp	L1@0.08	Ea	12.60	3.28	15.88
200 amp	L1@0.08	Ea	8.24	3.28	11.52
225 amp	L1@0.10	Ea	24.20	4.10	28.30
250 amp	L1@0.10	Ea	23.10	4.10	27.20
300 amp	L1@0.10	Ea	23.10	4.10	27.20
350 amp	L1@0.10	Ea	23.10	4.10	27.20
400 amp	L1@0.15	Ea	21.70	6.15	27.85
450 amp	L1@0.15	Ea	36.80	6.15	42.95
500 amp	L1@0.15	Ea	36.80	6.15	42.95
600 amp	L1@0.15	Ea	35.20	6.15	41.35

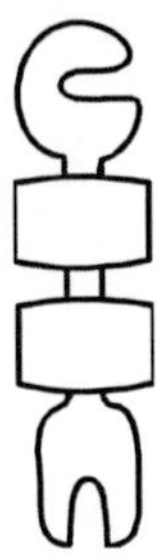

Use these figures to estimate the cost of fuses installed in safety switches and combination starters under the conditions described on pages 5 and 6. Costs listed are for each fuse installed. The crew size is one electrician working at a labor cost of $41.03 per manhour. These costs include layout, material handling, and normal waste. Add for sales tax, delivery supervision, mobilization, demobilization, cleanup, overhead and profit.

Material	Craft@Hrs	Unit	Material Cost	Labor Cost	Installed Cost
Ferrule type renewable 600 volt class H cartridge fuses with links					
1 amp	L1@0.06	Ea	62.60	2.46	65.06
2 amp	L1@0.06	Ea	62.60	2.46	65.06
3 amp	L1@0.06	Ea	62.60	2.46	65.06
4 amp	L1@0.06	Ea	62.60	2.46	65.06
5 amp	L1@0.06	Ea	62.60	2.46	65.06
6 amp	L1@0.06	Ea	58.60	2.46	61.06
8 amp	L1@0.06	Ea	62.60	2.46	65.06
10 amp	L1@0.06	Ea	55.20	2.46	57.66
12 amp	L1@0.06	Ea	56.10	2.46	58.56
15 amp	L1@0.06	Ea	50.10	2.46	52.56
20 amp	L1@0.06	Ea	43.20	2.46	45.66
25 amp	L1@0.06	Ea	53.70	2.46	56.16
30 amp	L1@0.06	Ea	37.60	2.46	40.06
35 amp	L1@0.06	Ea	84.50	2.46	86.96
40 amp	L1@0.06	Ea	84.50	2.46	86.96
45 amp	L1@0.06	Ea	88.30	2.46	90.76
50 amp	L1@0.06	Ea	84.50	2.46	86.96
60 amp	L1@0.06	Ea	59.10	2.46	61.56

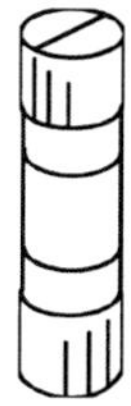

Material	Craft@Hrs	Unit	Material Cost	Labor Cost	Installed Cost
Blade type renewable 600 volt class H cartridge fuses with links					
70 amp	L1@0.08	Ea	192.00	3.28	195.28
80 amp	L1@0.08	Ea	192.00	3.28	195.28
90 amp	L1@0.08	Ea	199.00	3.28	202.28
100 amp	L1@0.08	Ea	134.00	3.28	137.28
110 amp	L1@0.08	Ea	387.00	3.28	390.28
125 amp	L1@0.08	Ea	376.00	3.28	379.28
150 amp	L1@0.08	Ea	376.00	3.28	379.28
175 amp	L1@0.08	Ea	387.00	3.28	390.28
200 amp	L1@0.08	Ea	260.00	3.28	263.28
225 amp	L1@0.10	Ea	785.00	4.10	789.10
250 amp	L1@0.10	Ea	785.00	4.10	789.10
300 amp	L1@0.10	Ea	785.00	4.10	789.10
350 amp	L1@0.10	Ea	785.00	4.10	789.10
400 amp	L1@0.15	Ea	603.00	6.15	609.15
450 amp	L1@0.15	Ea	624.00	6.15	630.15
500 amp	L1@0.15	Ea	1,110.00	6.15	1,116.15
600 amp	L1@0.15	Ea	997.00	6.15	1,003.15

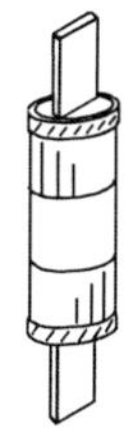

Use these figures to estimate the cost of fuses installed in safety switches and combination starters under the conditions described on pages 5 and 6. Costs listed are for each fuse installed. The crew size is one electrician working at a labor cost of $41.03 per manhour. These costs include layout, material handling, and normal waste. Add for sales tax, delivery supervision, mobilization, demobilization, cleanup, overhead and profit.

Cartridge Fuses Links, 600 Volt, Class H

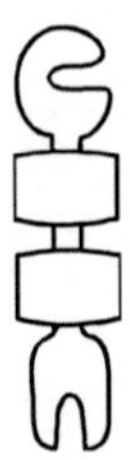

Links only for 600 volt class H cartridge fuses

Material	Craft@Hrs	Unit	Material Cost	Labor Cost	Installed Cost
1 amp	L1@0.06	Ea	8.03	2.46	10.49
3 amp	L1@0.06	Ea	8.03	2.46	10.49
4 amp	L1@0.06	Ea	8.03	2.46	10.49
5 amp	L1@0.06	Ea	8.03	2.46	10.49
6 amp	L1@0.06	Ea	8.03	2.46	10.49
8 amp	L1@0.06	Ea	8.03	2.46	10.49
10 amp	L1@0.06	Ea	8.03	2.46	10.49
12 amp	L1@0.06	Ea	7.26	2.46	9.72
15 amp	L1@0.06	Ea	3.21	2.46	5.67
20 amp	L1@0.06	Ea	2.19	2.46	4.65
25 amp	L1@0.06	Ea	2.19	2.46	4.65
30 amp	L1@0.06	Ea	3.21	2.46	5.67
35 amp	L1@0.06	Ea	2.19	2.46	4.65
40 amp	L1@0.06	Ea	5.95	2.46	8.41
45 amp	L1@0.06	Ea	5.95	2.46	8.41
50 amp	L1@0.06	Ea	5.95	2.46	8.41
60 amp	L1@0.06	Ea	5.95	2.46	8.41
70 amp	L1@0.08	Ea	12.40	3.28	15.68
80 amp	L1@0.08	Ea	12.40	3.28	15.68
90 amp	L1@0.08	Ea	12.40	3.28	15.68
100 amp	L1@0.08	Ea	8.24	3.28	11.52
110 amp	L1@0.08	Ea	22.60	3.28	25.88
125 amp	L1@0.08	Ea	22.60	3.28	25.88
150 amp	L1@0.08	Ea	22.60	3.28	25.88
175 amp	L1@0.08	Ea	22.60	3.28	25.88
200 amp	L1@0.08	Ea	17.20	3.28	20.48
225 amp	L1@0.10	Ea	42.70	4.10	46.80
250 amp	L1@0.10	Ea	42.70	4.10	46.80
300 amp	L1@0.10	Ea	35.30	4.10	39.40
350 amp	L1@0.10	Ea	35.30	4.10	39.40
400 amp	L1@0.15	Ea	40.90	6.15	47.05
450 amp	L1@0.15	Ea	59.10	6.15	65.25
500 amp	L1@0.15	Ea	59.10	6.15	65.25
600 amp	L1@0.15	Ea	59.10	6.15	65.25

Use these figures to estimate the cost of fuses installed in safety switches and combination starters under the conditions described on pages 5 and 6. Costs listed are for each fuse installed. The crew size is one electrician working at a labor cost of $41.03 per manhour. These costs include layout, material handling, and normal waste. Add for sales tax, delivery supervision, mobilization, demobilization, cleanup, overhead and profit.

Cartridge Fuses, Non-Time Delay, 600 Volt, Current Limiting, Class J

Ferrule type non-time delay 600 volt current limiting class J cartridge fuses

Material	Craft@Hrs	Unit	Material Cost	Labor Cost	Installed Cost
1 amp	L1@0.06	Ea	48.50	2.46	50.96
3 amp	L1@0.06	Ea	46.90	2.46	49.36
4 amp	L1@0.06	Ea	48.50	2.46	50.96
5 amp	L1@0.06	Ea	48.50	2.46	50.96
6 amp	L1@0.06	Ea	48.50	2.46	50.96
10 amp	L1@0.06	Ea	37.60	2.46	40.06
15 amp	L1@0.06	Ea	37.60	2.46	40.06
20 amp	L1@0.06	Ea	37.60	2.46	40.06
25 amp	L1@0.06	Ea	46.90	2.46	49.36
30 amp	L1@0.06	Ea	32.00	2.46	34.46
35 amp	L1@0.06	Ea	76.00	2.46	78.46
40 amp	L1@0.06	Ea	61.00	2.46	63.46
45 amp	L1@0.06	Ea	76.20	2.46	78.66
50 amp	L1@0.06	Ea	61.00	2.46	63.46
60 amp	L1@0.06	Ea	52.10	2.46	54.56

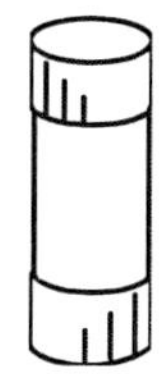

Blade type non-time delay 600 volt current limiting class J cartridge fuses

Material	Craft@Hrs	Unit	Material Cost	Labor Cost	Installed Cost
70 amp	L1@0.08	Ea	99.70	3.28	102.98
80 amp	L1@0.08	Ea	115.00	3.28	118.28
90 amp	L1@0.08	Ea	122.00	3.28	125.28
100 amp	L1@0.08	Ea	79.00	3.28	82.28
110 amp	L1@0.08	Ea	223.00	3.28	226.28
125 amp	L1@0.08	Ea	214.00	3.28	217.28
150 amp	L1@0.08	Ea	185.00	3.28	188.28
175 amp	L1@0.08	Ea	214.00	3.28	217.28
200 amp	L1@0.08	Ea	147.00	3.28	150.28
225 amp	L1@0.10	Ea	475.00	4.10	479.10
250 amp	L1@0.10	Ea	475.00	4.10	479.10
300 amp	L1@0.10	Ea	376.00	4.10	380.10
350 amp	L1@0.10	Ea	475.00	4.10	479.10
400 amp	L1@0.15	Ea	326.00	6.15	332.15
450 amp	L1@0.15	Ea	690.00	6.15	696.15
500 amp	L1@0.15	Ea	669.00	6.15	675.15
600 amp	L1@0.15	Ea	460.00	6.15	466.15

Use these figures to estimate the cost of fuses installed in safety switches and combination starters under the conditions described on pages 5 and 6. Costs listed are for each fuse installed. The crew size is one electrician working at a labor cost of $41.03 per manhour. These costs include layout, material handling, and normal waste. Add for sales tax, delivery supervision, mobilization, demobilization, cleanup, overhead and profit.

Cartridge Fuses, Time Delay, 600 Volt, Current Limiting, Class J

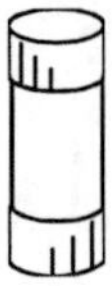

Material	Craft@Hrs	Unit	Material Cost	Labor Cost	Installed Cost
Ferrule type time delay 600 volt current limiting class J cartridge fuses					
1 amp	L1@0.06	Ea	26.20	2.46	28.66
3 amp	L1@0.06	Ea	26.20	2.46	28.66
6 amp	L1@0.06	Ea	26.20	2.46	28.66
10 amp	L1@0.06	Ea	26.20	2.46	28.66
15 amp	L1@0.06	Ea	26.20	2.46	28.66
20 amp	L1@0.06	Ea	26.20	2.46	28.66
25 amp	L1@0.06	Ea	26.20	2.46	28.66
30 amp	L1@0.06	Ea	26.20	2.46	28.66
35 amp	L1@0.06	Ea	45.40	2.46	47.86
40 amp	L1@0.06	Ea	45.40	2.46	47.86
45 amp	L1@0.06	Ea	45.40	2.46	47.86
50 amp	L1@0.06	Ea	45.40	2.46	47.86
60 amp	L1@0.06	Ea	45.40	2.46	47.86
Blade type time delay 600 volt current limiting class J cartridge fuses					
70 amp	L1@0.08	Ea	92.90	3.28	96.18
80 amp	L1@0.08	Ea	92.90	3.28	96.18
90 amp	L1@0.08	Ea	112.00	3.28	115.28
100 amp	L1@0.08	Ea	92.90	3.28	96.18
110 amp	L1@0.08	Ea	226.00	3.28	229.28
125 amp	L1@0.08	Ea	186.00	3.28	189.28
150 amp	L1@0.08	Ea	186.00	3.28	189.28
175 amp	L1@0.08	Ea	186.00	3.28	189.28
200 amp	L1@0.08	Ea	186.00	3.28	189.28
225 amp	L1@0.10	Ea	456.00	4.10	460.10
250 amp	L1@0.10	Ea	371.00	4.10	375.10
300 amp	L1@0.10	Ea	371.00	4.10	375.10
350 amp	L1@0.10	Ea	456.00	4.10	460.10
400 amp	L1@0.15	Ea	371.00	6.15	377.15
450 amp	L1@0.15	Ea	747.00	6.15	753.15
500 amp	L1@0.15	Ea	747.00	6.15	753.15
600 amp	L1@0.15	Ea	645.00	6.15	651.15

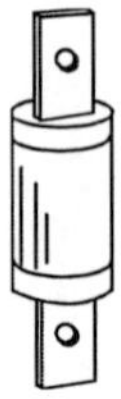

Use these figures to estimate the cost of fuses installed in safety switches and combination starters under the conditions described on pages 5 and 6. Costs listed are for each fuse installed. The crew size is one electrician working at a labor cost of $41.03 per manhour. These costs include layout, material handling, and normal waste. Add for sales tax, delivery supervision, mobilization, demobilization, cleanup, overhead and profit.

Bolt-on Type Cartridge Fuses, Non-Time Delay, 600 Volt, Current Limiting, Class L

Material	Craft@Hrs	Unit	Material Cost	Labor Cost	Installed Cost
Non-time delay 600 volt current limiting class L cartridge fuses, bolt-on type					
600 amp	L1@0.40	Ea	1,050.00	16.40	1,066.40
650 amp	L1@0.40	Ea	1,510.00	16.40	1,526.40
700 amp	L1@0.40	Ea	1,050.00	16.40	1,066.40
800 amp	L1@0.40	Ea	863.00	16.40	879.40
1000 amp	L1@0.40	Ea	863.00	16.40	879.40
1200 amp	L1@0.50	Ea	863.00	20.50	883.50
1350 amp	L1@0.50	Ea	1,960.00	20.50	1,980.50
1500 amp	L1@0.50	Ea	1,960.00	20.50	1,980.50
1600 amp	L1@0.50	Ea	1,100.00	20.50	1,120.50
1800 amp	L1@0.50	Ea	2,600.00	20.50	2,620.50
2000 amp	L1@0.60	Ea	1,470.00	24.60	1,494.60
2500 amp	L1@0.60	Ea	1,960.00	24.60	1,984.60
3000 amp	L1@0.60	Ea	2,260.00	24.60	2,284.60
3500 amp	L1@0.60	Ea	4,600.00	24.60	4,624.60
4000 amp	L1@0.75	Ea	3,090.00	30.80	3,120.80
4500 amp	L1@0.75	Ea	6,800.00	30.80	6,830.80
5000 amp	L1@0.75	Ea	4,870.00	30.80	4,900.80
6000 amp	L1@0.75	Ea	8,080.00	30.80	8,110.80

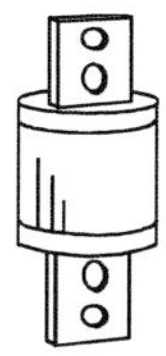

Time delay 600 volt current limiting class L cartridge fuses, bolt-on type

600 amp	L1@0.40	Ea	1,380.00	16.40	1,396.40
650 amp	L1@0.40	Ea	1,430.00	16.40	1,446.40
700 amp	L1@0.40	Ea	1,380.00	16.40	1,396.40
750 amp	L1@0.40	Ea	1,460.00	16.40	1,476.40
800 amp	L1@0.40	Ea	876.00	16.40	892.40
900 amp	L1@0.50	Ea	1,340.00	20.50	1,360.50
1000 amp	L1@0.50	Ea	1,100.00	20.50	1,120.50
1200 amp	L1@0.50	Ea	1,100.00	20.50	1,120.50
1350 amp	L1@0.50	Ea	1,920.00	20.50	1,940.50
1400 amp	L1@0.50	Ea	1,920.00	20.50	1,940.50
1500 amp	L1@0.60	Ea	1,920.00	24.60	1,944.60
1600 amp	L1@0.60	Ea	1,490.00	24.60	1,514.60
1800 amp	L1@0.60	Ea	2,570.00	24.60	2,594.60
2000 amp	L1@0.60	Ea	2,280.00	24.60	2,304.60
2500 amp	L1@0.60	Ea	3,260.00	24.60	3,284.60
3000 amp	L1@0.60	Ea	3,530.00	24.60	3,554.60
3500 amp	L1@0.75	Ea	5,400.00	30.80	5,430.80
4000 amp	L1@0.75	Ea	4,540.00	30.80	4,570.80
4500 amp	L1@0.75	Ea	6,560.00	30.80	6,590.80
5000 amp	L1@0.75	Ea	6,560.00	30.80	6,590.80
6000 amp	L1@0.55	Ea	8,280.00	22.60	8,302.60

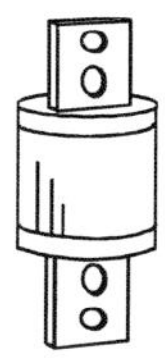

Use these figures to estimate the cost of fuses installed in safety switches and combination starters under the conditions described on pages 5 and 6. Costs listed are for each fuse installed. The crew size is one electrician working at a labor cost of $41.03 per manhour. These costs include layout, material handling, and normal waste. Add for sales tax, delivery supervision, mobilization, demobilization, cleanup, overhead and profit.

Cartridge Fuses, Non-Time Delay, 250 Volt, Current Limiting, Class RK1

Material	Craft@Hrs	Unit	Material Cost	Labor Cost	Installed Cost
Ferrule type non-time delay 250 volt current limiting class RK1 cartridge fuses					
1 amp	L1@0.06	Ea	37.10	2.46	39.56
3 amp	L1@0.06	Ea	37.10	2.46	39.56
5 amp	L1@0.06	Ea	37.10	2.46	39.56
6 amp	L1@0.06	Ea	37.10	2.46	39.56
10 amp	L1@0.06	Ea	37.10	2.46	39.56
15 amp	L1@0.06	Ea	35.40	2.46	37.86
20 amp	L1@0.06	Ea	35.40	2.46	37.86
25 amp	L1@0.06	Ea	37.10	2.46	39.56
30 amp	L1@0.06	Ea	35.40	2.46	37.86
35 amp	L1@0.06	Ea	84.00	2.46	86.46
40 amp	L1@0.06	Ea	87.80	2.46	90.26
45 amp	L1@0.06	Ea	87.80	2.46	90.26
50 amp	L1@0.06	Ea	87.80	2.46	90.26
60 amp	L1@0.06	Ea	78.10	2.46	80.56

Material	Craft@Hrs	Unit	Material Cost	Labor Cost	Installed Cost
Blade type non-time delay 250 volt current limiting class RK1 cartridge fuses					
70 amp	L1@0.08	Ea	180.00	3.28	183.28
80 amp	L1@0.08	Ea	186.00	3.28	189.28
90 amp	L1@0.08	Ea	186.00	3.28	189.28
100 amp	L1@0.08	Ea	166.00	3.28	169.28
110 amp	L1@0.08	Ea	371.00	3.28	374.28
125 amp	L1@0.08	Ea	354.00	3.28	357.28
150 amp	L1@0.08	Ea	306.00	3.28	309.28
175 amp	L1@0.08	Ea	354.00	3.28	357.28
200 amp	L1@0.08	Ea	287.00	3.28	290.28
225 amp	L1@0.10	Ea	710.00	4.10	714.10
250 amp	L1@0.10	Ea	710.00	4.10	714.10
300 amp	L1@0.10	Ea	710.00	4.10	714.10
350 amp	L1@0.10	Ea	737.00	4.10	741.10
400 amp	L1@0.15	Ea	664.00	6.15	670.15
450 amp	L1@0.15	Ea	963.00	6.15	969.15
500 amp	L1@0.15	Ea	963.00	6.15	969.15
600 amp	L1@0.15	Ea	860.00	6.15	866.15

Use these figures to estimate the cost of fuses installed in safety switches and combination starters under the conditions described on pages 5 and 6. Costs listed are for each fuse installed. The crew size is one electrician working at a labor cost of $41.03 per manhour. These costs include layout, material handling, and normal waste. Add for sales tax, delivery supervision, mobilization, demobilization, cleanup, overhead and profit.

Material	Craft@Hrs	Unit	Material Cost	Labor Cost	Installed Cost
Ferrule type time delay 250 volt current limiting class RK1 cartridge fuses					
1/10 amp	L1@0.06	Ea	29.30	2.46	31.76
15/100 amp	L1@0.06	Ea	29.30	2.46	31.76
2/10 amp	L1@0.06	Ea	29.30	2.46	31.76
3/10 amp	L1@0.06	Ea	30.00	2.46	32.46
4/10 amp	L1@0.06	Ea	29.30	2.46	31.76
1/2 amp	L1@0.06	Ea	22.30	2.46	24.76
6/10 amp	L1@0.06	Ea	25.70	2.46	28.16
8/10 amp	L1@0.06	Ea	27.50	2.46	29.96
1 amp	L1@0.06	Ea	18.70	2.46	21.16
1-1/8 amp	L1@0.06	Ea	29.30	2.46	31.76
1-1/4 amp	L1@0.06	Ea	27.50	2.46	29.96
1-4/10 amp	L1@0.06	Ea	29.30	2.46	31.76
1-6/10 amp	L1@0.06	Ea	19.20	2.46	21.66
2 amp	L1@0.06	Ea	29.30	2.46	31.76
2-1/4 amp	L1@0.06	Ea	15.50	2.46	17.96
2-1/2 amp	L1@0.06	Ea	29.30	2.46	31.76
2-8/10 amp	L1@0.06	Ea	21.80	2.46	24.26
3 amp	L1@0.06	Ea	17.20	2.46	19.66
3-2/10 amp	L1@0.06	Ea	17.20	2.46	19.66
3-1/2 amp	L1@0.06	Ea	19.90	2.46	22.36
4 amp	L1@0.06	Ea	17.20	2.46	19.66
4-1/2 amp	L1@0.06	Ea	19.90	2.46	22.36
5 amp	L1@0.06	Ea	14.30	2.46	16.76
5-6/10 amp	L1@0.06	Ea	24.60	2.46	27.06
6 amp	L1@0.06	Ea	17.20	2.46	19.66
6-1/4 amp	L1@0.06	Ea	19.80	2.46	22.26
7 amp	L1@0.06	Ea	19.80	2.46	22.26
8 amp	L1@0.06	Ea	17.20	2.46	19.66
9 amp	L1@0.06	Ea	23.30	2.46	25.76
10 amp	L1@0.06	Ea	14.30	2.46	16.76
12 amp	L1@0.06	Ea	17.20	2.46	19.66
15 amp	L1@0.06	Ea	11.60	2.46	14.06
17-1/2 amp	L1@0.06	Ea	14.30	2.46	16.76

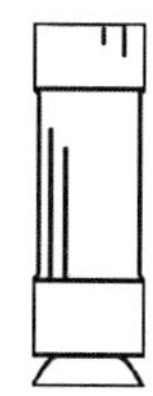

Use these figures to estimate the cost of fuses installed in safety switches and combination starters under the conditions described on pages 5 and 6. Costs listed are for each fuse installed. The crew size is one electrician working at a labor cost of $41.03 per manhour. These costs include layout, material handling, and normal waste. Add for sales tax, delivery supervision, mobilization, demobilization, cleanup, overhead and profit.

Cartridge Fuses, Time Delay, 250 Volt, Current Limiting, Class RK1

Material	Craft@Hrs	Unit	Material Cost	Labor Cost	Installed Cost
Ferrule type time delay 250 volt current limiting class RK1 cartridge fuses					
20 amp	L1@0.06	Ea	11.60	2.46	14.06
25 amp	L1@0.06	Ea	11.60	2.46	14.06
30 amp	L1@0.06	Ea	11.60	2.46	14.06
35 amp	L1@0.06	Ea	21.20	2.46	23.66
40 amp	L1@0.06	Ea	21.20	2.46	23.66
45 amp	L1@0.06	Ea	21.20	2.46	23.66
50 amp	L1@0.06	Ea	21.20	2.46	23.66
60 amp	L1@0.06	Ea	21.20	2.46	23.66

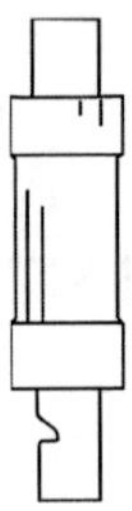

Material	Craft@Hrs	Unit	Material Cost	Labor Cost	Installed Cost
Blade type time delay 250 volt current limiting class RK1 cartridge fuses					
70 amp	L1@0.08	Ea	47.80	3.28	51.08
80 amp	L1@0.08	Ea	47.80	3.28	51.08
90 amp	L1@0.08	Ea	57.90	3.28	61.18
100 amp	L1@0.08	Ea	47.70	3.28	50.98
110 amp	L1@0.08	Ea	150.00	3.28	153.28
125 amp	L1@0.08	Ea	105.00	3.28	108.28
150 amp	L1@0.08	Ea	105.00	3.28	108.28
175 amp	L1@0.08	Ea	128.00	3.28	131.28
200 amp	L1@0.08	Ea	105.00	3.28	108.28
225 amp	L1@0.10	Ea	228.00	4.10	232.10
250 amp	L1@0.10	Ea	228.00	4.10	232.10
300 amp	L1@0.10	Ea	228.00	4.10	232.10
350 amp	L1@0.10	Ea	233.00	4.10	237.10
400 amp	L1@0.15	Ea	189.00	6.15	195.15
450 amp	L1@0.15	Ea	444.00	6.15	450.15
500 amp	L1@0.15	Ea	356.00	6.15	362.15
600 amp	L1@0.15	Ea	289.00	6.15	295.15

Use these figures to estimate the cost of fuses installed in safety switches and combination starters under the conditions described on pages 5 and 6. Costs listed are for each fuse installed. The crew size is one electrician working at a labor cost of $41.03 per manhour. These costs include layout, material handling, and normal waste. Add for sales tax, delivery supervision, mobilization, demobilization, cleanup, overhead and profit.

Material	Craft@Hrs	Unit	Material Cost	Labor Cost	Installed Cost
Ferrule type non-time delay 600 volt current limiting class RK1 cartridge fuses					
1 amp	L1@0.06	Ea	46.90	2.46	49.36
3 amp	L1@0.06	Ea	46.90	2.46	49.36
4 amp	L1@0.06	Ea	46.90	2.46	49.36
5 amp	L1@0.06	Ea	46.90	2.46	49.36
6 amp	L1@0.06	Ea	46.90	2.46	49.36
10 amp	L1@0.06	Ea	46.90	2.46	49.36
15 amp	L1@0.06	Ea	44.80	2.46	47.26
20 amp	L1@0.06	Ea	44.80	2.46	47.26
25 amp	L1@0.06	Ea	46.90	2.46	49.36
30 amp	L1@0.06	Ea	44.80	2.46	47.26
35 amp	L1@0.06	Ea	101.00	2.46	103.46
40 amp	L1@0.06	Ea	96.30	2.46	98.76
45 amp	L1@0.06	Ea	101.00	2.46	103.46
50 amp	L1@0.06	Ea	101.00	2.46	103.46
60 amp	L1@0.06	Ea	96.30	2.46	98.76

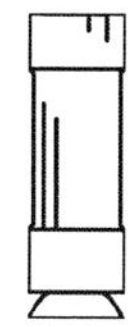

Material	Craft@Hrs	Unit	Material Cost	Labor Cost	Installed Cost
Blade type non-time delay 600 volt current limiting class RK1 cartridge fuses					
70 amp	L1@0.08	Ea	209.00	3.28	212.28
80 amp	L1@0.08	Ea	209.00	3.28	212.28
90 amp	L1@0.08	Ea	217.00	3.28	220.28
100 amp	L1@0.08	Ea	166.00	3.28	169.28
110 amp	L1@0.08	Ea	407.00	3.28	410.28
125 amp	L1@0.08	Ea	394.00	3.28	397.28
150 amp	L1@0.08	Ea	358.00	3.28	361.28
175 amp	L1@0.08	Ea	394.00	3.28	397.28
200 amp	L1@0.08	Ea	309.00	3.28	312.28
225 amp	L1@0.10	Ea	780.00	4.10	784.10
250 amp	L1@0.10	Ea	780.00	4.10	784.10
300 amp	L1@0.10	Ea	726.00	4.10	730.10
350 amp	L1@0.10	Ea	780.00	4.10	784.10
400 amp	L1@0.15	Ea	726.00	6.15	732.15
450 amp	L1@0.15	Ea	1,060.00	6.15	1,066.15
500 amp	L1@0.15	Ea	1,060.00	6.15	1,066.15
600 amp	L1@0.15	Ea	987.00	6.15	993.15

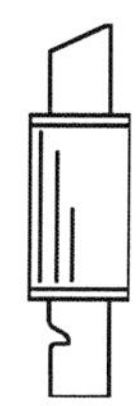

Use these figures to estimate the cost of fuses installed in safety switches and combination starters under the conditions described on pages 5 and 6. Costs listed are for each fuse installed. The crew size is one electrician working at a labor cost of $41.03 per manhour. These costs include layout, material handling, and normal waste. Add for sales tax, delivery supervision, mobilization, demobilization, cleanup, overhead and profit.

Cartridge Fuses, Time Delay, 600 Volt, Current Limiting, Class RK1

Material	Craft@Hrs	Unit	Material Cost	Labor Cost	Installed Cost
Ferrule type time delay 600 volt current limiting class RK1 cartridge fuses					
1/10 amp	L1@0.06	Ea	55.80	2.46	58.26
15/100 amp	L1@0.06	Ea	44.00	2.46	46.46
2/10 amp	L1@0.06	Ea	55.80	2.46	58.26
3/10 amp	L1@0.06	Ea	55.80	2.46	58.26
4/10 amp	L1@0.06	Ea	55.80	2.46	58.26
1/2 amp	L1@0.06	Ea	52.30	2.46	54.76
6/10 amp	L1@0.06	Ea	52.30	2.46	54.76
8/10 amp	L1@0.06	Ea	52.30	2.46	54.76
1 amp	L1@0.06	Ea	53.50	2.46	55.96
1-1/8 amp	L1@0.06	Ea	37.10	2.46	39.56
1-1/4 amp	L1@0.06	Ea	53.50	2.46	55.96
1-4/10 amp	L1@0.06	Ea	37.00	2.46	39.46
1-6/10 amp	L1@0.06	Ea	37.00	2.46	39.46
1-8/10 amp	L1@0.06	Ea	49.30	2.46	51.76
2 amp	L1@0.06	Ea	30.10	2.46	32.56
2-1/4 amp	L1@0.06	Ea	37.10	2.46	39.56
2-1/2 amp	L1@0.06	Ea	37.00	2.46	39.46
2-8/10 amp	L1@0.06	Ea	42.60	2.46	45.06
3 amp	L1@0.06	Ea	27.90	2.46	30.36
3-1/2 amp	L1@0.06	Ea	34.00	2.46	36.46
4 amp	L1@0.06	Ea	27.90	2.46	30.36
4-1/2 amp	L1@0.06	Ea	40.00	2.46	42.46
5 amp	L1@0.06	Ea	27.90	2.46	30.36
5-6/10 amp	L1@0.06	Ea	40.00	2.46	42.46
6 amp	L1@0.06	Ea	27.90	2.46	30.36
6-1/4 amp	L1@0.06	Ea	34.00	2.46	36.46
7 amp	L1@0.06	Ea	34.00	2.46	36.46
8 amp	L1@0.06	Ea	27.90	2.46	30.36
9 amp	L1@0.06	Ea	34.00	2.46	36.46
10 amp	L1@0.06	Ea	27.90	2.46	30.36
12 amp	L1@0.06	Ea	27.90	2.46	30.36
15 amp	L1@0.06	Ea	24.80	2.46	27.26
17-1/2 amp	L1@0.06	Ea	24.80	2.46	27.26

Use these figures to estimate the cost of fuses installed in safety switches and combination starters under the conditions described on pages 5 and 6. Costs listed are for each fuse installed. The crew size is one electrician working at a labor cost of $41.03 per manhour. These costs include layout, material handling, and normal waste. Add for sales tax, delivery supervision, mobilization, demobilization, cleanup, overhead and profit.

Material	Craft@Hrs	Unit	Material Cost	Labor Cost	Installed Cost
Ferrule type time delay 600 volt current limiting class RK1 cartridge fuses					
20 amp	L1@0.06	Ea	24.80	2.46	27.26
25 amp	L1@0.06	Ea	24.80	2.46	27.26
30 amp	L1@0.06	Ea	24.80	2.46	27.26
35 amp	L1@0.06	Ea	42.70	2.46	45.16
40 amp	L1@0.06	Ea	42.70	2.46	45.16
45 amp	L1@0.06	Ea	42.70	2.46	45.16
50 amp	L1@0.06	Ea	42.70	2.46	45.16
60 amp	L1@0.06	Ea	42.70	2.46	45.16
Blade type time delay 600 volt current limiting class RK1 cartridge fuses					
70 amp	L1@0.08	Ea	88.30	3.28	91.58
80 amp	L1@0.08	Ea	88.30	3.28	91.58
90 amp	L1@0.08	Ea	88.30	3.28	91.58
100 amp	L1@0.08	Ea	88.30	3.28	91.58
110 amp	L1@0.08	Ea	178.00	3.28	181.28
125 amp	L1@0.08	Ea	178.00	3.28	181.28
150 amp	L1@0.08	Ea	178.00	3.28	181.28
200 amp	L1@0.08	Ea	178.00	3.28	181.28
225 amp	L1@0.10	Ea	353.00	4.10	357.10
250 amp	L1@0.10	Ea	353.00	4.10	357.10
300 amp	L1@0.10	Ea	353.00	4.10	357.10
350 amp	L1@0.10	Ea	408.00	4.10	412.10
400 amp	L1@0.15	Ea	353.00	6.15	359.15
450 amp	L1@0.15	Ea	608.00	6.15	614.15
500 amp	L1@0.15	Ea	594.00	6.15	600.15
600 amp	L1@0.15	Ea	507.00	6.15	513.15

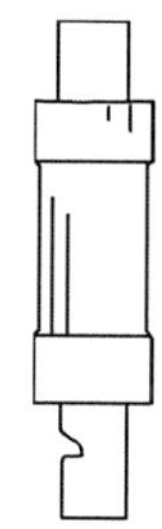

Use these figures to estimate the cost of fuses installed in safety switches and combination starters under the conditions described on pages 5 and 6. Costs listed are for each fuse installed. The crew size is one electrician working at a labor cost of $41.03 per manhour. These costs include layout, material handling, and normal waste. Add for sales tax, delivery supervision, mobilization, demobilization, cleanup, overhead and profit.

Cartridge Fuses, Time Delay, 250 Volt, Current Limiting, Class RK5

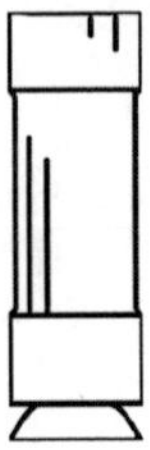

Material	Craft@Hrs	Unit	Material Cost	Labor Cost	Installed Cost
Ferrule type time delay 250 volt current limiting class RK5 cartridge fuses					
1/10 amp	L1@0.06	Ea	20.90	2.46	23.36
15/100 amp	L1@0.06	Ea	20.90	2.46	23.36
2/10 amp	L1@0.06	Ea	20.90	2.46	23.36
1/4 amp	L1@0.06	Ea	20.90	2.46	23.36
3/10 amp	L1@0.06	Ea	20.90	2.46	23.36
4/10 amp	L1@0.06	Ea	20.90	2.46	23.36
1/2 amp	L1@0.06	Ea	15.90	2.46	18.36
6/10 amp	L1@0.06	Ea	18.50	2.46	20.96
8/10 amp	L1@0.06	Ea	19.60	2.46	22.06
1 amp	L1@0.06	Ea	13.40	2.46	15.86
1-1/8 amp	L1@0.06	Ea	20.90	2.46	23.36
1-1/4 amp	L1@0.06	Ea	19.60	2.46	22.06
1-4/10 amp	L1@0.06	Ea	20.90	2.46	23.36
1-6/10 amp	L1@0.06	Ea	13.70	2.46	16.16
1-8/10 amp	L1@0.06	Ea	20.90	2.46	23.36
2 amp	L1@0.06	Ea	11.10	2.46	13.56
2-1/4 amp	L1@0.06	Ea	20.90	2.46	23.36
2-1/2 amp	L1@0.06	Ea	15.50	2.46	17.96
2-8/10 amp	L1@0.06	Ea	19.60	2.46	22.06
3 amp	L1@0.06	Ea	12.40	2.46	14.86
3-2/10 amp	L1@0.06	Ea	12.40	2.46	14.86
3-1/2 amp	L1@0.06	Ea	14.30	2.46	16.76
4 amp	L1@0.06	Ea	12.40	2.46	14.86
4-1/2 amp	L1@0.06	Ea	14.30	2.46	16.76
5 amp	L1@0.06	Ea	10.20	2.46	12.66
5-6/10 amp	L1@0.06	Ea	18.00	2.46	20.46
6 amp	L1@0.06	Ea	13.30	2.46	15.76
6-1/4 amp	L1@0.06	Ea	14.30	2.46	16.76
7 amp	L1@0.06	Ea	14.40	2.46	16.86
8 amp	L1@0.06	Ea	13.30	2.46	15.76
9 amp	L1@0.06	Ea	16.70	2.46	19.16
10 amp	L1@0.06	Ea	10.20	2.46	12.66
12 amp	L1@0.06	Ea	13.30	2.46	15.76
15 amp	L1@0.06	Ea	8.28	2.46	10.74
17-1/2 amp	L1@0.06	Ea	10.10	2.46	12.56

Use these figures to estimate the cost of fuses installed in safety switches and combination starters under the conditions described on pages 5 and 6. Costs listed are for each fuse installed. The crew size is one electrician working at a labor cost of $41.03 per manhour. These costs include layout, material handling, and normal waste. Add for sales tax, delivery supervision, mobilization, demobilization, cleanup, overhead and profit.

Material	Craft@Hrs	Unit	Material Cost	Labor Cost	Installed Cost
Ferrule type time delay 250 volt current limiting class RK5 cartridge fuses					
20 amp	L1@0.06	Ea	8.28	2.46	10.74
25 amp	L1@0.06	Ea	8.28	2.46	10.74
30 amp	L1@0.06	Ea	8.28	2.46	10.74
35 amp	L1@0.06	Ea	15.30	2.46	17.76
40 amp	L1@0.06	Ea	15.30	2.46	17.76
45 amp	L1@0.06	Ea	15.30	2.46	17.76
50 amp	L1@0.06	Ea	15.30	2.46	17.76
60 amp	L1@0.06	Ea	15.30	2.46	17.76
Blade type time delay 250 volt current limiting class RK5 cartridge fuses					
70 amp	L1@0.08	Ea	33.80	3.28	37.08
75 amp	L1@0.08	Ea	50.70	3.28	53.98
80 amp	L1@0.08	Ea	33.80	3.28	37.08
90 amp	L1@0.08	Ea	40.90	3.28	44.18
100 amp	L1@0.08	Ea	33.80	3.28	37.08
110 amp	L1@0.08	Ea	107.00	3.28	110.28
125 amp	L1@0.08	Ea	74.80	3.28	78.08
150 amp	L1@0.08	Ea	74.80	3.28	78.08
175 amp	L1@0.08	Ea	91.10	3.28	94.38
200 amp	L1@0.08	Ea	74.80	3.28	78.08
225 amp	L1@0.10	Ea	164.00	4.10	168.10
250 amp	L1@0.10	Ea	164.00	4.10	168.10
300 amp	L1@0.10	Ea	164.00	4.10	168.10
350 amp	L1@0.10	Ea	167.00	4.10	171.10
400 amp	L1@0.15	Ea	134.00	6.15	140.15
450 amp	L1@0.15	Ea	320.00	6.15	326.15
500 amp	L1@0.15	Ea	257.00	6.15	263.15
600 amp	L1@0.15	Ea	209.00	6.15	215.15

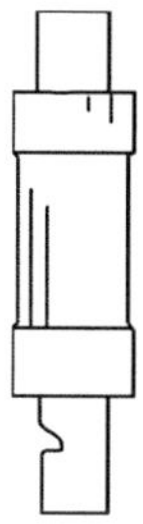

Use these figures to estimate the cost of fuses installed in safety switches and combination starters under the conditions described on pages 5 and 6. Costs listed are for each fuse installed. The crew size is one electrician working at a labor cost of $41.03 per manhour. These costs include layout, material handling, and normal waste. Add for sales tax, delivery supervision, mobilization, demobilization, cleanup, overhead and profit.

Cartridge Fuses, Time Delay, 600 Volt, Current Limiting, Class RK5

Material	Craft@Hrs	Unit	Material Cost	Labor Cost	Installed Cost
Ferrule type time delay 600 volt current limiting class RK5 cartridge fuses					
1/10 amp	L1@0.06	Ea	40.90	2.46	43.36
15/100 amp	L1@0.06	Ea	40.90	2.46	43.36
2/10 amp	L1@0.06	Ea	40.90	2.46	43.36
3/10 amp	L1@0.06	Ea	40.90	2.46	43.36
4/10 amp	L1@0.06	Ea	40.90	2.46	43.36
1/2 amp	L1@0.06	Ea	38.40	2.46	40.86
6/10 amp	L1@0.06	Ea	38.40	2.46	40.86
8/10 amp	L1@0.06	Ea	38.40	2.46	40.86
1 amp	L1@0.06	Ea	27.30	2.46	29.76
1-1/8 amp	L1@0.06	Ea	39.40	2.46	41.86
1-1/4 amp	L1@0.06	Ea	27.30	2.46	29.76
1-4/10 amp	L1@0.06	Ea	39.40	2.46	41.86
1-6/10 amp	L1@0.06	Ea	27.30	2.46	29.76
1-8/10 amp	L1@0.06	Ea	38.20	2.46	40.66
2 amp	L1@0.06	Ea	22.00	2.46	24.46
2-1/4 amp	L1@0.06	Ea	27.40	2.46	29.86
2-1/2 amp	L1@0.06	Ea	27.30	2.46	29.76
2-8/10 amp	L1@0.06	Ea	30.90	2.46	33.36
3 amp	L1@0.06	Ea	20.90	2.46	23.36
3-2/10 amp	L1@0.06	Ea	25.40	2.46	27.86
3-1/2 amp	L1@0.06	Ea	25.40	2.46	27.86
4 amp	L1@0.06	Ea	21.00	2.46	23.46
4-1/2 amp	L1@0.06	Ea	29.60	2.46	32.06
5 amp	L1@0.06	Ea	20.90	2.46	23.36
5-6/10 amp	L1@0.06	Ea	29.60	2.46	32.06
6 amp	L1@0.06	Ea	20.90	2.46	23.36
6-1/4 amp	L1@0.06	Ea	25.40	2.46	27.86
7 amp	L1@0.06	Ea	25.40	2.46	27.86
8 amp	L1@0.06	Ea	21.00	2.46	23.46
9 amp	L1@0.06	Ea	25.40	2.46	27.86
10 amp	L1@0.06	Ea	20.90	2.46	23.36
12 amp	L1@0.06	Ea	20.90	2.46	23.36
15 amp	L1@0.06	Ea	18.30	2.46	20.76
17-1/2 amp	L1@0.06	Ea	18.30	2.46	20.76

Use these figures to estimate the cost of fuses installed in safety switches and combination starters under the conditions described on pages 5 and 6. Costs listed are for each fuse installed. The crew size is one electrician working at a labor cost of $41.03 per manhour. These costs include layout, material handling, and normal waste. Add for sales tax, delivery supervision, mobilization, demobilization, cleanup, overhead and profit.

Cartridge Fuses, Time Delay, 600 Volt, Current Limiting, Class RK5

Material	Craft@Hrs	Unit	Material Cost	Labor Cost	Installed Cost
Ferrule type time delay 600 volt current limiting class RK5 cartridge fuses					
20 amp	L1@0.06	Ea	18.30	2.46	20.76
25 amp	L1@0.06	Ea	18.30	2.46	20.76
30 amp	L1@0.06	Ea	18.30	2.46	20.76
35 amp	L1@0.06	Ea	31.80	2.46	34.26
40 amp	L1@0.06	Ea	31.80	2.46	34.26
45 amp	L1@0.06	Ea	31.80	2.46	34.26
50 amp	L1@0.06	Ea	31.80	2.46	34.26
60 amp	L1@0.06	Ea	31.80	2.46	34.26
Blade type time delay 600 volt current limiting class RK5 cartridge fuses					
70 amp	L1@0.08	Ea	65.00	3.28	68.28
75 amp	L1@0.08	Ea	73.00	3.28	76.28
80 amp	L1@0.08	Ea	65.00	3.28	68.28
90 amp	L1@0.08	Ea	65.00	3.28	68.28
100 amp	L1@0.08	Ea	65.00	3.28	68.28
110 amp	L1@0.08	Ea	130.00	3.28	133.28
125 amp	L1@0.08	Ea	130.00	3.28	133.28
150 amp	L1@0.08	Ea	130.00	3.28	133.28
175 amp	L1@0.08	Ea	130.00	3.28	133.28
200 amp	L1@0.08	Ea	130.00	3.28	133.28
225 amp	L1@0.10	Ea	260.00	4.10	264.10
250 amp	L1@0.10	Ea	260.00	4.10	264.10
300 amp	L1@0.10	Ea	260.00	4.10	264.10
350 amp	L1@0.10	Ea	260.00	4.10	264.10
400 amp	L1@0.15	Ea	260.00	6.15	266.15
450 amp	L1@0.15	Ea	384.00	6.15	390.15
500 amp	L1@0.15	Ea	376.00	6.15	382.15
600 amp	L1@0.15	Ea	376.00	6.15	382.15

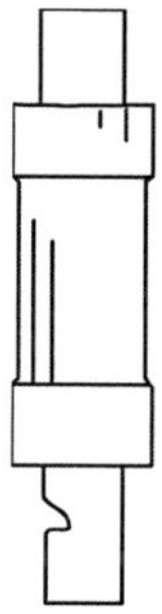

Use these figures to estimate the cost of fuses installed in safety switches and combination starters under the conditions described on pages 5 and 6. Costs listed are for each fuse installed. The crew size is one electrician working at a labor cost of $41.03 per manhour. These costs include layout, material handling, and normal waste. Add for sales tax, delivery supervision, mobilization, demobilization, cleanup, overhead and profit.

Cartridge Fuses, Fast Acting, 300 Volt, Current Limiting, Class T

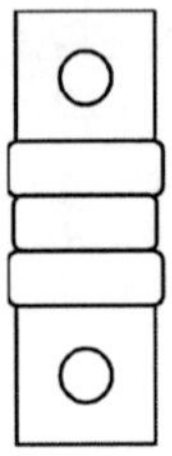

Material	Craft@Hrs	Unit	Material Cost	Labor Cost	Installed Cost
Blade type fast acting 300 volt current limiting class T cartridge fuses					
1 amp	L1@0.06	Ea	30.60	2.46	33.06
2 amp	L1@0.06	Ea	30.60	2.46	33.06
3 amp	L1@0.06	Ea	30.60	2.46	33.06
6 amp	L1@0.06	Ea	30.60	2.46	33.06
10 amp	L1@0.06	Ea	30.60	2.46	33.06
15 amp	L1@0.06	Ea	30.60	2.46	33.06
20 amp	L1@0.06	Ea	30.60	2.46	33.06
25 amp	L1@0.06	Ea	30.60	2.46	33.06
30 amp	L1@0.06	Ea	30.60	2.46	33.06
35 amp	L1@0.06	Ea	32.50	2.46	34.96
40 amp	L1@0.06	Ea	30.70	2.46	33.16
45 amp	L1@0.06	Ea	32.50	2.46	34.96
50 amp	L1@0.06	Ea	32.50	2.46	34.96
60 amp	L1@0.06	Ea	22.80	2.46	25.26
70 amp	L1@0.08	Ea	37.70	3.28	40.98
80 amp	L1@0.08	Ea	46.00	3.28	49.28
90 amp	L1@0.08	Ea	46.00	3.28	49.28
100 amp	L1@0.08	Ea	35.40	3.28	38.68
110 amp	L1@0.08	Ea	69.80	3.28	73.08
125 amp	L1@0.08	Ea	67.20	3.28	70.48
150 amp	L1@0.08	Ea	57.60	3.28	60.88
175 amp	L1@0.08	Ea	69.80	3.28	73.08
200 amp	L1@0.08	Ea	45.90	3.28	49.18
225 amp	L1@0.10	Ea	154.00	4.10	158.10
250 amp	L1@0.10	Ea	154.00	4.10	158.10
300 amp	L1@0.10	Ea	147.00	4.10	151.10
350 amp	L1@0.10	Ea	118.00	4.10	122.10
400 amp	L1@0.15	Ea	82.80	6.15	88.95
450 amp	L1@0.15	Ea	155.00	6.15	161.15
500 amp	L1@0.15	Ea	199.00	6.15	205.15
600 amp	L1@0.15	Ea	146.00	6.15	152.15

Use these figures to estimate the cost of fuses installed in safety switches and combination starters under the conditions described on pages 5 and 6. Costs listed are for each fuse installed. The crew size is one electrician working at a labor cost of $41.03 per manhour. These costs include layout, material handling, and normal waste. Add for sales tax, delivery supervision, mobilization, demobilization, cleanup, overhead and profit.

Material	Craft@Hrs	Unit	Material Cost	Labor Cost	Installed Cost
Blade type fast acting 600 volt current limiting class T cartridge fuses					
1 amp	L1@0.06	Ea	30.90	2.46	33.36
2 amp	L1@0.06	Ea	32.50	2.46	34.96
3 amp	L1@0.06	Ea	30.90	2.46	33.36
6 amp	L1@0.06	Ea	24.80	2.46	27.26
10 amp	L1@0.06	Ea	24.80	2.46	27.26
15 amp	L1@0.06	Ea	24.80	2.46	27.26
20 amp	L1@0.06	Ea	27.20	2.46	29.66
25 amp	L1@0.06	Ea	24.80	2.46	27.26
30 amp	L1@0.06	Ea	17.40	2.46	19.86
35 amp	L1@0.06	Ea	37.60	2.46	40.06
40 amp	L1@0.06	Ea	40.10	2.46	42.56
45 amp	L1@0.06	Ea	46.90	2.46	49.36
50 amp	L1@0.06	Ea	37.60	2.46	40.06
60 amp	L1@0.06	Ea	26.20	2.46	28.66
70 amp	L1@0.08	Ea	49.30	3.28	52.58
80 amp	L1@0.08	Ea	61.00	3.28	64.28
90 amp	L1@0.08	Ea	77.80	3.28	81.08
100 amp	L1@0.08	Ea	64.10	3.28	67.38
110 amp	L1@0.08	Ea	89.30	3.28	92.58
125 amp	L1@0.08	Ea	85.50	3.28	88.78
150 amp	L1@0.08	Ea	73.70	3.28	76.98
175 amp	L1@0.08	Ea	89.30	3.28	92.58
200 amp	L1@0.08	Ea	85.50	3.28	88.78
225 amp	L1@0.10	Ea	209.00	4.10	213.10
250 amp	L1@0.10	Ea	181.00	4.10	185.10
300 amp	L1@0.10	Ea	196.00	4.10	200.10
350 amp	L1@0.10	Ea	231.00	4.10	235.10
400 amp	L1@0.15	Ea	212.00	6.15	218.15
450 amp	L1@0.15	Ea	459.00	6.15	465.15
500 amp	L1@0.15	Ea	477.00	6.15	483.15
600 amp	L1@0.15	Ea	510.00	6.15	516.15

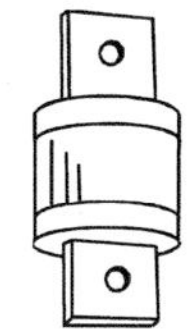

Use these figures to estimate the cost of fuses installed in safety switches and combination starters under the conditions described on pages 5 and 6. Costs listed are for each fuse installed. The crew size is one electrician working at a labor cost of $41.03 per manhour. These costs include layout, material handling, and normal waste. Add for sales tax, delivery supervision, mobilization, demobilization, cleanup, overhead and profit.

Plug-in Circuit Breakers

120/240 volt AC plug-in circuit breakers with 10,000 amp interrupt capacity

Material		Craft@Hrs	Unit	Material Cost	Labor Cost	Installed Cost
1 pole	10A	L1@0.10	Ea	20.80	4.10	24.90
1 pole	15A	L1@0.10	Ea	20.80	4.10	24.90
1 pole	20A	L1@0.10	Ea	20.80	4.10	24.90
1 pole	25A	L1@0.10	Ea	20.80	4.10	24.90
1 pole	30A	L1@0.10	Ea	20.80	4.10	24.90
1 pole	35A	L1@0.10	Ea	20.80	4.10	24.90
1 pole	40A	L1@0.10	Ea	20.80	4.10	24.90
1 pole	45A	L1@0.10	Ea	20.80	4.10	24.90
1 pole	50A	L1@0.10	Ea	20.80	4.10	24.90
1 pole	60A	L1@0.10	Ea	20.80	4.10	24.90
1 pole	70A	L1@0.10	Ea	46.40	4.10	50.50
2 pole	15A	L1@0.15	Ea	46.40	6.15	52.55
2 pole	20A	L1@0.15	Ea	46.40	6.15	52.55
2 pole	25A	L1@0.15	Ea	46.40	6.15	52.55
2 pole	30A	L1@0.15	Ea	46.40	6.15	52.55
2 pole	35A	L1@0.15	Ea	46.40	6.15	52.55
2 pole	40A	L1@0.15	Ea	46.40	6.15	52.55
2 pole	45A	L1@0.15	Ea	46.40	6.15	52.55
2 pole	50A	L1@0.15	Ea	46.40	6.15	52.55
2 pole	60A	L1@0.15	Ea	46.40	6.15	52.55
2 pole	70A	L1@0.15	Ea	96.90	6.15	103.05
2 pole	90A	L1@0.15	Ea	134.00	6.15	140.15
2 pole	100A	L1@0.15	Ea	134.00	6.15	140.15
3 pole	15A	L1@0.20	Ea	166.00	8.21	174.21
3 pole	20A	L1@0.20	Ea	166.00	8.21	174.21
3 pole	25A	L1@0.20	Ea	166.00	8.21	174.21
3 pole	30A	L1@0.20	Ea	166.00	8.21	174.21
3 pole	35A	L1@0.20	Ea	166.00	8.21	174.21
3 pole	40A	L1@0.20	Ea	166.00	8.21	174.21
3 pole	45A	L1@0.20	Ea	166.00	8.21	174.21
3 pole	50A	L1@0.20	Ea	166.00	8.21	174.21
3 pole	60A	L1@0.20	Ea	166.00	8.21	174.21
3 pole	70A	L1@0.20	Ea	214.00	8.21	222.21
3 pole	90A	L1@0.20	Ea	251.00	8.21	259.21
3 pole	100A	L1@0.20	Ea	251.00	8.21	259.21

Use these figures to estimate the cost of circuit breakers installed in enclosures under the conditions described on pages 5 and 6. Costs listed are for each circuit breaker installed. The crew is one electrician working at a labor cost of $41.03 per manhour. These costs include layout, material handling, and normal waste. Add for the enclosure, support, sales tax, delivery, supervision, mobilization, demobilization, cleanup, overhead and profit. The labor required for wire termination is not included in these figures. Add the labor cost for making terminations according to the wire size and the size of the enclosure. Note: Be sure to select the right mounting style and amp interrupt capacity (A.I.C.). Two and three pole circuit breakers have a common operating handle. Some multi-pole breakers have a common tie bar that holds several operating handles together. Most engineered systems require one operating handle on each circuit breaker.

Material		Craft@Hrs	Unit	Material Cost	Labor Cost	Installed Cost
Tandem 120/240 volt circuit breakers in single pole space						
15A - 15A		L1@0.20	Ea	36.10	8.21	44.31
15A - 20A		L1@0.20	Ea	36.10	8.21	44.31
20A - 20A		L1@0.20	Ea	35.60	8.21	43.81
120/240 volt bolt-on circuit breakers, 10,000 A.I.C.						
1 pole	15A	L1@0.15	Ea	26.50	6.15	32.65
1 pole	20A	L1@0.15	Ea	26.50	6.15	32.65
1 pole	25A	L1@0.15	Ea	26.50	6.15	32.65
1 pole	30A	L1@0.15	Ea	26.50	6.15	32.65
1 pole	40A	L1@0.15	Ea	26.50	6.15	32.65
1 pole	50A	L1@0.15	Ea	26.50	6.15	32.65
2 pole	15A	L1@0.20	Ea	59.40	8.21	67.61
2 pole	20A	L1@0.20	Ea	59.40	8.21	67.61
2 pole	30A	L1@0.20	Ea	59.40	8.21	67.61
2 pole	40A	L1@0.20	Ea	59.40	8.21	67.61
2 pole	50A	L1@0.20	Ea	59.40	8.21	67.61
2 pole	60A	L1@0.20	Ea	59.40	8.21	67.61
2 pole	70A	L1@0.20	Ea	59.40	8.21	67.61
3 pole	15A	L1@0.25	Ea	194.00	10.30	204.30
3 pole	20A	L1@0.25	Ea	194.00	10.30	204.30
3 pole	30A	L1@0.25	Ea	194.00	10.30	204.30
3 pole	40A	L1@0.25	Ea	194.00	10.30	204.30
3 pole	50A	L1@0.25	Ea	194.00	10.30	204.30
3 pole	60A	L1@0.25	Ea	194.00	10.30	204.30
3 pole	70A	L1@0.25	Ea	194.00	10.30	204.30
3 pole	80A	L1@0.25	Ea	245.00	10.30	255.30
3 pole	90A	L1@0.25	Ea	278.00	10.30	288.30
3 pole	100A	L1@0.25	Ea	278.00	10.30	288.30

Use these figures to estimate the cost of circuit breakers installed in enclosures under the conditions described on pages 5 and 6. Costs listed are for each circuit breaker installed. The crew is one electrician working at a labor cost of $41.03 per manhour. These costs include layout, material handling, and normal waste. Add for the enclosure, support, sales tax, delivery, supervision, mobilization, demobilization, cleanup, overhead and profit. The labor required for wire termination is not included in these figures. Add the labor cost for making terminations according to the wire size and the size of the enclosure. Note: Be sure to select the right mounting style and amp interrupt capacity (A.I.C.). Two and three pole circuit breakers have a common operating handle. Some multi-pole breakers have a common tie bar that holds several operating handles together. Most engineered systems require one operating handle on each circuit breaker.

120/240 Volt AC Thermal Magnetic Breakers

Material		Craft@Hrs	Unit	Material Cost	Labor Cost	Installed Cost
120/240 volt AC thermal magnetic bolt-on breakers with 100 amp frame						
1 pole	15A	L1@0.20	Ea	132.00	8.21	140.21
1 pole	20A	L1@0.20	Ea	132.00	8.21	140.21
1 pole	30A	L1@0.20	Ea	132.00	8.21	140.21
1 pole	40A	L1@0.20	Ea	132.00	8.21	140.21
1 pole	50A	L1@0.20	Ea	132.00	8.21	140.21
1 pole	60A	L1@0.20	Ea	132.00	8.21	140.21
1 pole	70A	L1@0.20	Ea	177.00	8.21	185.21
1 pole	90A	L1@0.20	Ea	177.00	8.21	185.21
1 pole	100A	L1@0.20	Ea	177.00	8.21	185.21
2 pole	15A	L1@0.25	Ea	224.00	10.30	234.30
2 pole	20A	L1@0.25	Ea	224.00	10.30	234.30
2 pole	30A	L1@0.25	Ea	224.00	10.30	234.30
2 pole	40A	L1@0.25	Ea	224.00	10.30	234.30
2 pole	50A	L1@0.25	Ea	224.00	10.30	234.30
2 pole	60A	L1@0.25	Ea	224.00	10.30	234.30
2 pole	70A	L1@0.25	Ea	363.00	10.30	373.30
2 pole	90A	L1@0.25	Ea	363.00	10.30	373.30
2 pole	100A	L1@0.25	Ea	363.00	10.30	373.30
3 pole	15A	L1@0.30	Ea	328.00	12.30	340.30
3 pole	20A	L1@0.30	Ea	328.00	12.30	340.30
3 pole	30A	L1@0.30	Ea	328.00	12.30	340.30
3 pole	40A	L1@0.30	Ea	328.00	12.30	340.30
3 pole	50A	L1@0.30	Ea	328.00	12.30	340.30
3 pole	60A	L1@0.30	Ea	328.00	12.30	340.30
3 pole	70A	L1@0.30	Ea	473.00	12.30	485.30
3 pole	90A	L1@0.30	Ea	473.00	12.30	485.30
3 pole	100A	L1@0.30	Ea	473.00	12.30	485.30

Use these figures to estimate the cost of circuit breakers installed in enclosures under the conditions described on pages 5 and 6. Costs listed are for each circuit breaker installed. The crew is one electrician working at a labor cost of $41.03 per manhour. These costs include layout, material handling, and normal waste. Add for the enclosure, support, sales tax, delivery, supervision, mobilization, demobilization, cleanup, overhead and profit. The labor required for wire termination is not included in these figures. Add the labor cost for making terminations according to the wire size and the size of the enclosure. Note: Be sure to select the right mounting style and amp interrupt capacity (A.I.C.). Two and three pole circuit breakers have a common operating handle. Some multi-pole breakers have a common tie bar that holds several operating handles together. Most engineered systems require one operating handle on each circuit breaker.

277/480 volt AC thermal magnetic breakers with 100 amp frame

Material		Craft@Hrs	Unit	Material Cost	Labor Cost	Installed Cost
1 pole	15A	L1@0.30	Ea	166.00	12.30	178.30
1 pole	20A	L1@0.30	Ea	166.00	12.30	178.30
1 pole	30A	L1@0.30	Ea	166.00	12.30	178.30
1 pole	40A	L1@0.30	Ea	166.00	12.30	178.30
1 pole	50A	L1@0.30	Ea	166.00	12.30	178.30
1 pole	60A	L1@0.30	Ea	167.00	12.30	179.30
1 pole	70A	L1@0.30	Ea	209.00	12.30	221.30
1 pole	90A	L1@0.30	Ea	209.00	12.30	221.30
1 pole	100A	L1@0.30	Ea	209.00	12.30	221.30
2 pole	15A	L1@0.40	Ea	408.00	16.40	424.40
2 pole	20A	L1@0.40	Ea	408.00	16.40	424.40
2 pole	30A	L1@0.40	Ea	408.00	16.40	424.40
2 pole	40A	L1@0.40	Ea	408.00	16.40	424.40
2 pole	50A	L1@0.40	Ea	408.00	16.40	424.40
2 pole	60A	L1@0.40	Ea	408.00	16.40	424.40
2 pole	70A	L1@0.40	Ea	524.00	16.40	540.40
2 pole	90A	L1@0.40	Ea	524.00	16.40	540.40
2 pole	100A	L1@0.40	Ea	524.00	16.40	540.40
3 pole	15A	L1@0.50	Ea	523.00	20.50	543.50
3 pole	20A	L1@0.50	Ea	523.00	20.50	543.50
3 pole	30A	L1@0.50	Ea	523.00	20.50	543.50
3 pole	40A	L1@0.50	Ea	523.00	20.50	543.50
3 pole	50A	L1@0.50	Ea	523.00	20.50	543.50
3 pole	60A	L1@0.50	Ea	523.00	20.50	543.50
3 pole	70A	L1@0.50	Ea	619.00	20.50	639.50
3 pole	90A	L1@0.50	Ea	619.00	20.50	639.50
3 pole	100A	L1@0.50	Ea	619.00	20.50	639.50

600 volt AC, 250 volt DC thermal magnetic breakers with 100 amp frame

Material		Craft@Hrs	Unit	Material Cost	Labor Cost	Installed Cost
2 pole	15A	L1@0.70	Ea	473.00	28.70	501.70
2 pole	20A	L1@0.70	Ea	473.00	28.70	501.70
2 pole	30A	L1@0.70	Ea	473.00	28.70	501.70
2 pole	40A	L1@0.70	Ea	473.00	28.70	501.70
2 pole	50A	L1@0.70	Ea	473.00	28.70	501.70
2 pole	60A	L1@0.70	Ea	473.00	28.70	501.70
2 pole	70A	L1@0.70	Ea	595.00	28.70	623.70
2 pole	90A	L1@0.70	Ea	595.00	28.70	623.70
2 pole	100A	L1@0.70	Ea	595.00	28.70	623.70

Use these figures to estimate the cost of circuit breakers installed in enclosures under the conditions described on pages 5 and 6. Costs listed are for each circuit breaker installed. The crew is one electrician working at a labor cost of $41.03 per manhour. These costs include layout, material handling, and normal waste. Add for the enclosure, supports, sales tax, delivery, supervision, mobilization, demobilization, cleanup, overhead and profit. The labor required for wire termination is not included in these figures. Add the labor cost for making terminations according to the wire size and the size of the enclosure. Note: Be sure to select the right mounting style and amp interrupt capacity (A.I.C.). Two and three pole circuit breakers have a common operating handle. Some multi-pole breakers have a common tie bar that holds several operating handles together. Most engineered systems require one operating handle on each circuit breaker.

Thermal Magnetic Breakers

600 volt AC, 250 volt DC thermal magnetic breakers with 100 amp frame

Material		Craft@Hrs	Unit	Material Cost	Labor Cost	Installed Cost
3 pole	15A	L1@0.75	Ea	602.00	30.80	632.80
3 pole	20A	L1@0.75	Ea	602.00	30.80	632.80
3 pole	30A	L1@0.75	Ea	602.00	30.80	632.80
3 pole	40A	L1@0.75	Ea	602.00	30.80	632.80
3 pole	50A	L1@0.75	Ea	602.00	30.80	632.80
3 pole	60A	L1@0.75	Ea	602.00	30.80	632.80
3 pole	70A	L1@0.75	Ea	747.00	30.80	777.80
3 pole	90A	L1@0.75	Ea	747.00	30.80	777.80
3 pole	100A	L1@0.75	Ea	747.00	30.80	777.80

600 volt AC, 250 volt DC thermal magnetic breakers with 250 amp frame

Material		Craft@Hrs	Unit	Material Cost	Labor Cost	Installed Cost
2 pole	70A	L1@1.00	Ea	1,330.00	41.00	1,371.00
2 pole	80A	L1@1.00	Ea	1,330.00	41.00	1,371.00
2 pole	90A	L1@1.00	Ea	1,330.00	41.00	1,371.00
2 pole	100A	L1@1.00	Ea	1,330.00	41.00	1,371.00
2 pole	110A	L1@1.00	Ea	1,330.00	41.00	1,371.00
2 pole	125A	L1@1.00	Ea	1,330.00	41.00	1,371.00
2 pole	150A	L1@1.00	Ea	1,330.00	41.00	1,371.00
2 pole	175A	L1@1.00	Ea	1,330.00	41.00	1,371.00
2 pole	200A	L1@1.00	Ea	1,330.00	41.00	1,371.00
2 pole	225A	L1@1.00	Ea	1,330.00	41.00	1,371.00
3 pole	70A	L1@1.25	Ea	1,660.00	51.30	1,711.30
3 pole	80A	L1@1.25	Ea	1,660.00	51.30	1,711.30
3 pole	90A	L1@1.25	Ea	1,660.00	51.30	1,711.30
3 pole	100A	L1@1.25	Ea	1,660.00	51.30	1,711.30
3 pole	110A	L1@1.25	Ea	1,660.00	51.30	1,711.30
3 pole	125A	L1@1.25	Ea	1,660.00	51.30	1,711.30
3 pole	150A	L1@1.25	Ea	1,660.00	51.30	1,711.30
3 pole	175A	L1@1.25	Ea	1,660.00	51.30	1,711.30
3 pole	200A	L1@1.25	Ea	1,660.00	51.30	1,711.30
3 pole	225A	L1@1.25	Ea	1,660.00	51.30	1,711.30

600 volt AC, 250 volt DC thermal magnetic breakers with 400 amp frame

Material		Craft@Hrs	Unit	Material Cost	Labor Cost	Installed Cost
2 pole	125A	L1@1.50	Ea	2,440.00	61.50	2,501.50
2 pole	150A	L1@1.50	Ea	2,440.00	61.50	2,501.50
2 pole	175A	L1@1.50	Ea	2,440.00	61.50	2,501.50
2 pole	200A	L1@1.50	Ea	2,440.00	61.50	2,501.50
2 pole	225A	L1@1.50	Ea	2,440.00	61.50	2,501.50
2 pole	250A	L1@1.50	Ea	2,440.00	61.50	2,501.50

Use these figures to estimate the cost of circuit breakers installed in enclosures under the conditions described on pages 5 and 6. Costs listed are for each circuit breaker installed. The crew is one electrician working at a labor cost of $41.03 per manhour. These costs include layout, material handling, and normal waste. Add for the enclosure, support, sales tax, delivery, supervision, mobilization, demobilization, cleanup, overhead and profit. The labor required for wire termination is not included in these figures. Add the labor cost for making terminations according to the wire size and the size of the enclosure. Note: Be sure to select the right mounting style and amp interrupt capacity (A.I.C.). Two and three pole circuit breakers have a common operating handle. Some multi-pole breakers have a common tie bar that holds several operating handles together. Most engineered systems require one operating handle on each circuit breaker.

600 volt AC, 250 volt DC thermal magnetic breakers with 400 amp frame

Material		Craft@Hrs	Unit	Material Cost	Labor Cost	Installed Cost
2 pole	300A	L1@1.50	Ea	2,440.00	61.50	2,501.50
2 pole	350A	L1@1.50	Ea	2,440.00	61.50	2,501.50
2 pole	400A	L1@1.50	Ea	2,440.00	61.50	2,501.50
3 pole	125A	L1@1.75	Ea	2,940.00	71.80	3,011.80
3 pole	150A	L1@1.75	Ea	2,940.00	71.80	3,011.80
3 pole	175A	L1@1.75	Ea	2,940.00	71.80	3,011.80
3 pole	200A	L1@1.75	Ea	2,940.00	71.80	3,011.80
3 pole	225A	L1@1.75	Ea	2,940.00	71.80	3,011.80
3 pole	250A	L1@1.75	Ea	2,940.00	71.80	3,011.80
3 pole	300A	L1@1.75	Ea	2,940.00	71.80	3,011.80
3 pole	350A	L1@1.75	Ea	2,940.00	71.80	3,011.80
3 pole	400A	L1@1.75	Ea	2,940.00	71.80	3,011.80

600 volt AC, 250 volt DC thermal magnetic breakers with 1,000 amp frame

Material		Craft@Hrs	Unit	Material Cost	Labor Cost	Installed Cost
2 pole	125A	L1@1.75	Ea	3,840.00	71.80	3,911.80
2 pole	150A	L1@1.75	Ea	3,840.00	71.80	3,911.80
2 pole	175A	L1@1.75	Ea	3,840.00	71.80	3,911.80
2 pole	200A	L1@1.75	Ea	3,840.00	71.80	3,911.80
2 pole	225A	L1@1.75	Ea	3,840.00	71.80	3,911.80
2 pole	250A	L1@1.75	Ea	3,840.00	71.80	3,911.80
2 pole	300A	L1@1.75	Ea	3,840.00	71.80	3,911.80
2 pole	350A	L1@1.75	Ea	3,840.00	71.80	3,911.80
2 pole	400A	L1@1.75	Ea	3,840.00	71.80	3,911.80
2 pole	450A	L1@1.75	Ea	3,840.00	71.80	3,911.80
2 pole	500A	L1@1.75	Ea	3,840.00	71.80	3,911.80
2 pole	600A	L1@1.75	Ea	3,840.00	71.80	3,911.80
2 pole	700A	L1@1.75	Ea	4,950.00	71.80	5,021.80
2 pole	800A	L1@1.75	Ea	4,950.00	71.80	5,021.80
2 pole	900A	L1@1.75	Ea	7,060.00	71.80	7,131.80
2 pole	1000A	L1@1.75	Ea	7,060.00	71.80	7,131.80
3 pole	125A	L1@2.00	Ea	4,880.00	82.10	4,962.10
3 pole	150A	L1@2.00	Ea	4,880.00	82.10	4,962.10
3 pole	175A	L1@2.00	Ea	4,880.00	82.10	4,962.10
3 pole	200A	L1@2.00	Ea	4,880.00	82.10	4,962.10
3 pole	225A	L1@2.00	Ea	4,880.00	82.10	4,962.10
3 pole	250A	L1@2.00	Ea	4,880.00	82.10	4,962.10
3 pole	300A	L1@2.00	Ea	4,880.00	82.10	4,962.10
3 pole	350A	L1@2.00	Ea	4,880.00	82.10	4,962.10
3 pole	400A	L1@2.00	Ea	4,880.00	82.10	4,962.10
3 pole	450A	L1@2.00	Ea	4,880.00	82.10	4,962.10

Use these figures to estimate the cost of circuit breakers installed in enclosures under the conditions described on pages 5 and 6. Costs listed are for each circuit breaker installed. The crew is one electrician working at a labor cost of $41.03 per manhour. These costs include layout, material handling, and normal waste. Add for the enclosure, support, sales tax, delivery, supervision, mobilization, demobilization, cleanup, overhead and profit. The labor required for wire termination is not included in these figures. Add the labor cost for making terminations according to the wire size and the size of the enclosure. Note: Be sure to select the right mounting style and amp interrupt capacity (A.I.C.). Two and three pole circuit breakers have a common operating handle. Some multi-pole breakers have a common tie bar that holds several operating handles together. Most engineered systems require one operating handle on each circuit breaker.

Thermal Magnetic Breakers

600 volt AC, 250 volt DC thermal magnetic breakers with 1,000 amp frame

Material		Craft@Hrs	Unit	Material Cost	Labor Cost	Installed Cost
3 pole	500A	L1@2.00	Ea	4,880.00	82.10	4,962.10
3 pole	600A	L1@2.00	Ea	4,880.00	82.10	4,962.10
3 pole	700A	L1@2.00	Ea	6,370.00	82.10	6,452.10
3 pole	800A	L1@2.00	Ea	6,420.00	82.10	6,502.10
3 pole	900A	L1@2.00	Ea	8,180.00	82.10	8,262.10
3 pole	1000A	L1@2.00	Ea	8,180.00	82.10	8,262.10

600 volt AC, 250 volt DC thermal magnetic breakers with 1,200 amp frame

Material		Craft@Hrs	Unit	Material Cost	Labor Cost	Installed Cost
2 pole	600A	L1@2.00	Ea	11,200.00	82.10	11,282.10
2 pole	700A	L1@2.00	Ea	11,200.00	82.10	11,282.10
2 pole	800A	L1@2.00	Ea	11,200.00	82.10	11,282.10
2 pole	1000A	L1@2.00	Ea	11,200.00	82.10	11,282.10
2 pole	1200A	L1@2.00	Ea	11,200.00	82.10	11,282.10
3 pole	600A	L1@2.50	Ea	12,400.00	103.00	12,503.00
3 pole	700A	L1@2.50	Ea	12,400.00	103.00	12,503.00
3 pole	800A	L1@2.50	Ea	12,400.00	103.00	12,503.00
3 pole	1000A	L1@2.50	Ea	12,400.00	103.00	12,503.00
3 pole	1200A	L1@2.50	Ea	12,400.00	103.00	12,503.00

277 volt AC plug-on breakers with 100 amp frame, 14,000 A.I.C.

Material		Craft@Hrs	Unit	Material Cost	Labor Cost	Installed Cost
1 pole	15A	L1@0.25	Ea	88.90	10.30	99.20
1 pole	20A	L1@0.25	Ea	88.90	10.30	99.20
1 pole	25A	L1@0.25	Ea	88.90	10.30	99.20
1 pole	30A	L1@0.25	Ea	88.90	10.30	99.20
1 pole	40A	L1@0.25	Ea	88.90	10.30	99.20
1 pole	50A	L1@0.25	Ea	88.90	10.30	99.20
1 pole	60A	L1@0.25	Ea	88.90	10.30	99.20

277 volt AC bolt-on breakers with 100 amp frame, 14,000 A.I.C.

Material		Craft@Hrs	Unit	Material Cost	Labor Cost	Installed Cost
1 pole	15A	L1@0.30	Ea	102.00	12.30	114.30
1 pole	20A	L1@0.30	Ea	102.00	12.30	114.30
1 pole	25A	L1@0.30	Ea	102.00	12.30	114.30
1 pole	30A	L1@0.30	Ea	102.00	12.30	114.30
1 pole	40A	L1@0.30	Ea	102.00	12.30	114.30
1 pole	50A	L1@0.30	Ea	102.00	12.30	114.30
1 pole	60A	L1@0.30	Ea	102.00	12.30	114.30

Use these figures to estimate the cost of circuit breakers installed in enclosures under the conditions described on pages 5 and 6. Costs listed are for each circuit breaker installed. The crew is one electrician working at a labor cost of $41.03 per manhour. These costs include layout, material handling, and normal waste. Add for the enclosure, support, sales tax, delivery, supervision, mobilization, demobilization, cleanup, overhead and profit. The labor required for wire termination is not included in these figures. Add the labor cost for making terminations according to the wire size and the size of the enclosure. Note: Be sure to select the right mounting style and amp interrupt capacity (A.I.C.). Two and three pole circuit breakers have a common operating handle. Some multi-pole breakers have a common tie bar that holds several operating handles together. Most engineered systems require one operating handle on each circuit breaker.

277 volt AC plug-on breakers with 100 amp frame, 14,000 A.I.C.

Material		Craft@Hrs	Unit	Material Cost	Labor Cost	Installed Cost
2 pole	15A	L1@0.40	Ea	255.00	16.40	271.40
2 pole	20A	L1@0.40	Ea	255.00	16.40	271.40
2 pole	25A	L1@0.40	Ea	255.00	16.40	271.40
2 pole	30A	L1@0.40	Ea	255.00	16.40	271.40
2 pole	40A	L1@0.40	Ea	255.00	16.40	271.40
2 pole	50A	L1@0.40	Ea	255.00	16.40	271.40
2 pole	60A	L1@0.40	Ea	255.00	16.40	271.40

277 volt AC bolt-on breakers with 100 amp frame, 14,000 A.I.C.

Material		Craft@Hrs	Unit	Material Cost	Labor Cost	Installed Cost
2 pole	15A	L1@0.50	Ea	274.00	20.50	294.50
2 pole	20A	L1@0.50	Ea	274.00	20.50	294.50
2 pole	25A	L1@0.50	Ea	274.00	20.50	294.50
2 pole	30A	L1@0.50	Ea	274.00	20.50	294.50
2 pole	40A	L1@0.50	Ea	274.00	20.50	294.50
2 pole	50A	L1@0.50	Ea	274.00	20.50	294.50
2 pole	60A	L1@0.50	Ea	274.00	20.50	294.50

277 volt AC plug-on breakers with 100 amp frame, 14,000 A.I.C.

Material		Craft@Hrs	Unit	Material Cost	Labor Cost	Installed Cost
3 pole	15A	L1@0.60	Ea	520.00	24.60	544.60
3 pole	20A	L1@0.60	Ea	520.00	24.60	544.60
3 pole	25A	L1@0.60	Ea	520.00	24.60	544.60
3 pole	30A	L1@0.60	Ea	520.00	24.60	544.60
3 pole	40A	L1@0.60	Ea	520.00	24.60	544.60
3 pole	50A	L1@0.60	Ea	520.00	24.60	544.60
3 pole	60A	L1@0.60	Ea	520.00	24.60	544.60

277 volt AC bolt-on breakers with 100 amp frame, 14,000 A.I.C.

Material		Craft@Hrs	Unit	Material Cost	Labor Cost	Installed Cost
3 pole	15A	L1@0.70	Ea	444.00	28.70	472.70
3 pole	20A	L1@0.70	Ea	444.00	28.70	472.70
3 pole	25A	L1@0.70	Ea	444.00	28.70	472.70
3 pole	30A	L1@0.70	Ea	444.00	28.70	472.70
3 pole	40A	L1@0.70	Ea	444.00	28.70	472.70
3 pole	50A	L1@0.70	Ea	444.00	28.70	472.70
3 pole	60A	L1@0.70	Ea	444.00	28.70	472.70

Use these figures to estimate the cost of circuit breakers installed in enclosures under the conditions described on pages 5 and 6. Costs listed are for each circuit breaker installed. The crew is one electrician working at a labor cost of $41.03 per manhour. These costs include layout, material handling, and normal waste. Add for the enclosure, support, sales tax, delivery, supervision, mobilization, demobilization, cleanup, overhead and profit. The labor required for wire termination is not included in these figures. Add the labor cost for making terminations according to the wire size and the size of the enclosure. Note: Be sure to select the right mounting style and amp interrupt capacity (A.I.C.). Two and three pole circuit breakers have a common operating handle. Some multi-pole breakers have a common tie bar that holds several operating handles together. Most engineered systems require one operating handle on each circuit breaker.

Thermal Magnetic Breakers

600 volt AC, 250 volt DC breakers with 225 amp frame, 10,000 A.I.C.

Material		Craft@Hrs	Unit	Material Cost	Labor Cost	Installed Cost
3 pole	125A	L1@1.50	Ea	1,460.00	61.50	1,521.50
3 pole	150A	L1@1.50	Ea	1,460.00	61.50	1,521.50
3 pole	175A	L1@1.50	Ea	1,460.00	61.50	1,521.50
3 pole	200A	L1@1.50	Ea	1,460.00	61.50	1,521.50
3 pole	225A	L1@1.50	Ea	1,460.00	61.50	1,521.50

600 volt AC, 250 volt DC breakers with 400 amp frame, 22,000 A.I.C.

Material		Craft@Hrs	Unit	Material Cost	Labor Cost	Installed Cost
2 pole	125A	L1@1.30	Ea	2,730.00	53.30	2,783.30
2 pole	150A	L1@1.30	Ea	2,730.00	53.30	2,783.30
2 pole	175A	L1@1.30	Ea	2,730.00	53.30	2,783.30
2 pole	200A	L1@1.30	Ea	2,730.00	53.30	2,783.30
2 pole	225A	L1@1.30	Ea	2,730.00	53.30	2,783.30
2 pole	250A	L1@1.30	Ea	2,730.00	53.30	2,783.30
2 pole	300A	L1@1.30	Ea	2,730.00	53.30	2,783.30
2 pole	350A	L1@1.30	Ea	2,730.00	53.30	2,783.30
2 pole	400A	L1@1.30	Ea	2,730.00	53.30	2,783.30
3 pole	125A	L1@1.70	Ea	3,300.00	69.80	3,369.80
3 pole	150A	L1@1.70	Ea	3,300.00	69.80	3,369.80
3 pole	175A	L1@1.70	Ea	3,300.00	69.80	3,369.80
3 pole	200A	L1@1.70	Ea	3,300.00	69.80	3,369.80
3 pole	225A	L1@1.70	Ea	3,300.00	69.80	3,369.80
3 pole	250A	L1@1.70	Ea	3,300.00	69.80	3,369.80
3 pole	300A	L1@1.70	Ea	3,300.00	69.80	3,369.80
3 pole	350A	L1@1.70	Ea	3,300.00	69.80	3,369.80
3 pole	400A	L1@1.70	Ea	3,300.00	69.80	3,369.80

600 volt AC, 250 volt DC breakers with 1,000 amp frame, 30,000 A.I.C.

Material		Craft@Hrs	Unit	Material Cost	Labor Cost	Installed Cost
2 pole	125A	L1@1.50	Ea	3,990.00	61.50	4,051.50
2 pole	150A	L1@1.50	Ea	3,990.00	61.50	4,051.50
2 pole	175A	L1@1.50	Ea	3,990.00	61.50	4,051.50
2 pole	200A	L1@1.50	Ea	3,990.00	61.50	4,051.50
2 pole	225A	L1@1.50	Ea	3,990.00	61.50	4,051.50
2 pole	250A	L1@1.50	Ea	3,990.00	61.50	4,051.50
2 pole	300A	L1@1.50	Ea	3,990.00	61.50	4,051.50
2 pole	350A	L1@1.50	Ea	3,990.00	61.50	4,051.50
2 pole	400A	L1@1.50	Ea	3,990.00	61.50	4,051.50

Use these figures to estimate the cost of circuit breakers installed in enclosures under the conditions described on pages 5 and 6. Costs listed are for each circuit breaker installed. The crew is one electrician working at a labor cost of $41.03 per manhour. These costs include layout, material handling, and normal waste. Add for the enclosure, support, sales tax, delivery, supervision, mobilization, demobilization, cleanup, overhead and profit. The labor required for wire termination is not included in these figures. Add the labor cost for making terminations according to the wire size and the size of the enclosure. Note: Be sure to select the right mounting style and amp interrupt capacity (A.I.C.). Two and three pole circuit breakers have a common operating handle. Some multi-pole breakers have a common tie bar that holds several operating handles together. Most engineered systems require one operating handle on each circuit breaker.

600 volt AC, 250 volt DC breakers with 1,000 amp frame, 30,000 A.I.C.

Material		Craft@Hrs	Unit	Material Cost	Labor Cost	Installed Cost
2 pole	450A	L1@1.50	Ea	3,990.00	61.50	4,051.50
2 pole	500A	L1@1.50	Ea	3,990.00	61.50	4,051.50
2 pole	600A	L1@1.50	Ea	3,990.00	61.50	4,051.50
2 pole	700A	L1@1.50	Ea	5,060.00	61.50	5,121.50
2 pole	800A	L1@1.50	Ea	5,160.00	61.50	5,221.50
2 pole	900A	L1@1.50	Ea	7,340.00	61.50	7,401.50
2 pole	1000A	L1@1.50	Ea	7,340.00	61.50	7,401.50
3 pole	125A	L1@2.00	Ea	5,070.00	82.10	5,152.10
3 pole	150A	L1@2.00	Ea	5,070.00	82.10	5,152.10
3 pole	175A	L1@2.00	Ea	5,070.00	82.10	5,152.10
3 pole	200A	L1@2.00	Ea	5,070.00	82.10	5,152.10
3 pole	225A	L1@2.00	Ea	5,070.00	82.10	5,152.10
3 pole	250A	L1@2.00	Ea	5,070.00	82.10	5,152.10
3 pole	300A	L1@2.00	Ea	5,070.00	82.10	5,152.10
3 pole	350A	L1@2.00	Ea	5,070.00	82.10	5,152.10
3 pole	400A	L1@2.00	Ea	5,070.00	82.10	5,152.10
3 pole	450A	L1@2.00	Ea	5,070.00	82.10	5,152.10
3 pole	500A	L1@2.00	Ea	5,070.00	82.10	5,152.10
3 pole	600A	L1@2.00	Ea	5,070.00	82.10	5,152.10
3 pole	700A	L1@2.00	Ea	6,680.00	82.10	6,762.10
3 pole	800A	L1@2.00	Ea	6,680.00	82.10	6,762.10
3 pole	900A	L1@2.00	Ea	8,490.00	82.10	8,572.10
3 pole	1000A	L1@2.00	Ea	8,490.00	82.10	8,572.10

NEMA 1 flush mounted circuit breaker enclosures

Material		Craft@Hrs	Unit	Material Cost	Labor Cost	Installed Cost
100A	120/240V	L1@0.30	Ea	187.00	12.30	199.30
100A	277/480V	L1@0.30	Ea	241.00	12.30	253.30
225A	120/240V	L1@0.35	Ea	221.00	14.40	235.40
225A	600V	L1@0.35	Ea	181.00	14.40	195.40
400A	600V	L1@0.50	Ea	324.00	20.50	344.50
1000A	600V	L1@0.70	Ea	583.00	28.70	611.70

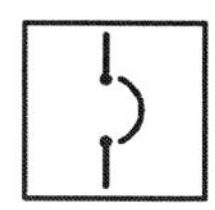

NEMA 1 surface mounted circuit breaker enclosures

Material		Craft@Hrs	Unit	Material Cost	Labor Cost	Installed Cost
100A	120/240V	L1@0.30	Ea	187.00	12.30	199.30
100A	277/480V	L1@0.30	Ea	241.00	12.30	253.30
225A	120/240V	L1@0.35	Ea	221.00	14.40	235.40
225A	277/480V	L1@0.35	Ea	181.00	14.40	195.40

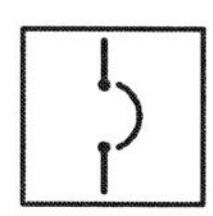

Use these figures to estimate the cost of circuit breakers installed in enclosures and circuit breaker enclosures installed in buildings under the conditions described on pages 5 and 6. Costs listed are for each circuit breaker or circuit breaker enclosure installed. The crew is one electrician working at a labor cost of $41.03 per manhour. These costs include layout, material handling, and normal waste. Add for the enclosure (except as noted above), support, sales tax, delivery, supervision, mobilization, demobilization, cleanup, overhead and profit. The labor required for wire termination is not included in these figures. Add the labor cost for making terminations according to the wire size and the size of the enclosure. Note: Be sure to select the right mounting style and amp interrupt capacity (A.I.C.). Two and three pole circuit breakers have a common operating handle. Some multi-pole breakers have a common tie bar that holds several operating handles together. Most engineered systems require one operating handle on each circuit breaker.

Surface Mounted Circuit Breaker Enclosures

NEMA 1 surface mounted circuit breaker enclosures

Material		Craft@Hrs	Unit	Material Cost	Labor Cost	Installed Cost
225A	600V	L1@0.35	Ea	180.00	14.40	194.40
400A	600V	L1@0.50	Ea	324.00	20.50	344.50
1000A	600V	L1@0.70	Ea	583.00	28.70	611.70
1200A	600V	L1@1.00	Ea	1,030.00	41.00	1,071.00

NEMA 3R surface mounted circuit breaker enclosures

3R

Material		Craft@Hrs	Unit	Material Cost	Labor Cost	Installed Cost
100A	120/240	L1@0.30	Ea	504.00	12.30	516.30
100A	277/480V	L1@0.30	Ea	561.00	12.30	573.30
225A	120/240V	L1@0.35	Ea	387.00	14.40	401.40
225A	277/480V	L1@0.35	Ea	434.00	14.40	448.40
225A	600V	L1@0.35	Ea	738.00	14.40	752.40
400A	600V	L1@0.50	Ea	1,650.00	20.50	1,670.50
1000A	600V	L1@0.70	Ea	2,160.00	28.70	2,188.70
1200A	600V	L1@1.00	Ea	2,680.00	41.00	2,721.00

NEMA 4 & 5 surface mounted circuit breaker enclosures

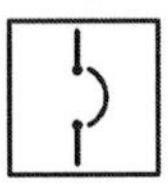
4/5

Material		Craft@Hrs	Unit	Material Cost	Labor Cost	Installed Cost
100A	277/480V	L1@0.50	Ea	1,450.00	20.50	1,470.50
225A	600V	L1@0.70	Ea	2,930.00	28.70	2,958.70
400A	600V	L1@1.00	Ea	5,690.00	41.00	5,731.00
1000A	600V	L1@1.50	Ea	10,200.00	61.50	10,261.50

NEMA 12 surface mounted circuit breaker enclosures

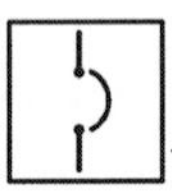
12

Material		Craft@Hrs	Unit	Material Cost	Labor Cost	Installed Cost
100A	277/480V	L1@0.35	Ea	308.00	14.40	322.40
225A	600V	L1@0.50	Ea	518.00	20.50	538.50
400A	600V	L1@0.70	Ea	880.00	28.70	908.70
1000A	600V	L1@1.00	Ea	1,650.00	41.00	1,691.00

NEMA 7 aluminum surface mounted circuit breaker enclosures

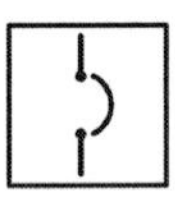
7

Material		Craft@Hrs	Unit	Material Cost	Labor Cost	Installed Cost
100A	277/480V	L1@0.50	Ea	2,000.00	20.50	2,020.50
225A	600V	L1@0.70	Ea	4,130.00	28.70	4,158.70
400A	600V	L1@1.00	Ea	9,620.00	41.00	9,661.00
800A	600V	L1@1.50	Ea	14,000.00	61.50	14,061.50

Use these figures to estimate the cost of circuit breaker enclosures installed in buildings under the conditions described on pages 5 and 6. Costs listed are for each enclosure installed. The crew is one electrician working at a labor cost of $41.03 per manhour. These costs include installation screws, layout, material handling, and normal waste. Add for circuit breakers, supports, sales tax, delivery, supervision, mobilization, demobilization, cleanup, overhead and profit. Note: Be sure to select the right enclosure for the style of circuit breaker required and for the installation location.

NEMA 3R 120/240 volt surface mounted meter sockets for overhead service

Material	Craft@Hrs	Unit	Material Cost	Labor Cost	Installed Cost
100A ring	L1@0.30	Ea	70.60	12.30	82.90
100A ringless	L1@0.30	Ea	70.60	12.30	82.90
150A ring	L1@0.40	Ea	106.00	16.40	122.40
150A ringless	L1@0.40	Ea	106.00	16.40	122.40
200A ring	L1@0.50	Ea	106.00	20.50	126.50
200A ringless	L1@0.50	Ea	106.00	20.50	126.50

NEMA 3R 120/240 volt surface mounted meter sockets for underground service

Material	Craft@Hrs	Unit	Material Cost	Labor Cost	Installed Cost
200A ring	L1@0.40	Ea	106.00	16.40	122.40
200A ringless	L1@0.40	Ea	106.00	16.40	122.40

NEMA 3R 120/240 volt surface mounted meter socket & main breaker for overhead service

Material	Craft@Hrs	Unit	Material Cost	Labor Cost	Installed Cost
100A main	L1@0.60	Ea	505.00	24.60	529.60
125A main	L1@0.60	Ea	779.00	24.60	803.60
150A main	L1@0.70	Ea	1,410.00	28.70	1,438.70
200A main	L1@0.70	Ea	1,410.00	28.70	1,438.70

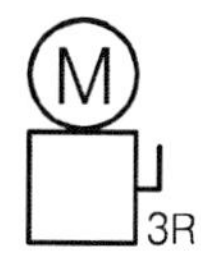

NEMA 3R 120/240 volt semi-flush meter socket & main, overhead service

Material	Craft@Hrs	Unit	Material Cost	Labor Cost	Installed Cost
100A main	L1@0.70	Ea	953.00	28.70	981.70
125A main	L1@0.70	Ea	1,950.00	28.70	1,978.70
150A main	L1@0.90	Ea	1,950.00	36.90	1,986.90
200A main	L1@0.90	Ea	1,950.00	36.90	1,986.90

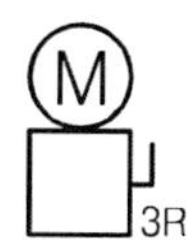

NEMA 3R 120/240 volt surface meter socket & main, underground service

Material	Craft@Hrs	Unit	Material Cost	Labor Cost	Installed Cost
100A main	L1@0.60	Ea	861.00	24.60	885.60
125A main	L1@0.60	Ea	1,150.00	24.60	1,174.60
150A main	L1@0.70	Ea	1,580.00	28.70	1,608.70
200A main	L1@0.70	Ea	1,580.00	28.70	1,608.70

NEMA 3R 120/240 volt semi-flush meter socket & main, underground service

Material	Craft@Hrs	Unit	Material Cost	Labor Cost	Installed Cost
100A main	L1@0.70	Ea	861.00	28.70	889.70
125A main	L1@0.70	Ea	1,150.00	28.70	1,178.70
150A main	L1@0.90	Ea	1,580.00	36.90	1,616.90
200A main	L1@0.90	Ea	1,580.00	36.90	1,616.90

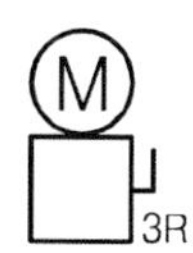

NEMA 3R 120/240 volt meter-breaker combination, overhead/underground service with main breaker, surface or semi-flush, add for branch breakers

Material	Craft@Hrs	Unit	Material Cost	Labor Cost	Installed Cost
100A, 8 to 12 poles	L1@1.00	Ea	303.00	41.00	344.00
125A 12 to 24 poles	L1@1.25	Ea	542.00	51.30	593.30
150A, 20 to 40 poles	L1@1.50	Ea	789.00	61.50	850.50

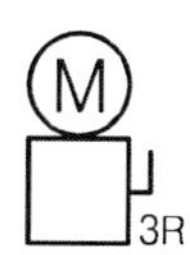

Use these figures to estimate the cost of meter sockets installed in buildings under the conditions described on pages 5 and 6. Costs listed are for each meter socket installed. The crew is one electrician working at a labor cost of $41.03 per manhour. These costs include sealing rings, layout, material handling, and normal waste. Add for supports, top conduit hubs, sales tax, delivery, supervision, mobilization, demobilization, cleanup, overhead and profit. Note: Top conduit hubs cost extra. Order the size appropriate for the size of conduit to be used. Blank caps are also available.

Safety Sockets and Meter Centers

NEMA 3R surface mounted safety sockets with test blocks

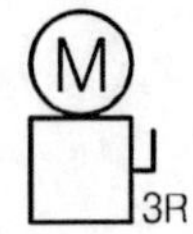

Material	Craft@Hrs	Unit	Material Cost	Labor Cost	Installed Cost
100A 5 jaw 240V	L1@0.50	Ea	587.00	20.50	607.50
100A 5 jaw 208V	L1@0.50	Ea	587.00	20.50	607.50
100A 5 jaw 480V	L1@0.50	Ea	735.00	20.50	755.50
100A 7 jaw 240V	L1@0.50	Ea	786.00	20.50	806.50
100A 7 jaw 480V	L1@0.50	Ea	786.00	20.50	806.50
200A 4 jaw 240V	L1@0.60	Ea	1,050.00	24.60	1,074.60
200A 5 jaw 208V	L1@0.60	Ea	1,050.00	24.60	1,074.60
200A 5 jaw 480V	L1@0.60	Ea	1,230.00	24.60	1,254.60
200A 7 jaw 240V	L1@0.60	Ea	1,370.00	24.60	1,394.60
200A 7 jaw 480V	L1@0.60	Ea	1,370.00	24.60	1,394.60

120/240 volt meter center & main, 800A bus, 4 jaw sockets, 125A max., 2 pole branch, indoor

Material	Craft@Hrs	Unit	Material Cost	Labor Cost	Installed Cost
3 meters & mains	L1@1.25	Ea	1,330.00	51.30	1,381.30
4 meters & mains	L1@1.50	Ea	1,760.00	61.50	1,821.50
5 meters & mains	L1@1.75	Ea	2,190.00	71.80	2,261.80
6 meters & mains	L1@2.00	Ea	2,550.00	82.10	2,632.10
7 meters & mains	L1@2.25	Ea	3,420.00	92.30	3,512.30
8 meters & mains	L1@2.50	Ea	3,420.00	103.00	3,523.00
10 meters & mains	L1@3.00	Ea	4,250.00	123.00	4,373.00

120/240 volt meter center, 1,200A bus, 4 jaw sockets, 200A max., 2 pole branch, indoor

Material	Craft@Hrs	Unit	Material Cost	Labor Cost	Installed Cost
3 meters & mains	L1@1.25	Ea	2,590.00	51.30	2,641.30
4 meters & mains	L1@1.50	Ea	3,510.00	61.50	3,571.50
6 meters & mains	L1@2.00	Ea	4,840.00	82.10	4,922.10
7 meters & mains	L1@2.25	Ea	5,090.00	92.30	5,182.30
8 meters & mains	L1@2.50	Ea	6,790.00	103.00	6,893.00

120/208 volt meter center, 800A bus, 5 jaw sockets, 125A max., 2 pole branch, indoor

Material	Craft@Hrs	Unit	Material Cost	Labor Cost	Installed Cost
3 meters & mains	L1@1.50	Ea	1,420.00	61.50	1,481.50
4 meters & mains	L1@1.75	Ea	1,900.00	71.80	1,971.80
5 meters & mains	L1@2.00	Ea	2,380.00	82.10	2,462.10
6 meters & mains	L1@2.25	Ea	2,750.00	92.30	2,842.30
7 meters & mains	L1@2.50	Ea	3,710.00	103.00	3,813.00
8 meters & mains	L1@2.75	Ea	3,710.00	113.00	3,823.00
10 meters & mains	L1@3.25	Ea	4,660.00	133.00	4,793.00

Use these figures to estimate the cost of meter centers installed in buildings under the conditions described on pages 5 and 6. Costs listed are for each meter socket installed. The crew is one electrician working at a labor cost of $41.03 per manhour. These costs include sealing rings, layout, material handling, and normal waste. Add for supports, top conduit hubs, sales tax, delivery, supervision, mobilization, demobilization, cleanup, overhead and profit. Note: Top conduit hubs cost extra. Order the size appropriate for the size of conduit to be used. Blank caps are also available.

120/208 volt meter centers, 1,200A bus, 5 jaw sockets, 10,000 A.I.C., 2 pole, indoor

Material	Craft@Hrs	Unit	Material Cost	Labor Cost	Installed Cost
3 meters & mains	L1@1.50	Ea	2,740.00	61.50	2,801.50
4 meters & mains	L1@1.75	Ea	3,640.00	71.80	3,711.80
6 meters & mains	L1@2.25	Ea	5,320.00	92.30	5,412.30
7 meters & mains	L1@2.50	Ea	5,320.00	103.00	5,423.00
8 meters & mains	L1@2.75	Ea	7,120.00	113.00	7,233.00

120/240 volt meter centers, 800A bus, 4 jaw sockets, 42,000 A.I.C., 2 pole, indoor

Material	Craft@Hrs	Unit	Material Cost	Labor Cost	Installed Cost
3 meters & mains	L1@1.75	Ea	2,060.00	71.80	2,131.80
4 meters & mains	L1@2.00	Ea	2,750.00	82.10	2,832.10
5 meters & mains	L1@2.25	Ea	3,430.00	92.30	3,522.30
6 meters & mains	L1@2.50	Ea	4,040.00	103.00	4,143.00
7 meters & mains	L1@2.75	Ea	5,170.00	113.00	5,283.00
8 meters & mains	L1@3.00	Ea	5,420.00	123.00	5,543.00
10 meters & mains	L1@3.50	Ea	6,760.00	144.00	6,904.00

120/240 volt meter centers, 1,200A bus, 4 jaw sockets, 42,000 A.I.C., 2 pole, indoor

Material	Craft@Hrs	Unit	Material Cost	Labor Cost	Installed Cost
3 meters & mains	L1@1.75	Ea	3,200.00	71.80	3,271.80
4 meters & mains	L1@2.00	Ea	4,470.00	82.10	4,552.10
6 meters & mains	L1@2.50	Ea	6,360.00	103.00	6,463.00
7 meters & mains	L1@2.75	Ea	6,790.00	113.00	6,903.00
8 meters & mains	L1@3.00	Ea	8,750.00	123.00	8,873.00

120/208 volt meter centers, 800A bus, 5 jaw sockets, 42,000 A.I.C., 2 pole, indoor

Material	Craft@Hrs	Unit	Material Cost	Labor Cost	Installed Cost
3 meters & mains	L1@2.00	Ea	2,140.00	82.10	2,222.10
4 meters & mains	L1@2.25	Ea	2,890.00	92.30	2,982.30
5 meters & mains	L1@2.50	Ea	3,570.00	103.00	3,673.00
6 meters & mains	L1@2.75	Ea	4,220.00	113.00	4,333.00
7 meters & mains	L1@3.00	Ea	5,420.00	123.00	5,543.00
8 meters & mains	L1@3.25	Ea	5,620.00	133.00	5,753.00
10 meters & mains	L1@3.75	Ea	6,890.00	154.00	7,044.00

Use these figures to estimate the cost of meter centers installed in buildings under the conditions described on pages 5 and 6. Costs listed are for each meter socket installed. The crew is one electrician working at a labor cost of $41.03 per manhour. These costs include sealing rings, layout, material handling, and normal waste. Add for supports, top conduit hubs, sales tax, delivery, supervision, mobilization, demobilization, cleanup, overhead and profit. Note: Top conduit hubs cost extra. Order the size appropriate for the size of conduit to be used. Blank caps are also available.

Meter Centers

120/208 volt meter centers, 1,200A bus, 5 jaw sockets, 42,000 A.I.C., 2 pole, indoor

Material	Craft@Hrs	Unit	Material Cost	Labor Cost	Installed Cost
3 meters & mains	L1@2.00	Ea	3,510.00	82.10	3,592.10
4 meters & mains	L1@2.25	Ea	4,670.00	92.30	4,762.30
6 meters & mains	L1@2.75	Ea	6,790.00	113.00	6,903.00
7 meters & mains	L1@3.00	Ea	7,060.00	123.00	7,183.00
8 meters & mains	L1@3.25	Ea	8,590.00	133.00	8,723.00

120/240 volt meter centers, 800A bus, 4 jaw sockets, 10,000 A.I.C., 2 pole, NEMA 3R

Material	Craft@Hrs	Unit	Material Cost	Labor Cost	Installed Cost
3 meters & mains	L1@1.25	Ea	1,320.00	51.30	1,371.30
4 meters & mains	L1@1.50	Ea	1,710.00	61.50	1,771.50
6 meters & mains	L1@2.00	Ea	2,490.00	82.10	2,572.10
7 meters & mains	L1@2.25	Ea	3,120.00	92.30	3,212.30
8 meters & mains	L1@2.50	Ea	4,210.00	103.00	4,313.00

120/240 volt meter centers, 1,200A bus, 4 jaw sockets, 10,000 A.I.C., 2 pole, NEMA 3R

Material	Craft@Hrs	Unit	Material Cost	Labor Cost	Installed Cost
3 meters & mains	L1@1.25	Ea	2,450.00	51.30	2,501.30
4 meters & mains	L1@1.50	Ea	3,260.00	61.50	3,321.50
6 meters & mains	L1@2.00	Ea	4,860.00	82.10	4,942.10
7 meters & mains	L1@2.25	Ea	4,860.00	92.30	4,952.30
8 meters & mains	L1@2.50	Ea	6,660.00	103.00	6,763.00

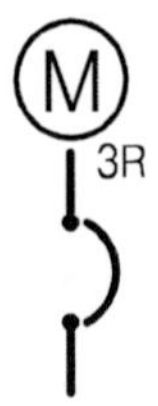

120/208 volt meter centers, 800A bus, 5 jaw sockets, 10,000 A.I.C., 2 pole, NEMA 3R

Material	Craft@Hrs	Unit	Material Cost	Labor Cost	Installed Cost
3 meters & mains	L1@1.50	Ea	1,420.00	61.50	1,481.50
4 meters & mains	L1@1.75	Ea	1,900.00	71.80	1,971.80
6 meters & mains	L1@2.25	Ea	2,690.00	92.30	2,782.30
7 meters & mains	L1@2.50	Ea	3,600.00	103.00	3,703.00
8 meters & mains	L1@2.75	Ea	3,600.00	113.00	3,713.00

Use these figures to estimate the cost of meter centers installed in buildings under the conditions described on pages 5 and 6. Costs listed are for each meter socket installed. The crew is one electrician working at a labor cost of $41.03 per manhour. These costs include sealing rings, layout, material handling, and normal waste. Add for supports, top conduit hubs, sales tax, delivery, supervision, mobilization, demobilization, cleanup, overhead and profit. Note: Main circuit breakers are included in these meter centers. But meters are not included. They are usually furnished by the electrical utility. Top conduit hubs cost extra. Order the size appropriate for the size of conduit to be used. Blank caps are also available.

120/208 volt raintight meter centers, 1,200A bus, 5 jaw sockets, 10,000 A.I.C., 2 pole, NEMA 3R

Material	Craft@Hrs	Unit	Material Cost	Labor Cost	Installed Cost
3 meters & mains	L1@1.50	Ea	2,690.00	61.50	2,751.50
4 meters & mains	L1@1.75	Ea	3,580.00	71.80	3,651.80
6 meters & mains	L1@2.25	Ea	5,320.00	92.30	5,412.30
7 meters & mains	L1@2.50	Ea	5,320.00	103.00	5,423.00
8 meters & mains	L1@2.75	Ea	6,970.00	113.00	7,083.00

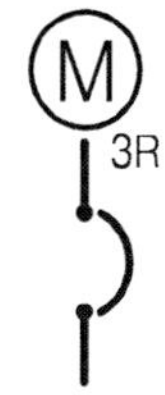

120/240 volt raintight meter centers, 800A bus, 4 jaw sockets, 42,000 A.I.C., 2 pole, NEMA 3R

Material	Craft@Hrs	Unit	Material Cost	Labor Cost	Installed Cost
3 meters & mains	L1@1.75	Ea	2,060.00	71.80	2,131.80
4 meters & mains	L1@2.00	Ea	2,750.00	82.10	2,832.10
6 meters & mains	L1@2.50	Ea	3,980.00	103.00	4,083.00
7 meters & mains	L1@2.75	Ea	5,140.00	113.00	5,253.00
8 meters & mains	L1@3.00	Ea	5,270.00	123.00	5,393.00

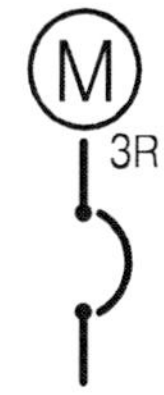

120/240 volt raintight meter centers, 1,200A bus, 4 jaw sockets, 42,000 A.I.C., 2 pole, NEMA 3R

Material	Craft@Hrs	Unit	Material Cost	Labor Cost	Installed Cost
3 meters & mains	L1@1.75	Ea	3,200.00	71.80	3,271.80
4 meters & mains	L1@2.00	Ea	4,410.00	82.10	4,492.10
6 meters & mains	L1@2.50	Ea	6,220.00	103.00	6,323.00
7 meters & mains	L1@2.75	Ea	6,420.00	113.00	6,533.00
8 meters & mains	L1@3.00	Ea	8,590.00	123.00	8,713.00

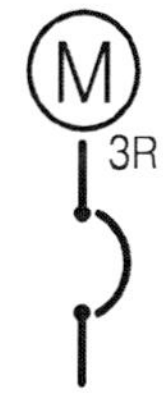

120/208 volt raintight meter centers, 800A bus, 5 jaw sockets, 42,000 A.I.C., 2 pole, NEMA 3R

Material	Craft@Hrs	Unit	Material Cost	Labor Cost	Installed Cost
3 meters & mains	L1@2.00	Ea	2,140.00	82.10	2,222.10
4 meters & mains	L1@2.25	Ea	2,870.00	92.30	2,962.30
6 meters & mains	L1@2.75	Ea	4,200.00	113.00	4,313.00
7 meters & mains	L1@3.00	Ea	5,290.00	123.00	5,413.00
8 meters & mains	L1@3.25	Ea	5,590.00	133.00	5,723.00

120/208 volt raintight meter centers, 1,200A bus, 5 jaw sockets, 42,000 A.I.C. , 2 pole, NEMA 3R

Material	Craft@Hrs	Unit	Material Cost	Labor Cost	Installed Cost
3 meters & mains	L1@2.00	Ea	3,510.00	82.10	3,592.10
4 meters & mains	L1@2.25	Ea	4,510.00	92.30	4,602.30
6 meters & mains	L1@2.75	Ea	6,650.00	113.00	6,763.00
7 meters & mains	L1@3.00	Ea	6,900.00	123.00	7,023.00
8 meters & mains	L1@3.25	Ea	8,600.00	133.00	8,733.00

Use these figures to estimate the cost of meter centers installed in buildings under the conditions described on pages 5 and 6. Costs listed are for each meter socket installed. The crew is one electrician working at a labor cost of $41.03 per manhour. These costs include sealing rings, layout, material handling, and normal waste. Add for supports, top conduit hubs, sales tax, delivery, supervision, mobilization, demobilization, cleanup, overhead and profit. Note: Main circuit breakers are included in these meter centers. But meters are not included. They are usually furnished by the electrical utility. Top conduit hubs cost extra. Order the size appropriate for the size of conduit to be used. Blank caps are also available.

120/240 Volt Loadcenters

Indoor 120/240 volt loadcenters with main lugs only, no breakers

Material		Craft@Hrs	Unit	Material Cost	Labor Cost	Installed Cost
30A	2 spaces	L1@0.35	Ea	26.30	14.40	40.70
40A	2 spaces	L1@0.35	Ea	75.60	14.40	90.00
70A	4 spaces	L1@0.40	Ea	86.00	16.40	102.40
100A	6 spaces	L1@0.50	Ea	52.10	20.50	72.60
100A	8 spaces	L1@0.60	Ea	86.00	24.60	110.60
100A	12 spaces	L1@0.70	Ea	153.00	28.70	181.70
125A	16 spaces	L1@0.75	Ea	120.00	30.80	150.80
125A	20 spaces	L1@0.80	Ea	150.00	32.80	182.80
125A	24 spaces	L1@0.90	Ea	216.00	36.90	252.90
150A	12 spaces	L1@0.75	Ea	196.00	30.80	226.80
150A	16 spaces	L1@0.80	Ea	196.00	32.80	228.80
150A	24 spaces	L1@1.00	Ea	220.00	41.00	261.00
150A	30 spaces	L1@1.20	Ea	220.00	49.20	269.20
200A	8 spaces	L1@0.70	Ea	252.00	28.70	280.70
200A	12 spaces	L1@0.80	Ea	252.00	32.80	284.80
200A	16 spaces	L1@0.95	Ea	252.00	39.00	291.00
200A	24 spaces	L1@1.10	Ea	252.00	45.10	297.10
200A	30 spaces	L1@1.30	Ea	252.00	53.30	305.30
225A	42 spaces	L1@1.50	Ea	459.00	61.50	520.50
400A	30 spaces	L1@1.25	Ea	970.00	51.30	1,021.30
400A	42 spaces	L1@1.75	Ea	970.00	71.80	1,041.80

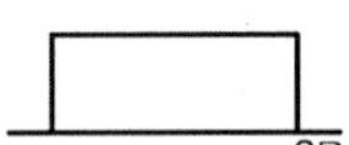

3R

Raintight 120/240 volt loadcenters with main lugs only, no breakers

Material		Craft@Hrs	Unit	Material Cost	Labor Cost	Installed Cost
40A	2 spaces	L1@0.35	Ea	86.00	14.40	100.40
70A	4 spaces	L1@0.40	Ea	91.30	16.40	107.70
100A	6 spaces	L1@0.50	Ea	124.00	20.50	144.50
100A	8 spaces	L1@0.60	Ea	124.00	24.60	148.60
100A	12 spaces	L1@0.70	Ea	124.00	28.70	152.70
125A	12 spaces	L1@0.70	Ea	200.00	28.70	228.70
125A	16 spaces	L1@0.75	Ea	243.00	30.80	273.80
125A	20 spaces	L1@0.80	Ea	284.00	32.80	316.80
125A	24 spaces	L1@0.90	Ea	284.00	36.90	320.90
150A	12 spaces	L1@0.75	Ea	375.00	30.80	405.80
150A	16 spaces	L1@0.80	Ea	375.00	32.80	407.80
150A	24 spaces	L1@1.00	Ea	375.00	41.00	416.00
150A	30 spaces	L1@1.25	Ea	704.00	51.30	755.30
200A	8 spaces	L1@0.70	Ea	375.00	28.70	403.70
200A	12 spaces	L1@0.80	Ea	375.00	32.80	407.80
200A	16 spaces	L1@0.95	Ea	375.00	39.00	414.00
200A	24 spaces	L1@1.10	Ea	375.00	45.10	420.10
200A	30 spaces	L1@1.30	Ea	704.00	53.30	757.30
225A	42 spaces	L1@1.50	Ea	1,010.00	61.50	1,071.50

Use these figures to estimate the cost of loadcenters installed in buildings under the conditions described on pages 5 and 6. Costs listed are for each loadcenter installed. The crew is one electrician working at a labor cost of $41.03 per manhour. These costs include the cover, bus, neutral bar, layout, material handling, and normal waste. Add for circuit breakers, supports, sales tax, delivery, supervision, mobilization, demobilization, cleanup, overhead and profit. Note: Be careful to select the right type of enclosure and voltage.

Indoor 120/208 volt loadcenters with main lugs only, no breakers

Material		Craft@Hrs	Unit	Material Cost	Labor Cost	Installed Cost
100A	12 spaces	L1@0.70	Ea	324.00	28.70	352.70
100A	16 spaces	L1@0.75	Ea	324.00	30.80	354.80
100A	20 spaces	L1@0.80	Ea	324.00	32.80	356.80
125A	24 spaces	L1@0.90	Ea	413.00	36.90	449.90
150A	16 spaces	L1@0.80	Ea	413.00	32.80	445.80
150A	20 spaces	L1@1.00	Ea	413.00	41.00	454.00
150A	24 spaces	L1@1.20	Ea	413.00	49.20	462.20
150A	30 spaces	L1@1.25	Ea	413.00	51.30	464.30
200A	16 spaces	L1@0.95	Ea	411.00	39.00	450.00
200A	20 spaces	L1@1.00	Ea	411.00	41.00	452.00
200A	24 spaces	L1@1.10	Ea	411.00	45.10	456.10
200A	30 spaces	L1@1.20	Ea	411.00	49.20	460.20
200A	40 spaces	L1@1.30	Ea	558.00	53.30	611.30
300A	42 spaces	L1@1.50	Ea	1,740.00	61.50	1,801.50
400A	42 spaces	L1@1.75	Ea	1,740.00	71.80	1,811.80

Raintight 120/208 volt loadcenters with main lugs only, no breakers

Material		Craft@Hrs	Unit	Material Cost	Labor Cost	Installed Cost
100A	12 spaces	L1@0.70	Ea	290.00	28.70	318.70
100A	16 spaces	L1@0.75	Ea	375.00	30.80	405.80
100A	20 spaces	L1@0.80	Ea	477.00	32.80	509.80
125A	24 spaces	L1@0.90	Ea	477.00	36.90	513.90
150A	16 spaces	L1@0.80	Ea	375.00	32.80	407.80
150A	20 spaces	L1@1.00	Ea	544.00	41.00	585.00
150A	24 spaces	L1@1.20	Ea	477.00	49.20	526.20
150A	30 spaces	L1@1.25	Ea	477.00	51.30	528.30
200A	16 spaces	L1@0.95	Ea	436.00	39.00	475.00
200A	20 spaces	L1@1.00	Ea	418.00	41.00	459.00

Use these figures to estimate the cost of loadcenters installed in buildings under the conditions described on pages 5 and 6. Costs listed are for each loadcenter installed. The crew is one electrician working at a labor cost of $41.03 per manhour. These costs include the cover, bus, neutral bar, layout, material handling, and normal waste. Add for circuit breakers, supports, sales tax, delivery, supervision, mobilization, demobilization, cleanup, overhead and profit. Note: Be careful to select the right type of enclosure and voltage.

Loadcenters and Panelboards

Raintight 120/208 volt C.B. loadcenters with main lugs but no other breakers

3R

Material		Craft@Hrs	Unit	Material Cost	Labor Cost	Installed Cost
200A	24 spaces	L1@1.10	Ea	566.00	45.10	611.10
200A	30 spaces	L1@1.20	Ea	566.00	49.20	615.20
200A	40 spaces	L1@1.30	Ea	566.00	53.30	619.30
225A	30 spaces	L1@1.40	Ea	1,020.00	57.40	1,077.40
225A	40 spaces	L1@1.50	Ea	1,020.00	61.50	1,081.50

Indoor 120/240 volt C.B. panelboards (NQO) with plug-in breakers to 60 amps, main lugs only

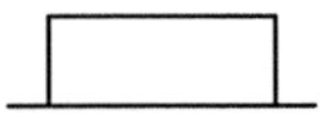

125A	8 poles	L1@3.40	Ea	720.00	140.00	860.00
125A	10 poles	L1@4.00	Ea	745.00	164.00	909.00
125A	12 poles	L1@4.60	Ea	773.00	189.00	962.00
125A	14 poles	L1@5.70	Ea	802.00	234.00	1,036.00
125A	16 poles	L1@6.30	Ea	829.00	258.00	1,087.00
125A	18 poles	L1@6.90	Ea	858.00	283.00	1,141.00
125A	20 poles	L1@7.50	Ea	886.00	308.00	1,194.00
225A	22 poles	L1@8.10	Ea	938.00	332.00	1,270.00
225A	24 poles	L1@8.70	Ea	970.00	357.00	1,327.00
225A	26 poles	L1@9.80	Ea	992.00	402.00	1,394.00
225A	28 poles	L1@10.4	Ea	1,020.00	427.00	1,447.00
225A	30 poles	L1@11.0	Ea	1,040.00	451.00	1,491.00
225A	32 poles	L1@11.6	Ea	1,080.00	476.00	1,556.00
225A	34 poles	L1@12.2	Ea	1,110.00	501.00	1,611.00
225A	36 poles	L1@12.8	Ea	1,140.00	525.00	1,665.00
225A	38 poles	L1@13.9	Ea	1,160.00	570.00	1,730.00
225A	40 poles	L1@14.5	Ea	1,190.00	595.00	1,785.00
225A	42 poles	L1@15.1	Ea	1,350.00	620.00	1,970.00
400A	38 poles	L1@14.4	Ea	1,390.00	591.00	1,981.00
400A	40 poles	L1@15.0	Ea	1,480.00	615.00	2,095.00
400A	42 poles	L1@15.6	Ea	1,500.00	640.00	2,140.00

Indoor 120/208 volt C.B. panelboards (NQO) with plug-in breakers to 60 amps and main breaker

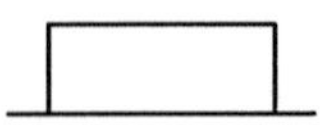

100A	8 poles	L1@3.40	Ea	1,500.00	140.00	1,640.00
100A	12 poles	L1@4.00	Ea	1,630.00	164.00	1,794.00
100A	16 poles	L1@4.60	Ea	1,800.00	189.00	1,989.00

Use these figures to estimate the cost of loadcenters and panelboards installed in buildings under the conditions described on pages 5 and 6. Costs listed are for each panelboard installed. The crew is one electrician working at a labor cost of $41.03 per manhour. These costs include the cover, bus, neutral bar, circuit breakers, layout, material handling, and normal waste. Labor costs include mounting the panel can, installing the interior section, making the circuit connections for each breaker and the main breaker, and hanging the door or cover. Reduce the labor cost for each spare breaker not connected to a circuit. Add for supports, sales tax, delivery, supervision, mobilization, demobilization, cleanup, overhead and profit. Note: Be sure to select the loadcenter or panelboard with the right voltage and type of enclosure.

Indoor 120/208 volt C.B. panelboards (NQO) with plug-in breakers to 60 amps and main breaker

Material		Craft@Hrs	Unit	Material Cost	Labor Cost	Installed Cost
100A	18 poles	L1@6.90	Ea	1,880.00	283.00	2,163.00
100A	20 poles	L1@7.50	Ea	1,940.00	308.00	2,248.00
225A	22 poles	L1@8.10	Ea	2,730.00	332.00	3,062.00
225A	24 poles	L1@8.70	Ea	2,760.00	357.00	3,117.00
225A	26 poles	L1@9.80	Ea	2,810.00	402.00	3,212.00
225A	28 poles	L1@10.4	Ea	2,940.00	427.00	3,367.00
225A	30 poles	L1@11.0	Ea	2,770.00	451.00	3,221.00
225A	32 poles	L1@11.6	Ea	3,190.00	476.00	3,666.00
225A	34 poles	L1@12.2	Ea	3,250.00	501.00	3,751.00
225A	36 poles	L1@12.8	Ea	3,290.00	525.00	3,815.00
225A	38 poles	L1@13.9	Ea	3,410.00	570.00	3,980.00
225A	40 poles	L1@14.5	Ea	3,420.00	595.00	4,015.00
225A	42 poles	L1@15.1	Ea	3,450.00	620.00	4,070.00

Indoor 120/240 volt C.B. panelboards (NQOB) with breakers to 60 amps, with main lugs only

Material		Craft@Hrs	Unit	Material Cost	Labor Cost	Installed Cost
100A	8 poles	L1@3.40	Ea	1,080.00	140.00	1,220.00
100A	10 poles	L1@4.00	Ea	1,140.00	164.00	1,304.00
100A	12 poles	L1@4.60	Ea	1,190.00	189.00	1,379.00
100A	14 poles	L1@5.70	Ea	1,380.00	234.00	1,614.00
100A	16 poles	L1@6.30	Ea	1,450.00	258.00	1,708.00
100A	18 poles	L1@6.90	Ea	1,490.00	283.00	1,773.00
100A	20 poles	L1@7.50	Ea	1,540.00	308.00	1,848.00
225A	22 poles	L1@8.10	Ea	1,790.00	332.00	2,122.00
225A	24 poles	L1@8.70	Ea	1,830.00	357.00	2,187.00
225A	26 poles	L1@9.80	Ea	1,910.00	402.00	2,312.00
225A	28 poles	L1@10.4	Ea	1,960.00	427.00	2,387.00
225A	30 poles	L1@11.0	Ea	2,000.00	451.00	2,451.00
225A	32 poles	L1@11.6	Ea	2,200.00	476.00	2,676.00
225A	34 poles	L1@12.2	Ea	2,150.00	501.00	2,651.00
225A	36 poles	L1@12.8	Ea	2,320.00	525.00	2,845.00
225A	38 poles	L1@13.9	Ea	2,420.00	570.00	2,990.00
225A	40 poles	L1@14.5	Ea	2,490.00	595.00	3,085.00
225A	42 poles	L1@15.1	Ea	2,500.00	620.00	3,120.00

Use these figures to estimate the cost of panelboards installed in buildings under the conditions described on pages 5 and 6. Costs listed are for each panelboard installed. The crew is one electrician working at a labor cost of $41.03 per manhour. These costs include the cover, bus, neutral bar, circuit breakers, layout, material handling, and normal waste. Labor costs include mounting the panel can, installing the interior section, making the circuit connections for each breaker and the main breaker, and hanging the door or cover. Reduce the labor cost for each spare breaker not connected to a circuit. Add for supports, sales tax, delivery, supervision, mobilization, demobilization, cleanup, overhead and profit. Note: Be sure to select the panelboard with the right voltage and type of enclosure.

Panelboards Including Bolt-on Breakers to 60 Amps

Indoor 120/208 volt C.B. panelboards with bolt-on breakers to 60 amps, with main lugs only

Material		Craft@Hrs	Unit	Material Cost	Labor Cost	Installed Cost
100A	8 poles	L1@3.40	Ea	1,100.00	140.00	1,240.00
100A	10 poles	L1@4.00	Ea	1,150.00	164.00	1,314.00
100A	12 poles	L1@4.60	Ea	1,200.00	189.00	1,389.00
100A	14 poles	L1@5.70	Ea	1,380.00	234.00	1,614.00
100A	16 poles	L1@6.30	Ea	1,450.00	258.00	1,708.00
100A	18 poles	L1@6.90	Ea	1,490.00	283.00	1,773.00
100A	20 poles	L1@7.50	Ea	1,540.00	308.00	1,848.00
100A	22 poles	L1@8.10	Ea	1,790.00	332.00	2,122.00
100A	24 poles	L1@8.70	Ea	1,830.00	357.00	2,187.00
225A	26 poles	L1@9.80	Ea	1,910.00	402.00	2,312.00
225A	28 poles	L1@10.4	Ea	1,960.00	427.00	2,387.00
225A	30 poles	L1@11.0	Ea	2,000.00	451.00	2,451.00
225A	32 poles	L1@11.6	Ea	2,940.00	476.00	3,416.00
225A	34 poles	L1@12.2	Ea	2,960.00	501.00	3,461.00
225A	36 poles	L1@12.8	Ea	3,030.00	525.00	3,555.00
225A	38 poles	L1@13.9	Ea	3,130.00	570.00	3,700.00
225A	40 poles	L1@14.5	Ea	3,190.00	595.00	3,785.00
225A	42 poles	L1@15.1	Ea	5,060.00	620.00	5,680.00

Indoor 120/240 volt C.B. panelboards with bolt-on breakers to 60 amps and main breaker

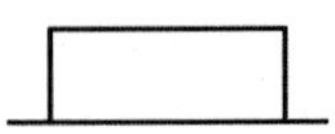

Material		Craft@Hrs	Unit	Material Cost	Labor Cost	Installed Cost
100A	8 poles	L1@3.40	Ea	1,500.00	140.00	1,640.00
100A	10 poles	L1@4.00	Ea	1,540.00	164.00	1,704.00
100A	12 poles	L1@4.60	Ea	1,630.00	189.00	1,819.00
100A	14 poles	L1@5.70	Ea	1,740.00	234.00	1,974.00
100A	16 poles	L1@6.30	Ea	1,800.00	258.00	2,058.00
100A	18 poles	L1@6.90	Ea	1,880.00	283.00	2,163.00
100A	20 poles	L1@7.50	Ea	1,940.00	308.00	2,248.00
225A	22 poles	L1@8.10	Ea	2,730.00	332.00	3,062.00
225A	24 poles	L1@8.70	Ea	2,760.00	357.00	3,117.00
225A	26 poles	L1@9.80	Ea	2,810.00	402.00	3,212.00
225A	28 poles	L1@10.4	Ea	2,940.00	427.00	3,367.00
225A	30 poles	L1@11.0	Ea	2,950.00	451.00	3,401.00
225A	32 poles	L1@11.6	Ea	3,190.00	476.00	3,666.00
225A	34 poles	L1@12.2	Ea	3,250.00	501.00	3,751.00
225A	36 poles	L1@12.8	Ea	3,290.00	525.00	3,815.00
225A	38 poles	L1@13.9	Ea	3,410.00	570.00	3,980.00
225A	40 poles	L1@14.5	Ea	3,420.00	595.00	4,015.00
225A	42 poles	L1@15.1	Ea	3,450.00	620.00	4,070.00

Use these figures to estimate the cost of panelboards installed in buildings under the conditions described on pages 5 and 6. Costs listed are for each panelboard installed. The crew is one electrician working at a labor cost of $41.03 per manhour. These costs include the cover, bus, neutral bar, circuit breakers, layout, material handling, and normal waste. Labor costs include mounting the panel can, installing the interior section, making the circuit connections for each breaker and the main breaker, and hanging the door or cover. Reduce the labor cost for each spare breaker not connected to a circuit. Add for supports, sales tax, delivery, supervision, mobilization, demobilization, cleanup, overhead and profit. Note: Be sure to select the panelboard with the right voltage and type of enclosure.

120/240 volt 14" wide C.B. panelboards with bolt-on breakers up to 60 amps and main lugs only

Material		Craft@Hrs	Unit	Material Cost	Labor Cost	Installed Cost
100A	8 poles	L1@3.40	Ea	1,410.00	140.00	1,550.00
100A	10 poles	L1@4.00	Ea	1,460.00	164.00	1,624.00
100A	12 poles	L1@4.60	Ea	1,500.00	189.00	1,689.00
100A	14 poles	L1@5.70	Ea	1,590.00	234.00	1,824.00
100A	16 poles	L1@6.30	Ea	1,630.00	258.00	1,888.00
100A	18 poles	L1@6.90	Ea	1,770.00	283.00	2,053.00
100A	20 poles	L1@7.50	Ea	1,830.00	308.00	2,138.00
225A	22 poles	L1@8.10	Ea	1,910.00	332.00	2,242.00
225A	24 poles	L1@8.70	Ea	1,960.00	357.00	2,317.00
225A	26 poles	L1@9.80	Ea	2,000.00	402.00	2,402.00
225A	28 poles	L1@10.4	Ea	2,070.00	427.00	2,497.00
225A	30 poles	L1@11.0	Ea	2,120.00	451.00	2,571.00
225A	32 poles	L1@11.6	Ea	2,400.00	476.00	2,876.00
225A	34 poles	L1@12.2	Ea	2,470.00	501.00	2,971.00
225A	36 poles	L1@12.8	Ea	2,500.00	525.00	3,025.00
225A	38 poles	L1@13.9	Ea	2,580.00	570.00	3,150.00
225A	40 poles	L1@14.5	Ea	2,620.00	595.00	3,215.00
225A	42 poles	L1@15.1	Ea	2,720.00	620.00	3,340.00

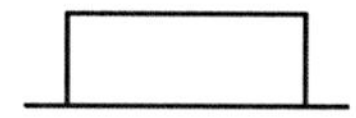

NEMA 1 telephone & signal terminal cabinets with keyed door lock, wood backing

Material	Craft@Hrs	Unit	Material Cost	Labor Cost	Installed Cost
12"W x 12"H x 4"D	L1@0.45	Ea	196.00	18.50	214.50
12"W x 16"H x 4"D	L1@0.60	Ea	196.00	24.60	220.60
12"W x 16"H x 6"D	L1@0.70	Ea	272.00	28.70	300.70
12"W x 18"H x 4"D	L1@0.90	Ea	236.00	36.90	272.90
12"W x 18"H x 6"D	L1@1.00	Ea	285.00	41.00	326.00
12"W x 24"H x 4"D	L1@1.20	Ea	255.00	49.20	304.20
12"W x 24"H x 6"D	L1@1.30	Ea	400.00	53.30	453.30
18"W x 18"H x 4"D	L1@1.20	Ea	263.00	49.20	312.20
18"W x 18"H x 6"D	L1@1.30	Ea	305.00	53.30	358.30
18"W x 24"H x 4"D	L1@1.50	Ea	352.00	61.50	413.50
18"W x 24"H x 6"D	L1@1.60	Ea	437.00	65.60	502.60
18"W x 30"H x 4"D	L1@1.80	Ea	400.00	73.90	473.90
18"W x 30"H x 6"D	L1@1.90	Ea	460.00	78.00	538.00
24"W x 24"H x 4"D	L1@2.10	Ea	425.00	86.20	511.20
24"W x 24"H x 6"D	L1@2.20	Ea	493.00	90.30	583.30
24"W x 30"H x 4"D	L1@2.50	Ea	538.00	103.00	641.00
24"W x 30"H x 6"D	L1@2.60	Ea	591.00	107.00	698.00
24"W x 36"H x 4"D	L1@3.00	Ea	622.00	123.00	745.00
24"W x 36"H x 6"D	L1@3.10	Ea	687.00	127.00	814.00

Use these figures to estimate the cost of panelboards and terminal cabinets installed in buildings under the conditions described on pages 5 and 6. Costs listed are for each panelboard or terminal cabinet installed. The crew is one electrician working at a labor cost of $41.03 per manhour. These costs include the cover, bus, neutral bar, circuit breakers, layout, material handling, and normal waste. Labor costs include mounting the panel can, installing the interior section, making the circuit connections for each breaker and the main breaker, and hanging the door or cover. Reduce the labor cost for each spare breaker not connected to a circuit. Add for supports, sales tax, delivery, supervision, mobilization, demobilization, cleanup, overhead and profit. Note: Be sure to select the panelboard with the right voltage and type of enclosure.

Signal Cabinets and Wireway

NEMA 1 telephone & signal terminal cabinets with keyed door lock, wood backing

Material	Craft@Hrs	Unit	Material Cost	Labor Cost	Installed Cost
30"W x 30"H x 4"D	L1@3.00	Ea	853.00	123.00	976.00
30"W x 30"H x 6"D	L1@3.10	Ea	918.00	127.00	1,045.00
30"W x 36"H x 4"D	L1@3.30	Ea	1,040.00	135.00	1,175.00
30"W x 36"H x 6"D	L1@3.40	Ea	1,130.00	140.00	1,270.00
36"W x 48"H x 4"D	L1@3.60	Ea	1,900.00	148.00	2,048.00
36"W x 48"H x 6"D	L1@3.70	Ea	2,180.00	152.00	2,332.00

NEMA 1 screw cover wireway

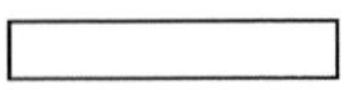

Material	Craft@Hrs	Unit	Material Cost	Labor Cost	Installed Cost
3" x 3" x 12"	L1@0.40	Ea	15.90	16.40	32.30
3" x 3" x 18"	L1@0.45	Ea	24.00	18.50	42.50
3" x 3" x 24"	L1@0.50	Ea	26.60	20.50	47.10
3" x 3" x 36"	L1@0.60	Ea	40.20	24.60	64.80
3" x 3" x 48"	L1@0.70	Ea	54.30	28.70	83.00
3" x 3" x 60"	L1@0.80	Ea	61.90	32.80	94.70
3" x 3" x 72"	L1@1.00	Ea	76.50	41.00	117.50
3" x 3" x 120"	L1@1.15	Ea	127.00	47.20	174.20
4" x 4" x 12"	L1@0.50	Ea	17.10	20.50	37.60
4" x 4" x 18"	L1@0.60	Ea	26.60	24.60	51.20
4" x 4" x 24"	L1@0.70	Ea	27.30	28.70	56.00
4" x 4" x 36"	L1@0.80	Ea	42.40	32.80	75.20
4" x 4" x 48"	L1@1.00	Ea	55.60	41.00	96.60
4" x 4" x 60"	L1@1.20	Ea	60.00	49.20	109.20
4" x 4" x 72"	L1@1.50	Ea	83.30	61.50	144.80
4" x 4" x 120"	L1@1.75	Ea	129.00	71.80	200.80
4" x 6" x 12"	L1@0.55	Ea	31.60	22.60	54.20
4" x 6" x 18"	L1@0.65	Ea	38.60	26.70	65.30
4" x 6" x 24"	L1@0.75	Ea	44.10	30.80	74.90
4" x 6" x 36"	L1@0.95	Ea	62.20	39.00	101.20
4" x 6" x 48"	L1@1.05	Ea	79.40	43.10	122.50
4" x 6" x 60"	L1@1.25	Ea	98.00	51.30	149.30
4" x 6" x 72"	L1@1.55	Ea	105.00	63.60	168.60
6" x 6" x 12"	L1@0.60	Ea	42.10	24.60	66.70
6" x 6" x 18"	L1@0.70	Ea	48.40	28.70	77.10
6" x 6" x 24"	L1@0.80	Ea	47.10	32.80	79.90
6" x 6" x 36"	L1@1.00	Ea	63.00	41.00	104.00
6" x 6" x 48"	L1@1.20	Ea	84.50	49.20	133.70
6" x 6" x 60"	L1@1.30	Ea	92.60	53.30	145.90
6" x 6" x 72"	L1@1.60	Ea	119.00	65.60	184.60
6" x 6" x 120"	L1@1.75	Ea	219.00	71.80	290.80

Use these figures to estimate the cost of terminal cabinets and wireway installed in buildings under the conditions described on pages 5 and 6. Costs listed are for each cabinet and wireway installed. The crew is one electrician working at a labor cost of $41.03 per manhour. These costs include the cover, layout, material handling, and normal waste. Add for wireway ends, couplings, wireway fittings, supports, sales tax, delivery, supervision, mobilization, demobilization, cleanup, overhead and profit. Note: Wireway size depends on the number of wires and the equipment that will be attached.

Material	Craft@Hrs	Unit	Material Cost	Labor Cost	Installed Cost

NEMA 1 screw cover wireway

Material	Craft@Hrs	Unit	Material Cost	Labor Cost	Installed Cost
8" x 8" x 12"	L1@0.65	Ea	80.70	26.70	107.40
8" x 8" x 24"	L1@0.85	Ea	89.20	34.90	124.10
8" x 8" x 36"	L1@1.05	Ea	134.00	43.10	177.10
8" x 8" x 48"	L1@1.25	Ea	159.00	51.30	210.30
8" x 8" x 60"	L1@1.35	Ea	185.00	55.40	240.40
8" x 8" x 72"	L1@1.65	Ea	252.00	67.70	319.70
8" x 8" x 120"	L1@1.80	Ea	384.00	73.90	457.90
10" x 10" x 24"	L1@0.70	Ea	137.00	28.70	165.70
10" x 10" x 36"	L1@1.15	Ea	192.00	47.20	239.20
10" x 10" x 48"	L1@1.35	Ea	228.00	55.40	283.40
10" x 10" x 60"	L1@1.45	Ea	277.00	59.50	336.50
10" x 10" x 72"	L1@1.75	Ea	359.00	71.80	430.80
12" x 12" x 24"	L1@0.90	Ea	158.00	36.90	194.90
12" x 12" x 36"	L1@1.25	Ea	228.00	51.30	279.30
12" x 12" x 48"	L1@1.45	Ea	278.00	59.50	337.50
12" x 12" x 60"	L1@1.55	Ea	322.00	63.60	385.60
12" x 12" x 72"	L1@1.85	Ea	404.00	75.90	479.90

NEMA 1 hinged cover wireway

Material	Craft@Hrs	Unit	Material Cost	Labor Cost	Installed Cost
4" x 4" x 12"	L1@0.50	Ea	20.90	20.50	41.40
4" x 4" x 24"	L1@0.70	Ea	30.10	28.70	58.80
4" x 4" x 36"	L1@0.80	Ea	38.70	32.80	71.50
4" x 4" x 48"	L1@1.00	Ea	57.60	41.00	98.60
4" x 4" x 60"	L1@1.20	Ea	69.10	49.20	118.30
4" x 4" x 120"	L1@1.75	Ea	158.00	71.80	229.80
6" x 6" x 12"	L1@0.60	Ea	42.40	24.60	67.00
6" x 6" x 24"	L1@0.80	Ea	54.80	32.80	87.60
6" x 6" x 36"	L1@1.00	Ea	73.50	41.00	114.50
6" x 6" x 48"	L1@1.20	Ea	101.00	49.20	150.20
6" x 6" x 60"	L1@1.30	Ea	109.00	53.30	162.30
6" x 6" x 72"	L1@1.60	Ea	136.00	65.60	201.60
6" x 6" x 120"	L1@1.75	Ea	260.00	71.80	331.80
8" x 8" x 12"	L1@0.65	Ea	69.40	26.70	96.10
8" x 8" x 24"	L1@0.85	Ea	105.00	34.90	139.90
8" x 8" x 36"	L1@1.05	Ea	161.00	43.10	204.10
8" x 8" x 48"	L1@1.25	Ea	187.00	51.30	238.30
8" x 8" x 60"	L1@1.35	Ea	219.00	55.40	274.40
8" x 8" x 120"	L1@1.80	Ea	458.00	73.90	531.90

Use these figures to estimate the cost of wireway installed in buildings under the conditions described on pages 5 and 6. Costs listed are for each wireway installed. The crew is one electrician working at a labor cost of $41.03 per manhour. These costs include the cover, layout, material handling, and normal waste. Add for supports, wireway ends, couplings, wireway fittings, sales tax, delivery, supervision, mobilization, demobilization, cleanup, overhead and profit. Note: Wireway size depends on the number of wires and the equipment that will be attached.

Wireway Fittings

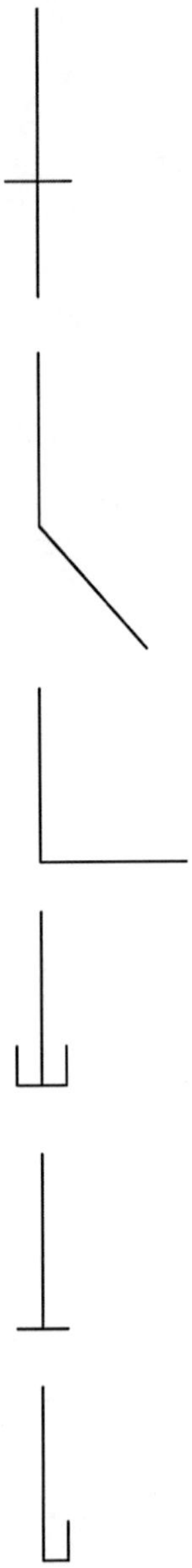

Material	Craft@Hrs	Unit	Material Cost	Labor Cost	Installed Cost
NEMA 1 wireway fittings					
3" x 3" couplings	L1@0.10	Ea	5.76	4.10	9.86
4" x 4" couplings	L1@0.10	Ea	6.90	4.10	11.00
4" x 6" couplings	L1@0.10	Ea	10.10	4.10	14.20
6" x 6" couplings	L1@0.10	Ea	10.30	4.10	14.40
8" x 8" couplings	L1@0.15	Ea	13.80	6.15	19.95
10" x 10" couplings	L1@0.20	Ea	21.70	8.21	29.91
12" x 12" couplings	L1@0.25	Ea	28.90	10.30	39.20
3" x 3" elbows 45	L1@0.20	Ea	46.80	8.21	55.01
4" x 4" elbows 45	L1@0.20	Ea	56.00	8.21	64.21
4" x 6" elbows 45	L1@0.20	Ea	59.70	8.21	67.91
6" x 6" elbows 45	L1@0.20	Ea	65.00	8.21	73.21
8" x 8" elbows 45	L1@0.30	Ea	101.00	12.30	113.30
3" x 3" elbows 90	L1@0.20	Ea	46.80	8.21	55.01
4" x 4" elbows 90	L1@0.20	Ea	44.30	8.21	52.51
4" x 6" elbows 90	L1@0.20	Ea	49.80	8.21	58.01
6" x 6" elbows 90	L1@0.20	Ea	52.90	8.21	61.11
8" x 8" elbows 90	L1@0.30	Ea	84.80	12.30	97.10
10" x 10" elbows 90	L1@0.40	Ea	140.00	16.40	156.40
12" x 12" elbows 90	L1@0.50	Ea	168.00	20.50	188.50
3" x 3" ends	L1@0.10	Ea	5.78	4.10	9.88
4" x 4" ends	L1@0.10	Ea	6.90	4.10	11.00
4" x 6" ends	L1@0.10	Ea	8.73	4.10	12.83
6" x 6" ends	L1@0.10	Ea	10.10	4.10	14.20
8" x 8" ends	L1@0.15	Ea	13.60	6.15	19.75
10" x 10" ends	L1@0.20	Ea	21.30	8.21	29.51
12" x 12" ends	L1@0.25	Ea	30.40	10.30	40.70
3" x 3" flanges	L1@0.20	Ea	17.70	8.21	25.91
4" x 4" flanges	L1@0.20	Ea	21.30	8.21	29.51
4" x 6" flanges	L1@0.20	Ea	27.30	8.21	35.51
6" x 6" flanges	L1@0.30	Ea	29.70	12.30	42.00
8" x 8" flanges	L1@0.40	Ea	42.10	16.40	58.50
10" x 10" flanges	L1@0.50	Ea	63.90	20.50	84.40
12" x 12" flanges	L1@0.70	Ea	85.30	28.70	114.00
3" x 3" hangers	L1@0.25	Ea	12.50	10.30	22.80
4" x 4" hangers	L1@0.25	Ea	17.10	10.30	27.40
6" x 6" hangers	L1@0.30	Ea	22.20	12.30	34.50
8" x 8" hangers	L1@0.40	Ea	27.40	16.40	43.80

Use these figures to estimate the cost of wireway fittings installed in buildings under the conditions described on pages 5 and 6. Costs listed are for each fitting installed. The crew is one electrician working at a labor cost of $41.03 per manhour. These costs include connecting screws, layout, material handling, and normal waste. Add for wireway, supports, sales tax, delivery, supervision, mobilization, demobilization, cleanup, overhead and profit.

NEMA 1 wireway fittings

Material	Craft@Hrs	Unit	Material Cost	Labor Cost	Installed Cost
4" - 3" reducers	L1@0.15	Ea	34.70	6.15	40.85
6" - 4" reducers	L1@0.20	Ea	65.70	8.21	73.91
8" - 6" reducers	L1@0.25	Ea	71.40	10.30	81.70
3" x 3" tees	L1@0.25	Ea	92.00	10.30	102.30
4" x 4" tees	L1@0.25	Ea	63.70	10.30	74.00
4" x 6" tees	L1@0.30	Ea	101.00	12.30	113.30
6" x 6" tees	L1@0.30	Ea	123.00	12.30	135.30
8" x 8" tees	L1@0.40	Ea	217.00	16.40	233.40
10" x 10" tees	L1@0.50	Ea	324.00	20.50	344.50
12" x 12" tees	L1@0.70	Ea	405.00	28.70	433.70
3" x 3" X's	L1@0.30	Ea	54.40	12.30	66.70
4" x 4" X's	L1@0.30	Ea	63.70	12.30	76.00
6" x 6" X's	L1@0.40	Ea	76.50	16.40	92.90
8" x 8" X's	L1@0.50	Ea	120.00	20.50	140.50

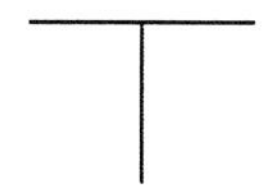
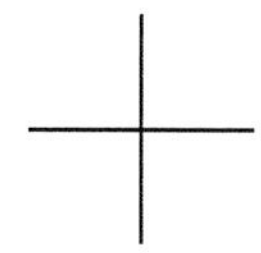

NEMA 3R wireway

Material	Craft@Hrs	Unit	Material Cost	Labor Cost	Installed Cost
4" x 4" x 12"	L1@0.30	Ea	21.50	12.30	33.80
4" x 4" x 24"	L1@0.40	Ea	26.60	16.40	43.00
4" x 4" x 36"	L1@0.50	Ea	42.40	20.50	62.90
4" x 4" x 48"	L1@0.60	Ea	54.40	24.60	79.00
4" x 4" x 60"	L1@0.70	Ea	63.70	28.70	92.40
4" x 4" x 120"	L1@1.00	Ea	126.00	41.00	167.00
6" x 6" x 12"	L1@0.40	Ea	37.80	16.40	54.20
6" x 6" x 24"	L1@0.50	Ea	49.40	20.50	69.90
6" x 6" x 36"	L1@0.60	Ea	60.90	24.60	85.50
6" x 6" x 48"	L1@0.70	Ea	83.30	28.70	112.00
6" x 6" x 60"	L1@0.80	Ea	97.40	32.80	130.20
6" x 6" x 120"	L1@1.10	Ea	213.00	45.10	258.10
8" x 8" x 12"	L1@0.50	Ea	60.90	20.50	81.40
8" x 8" x 24"	L1@0.60	Ea	87.30	24.60	111.90
8" x 8" x 36"	L1@0.70	Ea	125.00	28.70	153.70
8" x 8" x 48"	L1@0.80	Ea	158.00	32.80	190.80
8" x 8" x 60"	L1@0.90	Ea	181.00	36.90	217.90
8" x 8" x 120"	L1@1.20	Ea	374.00	49.20	423.20

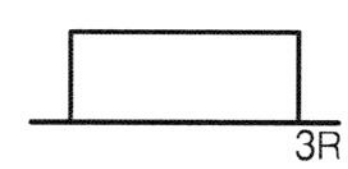

Use these figures to estimate the cost of wireway fittings installed in buildings under the conditions described on pages 5 and 6. See the footnote on the previous page for notes on wireway fittings. The crew is one electrician. The cost per manhour is $41.03. These costs include layout, material handling, and normal waste. Add for sales tax, delivery, supervision, mobilization, demobilization, cleanup, overhead and profit.

Dry Type Transformers, Indoor/Outdoor

240/480 volt primary, 120/240 volt secondary

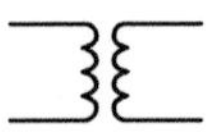

Material		Craft@Hrs	Unit	Material Cost	Labor Cost	Installed Cost
.050	KVA	L1@0.25	Ea	168.00	10.30	178.30
.075	KVA	L1@0.25	Ea	187.00	10.30	197.30
.100	KVA	L1@0.25	Ea	195.00	10.30	205.30
.150	KVA	L1@0.30	Ea	231.00	12.30	243.30
.250	KVA	L1@0.30	Ea	243.00	12.30	255.30
.500	KVA	L1@0.35	Ea	347.00	14.40	361.40
.750	KVA	L1@0.40	Ea	414.00	16.40	430.40
1	KVA	L1@0.50	Ea	547.00	20.50	567.50
1.5	KVA	L1@0.55	Ea	655.00	22.60	677.60
2	KVA	L1@0.60	Ea	811.00	24.60	835.60
3	KVA	L1@0.70	Ea	1,040.00	28.70	1,068.70

120/240 volt primary, 120/240 volt secondary

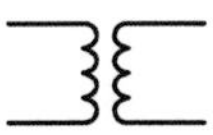

Material		Craft@Hrs	Unit	Material Cost	Labor Cost	Installed Cost
.050	KVA	L1@0.25	Ea	222.00	10.30	232.30
.100	KVA	L1@0.25	Ea	264.00	10.30	274.30
.150	KVA	L1@0.25	Ea	322.00	10.30	332.30
.250	KVA	L1@0.30	Ea	380.00	12.30	392.30
.500	KVA	L1@0.30	Ea	530.00	12.30	542.30
.750	KVA	L1@0.35	Ea	668.00	14.40	682.40
1	KVA	L1@0.50	Ea	849.00	20.50	869.50
1.5	KVA	L1@0.55	Ea	1,040.00	22.60	1,062.60
2	KVA	L1@0.60	Ea	1,260.00	24.60	1,284.60
3	KVA	L1@0.70	Ea	1,530.00	28.70	1,558.70
5	KVA	L1@1.00	Ea	2,360.00	41.00	2,401.00
7.5	KVA	L1@1.25	Ea	3,280.00	51.30	3,331.30
10	KVA	L2@1.50	Ea	4,200.00	61.50	4,261.50
15	KVA	L2@1.75	Ea	5,510.00	71.80	5,581.80
25	KVA	L2@2.00	Ea	7,400.00	82.10	7,482.10

120/240 volt primary, 12/24 volt secondary

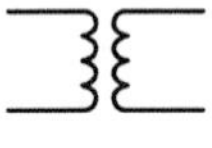

Material		Craft@Hrs	Unit	Material Cost	Labor Cost	Installed Cost
.050	KVA	L1@0.25	Ea	208.00	10.30	218.30
.075	KVA	L1@0.25	Ea	231.00	10.30	241.30
.100	KVA	L1@0.25	Ea	247.00	10.30	257.30
.150	KVA	L1@0.30	Ea	303.00	12.30	315.30
.250	KVA	L1@0.30	Ea	364.00	12.30	376.30
.500	KVA	L1@0.35	Ea	496.00	14.40	510.40
.750	KVA	L1@0.40	Ea	638.00	16.40	654.40
1	KVA	L1@0.50	Ea	796.00	20.50	816.50
1.5	KVA	L1@0.55	Ea	973.00	22.60	995.60
2	KVA	L1@0.60	Ea	1,180.00	24.60	1,204.60
3	KVA	L1@0.70	Ea	1,580.00	28.70	1,608.70

Use these figures to estimate the cost of transformers installed in buildings under the conditions described on pages 5 and 6. Costs listed are for each transformer installed. The crew is one electrician for transformers to 7.5 KVA and two electricians for transformers over 7.5 KVA. The cost per manhour is $41.03. These costs include layout, material handling, and normal waste. Add for protective devices, supports, lifting equipment (if needed), sales tax, delivery, supervision, mobilization, demobilization, cleanup, overhead and profit. Note: Transformers with a 120/240 volt secondary are single phase. Transformers with a 208/120 volt secondary are three phase. Taps on transformers usually permit 2-1/2% adjustments in the output voltage. Generally two taps are available for raising the voltage and two are available for lowering the voltage.

120/240 volt primary, 16/32 volt secondary

Material		Craft@Hrs	Unit	Material Cost	Labor Cost	Installed Cost
.050	KVA	L1@0.25	Ea	208.00	10.30	218.30
.075	KVA	L1@0.25	Ea	231.00	10.30	241.30
.100	KVA	L1@0.25	Ea	247.00	10.30	257.30
.150	KVA	L1@0.30	Ea	456.00	12.30	468.30
.250	KVA	L1@0.30	Ea	364.00	12.30	376.30
.500	KVA	L1@0.35	Ea	496.00	14.40	510.40
.750	KVA	L1@0.40	Ea	638.00	16.40	654.40
1	KVA	L1@0.50	Ea	796.00	20.50	816.50
1.5	KVA	L1@0.55	Ea	973.00	22.60	995.60
2	KVA	L1@0.60	Ea	1,180.00	24.60	1,204.60
3	KVA	L1@0.70	Ea	1,580.00	28.70	1,608.70

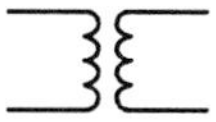

240/480 volt primary, 120/240 volt secondary

Material		Craft@Hrs	Unit	Material Cost	Labor Cost	Installed Cost
5	KVA	L1@1.00	Ea	1,350.00	41.00	1,391.00
7.5	KVA	L1@1.25	Ea	1,670.00	51.30	1,721.30
10	KVA	L2@1.50	Ea	2,080.00	61.50	2,141.50
15	KVA	L2@1.75	Ea	2,410.00	71.80	2,481.80
25	KVA	L2@2.00	Ea	3,220.00	82.10	3,302.10

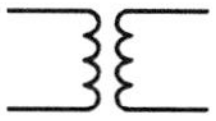

240/480 volt primary, 120/240 volt secondary, with taps

Material		Craft@Hrs	Unit	Material Cost	Labor Cost	Installed Cost
37.5	KVA	L2@2.50	Ea	4,060.00	103.00	4,163.00
50	KVA	L2@3.00	Ea	4,810.00	123.00	4,933.00
75	KVA	L2@3.50	Ea	6,850.00	144.00	6,994.00
100	KVA	L2@4.00	Ea	8,720.00	164.00	8,884.00
167	KVA	L2@4.50	Ea	17,000.00	185.00	17,185.00

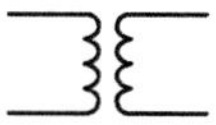

480 volt primary, 120/240 volt secondary, with taps

Material		Craft@Hrs	Unit	Material Cost	Labor Cost	Installed Cost
5	KVA	L1@1.00	Ea	2,210.00	41.00	2,251.00
7.5	KVA	L1@1.25	Ea	3,060.00	51.30	3,111.30
10	KVA	L2@1.50	Ea	3,880.00	61.50	3,941.50
15	KVA	L2@1.75	Ea	5,310.00	71.80	5,381.80
25	KVA	L2@2.00	Ea	7,800.00	82.10	7,882.10

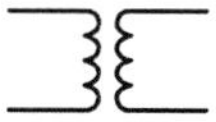

Use these figures to estimate the cost of transformers installed in buildings under the conditions described on pages 5 and 6. See the footnote on page 330 for notes on wireway fittings. Costs listed are for each transformer installed. The crew is one electrician for transformers to 7.5 KVA and two electricians for transformers over 7.5 KVA. The cost per manhour is $41.03. These costs include layout, material handling, and normal waste. Add for protective devices, supports, lifting equipment (if needed), sales tax, delivery, supervision, mobilization, demobilization, cleanup, overhead and profit. Note: Transformers with a 120/240 volt secondary are single phase. Transformers with a 208/120 volt secondary are three phase. Taps on transformers usually permit 2-1/2% adjustments in the output voltage. Generally two taps are available for raising the voltage and two are available for lowering the voltage.

Dry Type Transformers, Indoor/Outdoor

480 volt primary, 208/120 volt secondary

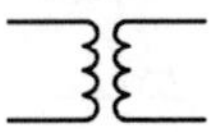

Material	Craft@Hrs	Unit	Material Cost	Labor Cost	Installed Cost
3 KVA	L2@0.75	Ea	1,990.00	30.80	2,020.80
6 KVA	L2@1.00	Ea	2,310.00	41.00	2,351.00
9 KVA	L2@1.40	Ea	3,060.00	57.40	3,117.40
15 KVA	L2@1.60	Ea	4,140.00	65.60	4,205.60
30 KVA	L2@2.25	Ea	6,200.00	92.30	6,292.30
45 KVA	L2@2.75	Ea	7,400.00	113.00	7,513.00
75 KVA	L2@3.50	Ea	11,200.00	144.00	11,344.00
112.5 KVA	L2@4.25	Ea	14,700.00	174.00	14,874.00
150 KVA	L2@4.25	Ea	19,500.00	174.00	19,674.00
225 KVA	L2@5.00	Ea	26,100.00	205.00	26,305.00
300 KVA	L2@5.50	Ea	33,400.00	226.00	33,626.00

600 volt primary, 120/240 volt secondary

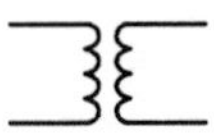

Material	Craft@Hrs	Unit	Material Cost	Labor Cost	Installed Cost
5 KVA	L1@1.00	Ea	2,190.00	41.00	2,231.00
7.5 KVA	L1@1.25	Ea	2,900.00	51.30	2,951.30
10 KVA	L2@1.50	Ea	3,730.00	61.50	3,791.50
15 KVA	L2@1.75	Ea	5,310.00	71.80	5,381.80
25 KVA	L2@2.00	Ea	7,510.00	82.10	7,592.10

600 volt primary, 120/240 volt secondary, with taps

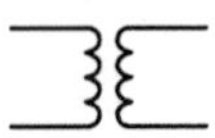

Material	Craft@Hrs	Unit	Material Cost	Labor Cost	Installed Cost
5 KVA	L1@1.00	Ea	2,290.00	41.00	2,331.00
7.5 KVA	L1@1.25	Ea	3,120.00	51.30	3,171.30
10 KVA	L2@1.50	Ea	3,640.00	61.50	3,701.50
15 KVA	L2@1.75	Ea	5,470.00	71.80	5,541.80
25 KVA	L2@2.00	Ea	8,200.00	82.10	8,282.10

Use these figures to estimate the cost of transformers installed in buildings under the conditions described on pages 5 and 6. Costs listed are for each transformer installed. The crew is one electrician for transformers to 7.5 KVA and two electricians for transformers over 7.5 KVA. The cost per manhour is $41.03. These costs include layout, material handling, and normal waste. Add for protective devices, supports, lifting equipment (if needed), sales tax, delivery, supervision, mobilization, demobilization, cleanup, overhead and profit. Note: Transformers with a 120/240 volt secondary are single phase. Transformers with a 208/120 volt secondary are three phase. Taps on transformers usually permit 2-1/2% adjustments in the output voltage. Generally two taps are available for raising the voltage and two are available for lowering the voltage.

Section 7: Underfloor Raceway

Most residential buildings have electrical requirements that can be expected to change very little during the normal life of the building. In most cases, you can wire it and forget it. Larger office buildings are different. They need electrical systems that offer more flexibility. They need a wiring system that adapts easily each time a new tenant moves in, and that is easy to expand as the needs of each occupant change. That's why owners of many office and industrial buildings install underfloor raceway systems.

Underfloor raceway permits major changes in the size and number of conductors or signal cables during the life of the building, without compromise in fire resistance — install a major piece of new equipment; add a new row of desks down the middle of a room; put computer terminals on each desk and connect them with the computer located in another room. Changes like this can be expected in an office building. Each may require major electrical work — unless the building was equipped with an underfloor raceway system.

Underfloor raceway is used principally in fire-resistant construction. The raceway provides an open channel in or under the floor slab for power and communications lines. Wiring devices can be installed under every work space in an office or plant. Figure 7-1 shows a typical underfloor duct system.

Duct is usually set in the floor before the concrete deck is poured. Concrete surrounds the duct and finishes flush with each duct insert. Outlet fixtures are installed in the inserts wherever electrical and communications access will be needed. This provides power and signal lines to each desk and work space, even to areas located away from wall outlets.

Courtesy: Walkerduct

Figure 7-1
Underfloor Duct System

The network of duct lines are joined by junction boxes that make the system more rigid and keep raceway laterals in place and at right angles. Junction boxes don't provide outlets for the system. They're used only when pulling wire through the duct. Junction boxes should be placed close enough together to make it easy to fish wire or cable through the cells. It's good practice to provide a regular pattern of junction boxes that can be rewired easily later when additional circuits or cables are needed.

Junction boxes are used at right angle turns, at tee connections, at cross connections and at feed-in positions. The junction box finish ring should be set at screed level so the finished concrete floor is level with the junction box lid.

Minor adjustment is pos.sible in junction box cover rings. The ring can be raised or lowered slightly after floor tile is laid. When the floor is carpeted, a special cover is used to bring the junction box cover up to the level of the carpet. This cover is called a ***carpet pan***. When the finished floor is terrazzo, a similar ring can be used.

Underfloor Duct Materials

The underfloor duct is made in two common sizes. The Type 1 duct is usually used for power circuits. Cell size is 1$^3/_8$" x 3$^1/_8$" x 10'. The Type 2 duct is commonly used for communications cables such as telephone, alarm, and computer systems. Type 2 duct measures 1$^3/_8$" x 7$^1/_4$" x 10'. Both types of duct are made in blank sections for feeder lines and with insert access holes placed 24 inches on center along one face of the duct. The first insert can be set 6 inches from one end and the last is 18 inches from the other end. Other spacing is available from some manufacturers on special order.

Cell insert height can vary from $^7/_8$" to 3$^3/_8$". But note that inserts aren't usually interchangeable. Each must be ordered to fit the selected type of duct. When the junction box is set level with the screed line, the duct inserts must also be at the screed line.

Duct Supports

You'll hear duct supports called ***chairs***. Supports hold the duct in place and at the correct height while the concrete floor is being placed. Once the concrete has hardened, supports serve no useful purpose. A special type of support serves both as a support and as a coupling at duct joints. It replaces the common duct coupling.

Install enough supports to hold the duct in place as the deck is being poured. For planning purposes, allow one support every 5 feet and at the end of each duct run.

Some duct systems use more than one level of duct in the same floor. For example, the system may combine Type 1 duct for electrical lines and Type 2 duct for communication lines. When that's the case, use duct supports made for combination duct runs and combination junction boxes. Use standard supports and duct couplings at each duct joint.

All duct supports have leveling screws so you can level access rings at the screed line between junction boxes. Leveling legs are available in any length to meet any slab thicknesses.

Duct Elbows

Duct elbows are used to make vertical or horizontal turns. Vertical elbows turn the duct above the level of the finish floor. That's necessary when duct has to be connected directly to a terminal cabinet in a wall, for example. Use blank feeder duct in an exposed vertical section above an elbow. Vertical elbows form right angles, but have a fairly long radius to make pulling wire or cable easier.

Horizontal elbows are used to change direction without leaving the concrete deck. They come in 90-degree bends, 45-degree bends and adjustable 15- to 30-degree bends. All of the bends have a radius long enough to make pulling wire relatively easy.

A special offset elbow is made to simplify running duct under obstacles or other underfloor duct. The offset elbow is available for both Type 1 and Type 2 duct and is a one-piece unit. Notice that it takes two offset elbows to go under an obstacle and return to the original level.

Fittings

The fittings made for underfloor duct will meet any need. For the estimator, the only hard part is counting all the fittings you'll need. Here are the most common fittings:

Wyes are used when one cell divides into two runs. Both 30- and 45-degree wyes are made for both sizes of duct.

Plugs are used to fill any unused openings in junction boxes so liquid concrete doesn't flow into the box. The plugs come in both duct sizes. Corners in junction boxes have entrances for 2 inch conduit. You'll also need plugs to fill any openings that aren't used for conduit.

There are special duct plugs for capping the ends of duct runs that terminate without a junction box. The caps go into the end of the duct to keep concrete out of the cell.

Adapters are sometimes needed when conduit joins an underfloor duct system. When the electrical feed to a duct system comes from conduit in the slab, the conduit usually enters at a corner of a junction box. Entrance hubs at the junction box corner are usually for 2 inch conduit. If the conduit feed isn't 2 inches, an adapter will be needed to make a concrete-tight connection. Adapter sizes range from ½" to 1¼".

Adapters are also available for joining the end of both Type 1 and Type 2 duct to conduit. Insert the right size adapter into the end of the duct and connect the conduit.

Insert caps are installed in the duct inserts at the factory. Leave the caps in place until it's time to install the service fittings. If you need extra insert caps, they're available separately. There are two common types of inserts. The most common has a 2 inch diameter. Another has an ellipsoid shape. Ellipsoid insert caps are also available.

Cabinet connectors are used to attach the duct to a service panel or cabinet. The connector provides ground continuity within the metallic raceway system. It's attached with machine screws and has a protective bushing to keep wire insulation from getting stripped off as it's pulled into the duct.

Electrical outlet receptacles are installed in the service fittings. Note that the fittings listed in this section include the cost of labor and material for the basic housing only — not the receptacle itself. Add the cost of buying and installing receptacles separately. Job specs will identify the type and grade of receptacles to use.

Abandon plugs are used when a service fitting has to be relocated. The opening that remains is plugged with a special abandon plug. They're made either of brass or aluminum. Install the plug immediately after the service fitting has been removed to keep debris out of the duct. The plug has a toggle-type bar that extends into the duct through the insert. When the screw is tightened, the plug seals the hole.

The abandon plug can be removed later so the service fitting can be reinstalled.

When a service fitting is abandoned, the wire or cable can either be removed or taped off. If it's left in place, tape the end carefully and tuck it back into the insert.

Finding Caps and Installing Fittings

Sometimes underfloor raceway gets buried a little too deep in the concrete deck. That makes finding the inserts more difficult. To locate "lost" inserts, measure from the last junction box or the last service fitting previously installed. Remember, standard insert spacing is 24 inches on center and the first insert is 6 inches from the box. Be careful to avoid damaging the floor when searching for the insert. Don't start chipping concrete until you know where the insert is.

The further away from the junction box, the harder it is to find inserts. To simplify locating inserts, put a marking screw in the last insert in every duct run. The screw should extend ½ inch above the surface of the finished floor. Snap a chalk line between the marking screw and the junction box. Inserts will be on that line. Find the first insert from the junction box and measure in 2 foot increments along the line to the next service fitting. Mark each insert location and then remove the marking screw.

When each insert is located in the floor, use a chipping gun or chisel to remove the small amount of concrete cover. Then vacuum up the concrete dust and chips before opening the insert cap. That keeps debris out of the duct. Then fish wire or cable to the new service fitting location and install the service fitting.

Service fittings go on over the duct insert and are mechanically connected with a chase nipple-type fitting. Tighten the nipple to hold the service fitting in place and ground the fitting to the duct.

Labor for Underfloor Duct

The labor units listed in this section assume no unusual conditions, and good access. Be alert for other situations. Start by visualizing what the site will look like on installation day. A congested site will add to installation time and increase the labor cost. A tight installation schedule will increase conflicts with other trades. A deeper slab will usually extend the installation time. If the floor carries several other duct runs, expect some problems. Several extra offset fittings may be needed to avoid other lines.

If it looks like the system will be hard to lay out, hard to get to, will share the slab with grade beams or structural beams, if there's too much conduit or piping in the area, and if the job involves less than 100 linear feet of duct, expect labor time to be 25 to 50 percent higher than listed in this section.

Installations over 100 feet may take less time, especially if the duct runs are long. If you're installing over 200 linear feet of duct, labor time can be 10 to 20 percent less. Even an inexperienced crew will have picked up enough about duct installation to increase their productivity after running the first 100 feet.

Doing the Take-off

Your take-off of underfloor duct begins with a lined worksheet just like the other estimating worksheets you've used. Draw a vertical line down the left side about 1½ inches from the left edge. At the top of the page, on the right side of the vertical line, head the column with the plan page number. In the left column, list the material descriptions.

First, list the junction boxes for single duct runs. Start the junction boxes for Type 1 duct. On the next line list junction boxes for Type 2 duct, if there are any. Then do the same for combination duct runs.

Next measure the lengths of Type 1 and Type 2 duct. List all elbows and offsets as you find them on the plan. After measuring all of the duct, calculate the number of duct supports of each type.

Next, locate all conduit that feeds the duct system. Determine the conduit size and list the adapter size needed for each connection. Watch for duct couplings or combination coupling and support pieces. There won't be too many couplings. Underfloor duct comes in 10-foot lengths.

Watch for blank duct. Blank duct doesn't have inserts. It will usually be installed in the floor of rooms or corridors that don't require service fittings. The plans should show where blank duct is to be used.

Blank duct is also used in risers from the underfloor system to panels or cabinets. It can be exposed or concealed in the wall. When a riser is to be connected to a panel or cabinet, list an adapter for that duct size.

Next, identify and count all the service fittings that will be required. Check the job specifications and see if additional service fittings are to be delivered to the owner for future installation. These extra service fittings won't show up on the plans.

Using Duct Systems

Most underfloor duct jobs require many cut pieces of duct. The runs between junction boxes are seldom spaced the full length of a duct section. Usually you'll lay two or three full-duct sections and have to cut the next section to complete the run to the next junction box.

Cutting duct by hand with a hacksaw is slow work. A power hacksaw will both save time and produce straighter cuts. Be sure to ream the inside edge with a file to remove the cutting burr.

Learn From Your Mistakes

My most memorable underfloor duct job was a large office building for a state Department of Motor Vehicles. The plans called for both Type 1 and Type 2 duct in one large room.

The construction schedule was tight. Close coordination with the general contractor was essential. Notice to proceed with construction was issued before the subcontracts were completed. The contractor wanted to start the job right away, but full descriptions of all key materials, including the underfloor duct, had to be submitted for approval before we could place orders.

We worked quickly with suppliers to prepare shop drawings and catalog cuts for the submittals. It took about two weeks to gather all the information that was required. We prepared a material list and packaged it with the catalog cuts for the submittal. The shop drawings for switchboards, motor control centers and panels would be late, so we planned to submit them separately at a later date.

The final submittal package included ten submittal brochures. We rushed these to the general contractor and he passed them immediately to the owner. Along with the brochures went a request for prompt review so we could stay on the construction schedule.

The approved brochures were returned in three weeks — not exactly record time, but pretty good for a state agency. By then the contractor had graded the site and excavated for the footings. The plans called for wire mesh in the floor slabs and the mesh was already on the job site. The contractor was ready to form the floor slab. But the underfloor duct had to go in that slab! Construction would have to wait until we could set the duct.

We notified the supplier that the formwork was about to start. Our supplier put pressure on the manufacturer to rush the underfloor duct and fittings. This was probably a mistake, but we didn't realize it at the time. The duct arrived about a week later. The forms were in place for the slab and concrete delivery had been scheduled. Our underfloor duct was right on the critical path, but everything was looking good at this point. We were already congratulating ourselves on a job well done. As it turned out, that was premature.

Our tradesmen began to set duct at one end of the building. The slab crew went to work beside them setting screed pins for leveling the concrete to a close tolerance. The junction boxes were spotted and set in small batches of bagged concrete mix. The concrete would anchor the junction box firmly in place until the slab was poured. The top of each box was set at screed level.

As the slab excavation crew would screed the underfloor material, fill dirt was moved back across the floor area. Screeding was done in patterns because the floor was going to be poured in a checkerboard pattern, A 2 x 4 header marked the boundary of each pour.

With the junction boxes in place, we began to unload the duct sections. It was at this time that one of our electricians noticed a problem. A seam on one length of duct wasn't welded closed. We quickly started checking other duct. Out of the first few dozen sections checked, about one-third had the same flaw! Now we were in trouble.

We immediately notified the contractor of the problem. Next, we phoned our supplier. Based on our guess that 30 percent was going to be rejected, the manufacturer immediately shipped 30 percent more duct. The project waited until the duct arrived, of course. But we were able to finish on time by compressing other parts of the job.

It's always embarrassing to be the cause of a delay on a project — especially when everyone on the job is working against a tight deadline. But in truth, there wasn't much we could have done to avoid the problem. And it could have been worse. Think how embarrassed we would have been if that defective duct had been in our warehouse for two weeks before we started unloading it on the job.

We learned one other lesson on this job. Most of the duct was staked in place right on the ground while the slab crew poured concrete around it. This can lead to trouble. When the laborers placed, screeded and tamped the concrete, they walked on and stumbled over the underfloor duct. That was unavoidable. The stakes tended to hold the duct in place, of course. But they also tended to hold it wherever it was moved during the pour. We found that some of the inserts ended up deeper than others.

The duct that wasn't staked survived the pour better. The laborers walked on the unstaked duct too, of course. And the unstaked duct would move when it was disturbed. But it would also return closer to its intended position after being disturbed because no stakes were holding it down.

On the whole, this job was both successful and profitable, both for the general contractor and my company. And the state got the facility it paid for. But there were more than a few anxious moments for the electrical crew that day when we were setting seamless underfloor duct.

Material	Craft@Hrs	Unit	Material Cost	Labor Cost	Installed Cost
Underfloor raceway 14 gauge power duct, 3-1/4" wide, 10' long, Type 1 duct					
Blank duct	L2@0.30	Ea	139.00	12.30	151.30
7/8" insert height	L2@0.30	Ea	182.00	12.30	194.30
1-3/8" insert height	L2@0.30	Ea	182.00	12.30	194.30
1-7/8" insert height	L2@0.30	Ea	182.00	12.30	194.30
2-3/8" insert height	L2@0.30	Ea	182.00	12.30	194.30
3-3/8" insert height	L2@0.30	Ea	182.00	12.30	194.30
Underfloor raceway 14 gauge communications duct, 7-1/4" wide, 10' long, Type 2 duct					
Blank duct	L2@0.40	Ea	276.00	16.40	292.40
7/8" insert height	L2@0.40	Ea	320.00	16.40	336.40
1-3/8" insert height	L2@0.40	Ea	320.00	16.40	336.40
1-7/8" insert height	L2@0.40	Ea	320.00	16.40	336.40
2-3/8" insert height	L2@0.40	Ea	320.00	16.40	336.40
3-3/8" insert height	L2@0.40	Ea	320.00	16.40	336.40
Underfloor raceway one level junction boxes, 7/8" insert height					
4 #1 ducts	L2@0.50	Ea	623.00	20.50	643.50
8 #1 ducts	L2@0.60	Ea	623.00	24.60	647.60
12 #1 ducts	L2@0.75	Ea	1,240.00	30.80	1,270.80
4 #2 ducts	L2@0.60	Ea	647.00	24.60	671.60
8 #2 ducts	L2@1.00	Ea	1,620.00	41.00	1,661.00
4 #1 & 4 #2 ducts	L2@0.80	Ea	1,240.00	32.80	1,272.80
Underfloor raceway one level junction boxes, 1-3/8" insert height					
4 #1 ducts	L2@0.50	Ea	623.00	20.50	643.50
8 #1 ducts	L2@0.60	Ea	623.00	24.60	647.60
12 #1 ducts	L2@0.75	Ea	1,240.00	30.80	1,270.80
4 #2 ducts	L2@0.60	Ea	647.00	24.60	671.60
8 #2 ducts	L2@1.00	Ea	1,610.00	41.00	1,651.00
4 #1 & 4 #2 ducts	L2@0.80	Ea	1,230.00	32.80	1,262.80

J

J

Use these figures to estimate the cost of underfloor raceway installed in building floors under the conditions described on pages 5 and 6. Costs listed are for each length installed. The crew is two electricians working at a labor cost of $41.03 per manhour. These costs include leveling, layout, material handling, and normal waste. Add for supports, sales tax, delivery, supervision, mobilization, demobilization, cleanup, overhead and profit. Labor costs will be higher when installing duct in slabs on grade. Extra care is needed to anchor the duct properly and prevent it from floating as the concrete is poured. Some installers place all the concrete that's poured around duct to be sure the duct isn't disturbed. When duct is installed on wood deck forms, the junction boxes and duct supports can be fastened to the form to prevent floating. No matter where the duct is placed, be sure to seal all openings in the duct before concrete is poured. In the table above, No. 1 duct is power duct and No. 2 duct is communications duct. Junction boxes have one, two or three duct entrances in each side. When a dead end, straight through, or tee junction is needed, fill the unused duct outlets with blank plugs.

Underfloor Raceway

Underfloor raceway one level junction boxes, 1-7/8" insert height

Material	Craft@Hrs	Unit	Material Cost	Labor Cost	Installed Cost
4 #1 ducts	L2@0.50	Ea	423.00	20.50	443.50
8 #1 ducts	L2@0.60	Ea	544.00	24.60	568.60
12 #1 ducts	L2@0.75	Ea	889.00	30.80	919.80
4 #2 ducts	L2@0.60	Ea	544.00	24.60	568.60
8 #2 ducts	L2@1.00	Ea	1,310.00	41.00	1,351.00
4 #1 & 4 #2 ducts	L2@0.80	Ea	889.00	32.80	921.80

Underfloor raceway one level junction boxes, 2-3/8" insert height

Material	Craft@Hrs	Unit	Material Cost	Labor Cost	Installed Cost
4 #1 ducts	L2@0.50	Ea	424.00	20.50	444.50
8 #1 ducts	L2@0.60	Ea	544.00	24.60	568.60
12 #1 ducts	L2@0.75	Ea	889.00	30.80	919.80
4 #2 ducts	L2@0.60	Ea	485.00	24.60	509.60
8 #2 ducts	L2@1.00	Ea	1,310.00	41.00	1,351.00
4 #1 & 4 #2 ducts	L2@0.80	Ea	889.00	32.80	921.80

Underfloor raceway one level junction boxes, 3-3/8" insert height

Material	Craft@Hrs	Unit	Material Cost	Labor Cost	Installed Cost
4 #1 ducts	L2@0.50	Ea	423.00	20.50	443.50
8 #1 ducts	L2@0.60	Ea	544.00	24.60	568.60
12 #1 ducts	L2@0.75	Ea	889.00	30.80	919.80
4 #2 ducts	L2@0.60	Ea	485.00	24.60	509.60
8 #2 ducts	L2@1.00	Ea	1,310.00	41.00	1,351.00
4 #1 & 4 #2 ducts	L2@0.80	Ea	1,310.00	32.80	1,342.80

Underfloor raceway duct supports with leveling screws

Material	Craft@Hrs	Unit	Material Cost	Labor Cost	Installed Cost
1 #1 duct	L2@0.15	Ea	45.70	6.15	51.85
2 #1 ducts	L2@0.20	Ea	45.70	8.21	53.91
3 #1 ducts	L2@0.25	Ea	45.70	10.30	56.00
1 #2 ducts	L2@0.20	Ea	45.70	8.21	53.91
2 #2 ducts	L2@0.25	Ea	45.70	10.30	56.00
1 #1 & 1 #2 ducts	L2@0.25	Ea	45.70	10.30	56.00
2 #1 & 1 #2 ducts	L2@0.25	Ea	45.70	10.30	56.00
1 #1 & 2 #2 ducts	L2@0.30	Ea	45.70	12.30	58.00

Use these figures to estimate the cost of underfloor raceway installed in building floors under the conditions described on pages 5 and 6. Costs listed are for each length installed. The crew is two electricians working at a labor cost of $41.03 per manhour. These costs include leveling, layout, material handling, and normal waste. Add for supports, sales tax, delivery, supervision, mobilization, demobilization, cleanup, overhead and profit. Labor costs will be higher when installing duct in slabs on grade. Extra care is needed to anchor the duct properly and prevent it from floating as the concrete is poured. Some installers place all the concrete that's poured around duct to be sure the duct isn't disturbed. When duct is installed on wood deck forms, the junction boxes and duct supports can be fastened to the form to prevent floating. No matter where the duct is placed, be sure to seal all openings in the duct before concrete is poured. In the table above, No. 1 duct is power duct and No. 2 duct is communications duct. Junction boxes have one, two or three duct entrances in each side. When a dead end, straight through, or tee junction is needed, fill the unused duct outlets with blank plugs.

Material	Craft@Hrs	Unit	Material Cost	Labor Cost	Installed Cost
Fittings for underfloor raceway					
#1 couplings	L1@0.05	Ea	63.30	2.05	65.35
#2 couplings	L1@0.06	Ea	80.60	2.46	83.06
#1 vertical elbows	L1@0.20	Ea	162.00	8.21	170.21
#2 vertical elbows	L1@0.25	Ea	162.00	10.30	172.30
#1 horizontal elbows	L1@0.20	Ea	207.00	8.21	215.21
#2 horizontal elbows	L1@0.25	Ea	207.00	10.30	217.30
#1 horz. elbows, 30	L1@0.20	Ea	207.00	8.21	215.21
#2 horz. elbows, 30	L1@0.25	Ea	207.00	10.30	217.30
#1 horz. elbows, 45	L1@0.20	Ea	207.00	8.21	215.21
#2 horz. elbows, 45	L1@0.25	Ea	207.00	10.30	217.30
#1 offset elbows	L1@0.20	Ea	207.00	8.21	215.21
#2 offset elbows	L1@0.25	Ea	207.00	10.30	217.30
#1 box plugs	L1@0.05	Ea	10.70	2.05	12.75
#2 box plugs	L1@0.06	Ea	10.70	2.46	13.16
#1 duct plugs	L1@0.05	Ea	10.70	2.05	12.75
#2 duct plugs	L1@0.06	Ea	10.70	2.46	13.16
1/2" pipe adapters	L1@0.05	Ea	80.60	2.05	82.65
3/4" pipe adapters	L1@0.06	Ea	80.60	2.46	83.06
1" pipe adapters	L1@0.08	Ea	80.60	3.28	83.88
1-1/4" pipe adapters	L1@0.10	Ea	80.60	4.10	84.70
2" pipe adapters	L1@0.15	Ea	80.60	6.15	86.75
2"-1-1/4" pipe adapters	L1@0.20	Ea	80.60	8.21	88.81
3"-1-1/4" pipe adapters	L1@0.25	Ea	80.60	10.30	90.90
1/2" pipe inserts	L1@0.05	Ea	53.80	2.05	55.85
3/4" pipe inserts	L1@0.06	Ea	53.80	2.46	56.26
1" pipe inserts	L1@0.08	Ea	53.80	3.28	57.08
1-1/4" pipe inserts	L1@0.10	Ea	53.80	4.10	57.90
1-1/2" pipe inserts	L1@0.10	Ea	53.80	4.10	57.90
1-1/4"-3/4" re bush	L1@0.10	Ea	53.80	4.10	57.90
1-1/4"-1" re bush	L1@0.10	Ea	53.80	4.10	57.90
2"-1-1/2" re bush	L1@0.15	Ea	80.60	6.15	86.75
Insert cap	L1@0.05	Ea	10.70	2.05	12.75
Marker screw	L1@0.25	Ea	21.50	10.30	31.80
#1 cabinet connectors	L1@0.30	Ea	60.40	12.30	72.70
#2 cabinet connectors	L1@0.50	Ea	60.40	20.50	80.90

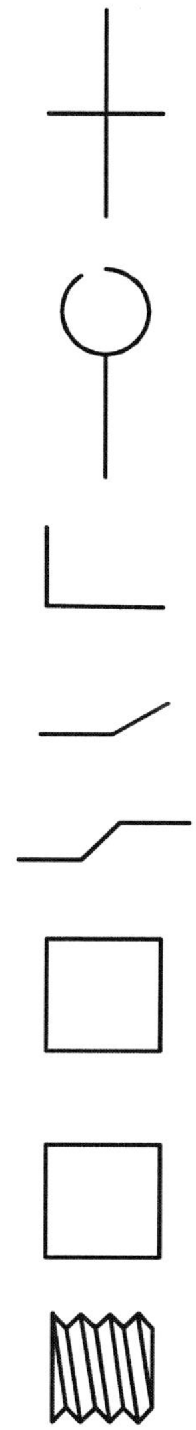

Use these figures to estimate the cost of underfloor raceway fittings installed in building floors under the conditions described on pages 5 and 6. Costs listed are for each fitting installed. The crew is one electrician working at a labor cost of $41.03 per manhour. These costs include leveling, mounting accessories, layout, material handling, and normal waste. Add for wiring devices, sales tax, delivery, supervision, mobilization, demobilization, cleanup, overhead and profit.

Underfloor Raceway Fittings

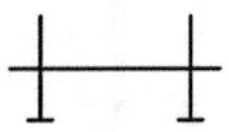

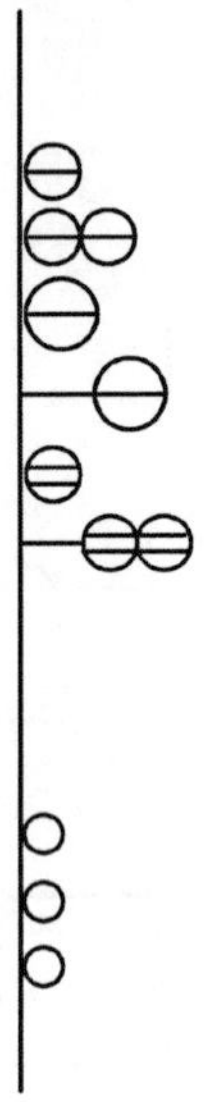

Material	Craft@Hrs	Unit	Material Cost	Labor Cost	Installed Cost
Fittings for underfloor raceway					
#1 wye coupling	L1@0.25	Ea	134.00	10.30	144.30
#2 wye connectors	L1@0.30	Ea	134.00	12.30	146.30
2" leveling legs	L1@0.05	Ea	8.05	2.05	10.10
3" leveling legs	L1@0.06	Ea	8.05	2.46	10.51
4" leveling legs	L1@0.08	Ea	8.05	3.28	11.33
6" leveling legs	L1@0.10	Ea	8.05	4.10	12.15
8" leveling legs	L1@0.12	Ea	9.41	4.92	14.33
10" leveling legs	L1@0.15	Ea	9.41	6.15	15.56
12" leveling legs	L1@0.20	Ea	9.41	8.21	17.62
14" leveling legs	L1@0.25	Ea	9.41	10.30	19.71
Service fittings without receptacles for underfloor raceway					
1 single recept. 20A	L1@0.25	Ea	107.00	10.30	117.30
2 single recept. 20A	L1@0.25	Ea	107.00	10.30	117.30
1 single recept. 30A	L1@0.25	Ea	107.00	10.30	117.30
1 single recept. 50A	L1@0.25	Ea	107.00	10.30	117.30
1 duplex receptacle	L1@0.25	Ea	107.00	10.30	117.30
2 duplex receptacles	L1@0.25	Ea	107.00	10.30	117.30
1 1" insulated bushing	L1@0.25	Ea	107.00	10.30	117.30
2 1" insulated bushing	L1@0.25	Ea	107.00	10.30	117.30
3/4" x 3" standpipes	L1@0.20	Ea	101.00	8.21	109.21
1" x 3" standpipes	L1@0.20	Ea	134.00	8.21	142.21
2" x 3" standpipes	L1@0.25	Ea	134.00	10.30	144.30
Brass abandon plugs	L1@0.15	Ea	53.80	6.15	59.95
Alum. abandon plugs	L1@0.15	Ea	53.80	6.15	59.95

Use these figures to estimate the cost of service fittings installed in building floors under the conditions described on pages 5 and 6. Costs listed are for each fitting installed. The crew is one electrician working at a labor cost of $41.03 per manhour. These costs include mounting accessories, layout, material handling, and normal waste. Add for wiring devices, sales tax, delivery, supervision, mobilization, demobilization, cleanup, overhead and profit. Note: Many other underfloor raceway fittings are available. Supplier catalogs list other service fittings, and fittings for special applications.

Section 8: Bus Duct

Bus duct is used in place of wire to distribute high amperage current within a building. It's made of copper or aluminum bus bars mounted in a protective sheet metal enclosure. Bus duct comes in prefabricated sections from one to 10 feet long and can't be cut to length or modified on the job.

Bus duct is common in high-rise commercial and apartment buildings. It reduces installation cost when long runs of high-amperage feeder or branch circuit are required.

Several types of bus duct are made. Each serves a different purpose. ***Feeder bus duct*** is used for large feeder circuits. ***Plug-in bus duct*** is used where multiple-power take-offs are needed.

Feeder Bus Duct

Feeder bus duct is rated as either indoor (NEMA 1) or raintight (NEMA 3R). The capacity of the bus duct usually begins at 225 amps and ranges up to several thousands of amps. Standard voltage ratings run from 250 volts to 600.

The feeder duct is used primarily for carrying large circuit loads from primary distribution centers or switchboards to other distribution centers or switchboards. Feeder bus duct is also used in high-rise buildings to carry the power feed vertically to the power center on each floor. It replaces large multiple conduit and wire feeders.

Many fittings are available for feeder duct: 90-degree flat elbows, 45-degree flat elbows, special bend flat elbows, tees, crosses or X's, flange adapters for switchboard connections, hangers, supports, and expansion joints.

A special transition section is needed to connect a run of outdoor raintight NEMA class 3R bus duct to an extension of indoor NEMA class 1 bus duct. Another type of transition section may be needed when extending bus duct from an area with one fire or service rating to an area with a different fire or service rating.

Feeder bus duct is available with either aluminum or copper bus bars. Job specs usually dictate which is to be used. Aluminum bus duct is lighter than copper duct. Most inspection authorities will not permit mixing the two types together in the same system.

Plug-In Bus Duct

Many machine shops uses plug-in bus duct to supply power to heavy electrical equipment. The duct is run overhead with individual feeds plugged into the bus where needed. Each service location is equipped with a plug-in circuit breaker.

The bus duct has ports placed along the face of the housing. Wherever power is needed, the port cover is moved aside so a plug-in device can be installed. After the device is secured in place, conduit and wire are extended to the machine's disconnect switch.

The plug-in device is operated from the floor with a ***hook stick***. The hook stick switches power off and on.

Fittings for plug-in bus duct include elbows, tees, crosses, hangers, ends, feed-in cable tap boxes and expansion joints. Bus duct reducers are used down the line when amperage requirements have dropped and smaller duct is appropriate.

Plug-In Devices

Plug-in devices are made to fit every need. When selecting the appropriate device, be sure the voltage rating and conductor configuration matches the bus duct. For example, bus duct might be 3-phase, 3-wire service or 3-phase, 4-wire service. The device should be the same.

The plug-in device is secured directly to the bus duct housing with screw-type clamps. Installation is usually very simple. But be careful when inserting the plug-in device. Line up the device squarely with the bus duct. Insert the contacts through the installation ports on the face of the bus duct.

Hangers

All bus duct manufacturers provide instructions for installing their duct. Be especially careful to follow the instructions for duct supports. Placement and spacing of hangers are important. Be sure the structure the bus duct is attached to is strong enough to support the duct weight.

Generally, you'll place hangers at least every 10 feet. The size of the hanger rod should be a minimum of 3/8 inch for the small size duct and increased proportionately for heavier duct. Usually

Plug-in Bus Duct

Conventional Type
225 Amp Through 1000 Amp

Materials For Run # 1

Quantity	Cat. No.	Description
1	LF1003	1000A 3 wire left flanged end with bus extension
1	DE1003	1000A 3 wire downward elbow
2	FST1003-2	1000A 3 wire feeder 2 ft. straight lengths
1	EC1003	1000A 3 wire edgewise cross
5	FST1003	1000A 3 wire feeder 10 ft. straight lengths
1	FST1003-8	1000A 3 wire feeder 8 ft. straight length
2	FE1003	1000A 3 wire forward elbows
2	RR1043	1000/400A 3 wire right unfused reducer adapters
1	LR1043	1000/400A 3 wire left unfused reducer adapter
12	ST403	400A 3 wire plug-in 10 ft. straight lengths
3	ST403-2'-6"	400A 3 wire plug-in 2'-6" straight lengths
2	RN403	400A 3 wire right end closers
1	LN403	400A 3 wire left end closer

Materials For Run # 2

Quantity	Cat. No.	Description
1	RF1003	1000A 3 wire right flanged end with bus extension
1	DE1003	1000A 3 wire downward elbow
1	FST1003-2	1000A 3 wire feeder 2 ft. straight length
10	FST1003	1000A 3 wire feeder 10 ft. straight lengths
2	FST1003-9	1000A 3 wire feeder 9 ft. straight lengths
2	FT1003	1000A 3 wire forward tees
1	FE1003	1000A 3 wire forward elbow
2	RR1043	1000/400A 3 wire right unfused reducer adapters
1	LR1043	1000/400A 3 wire left unfused reducer adapter
12	ST403	400A 3 wire plug-in 10 ft. straight lengths
3	ST403-7'-6"	400A 3 wire plug-in 7'-6" straight lengths
2	RN403	400A 3 wire right end closers
1	LN403	400A 3 wire left end closer

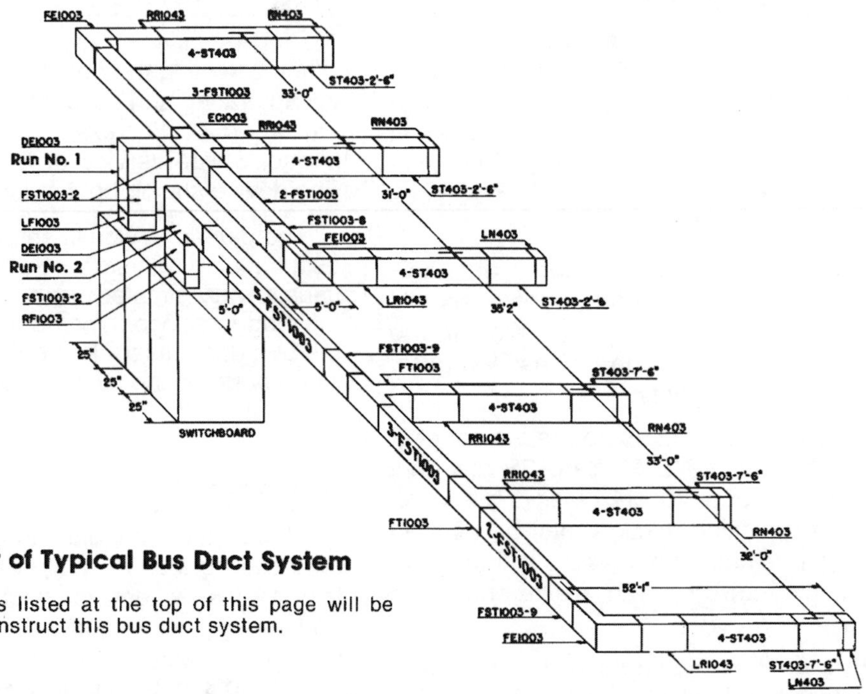

Layout of Typical Bus Duct System

The materials listed at the top of this page will be needed to construct this bus duct system.

Figure 8-1
Plug-in Bus Duct

two hangers are used, one on each side of the duct. A trapeze bar or channel is used in some cases to seat the duct. Rods carry the trapeze bar.

When the bus duct is run vertically through a floor, a special collar-like support is used at the floor line to carry the weight of the duct.

When laying out a bus duct system, be careful to avoid obstructions. Routing duct around a column or overhead crane can add four elbows and several hundred dollars to the cost. It's hard to see some of these potential conflicts on the plans. Consider what an elevation view would look like, even if you have only a plan view of the electrical and mechanical system.

Before beginning installation, be sure all the right fittings and devices are on hand. If you haven't estimated or installed the particular type of bus duct before, get help from a manufacturer's representative.

Estimating Bus Duct

Start your estimate by finding the plan sheet details and sections in the specs that cover bus duct. Determine the amperage and voltage required. Locate the run or runs of bus duct on the plans. All bus duct on most jobs will be the same size and be rated at the same voltage. But this isn't always the case.

Study the floor plan and details. Figure out what purpose the bus duct serves and try to visualize the way it should be installed. Then scale off the length of duct as shown on the floor plan. Calculate the length of vertical risers. List these measurements on your worksheet. Then look for the fittings you'll need. List the elbows as either vertical or horizontal and by the degree of bend, either 15, 30, 45, 60 or 90 degrees. List the tees, crosses and offsets. List flanges needed for switchboard connections. When duct penetrates a fire wall, there will be a fire stop fitting.

After all fittings have been listed, check for hangers. The hangers will probably need beam clamps, nuts, rods, and trapeze racks. Bracing to eliminate sway movement may be required. List all block-outs or chases that must be made in structural walls or floors.

It's easiest to install bus duct in open spaces where overhead structural support is about 12 feet above the floor. This permits easier handling and hoisting. Power bus duct made with copper bus bars is both heavy and difficult to handle.

Some types of bus duct have simple through-bolt connections; others require bolting each bus bar together. Bolts have to be tightened with a torque wrench. A cover assembly is then placed over the joint. Connections are needed between each duct section and at each fitting.

Aluminum bus duct is much lighter and is easier to handle. Reduce the labor cost by about 30 percent when installing aluminum bus duct and fittings. But note that installing the hangers will take about the same time as copper bus duct.

If the installation area is cluttered with obstructions and if runs are short, increase the labor units by 15 or 20 percent.

You may need lifting equipment to move bus duct from the delivery truck to the storage area. Be sure to protect the duct sections from moisture and dust while they're in storage. Keep the ends covered until it's time to install the system.

Bus Duct, Aluminum

Material	Craft@Hrs	Unit	Material Cost	Labor Cost	Installed Cost
Aluminum feeder bus duct, 600 volt, 3-wire					
800A 12"	L4@0.30	Ea	253.00	12.30	265.30
1000A 12"	L4@0.40	Ea	280.00	16.40	296.40
1200A 12"	L4@0.50	Ea	372.00	20.50	392.50
1350A 12"	L4@0.60	Ea	534.00	24.60	558.60
1600A 12"	L4@0.70	Ea	648.00	28.70	676.70
2000A 12"	L4@0.80	Ea	788.00	32.80	820.80
2500A 12"	L4@0.90	Ea	886.00	36.90	922.90
3000A 12"	L4@1.00	Ea	1,150.00	41.00	1,191.00
4000A 12"	L4@1.25	Ea	1,630.00	51.30	1,681.30
800A 18"	L4@0.50	Ea	506.00	20.50	526.50
1000A 18"	L4@0.60	Ea	576.00	24.60	600.60
1200A 18"	L4@0.70	Ea	773.00	28.70	801.70
1600A 18"	L4@0.90	Ea	1,070.00	36.90	1,106.90
2000A 18"	L4@1.00	Ea	1,300.00	41.00	1,341.00
800A 24"	L4@0.60	Ea	506.00	24.60	530.60
1000A 24"	L4@0.70	Ea	604.00	28.70	632.70
1200A 24"	L4@0.80	Ea	773.00	32.80	805.80
1600A 24"	L4@1.00	Ea	1,080.00	41.00	1,121.00
2000A 24"	L4@1.10	Ea	1,300.00	45.10	1,345.10
800A 72"	L4@1.50	Ea	1,510.00	61.50	1,571.50
1000A 72"	L4@1.75	Ea	1,670.00	71.80	1,741.80
1200A 72"	L4@2.00	Ea	2,270.00	82.10	2,352.10
1600A 72"	L4@2.50	Ea	3,220.00	103.00	3,323.00
2000A 72"	L4@2.75	Ea	3,840.00	113.00	3,953.00
800A 84"	L4@1.75	Ea	1,760.00	71.80	1,831.80
1000A 84"	L4@2.00	Ea	1,950.00	82.10	2,032.10
1200A 84"	L4@2.25	Ea	2,600.00	92.30	2,692.30
1600A 84"	L4@2.75	Ea	3,700.00	113.00	3,813.00
2000A 84"	L4@3.00	Ea	4,430.00	123.00	4,553.00
800A 120"	L4@2.00	Ea	2,520.00	82.10	2,602.10
1000A 120"	L4@2.25	Ea	2,790.00	92.30	2,882.30
1200A 120"	L4@2.50	Ea	3,770.00	103.00	3,873.00
1350A 120"	L4@2.75	Ea	4,400.00	113.00	4,513.00
1600A 120"	L4@3.00	Ea	5,340.00	123.00	5,463.00
2000A 120"	L4@3.25	Ea	6,350.00	133.00	6,483.00
2500A 120"	L4@3.50	Ea	7,910.00	144.00	8,054.00
3000A 120"	L4@3.75	Ea	8,880.00	154.00	9,034.00
4000A 120"	L4@4.00	Ea	12,200.00	164.00	12,364.00

Use these figures to estimate the cost of indoor bus duct installed in buildings under the conditions described on pages 5 and 6. Costs listed are for each length of duct installed. The crew is four electricians working at a labor cost of $41.03 per manhour. These costs include connecting hardware, layout, material handling, and normal waste. Add for supports, service devices, sales tax, delivery, supervision, mobilization, demobilization, cleanup, overhead and profit. Note: Never mix copper and aluminum duct or fittings in a single system. Dissimilar metals will lead to a failure in a very short time. Using contact compounds won't overcome this problem. Some bus duct and fittings are made with quick connects. The duct slips together and is held in place with a tightening stud. Use a torque wrench to apply the specified pressure to the stud. All bus duct must be supported according to the manufacturer's recommendations.

Material		Craft@Hrs	Unit	Material Cost	Labor Cost	Installed Cost
Aluminum feeder bus duct, 600 volt, 4-wire						
800A	12"	L4@0.40	Ea	293.00	16.40	309.40
1000A	12"	L4@0.50	Ea	367.00	20.50	387.50
1200A	12"	L4@0.60	Ea	454.00	24.60	478.60
1350A	12"	L4@0.70	Ea	521.00	28.70	549.70
1600A	12"	L4@0.80	Ea	632.00	32.80	664.80
2000A	12"	L4@0.90	Ea	773.00	36.90	809.90
2500A	12"	L4@1.00	Ea	948.00	41.00	989.00
3000A	12"	L4@1.10	Ea	1,100.00	45.10	1,145.10
4000A	12"	L4@1.20	Ea	1,470.00	49.20	1,519.20
800A	18"	L4@0.50	Ea	584.00	20.50	604.50
1000A	18"	L4@0.60	Ea	725.00	24.60	749.60
1200A	18"	L4@0.70	Ea	906.00	28.70	934.70
1600A	18"	L4@0.90	Ea	1,260.00	36.90	1,296.90
2000A	18"	L4@1.00	Ea	1,540.00	41.00	1,581.00
800A	24"	L4@0.60	Ea	584.00	24.60	608.60
1000A	24"	L4@0.70	Ea	725.00	28.70	753.70
1200A	24"	L4@0.80	Ea	906.00	32.80	938.80
1600A	24"	L4@1.00	Ea	1,260.00	41.00	1,301.00
2000A	24"	L4@1.10	Ea	1,540.00	45.10	1,585.10
800A	72"	L4@1.75	Ea	1,760.00	71.80	1,831.80
1000A	72"	L4@2.00	Ea	2,200.00	82.10	2,282.10
1200A	72"	L4@2.25	Ea	2,730.00	92.30	2,822.30
1600A	72"	L4@2.75	Ea	3,760.00	113.00	3,873.00
2000A	72"	L4@3.00	Ea	4,510.00	123.00	4,633.00
800A	84"	L4@2.00	Ea	2,030.00	82.10	2,112.10
1000A	84"	L4@2.25	Ea	2,550.00	92.30	2,642.30
1200A	84"	L4@2.50	Ea	3,200.00	103.00	3,303.00
1600A	84"	L4@3.00	Ea	4,430.00	123.00	4,553.00
2000A	84"	L4@3.25	Ea	5,390.00	133.00	5,523.00
800A	120"	L4@2.25	Ea	2,930.00	92.30	3,022.30
1000A	120"	L4@2.50	Ea	3,670.00	103.00	3,773.00
1200A	120"	L4@2.75	Ea	4,760.00	113.00	4,873.00
1350A	120"	L4@3.00	Ea	5,230.00	123.00	5,353.00
1600A	120"	L4@3.25	Ea	6,320.00	133.00	6,453.00
2000A	120"	L4@3.50	Ea	7,720.00	144.00	7,864.00
2500A	120"	L4@3.75	Ea	9,480.00	154.00	9,634.00
3000A	120"	L4@4.00	Ea	11,000.00	164.00	11,164.00
4000A	120"	L4@4.25	Ea	14,700.00	174.00	14,874.00

F

F

Use these figures to estimate the cost of indoor bus duct installed in buildings under the conditions described on pages 5 and 6. Costs listed are for each length of duct installed. The crew is four electricians working at a labor cost of $41.03 per manhour. These costs include connecting hardware, layout, material handling, and normal waste. Add for supports, service devices, sales tax, delivery, supervision, mobilization, demobilization, cleanup, overhead and profit. Note: Never mix copper and aluminum duct or fittings in a single system. Dissimilar metals will lead to a failure in a very short time. Using contact compounds won't overcome this problem. Some bus duct and fittings are made with quick connects. The duct slips together and is held in place with a tightening stud. Use a torque wrench to apply the specified pressure to the stud. All bus duct must be supported according to the manufacturer's recommendations.

Bus Duct, Aluminum

Material	Craft@Hrs	Unit	Material Cost	Labor Cost	Installed Cost
Aluminum plug-in bus duct, 600 volt, 3-phase, 3-wire, 6' length					
225A	L4@1.00	Ea	722.00	41.00	763.00
400A	L4@1.25	Ea	881.00	51.30	932.30
600A	L4@1.50	Ea	1,100.00	61.50	1,161.50
Aluminum plug-in bus duct, 600 volt, 3-phase, 3-wire, 10' length					
225A	L4@1.50	Ea	1,200.00	61.50	1,261.50
400A	L4@1.75	Ea	1,460.00	71.80	1,531.80
600A	L4@2.00	Ea	1,840.00	82.10	1,922.10
Aluminum plug-in bus duct, 600 volt, 3-phase, 4-wire, 6' length					
225A	L4@1.00	Ea	881.00	41.00	922.00
400A	L4@1.25	Ea	1,100.00	51.30	1,151.30
600A	L4@1.50	Ea	1,520.00	61.50	1,581.50
Aluminum plug-in bus duct, 600 volt, 3-phase, 4-wire, 10' length					
225A	L4@1.50	Ea	1,460.00	61.50	1,521.50
400A	L4@1.75	Ea	1,840.00	71.80	1,911.80
600A	L4@2.00	Ea	2,520.00	82.10	2,602.10

Use these figures to estimate the cost of indoor bus duct installed in buildings under the conditions described on pages 5 and 6. Costs listed are for each length of duct installed. The crew is four electricians working at a labor cost of $41.03 per manhour. These costs include connecting hardware, layout, material handling, and normal waste. Add for supports, service devices, sales tax, delivery, supervision, mobilization, demobilization, cleanup, overhead and profit. Note: Never mix copper and aluminum duct or fittings in a single system. Dissimilar metals will lead to a failure in a very short time. Using contact compounds won't overcome this problem. Some bus duct and fittings are made with quick connects. The duct slips together and is held in place with a tightening stud. Use a torque wrench to apply the specified pressure to the stud. All bus duct must be supported according to the manufacturer's recommendations.

Material	Craft@Hrs	Unit	Material Cost	Labor Cost	Installed Cost
Aluminum plug-in bus duct, 600 volt, 3-phase, 3-wire, 6' length with integral ground bus					
225A	L4@1.00	Ea	992.00	41.00	1,033.00
400A	L4@1.25	Ea	1,150.00	51.30	1,201.30
600A	L4@1.50	Ea	1,400.00	61.50	1,461.50
Aluminum plug-in bus duct, 600 volt, 3-phase, 3-wire, 10' length with integral ground bus					
225A	L4@1.50	Ea	1,640.00	61.50	1,701.50
400A	L4@1.75	Ea	1,920.00	71.80	1,991.80
600A	L4@2.00	Ea	2,350.00	82.10	2,432.10
Aluminum plug-in bus duct, 600 volt, 3-phase, 4-wire, 6' length with integral ground bus					
225A	L4@1.00	Ea	1,150.00	41.00	1,191.00
400A	L4@1.25	Ea	1,390.00	51.30	1,441.30
600A	L4@1.50	Ea	1,830.00	61.50	1,891.50
Aluminum plug-in bus duct, 600 volt, 3-phase, 4-wire, 10' length with integral ground bus					
225A	L4@1.50	Ea	1,920.00	61.50	1,981.50
400A	L4@1.75	Ea	2,320.00	71.80	2,391.80
600A	L4@2.00	Ea	3,050.00	82.10	3,132.10

P

P

Use these figures to estimate the cost of indoor bus duct installed in buildings under the conditions described on pages 5 and 6. Costs listed are for each length of duct installed. The crew is four electricians working at a labor cost of $41.03 per manhour. These costs include connecting hardware, layout, material handling, and normal waste. Add for supports, service devices, sales tax, delivery, supervision, mobilization, demobilization, cleanup, overhead and profit. Note: Never mix copper and aluminum duct or fittings in a single system. Dissimilar metals will lead to a failure in a very short time. Using contact compounds won't overcome this problem. Some bus duct and fittings are made with quick connects. The duct slips together and is held in place with a tightening stud. Use a torque wrench to apply the specified pressure to the stud. All bus duct must be supported according to the manufacturer's recommendations.

Bus Duct, Copper

Material	Craft@Hrs	Unit	Material Cost	Labor Cost	Installed Cost
Copper feeder bus duct, 600 volt, 3-wire					
800A 12"	L4@0.50	Ea	488.00	20.50	508.50
1000A 12"	L4@0.60	Ea	517.00	24.60	541.60
1200A 12"	L4@0.70	Ea	675.00	28.70	703.70
1350A 12"	L4@0.80	Ea	781.00	32.80	813.80
1600A 12"	L4@0.90	Ea	906.00	36.90	942.90
2000A 12"	L4@1.00	Ea	1,150.00	41.00	1,191.00
2500A 12"	L4@1.10	Ea	1,450.00	45.10	1,495.10
3000A 12"	L4@1.20	Ea	1,730.00	49.20	1,779.20
4000A 12"	L4@1.30	Ea	2,250.00	53.30	2,303.30
800A 18"	L4@0.60	Ea	973.00	24.60	997.60
1000A 18"	L4@0.70	Ea	1,040.00	28.70	1,068.70
1200A 18"	L4@0.80	Ea	1,340.00	32.80	1,372.80
1600A 18"	L4@1.00	Ea	1,810.00	41.00	1,851.00
2000A 18"	L4@1.10	Ea	2,310.00	45.10	2,355.10
800A 24"	L4@0.80	Ea	973.00	32.80	1,005.80
1000A 24"	L4@0.90	Ea	1,040.00	36.90	1,076.90
1200A 24"	L4@1.00	Ea	1,340.00	41.00	1,381.00
1600A 24"	L4@1.20	Ea	1,810.00	49.20	1,859.20
2000A 24"	L4@1.30	Ea	2,310.00	53.30	2,363.30
800A 72"	L4@3.00	Ea	2,940.00	123.00	3,063.00
1000A 72"	L4@3.50	Ea	3,070.00	144.00	3,214.00
1200A 72"	L4@3.70	Ea	4,050.00	152.00	4,202.00
1600A 72"	L4@4.10	Ea	5,390.00	168.00	5,558.00
2000A 72"	L4@4.25	Ea	6,900.00	174.00	7,074.00
800A 84"	L4@3.30	Ea	3,450.00	135.00	3,585.00
1000A 84"	L4@3.85	Ea	3,610.00	158.00	3,768.00
1200A 84"	L4@4.00	Ea	4,700.00	164.00	4,864.00
1600A 84"	L4@4.40	Ea	6,310.00	181.00	6,491.00
2000A 84"	L4@4.60	Ea	8,060.00	189.00	8,249.00
800A 120"	L4@3.75	Ea	4,870.00	154.00	5,024.00
1000A 120"	L4@4.25	Ea	5,150.00	174.00	5,324.00
1200A 120"	L4@4.40	Ea	6,680.00	181.00	6,861.00
1350A 120"	L4@4.60	Ea	8,000.00	189.00	8,189.00
1600A 120"	L4@4.80	Ea	8,960.00	197.00	9,157.00
2000A 120"	L4@5.00	Ea	11,300.00	205.00	11,505.00
2500A 120"	L4@5.20	Ea	14,300.00	213.00	14,513.00
3000A 120"	L4@5.40	Ea	17,200.00	222.00	17,422.00
4000A 120"	L4@5.60	Ea	22,300.00	230.00	22,530.00

F

F

Use these figures to estimate the cost of indoor bus duct installed in buildings under the conditions described on pages 5 and 6. Costs listed are for each length of duct installed. The crew is four electricians working at a labor cost of $41.03 per manhour. These costs include connecting hardware, layout, material handling, and normal waste. Add for supports, service devices, sales tax, delivery, supervision, mobilization, demobilization, cleanup, overhead and profit. Note: Never mix copper and aluminum duct or fittings in a single system. Dissimilar metals will lead to a failure in a very short time. Using contact compounds won't overcome this problem. Some bus duct and fittings are made with quick connects. The duct slips together and is held in place with a tightening stud. Use a torque wrench to apply the specified pressure to the stud. All bus duct must be supported according to the manufacturer's recommendations.

Copper feeder bus duct, 600 volt, 4-wire

Material	Craft@Hrs	Unit	Material Cost	Labor Cost	Installed Cost
800A 12"	L4@0.60	Ea	630.00	24.60	654.60
1000A 12"	L4@0.70	Ea	654.00	28.70	682.70
1200A 12"	L4@0.80	Ea	838.00	32.80	870.80
1350A 12"	L4@0.90	Ea	965.00	36.90	1,001.90
1600A 12"	L4@1.00	Ea	1,130.00	41.00	1,171.00
2000A 12"	L4@1.10	Ea	1,410.00	45.10	1,455.10
2500A 12"	L4@1.20	Ea	1,760.00	49.20	1,809.20
3000A 12"	L4@1.30	Ea	2,100.00	53.30	2,153.30
4000A 12"	L4@1.40	Ea	2,700.00	57.40	2,757.40
800A 18"	L4@0.70	Ea	1,280.00	28.70	1,308.70
1000A 18"	L4@0.80	Ea	1,310.00	32.80	1,342.80
1200A 18"	L4@0.90	Ea	1,680.00	36.90	1,716.90
1600A 18"	L4@1.10	Ea	2,310.00	45.10	2,355.10
2000A 18"	L4@1.20	Ea	2,880.00	49.20	2,929.20
800A 24"	L4@0.90	Ea	1,280.00	36.90	1,316.90
1000A 24"	L4@1.00	Ea	1,310.00	41.00	1,351.00
1200A 24"	L4@1.10	Ea	1,680.00	45.10	1,725.10
1600A 24"	L4@1.30	Ea	2,310.00	53.30	2,363.30
2000A 24"	L4@1.40	Ea	2,880.00	57.40	2,937.40
800A 72"	L4@3.10	Ea	3,800.00	127.00	3,927.00
1000A 72"	L4@3.60	Ea	3,970.00	148.00	4,118.00
1200A 72"	L4@3.80	Ea	5,020.00	156.00	5,176.00
1600A 72"	L4@4.20	Ea	6,910.00	172.00	7,082.00
2000A 72"	L4@4.40	Ea	7,710.00	181.00	7,891.00
800A 84"	L4@3.40	Ea	4,500.00	140.00	4,640.00
1000A 84"	L4@3.95	Ea	4,630.00	162.00	4,792.00
1200A 84"	L4@4.10	Ea	5,910.00	168.00	6,078.00
1600A 84"	L4@4.50	Ea	8,060.00	185.00	8,245.00
2000A 84"	L4@4.70	Ea	10,000.00	193.00	10,193.00
800A 120"	L4@4.00	Ea	6,290.00	164.00	6,454.00
1000A 120"	L4@4.50	Ea	7,450.00	185.00	7,635.00
1200A 120"	L4@4.75	Ea	8,900.00	195.00	9,095.00
1350A 120"	L4@5.00	Ea	10,200.00	205.00	10,405.00
1600A 120"	L4@5.25	Ea	12,000.00	215.00	12,215.00
2000A 120"	L4@5.50	Ea	14,600.00	226.00	14,826.00
2500A 120"	L4@5.75	Ea	18,100.00	236.00	18,336.00
3000A 120"	L4@6.00	Ea	23,300.00	246.00	23,546.00
4000A 120"	L4@6.25	Ea	29,200.00	256.00	29,456.00

Use these figures to estimate the cost of indoor bus duct installed in buildings under the conditions described on pages 5 and 6. Costs listed are for each length of duct installed. The crew is four electricians working at a labor cost of $41.03 per manhour. These costs include connecting hardware, layout, material handling, and normal waste. Add for supports, service devices, sales tax, delivery, supervision, mobilization, demobilization, cleanup, overhead and profit. Note: Never mix copper and aluminum duct or fittings in a single system. Dissimilar metals will lead to a failure in a very short time. Using contact compounds won't overcome this problem. Some bus duct and fittings are made with quick connects. The duct slips together and is held in place with a tightening stud. Use a torque wrench to apply the specified pressure to the stud. All bus duct must be supported according to the manufacturer's recommendations.

Bus Duct, Copper

Material	Craft@Hrs	Unit	Material Cost	Labor Cost	Installed Cost
Copper plug-in bus duct, 600 volt, 3-phase, 3-wire, 6'- 0" length					
225A	L4@1.50	Ea	1,030.00	61.50	1,091.50
400A	L4@1.75	Ea	1,590.00	71.80	1,661.80
600A	L4@2.00	Ea	2,000.00	82.10	2,082.10
Copper plug-in bus duct, 600 volt, 3-phase, 3-wire, 10'- 0" length					
225A	L4@2.00	Ea	1,720.00	82.10	1,802.10
400A	L4@2.50	Ea	2,630.00	103.00	2,733.00
600A	L4@3.00	Ea	3,350.00	123.00	3,473.00
Copper plug-in bus duct, 600 volt, 3-phase, 4-wire, 6'- 0" length					
225A	L4@1.50	Ea	1,390.00	61.50	1,451.50
400A	L4@1.75	Ea	2,310.00	71.80	2,381.80
600A	L4@2.00	Ea	2,630.00	82.10	2,712.10
Copper plug-in bus duct, 600 volt, 3-phase, 4-wire, 10'- 0" length					
225A	L4@2.00	Ea	2,330.00	82.10	2,412.10
400A	L4@2.50	Ea	3,850.00	103.00	3,953.00
600A	L4@3.00	Ea	4,360.00	123.00	4,483.00

P

P

Use these figures to estimate the cost of indoor bus duct installed in buildings under the conditions described on pages 5 and 6. Costs listed are for each length of duct installed. The crew is four electricians working at a labor cost of $41.03 per manhour. These costs include connecting hardware, layout, material handling, and normal waste. Add for supports, service devices, sales tax, delivery, supervision, mobilization, demobilization, cleanup, overhead and profit. Note: Never mix copper and aluminum duct or fittings in a single system. Dissimilar metals will lead to a failure in a very short time. Using contact compounds won't overcome this problem. Some bus duct and fittings are made with quick connects. The duct slips together and is held in place with a tightening stud. Use a torque wrench to apply the specified pressure to the stud. All bus duct must be supported according to the manufacturer's recommendations.

Material	Craft@Hrs	Unit	Material Cost	Labor Cost	Installed Cost

Copper plug-in bus duct, 600 volt, 3-phase, 3-wire, 6'- 0" length with integral ground bus

Material	Craft@Hrs	Unit	Material Cost	Labor Cost	Installed Cost
225A	L4@1.50	Ea	1,490.00	61.50	1,551.50
400A	L4@1.75	Ea	2,040.00	71.80	2,111.80
600A	L4@2.00	Ea	2,510.00	82.10	2,592.10

Copper plug-in bus duct, 600 volt, 3-phase, 3-wire, 10'- 0" length with integral ground bus

Material	Craft@Hrs	Unit	Material Cost	Labor Cost	Installed Cost
225A	L4@2.00	Ea	2,460.00	82.10	2,542.10
400A	L4@2.50	Ea	3,340.00	103.00	3,443.00
600A	L4@3.00	Ea	4,130.00	123.00	4,253.00

P

Copper plug-in bus duct, 600 volt, 3-phase, 4-wire, 6'- 0" length with integral ground bus

Material	Craft@Hrs	Unit	Material Cost	Labor Cost	Installed Cost
225A	L4@1.50	Ea	1,860.00	61.50	1,921.50
400A	L4@1.75	Ea	2,760.00	71.80	2,831.80
600A	L4@2.00	Ea	3,130.00	82.10	3,212.10

Copper plug-in bus duct, 600 volt, 3-phase, 4-wire, 10'- 0" length with integral ground bus

Material	Craft@Hrs	Unit	Material Cost	Labor Cost	Installed Cost
225A	L4@2.00	Ea	3,070.00	82.10	3,152.10
400A	L4@2.50	Ea	4,630.00	103.00	4,733.00
600A	L4@3.00	Ea	5,170.00	123.00	5,293.00

P

Use these figures to estimate the cost of indoor bus duct installed in buildings under the conditions described on pages 5 and 6. Costs listed are for each length of duct installed. The crew is four electricians working at a labor cost of $41.03 per manhour. These costs include connecting hardware, layout, material handling, and normal waste. Add for supports, service devices, sales tax, delivery, supervision, mobilization, demobilization, cleanup, overhead and profit. Note: Never mix copper and aluminum duct or fittings in a single system. Dissimilar metals will lead to a failure in a very short time. Using contact compounds won't overcome this problem. Some bus duct and fittings are made with quick connects. The duct slips together and is held in place with a tightening stud. Use a torque wrench to apply the specified pressure to the stud. All bus duct must be supported according to the manufacturer's recommendations.

Bus Duct Fittings

Flat elbows for bus duct, 600 volt, 3- or 4-wire

Material	Craft@Hrs	Unit	Material Cost	Labor Cost	Installed Cost
800A	L4@1.50	Ea	2,000.00	61.50	2,061.50
1000A	L4@1.75	Ea	2,230.00	71.80	2,301.80
1200A	L4@2.00	Ea	2,960.00	82.10	3,042.10
1350A	L4@2.25	Ea	3,220.00	92.30	3,312.30
1600A	L4@2.50	Ea	3,520.00	103.00	3,623.00
2000A	L4@2.75	Ea	4,200.00	113.00	4,313.00
2500A	L4@3.00	Ea	5,080.00	123.00	5,203.00
3000A	L4@3.25	Ea	5,780.00	133.00	5,913.00
4000A	L4@3.50	Ea	8,690.00	144.00	8,834.00

Tee for bus duct, 600 volt, 3- or 4-wire

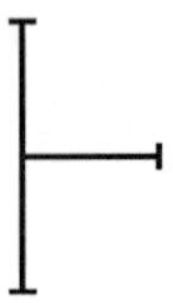

Material	Craft@Hrs	Unit	Material Cost	Labor Cost	Installed Cost
800A	L4@2.25	Ea	1,590.00	92.30	1,682.30
1000A	L4@2.50	Ea	1,590.00	103.00	1,693.00
1200A	L4@2.75	Ea	2,210.00	113.00	2,323.00
1350A	L4@3.00	Ea	2,210.00	123.00	2,333.00
1600A	L4@3.25	Ea	2,210.00	133.00	2,343.00
2000A	L4@3.50	Ea	2,210.00	144.00	2,354.00
2500A	L4@3.75	Ea	2,210.00	154.00	2,364.00
3000A	L4@4.00	Ea	2,210.00	164.00	2,374.00
4000A	L4@4.25	Ea	2,600.00	174.00	2,774.00

Cross for bus duct, 600 volt, 3- or 4-wire

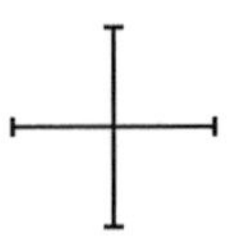

Material	Craft@Hrs	Unit	Material Cost	Labor Cost	Installed Cost
800A	L4@2.50	Ea	3,170.00	103.00	3,273.00
1000A	L4@2.75	Ea	3,170.00	113.00	3,283.00
1200A	L4@3.00	Ea	4,410.00	123.00	4,533.00
1350A	L4@3.25	Ea	4,410.00	133.00	4,543.00
1600A	L4@3.50	Ea	4,410.00	144.00	4,554.00
2000A	L4@3.75	Ea	4,410.00	154.00	4,564.00
2500A	L4@4.00	Ea	4,410.00	164.00	4,574.00
3000A	L4@4.25	Ea	4,410.00	174.00	4,584.00
4000A	L4@4.50	Ea	5,150.00	185.00	5,335.00

Flanged end for bus duct, 600 volt, 3- or 4-wire

Material	Craft@Hrs	Unit	Material Cost	Labor Cost	Installed Cost
800A	L4@1.50	Ea	1,220.00	61.50	1,281.50
1000A	L4@1.75	Ea	1,550.00	71.80	1,621.80
1200A	L4@2.00	Ea	1,550.00	82.10	1,632.10
1350A	L4@2.25	Ea	1,620.00	92.30	1,712.30
1600A	L4@2.50	Ea	1,880.00	103.00	1,983.00
2000A	L4@2.75	Ea	2,180.00	113.00	2,293.00
2500A	L4@3.00	Ea	2,640.00	123.00	2,763.00
3000A	L4@3.25	Ea	3,130.00	133.00	3,263.00
4000A	L4@3.50	Ea	3,870.00	144.00	4,014.00

Use these figures to estimate the cost of indoor bus duct fittings installed in buildings under the conditions described on pages 5 and 6. Costs listed are for each fitting installed. The crew is four electricians working at a labor cost of $41.03 per manhour. These costs include connecting hardware, layout, material handling, and normal waste. Add for supports, service devices, sales tax, delivery, supervision, mobilization, demobilization, cleanup, overhead and profit. Note: Never mix copper and aluminum duct or fittings in a single system. Dissimilar metals will lead to a failure in a very short time. Using contact compounds won't overcome this problem. Material and labor costs for the elbows listed above will be about the same whether the elbow is vertical or horizontal. But horizontal and vertical elbows are not interchangeable. Order the type needed on your job. Some bus duct and fittings are made with quick connects. The duct slips together and is held in place with a tightening stud. Use a torque wrench to apply the specified pressure to the stud. All bus duct must be supported according to the manufacturer's recommendations.

Tap box for bus duct, 600 volt, 3- or 4-wire

Material	Craft@Hrs	Unit	Material Cost	Labor Cost	Installed Cost
800A	L4@1.50	Ea	4,070.00	61.50	4,131.50
1000A	L4@1.75	Ea	4,310.00	71.80	4,381.80
1200A	L4@2.00	Ea	4,560.00	82.10	4,642.10
1350A	L4@2.25	Ea	4,760.00	92.30	4,852.30
1600A	L4@2.50	Ea	5,060.00	103.00	5,163.00
2000A	L4@2.75	Ea	5,610.00	113.00	5,723.00
2500A	L4@3.00	Ea	6,800.00	123.00	6,923.00
3000A	L4@3.25	Ea	7,510.00	133.00	7,643.00
4000A	L4@3.50	Ea	8,750.00	144.00	8,894.00

Reducer for bus duct, 600 volt, 3- or 4-wire

Material	Craft@Hrs	Unit	Material Cost	Labor Cost	Installed Cost
800A	L4@1.75	Ea	1,300.00	71.80	1,371.80
1000A	L4@2.00	Ea	1,540.00	82.10	1,622.10
1200A	L4@2.25	Ea	2,620.00	92.30	2,712.30
1350A	L4@2.50	Ea	3,320.00	103.00	3,423.00
1600A	L4@2.75	Ea	3,600.00	113.00	3,713.00
2000A	L4@3.00	Ea	4,850.00	123.00	4,973.00
2500A	L4@3.25	Ea	5,980.00	133.00	6,113.00
3000A	L4@3.50	Ea	7,200.00	144.00	7,344.00
4000A	L4@3.75	Ea	10,700.00	154.00	10,854.00

Expansion joint for bus duct, 600 volt, 3- or 4-wire

Material	Craft@Hrs	Unit	Material Cost	Labor Cost	Installed Cost
800A	L4@1.75	Ea	3,290.00	71.80	3,361.80
1000A	L4@2.00	Ea	3,740.00	82.10	3,822.10
1200A	L4@2.25	Ea	4,490.00	92.30	4,582.30
1350A	L4@2.50	Ea	4,940.00	103.00	5,043.00
1600A	L4@2.75	Ea	5,990.00	113.00	6,103.00
2000A	L4@3.00	Ea	6,370.00	123.00	6,493.00
2500A	L4@3.25	Ea	7,020.00	133.00	7,153.00
3000A	L4@3.50	Ea	9,020.00	144.00	9,164.00
4000A	L4@3.75	Ea	10,400.00	154.00	10,554.00

End closures for bus duct, 3- or 4-wire

Material	Craft@Hrs	Unit	Material Cost	Labor Cost	Installed Cost
800A	L4@0.25	Ea	335.00	10.30	345.30
1000A	L4@0.30	Ea	335.00	12.30	347.30
1200A	L4@0.35	Ea	335.00	14.40	349.40
1350A	L4@0.40	Ea	335.00	16.40	351.40
1600A	L4@0.45	Ea	335.00	18.50	353.50
2000A	L4@0.50	Ea	436.00	20.50	456.50
2500A	L4@0.55	Ea	436.00	22.60	458.60
3000A	L4@0.60	Ea	436.00	24.60	460.60
4000A	L4@0.70	Ea	555.00	28.70	583.70

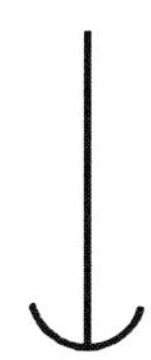

Use these figures to estimate the cost of indoor bus duct fittings installed in buildings under the conditions described on pages 5 and 6. Costs listed are for each fitting installed. The crew is four electricians working at a labor cost of $41.03 per manhour. These costs include connecting hardware, layout, material handling, and normal waste. Add for supports, service devices, sales tax, delivery, supervision, mobilization, demobilization, cleanup, overhead and profit. Note: Never mix copper and aluminum duct or fittings in a single system. Dissimilar metals will lead to a failure in a very short time. Using contact compounds won't overcome this problem. Some bus duct and fittings are made with quick connects. The duct slips together and is held in place with a tightening stud. Use a torque wrench to apply the specified pressure to the stud. All bus duct must be supported according to the manufacturer's recommendations.

Bus Duct Plug-in Units

Material	Craft@Hrs	Unit	Material Cost	Labor Cost	Installed Cost
Fusible plug-in switches for bus duct					
30A, 3P, 240V	L2@0.40	Ea	747.00	16.40	763.40
60A, 3P, 240V	L2@0.50	Ea	798.00	20.50	818.50
100A, 3P, 240V	L2@0.75	Ea	1,120.00	30.80	1,150.80
200A, 3P, 240V	L2@1.00	Ea	1,880.00	41.00	1,921.00
400A, 3P, 240V	L2@1.50	Ea	4,900.00	61.50	4,961.50
600A, 3P, 240V	L2@2.00	Ea	7,060.00	82.10	7,142.10
30A, 4P, 240V	L2@0.50	Ea	869.00	20.50	889.50
60A, 4P, 240V	L2@0.75	Ea	902.00	30.80	932.80
100A, 4P, 240V	L2@1.00	Ea	1,230.00	41.00	1,271.00
200A, 4P, 240V	L2@1.50	Ea	2,080.00	61.50	2,141.50
400A, 4P, 240V	L2@2.00	Ea	5,350.00	82.10	5,432.10
600A, 3P, 240V	L2@2.50	Ea	7,760.00	103.00	7,863.00
30A, 3P, 480V	L2@0.50	Ea	902.00	20.50	922.50
60A, 3P, 480V	L2@0.75	Ea	943.00	30.80	973.80
100A, 3P, 480V	L2@1.00	Ea	1,330.00	41.00	1,371.00
200A, 3P, 480V	L2@1.50	Ea	2,200.00	61.50	2,261.50
400A, 3P, 480V	L2@2.00	Ea	5,350.00	82.10	5,432.10
600A, 3P, 480V	L2@2.50	Ea	7,760.00	103.00	7,863.00
30A, 3P, 600V	L2@0.50	Ea	798.00	20.50	818.50
60A, 3P, 600V	L2@0.75	Ea	856.00	30.80	886.80
100A, 3P, 600V	L2@1.00	Ea	1,150.00	41.00	1,191.00
200A, 3P, 600V	L2@1.50	Ea	1,970.00	61.50	2,031.50
400A, 3P, 600V	L2@2.00	Ea	4,900.00	82.10	4,982.10
600A, 3P, 600V	L2@2.50	Ea	7,080.00	103.00	7,183.00
Circuit breaker plug-in switches for bus duct					
15A, 3P, 240V	L2@0.50	Ea	1,100.00	20.50	1,120.50
20A, 3P, 240V	L2@0.50	Ea	1,100.00	20.50	1,120.50
30A, 3P, 240V	L2@0.50	Ea	1,100.00	20.50	1,120.50
40A, 3P, 240V	L2@0.50	Ea	1,100.00	20.50	1,120.50
50A, 3P, 240V	L2@0.50	Ea	1,100.00	20.50	1,120.50
60A, 3P, 240V	L2@0.50	Ea	1,100.00	20.50	1,120.50
70A, 3P, 240V	L2@0.75	Ea	1,290.00	30.80	1,320.80
90A, 3P, 240V	L2@0.75	Ea	1,290.00	30.80	1,320.80
100A, 3P, 240V	L2@0.75	Ea	1,290.00	30.80	1,320.80
15A, 3P, 240V, S/N	L2@0.60	Ea	1,250.00	24.60	1,274.60
20A, 3P, 240V, S/N	L2@0.60	Ea	1,250.00	24.60	1,274.60
30A, 3P, 240V, S/N	L2@0.60	Ea	1,250.00	24.60	1,274.60
40A, 3P, 240V, S/N	L2@0.60	Ea	1,250.00	24.60	1,274.60
50A, 3P, 240V, S/N	L2@0.60	Ea	1,250.00	24.60	1,274.60
60A, 3P, 240V, S/N	L2@0.60	Ea	1,250.00	24.60	1,274.60
70A, 3P, 240V, S/N	L2@0.90	Ea	1,420.00	36.90	1,456.90

Use these figures to estimate the cost of plug-in switches installed in bus duct under the conditions described on pages 5 and 6. Costs listed are for each switch installed. The crew is two electricians working at a labor cost of $41.03 per manhour. These costs include connecting hardware, layout, material handling, and normal waste. Add for supports, service devices, sales tax, delivery, supervision, mobilization, demobilization, cleanup, overhead and profit. Note: Bus duct made as feeder duct only will not accept plug-in breakers.

Material	Craft@Hrs	Unit	Material Cost	Labor Cost	Installed Cost

Circuit breaker plug-in switches for bus duct

90A, 3P, 240V, S/N	L2@1.00	Ea	1,420.00	41.00	1,461.00
100A, 3P, 240V, S/N	L2@1.25	Ea	1,420.00	51.30	1,471.30
15A, 3P, 480V	L2@0.50	Ea	1,340.00	20.50	1,360.50
20A, 3P, 480V	L2@0.50	Ea	1,340.00	20.50	1,360.50
30A, 3P, 480V	L2@0.50	Ea	1,340.00	20.50	1,360.50
40A, 3P, 480V	L2@0.50	Ea	1,340.00	20.50	1,360.50
50A, 3P, 480V	L2@0.50	Ea	1,340.00	20.50	1,360.50
60A, 3P, 480V	L2@0.50	Ea	1,340.00	20.50	1,360.50
70A, 3P, 480V	L2@0.75	Ea	1,490.00	30.80	1,520.80
90A, 3P, 480V	L2@0.75	Ea	1,490.00	30.80	1,520.80
100A, 3P, 480V	L2@0.75	Ea	1,410.00	30.80	1,440.80
15A, 3P, 480V, S/N	L2@0.60	Ea	1,490.00	24.60	1,514.60
20A, 3P, 480V, S/N	L2@0.60	Ea	1,490.00	24.60	1,514.60
30A, 3P, 480V, S/N	L2@0.60	Ea	1,490.00	24.60	1,514.60
40A, 3P, 480V, S/N	L2@0.60	Ea	1,490.00	24.60	1,514.60
50A, 3P, 480V, S/N	L2@0.60	Ea	1,490.00	24.60	1,514.60
60A, 3P, 480V, S/N	L2@0.60	Ea	1,490.00	24.60	1,514.60
70A, 3P, 480V, S/N	L2@0.90	Ea	1,610.00	36.90	1,646.90
90A, 3P, 480V, S/N	L2@0.90	Ea	1,610.00	36.90	1,646.90
100A, 3P, 480V, S/N	L2@0.90	Ea	1,610.00	36.90	1,646.90
15A, 3P, 600V	L2@0.50	Ea	1,450.00	20.50	1,470.50
20A, 3P, 600V	L2@0.50	Ea	1,450.00	20.50	1,470.50
30A, 3P, 600V	L2@0.50	Ea	1,450.00	20.50	1,470.50
40A, 3P, 600V	L2@0.50	Ea	1,450.00	20.50	1,470.50
50A, 3P, 600V	L2@0.50	Ea	1,450.00	20.50	1,470.50
60A, 3P, 600V	L2@0.50	Ea	1,450.00	20.50	1,470.50
70A, 3P, 600V	L2@0.75	Ea	1,610.00	30.80	1,640.80
90A, 3P, 600V	L2@0.75	Ea	1,610.00	30.80	1,640.80
100A, 3P, 600V	L2@0.75	Ea	1,610.00	30.80	1,640.80

Combination starter/fusible switch plug-in unit for bus duct

Size 0, 3P, 240V	L2@1.50	Ea	2,310.00	61.50	2,371.50
Size 1, 3P, 240V	L2@1.75	Ea	2,490.00	71.80	2,561.80
Size 2, 3P, 240V	L2@2.00	Ea	2,900.00	82.10	2,982.10
Size 3, 3P, 240V	L2@3.00	Ea	5,020.00	123.00	5,143.00

Combination starter/circuit breaker plug-in unit for bus duct

Size 0, 3P, 240V	L2@1.50	Ea	2,420.00	61.50	2,481.50
Size 1, 3P, 240V	L2@1.75	Ea	2,490.00	71.80	2,561.80
Size 2, 3P, 240V	L2@2.00	Ea	3,560.00	82.10	3,642.10
Size 3, 3P, 240V	L2@3.00	Ea	4,630.00	123.00	4,753.00

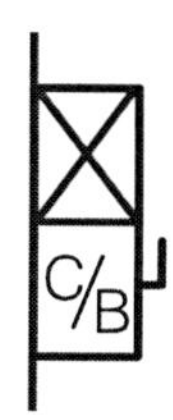

Use these figures to estimate the cost of plug-in switches or breakers installed in bus duct under the conditions described on pages 5 and 6. Costs listed are for each switch or breaker installed. The crew is two electricians working at a labor cost of $41.03 per manhour. These costs include connecting hardware, layout, material handling, and normal waste. Add for supports, service devices, sales tax, delivery, supervision, mobilization, demobilization, cleanup, overhead and profit. Note: Bus duct made as feeder duct only will not accept plug-in breakers.

Section 9: Cable Tray

Cable tray has been installed in industrial and commercial buildings for many years. In fact, the first cable tray was developed when large buildings were first wired for electric power. Since then, there have been many improvements in design, carrying capacities and materials.

Cable tray in its simplest form is just a continuous tray that carries cable. The types of cable tray available include the following:

Ladder tray has two side rails connected by uniformly-spaced rungs.

Trough tray has a ventilated bottom and closely-spaced supports that are carried by side rails.

Channel tray has a one-piece bottom and side rails.

No matter what type of tray is used, cable is laid in the tray bottom. Sides of the tray keep the cable from shifting out of place. Figure 9-1 shows a typical installation.

This section covers the most common sizes and types of cable tray systems. The sizes listed are the sizes used most commonly today. Some manufacturers will fabricate special sizes to meet unusual job requirements. Your suppliers can provide information on custom orders.

Most cable tray is open to permit free air circulation. But some cable tray comes with a cover to enclose the system. A cover helps protect cable run in areas where it could be damaged, and reduces the chance of developing dust problems in a dusty environment.

Angle fittings are made for routing the cable tray over and under obstructions or to provide a lateral offset.

Most cable tray catalogs don't list prices. Pricing is usually done by a factory representative from a material takeoff done by the electrical contractor. Some factory reps will do the takeoff from your plans and then route the price quotation to your supplier.

Hanger spacing is critical with cable tray. Be sure that the hangers are spaced to carry the weight of the tray and the weight of all cables that can be installed. Some catalogs offer guidelines for calculating hanger spacing. Cable tray class is a guide to hanger spacing. The National Electrical Manufacturer's Association has established four classes of cable tray. Class 1 can span 12 feet between hangers, and supports up to 35 pounds of cable per linear foot. Class 2 can also span 12 feet, but supports up to 50 pounds per linear foot. Class 3 spans 20 feet and supports up to 45 pounds. Class 4 spans 20 feet and supports up to 75 pounds.

Be careful when attaching hangers to structural members. What you're attaching to has to be strong enough to support a lot of weight. In some cases you'll need bracing to spread the load over a wider area.

The most common support for cable tray is threaded hanger rod attached to overhead beams with beam clamps or angle brackets. Rod is usually attached to a metal channel that forms a trapeze support for the tray. The size of the rod depends on the gross weight. The smallest rod to use is $^3/_8$" in diameter. It should be installed on each side of the tray. If you use larger diameter rod, a single rod hanger running through the middle of the tray may be enough.

When centered single hangers are used, the tray may not hang level if the cable load isn't distributed evenly. One easy way to remedy this problem is to install a piece of conduit over nearly the full length of rod. Be sure to cut the ends of the conduit square. When the rod nuts are tightened, the conduit becomes a spacer that stiffens the hanger. With a conduit spacer, even a moderately unbalanced load won't affect the tray.

It's common to install cable tray but no cable. Tray is installed when the building is constructed. Cable will be run later when the building is occupied by a tenant who needs power for high amperage loads or when the phone company installs a phone system.

Planning the layout is a very important part of every cable tray job. Do detailed planning before ordering materials. Study the drawings carefully. Ask yourself some questions: What is the elevation of the cable tray? What obstruction will also be at that level? What's the best way to avoid potential obstructions? Can you improve the layout to eliminate any unnecessary turns or offsets? Special fittings are expensive, both to buy from the manufacturer and to install. Discovering that you need a special fitting after installation has begun can delay completion for days or weeks. Try to plan the simplest layout possible and anticipate all of the potential problems.

P-W Cable Tray Systems

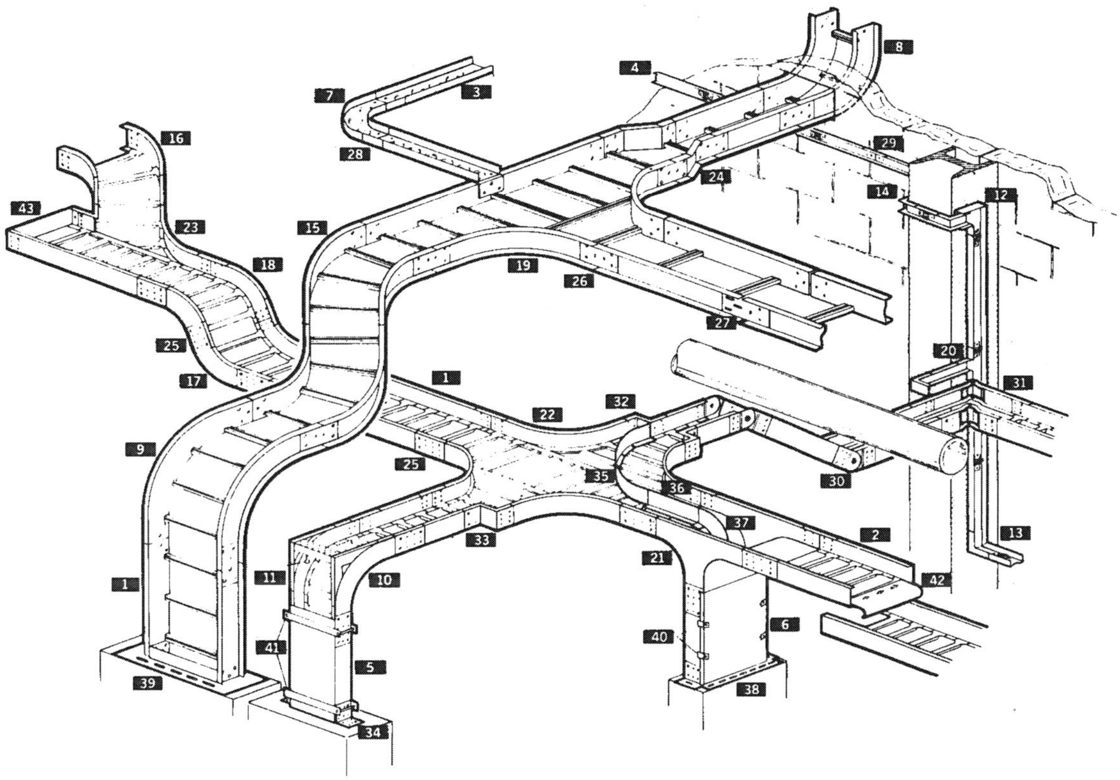

1. Straight Length
2. Straight Length
3. Straight Length
4. Straight Length
5. Flat Covers
6. Flanged Covers
7. Horizontal Fittings
8. Inside 90° Vertical Elbow
9. Outside 90° Verticle Elbow
10. Cable Support Elbow
11. "J" Hooks for hanging Cable
12. 90° Raceway Connectors
13. 90°Inside Elbow Raceway
14. 90° Outside Elbow, Raceway
15. 45° Horizontal Elbow
16. 45°Horizontal Elbow
17. 45° Inside Vertical Elbow
18. 45° Outside Vertical Elbow
19. Horizontal Tee
20. Tee Connector Raceway
21. Veritcal Tee
22. Horizontal Cross
23. 45° Horizontal Wye Branch
24. Reducers
25. Channel Connector
26. Plate Connector
27. Expansion Connector
28. 4" Channel Connector
29. Straight Raceway Connector
30. Adj. Vertical Connector
31. Adj. Horizontal Connector
32. Straight Reducer
33. Offset Reducer
34. Angle Connector
35. Adjustable Horizontal Divider
36. Straight Divider
37. Vertical Elbow Divider
38. Trough-to-Box Connector
39. Trough-to-Box Connector
40. Cover Connector Clip
41. Cover Connector Strap
42. Dropout
43. Blind End

Courtesy: P-W Cable Tray Systems

Figure 9-1
Cable Tray

Most tray material is lightweight and easy to handle. But the 12-foot sections are hard to install in an existing drop ceiling. Allow additional time for this type of work and for any retrofit job that involves cable tray.

Installation in an existing building usually requires working over and around machinery, office furniture, drop ceilings or in confined areas. There may even be a problem with storage space on site. Sometimes only small bundles of material can be stockpiled on site at one time. Allow extra labor if you anticipate conditions like this. With cable tray, even an apparently inconsequential discrepancy can turn normal installation into a nightmare. That turns a good estimate into a major loss.

The labor units listed in this section apply on standard overhead installations at heights up to 12 feet above the floor. Higher installations will take longer.

Estimating Cable Tray

Cable tray take-off is very similar to bus duct take-off. First, study the plans and specifications. Find the type and size of cable tray to be used. Study the area where the system will be installed.

The ideal location is in an open area about 12 feet above the floor without any significant obstructions. That's an easy job, especially if the plans indicate very clearly all bends and tray drops required. Most cable tray jobs aren't that easy. The hardest cable tray job would be in an existing building, where your electricians have to work around furniture or machinery. If you have to run tray in a drop ceiling that's already filled with air conditioning duct, fire sprinkler piping, electrical conduit and recessed lighting fixtures, installation time will be much longer than the tables in this section indicate.

In some cases it may be nearly impossible to install cable tray as specified. On any job over a dropped ceiling, remember that cable tray comes in 12-foot lengths. Most tee-bar ceiling grid has 2 foot by 4 foot openings. It will take more than grease to pass tray through that grid. Some of the ceiling will have to be removed to install the tray. Your electricians shouldn't have to remove and reset the ceiling grid. If removal and replacement aren't covered separately in the specs, get a quote on the gridwork from a ceiling contractor.

The labor cost for installing cable tray in some buildings may be double the figures listed in the tables that follow. Don't be reluctant to increase the labor cost when you anticipate access problems or note a lot of possible obstructions. Remember that a greater installation height will increase your labor cost. Figure that cable tray installed 18 feet above the floor may require 15 percent more labor than the same tray installed 12 feet above the floor.

On many jobs there will be conflicts with fire sprinkler piping, air ducts and light fixtures. This is the most common problem. A slight change in elevation might help. But remember that tray that isn't accessible for installation of cable has little value.

16 gauge galvanized cable tray with louvered openings

Material	Craft@Hrs	Unit	Material Cost	Labor Cost	Installed Cost
12" wide	L2@1.25	Ea	523.00	51.30	574.30
18" wide	L2@1.50	Ea	574.00	61.50	635.50
24" wide	L2@1.75	Ea	638.00	71.80	709.80

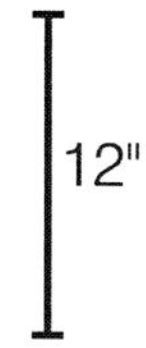

12" radius elbows for louvered cable tray

Material	Craft@Hrs	Unit	Material Cost	Labor Cost	Installed Cost
12" 45 degree, flat	L2@0.50	Ea	319.00	20.50	339.50
18" 45 degree, flat	L2@0.60	Ea	382.00	24.60	406.60
24" 45 degree, flat	L2@0.75	Ea	439.00	30.80	469.80
12" 90 degree, flat	L2@0.60	Ea	408.00	24.60	432.60
18" 90 degree, flat	L2@0.75	Ea	435.00	30.80	465.80
24" 90 degree, flat	L2@1.00	Ea	561.00	41.00	602.00
12" 90 degree, vert.	L2@0.60	Ea	346.00	24.60	370.60
18" 90 degree, vert.	L2@0.75	Ea	417.00	30.80	447.80
24" 90 degree, vert.	L2@1.00	Ea	361.00	41.00	402.00

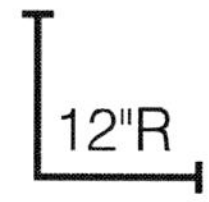

12" radius tees for louvered cable tray

Material	Craft@Hrs	Unit	Material Cost	Labor Cost	Installed Cost
12" flat	L2@0.75	Ea	561.00	30.80	591.80
18" flat	L2@1.00	Ea	793.00	41.00	834.00
24" flat	L2@1.25	Ea	912.00	51.30	963.30

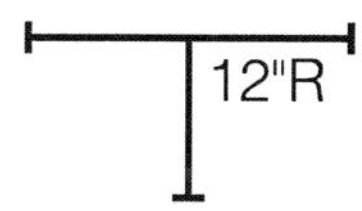

Panel or box adapter for louvered cable tray

Material	Craft@Hrs	Unit	Material Cost	Labor Cost	Installed Cost
12"	L2@0.30	Ea	57.10	12.30	69.40
18"	L2@0.30	Ea	57.10	12.30	69.40
24"	L2@0.30	Ea	57.10	12.30	69.40

End caps for louvered cable tray

Material	Craft@Hrs	Unit	Material Cost	Labor Cost	Installed Cost
12"	L2@0.10	Ea	72.00	4.10	76.10
18"	L2@0.15	Ea	72.00	6.15	78.15
24"	L2@0.20	Ea	72.00	8.21	80.21

Use these figures to estimate the cost of cable tray and fittings installed in buildings under the conditions described on pages 5 and 6. Costs listed are for each tray section or fitting installed. The crew is two electricians working at a labor cost of $41.03 per manhour. These costs include layout, material handling, and normal waste. Add for supports, sales tax, delivery, supervision, mobilization, demobilization, cleanup, overhead and profit. Note: Covers are available for cable tray to keep dust away from the conductors. Order special extra-wide sections or reducing sections when changing from one width to another. Be sure to use the right size supports. Cable tray can be suspended from single support hangers. The hanger attaches to the center of the tray. A stiffener bar goes under the tray and the rod hanger extends up through the tray to the building structure above. If you use single hangers, be sure the hanger is sized for the tray and cable load. Spacing of hangers can't exceed 10' and must be closer for higher carrying capacities. The manufacturer's catalog will list approved spacing and other fittings that may be needed for special applications.

Fittings for 16 Gauge Louvered Cable Tray

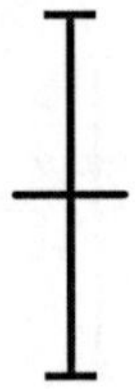

Material	Craft@Hrs	Unit	Material Cost	Labor Cost	Installed Cost
Hangers and joiners for 16 gauge louvered cable tray					
Hangers, pair	L2@0.50	Ea	39.70	20.50	60.20
Joiners, standard, pair	L2@0.30	Ea	39.70	12.30	52.00
Joiners, adj., flat, pair	L2@0.40	Ea	141.00	16.40	157.40
Joiners, adj., vert, pair	L2@0.30	Ea	95.80	12.30	108.10
Dropouts for 16 gauge louvered cable tray					
12"	L2@0.25	Ea	48.30	10.30	58.60
18"	L2@0.30	Ea	51.40	12.30	63.70
24"	L2@0.35	Ea	57.10	14.40	71.50

Use these figures to estimate the cost of cable tray and fittings installed in bus duct under the conditions described on pages 5 and 6. Costs listed are for each tray section or fitting installed. The crew is two electricians working at a labor cost of $41.03 per manhour. These costs include layout, material handling, and normal waste. Add for supports, sales tax, delivery, supervision, mobilization, demobilization, cleanup, overhead and profit. Note: Covers are available for cable tray to keep dust away from the conductors. Order special extra-wide sections or reducing sections when changing form one width to another. Be sure to use the right size supports. Cable tray can be suspended from single support hangers. The hanger attaches to the center of the tray. A stiffener bar goes under the tray and the rod hanger extends up through the tray to the building structure above. If you use single hangers, be sure the hanger is sized for the tray and cable load. Spacing of hangers can't exceed 10' and must be closer for higher carrying capacities. The manufacturer's catalog will list approved spacing and other fittings that may be needed for special applications.

12' long aluminum ladder cable tray

Material	Craft@Hrs	Unit	Material Cost	Labor Cost	Installed Cost
12" wide	L2@1.00	Ea	639.00	41.00	680.00
18" wide	L2@1.25	Ea	699.00	51.30	750.30
24" wide	L2@1.50	Ea	759.00	61.50	820.50

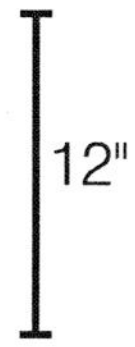

12" radius elbows for aluminum ladder cable tray

Material	Craft@Hrs	Unit	Material Cost	Labor Cost	Installed Cost
12" 45 degree, flat	L2@0.40	Ea	404.00	16.40	420.40
18" 45 degree, flat	L2@0.50	Ea	424.00	20.50	444.50
24" 45 degree, flat	L2@0.60	Ea	450.00	24.60	474.60
12" 90 degree, flat	L2@0.50	Ea	414.00	20.50	434.50
18" 90 degree, flat	L2@0.60	Ea	493.00	24.60	517.60
24" 90 degree, flat	L2@0.75	Ea	504.00	30.80	534.80
12" 90 degree, vert.	L2@0.50	Ea	384.00	20.50	404.50
18" 90 degree, vert.	L2@0.60	Ea	399.00	24.60	423.60
24" 90 degree, vert.	L2@0.75	Ea	406.00	30.80	436.80

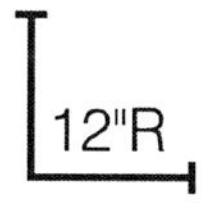

12" radius tees for aluminum ladder cable tray

Material	Craft@Hrs	Unit	Material Cost	Labor Cost	Installed Cost
12"	L2@0.70	Ea	837.00	28.70	865.70
18"	L2@0.90	Ea	896.00	36.90	932.90
24"	L2@1.20	Ea	908.00	49.20	957.20

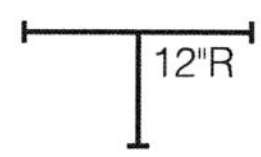

Panel or box adapters for aluminum ladder cable tray

Material	Craft@Hrs	Unit	Material Cost	Labor Cost	Installed Cost
12"	L2@0.30	Ea	53.30	12.30	65.60
18"	L2@0.30	Ea	53.30	12.30	65.60
24"	L2@0.30	Ea	53.30	12.30	65.60

End caps for aluminum ladder cable tray

Material	Craft@Hrs	Unit	Material Cost	Labor Cost	Installed Cost
12"	L2@0.10	Ea	73.00	4.10	77.10
18"	L2@0.15	Ea	73.00	6.15	79.15
24"	L2@0.20	Ea	73.00	8.21	81.21

Use these figures to estimate the cost of cable tray and fittings installed in bus duct under the conditions described on pages 5 and 6. Costs listed are for each tray section or fitting installed. The crew is two electricians working at a labor cost of $41.03 per manhour. These costs include layout, material handling, and normal waste. Add for supports, sales tax, delivery, supervision, mobilization, demobilization, cleanup, overhead and profit. Note: Covers are available for cable tray to keep dust away from the conductors. Order special extra-wide sections or reducing sections when changing form one width to another. Be sure to use the right size supports. Cable tray can be suspended from single support hangers. The hanger attaches to the center of the tray. A stiffener bar goes under the tray and the rod hanger extends up through the tray to the building structure above. If you use single hangers, be sure the hanger is sized for the tray and cable load. Spacing of hangers can't exceed 10' and must be closer for higher carrying capacities. The manufacturer's catalog will list approved spacing and other fittings that may be needed for special applications.

Aluminum Ladder Cable Tray Fittings

Material	Craft@Hrs	Unit	Material Cost	Labor Cost	Installed Cost
Dropouts for aluminum ladder cable tray					
12"	L2@0.25	Ea	48.10	10.30	58.40
18"	L2@0.30	Ea	67.50	12.30	79.80
24"	L2@0.35	Ea	73.00	14.40	87.40
Joiners for aluminum ladder cable tray					
Standard, pairs	L2@0.30	Ea	50.50	12.30	62.80
Adjustable, flat	L2@0.40	Ea	198.00	16.40	214.40
Adjustable, vert.	L2@0.30	Ea	95.60	12.30	107.90
Hangers in pairs for aluminum ladder cable tray					
Hangers, pairs	L2@0.50	Ea	36.80	20.50	57.30
10' wire basket cable tray					
4" X 2" deep	L2@0.05	Ea	113.00	2.05	115.05
4" X 4" deep	L2@0.05	Ea	122.00	2.05	124.05
6" X 2" deep	L2@0.10	Ea	127.00	4.10	131.10
12" X 4" deep	L2@0.10	Ea	175.00	4.10	179.10
12" X 6" deep	L2@0.10	Ea	192.00	4.10	196.10
18" X 4" deep	L2@0.10	Ea	209.00	4.10	213.10

Use these figures to estimate the cost of cable tray and fittings installed in bus duct under the conditions described on pages 5 and 6. Costs listed are for each tray section or fitting installed. The crew is two electricians working at a labor cost of $41.03 per manhour. These costs include layout, material handling, and normal waste. Add for supports, sales tax, delivery, supervision, mobilization, demobilization, cleanup, overhead and profit. Note: Covers are available for cable tray to keep dust away from the conductors. Order special extra-wide sections or reducing sections when changing form one width to another. Be sure to use the right size supports. Cable tray can be suspended from single support hangers. The hanger attaches to the center of the tray. A stiffener bar goes under the tray and the rod hanger extends up through the tray to the building structure above. If you use single hangers, be sure the hanger is sized for the tray and cable load. Spacing of hangers can't exceed 10' and must be closer for higher carrying capacities. The manufacturer's catalog will list approved spacing and other fittings that may be needed for special applications.

Section 10: Signal Systems

Most buildings have some type of signal system. It could be anything from a simple bell circuit to a complex building management and control system. Many of the more sophisticated systems are installed by specialists. If you don't want to tackle the work yourself, get a quote from a company that handles the work required. If the project includes a sophisticated signal system, the designers probably wrote a separate specification section that covers signal work. Detach that portion of the plans and specs and send it out for bid.

Buildings with high-tech management systems are sometimes called "smart" buildings. Building management systems are designed to control all energy consumption: heating, cooling, ventilation, and lighting. Some management systems can detect when a room isn't being used and will shut off the lights and adjust the room temperature to conserve energy. Devices in the ceiling or in the lighting fixtures can sense the amount of lighting needed throughout the work day. When daylight is available, artificial lighting is reduced. Other sensors control the demand for electrical service during peak consumption hours by shedding nonessential loads.

Building management systems usually have security circuits and sensors to detect unauthorized entry. Security devices may include video cameras, door and window alarms, coded entry stations, automatic parking entry control, building emergency evacuation systems, standby electrical generator units, standby battery-powered centers, and uninterruptable power supplies, to name a few.

Bells

Bells are the most common signal device. In fact, door bells have been in common use longer than electric lighting. School jobs nearly always include bell systems to signal the beginning and ending of class periods.

Bells are intended for either indoor or outdoor use and may be either single stroke or the vibrating type. Low-voltage bells are powered by a small transformer that's activated by pressing a push button. Low-voltage bell wiring uses twisted pair conductors. No conduit is needed unless the conductors go through masonry or concrete.

Buzzers

Buzzer wiring is like bell wiring. Because buzzers are usually smaller than bells, the entire buzzer may be mounted in a flush single-gang box. Buzzers are made in a variety of sizes and voltages. Some are adjustable so the sound can be controlled.

Sirens

Sirens are warning devices and emergency alarms. For example, a siren would be required in a noisy engine room if an automatic halon or carbon dioxide fire-extinguishing system is used.

Sirens are made in many sizes and voltages and are made either for indoor or outdoor use.

Horns

Horns are speakers used for amplification, usually of the human voice. For example, a horn might be required as part of a fire-protection system to signal occupants to evacuate a building. Horns are also used in paging and public address systems.

Horns are designed for either indoor or outdoor use. Some have grilles. Others have double projection cones for use in corridors where the sound has to be aimed in two directions.

Beacons

Beacons provide a visual signal. The light can be steady, flashing, or rotating to give the effect of a flashing light. Beacons are required where it's important to draw attention to some hazard or to mark the boundary of a hazard.

Beacons are designed either for indoor or outdoor use. Navigation hazard beacons usually flash at intervals that are set by rules of navigation.

Chimes

Some chimes can produce several bars of music. The most common application is in place of a door bell.

Chimes are designed either for indoor or outdoor use. They're made in several sizes and voltages and may have an adjustable volume level.

Push Buttons

Push buttons operate signal devices. They can be either surface or flush mounted with either a plain or decorative exterior.

Note that push buttons are rated by voltage, from simple residential types to heavy-duty industrial models. Some push buttons have lighted buttons. Even color-coded buttons are available. Some heavy-duty push buttons are rated as oiltight.

Signal Transformers

Every low-voltage signal device needs a small transformer to reduce the line voltage. Signal transformers are usually the dry type and are intended for continuous service.

Signal transformers are made in many sizes and voltages. Some have multiple taps so you can select the voltage that's needed.

Clocks

Commercial wall clocks are available for use with most of the standard voltages. Some are centrally operated. A main program control panel transmits impulses that advance all clocks in the system, keeping all clocks synchronized. All clocks in the system have to be connected to the central control panel.

Where precise control of many clocks is needed, a frequency generator can be used to superimpose a precise frequency on a building's electrical supply. That way every clock in the building is synchronized with the control center's impulse frequency — without any special wiring.

Smoke Detectors

Some fire alarm systems use both smoke detectors and manual pull stations. In an office, detectors mount in the drop ceiling cavity. The code also requires smoke detectors in some larger air conditioning units. Detectors can be wired so they both sound alarms and shut down the equipment that they monitor.

There are ionization detectors, fixed-temperature detectors and rate-of-rise detectors. The local fire marshal will explain which type is needed for your building. You'll probably have to submit plans for approval by the fire-protection authority if smoke or fire detectors are required.

Intrusion Detectors

There's an intrusion detection device made to meet every need. Some are very simple. Others have silent alarms, video filming, general alarms, visual or audible alarms and extensive communication ability.

Some of the more common intrusion detecting devices are motion detectors, magnetic window or door switches, floor mats with pressure switches built in, door releases, door openers, door closers and cameras.

Many larger apartment buildings have entry control systems. Each apartment has a microphone and speaker that permits communication with selected other stations. Occupants can respond to someone at the entry without leaving their living unit and without permitting entry.

Material		Craft@Hrs	Unit	Material Cost	Labor Cost	Installed Cost

Bells

4"	12 VAC	L1@0.25	Ea	72.10	10.30	82.40
4"	16 VAC	L1@0.25	Ea	72.10	10.30	82.40
4"	18 VAC	L1@0.30	Ea	71.10	12.30	83.40
4"	24 VAC	L1@0.30	Ea	71.10	12.30	83.40
4"	120 VAC	L1@0.30	Ea	71.10	12.30	83.40
4"	240 VAC	L1@0.30	Ea	72.10	12.30	84.40
6"	12 VAC	L1@0.30	Ea	86.40	12.30	98.70
6"	16 VAC	L1@0.35	Ea	86.40	14.40	100.80
6"	18 VAC	L1@0.35	Ea	84.70	14.40	99.10
6"	24 VAC	L1@0.35	Ea	84.70	14.40	99.10
6"	120 VAC	L1@0.35	Ea	84.70	14.40	99.10
6"	240 VAC	L1@0.45	Ea	86.40	18.50	104.90
10"	120 VAC	L1@0.45	Ea	100.00	18.50	118.50
10"	240 VAC	L1@0.45	Ea	103.00	18.50	121.50

BV

Single stroke bells

2-1/2"	24 VAC	L1@0.30	Ea	16.60	12.30	28.90
2-1/2"	120 VAC	L1@0.30	Ea	17.60	12.30	29.90
4"	24 VAC	L1@0.35	Ea	80.20	14.40	94.60
4"	120 VAC	L1@0.35	Ea	80.20	14.40	94.60
6"	24 VAC	L1@0.40	Ea	95.30	16.40	111.70
6"	120 VAC	L1@0.40	Ea	95.30	16.40	111.70
10"	24 VAC	L1@0.45	Ea	105.00	18.50	123.50
10"	120 VAC	L1@0.45	Ea	105.00	18.50	123.50

BS

Surface mounted buzzers

8 VAC	71.5 db	L1@0.30	Ea	22.50	12.30	34.80
24 VAC	71.5 db	L1@0.30	Ea	22.50	12.30	34.80
8 VAC	74 db	L1@0.30	Ea	18.60	12.30	30.90
24 VAC	74 db	L1@0.30	Ea	18.60	12.30	30.90
8 VAC	75 db	L1@0.30	Ea	19.60	12.30	31.90
24 VAC	75 db	L1@0.30	Ea	19.60	12.30	31.90
24 VAC	80.5 db	L1@0.30	Ea	20.60	12.30	32.90

BZ

Sirens

12 VAC		L1@0.50	Ea	388.00	20.50	408.50
24 VAC		L1@0.50	Ea	388.00	20.50	408.50
120 VAC		L1@0.50	Ea	388.00	20.50	408.50
240 VAC		L1@0.50	Ea	407.00	20.50	427.50
12 VDC		L1@0.50	Ea	388.00	20.50	408.50
24 VDC		L1@0.50	Ea	388.00	20.50	408.50
125 VDC		L1@0.50	Ea	388.00	20.50	408.50
250 VDC		L1@0.50	Ea	407.00	20.50	427.50

SN

Use these figures to estimate the cost of signal devices installed in buildings under the conditions described on pages 5 and 6. Costs listed are for each device installed. The crew is one electrician working at a labor cost of $41.03 per manhour. These costs include mounting screws, layout, material handling, and normal waste. Add for the outlet box, supports, sales tax, delivery, supervision, mobilization, demobilization, cleanup, overhead and profit.

Sirens and Horns

Material	Craft@Hrs	Unit	Material Cost	Labor Cost	Installed Cost
Horn sirens, explosion proof					
12 VAC or DC	L1@0.50	Ea	1,040.00	20.50	1,060.50
24 VAC or DC	L1@0.50	Ea	1,040.00	20.50	1,060.50
120 VAC or 125 VDC	L1@0.50	Ea	1,040.00	20.50	1,060.50
240 VAC	L1@0.50	Ea	1,040.00	20.50	1,060.50
250 VDC	L1@0.50	Ea	1,040.00	20.50	1,060.50
Horns, AC					
24 VAC, flush	L1@0.35	Ea	73.10	14.40	87.50
120 VAC, flush	L1@0.35	Ea	66.20	14.40	80.60
240 VAC, flush	L1@0.35	Ea	76.00	14.40	90.40
24 VAC, projector	L1@0.40	Ea	77.10	16.40	93.50
120 VAC, projector	L1@0.40	Ea	77.10	16.40	93.50
240 VAC, projector	L1@0.40	Ea	80.20	16.40	96.60
24 VAC, grille	L1@0.40	Ea	65.20	16.40	81.60
120 VAC, grille	L1@0.40	Ea	65.20	16.40	81.60
240 VAC, grille	L1@0.40	Ea	68.20	16.40	84.60
24 VAC, weatherproof	L1@0.40	Ea	87.20	16.40	103.60
120 VAC, weatherproof	L1@0.40	Ea	87.20	16.40	103.60
240 VAC, weatherproof	L1@0.40	Ea	90.40	16.40	106.80
Horns, DC					
24 VDC, flush	L1@0.40	Ea	73.10	16.40	89.50
125 VDC, flush	L1@0.40	Ea	73.10	16.40	89.50
24 VDC, projector	L1@0.40	Ea	24.40	16.40	40.80
125 VDC, projector	L1@0.40	Ea	24.40	16.40	40.80
12 VDC, grille	L1@0.40	Ea	73.10	16.40	89.50
24 VDC, grille	L1@0.40	Ea	73.10	16.40	89.50
32 VDC, grille	L1@0.40	Ea	73.10	16.40	89.50
125 VDC, grille	L1@0.40	Ea	73.10	16.40	89.50
250 VDC, grille	L1@0.40	Ea	73.10	16.40	89.50

Use these figures to estimate the cost of signal devices installed in buildings under the conditions described on pages 5 and 6. Costs listed are for each device installed. The crew is one electrician working at a labor cost of $41.03 per manhour. These costs include mounting screws, layout, material handling, and normal waste. Add for the outlet box, supports, sales tax, delivery, supervision, mobilization, demobilization, cleanup, overhead and profit.

Material	Craft@Hrs	Unit	Material Cost	Labor Cost	Installed Cost
Beacons					
Amber, 12VDC	L1@0.50	Ea	92.90	20.50	113.40
Blue, 12VDC	L1@0.50	Ea	92.90	20.50	113.40
Red, 12VDC	L1@0.50	Ea	92.90	20.50	113.40
Clear, 12VDC	L1@0.50	Ea	92.90	20.50	113.40
Flashing beacons					
Amber, 120VAC	L1@0.50	Ea	160.00	20.50	180.50
Blue, 120VAC	L1@0.50	Ea	160.00	20.50	180.50
Red, 120VAC	L1@0.50	Ea	160.00	20.50	180.50
Clear, 120VAC	L1@0.50	Ea	160.00	20.50	180.50
Rotating beacons					
Amber, 120VAC	L1@0.50	Ea	137.00	20.50	157.50
Blue, 120VAC	L1@0.50	Ea	137.00	20.50	157.50
Red, 120VAC	L1@0.50	Ea	137.00	20.50	157.50
Clear, 120VAC	L1@0.50	Ea	137.00	20.50	157.50
Amber, 12VDC	L1@0.50	Ea	104.00	20.50	124.50
Blue, 12VDC	L1@0.50	Ea	104.00	20.50	124.50
Red, 12VDC	L1@0.50	Ea	104.00	20.50	124.50
Clear, 12VDC	L1@0.50	Ea	104.00	20.50	124.50
Chimes					
One entrance, white	L1@0.40	Ea	15.60	16.40	32.00
One entrance, beige	L1@0.40	Ea	15.60	16.40	32.00
Two entrance, white	L1@0.40	Ea	11.60	16.40	28.00
Two entrance, beige	L1@0.40	Ea	11.60	16.40	28.00
Bar chimes					
White	L1@0.40	Ea	50.60	16.40	67.00
Maple	L1@0.40	Ea	50.60	16.40	67.00
Light oak	L1@0.40	Ea	63.30	16.40	79.70
Maple, with clock	L1@0.40	Ea	202.00	16.40	218.40

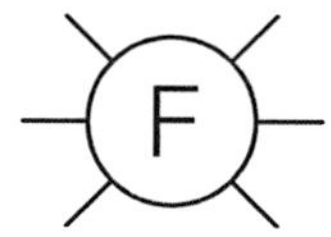

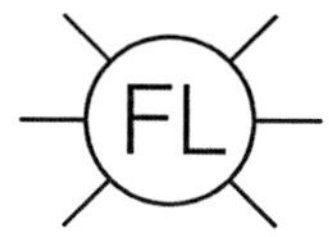

Use these figures to estimate the cost of signal devices installed in buildings under the conditions described on pages 5 and 6. Costs listed are for each device installed. The crew is one electrician working at a labor cost of $41.03 per manhour. These costs include mounting screws, layout, material handling, and normal waste. Add for the outlet box, supports, sales tax, delivery, supervision, mobilization, demobilization, cleanup, overhead and profit.

Signal Systems

Material	Craft@Hrs	Unit	Material Cost	Labor Cost	Installed Cost
Corridor dome lights for signal systems					
24 VAC, 1-lamp	L1@0.40	Ea	34.70	16.40	51.10
120 VAC, 1-lamp	L1@0.40	Ea	37.00	16.40	53.40
24 VAC, 2-lamp	L1@0.40	Ea	44.50	16.40	60.90
120 VAC, 2-lamp	L1@0.40	Ea	44.50	16.40	60.90
Surface mounted push buttons for signal systems					
Round, plain	L1@0.20	Ea	3.24	8.21	11.45
Round, gold	L1@0.20	Ea	3.24	8.21	11.45
Oblong, brown	L1@0.20	Ea	2.73	8.21	10.94
Oblong, ivory	L1@0.20	Ea	2.73	8.21	10.94
Flush mounted push buttons for signal systems					
5/8" chrome	L1@0.25	Ea	2.32	10.30	12.62
5/8" brass	L1@0.25	Ea	2.32	10.30	12.62
5/8" lighted, plain	L1@0.25	Ea	3.96	10.30	14.26
5/8" lighted, brass	L1@0.25	Ea	3.96	10.30	14.26
Signal transformers					
5W 120/10V	L1@0.30	Ea	11.10	12.30	23.40
5W 240/10V	L1@0.30	Ea	14.50	12.30	26.80
10W 120/16V	L1@0.30	Ea	11.10	12.30	23.40
15W 120/6-12-18V	L1@0.30	Ea	13.10	12.30	25.40
20W 120/8-16-24V	L1@0.35	Ea	14.50	14.40	28.90
20W 240/8-16-24V	L1@0.35	Ea	16.60	14.40	31.00
30W 120/8-16-24V	L1@0.35	Ea	16.60	14.40	31.00
30W 240/8-16-24V	L1@0.35	Ea	18.80	14.40	33.20
40W 120/24V	L1@0.35	Ea	16.60	14.40	31.00
Adapter plate	L1@0.10	Ea	8.83	4.10	12.93
Commercial grade clocks					
8" dia. silver	L1@0.50	Ea	40.90	20.50	61.40
8" dia. brown	L1@0.50	Ea	40.90	20.50	61.40
12" dia. silver	L1@0.50	Ea	43.00	20.50	63.50
12" dia. brown	L1@0.50	Ea	43.00	20.50	63.50
15" dia. silver	L1@0.50	Ea	86.40	20.50	106.90
15" dia. brown	L1@0.50	Ea	86.40	20.50	106.90

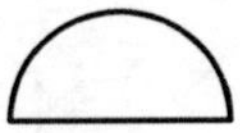

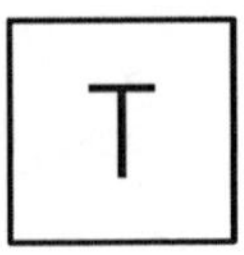

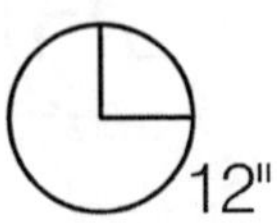

Use these figures to estimate the cost of signal devices installed in buildings under the conditions described on pages 5 and 6. Costs listed are for each device installed. The crew is one electrician working at a labor cost of $41.03 per manhour. These costs include mounting screws, layout, material handling, and normal waste. Add for the outlet box, supports, sales tax, delivery, supervision, mobilization, demobilization, cleanup, overhead and profit.

Material	Craft@Hrs	Unit	Material Cost	Labor Cost	Installed Cost
Battery type smoke detectors					
No LED indicator	L1@0.30	Ea	9.45	12.30	21.75
Blinking red LED	L1@0.30	Ea	10.40	12.30	22.70
120 volt smoke detectors					
Continuous red LED	L1@0.50	Ea	11.20	20.50	31.70
Multi or single station, 120 volt, smoke detectors					
Up to 6 units	L1@0.50	Ea	11.60	20.50	32.10
Up to 12 units	L1@0.50	Ea	12.70	20.50	33.20
Temperature detectors					
Rise heat, 135 degree	L1@0.50	Ea	18.00	20.50	38.50
Rise heat, 200 degree	L1@0.50	Ea	18.00	20.50	38.50
Fixed temp, 135 degree	L1@0.50	Ea	18.00	20.50	38.50
Fixed temp, 190 degree	L1@0.50	Ea	18.00	20.50	38.50
Intrusion detectors					
Magnetic detector	L1@0.30	Ea	4.75	12.30	17.05
Floor mat, 17" x 23"	L1@0.25	Ea	38.80	10.30	49.10
Roller/plunger det	L1@0.25	Ea	8.39	10.30	18.69
Door switch	L1@0.25	Ea	9.45	10.30	19.75
Door opener, 3-6VDC	L1@0.50	Ea	35.60	20.50	56.10
Door opener, 16VAC	L1@0.50	Ea	35.60	20.50	56.10
Door opener, 24VAC	L1@0.50	Ea	44.30	20.50	64.80
Door opener, 24VDC	L1@0.50	Ea	47.40	20.50	67.90
Door trip	L1@0.25	Ea	11.90	10.30	22.20
Desk push button	L1@0.25	Ea	20.10	10.30	30.40
Door switch, closed cir.	L1@0.40	Ea	9.69	16.40	26.09

Use these figures to estimate the cost of signal devices installed in buildings under the conditions described on pages 5 and 6. Costs listed are for each device installed. The crew is one electrician working at a labor cost of $41.03 per manhour. These costs include mounting screws, layout, material handling, and normal waste. Add for the outlet box, supports, sales tax, delivery, supervision, mobilization, demobilization, cleanup, overhead and profit.

Apartment Entry Control

Material	Craft@Hrs	Unit	Material Cost	Labor Cost	Installed Cost
Apartment entry control with speakers and flush mounted push buttons					
4 buttons, brass	L1@0.75	Ea	104.00	30.80	134.80
4 buttons, silver	L1@0.75	Ea	104.00	30.80	134.80
* 6 buttons, brass	L1@1.00	Ea	114.00	41.00	155.00
6 buttons, silver	L1@1.00	Ea	114.00	41.00	155.00
8 buttons, brass	L1@1.25	Ea	126.00	51.30	177.30
8 buttons, silver	L1@1.25	Ea	126.00	51.30	177.30
12 buttons, brass	L1@1.50	Ea	147.00	61.50	208.50
12 buttons, silver	L1@1.50	Ea	147.00	61.50	208.50
16 buttons, brass	L1@1.75	Ea	165.00	71.80	236.80
16 buttons, silver	L1@1.75	Ea	165.00	71.80	236.80
* 20 buttons, brass	L1@2.00	Ea	188.00	82.10	270.10
20 buttons, silver	L1@2.00	Ea	188.00	82.10	270.10
Apartment speakers	L1@0.30	Ea	71.80	12.30	84.10
Entry release switch	L1@0.20	Ea	69.10	8.21	77.31
Timed release switch	L1@0.20	Ea	58.30	8.21	66.51

Use these figures to estimate the cost of signal devices installed in buildings under the conditions described on pages 5 and 6. Costs listed are for each device installed. The crew is one electrician working at a labor cost of $41.03 per manhour. These costs include mounting screws, layout, material handling, and normal waste. Add for outlet box, supports, sales tax, delivery, supervision, mobilization, demobilization, cleanup, overhead and profit.

Section 11: Precast Concrete Access Boxes

This section covers precast concrete utility boxes, handholes, pull boxes and manholes. These boxes provide access to underground electrical cable during construction and after installation is complete. Of course, it's possible to form and pour your own access boxes. But it's nearly always cheaper to buy a precast box if a box of the right size and shape is available.

Note that the boxes listed in this section are typical sizes. Your local electric company probably dictates the exact size required. Be sure the box you price will meet job specs.

Access boxes usually (but not always) come with the cover that's required. Covers may be precast concrete, flat metal or cast iron. The location of the precast box dictates the type of cover to use. A concrete cover is adequate if it's placed away from vehicle traffic. Cast iron and flat steel covers are used in roadways and parking areas. Traffic covers are rated by the anticipated wheel load. Watch for these requirements in the job specs.

The steel covers are usually galvanized and have a checker-plate surface. Larger sizes intended for traffic areas have structural reinforcing under the cover.

Handholes

Handholes comes in several standard sizes. The size to use depends on the wire diameter and intended service. Most handholes are composed of a base section, a body section and a cover. Some have open bottoms. Others have a cast concrete bottom. Handholes have knockouts that are removed where conduit is to enter the box.

The cover will have a word that identifies the type of service. Standard markings are ***Electric, Telephone, CATV, Street Lighting, Cathodic Protection*** and ***Signal***. Many suppliers can provide other wording on request.

Precast Pull Boxes and Manholes

Precast pull boxes and manholes usually come in sections. The larger sizes have a bottom section, midsection and top section. Large tunnel vaults are usually made in vertical sections that can be joined to extend the tunnel to any length.

Some specs require mastic in the joints between sections. This mastic comes in long, thin lengths that are laid in the tongue and groove of each joint. Mastic keeps moisture out of the vault.

Some specs also require an exterior waterseal. Most suppliers can provide precast sections with the waterseal already on the outer surface.

Most boxes, manholes and vaults have knockouts in the ends and sides. Placement of knockouts and the number of knockouts depends on the size of the box. Remove knockouts with a hammer.

If a duct system is installed by direct burial, attach an end bell to the first piece of duct. Insert the bell into the knockout opening and grout it into place with masonry grout.

The procedure for concrete-encased duct is different. Again, attach an end bell to the first length of duct and insert the bell in a knockout. Use a short length of wood placed inside the box as a brace to hold the bell in place. When encasing the duct, be sure to ***rod*** the concrete around the end bell. That makes the concrete flow completely around the end bell and creates a good seal. Sometimes a little grout is needed to completely seal the end bell in place.

Larger pull boxes, manholes and tunnels have steel ladders installed from the cover to the bottom. The ladder must be installed in inserts when the box is placed. Inserts are cast into the box by the manufacturer.

Some boxes, manholes and tunnels require cable racks for holding cable that's to be installed later on. If cable racks are needed, be sure to order boxes with inserts or flush channels for cable racks. The specs should indicate spacing of racks in the box.

Most precast boxes and vaults have sumps in the bottom. The sump may be a simple block-out for a sump pump suction hose. Some specifications require that a drain line be installed from the sump so any accumulation of water is drained away. Other specifications require a sump pump in the box.

Excavation for the box usually has to be several feet larger than the box itself. That allows room to compact backfill around the box. The amount of overexcavation depends on the compaction equipment to be used. Minimum over excavation is usually 2 feet on each side.

The depth of the excavation is determined by the overall height of the box, plus the neck and cover. The finish ring should be at least ½ inch above the finish grade or surface. Place sand or other fine soil at the bottom of the pit so it's easy to make small final adjustments in the box position.

Most underground installers set the precast box first and then place the duct. Usually the box is completely installed and backfilled before duct trenching even begins. Duct is then run from point to point or box to box.

Precast Pad Mount Transformer Slabs

Pad mount transformers are installed on concrete slabs, either poured and finished or precast. Precast slabs are often cheaper because there is no forming, pouring or finishing required. And precast slabs come with a block-out where the transformer will be placed. That makes it easy to stub conduit up into the transformer entrance section. Pads come with inserts already installed for anchoring the transformer in place.

Many sizes of slabs are available. Some come with underground pull boxes for cable entrance to the transformer. The size of the pull box has to conform with utility company requirements.

Note that pad mounted transformers must be grounded. Grounding has to be installed before the slab is set. Your power company will describe grounding requirements for the transformer you're placing.

Delivery Costs

Delivery cost is usually a major part of the installed cost of precast boxes, handholes and slabs. Don't settle for a price quote F.O.B. the manufacturer's yard. Get a price that includes delivery to the site.

Most precast concrete companies charge extra for setting a box or pad in place. They'll quote setting prices, usually based upon a flat rate per item. Some will only quote hourly costs.

Some vaults are too large to be set with the precast company's equipment. In that case, you'll need hoisting equipment. Some large manholes require a large motor crane to move sections from the delivery truck to the pit or trench.

Minimize setting costs with careful planning and scheduling. Be sure the pit is ready before the delivery truck arrives. Have someone meet the delivery truck and direct it to the point of installation. The area around the pit should be clear so unloading isn't obstructed.

Pricing Precast Concrete Products

Get at least three quotes for major items like large boxes and vaults. Call for quotes early enough so the yard can submit a written quotation. Some precast companies will include all of the hardware, cable racks, hooks and insulators in their quote.

Most manufacturers of precast boxes and vaults offer standard products that meet the requirements of the local power company. But box and vault sizes can vary slightly from one manufacturer to another. Protect yourself. Get written quotes for each job.

Material	Craft@Hrs	Unit	Material Cost	Labor Cost	Installed Cost
Precast concrete handholes for underground systems					
14" x 27"	L2@0.50	Ea	225.00	20.50	245.50
18" x 30"	L2@0.70	Ea	294.00	28.70	322.70

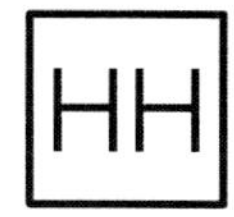

Material	Craft@Hrs	Unit	Material Cost	Labor Cost	Installed Cost
Precast concrete pull boxes					
24" x 36" x 24"	L2@1.00	Ea	1,300.00	41.00	1,341.00
24" x 36" x 36"	L2@1.25	Ea	1,680.00	51.30	1,731.30
30" x 48" x 36"	L2@1.50	Ea	2,030.00	61.50	2,091.50
36" x 72" x 24"	L2@2.25	Ea	3,720.00	92.30	3,812.30
36" x 60" x 36"	L2@2.50	Ea	4,090.00	103.00	4,193.00
48" x 48" x 36"	L2@2.50	Ea	4,480.00	103.00	4,583.00
48" x 60" x 48"	L2@2.75	Ea	5,600.00	113.00	5,713.00

Material	Craft@Hrs	Unit	Material Cost	Labor Cost	Installed Cost
Precast concrete manholes for underground systems					
48" x 90" x 84"	L2@3.50	Ea	4,480.00	144.00	4,624.00
48" x 48" x 72"	L2@3.00	Ea	4,780.00	123.00	4,903.00
48" x 48" x 84"	L2@3.00	Ea	5,150.00	123.00	5,273.00
48" x 78" x 86"	L2@3.50	Ea	4,780.00	144.00	4,924.00
54" x 102" x 78"	L2@4.00	Ea	5,940.00	164.00	6,104.00
72" x 72" x 72"	L2@4.00	Ea	6,300.00	164.00	6,464.00
72" x 72" x 60"	L2@3.75	Ea	5,600.00	154.00	5,754.00
72" x 120" x 84"	L2@4.50	Ea	8,470.00	185.00	8,655.00
96" x 120" x 72"	L2@5.00	Ea	8,180.00	205.00	8,385.00
60" x 120" x 72"	L2@4.50	Ea	7,420.00	185.00	7,605.00
60" x 120" x 86"	L2@4.50	Ea	7,780.00	185.00	7,965.00
72" x 180" x 108"	L2@6.00	Ea	11,100.00	246.00	11,346.00
72" x 180" x 144"	L2@6.50	Ea	11,100.00	267.00	11,367.00
96" x 168" x 100"	L2@8.00	Ea	23,000.00	328.00	23,328.00
72" x 228" x 108"	L2@9.00	Ea	18,300.00	369.00	18,669.00
96" x 168" x 112"	L2@10.0	Ea	14,800.00	410.00	15,210.00
96" x 240" x 112"	L2@12.0	Ea	20,300.00	492.00	20,792.00
96" x 312" x 112"	L2@14.0	Ea	25,900.00	574.00	26,474.00
84" x 264" x 96"	L2@12.0	Ea	18,300.00	492.00	18,792.00
120" x 180" x 120"	L2@16.0	Ea	37,200.00	656.00	37,856.00

Use these figures to estimate the cost of precast concrete products installed under the conditions described on pages 5 and 6. Costs listed are for each item installed. The crew is two electricians working at a labor cost of $41.03 per manhour. These costs include unloading and setting, layout, material handling, and normal waste. Add for excavation, backfill, hoisting equipment (when needed), surface patching, mastic water sealing around the item set, interior hardware, grounding, sales tax, delivery, supervision, mobilization, demobilization, cleanup, overhead and profit.

Precast Concrete Products

Material	Craft@Hrs	Unit	Material Cost	Labor Cost	Installed Cost
Precast concrete manhole necking for underground systems					
Concrete cone 27" dia	L2@0.30	Ea	183.00	12.30	195.30
Concrete grade ring	L2@0.30	Ea	153.00	12.30	165.30
Iron ring & cover	L2@1.00	Ea	929.00	41.00	970.00
48" square cover set	L2@1.00	Ea	2,940.00	41.00	2,981.00
Precast concrete pad mount transformer slabs for underground systems					
31" x 46"	L2@0.50	Ea	244.00	20.50	264.50
32" x 38"	L2@0.50	Ea	183.00	20.50	203.50
42" x 49"	L2@0.70	Ea	183.00	28.70	211.70
44" x 46"	L2@0.70	Ea	281.00	28.70	309.70
48" x 54"	L2@1.00	Ea	372.00	41.00	413.00
57" x 76"	L2@1.25	Ea	656.00	51.30	707.30
72" x 66"	L2@1.50	Ea	929.00	61.50	990.50
73" x 75"	L2@1.50	Ea	929.00	61.50	990.50
94" x 72"	L2@1.50	Ea	1,010.00	61.50	1,071.50

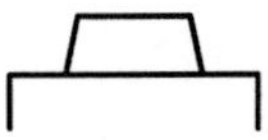

Use these figures to estimate the cost of precast concrete products installed under the conditions described on pages 5 and 6. Costs listed are for each item installed. The crew is two electricians working at a labor cost of $41.03 per manhour. These costs include unloading and setting, layout, material handling, and normal waste. Add for excavation, backfill, hoisting equipment (when needed), surface patching, mastic water sealing around the item set, interior hardware, grounding, sales tax, delivery, supervision, mobilization, demobilization, cleanup, overhead and profit.

Section 12: Equipment Hookup

This section covers the simple equipment hookups that are common on small commercial jobs. If you're handling more complex work or the hookup of heavy industrial equipment, there won't be much information here to help you. But what's found in this section covers nearly all the hookups most electrical contractors have to handle.

Equipment hookup isn't as easy as it may seem to the novice. My advice is to read every word in the specs and examine the plans very carefully any time hookup is required. Even small details can make a big difference in the labor and material cost. Study the type of equipment, connection details and the power requirements. Be sure to include a labor and material cost for each accessory. Make an allowance for difficult conditions when you anticipate problems.

Motors

The labor costs in this section for electric motor hookup assume that the motor is easily accessible, ready for wire connection and that there are no complications. Labor and material costs will be higher for any other type of work. Prices for hookup materials are assumed to be standard commodity grade. The motors are assumed to be standard frame, open or drip proof, and either single or three phase up to 480 volts.

Before connecting any electric motor, check for wiring instructions on or near the dataplate. Sometimes you'll find wiring instructions in the motor terminal box or pasted to the inside of the terminal box cover.

To make the connection, first determine what direction the motor must turn. If you're in doubt, the equipment supplier should be able to supply the answer. Look for rotation arrows around the outside of the motor case. Some types of motors are damaged if the motor turns the wrong direction. Most pumps are damaged if the rotation is wrong. The impeller inside the pump may even unscrew from the shaft.

To reverse the rotation of a standard single phase motor, open the terminal box and disconnect the line wiring. Look for two internal wires attached to the terminals where the line wires make contact. Reverse those two internal lead wires and reconnect the line wiring. That should reverse the direction of rotation.

If a single phase motor has four wires connected to the terminal posts, two on each side, reverse the top two internal lead wires and reconnect the line wiring. Single phase motors have two separate windings inside the stator. One starts the motor; the other is the power source after the motor is started. Both are energized when the motor is first turned on.

On larger single phase motors, there's a centrifugal switch built onto the rotor that energizes the starting coil until the rotation reaches a higher R.P.M. When the motor is at rest, the centrifugal switch is at rest and presses against a stationary contact device. The starting winding is connected to this stationary switch device. When the motor starts and gains speed, the centrifugal switch opens the starting circuit. The motor then runs completely on the main coil. To reverse rotation, swap the two starting leads. That starts rotation in the opposite direction. The running leads don't need to be changed.

A three phase motor is even simpler. Three line wires are connected to the motor. To reverse rotation, swap any two of the three line wires at either the motor, the disconnect switch or at the motor starter. The line wires are usually identified as phase A, B or C. Simply change A and B or B and C or A and C. Any swap will change the direction of rotation. There's no starting device in a three phase motor and there is no starting winding in the stator.

Some motors have thermal overload switches or built-in stator heaters that have to be connected also. Heaters are connected to a separate circuit that's turned on before the motor is started. Usually these connections are made at the motor terminal box.

All motors need overload protection to prevent motor damage. The overload protection is usually built in on smaller motors. Motor starters provide this protection on larger motors. The largest electric motors operate from motor control centers. That's the subject of the next section.

Manual motor starters are built so that overload heating elements can be inserted. When the current flow to the motor exceeds the design maximum, the heater element opens a bi-metallic switch to disconnect the line wiring.

Magnetic starters can also have overload heater elements. Many specifications require that all three phases be protected from overload. In that case all three line wires have to be connected to an overload device. When any one of the three overload heaters opens, the starter circuit opens and stops the motor.

Overload heater elements are sold in sizes that cover a small range of amperages. Each manufacturer offers heaters of a distinct style and shape. Generally heaters of different manufacturers are not interchangeable. To size the overload heater element, check the motor nameplate for the full load amperage rating (FLA). Then check the chart inside the starter for the corresponding heater catalog number for that FLA.

Don't install the wrong heater. If you use a heater with an FLA rating that's too low, the normal running load will activate the heater switch, dropping the starter coil out of the circuit and stopping the motor. If you install a heater with an FLA rating that's too high, an overload will probably damage the motor eventually.

There are many types of electric motors: two speed, variable speed, wound rotor, synchronous, and a variety of D.C. and special wound motors. Be sure to check the wiring diagrams before making any line hookup.

Mechanical Equipment Hookup

The mechanical equipment covered in this section is the type of equipment that you'll find in small and moderate size commercial buildings. The figures assume no special conditions or complicated control schemes. High capacity mechanical equipment will cost considerably more.

Equipment hookup can be very complex. Some mechanical equipment is controlled with electronic programmers or microprocessors. Other equipment has multiple sections that have to be interconnected either electrically or mechanically. Study the control diagrams for the equipment. Have a clear picture of what work is covered in your bid.

Most mechanical equipment controls are supplied by the mechanical contractor and installed or wired by the electrical contractor. You'll see these controls on the installation drawings and note them in the specs. Solenoid valves, stats, pressure switches and control switches are usually easy to hook up. Assume that hookup of these controls is your work until you are advised otherwise.

Kitchen Equipment

Commercial food handling equipment comes in many sizes. Smaller units are easy to connect to the power source. Larger pieces of equipment take more time, material, skill and planning. Consultants who specialize in food handling facilities usually do the complete kitchen layout. The electrical system will be specially designed to meet the needs of the equipment.

Only the more common commercial grade kitchen equipment is listed in this section. Note that costs will be higher if the plan calls for a special control system that interconnects several pieces of equipment.

Standby Engine-Generator Units

Standby engine-generators provide backup power in case of a general power failure. Larger generators are usually diesel powered. The fuel used can have an effect on the type of hookup you make. Special features and accessories on the generator can increase hookup costs considerably.

Emergency engine-generators have precise start-up controls that bring the system on line in the shortest time possible. Some engine-generators have block heaters to keep the engine oil heated — and sometimes circulating. They also have expensive transfer switches that cut in emergency power when that power is needed and available. Expect connecting the power transfer system to be fairly complex.

Engine-generators need special long-life and high capacity batteries for startup. The voltage needed determines the number of cells required. Some specifications require dual batteries. And, of course, each battery must have a charger — perhaps one of the more expensive solid state chargers.

Look for several auxiliary items in any emergency electrical system: fuel transfer pumps, cooling system heat exchanging units, oil purifiers and monitoring devices. Each will add to the cost of electrical hookup.

It's a good idea to discuss installation with the engine-generator supplier before bidding the job. Go over system requirements in detail. Make your contact as early as possible during the bidding period so your supplier can get answers from the manufacturer if necessary.

Other Electrical Hookup Items

Most commercial jobs include several miscellaneous hookup items. Some of this equipment may have been designed and built specially for the job at hand. If you don't know what type of hookup is required, get help. Experienced electrical estimators have learned not to be bashful about calling the design engineer. If he can't supply an answer, get a suggestion on where more information can be found. Usually the design engineer is both well-informed and anxious to be helpful.

Material		Craft@Hrs	Unit	Material Cost	Labor Cost	Installed Cost
Hook up single or three-phase motors to 600 volts						
Fractional	HP	L1@0.50	Ea	4.63	20.50	25.13
1	HP	L1@0.75	Ea	6.01	30.80	36.81
1-1/2	HP	L1@0.75	Ea	6.01	30.80	36.81
2	HP	L1@0.75	Ea	6.01	30.80	36.81
3	HP	L1@0.75	Ea	6.01	30.80	36.81
5	HP	L1@1.00	Ea	8.76	41.00	49.76
7-1/2	HP	L1@1.00	Ea	8.76	41.00	49.76
10	HP	L1@1.25	Ea	17.20	51.30	68.50
15	HP	L1@1.25	Ea	17.20	51.30	68.50
20	HP	L1@1.50	Ea	29.50	61.50	91.00
25	HP	L1@1.50	Ea	29.50	61.50	91.00
30	HP	L1@2.00	Ea	58.50	82.10	140.60
50	HP	L1@3.00	Ea	87.80	123.00	210.80
75	HP	L1@4.00	Ea	145.00	164.00	309.00
100	HP	L1@5.00	Ea	207.00	205.00	412.00

Material	Craft@Hrs	Unit	Material Cost	Labor Cost	Installed Cost
Hook up mechanical equipment					
Air compressor	L1@1.50	Ea	43.60	61.50	105.10
Air alternator	L1@1.25	Ea	29.50	51.30	80.80
Air handlers	L1@1.50	Ea	43.60	61.50	105.10
Aquastats	L1@0.50	Ea	145.00	20.50	165.50
Boilers	L1@8.00	Ea	295.00	328.00	623.00
Boiler control panels	L1@8.00	Ea	145.00	328.00	473.00
Bridge cranes	L1@16.0	Ea	368.00	656.00	1,024.00
Budget hoists	L1@2.00	Ea	117.00	82.10	199.10
Chiller, water	L1@24.0	Ea	878.00	985.00	1,863.00
Chiller control panels	L1@16.0	Ea	145.00	656.00	801.00
Conveyors, single dr	L1@16.0	Ea	368.00	656.00	1,024.00
Conveyor control panels	L1@16.0	Ea	145.00	656.00	801.00
Cooling tower, single	L1@10.0	Ea	368.00	410.00	778.00
Cooling tower, twin	L1@16.0	Ea	511.00	656.00	1,167.00
Duct heaters, 1-ph	L1@2.00	Ea	14.50	82.10	96.60
Duct heaters, 3-ph	L1@2.25	Ea	29.50	92.30	121.80
Elevators, small	L1@24.0	Ea	732.00	985.00	1,717.00
Evap coolers, small	L1@2.00	Ea	221.00	82.10	303.10
Exhaust fans, small	L1@1.50	Ea	43.60	61.50	105.10
Fan coil units	L1@2.00	Ea	43.60	82.10	125.70
Flow switches	L1@1.00	Ea	14.50	41.00	55.50
Furnaces, small	L1@2.50	Ea	74.10	103.00	177.10
Humidistat	L1@1.00	Ea	14.50	41.00	55.50

Use these figures to estimate the cost of equipment hookup installed in buildings under the conditions described on pages 5 and 6. Costs listed are for each piece of equipment connected. The crew is one electrician working at a labor cost of $41.03 per manhour. These costs include flexible connections, connectors, wire, phase and rotation identification, testing, layout, material handling, and normal waste. Add for junction boxes, safety switches, control devices, electric motors, disconnect switches, grounding, sales tax, delivery, supervision, mobilization, demobilization, cleanup, overhead and profit. Note: Some equipment items require sequence planning.

Equipment Hookup

Material	Craft@Hrs	Unit	Material Cost	Labor Cost	Installed Cost
Hook up mechanical equipment					
Mech control panel, small	L1@4.00	Ea	87.80	164.00	251.80
Modulating valves	L1@1.50	Ea	14.50	61.50	76.00
Monorail trolley, small	L1@8.00	Ea	145.00	328.00	473.00
Motorized valves	L1@1.50	Ea	14.50	61.50	76.00
Overhead door, res.	L1@2.50	Ea	87.80	103.00	190.80
Overhead door, comm.	L1@8.00	Ea	436.00	328.00	764.00
Pneumatic switches	L1@1.00	Ea	14.50	41.00	55.50
Pressure switches	L1@1.00	Ea	14.50	41.00	55.50
Pump control panels	L1@2.50	Ea	74.10	103.00	177.10
Refrigeration, small	L1@4.00	Ea	145.00	164.00	309.00
Sail switches	L1@1.25	Ea	14.50	51.30	65.80
Solenoid valves	L1@1.50	Ea	14.50	61.50	76.00
Stats, outside air	L1@1.25	Ea	14.50	51.30	65.80
Sump pumps, small	L1@2.00	Ea	29.40	82.10	111.50
Sump control panels	L1@2.50	Ea	29.40	103.00	132.40
Sump alternators	L1@1.25	Ea	14.50	51.30	65.80
Unit heaters, small	L1@2.00	Ea	76.40	82.10	158.50
Valves, 3-way	L1@1.50	Ea	14.50	61.50	76.00
Hook up kitchen equipment					
Booster water heater	L1@2.50	Ea	101.00	103.00	204.00
Dishwasher, small	L1@5.00	Ea	207.00	205.00	412.00
Elec hot tables	L1@2.00	Ea	43.60	82.10	125.70
Elec hot water tables	L1@1.50	Ea	43.60	61.50	105.10
Elec steam tables	L1@2.00	Ea	43.60	82.10	125.70
Garbage disposals	L1@1.25	Ea	29.50	51.30	80.80
Grill	L1@1.50	Ea	29.50	61.50	91.00
Grill hood fan	L1@2.00	Ea	29.50	82.10	111.60
Portable food warmer	L1@1.00	Ea	43.60	41.00	84.60
Ovens	L1@1.50	Ea	29.50	61.50	91.00
Stationary coffee urn	L1@1.25	Ea	14.50	51.30	65.80
Stationary food mixer	L1@1.25	Ea	14.50	51.30	65.80
Stationary refrig.	L1@1.50	Ea	14.50	61.50	76.00
Walk-in freezer	L1@8.00	Ea	221.00	328.00	549.00
Walk-in refigerator	L1@8.00	Ea	221.00	328.00	549.00

Use these figures to estimate the cost of equipment hookup installed in buildings under the conditions described on pages 5 and 6. Costs listed are for each piece of equipment connected. The crew is one electrician working at a labor cost of $41.03 per manhour. These costs include flexible connections, connectors, wire, phase and rotation identification, testing, layout, material handling, and normal waste. Add for junction boxes, safety switches, control devices, electric motors, disconnect switches, grounding, sales tax, delivery, supervision, mobilization, demobilization, cleanup, overhead and profit. Note: Some equipment items require sequence planning.

Hook up Standby Engine-Generators

Material	Craft@Hrs	Unit	Material Cost	Labor Cost	Installed Cost
Hook up standby engine-generators to 600 volts, no accessories included					
10KW, single phase	L1@1.50	Ea	106.00	61.50	167.50
15KW, single phase	L1@2.00	Ea	154.00	82.10	236.10
25KW, single phase	L1@4.00	Ea	306.00	164.00	470.00
50KW, single phase	L1@6.00	Ea	460.00	246.00	706.00
75KW, single phase	L1@8.00	Ea	533.00	328.00	861.00
100KW, single phase	L1@14.0	Ea	765.00	574.00	1,339.00
10KW, three phase	L1@2.00	Ea	154.00	82.10	236.10
15KW, three phase	L1@2.50	Ea	233.00	103.00	336.00
25KW, three phase	L1@5.00	Ea	460.00	205.00	665.00
50KW, three phase	L1@7.00	Ea	611.00	287.00	898.00
75KW, three phase	L1@10.0	Ea	765.00	410.00	1,175.00
100KW, three phase	L1@16.0	Ea	1,070.00	656.00	1,726.00

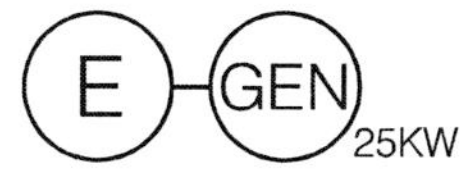

Use these figures to estimate the cost of equipment hookup installed in buildings under the conditions described on pages 5 and 6. Costs listed are for each piece of equipment connected. The crew is one electrician working at a labor cost of $41.03 per manhour. These costs include flexible connections, connectors, wire, phase and rotation identification, testing, layout, material handling, and normal waste. Add for junction boxes, safety switches, control devices, electric motors, disconnect switches, grounding, sales tax, delivery, supervision, mobilization, demobilization, cleanup, overhead and profit. Note: Some equipment items require sequence planning.

Section 13: Motor Control Equipment

Many commercial jobs include some kind of motor control equipment. It may be very simple or highly complex, with interlocking devices such as relays or contact blocks attached to magnetic starters for switching control circuits.

This section covers only some of the motor control equipment that's available. But each item listed here is representative of a larger group of motor control devices that your supplier sells.

The electrical designer will provide a schematic wiring diagram of the motor control system. It should show the control devices needed and how they're connected. It takes some experience to understand the sequence of control the designer intends. Sometimes it's hard enough just to follow a control circuit from its point of origin to its termination.

When doing a take-off of motor control equipment, start by checking the mechanical plans for mechanical equipment controls. Usually the designer will indicate which devices are furnished and installed by the electrical contractor and which are to be furnished by the mechanical contractor and installed by the electrical contractor. Some specs require that the mechanical contractor be responsible for all control systems that operate at less than 100 volts.

Some motor control systems aren't electric, although electric components may be included. For example, pneumatic controls are available for heating, ventilating and air conditioning systems. Pneumatic tubing controls thermostats, valves, pressure switches and other devices. No wiring is required. Air pressure is regulated by a pneumatic control panel.

Manual Motor Starters

Manual motor starters prevent motor overload damage. A single pole starter has a toggle switch handle that controls the ***on*** and ***off*** position. Included in the starter will be a bi-metallic switch with an overload heater element sized for the full motor load. If the current feeding the motor exceeds the value of the element, the element heats up and opens the switch. The toggle handle goes into a trip position. It has to be reset to ***off*** before it can be turned on again. Most overloads happen when the motor starts under stress. Some adjustment is needed to reduce that load.

Manual motor starters come with single pole, double pole and three pole contacts for various applications. Some have a pilot light which indicates when the circuit is on.

Manual motor starters are rated in the horsepower of the motor they're intended to control. Sizes range from fractional horsepower to 10 horsepower. Voltages vary from 115 to 600 volts. Your supplier's catalog will list the appropriate heater element size and the type of enclosure for each starter.

Magnetic Contactors

Magnetic contactors control high amperage loads but don't have overload relays for overcurrent protection. They're made in a wide range of sizes and are available with enclosures to meet any need. Contactors are rated by horsepower and are available in various voltages. The holding coil can be changed to accommodate any voltage needed.

Another type of contactor is the mechanically held contactor. It's operated with a momentary control switch. Once the contactor is closed, it stays closed until opened by a momentary control switch. The opening circuit is connected to another operating coil which causes the switch to jump open and release the contactor.

AC Magnetic Starters

The most common motor starter for single phase and poly-phase motors is the magnetic starter. The contacts are held in position electrically. The starter has overload relays that require overload heater elements. Size the relays to match the full load amp rating on the motor nameplate.

Enclosures for magnetic starters come in several classes: NEMA 1, NEMA 3R, NEMA 4, NEMA 4X, NEMA 7 and 9 and NEMA 12. They're made in all the common voltages and are rated by the horsepower of the motor controlled. Operating coils can be changed to match the control voltage. Most have a button on the face of the enclosure that resets the relay after it's been tripped by an overload.

Only the most common magnetic starters are listed in this section. Many more will be available from your supplier: reversing starters, reduced voltage starters, two speed starters, variable speed starters, special pump starters and multi-speed starters, among others. Each type of starter comes in a range of sizes and types appropriate to protect, control and operate the motor that's connected.

When magnetic starters are housed in a single enclosure, it's usually called a ***motor control center*** (MCC). These units are custom-made at the factory to meet individual job specs. The MCC's are assembled at the factory from standard modular components. The magnetic starter is usually mounted in a basket-type box which is inserted into an open module in the MCC. Power supply probes extend from the starter to the housing bus. A door assembly is then attached to enclose the starter. The door may have a reset push button, pilot lights, selector switches, operating handles and nameplates that indicate how the motor or equipment is to be controlled. Motor control centers usually stand directly on the floor.

Other devices can be installed in a MCC. Panelboards, step down transformers, circuit breakers, combination disconnect switch and magnetic starters, combination circuit breaker and magnetic starters, blank spaces and relay or control sections are all found in some motor control centers.

Control Devices

A few common control devices are listed in this section. These devices are usually assembled from standard components and installed in a single enclosure. They may use a common operating mechanism, standard contact blocks and standard enclosures. The assembly will include all of the switches needed for control purposes. Most can be ganged into a common enclosure or adapted to other enclosures, such as a motor control center or relay panel.

Manual Motor Starters Without Overload Relays

	Material	Craft@Hrs	Unit	Material Cost	Labor Cost	Installed Cost
	NEMA 1 manual motor starters without overload relays, toggle handle					
	2P standard	L1@0.35	Ea	75.30	14.40	89.70
	2P w/ pilot light, 115V	L1@0.35	Ea	111.00	14.40	125.40
ST	2P w/ pilot light, 230V	L1@0.35	Ea	111.00	14.40	125.40
	3P standard	L1@0.40	Ea	94.00	16.40	110.40
	3P w/ pilot light, 240V	L1@0.40	Ea	132.00	16.40	148.40
	3P w/ pilot light, 600V	L1@0.40	Ea	274.00	16.40	290.40
	NEMA 1 flush mounted manual motor starters without overload relays					
	2P standard	L1@0.35	Ea	72.40	14.40	86.80
	2P w/ pilot light, 115V	L1@0.35	Ea	130.00	14.40	144.40
ST	2P w/ pilot light, 230V	L1@0.35	Ea	130.00	14.40	144.40
	3P standard	L1@0.40	Ea	86.90	16.40	103.30
	3P w/ pilot light, 240V	L1@0.40	Ea	130.00	16.40	146.40
	3P w/ pilot light, 600V	L1@0.40	Ea	130.00	16.40	146.40
	NEMA 1 flush mounted manual motor starters, no overload relays, stainless cover					
	2P standard	L1@0.35	Ea	86.90	14.40	101.30
	2P w/ pilot light, 115V	L1@0.35	Ea	111.00	14.40	125.40
ST	2P w/ pilot light, 230V	L1@0.35	Ea	111.00	14.40	125.40
	3P standard	L1@0.40	Ea	130.00	16.40	146.40
	3P w/ pilot light, 240V	L1@0.40	Ea	158.00	16.40	174.40
	3P w/ pilot light, 600V	L1@0.40	Ea	158.00	16.40	174.40
	NEMA 4 & 5 manual motor starters without overload relays					
	2P standard	L1@0.50	Ea	274.00	20.50	294.50
	2P w/ pilot light, 115V	L1@0.50	Ea	377.00	20.50	397.50
$ST_{4/5}$	2P w/ pilot light, 230V	L1@0.50	Ea	377.00	20.50	397.50
	3P standard	L1@0.60	Ea	340.00	24.60	364.60
	3P w/ pilot light, 240V	L1@0.60	Ea	434.00	24.60	458.60
	3P w/ pilot light, 600V	L1@0.60	Ea	434.00	24.60	458.60

Use these figures to estimate the cost of manual motor starters installed in buildings under the conditions described on pages 5 and 6. Costs listed are for each starter installed. The crew size is one electrician working at a labor cost of $41.03 per manhour. These costs include layout, material handling, and normal waste. Add for thermal overload devices, sales tax, delivery, supervision, mobilization, demobilization, cleanup, overhead and profit. Overload heating elements are not included in the cost of these manual motor starters. Size of the overload relay is based on the full motor load in amps. Check the manufacturer's overload heater chart for the correct heater size. These manual motor starters will fit into a standard single gang switch box. They are used as disconnect switches for small machinery, appliances, pumps and fans.

Key operated manual motor starters without overload relays

Material	Craft@Hrs	Unit	Material Cost	Labor Cost	Installed Cost
2P standard	L1@0.35	Ea	111.00	14.40	125.40
2P w/ pilot light, 115V	L1@0.35	Ea	151.00	14.40	165.40
2P w/ pilot light, 230V	L1@0.35	Ea	151.00	14.40	165.40
3P standard	L1@0.40	Ea	155.00	16.40	171.40
3P w/ pilot light, 240V	L1@0.40	Ea	217.00	16.40	233.40
3P w/ pilot light, 600V	L1@0.40	Ea	217.00	16.40	233.40

STK

Open type manual motor starters without overload relays

Material	Craft@Hrs	Unit	Material Cost	Labor Cost	Installed Cost
2P standard	L1@0.30	Ea	73.80	12.30	86.10
2P w/ pilot light, 115V	L1@0.30	Ea	111.00	12.30	123.30
2P w/ pilot light, 230V	L1@0.30	Ea	111.00	12.30	123.30
3P standard	L1@0.40	Ea	117.00	16.40	133.40
3P w/ pilot light, 240V	L1@0.40	Ea	179.00	16.40	195.40
3P w/ pilot light, 600V	L1@0.40	Ea	179.00	16.40	195.40

STO

Manual motor starters with overload relays

Material	Craft@Hrs	Unit	Material Cost	Labor Cost	Installed Cost
1P standard	L1@0.30	Ea	86.90	12.30	99.20
1P w/ pilot light, 115V	L1@0.30	Ea	125.00	12.30	137.30
2P standard	L1@0.40	Ea	98.40	16.40	114.80
2P w/ pilot light, 115V	L1@0.40	Ea	137.00	16.40	153.40
2P w/ pilot light, 230V	L1@0.40	Ea	137.00	16.40	153.40

ST

Flush mounted manual motor starters with overload relays

Material	Craft@Hrs	Unit	Material Cost	Labor Cost	Installed Cost
1P standard	L1@0.30	Ea	80.80	12.30	93.10
1P w/ pilot light, 115V	L1@0.30	Ea	118.00	12.30	130.30
2P standard	L1@0.40	Ea	94.00	16.40	110.40
2P w/ pilot light, 115V	L1@0.40	Ea	130.00	16.40	146.40
2P w/ pilot light, 230V	L1@0.40	Ea	130.00	16.40	146.40

ST

Use these figures to estimate the cost of manual motor starters installed in buildings under the conditions described on pages 5 and 6. Costs listed are for each starter installed. The crew size is one electrician working at a labor cost of $41.03 per manhour. These costs include layout, material handling, and normal waste. Add for thermal overload devices, sales tax, delivery, supervision, mobilization, demobilization, cleanup, overhead and profit. Overload heating elements are not included in the cost of these manual motor starters. Size of the overload relay is based on the full motor load in amps. Check the manufacturer's overload heater chart for the correct heater size. These manual motor starters will fit into a standard single gang switch box. They are used as disconnect switches for small machinery, appliances, pumps and fans.

Manual Motor Starters With Overload Relays

	Material		Craft@Hrs	Unit	Material Cost	Labor Cost	Installed Cost
	NEMA 4 & 5 horsepower rated manual motor starters with overload relays						
ST4/5	1P standard		L1@0.40	Ea	273.00	16.40	289.40
	1P w/ pilot light, 115V		L1@0.40	Ea	386.00	16.40	402.40
	2P standard		L1@0.50	Ea	273.00	20.50	293.50
	2P w/ pilot light, 115V		L1@0.50	Ea	386.00	20.50	406.50
	2P w/ pilot light, 230V		L1@0.50	Ea	386.00	20.50	406.50
	NEMA 1 flush manual motor starters with overload relays, stainless steel cover						
ST	1P standard		L1@0.30	Ea	85.50	12.30	97.80
	1P w/ pilot light, 115V		L1@0.30	Ea	125.00	12.30	137.30
	2P standard		L1@0.40	Ea	111.00	16.40	127.40
	2P w/ pilot light, 115V		L1@0.40	Ea	137.00	16.40	153.40
	2P w/ pilot light, 230V		L1@0.40	Ea	137.00	16.40	153.40
	NEMA 7 & 9 horsepower rated manual motor starters with overload relays						
ST7/9	1P standard		L1@0.60	Ea	318.00	24.60	342.60
	2P standard		L1@0.70	Ea	329.00	28.70	357.70
	Open type horsepower rated manual motor starters with overload relays						
STO	2 pole	1 HP	L1@0.35	Ea	286.00	14.40	300.40
	2 pole	2 HP	L1@0.35	Ea	286.00	14.40	300.40
	2 pole	3 HP	L1@0.35	Ea	286.00	14.40	300.40
	2 pole	5 HP	L1@0.35	Ea	286.00	14.40	300.40
	3 pole	3 HP	L1@0.40	Ea	155.00	16.40	171.40
	3 pole	5 HP	L1@0.40	Ea	287.00	16.40	303.40
	3 pole	7.5 HP	L1@0.40	Ea	345.00	16.40	361.40
	3 pole	10 HP	L1@0.40	Ea	345.00	16.40	361.40

Use these figures to estimate the cost of manual motor starters installed in buildings under the conditions described on pages 5 and 6. Costs listed are for each starter installed. The crew size is one electrician working at a labor cost of $41.03 per manhour. These costs include layout, material handling, and normal waste. Add for thermal overload devices, sales tax, delivery, supervision, mobilization, demobilization, cleanup, overhead and profit. Overload heating elements are not included in the cost of these manual motor starters. Size of the overload relay is based on the full motor load in amps. Check the manufacturer's overload heater chart for the correct heater size.

Material		Craft@Hrs	Unit	Material Cost	Labor Cost	Installed Cost
NEMA 1 manual motor starters with overload relays						
2 pole	1 HP	L1@0.40	Ea	246.00	16.40	262.40
2 pole	2 HP	L1@0.40	Ea	246.00	16.40	262.40
2 pole	3 HP	L1@0.40	Ea	378.00	16.40	394.40
2 pole	5 HP	L1@0.40	Ea	444.00	16.40	460.40
3 pole	3 HP	L1@0.45	Ea	362.00	18.50	380.50
3 pole	5 HP	L1@0.45	Ea	362.00	18.50	380.50
3 pole	7.5 HP	L1@0.45	Ea	371.00	18.50	389.50
3 pole	10 HP	L1@0.45	Ea	371.00	18.50	389.50
NEMA 4 manual motor starters with overload relays						
2 pole	1 HP	L1@0.35	Ea	644.00	14.40	658.40
2 pole	2 HP	L1@0.35	Ea	644.00	14.40	658.40
2 pole	3 HP	L1@0.35	Ea	963.00	14.40	977.40
2 pole	5 HP	L1@0.35	Ea	963.00	14.40	977.40
3 pole	3 HP	L1@0.40	Ea	707.00	16.40	723.40
3 pole	5 HP	L1@0.40	Ea	707.00	16.40	723.40
3 pole	7.5 HP	L1@0.40	Ea	856.00	16.40	872.40
3 pole	10 HP	L1@0.40	Ea	856.00	16.40	872.40
NEMA 4X manual motor starters with overload relays						
2 pole	1 HP	L1@0.40	Ea	644.00	16.40	660.40
2 pole	2 HP	L1@0.40	Ea	644.00	16.40	660.40
2 pole	3 HP	L1@0.40	Ea	983.00	16.40	999.40
2 pole	5 HP	L1@0.40	Ea	983.00	16.40	999.40
3 pole	3 HP	L1@0.45	Ea	704.00	18.50	722.50
3 pole	5 HP	L1@0.45	Ea	704.00	18.50	722.50
3 pole	7.5 HP	L1@0.45	Ea	856.00	18.50	874.50
3 pole	10 HP	L1@0.45	Ea	856.00	18.50	874.50
NEMA 7 & 9 manual motor starters with overload relays						
2 pole	1 HP	L1@0.40	Ea	891.00	16.40	907.40
2 pole	2 HP	L1@0.40	Ea	891.00	16.40	907.40
2 pole	3 HP	L1@0.40	Ea	940.00	16.40	956.40
2 pole	5 HP	L1@0.40	Ea	940.00	16.40	956.40

ST

ST_4

ST_{4X}

$ST_{7/9}$

Use these figures to estimate the cost of manual motor starters installed in buildings under the conditions described on pages 5 and 6. Costs listed are for each starter installed. The crew size is one electrician working at a labor cost of $41.03 per manhour. These costs include layout, material handling, and normal waste. Add for thermal overload devices, sales tax, delivery, supervision, mobilization, demobilization, cleanup, overhead and profit. Overload heating elements are not included in the cost of these manual motor starters. Size of the overload relay is based on the full motor load in amps. Check the manufacturer's overload heater chart for the correct heater size. These manual motor starters will fit into a standard single gang switch box. They are used as disconnect switches for small machinery, appliances, pumps and fans.

HP Rated Manual Motor Starters with Overload Relays

	Material	Craft@Hrs	Unit	Material Cost	Labor Cost	Installed Cost
	NEMA 7 & 9 HP rated manual motor starters with overload relays					
ST7/9	3 pole 3 HP	L1@0.45	Ea	959.00	18.50	977.50
	3 pole 5 HP	L1@0.45	Ea	959.00	18.50	977.50
	3 pole 7.5 HP	L1@0.45	Ea	1,120.00	18.50	1,138.50
	3 pole 10 HP	L1@0.45	Ea	1,120.00	18.50	1,138.50
	NEMA 1 reversing HP rated manual motor starters with overload relays					
ST	3 pole 3 HP	L1@1.25	Ea	893.00	51.30	944.30
	3 pole 5 HP	L1@1.25	Ea	893.00	51.30	944.30
	3 pole 7.5 HP	L1@1.50	Ea	1,070.00	61.50	1,131.50
	3 pole 10 HP	L1@1.50	Ea	1,070.00	61.50	1,131.50
	NEMA 1 two speed HP rated manual motor starters with overload relays					
ST	3 pole 3 HP	L1@1.00	Ea	893.00	41.00	934.00
	3 pole 5 HP	L1@1.00	Ea	893.00	41.00	934.00
	3 pole 7.5 HP	L1@1.25	Ea	1,070.00	51.30	1,121.30
	3 pole 10 HP	L1@1.25	Ea	1,070.00	51.30	1,121.30
	NEMA 1 reversing HP rated manual starters, overload relays, low voltage protection					
ST	3 pole 3 HP, 208V	L1@1.25	Ea	1,180.00	51.30	1,231.30
	3 pole 3 HP, 240V	L1@1.25	Ea	1,180.00	51.30	1,231.30
	3 pole 5 HP, 480V	L1@1.25	Ea	1,180.00	51.30	1,231.30
	3 pole 5 HP, 600V	L1@1.25	Ea	1,180.00	51.30	1,231.30
	3 pole 7.5 HP, 208V	L1@1.50	Ea	1,370.00	61.50	1,431.50
	3 pole 7.5 HP, 240V	L1@1.50	Ea	1,370.00	61.50	1,431.50
	3 pole 10 HP, 480V	L1@1.50	Ea	1,370.00	61.50	1,431.50
	3 pole 10 HP, 600V	L1@1.50	Ea	1,370.00	61.50	1,431.50
	NEMA 1 two speed HP rated manual starters, overload relays, low voltage protection					
ST	3 pole 3 HP, 208V	L1@1.25	Ea	1,180.00	51.30	1,231.30
	3 pole 3 HP, 240V	L1@1.25	Ea	1,180.00	51.30	1,231.30
	3 pole 5 HP, 480V	L1@1.25	Ea	1,180.00	51.30	1,231.30
	3 pole 5 HP, 600V	L1@1.25	Ea	1,180.00	51.30	1,231.30
	3 pole 7.5 HP, 208V	L1@1.50	Ea	1,370.00	61.50	1,431.50
	3 pole 7.5 HP, 240V	L1@1.50	Ea	1,370.00	61.50	1,431.50
	3 pole 10 HP, 480V	L1@1.50	Ea	1,370.00	61.50	1,431.50
	3 pole 10 HP, 600V	L1@1.50	Ea	1,370.00	61.50	1,431.50

Use these figures to estimate the cost of manual motor starters installed in buildings under the conditions described on pages 5 and 6. Costs listed are for each starter installed. The crew size is one electrician working at a labor cost of $41.03 per manhour. These costs include layout, material handling, and normal waste. Add for thermal overload devices, sales tax, delivery, supervision, mobilization, demobilization, cleanup, overhead and profit. Overload heating elements are not included in the cost of these manual motor starters. Size of the overload relay is based on the full motor load in amps. Check the manufacturer's overload heater chart for the correct heater size.

Open type HP rated reversing two-speed manual starters, overload relays, low voltage protection

Material	Craft@Hrs	Unit	Material Cost	Labor Cost	Installed Cost
3 pole 3 HP, 208V	L1@1.00	Ea	1,100.00	41.00	1,141.00
3 pole 3 HP, 240V	L1@1.00	Ea	1,100.00	41.00	1,141.00
3 pole 5 HP, 480V	L1@1.00	Ea	1,100.00	41.00	1,141.00
3 pole 5 HP, 600V	L1@1.00	Ea	1,100.00	41.00	1,141.00
3 pole 7.5 HP, 208V	L1@1.25	Ea	1,300.00	51.30	1,351.30
3 pole 7.5 HP, 240V	L1@1.25	Ea	1,300.00	51.30	1,351.30
3 pole 10 HP, 480V	L1@1.25	Ea	1,300.00	51.30	1,351.30
3 pole 10 HP, 600V	L1@1.25	Ea	1,300.00	51.30	1,351.30

STO

NEMA 1 two pole AC magnetic contactors without overload relays

Material	Craft@Hrs	Unit	Material Cost	Labor Cost	Installed Cost
Size 00 1/3 HP, 115V	L1@0.50	Ea	188.00	20.50	208.50
Size 00 1 HP, 230V	L1@0.50	Ea	188.00	20.50	208.50
Size 0 1 HP, 115V	L1@0.60	Ea	244.00	24.60	268.60
Size 0 2 HP, 230V	L1@0.60	Ea	244.00	24.60	268.60
Size 1 2 HP, 115V	L1@0.75	Ea	290.00	30.80	320.80
Size 1 3 HP, 230V	L1@0.75	Ea	290.00	30.80	320.80
Size 2 3 HP, 115V	L1@1.00	Ea	540.00	41.00	581.00
Size 2 7.5 HP, 230V	L1@1.00	Ea	540.00	41.00	581.00
Size 3 7.5 HP, 115V	L1@1.25	Ea	930.00	51.30	981.30
Size 3 15 HP, 230V	L1@1.25	Ea	930.00	51.30	981.30

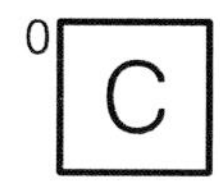

NEMA 4 two pole AC magnetic contactors without overload relays

Material	Craft@Hrs	Unit	Material Cost	Labor Cost	Installed Cost
Size 0 1 HP, 115V	L1@1.00	Ea	990.00	41.00	1,031.00
Size 0 2 HP, 230V	L1@1.00	Ea	990.00	41.00	1,031.00
Size 1 2 HP, 115V	L1@1.25	Ea	1,090.00	51.30	1,141.30
Size 1 3 HP, 230V	L1@1.25	Ea	1,090.00	51.30	1,141.30
Size 2 3 HP, 115V	L1@1.50	Ea	2,220.00	61.50	2,281.50
Size 2 7.5 HP, 230V	L1@1.50	Ea	2,220.00	61.50	2,281.50
Size 3 7.5 HP, 115V	L1@1.75	Ea	3,360.00	71.80	3,431.80
Size 3 15 HP, 230V	L1@1.75	Ea	3,360.00	71.80	3,431.80

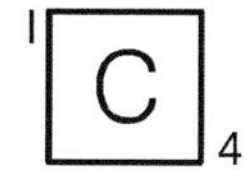

Use these figures to estimate the cost of manual motor starters and magnetic contactors installed in buildings under the conditions described on pages 5 and 6. Costs listed are for each starter or contactor installed. The crew size is one electrician working at a labor cost of $41.03 per manhour. These costs include layout, material handling, and normal waste. Add for supports, sales tax, delivery, supervision, mobilization, demobilization, cleanup, overhead and profit. Magnetic contactors do not provide overload protection. They are used as a relay device only.

Two Pole AC Magnetic Contactors Without Overload Relays

NEMA 4X two-pole AC magnetic contactors without overload relays

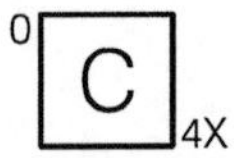

Material	Craft@Hrs	Unit	Material Cost	Labor Cost	Installed Cost
Size 0 1 HP, 115V	L1@1.00	Ea	1,100.00	41.00	1,141.00
Size 0 2 HP, 230V	L1@1.00	Ea	1,100.00	41.00	1,141.00
Size 1 2 HP, 115V	L1@1.25	Ea	1,230.00	51.30	1,281.30
Size 1 3 HP, 230V	L1@1.25	Ea	1,230.00	51.30	1,281.30
Size 2 3 HP, 115V	L1@1.50	Ea	2,460.00	61.50	2,521.50
Size 2 7.5 HP, 230V	L1@1.50	Ea	2,460.00	61.50	2,521.50

NEMA 7 & 9 two-pole AC magnetic contactors without overload relays

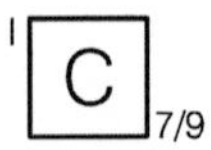

Material	Craft@Hrs	Unit	Material Cost	Labor Cost	Installed Cost
Size 0 1 HP, 115V	L1@1.25	Ea	2,460.00	51.30	2,511.30
Size 0 2 HP, 230V	L1@1.25	Ea	2,460.00	51.30	2,511.30
Size 1 2 HP, 115V	L1@1.50	Ea	2,570.00	61.50	2,631.50
Size 1 3 HP, 230V	L1@1.50	Ea	2,570.00	61.50	2,631.50
Size 2 3 HP, 115V	L1@1.75	Ea	4,190.00	71.80	4,261.80
Size 2 7.5 HP, 230V	L1@1.75	Ea	4,190.00	71.80	4,261.80
Size 3 7.5 HP, 115V	L1@2.00	Ea	6,250.00	82.10	6,332.10
Size 3 15 HP, 230V	L1@2.00	Ea	6,250.00	82.10	6,332.10

NEMA 12 two-pole AC magnetic contactors without overload relays

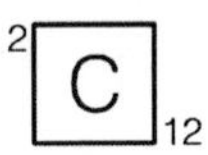

Material	Craft@Hrs	Unit	Material Cost	Labor Cost	Installed Cost
Size 0 1 HP, 115V	L1@0.50	Ea	705.00	20.50	725.50
Size 0 2 HP, 230V	L1@0.50	Ea	705.00	20.50	725.50
Size 1 2 HP, 115V	L1@0.60	Ea	791.00	24.60	815.60
Size 1 3 HP, 230V	L1@0.60	Ea	791.00	24.60	815.60
Size 2 3 HP, 115V	L1@0.75	Ea	1,570.00	30.80	1,600.80
Size 2 7.5 HP, 230V	L1@0.75	Ea	1,570.00	30.80	1,600.80
Size 3 7.5 HP, 115V	L1@1.00	Ea	2,410.00	41.00	2,451.00
Size 3 15 HP, 230V	L1@1.00	Ea	2,410.00	41.00	2,451.00

Open type two-pole AC magnetic contactors without overload relays

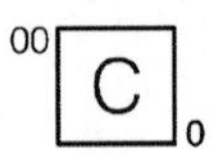

Material	Craft@Hrs	Unit	Material Cost	Labor Cost	Installed Cost
Size 00 1/3 HP, 115V	L1@0.40	Ea	345.00	16.40	361.40
Size 00 1 HP, 230V	L1@0.40	Ea	345.00	16.40	361.40
Size 0 1 HP, 115V	L1@0.50	Ea	454.00	20.50	474.50
Size 0 2 HP, 230V	L1@0.50	Ea	454.00	20.50	474.50

Use these figures to estimate the cost of magnetic contactors installed in buildings under the conditions described on pages 5 and 6. Costs listed are for each contactor installed. The crew size is one electrician working at a labor cost of $41.03 per manhour. These costs include layout, material handling, and normal waste. Add for supports, sales tax, delivery, supervision, mobilization, demobilization, cleanup, overhead and profit. These magnetic contactors do not provide overload protection. They are used as a relay device only. For motor overload protection, use magnetic starters with overload relays and heating elements. Other types of magnetic contactors are available for incandescent lamp loads and induction loads.

AC Magnetic Contactors Without Overload Relays

Two pole open type AC magnetic contactors without overload relays

Material	Craft@Hrs	Unit	Material Cost	Labor Cost	Installed Cost
Size 1 2 HP, 115V	L1@0.60	Ea	544.00	24.60	568.60
Size 1 3 HP, 230V	L1@0.60	Ea	544.00	24.60	568.60
Size 2 3 HP, 115V	L1@0.70	Ea	1,010.00	28.70	1,038.70
Size 2 7.5 HP, 230V	L1@0.70	Ea	1,010.00	28.70	1,038.70
Size 3 7.5 HP, 115V	L1@0.90	Ea	1,610.00	36.90	1,646.90
Size 3 15 HP, 230V	L1@0.90	Ea	1,610.00	36.90	1,646.90

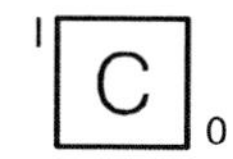

NEMA 1 three pole AC magnetic contactors without overload relays

Material	Craft@Hrs	Unit	Material Cost	Labor Cost	Installed Cost
Size 00 1.5 HP, 208V	L1@0.60	Ea	444.00	24.60	468.60
Size 00 2 HP, 480V	L1@0.60	Ea	444.00	24.60	468.60
Size 0 3 HP, 208V	L1@0.60	Ea	545.00	24.60	569.60
Size 0 5 HP, 480V	L1@0.60	Ea	545.00	24.60	569.60
Size 1 7.5 HP, 208V	L1@0.70	Ea	633.00	28.70	661.70
Size 1 10 HP, 480V	L1@0.70	Ea	633.00	28.70	661.70
Size 2 10 HP, 208V	L1@1.00	Ea	1,260.00	41.00	1,301.00
Size 2 25 HP, 480V	L1@1.00	Ea	1,260.00	41.00	1,301.00
Size 3 25 HP, 208V	L1@1.25	Ea	2,100.00	51.30	2,151.30
Size 3 50 HP, 480V	L1@1.25	Ea	2,100.00	51.30	2,151.30
Size 4 40 HP, 208V	L1@2.00	Ea	4,980.00	82.10	5,062.10
Size 4 100 HP, 480V	L1@2.00	Ea	4,980.00	82.10	5,062.10
Size 5 75 HP, 208V	L1@4.00	Ea	10,400.00	164.00	10,564.00
Size 5 200 HP, 480V	L1@4.00	Ea	10,400.00	164.00	10,564.00
Size 6 150 HP, 208V	L1@6.00	Ea	30,900.00	246.00	31,146.00
Size 6 400 HP, 480V	L1@6.00	Ea	30,900.00	246.00	31,146.00
Size 7 300 HP, 208V	L1@8.00	Ea	41,900.00	328.00	42,228.00
Size 7 600 HP, 480V	L1@8.00	Ea	41,900.00	328.00	42,228.00

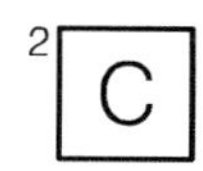

NEMA 4 three pole AC magnetic contactors without overload relays

Material	Craft@Hrs	Unit	Material Cost	Labor Cost	Installed Cost
Size 0 3 HP, 208V	L1@1.25	Ea	1,160.00	51.30	1,211.30
Size 0 5 HP, 480V	L1@1.25	Ea	1,160.00	51.30	1,211.30
Size 1 7.5 HP, 208V	L1@1.50	Ea	1,250.00	61.50	1,311.50
Size 1 10 HP, 480V	L1@1.50	Ea	1,250.00	61.50	1,311.50
Size 2 10 HP, 208V	L1@1.75	Ea	2,500.00	71.80	2,571.80
Size 2 25 HP, 480V	L1@1.75	Ea	2,500.00	71.80	2,571.80
Size 3 25 HP, 208V	L1@2.00	Ea	3,880.00	82.10	3,962.10
Size 3 50 HP, 480V	L1@2.00	Ea	3,880.00	82.10	3,962.10
Size 4 40 HP, 208V	L1@3.00	Ea	8,020.00	123.00	8,143.00
Size 4 100 HP, 480V	L1@3.00	Ea	8,020.00	123.00	8,143.00
Size 5 75 HP, 208V	L1@6.00	Ea	14,300.00	246.00	14,546.00
Size 5 200 HP, 480V	L1@6.00	Ea	14,300.00	246.00	14,546.00

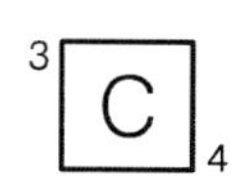

Use these figures to estimate the cost of magnetic contactors installed in buildings under the conditions described on pages 5 and 6. Costs listed are for each contactor installed. The crew size is one electrician working at a labor cost of $41.03 per manhour. These costs include layout, material handling, and normal waste. Add for supports, sales tax, delivery, supervision, mobilization, demobilization, cleanup, overhead and profit. These magnetic contactors do not provide overload protection. They are used as a relay device only. For motor overload protection, use magnetic starters with overload relays and heating elements. Other types of magnetic contactors are available for incandescent lamp loads and induction loads.

Three Pole AC Magnetic Contactors Without Overload Relays

Material	Craft@Hrs	Unit	Material Cost	Labor Cost	Installed Cost
NEMA 4 three pole AC magnetic contactors without overload relays					
Size 6 150 HP, 208V	L1@8.00	Ea	41,900.00	328.00	42,228.00
Size 6 400 HP, 480V	L1@8.00	Ea	41,900.00	328.00	42,228.00
Size 7 300 HP, 208V	L1@12.0	Ea	50,200.00	492.00	50,692.00
Size 7 600 HP, 480V	L1@12.0	Ea	50,200.00	492.00	50,692.00

NEMA 4X three pole AC magnetic contactors without overload relays

Size 0 3 HP, 208V	L1@1.25	Ea	1,170.00	51.30	1,221.30
Size 0 5 HP, 480V	L1@1.25	Ea	1,170.00	51.30	1,221.30
Size 1 7.5 HP, 208V	L1@1.50	Ea	1,260.00	61.50	1,321.50
Size 1 10 HP, 480V	L1@1.50	Ea	1,260.00	61.50	1,321.50
Size 2 10 HP, 208V	L1@1.75	Ea	2,500.00	71.80	2,571.80
Size 2 25 HP, 480V	L1@1.75	Ea	2,500.00	71.80	2,571.80
Size 3 25 HP, 208V	L1@2.00	Ea	4,840.00	82.10	4,922.10
Size 3 50 HP, 480V	L1@2.00	Ea	4,840.00	82.10	4,922.10
Size 4 40 HP, 208V	L1@3.00	Ea	9,990.00	123.00	10,113.00
Size 4 100 HP, 480V	L1@3.00	Ea	9,990.00	123.00	10,113.00

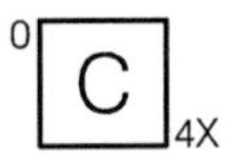

NEMA 7 & 9 three pole AC magnetic contactors without overload relays

Size 0 3 HP, 208V	L1@1.50	Ea	1,790.00	61.50	1,851.50
Size 0 5 HP, 480V	L1@1.50	Ea	1,790.00	61.50	1,851.50
Size 1 7.5 HP, 208V	L1@1.75	Ea	1,890.00	71.80	1,961.80
Size 1 10 HP, 480V	L1@1.75	Ea	1,890.00	71.80	1,961.80
Size 2 10 HP, 208V	L1@2.00	Ea	3,010.00	82.10	3,092.10
Size 2 25 HP, 480V	L1@2.00	Ea	3,010.00	82.10	3,092.10
Size 3 25 HP, 208V	L1@3.00	Ea	4,530.00	123.00	4,653.00
Size 3 50 HP, 480V	L1@3.00	Ea	4,530.00	123.00	4,653.00
Size 4 40 HP, 208V	L1@4.00	Ea	7,310.00	164.00	7,474.00
Size 4 100 HP, 480V	L1@4.00	Ea	7,310.00	164.00	7,474.00
Size 5 75 HP, 208V	L1@8.00	Ea	21,100.00	328.00	21,428.00
Size 5 200 HP, 480V	L1@8.00	Ea	21,100.00	328.00	21,428.00
Size 6 150 HP, 208V	L1@12.0	Ea	56,100.00	492.00	56,592.00
Size 6 400 HP, 480V	L1@12.0	Ea	56,100.00	492.00	56,592.00

Use these figures to estimate the cost of magnetic contactors installed in buildings under the conditions described on pages 5 and 6. Costs listed are for each contactor installed. The crew size is one electrician working at a labor cost of $41.03 per manhour. These costs include layout, material handling, and normal waste. Add for supports, sales tax, delivery, supervision, mobilization, demobilization, cleanup, overhead and profit. These magnetic contactors do not provide overload protection. They are used as a relay device only. For motor overload protection, use magnetic starters with overload relays and heating elements. Other types of magnetic contactors are available for incandescent lamp loads and induction loads.

Three Pole AC Magnetic Contactors Without Overload Relays

NEMA 12 three pole AC magnetic contactors without overload relays

Material	Craft@Hrs	Unit	Material Cost	Labor Cost	Installed Cost
Size 0 3 HP, 208V	L1@0.70	Ea	536.00	28.70	564.70
Size 0 5 HP, 480V	L1@0.70	Ea	536.00	28.70	564.70
Size 1 7.5 HP, 208V	L1@0.90	Ea	598.00	36.90	634.90
Size 1 10 HP, 480V	L1@0.90	Ea	598.00	36.90	634.90
Size 2 10 HP, 208V	L1@1.00	Ea	1,170.00	41.00	1,211.00
Size 2 25 HP, 480V	L1@1.00	Ea	1,170.00	41.00	1,211.00
Size 3 25 HP, 208V	L1@1.25	Ea	1,800.00	51.30	1,851.30
Size 3 50 HP, 480V	L1@1.25	Ea	1,800.00	51.30	1,851.30
Size 4 40 HP, 208V	L1@2.00	Ea	4,580.00	82.10	4,662.10
Size 4 100 HP, 480V	L1@2.00	Ea	4,580.00	82.10	4,662.10
Size 5 75 HP, 208V	L1@4.00	Ea	10,100.00	164.00	10,264.00
Size 5 200 HP, 480V	L1@4.00	Ea	10,100.00	164.00	10,264.00
Size 6 150 HP, 208V	L1@6.00	Ea	25,000.00	246.00	25,246.00
Size 6 400 HP, 480V	L1@6.00	Ea	25,000.00	246.00	25,246.00
Size 7 300 HP, 208V	L1@8.00	Ea	32,800.00	328.00	33,128.00
Size 7 600 HP, 480V	L1@8.00	Ea	32,800.00	328.00	33,128.00

0 C 12

Open type three pole AC magnetic contactors without overload relays

Material	Craft@Hrs	Unit	Material Cost	Labor Cost	Installed Cost
Size 00 1.5 HP, 208V	L1@0.50	Ea	287.00	20.50	307.50
Size 00 2 HP, 480V	L1@0.50	Ea	287.00	20.50	307.50
Size 0 3 HP, 208V	L1@0.50	Ea	358.00	20.50	378.50
Size 0 5 HP, 480V	L1@0.50	Ea	358.00	20.50	378.50
Size 1 7.5 HP, 208V	L1@0.60	Ea	421.00	24.60	445.60
Size 1 10 HP, 480V	L1@0.60	Ea	421.00	24.60	445.60
Size 2 10 HP, 208V	L1@0.75	Ea	738.00	30.80	768.80
Size 2 25 HP, 480V	L1@0.75	Ea	738.00	30.80	768.80
Size 3 25 HP, 208V	L1@1.00	Ea	1,250.00	41.00	1,291.00
Size 3 50 HP, 480V	L1@1.00	Ea	1,250.00	41.00	1,291.00
Size 4 40 HP, 208V	L1@1.75	Ea	2,970.00	71.80	3,041.80
Size 4 100 HP, 480V	L1@1.75	Ea	2,970.00	71.80	3,041.80
Size 5 75 HP, 208V	L1@3.50	Ea	6,470.00	144.00	6,614.00
Size 5 200 HP, 480V	L1@3.50	Ea	6,470.00	144.00	6,614.00
Size 6 150 HP, 208V	L1@5.50	Ea	17,600.00	226.00	17,826.00
Size 6 400 HP, 480V	L1@5.50	Ea	17,600.00	226.00	17,826.00
Size 7 300 HP, 208V	L1@7.50	Ea	25,000.00	308.00	25,308.00
Size 7 600 HP, 480V	L1@7.50	Ea	25,000.00	308.00	25,308.00

I C 0

Use these figures to estimate the cost of magnetic contactors installed in buildings under the conditions described on pages 5 and 6. Costs listed are for each contactor installed. The crew size is one electrician working at a labor cost of $41.03 per manhour. These costs include layout, material handling, and normal waste. Add for supports, sales tax, delivery, supervision, mobilization, demobilization, cleanup, overhead and profit. These magnetic contactors do not provide overload protection. They are used as a relay device only. For motor overload protection, use magnetic starters with overload relays and heating elements. Other types of magnetic contactors are available for incandescent lamp loads and induction loads.

Four Pole AC Magnetic Contactors Without Overload Relays

NEMA 1 four pole AC magnetic contactors without overload relays

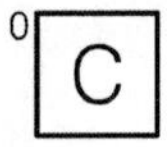

Material	Craft@Hrs	Unit	Material Cost	Labor Cost	Installed Cost
Size 0 3 HP, 208V	L1@0.70	Ea	488.00	28.70	516.70
Size 0 5 HP, 480V	L1@0.70	Ea	488.00	28.70	516.70
Size 1 7.5 HP, 208V	L1@0.90	Ea	550.00	36.90	586.90
Size 1 10 HP, 480V	L1@0.90	Ea	550.00	36.90	586.90
Size 2 10 HP, 208V	L1@1.25	Ea	1,110.00	51.30	1,161.30
Size 2 15 HP, 230V	L1@1.25	Ea	1,110.00	51.30	1,161.30
Size 2 25 HP, 480V	L1@1.25	Ea	1,110.00	51.30	1,161.30
Size 3 25 HP, 208V	L1@1.50	Ea	1,860.00	61.50	1,921.50
Size 3 30 HP, 230V	L1@1.50	Ea	1,860.00	61.50	1,921.50
Size 3 50 HP, 480V	L1@1.50	Ea	1,860.00	61.50	1,921.50
Size 4 40 HP, 208V	L1@1.75	Ea	4,660.00	71.80	4,731.80
Size 4 75 HP, 230V	L1@1.75	Ea	4,660.00	71.80	4,731.80
Size 4 100 HP, 480V	L1@1.75	Ea	4,660.00	71.80	4,731.80

NEMA 4 four pole AC magnetic contactors without overload relays

Material	Craft@Hrs	Unit	Material Cost	Labor Cost	Installed Cost
Size 0 3 HP, 208V	L1@1.50	Ea	930.00	61.50	991.50
Size 0 5 HP, 480V	L1@1.50	Ea	930.00	61.50	991.50
Size 1 7.5 HP, 208V	L1@1.75	Ea	996.00	71.80	1,067.80
Size 1 10 HP, 480V	L1@1.75	Ea	996.00	71.80	1,067.80
Size 2 10 HP, 208V	L1@2.00	Ea	2,370.00	82.10	2,452.10
Size 2 15 HP, 230V	L1@2.00	Ea	2,370.00	82.10	2,452.10
Size 2 25 HP, 480V	L1@2.00	Ea	2,370.00	82.10	2,452.10
Size 3 25 HP, 208V	L1@2.25	Ea	3,440.00	92.30	3,532.30
Size 3 30 HP, 230V	L1@2.25	Ea	3,440.00	92.30	3,532.30
Size 3 50 HP, 480V	L1@2.25	Ea	3,440.00	92.30	3,532.30
Size 4 40 HP, 208V	L1@2.75	Ea	7,700.00	113.00	7,813.00
Size 4 75 HP, 230V	L1@2.75	Ea	7,700.00	113.00	7,813.00
Size 4 100 HP, 480V	L1@2.75	Ea	7,700.00	113.00	7,813.00

NEMA 4X four pole AC magnetic contactors without overload relays

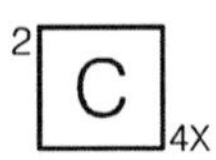

Material	Craft@Hrs	Unit	Material Cost	Labor Cost	Installed Cost
Size 0 3 HP, 208V	L1@1.50	Ea	930.00	61.50	991.50
Size 0 5 HP, 480V	L1@1.50	Ea	930.00	61.50	991.50
Size 1 7.5 HP, 208V	L1@1.75	Ea	996.00	71.80	1,067.80
Size 1 10 HP, 480V	L1@1.75	Ea	996.00	71.80	1,067.80
Size 2 10 HP, 208V	L1@2.00	Ea	2,370.00	82.10	2,452.10
Size 2 15 HP, 230V	L1@2.00	Ea	2,370.00	82.10	2,452.10
Size 2 25 HP, 480V	L1@2.00	Ea	2,370.00	82.10	2,452.10

Use these figures to estimate the cost of magnetic contactors installed in buildings under the conditions described on pages 5 and 6. Costs listed are for each contactor installed. The crew size is one electrician working at a labor cost of $41.03 per manhour. These costs include layout, material handling, and normal waste. Add for supports, sales tax, delivery, supervision, mobilization, demobilization, cleanup, overhead and profit. These magnetic contactors do not provide overload protection. They are used as a relay device only. For motor overload protection, use magnetic starters with overload relays and heating elements. Other types of magnetic contactors are available for incandescent lamp loads and induction loads.

Four Pole AC Magnetic Contactors Without Overload Relays

NEMA 7 & 9 four pole AC magnetic contactors without overload relays

Material		Craft@Hrs	Unit	Material Cost	Labor Cost	Installed Cost
Size 0	3 HP, 208V	L1@1.75	Ea	1,910.00	71.80	1,981.80
Size 0	5 HP, 480V	L1@1.75	Ea	1,910.00	71.80	1,981.80
Size 1	7.5 HP, 208V	L1@2.00	Ea	1,980.00	82.10	2,062.10
Size 1	10 HP, 480V	L1@2.00	Ea	1,980.00	82.10	2,062.10
Size 2	10 HP, 208V	L1@2.00	Ea	3,630.00	82.10	3,712.10
Size 2	15 HP, 230V	L1@2.25	Ea	3,630.00	92.30	3,722.30
Size 2	25 HP, 480V	L1@2.25	Ea	3,290.00	92.30	3,382.30
Size 3	25 HP, 208V	L1@2.50	Ea	4,930.00	103.00	5,033.00
Size 3	30 HP, 230V	L1@2.50	Ea	4,930.00	103.00	5,033.00
Size 3	50 HP, 480V	L1@2.50	Ea	4,930.00	103.00	5,033.00
Size 4	40 HP, 208V	L1@3.00	Ea	7,480.00	123.00	7,603.00
Size 4	75 HP, 230V	L1@3.00	Ea	7,480.00	123.00	7,603.00
Size 4	100 HP, 480V	L1@3.00	Ea	7,480.00	123.00	7,603.00

3 C 7/9

NEMA 12 four pole AC magnetic contactors without overload relays

Material		Craft@Hrs	Unit	Material Cost	Labor Cost	Installed Cost
Size 0	3 HP, 208V	L1@0.75	Ea	635.00	30.80	665.80
Size 0	5 HP, 480V	L1@0.75	Ea	635.00	30.80	665.80
Size 1	7.5 HP, 208V	L1@1.00	Ea	697.00	41.00	738.00
Size 1	10 HP, 480V	L1@1.00	Ea	697.00	41.00	738.00
Size 2	10 HP, 208V	L1@1.30	Ea	1,380.00	53.30	1,433.30
Size 2	15 HP, 230V	L1@1.30	Ea	1,380.00	53.30	1,433.30
Size 2	25 HP, 480V	L1@1.30	Ea	1,380.00	53.30	1,433.30
Size 3	25 HP, 208V	L1@1.60	Ea	2,160.00	65.60	2,225.60
Size 3	30 HP, 230V	L1@1.60	Ea	2,160.00	65.60	2,225.60
Size 3	50 HP, 480V	L1@1.60	Ea	2,160.00	65.60	2,225.60
Size 4	40 HP, 208V	L1@3.00	Ea	6,100.00	123.00	6,223.00
Size 4	75 HP, 230V	L1@3.00	Ea	6,100.00	123.00	6,223.00
Size 4	100 HP, 480V	L1@3.00	Ea	6,100.00	123.00	6,223.00

1 C 12

Use these figures to estimate the cost of magnetic contactors installed in buildings under the conditions described on pages 5 and 6. Costs listed are for each contactor installed. The crew size is one electrician working at a labor cost of $41.03 per manhour. These costs include layout, material handling, and normal waste. Add for supports, sales tax, delivery, supervision, mobilization, demobilization, cleanup, overhead and profit. These magnetic contactors do not provide overload protection. They are used as a relay device only. For motor overload protection, use magnetic starters with overload relays and heating elements. Other types of magnetic contactors are available for incandescent lamp loads and induction loads.

AC Magnetic Contactors Without Overload Relays

Open type four pole AC magnetic contactors without overload relays

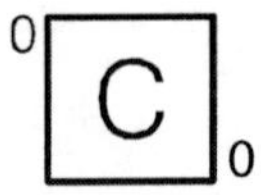

Material	Craft@Hrs	Unit	Material Cost	Labor Cost	Installed Cost
Size 0 3 HP, 208V	L1@0.60	Ea	458.00	24.60	482.60
Size 0 5 HP, 480V	L1@0.60	Ea	458.00	24.60	482.60
Size 1 7.5 HP, 208V	L1@0.75	Ea	519.00	30.80	549.80
Size 1 10 HP, 480V	L1@0.75	Ea	519.00	30.80	549.80
Size 2 10 HP, 208V	L1@1.00	Ea	989.00	41.00	1,030.00
Size 2 15 HP, 230V	L1@1.00	Ea	989.00	41.00	1,030.00
Size 2 25 HP, 480V	L1@1.00	Ea	989.00	41.00	1,030.00
Size 3 25 HP, 208V	L1@1.25	Ea	1,570.00	51.30	1,621.30
Size 3 30 HP, 230V	L1@1.25	Ea	1,570.00	51.30	1,621.30
Size 3 50 HP, 480V	L1@1.25	Ea	1,570.00	51.30	1,621.30
Size 4 40 HP, 208V	L1@1.50	Ea	4,110.00	61.50	4,171.50
Size 4 75 HP, 230V	L1@1.50	Ea	4,110.00	61.50	4,171.50
Size 4 100 HP, 480V	L1@1.50	Ea	4,110.00	61.50	4,171.50

NEMA 1 five pole AC magnetic contactors without overload relays

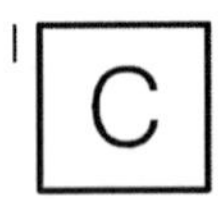

Material	Craft@Hrs	Unit	Material Cost	Labor Cost	Installed Cost
Size 0 3 HP, 208V	L1@1.00	Ea	624.00	41.00	665.00
Size 0 5 HP, 480V	L1@1.00	Ea	624.00	41.00	665.00
Size 1 7.5 HP, 208V	L1@1.25	Ea	683.00	51.30	734.30
Size 1 10 HP, 480V	L1@1.25	Ea	683.00	51.30	734.30
Size 2 10 HP, 208V	L1@1.50	Ea	1,610.00	61.50	1,671.50
Size 2 15 HP, 230V	L1@1.50	Ea	1,610.00	61.50	1,671.50
Size 2 25 HP, 480V	L1@1.50	Ea	1,610.00	61.50	1,671.50
Size 3 25 HP, 208V	L1@1.75	Ea	2,690.00	71.80	2,761.80
Size 3 30 HP, 230V	L1@1.75	Ea	2,690.00	71.80	2,761.80
Size 3 50 HP, 480V	L1@1.75	Ea	2,690.00	71.80	2,761.80
Size 4 40 HP, 208V	L1@2.00	Ea	6,240.00	82.10	6,322.10
Size 4 75 HP, 230V	L1@2.00	Ea	6,240.00	82.10	6,322.10
Size 4 100 HP, 480V	L1@2.00	Ea	6,240.00	82.10	6,322.10

Use these figures to estimate the cost of magnetic contactors installed in buildings under the conditions described on pages 5 and 6. Costs listed are for each contactor installed. The crew size is one electrician working at a labor cost of $41.03 per manhour. These costs include layout, material handling, and normal waste. Add for supports, sales tax, delivery, supervision, mobilization, demobilization, cleanup, overhead and profit. These magnetic contactors do not provide overload protection. They are used as a relay device only. For motor overload protection, use magnetic starters with overload relays and heating elements. Other types of magnetic contactors are available for incandescent lamp loads and induction loads.

NEMA 4 five pole AC magnetic contactors without overload relays

Material	Craft@Hrs	Unit	Material Cost	Labor Cost	Installed Cost
Size 0 3 HP, 208V	L1@1.75	Ea	1,070.00	71.80	1,141.80
Size 0 5 HP, 480V	L1@1.75	Ea	1,070.00	71.80	1,141.80
Size 1 7.5 HP, 208V	L1@2.00	Ea	1,090.00	82.10	1,172.10
Size 1 10 HP, 480V	L1@2.00	Ea	1,090.00	82.10	1,172.10
Size 2 10 HP, 208V	L1@2.25	Ea	2,870.00	92.30	2,962.30
Size 2 15 HP, 230V	L1@2.25	Ea	2,870.00	92.30	2,962.30
Size 2 25 HP, 480V	L1@2.25	Ea	2,870.00	92.30	2,962.30
Size 3 25 HP, 208V	L1@2.50	Ea	4,230.00	103.00	4,333.00
Size 3 30 HP, 230V	L1@2.50	Ea	4,230.00	103.00	4,333.00
Size 3 50 HP, 480V	L1@2.50	Ea	4,230.00	103.00	4,333.00
Size 4 40 HP, 208V	L1@2.75	Ea	9,270.00	113.00	9,383.00
Size 4 75 HP, 230V	L1@2.75	Ea	9,270.00	113.00	9,383.00
Size 4 100 HP, 480V	L1@2.75	Ea	9,270.00	113.00	9,383.00

Open type five pole AC magnetic contactors without overload relays

Material	Craft@Hrs	Unit	Material Cost	Labor Cost	Installed Cost
Size 0 3 HP, 208V	L1@0.75	Ea	593.00	30.80	623.80
Size 0 5 HP, 480V	L1@0.75	Ea	593.00	30.80	623.80
Size 1 7.5 HP, 208V	L1@1.00	Ea	656.00	41.00	697.00
Size 1 10 HP, 480V	L1@1.00	Ea	656.00	41.00	697.00
Size 2 10 HP, 208V	L1@1.25	Ea	1,510.00	51.30	1,561.30
Size 2 15 HP, 230V	L1@1.25	Ea	1,510.00	51.30	1,561.30
Size 2 25 HP, 480V	L1@1.25	Ea	1,510.00	51.30	1,561.30
Size 3 25 HP, 208V	L1@1.50	Ea	2,390.00	61.50	2,451.50
Size 3 30 HP, 230V	L1@1.50	Ea	2,390.00	61.50	2,451.50
Size 3 50 HP, 480V	L1@1.50	Ea	2,390.00	61.50	2,451.50
Size 4 40 HP, 208V	L1@2.00	Ea	5,720.00	82.10	5,802.10
Size 4 75 HP, 230V	L1@2.00	Ea	5,720.00	82.10	5,802.10
Size 4 100 HP, 480V	L1@2.00	Ea	5,720.00	82.10	5,802.10

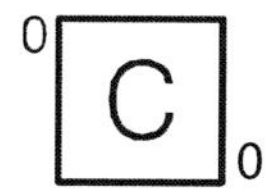

Use these figures to estimate the cost of magnetic contactors installed in buildings under the conditions described on pages 5 and 6. Costs listed are for each contactor installed. The crew size is one electrician working at a labor cost of $41.03 per manhour. These costs include layout, material handling, and normal waste. Add for supports, sales tax, delivery, supervision, mobilization, demobilization, cleanup, overhead and profit. These magnetic contactors do not provide overload protection. They are used as a relay device only. For motor overload protection, use magnetic starters with overload relays and heating elements. Other types of magnetic contactors are available for incandescent lamp loads and induction loads.

AC Magnetic Two Pole Starters with Overload Relays

NEMA 1 two pole AC magnetic starters with overload relays

Material	Craft@Hrs	Unit	Material Cost	Labor Cost	Installed Cost
Size 00 1/3 HP, 115V	L1@0.60	Ea	402.00	24.60	426.60
Size 00 1 HP, 230V	L1@0.60	Ea	402.00	24.60	426.60
Size 0 1 HP, 115V	L1@0.70	Ea	444.00	28.70	472.70
Size 0 2 HP, 230V	L1@0.70	Ea	444.00	28.70	472.70
Size 1 2 HP, 115V	L1@0.80	Ea	504.00	32.80	536.80
Size 1 3 HP, 230V	L1@0.80	Ea	504.00	32.80	536.80
Size 2 3 HP, 115V	L1@1.10	Ea	643.00	45.10	688.10
Size 2 7.5 HP, 230V	L1@1.10	Ea	643.00	45.10	688.10
Size 3 7.5 HP, 115V	L1@1.30	Ea	983.00	53.30	1,036.30
Size 3 15 HP, 230V	L1@1.30	Ea	983.00	53.30	1,036.30

NEMA 3R two pole AC magnetic starters with overload relays

Material	Craft@Hrs	Unit	Material Cost	Labor Cost	Installed Cost
Size 00 1/3 HP, 115V	L1@0.75	Ea	537.00	30.80	567.80
Size 00 1 HP, 230V	L1@0.75	Ea	537.00	30.80	567.80
Size 0 1 HP, 115V	L1@0.75	Ea	537.00	30.80	567.80
Size 0 2 HP, 230V	L1@0.75	Ea	537.00	30.80	567.80
Size 1 2 HP, 115V	L1@1.00	Ea	598.00	41.00	639.00
Size 1 3 HP, 230V	L1@1.00	Ea	598.00	41.00	639.00
Size 2 3 HP, 115V	L1@1.25	Ea	1,150.00	51.30	1,201.30
Size 2 7.5 HP, 230V	L1@1.25	Ea	1,150.00	51.30	1,201.30
Size 3 7.5 HP, 115V	L1@1.50	Ea	1,760.00	61.50	1,821.50
Size 3 15 HP, 230V	L1@1.50	Ea	1,760.00	61.50	1,821.50

NEMA 4 two pole AC magnetic starters with overload relays

Material	Craft@Hrs	Unit	Material Cost	Labor Cost	Installed Cost
Size 00 1/3 HP, 115V	L1@1.00	Ea	822.00	41.00	863.00
Size 00 1 HP, 230V	L1@1.00	Ea	822.00	41.00	863.00
Size 0 1 HP, 115V	L1@1.00	Ea	822.00	41.00	863.00
Size 0 2 HP, 230V	L1@1.00	Ea	822.00	41.00	863.00
Size 1 2 HP, 115V	L1@1.25	Ea	896.00	51.30	947.30
Size 1 3 HP, 230V	L1@1.25	Ea	896.00	51.30	947.30
Size 2 3 HP, 115V	L1@1.50	Ea	1,780.00	61.50	1,841.50
Size 2 7.5 HP, 230V	L1@1.50	Ea	1,780.00	61.50	1,841.50
Size 3 7.5 HP, 115V	L1@1.75	Ea	2,690.00	71.80	2,761.80
Size 3 15 HP, 230V	L1@1.75	Ea	2,690.00	71.80	2,761.80

Use these figures to estimate the cost of magnetic starters installed in buildings under the conditions described on pages 5 and 6. Costs listed are for each starter installed. The crew size is one electrician working at a labor cost of $41.03 per manhour. These costs include layout, material handling, and normal waste. Add for thermal overload devices, supports, sales tax, delivery, supervision, mobilization, demobilization, cleanup, overhead and profit. Note: Select the right size for load and voltage. To select the right overload relay element, find the full load amps rating opposite the appropriate voltage on the motor nameplate. Then check the manufacturer's overload element chart for the correct element or heater. Heater elements are inserted into the overload relay. If the running current exceeds the full load amperage rating, the heater element heats a bi-metallic switch which opens the control circuit and interrupts power to the motor.

AC Magnetic Two Pole Starters with Overload Relays

NEMA 4X two pole AC magnetic starters with overload relays

Material	Craft@Hrs	Unit	Material Cost	Labor Cost	Installed Cost
Size 00 1/3 HP, 115V	L1@1.00	Ea	822.00	41.00	863.00
Size 00 1 HP, 230V	L1@1.00	Ea	822.00	41.00	863.00
Size 0 1 HP, 115V	L1@1.00	Ea	822.00	41.00	863.00
Size 0 2 HP, 230V	L1@1.00	Ea	822.00	41.00	863.00
Size 1 2 HP, 115V	L1@1.25	Ea	896.00	51.30	947.30
Size 1 3 HP, 230V	L1@1.25	Ea	896.00	51.30	947.30
Size 2 3 HP, 115V	L1@1.50	Ea	1,780.00	61.50	1,841.50
Size 2 7.5 HP, 230V	L1@1.50	Ea	1,780.00	61.50	1,841.50
Size 3 7.5 HP, 115V	L1@1.75	Ea	2,690.00	71.80	2,761.80
Size 3 15 HP, 230V	L1@1.75	Ea	2,690.00	71.80	2,761.80

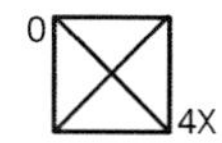

NEMA 7 & 9 two pole AC magnetic starters with overload relays

Material	Craft@Hrs	Unit	Material Cost	Labor Cost	Installed Cost
Size 00 1/3 HP, 115V	L1@1.25	Ea	1,790.00	51.30	1,841.30
Size 00 1 HP, 230V	L1@1.25	Ea	1,790.00	51.30	1,841.30
Size 0 1 HP, 115V	L1@1.25	Ea	1,790.00	51.30	1,841.30
Size 0 2 HP, 230V	L1@1.25	Ea	1,790.00	51.30	1,841.30
Size 1 2 HP, 115V	L1@1.50	Ea	1,880.00	61.50	1,941.50
Size 1 3 HP, 230V	L1@1.50	Ea	1,880.00	61.50	1,941.50
Size 2 3 HP, 115V	L1@1.75	Ea	2,990.00	71.80	3,061.80
Size 2 7.5 HP, 230V	L1@1.75	Ea	2,990.00	71.80	3,061.80

Open type two pole AC magnetic starters with overload relays

Material	Craft@Hrs	Unit	Material Cost	Labor Cost	Installed Cost
Size 00 1/3 HP, 115V	L1@0.50	Ea	288.00	20.50	308.50
Size 00 1 HP, 230V	L1@0.50	Ea	288.00	20.50	308.50
Size 0 1 HP, 115V	L1@0.50	Ea	361.00	20.50	381.50
Size 0 2 HP, 230V	L1@0.50	Ea	361.00	20.50	381.50
Size 1 2 HP, 115V	L1@0.70	Ea	421.00	28.70	449.70
Size 1 3 HP, 230V	L1@0.70	Ea	421.00	28.70	449.70
Size 2 3 HP, 115V	L1@1.00	Ea	756.00	41.00	797.00
Size 2 7.5 HP, 230V	L1@1.00	Ea	756.00	41.00	797.00
Size 3 7.5 HP, 115V	L1@1.25	Ea	1,170.00	51.30	1,221.30
Size 3 15 HP, 230V	L1@1.25	Ea	1,170.00	51.30	1,221.30

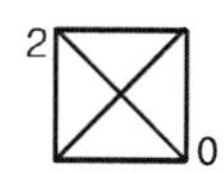

Use these figures to estimate the cost of magnetic starters installed in buildings under the conditions described on pages 5 and 6. Costs listed are for each starter installed. The crew size is one electrician working at a labor cost of $41.03 per manhour. These costs include layout, material handling, and normal waste. Add for thermal overload devices, supports, sales tax, delivery, supervision, mobilization, demobilization, cleanup, overhead and profit. Note: Select the right size for load and voltage. To select the right overload relay element, find the full load amps rating opposite the appropriate voltage on the motor nameplate. Then check the manufacturer's overload element chart for the correct element or heater. Heater elements are inserted into the overload relay. If the running current exceeds the full load amperage rating, the heater element heats a bi-metallic switch which opens the control circuit and interrupts power to the motor.

AC Magnetic Three Pole Starters with Overload Relays

NEMA 1 three pole AC magnetic starters with overload relays

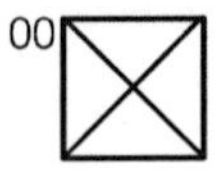

Material	Craft@Hrs	Unit	Material Cost	Labor Cost	Installed Cost
Size 00 1.5 HP, 208V	L1@0.70	Ea	368.00	28.70	396.70
Size 00 2 HP, 480V	L1@0.70	Ea	368.00	28.70	396.70
Size 0 3 HP, 208V	L1@0.80	Ea	413.00	32.80	445.80
Size 0 5 HP, 480V	L1@0.80	Ea	413.00	32.80	445.80
Size 1 7.5 HP, 208V	L1@1.00	Ea	476.00	41.00	517.00
Size 1 10 HP, 480V	L1@1.00	Ea	476.00	41.00	517.00
Size 2 10 HP, 208V	L1@1.25	Ea	822.00	51.30	873.30
Size 2 15 HP, 230V	L1@1.25	Ea	822.00	51.30	873.30
Size 2 25 HP, 480V	L1@1.25	Ea	822.00	51.30	873.30
Size 3 25 HP, 208V	L1@1.50	Ea	1,300.00	61.50	1,361.50
Size 3 30 HP, 230V	L1@1.50	Ea	1,300.00	61.50	1,361.50
Size 3 50 HP, 480V	L1@1.50	Ea	1,300.00	61.50	1,361.50
Size 4 40 HP, 208V	L1@2.25	Ea	3,010.00	92.30	3,102.30
Size 4 75 HP, 230V	L1@2.25	Ea	3,010.00	92.30	3,102.30
Size 4 100 HP, 480V	L1@2.25	Ea	3,010.00	92.30	3,102.30
Size 5 75 HP, 208V	L2@4.50	Ea	6,520.00	185.00	6,705.00
Size 5 100 HP, 230V	L2@4.50	Ea	6,520.00	185.00	6,705.00
Size 5 200 HP, 480V	L2@4.50	Ea	6,520.00	185.00	6,705.00
Size 6 150 HP, 208V	L2@6.50	Ea	17,600.00	267.00	17,867.00
Size 6 200 HP, 230V	L2@6.50	Ea	17,600.00	267.00	17,867.00
Size 6 400 HP, 480V	L2@6.50	Ea	17,600.00	267.00	17,867.00
Size 7 300 HP, 230V	L2@8.50	Ea	25,200.00	349.00	25,549.00
Size 7 600 HP, 480V	L2@8.50	Ea	25,200.00	349.00	25,549.00

NEMA 3R three pole AC magnetic starters with overload relays

Material	Craft@Hrs	Unit	Material Cost	Labor Cost	Installed Cost
Size 00 1.5 HP, 208V	L1@0.80	Ea	592.00	32.80	624.80
Size 00 2 HP, 480V	L1@0.80	Ea	592.00	32.80	624.80
Size 0 3 HP, 208V	L1@1.00	Ea	592.00	41.00	633.00
Size 0 5 HP, 480V	L1@1.00	Ea	592.00	41.00	633.00
Size 1 7.5 HP, 208V	L1@1.25	Ea	655.00	51.30	706.30
Size 1 10 HP, 480V	L1@1.25	Ea	655.00	51.30	706.30
Size 2 10 HP, 208V	L1@1.50	Ea	1,230.00	61.50	1,291.50
Size 2 15 HP, 230V	L1@1.50	Ea	1,230.00	61.50	1,291.50
Size 2 25 HP, 480V	L1@1.50	Ea	1,230.00	61.50	1,291.50
Size 3 25 HP, 208V	L1@1.75	Ea	1,880.00	71.80	1,951.80
Size 3 30 HP, 230V	L1@1.75	Ea	1,880.00	71.80	1,951.80
Size 3 50 HP, 480V	L1@1.75	Ea	1,880.00	71.80	1,951.80
Size 4 40 HP, 208V	L1@2.00	Ea	4,630.00	82.10	4,712.10
Size 4 75 HP, 230V	L1@2.00	Ea	4,630.00	82.10	4,712.10
Size 4 100 HP, 480V	L1@2.00	Ea	4,630.00	82.10	4,712.10

Use these figures to estimate the cost of magnetic starters installed in buildings under the conditions described on pages 5 and 6. Costs listed are for each starter installed. The crew size is one electrician for starters up to size 4 and two electricians for larger starters. The labor cost per manhour is $41.03. These costs include layout, material handling, and normal waste. Add for thermal overload devices, supports, sales tax, delivery, supervision, mobilization, demobilization, cleanup, overhead and profit. Note: Select the right size for load and voltage. To select the right overload relay element, find the full load amps rating opposite the appropriate voltage on the motor nameplate. Then check the manufacturer's overload element chart for the correct element or heater. Heater elements are inserted into the overload relay. If the running current exceeds the full load amperage rating, the heater element heats a bi-metallic switch which opens the control circuit and interrupts power to the motor.

AC Magnetic Three Pole Starters with Overload Relays

Material	Craft@Hrs	Unit	Material Cost	Labor Cost	Installed Cost
NEMA 4 three pole AC magnetic starters with overload relays					
Size 00 1.5 HP, 208V	L1@1.25	Ea	879.00	51.30	930.30
Size 00 2 HP, 480V	L1@1.25	Ea	879.00	51.30	930.30
Size 0 3 HP, 208V	L1@1.25	Ea	879.00	51.30	930.30
Size 0 5 HP, 480V	L1@1.25	Ea	879.00	51.30	930.30
Size 1 7.5 HP, 208V	L1@1.50	Ea	952.00	61.50	1,013.50
Size 1 10 HP, 480V	L1@1.50	Ea	952.00	61.50	1,013.50
Size 2 10 HP, 208V	L1@1.75	Ea	1,860.00	71.80	1,931.80
Size 2 15 HP, 230V	L1@1.75	Ea	1,860.00	71.80	1,931.80
Size 2 25 HP, 480V	L1@1.75	Ea	1,860.00	71.80	1,931.80
Size 3 25 HP, 208V	L1@2.00	Ea	3,490.00	82.10	3,572.10
Size 3 30 HP, 230V	L1@2.00	Ea	3,490.00	82.10	3,572.10
Size 3 50 HP, 480V	L1@2.00	Ea	3,490.00	82.10	3,572.10
Size 4 40 HP, 208V	L1@2.50	Ea	7,130.00	103.00	7,233.00
Size 4 75 HP, 230V	L1@2.50	Ea	7,130.00	103.00	7,233.00
Size 4 100 HP, 480V	L1@2.50	Ea	7,130.00	103.00	7,233.00
Size 5 75 HP, 208V	L2@5.50	Ea	10,200.00	226.00	10,426.00
Size 5 100 HP, 230V	L2@5.50	Ea	10,200.00	226.00	10,426.00
Size 5 200 HP, 480V	L2@5.50	Ea	10,200.00	226.00	10,426.00
Size 6 150 HP, 208V	L2@7.00	Ea	28,500.00	287.00	28,787.00
Size 6 200 HP, 230V	L2@7.00	Ea	28,500.00	287.00	28,787.00
Size 6 400 HP, 480V	L2@7.00	Ea	28,500.00	287.00	28,787.00
Size 7 300 HP, 230V	L2@9.00	Ea	35,500.00	369.00	35,869.00
Size 7 600 HP, 480V	L2@9.00	Ea	35,500.00	369.00	35,869.00
NEMA 4X three pole AC magnetic starters with overload relays					
Size 00 1.5 HP, 208V	L1@1.25	Ea	879.00	51.30	930.30
Size 00 2 HP, 480V	L1@1.25	Ea	879.00	51.30	930.30
Size 0 3 HP, 208V	L1@1.25	Ea	879.00	51.30	930.30
Size 0 5 HP, 480V	L1@1.25	Ea	879.00	51.30	930.30
Size 1 7.5 HP, 208V	L1@1.50	Ea	952.00	61.50	1,013.50
Size 1 10 HP, 480V	L1@1.50	Ea	952.00	61.50	1,013.50
Size 2 10 HP, 208V	L1@1.75	Ea	1,860.00	71.80	1,931.80
Size 2 15 HP. 230V	L1@1.75	Ea	1,860.00	71.80	1,931.80
Size 2 25 HP, 480V	L1@1.75	Ea	1,860.00	71.80	1,931.80

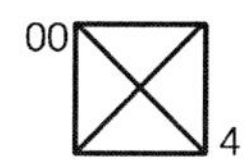

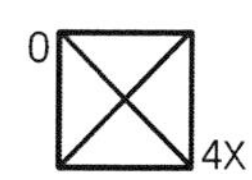

Use these figures to estimate the cost of magnetic starters installed in buildings under the conditions described on pages 5 and 6. Costs listed are for each starter installed. The crew size is one electrician for starters up to size 4 and two electricians for larger starters. The labor cost per manhour is $41.03. These costs include layout, material handling, and normal waste. Add for thermal overload devices, supports, sales tax, delivery, supervision, mobilization, demobilization, cleanup, overhead and profit. Note: Select the right size for load and voltage. To select the right overload relay element, find the full load amps rating opposite the appropriate voltage on the motor nameplate. Then check the manufacturer's overload element chart for the correct element or heater. Heater elements are inserted into the overload relay. If the running current exceeds the full load amperage rating, the heater element heats a bi-metallic switch which opens the control circuit and interrupts power to the motor.

AC Magnetic Three Pole Starters with Overload Relays

NEMA 4X three pole AC magnetic starters with overload relays

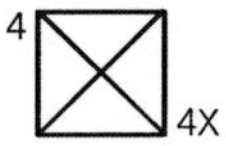

Material	Craft@Hrs	Unit	Material Cost	Labor Cost	Installed Cost
Size 3 25 HP, 208V	L1@2.00	Ea	3,490.00	82.10	3,572.10
Size 3 30 HP, 230V	L1@2.00	Ea	3,490.00	82.10	3,572.10
Size 3 50 HP, 480V	L1@2.00	Ea	3,490.00	82.10	3,572.10
Size 4 40 HP, 208V	L2@2.50	Ea	7,130.00	103.00	7,233.00
Size 4 75 HP, 230V	L2@2.50	Ea	7,130.00	103.00	7,233.00
Size 4 100 HP, 480V	L2@2.50	Ea	7,130.00	103.00	7,233.00

NEMA 7 & 9 three pole AC magnetic starters with overload relays

Material	Craft@Hrs	Unit	Material Cost	Labor Cost	Installed Cost
Size 00 1.5 HP, 208V	L1@1.50	Ea	1,870.00	61.50	1,931.50
Size 00 2 HP, 480V	L1@1.50	Ea	1,870.00	61.50	1,931.50
Size 0 3 HP, 208V	L1@1.50	Ea	1,870.00	61.50	1,931.50
Size 0 5 HP, 480V	L1@1.50	Ea	1,870.00	61.50	1,931.50
Size 1 7.5 HP, 208V	L1@1.75	Ea	1,950.00	71.80	2,021.80
Size 1 10 HP, 480V	L1@1.75	Ea	1,950.00	71.80	2,021.80
Size 2 10 HP, 208V	L1@2.00	Ea	3,090.00	82.10	3,172.10
Size 2 15 HP, 230V	L1@2.00	Ea	3,090.00	82.10	3,172.10
Size 2 25 HP, 480V	L1@2.00	Ea	3,090.00	82.10	3,172.10
Size 3 25 HP, 208V	L1@2.25	Ea	4,580.00	92.30	4,672.30
Size 3 30 HP, 230V	L1@2.25	Ea	4,580.00	92.30	4,672.30
Size 3 50 HP, 480V	L1@2.25	Ea	4,580.00	92.30	4,672.30
Size 4 40 HP, 208V	L1@3.00	Ea	7,370.00	123.00	7,493.00
Size 4 75 HP, 230V	L1@3.00	Ea	7,370.00	123.00	7,493.00
Size 4 100 HP, 480V	L1@3.00	Ea	7,370.00	123.00	7,493.00
Size 5 75 HP, 208V	L2@6.00	Ea	19,000.00	246.00	19,246.00
Size 5 100 HP, 230V	L2@6.00	Ea	19,000.00	246.00	19,246.00
Size 5 200 HP, 480V	L2@6.00	Ea	19,000.00	246.00	19,246.00
Size 6 150 HP, 208V	L2@8.00	Ea	46,700.00	328.00	47,028.00
Size 6 200 HP, 230V	L2@8.00	Ea	46,700.00	328.00	47,028.00
Size 6 400 HP, 480V	L2@8.00	Ea	46,700.00	328.00	47,028.00

NEMA 12 three pole AC magnetic starters with overload relays

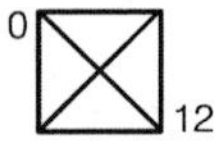

Material	Craft@Hrs	Unit	Material Cost	Labor Cost	Installed Cost
Size 00 1.5 HP, 208V	L1@0.80	Ea	592.00	32.80	624.80
Size 00 2 HP, 480V	L1@0.80	Ea	592.00	32.80	624.80
Size 0 3 HP, 208V	L1@1.00	Ea	592.00	41.00	633.00
Size 0 5 HP, 480V	L1@1.00	Ea	592.00	41.00	633.00
Size 1 7.5 HP, 208V	L1@1.25	Ea	655.00	51.30	706.30
Size 1 10 HP, 480V	L1@1.25	Ea	655.00	51.30	706.30

Use these figures to estimate the cost of magnetic starters installed in buildings under the conditions described on pages 5 and 6. Costs listed are for each starter installed. The crew size is one electrician for starters up to size 4 and two electricians for larger starters. The labor cost per manhour is $41.03. These costs include layout, material handling, and normal waste. Add for thermal overload devices, supports, sales tax, delivery, supervision, mobilization, demobilization, cleanup, overhead and profit. Note: Select the right size for load and voltage. To select the right overload relay element, find the full load amps rating opposite the appropriate voltage on the motor nameplate. Then check the manufacturer's overload element chart for the correct element or heater. Heater elements are inserted into the overload relay. If the running current exceeds the full load amperage rating, the heater element heats a bi-metallic switch which opens the control circuit and interrupts power to the motor.

AC Magnetic Three Pole Starters with Overload Relays

NEMA 12 three pole AC magnetic starters with overload relays

Material	Craft@Hrs	Unit	Material Cost	Labor Cost	Installed Cost
Size 2 10 HP, 208V	L1@1.50	Ea	1,230.00	61.50	1,291.50
Size 2 15 HP, 230V	L1@1.50	Ea	1,230.00	61.50	1,291.50
Size 2 25 HP, 480V	L1@1.50	Ea	1,230.00	61.50	1,291.50
Size 3 25 HP, 208V	L1@1.75	Ea	1,880.00	71.80	1,951.80
Size 3 30 HP, 230V	L1@1.75	Ea	1,880.00	71.80	1,951.80
Size 3 50 HP, 480V	L1@1.75	Ea	1,880.00	71.80	1,951.80
Size 4 40 HP, 208V	L1@2.00	Ea	4,630.00	82.10	4,712.10
Size 4 75 HP, 230V	L1@2.00	Ea	4,630.00	82.10	4,712.10
Size 4 100 HP, 480V	L1@2.00	Ea	4,630.00	82.10	4,712.10
Size 5 75 HP, 208V	L2@5.00	Ea	10,200.00	205.00	10,405.00
Size 5 100 HP, 230V	L2@5.00	Ea	10,200.00	205.00	10,405.00
Size 5 200 HP, 480V	L2@5.00	Ea	10,200.00	205.00	10,405.00
Size 6 150 HP, 208V	L2@7.00	Ea	25,000.00	287.00	25,287.00
Size 6 200 HP, 230V	L2@7.00	Ea	25,000.00	287.00	25,287.00
Size 6 400 HP, 480V	L2@7.00	Ea	25,000.00	287.00	25,287.00
Size 7 300 HP, 230V	L2@9.00	Ea	32,900.00	369.00	33,269.00
Size 7 600 HP, 480V	L2@9.00	Ea	32,900.00	369.00	33,269.00

Open type three pole AC magnetic starters with overload relays

Material	Craft@Hrs	Unit	Material Cost	Labor Cost	Installed Cost
Size 00 1.5 HP, 208V	L1@0.70	Ea	342.00	28.70	370.70
Size 00 2 HP, 480V	L1@0.70	Ea	342.00	28.70	370.70
Size 0 3 HP, 208V	L1@0.70	Ea	413.00	28.70	441.70
Size 0 5 HP, 480V	L1@0.70	Ea	413.00	28.70	441.70
Size 1 7.5 HP, 208V	L1@0.90	Ea	476.00	36.90	512.90
Size 1 10 HP, 480V	L1@0.90	Ea	476.00	36.90	512.90
Size 2 10 HP, 208V	L1@1.10	Ea	822.00	45.10	867.10
Size 2 15 HP, 230V	L1@1.10	Ea	822.00	45.10	867.10
Size 2 25 HP, 480V	L1@1.10	Ea	822.00	45.10	867.10
Size 3 25 HP, 208V	L1@1.20	Ea	1,300.00	49.20	1,349.20
Size 3 30 HP, 230V	L1@1.20	Ea	1,300.00	49.20	1,349.20
Size 3 50 HP, 480V	L1@1.20	Ea	1,300.00	49.20	1,349.20
Size 4 40 HP, 208V	L1@2.00	Ea	3,010.00	82.10	3,092.10
Size 4 75 HP, 230V	L1@2.00	Ea	3,010.00	82.10	3,092.10
Size 4 100 HP, 480V	L1@2.00	Ea	3,010.00	82.10	3,092.10
Size 5 75 HP, 208V	L2@4.25	Ea	6,520.00	174.00	6,694.00
Size 5 100 HP, 230V	L2@4.25	Ea	6,520.00	174.00	6,694.00
Size 5 200 HP, 480V	L2@4.25	Ea	6,520.00	174.00	6,694.00
Size 6 150 HP, 208V	L2@6.25	Ea	17,600.00	256.00	17,856.00
Size 6 200 HP, 230V	L2@6.25	Ea	17,600.00	256.00	17,856.00
Size 6 400 HP, 480V	L2@6.25	Ea	17,600.00	256.00	17,856.00
Size 7 300 HP, 230V	L2@8.25	Ea	25,200.00	338.00	25,538.00
Size 7 600 HP, 480V	L2@8.25	Ea	25,200.00	338.00	25,538.00

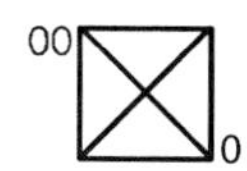

Use these figures to estimate the cost of magnetic starters installed in buildings under the conditions described on pages 5 and 6. Costs listed are for each starter installed. The crew size is one electrician for starters up to size 4 and two electricians for larger starters. The labor cost per manhour is $41.03. These costs include layout, material handling, and normal waste. Add for thermal overload devices, supports, sales tax, delivery, supervision, mobilization, demobilization, cleanup, overhead and profit. Note: Select the right size for load and voltage. To select the right overload relay element, find the full load amps rating opposite the appropriate voltage on the motor nameplate. Then check the manufacturer's overload element chart for the correct element or heater. Heater elements are inserted into the overload relay. If the running current exceeds the full load amperage rating, the heater element heats a bi-metallic switch which opens the control circuit and interrupts power to the motor.

Combination AC Magnetic Three Pole Starters with Non-fused Disconnect and Overload Relays

NEMA 1 combination three pole AC magnetic starters, non-fused disconnect

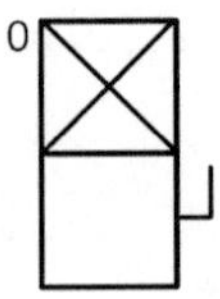

Material		Craft@Hrs	Unit	Material Cost	Labor Cost	Installed Cost
Size 0	3 HP, 208V	L1@1.25	Ea	1,180.00	51.30	1,231.30
Size 0	3 HP, 230V	L1@1.25	Ea	1,180.00	51.30	1,231.30
Size 0	5 HP, 480V	L1@1.25	Ea	1,180.00	51.30	1,231.30
Size 1	7.5 HP, 208V	L1@1.25	Ea	1,260.00	51.30	1,311.30
Size 1	7.5 HP, 230V	L1@1.25	Ea	1,260.00	51.30	1,311.30
Size 1	10 HP, 480V	L1@1.25	Ea	1,260.00	51.30	1,311.30
Size 2	10 HP, 208V	L1@1.50	Ea	1,950.00	61.50	2,011.50
Size 2	15 HP, 230V	L1@1.50	Ea	1,950.00	61.50	2,011.50
Size 2	25 HP, 480V	L1@1.50	Ea	1,950.00	61.50	2,011.50
Size 3	25 HP, 208V	L1@2.00	Ea	3,180.00	82.10	3,262.10
Size 3	30 HP, 230V	L1@2.00	Ea	3,180.00	82.10	3,262.10
Size 3	50 HP, 480V	L1@2.00	Ea	3,180.00	82.10	3,262.10
Size 4	40 HP, 208V	L1@3.00	Ea	6,100.00	123.00	6,223.00
Size 4	50 HP, 230V	L1@3.00	Ea	6,100.00	123.00	6,223.00
Size 4	100 HP, 480V	L1@3.00	Ea	6,100.00	123.00	6,223.00
Size 5	75 HP, 208V	L2@6.00	Ea	13,600.00	246.00	13,846.00
Size 5	100 HP, 230V	L2@6.00	Ea	13,600.00	246.00	13,846.00
Size 5	200 HP, 480V	L2@6.00	Ea	13,600.00	246.00	13,846.00
Size 6	150 HP, 208V	L2@8.00	Ea	35,800.00	328.00	36,128.00
Size 6	200 HP, 230V	L2@8.00	Ea	35,800.00	328.00	36,128.00
Size 6	400 HP, 480V	L2@8.00	Ea	35,800.00	328.00	36,128.00

NEMA 4 combination three pole AC magnetic starters, non-fused disconnect

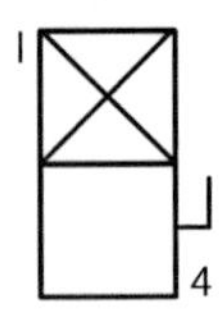

Material		Craft@Hrs	Unit	Material Cost	Labor Cost	Installed Cost
Size 0	3 HP, 208V	L1@1.50	Ea	2,320.00	61.50	2,381.50
Size 0	3 HP, 230V	L1@1.50	Ea	2,320.00	61.50	2,381.50
Size 0	5 HP, 480V	L1@1.50	Ea	2,320.00	61.50	2,381.50
Size 1	7.5 HP, 208V	L1@1.50	Ea	2,410.00	61.50	2,471.50
Size 1	7.5 HP, 230V	L1@1.50	Ea	2,410.00	61.50	2,471.50
Size 1	10 HP, 480V	L1@1.50	Ea	2,410.00	61.50	2,471.50
Size 2	10 HP, 208V	L1@2.00	Ea	3,720.00	82.10	3,802.10
Size 2	15 HP, 230V	L1@2.00	Ea	3,720.00	82.10	3,802.10
Size 2	25 HP, 480V	L1@2.00	Ea	3,720.00	82.10	3,802.10
Size 3	25 HP, 208V	L1@3.00	Ea	6,330.00	123.00	6,453.00
Size 3	30 HP, 230V	L1@3.00	Ea	6,330.00	123.00	6,453.00
Size 3	50 HP, 480V	L1@3.00	Ea	6,330.00	123.00	6,453.00
Size 4	40 HP, 208V	L1@4.00	Ea	10,100.00	164.00	10,264.00
Size 4	50 HP, 230V	L1@4.00	Ea	10,100.00	164.00	10,264.00
Size 4	100 HP, 480V	L1@4.00	Ea	10,100.00	164.00	10,264.00
Size 5	75 HP, 208V	L2@7.00	Ea	24,100.00	287.00	24,387.00
Size 5	100 HP, 230V	L2@7.00	Ea	24,100.00	287.00	24,387.00
Size 5	200 HP, 480V	L2@7.00	Ea	24,100.00	287.00	24,387.00
Size 6	150 HP, 208V	L2@10.0	Ea	45,700.00	410.00	46,110.00
Size 6	200 HP, 230V	L2@10.0	Ea	45,700.00	410.00	46,110.00
Size 6	400 HP, 480V	L2@10.0	Ea	45,700.00	410.00	46,110.00

Use these figures to estimate the cost of combination starters installed in buildings under the conditions described on pages 5 and 6. Costs listed are for each starter installed. The crew size is one electrician for starters up to size 4 and two electricians for starters over size 4. Cost per manhour is $41.03. These costs include layout, material handling, and normal waste. Add for overload relay heater elements, sales tax, delivery, supervision, mobilization, demobilization, cleanup, overhead and profit. Note: AC magnetic starters are non-reversing with three melting alloy overload relays for motor protection. The heating elements must be appropriate for the full load amp rating of the connected motor.

Combination AC Magnetic Three Pole Starters with Non-fused Disconnect and Overload Relays

NEMA 4X combination three pole AC magnetic starters, non-fused disconnect

Material		Craft@Hrs	Unit	Material Cost	Labor Cost	Installed Cost
Size 0	3 HP, 208V	L1@1.75	Ea	2,390.00	71.80	2,461.80
Size 0	3 HP, 230V	L1@1.75	Ea	2,390.00	71.80	2,461.80
Size 0	5 HP, 480V	L1@1.75	Ea	2,390.00	71.80	2,461.80
Size 1	7.5 HP, 208V	L1@1.75	Ea	2,440.00	71.80	2,511.80
Size 1	7.5 HP, 230V	L1@1.75	Ea	2,440.00	71.80	2,511.80
Size 1	10 HP, 480V	L1@1.75	Ea	2,440.00	71.80	2,511.80
Size 2	10 HP, 208V	L1@2.00	Ea	3,780.00	82.10	3,862.10
Size 2	15 HP, 230V	L1@2.00	Ea	3,780.00	82.10	3,862.10
Size 2	25 HP, 480V	L1@2.00	Ea	3,780.00	82.10	3,862.10
Size 3	25 HP, 208V	L1@2.50	Ea	6,390.00	103.00	6,493.00
Size 3	30 HP, 230V	L1@2.50	Ea	6,390.00	103.00	6,493.00
Size 3	50 HP, 480V	L1@2.50	Ea	6,390.00	103.00	6,493.00

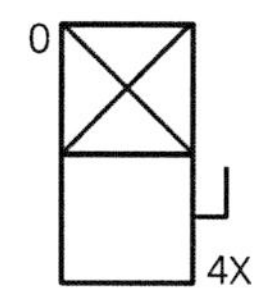

NEMA 12 combination three pole AC magnetic starters, non-fused disconnect

Material		Craft@Hrs	Unit	Material Cost	Labor Cost	Installed Cost
Size 0	3 HP, 208V	L1@1.50	Ea	1,510.00	61.50	1,571.50
Size 0	3 HP, 230V	L1@1.50	Ea	1,510.00	61.50	1,571.50
Size 0	5 HP, 480V	L1@1.50	Ea	1,510.00	61.50	1,571.50
Size 1	7.5 HP, 208V	L1@1.50	Ea	1,550.00	61.50	1,611.50
Size 1	7.5 HP, 230V	L1@1.50	Ea	1,550.00	61.50	1,611.50
Size 1	10 HP, 480V	L1@1.50	Ea	1,550.00	61.50	1,611.50
Size 2	10 HP, 208V	L1@2.00	Ea	2,370.00	82.10	2,452.10
Size 2	15 HP, 230V	L1@2.00	Ea	2,370.00	82.10	2,452.10
Size 2	25 HP, 480V	L1@2.00	Ea	2,370.00	82.10	2,452.10
Size 3	25 HP, 208V	L1@3.00	Ea	3,720.00	123.00	3,843.00
Size 3	30 HP, 230V	L1@3.00	Ea	3,720.00	123.00	3,843.00
Size 3	50 HP, 480V	L1@3.00	Ea	3,720.00	123.00	3,843.00
Size 4	40 HP, 208V	L1@4.00	Ea	7,620.00	164.00	7,784.00
Size 4	50 HP, 230V	L1@4.00	Ea	7,620.00	164.00	7,784.00
Size 4	100 HP, 480V	L1@4.00	Ea	7,620.00	164.00	7,784.00
Size 5	75 HP, 208V	L2@7.00	Ea	17,300.00	287.00	17,587.00
Size 5	100 HP, 230V	L2@7.00	Ea	17,300.00	287.00	17,587.00
Size 5	200 HP, 480V	L2@7.00	Ea	17,300.00	287.00	17,587.00
Size 6	150 HP, 208V	L2@10.0	Ea	39,800.00	410.00	40,210.00
Size 6	200 HP, 230V	L2@10.0	Ea	39,800.00	410.00	40,210.00
Size 6	400 HP, 480V	L2@10.0	Ea	39,800.00	410.00	40,210.00

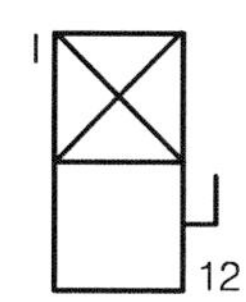

Use these figures to estimate the cost of combination starters installed in buildings under the conditions described on pages 5 and 6. Costs listed are for each starter installed. The crew size is one electrician for starters up to size 4 and two electricians for starters over size 4. Cost per manhour is $41.03. These costs include layout, material handling, and normal waste. Add for overload relay heater elements, sales tax, delivery, supervision, mobilization, demobilization, cleanup, overhead and profit. Note: AC magnetic starters are non-reversing with three melting alloy overload relays for motor protection. The heating elements must be appropriate for the full load amp rating of the connected motor.

Combination AC Magnetic Three Pole Starters with Fusible Disconnect and Overload Relays

NEMA 1 combination three pole AC magnetic starters, fusible disconnect

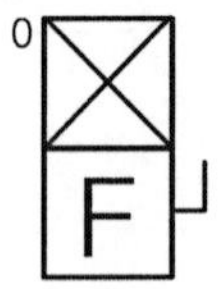

Material	Craft@Hrs	Unit	Material Cost	Labor Cost	Installed Cost
Size 0 3 HP, 208V	L1@1.25	Ea	1,230.00	51.30	1,281.30
Size 0 3 HP, 230V	L1@1.50	Ea	1,230.00	61.50	1,291.50
Size 0 5 HP, 480V	L1@1.50	Ea	1,230.00	61.50	1,291.50
Size 1 5 HP, 208V	L1@1.25	Ea	1,300.00	51.30	1,351.30
Size 1 7.5 HP, 208V	L1@1.25	Ea	1,300.00	51.30	1,351.30
Size 1 5 HP, 230V	L1@1.25	Ea	1,300.00	51.30	1,351.30
Size 1 7.5 HP, 230V	L1@1.25	Ea	1,300.00	51.30	1,351.30
Size 1 10 HP, 480V	L1@1.25	Ea	1,300.00	51.30	1,351.30
Size 2 10 HP, 208V	L1@1.50	Ea	1,990.00	61.50	2,051.50
Size 2 15 HP, 230V	L1@1.50	Ea	1,990.00	61.50	2,051.50
Size 2 25 HP, 480V	L1@1.50	Ea	1,990.00	61.50	2,051.50
Size 3 20 HP, 208V	L1@2.00	Ea	3,330.00	82.10	3,412.10
Size 3 25 HP, 208V	L1@2.00	Ea	3,600.00	82.10	3,682.10
Size 3 25 HP, 230V	L1@2.00	Ea	3,340.00	82.10	3,422.10
Size 3 50 HP, 480V	L1@2.00	Ea	3,330.00	82.10	3,412.10
Size 4 40 HP, 208V	L1@3.00	Ea	6,300.00	123.00	6,423.00
Size 4 50 HP, 230V	L1@3.00	Ea	6,300.00	123.00	6,423.00
Size 4 100 HP, 480V	L1@3.00	Ea	6,480.00	123.00	6,603.00
Size 5 75 HP, 208V	L2@4.00	Ea	13,900.00	164.00	14,064.00
Size 5 100 HP, 230V	L2@4.00	Ea	13,900.00	164.00	14,064.00
Size 5 200 HP, 480V	L2@4.00	Ea	13,900.00	164.00	14,064.00
Size 6 150 HP, 208V	L2@8.00	Ea	36,700.00	328.00	37,028.00
Size 6 200 HP, 230V	L2@8.00	Ea	36,700.00	328.00	37,028.00
Size 6 400 HP, 480V	L2@8.00	Ea	36,700.00	328.00	37,028.00

NEMA 4 combination three pole AC magnetic starters, fusible disconnect

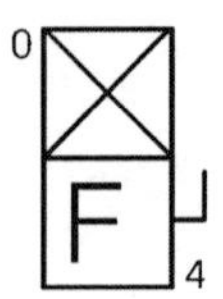

Material	Craft@Hrs	Unit	Material Cost	Labor Cost	Installed Cost
Size 0 3 HP, 208V	L1@1.50	Ea	2,430.00	61.50	2,491.50
Size 0 3 HP, 230V	L1@1.50	Ea	2,430.00	61.50	2,491.50
Size 0 5 HP, 480V	L1@1.50	Ea	2,430.00	61.50	2,491.50
Size 1 5 HP, 208V	L1@1.50	Ea	2,460.00	61.50	2,521.50
Size 1 7.5 HP, 208V	L1@1.50	Ea	2,490.00	61.50	2,551.50
Size 1 5 HP, 230V	L1@1.50	Ea	1,300.00	61.50	1,361.50
Size 1 7.5 HP, 230V	L1@1.50	Ea	1,300.00	61.50	1,361.50
Size 1 10 HP, 480V	L1@1.50	Ea	1,300.00	61.50	1,361.50
Size 2 10 HP, 208V	L1@2.00	Ea	1,990.00	82.10	2,072.10
Size 2 15 HP, 230V	L1@2.00	Ea	1,990.00	82.10	2,072.10
Size 2 25 HP, 480V	L1@2.00	Ea	2,030.00	82.10	2,112.10

Use these figures to estimate the cost of combination starters installed in buildings under the conditions described on pages 5 and 6. Costs listed are for each starter installed. The crew size is one electrician for starters up to size 4 and two electricians for starters over size 4. Cost per manhour is $41.03. These costs include layout, material handling, and normal waste. Add for overload relay heater elements, sales tax, delivery, supervision, mobilization, demobilization, cleanup, overhead and profit. Note: AC magnetic starters are non-reversing with three melting alloy overload relays for motor protection. The heating elements must be appropriate for the full load amp rating of the connected motor.

Fuses are on pages 283 to 305.

Combination AC Magnetic Three Pole Starters with Fusible Disconnect and Overload Relays

NEMA 4 combination three pole AC magnetic starters, fusible disconnect

Material		Craft@Hrs	Unit	Material Cost	Labor Cost	Installed Cost
Size 3	20 HP, 208V	L1@3.00	Ea	3,330.00	123.00	3,453.00
Size 3	25 HP, 208V	L1@3.00	Ea	3,540.00	123.00	3,663.00
Size 3	25 HP, 230V	L1@3.00	Ea	3,330.00	123.00	3,453.00
Size 3	30 HP, 230V	L1@3.00	Ea	3,600.00	123.00	3,723.00
Size 3	50 HP, 480V	L1@3.00	Ea	3,300.00	123.00	3,423.00
Size 4	40 HP, 208V	L1@4.00	Ea	6,300.00	164.00	6,464.00
Size 4	50 HP, 230V	L1@4.00	Ea	6,300.00	164.00	6,464.00
Size 4	100 HP, 480V	L1@4.00	Ea	6,350.00	164.00	6,514.00
Size 5	75 HP, 208V	L2@7.00	Ea	13,900.00	287.00	14,187.00
Size 5	100 HP, 230V	L2@7.00	Ea	13,900.00	287.00	14,187.00
Size 5	200 HP, 480V	L2@7.00	Ea	13,900.00	287.00	14,187.00
Size 6	150 HP, 208V	L2@10.0	Ea	36,700.00	410.00	37,110.00
Size 6	200 HP, 230V	L2@10.0	Ea	36,700.00	410.00	37,110.00
Size 6	400 HP, 480V	L2@10.0	Ea	36,700.00	410.00	37,110.00

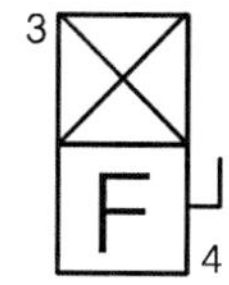

NEMA 4X combination three pole AC magnetic starters, fusible disconnect

Material		Craft@Hrs	Unit	Material Cost	Labor Cost	Installed Cost
Size 0	3 HP, 208V	L1@1.75	Ea	2,430.00	71.80	2,501.80
Size 0	3 HP, 230V	L1@1.75	Ea	2,430.00	71.80	2,501.80
Size 0	5 HP, 480V	L1@1.75	Ea	2,430.00	71.80	2,501.80
Size 1	5 HP, 208V	L1@1.75	Ea	2,540.00	71.80	2,611.80
Size 1	7.5 HP, 208V	L1@1.75	Ea	2,540.00	71.80	2,611.80
Size 1	5 HP, 230V	L1@1.75	Ea	1,300.00	71.80	1,371.80
Size 1	7.5 HP, 230V	L1@1.75	Ea	1,300.00	71.80	1,371.80
Size 1	10 HP, 480V	L1@1.75	Ea	1,300.00	71.80	1,371.80
Size 2	10 HP, 208V	L1@2.00	Ea	1,920.00	82.10	2,002.10
Size 2	15 HP, 230V	L1@2.00	Ea	1,920.00	82.10	2,002.10
Size 2	15 HP, 480V	L1@2.00	Ea	2,030.00	82.10	2,112.10
Size 2	25 HP, 480V	L1@2.00	Ea	2,030.00	82.10	2,112.10
Size 3	20 HP, 208V	L1@3.00	Ea	3,330.00	123.00	3,453.00
Size 3	25 HP, 230V	L1@3.00	Ea	3,600.00	123.00	3,723.00
Size 3	50 HP, 480V	L1@3.00	Ea	3,330.00	123.00	3,453.00

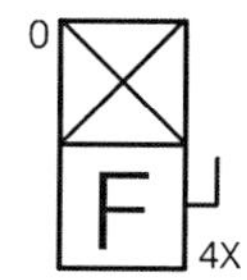

NEMA 12 combination three pole AC magnetic starters, fusible disconnect

Material		Craft@Hrs	Unit	Material Cost	Labor Cost	Installed Cost
Size 0	3 HP, 208V	L1@1.50	Ea	1,470.00	61.50	1,531.50
Size 0	3 HP, 230V	L1@1.50	Ea	1,470.00	61.50	1,531.50
Size 0	5 HP, 480V	L1@1.50	Ea	1,470.00	61.50	1,531.50

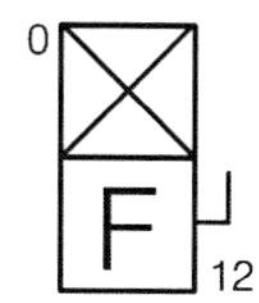

Use these figures to estimate the cost of combination starters installed in buildings under the conditions described on pages 5 and 6. Costs listed are for each starter installed. The crew size is one electrician for starters up to size 4 and two electricians for starters over size 4. Cost per manhour is $41.03. These costs include layout, material handling, and normal waste. Add for overload relay heater elements, sales tax, delivery, supervision, mobilization, demobilization, cleanup, overhead and profit. Note: AC magnetic starters are non-reversing with three melting alloy overload relays for motor protection. The heating elements must be appropriate for the full load amp rating of the connected motor.

Fuses are on pages 283 to 305.

Combination AC Magnetic Three Pole Starters with Fusible Disconnect and Overload Relays for Class R Fuses

NEMA 12 combination three pole AC magnetic starters with fusible disconnect

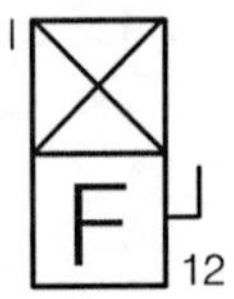

Material	Craft@Hrs	Unit	Material Cost	Labor Cost	Installed Cost
Size 1 5 HP, 208V	L1@1.50	Ea	1,540.00	61.50	1,601.50
Size 1 7.5 HP, 208V	L1@1.50	Ea	1,560.00	61.50	1,621.50
Size 1 5 HP, 230V	L1@1.50	Ea	1,540.00	61.50	1,601.50
Size 1 7.5 HP, 230V	L1@1.50	Ea	1,560.00	61.50	1,621.50
Size 1 10 HP, 480V	L1@1.50	Ea	1,560.00	61.50	1,621.50
Size 2 10 HP, 208V	L1@2.00	Ea	2,370.00	82.10	2,452.10
Size 2 15 HP, 230V	L1@2.00	Ea	2,370.00	82.10	2,452.10
Size 2 15 HP, 480V	L1@2.00	Ea	2,370.00	82.10	2,452.10
Size 2 25 HP, 480V	L1@2.00	Ea	2,390.00	82.10	2,472.10
Size 3 20 HP, 208V	L1@3.00	Ea	3,770.00	123.00	3,893.00
Size 3 25 HP, 208V	L1@3.00	Ea	4,020.00	123.00	4,143.00
Size 3 25 HP, 230V	L1@3.00	Ea	3,770.00	123.00	3,893.00
Size 3 30 HP, 230V	L1@3.00	Ea	4,020.00	123.00	4,143.00
Size 3 50 HP, 480V	L1@3.00	Ea	3,840.00	123.00	3,963.00
Size 4 40 HP, 208V	L1@4.00	Ea	7,620.00	164.00	7,784.00
Size 4 50 HP, 230V	L1@4.00	Ea	7,620.00	164.00	7,784.00
Size 4 100 HP, 480V	L1@4.00	Ea	7,620.00	164.00	7,784.00
Size 5 75 HP, 208V	L2@7.00	Ea	17,300.00	287.00	17,587.00
Size 5 100 HP, 230V	L2@7.00	Ea	17,300.00	287.00	17,587.00
Size 5 200 HP, 480V	L2@7.00	Ea	17,300.00	287.00	17,587.00
Size 6 150 HP, 208V	L2@10.0	Ea	40,100.00	410.00	40,510.00
Size 6 200 HP, 230V	L2@10.0	Ea	40,100.00	410.00	40,510.00
Size 6 400 HP, 480V	L2@10.0	Ea	40,100.00	410.00	40,510.00

NEMA 1 combination three pole AC magnetic starters with class R fuse disconnect

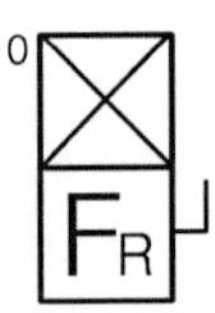

Material	Craft@Hrs	Unit	Material Cost	Labor Cost	Installed Cost
Size 0 3 HP, 208V	L1@1.25	Ea	1,220.00	51.30	1,271.30
Size 0 3 HP, 230V	L1@1.25	Ea	1,220.00	51.30	1,271.30
Size 0 5 HP, 480V	L1@1.25	Ea	1,240.00	51.30	1,291.30
Size 1 5 HP, 208V	L1@1.25	Ea	1,270.00	51.30	1,321.30
Size 1 7.5 HP, 208V	L1@1.25	Ea	1,300.00	51.30	1,351.30
Size 1 5 HP, 230V	L1@1.25	Ea	1,270.00	51.30	1,321.30
Size 1 7.5 HP, 230V	L1@1.25	Ea	1,300.00	51.30	1,351.30
Size 1 10 HP, 480V	L1@1.25	Ea	1,300.00	51.30	1,351.30
Size 2 10 HP, 208V	L1@1.50	Ea	1,970.00	61.50	2,031.50
Size 2 15 HP, 230V	L1@1.50	Ea	1,970.00	61.50	2,031.50
Size 2 15 HP, 480V	L1@1.50	Ea	1,980.00	61.50	2,041.50
Size 2 25 HP, 480V	L1@1.50	Ea	1,990.00	61.50	2,051.50
Size 3 20 HP, 208V	L1@2.00	Ea	3,280.00	82.10	3,362.10
Size 3 25 HP, 208V	L1@2.00	Ea	3,530.00	82.10	3,612.10

Use these figures to estimate the cost of combination starters installed in buildings under the conditions described on pages 5 and 6. Costs listed are for each starter installed. The crew size is one electrician for starters up to size 4 and two electricians for over size 4. Cost per manhour is $41.03. These costs include layout, material handling, and normal waste. Add for overload relay heater elements, sales tax, delivery, supervision, mobilization, demobilization, cleanup, overhead and profit. Note: AC magnetic starters are non-reversing with three melting alloy overload relays for motor protection. The heating elements must be appropriate for the full load amp rating of the connected motor.

Fuses are on pages 283 to 305.

Combination AC Magnetic Three Pole Starters with Fusible Disconnect and Overload Relays for Class R Fuses

NEMA 1 combination three pole AC magnetic starters with class R fuse disconnect

Material	Craft@Hrs	Unit	Material Cost	Labor Cost	Installed Cost
Size 3 25 HP, 230V	L1@2.00	Ea	3,280.00	82.10	3,362.10
Size 3 30 HP, 230V	L1@2.00	Ea	3,530.00	82.10	3,612.10
Size 3 50 HP, 480V	L1@2.00	Ea	3,330.00	82.10	3,412.10
Size 4 40 HP, 208V	L1@3.00	Ea	6,180.00	123.00	6,303.00
Size 4 50 HP, 230V	L1@3.00	Ea	6,180.00	123.00	6,303.00
Size 4 100 HP, 480V	L1@3.00	Ea	6,230.00	123.00	6,353.00
Size 5 75 HP, 208V	L2@4.00	Ea	13,100.00	164.00	13,264.00
Size 5 100 HP, 230V	L2@4.00	Ea	13,100.00	164.00	13,264.00
Size 5 200 HP, 480V	L2@4.00	Ea	13,100.00	164.00	13,264.00
Size 6 150 HP, 208V	L2@8.00	Ea	36,300.00	328.00	36,628.00
Size 6 200 HP, 230V	L2@8.00	Ea	36,300.00	328.00	36,628.00
Size 6 400 HP, 480V	L2@8.00	Ea	36,300.00	328.00	36,628.00

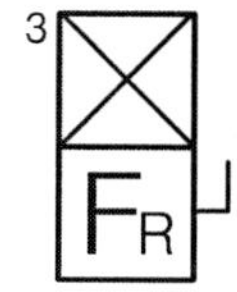

NEMA 4 combination three pole AC magnetic starters with class R fuse disconnect

Material	Craft@Hrs	Unit	Material Cost	Labor Cost	Installed Cost
Size 0 3 HP, 208V	L1@1.50	Ea	2,380.00	61.50	2,441.50
Size 0 3 HP, 230V	L1@1.50	Ea	2,380.00	61.50	2,441.50
Size 0 5 HP, 480V	L1@1.50	Ea	2,400.00	61.50	2,461.50
Size 1 5 HP, 208V	L1@1.50	Ea	2,430.00	61.50	2,491.50
Size 1 7.5 HP, 208V	L1@1.50	Ea	2,460.00	61.50	2,521.50
Size 1 5 HP, 230V	L1@1.50	Ea	2,430.00	61.50	2,491.50
Size 1 7.5 HP, 230V	L1@1.50	Ea	2,460.00	61.50	2,521.50
Size 1 10 HP, 480V	L1@1.50	Ea	2,460.00	61.50	2,521.50
Size 2 10 HP, 208V	L1@2.00	Ea	3,720.00	82.10	3,802.10
Size 2 15 HP, 230V	L1@2.00	Ea	3,720.00	82.10	3,802.10
Size 2 15 HP, 480V	L1@2.00	Ea	3,760.00	82.10	3,842.10
Size 2 25 HP, 480V	L1@2.00	Ea	3,760.00	82.10	3,842.10
Size 3 20 HP, 208V	L1@3.00	Ea	6,390.00	123.00	6,513.00
Size 3 25 HP, 208V	L1@3.00	Ea	6,650.00	123.00	6,773.00
Size 3 25 HP, 230V	L1@3.00	Ea	6,390.00	123.00	6,513.00
Size 3 30 HP, 480V	L1@3.00	Ea	6,430.00	123.00	6,553.00
Size 4 40 HP, 208V	L1@4.00	Ea	10,100.00	164.00	10,264.00
Size 4 50 HP, 230V	L1@4.00	Ea	10,100.00	164.00	10,264.00
Size 4 100 HP, 480V	L1@4.00	Ea	10,200.00	164.00	10,364.00
Size 5 75 HP, 208V	L2@7.00	Ea	24,000.00	287.00	24,287.00
Size 5 100 HP, 230V	L2@7.00	Ea	24,000.00	287.00	24,287.00
Size 5 200 HP, 480V	L2@7.00	Ea	24,000.00	287.00	24,287.00
Size 6 150 HP, 208V	L2@10.0	Ea	47,000.00	410.00	47,410.00
Size 6 200 HP, 230V	L2@10.0	Ea	47,000.00	410.00	47,410.00
Size 6 400 HP, 480V	L2@10.0	Ea	47,000.00	410.00	47,410.00

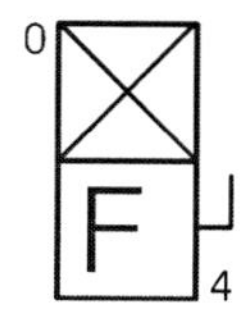

Use these figures to estimate the cost of combination starters installed in buildings under the conditions described on pages 5 and 6. Costs listed are for each starter installed. The crew size is one electrician for starters up to size 4 and two electricians for over size 4. Cost per manhour is $41.03. These costs include layout, material handling, and normal waste. Add for overload relay heater elements, sales tax, delivery, supervision, mobilization, demobilization, cleanup, overhead and profit. Note: AC magnetic starters are non-reversing with three melting alloy overload relays for motor protection. The heating elements must be appropriate for the full load amp rating of the connected motor.

Fuses are on pages 283 to 305.

Combination AC Magnetic Three Pole Starters with Fusible Disconnect and Overload Relays for Class R Fuses

NEMA 4X combination three pole AC magnetic starters with class R fuse disconnect

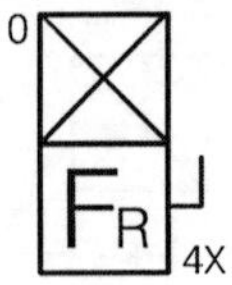

Material	Craft@Hrs	Unit	Material Cost	Labor Cost	Installed Cost
Size 0 3 HP, 208V	L1@1.75	Ea	2,710.00	71.80	2,781.80
Size 0 3 HP, 230V	L1@1.75	Ea	2,710.00	71.80	2,781.80
Size 0 5 HP, 480V	L1@1.75	Ea	2,710.00	71.80	2,781.80
Size 1 5 HP, 208V	L1@1.75	Ea	2,780.00	71.80	2,851.80
Size 1 7.5 HP, 230V	L1@1.75	Ea	2,780.00	71.80	2,851.80
Size 1 10 HP, 480V	L1@1.75	Ea	2,780.00	71.80	2,851.80
Size 2 10 HP, 208V	L1@2.00	Ea	4,110.00	82.10	4,192.10
Size 2 15 HP, 230V	L1@2.00	Ea	4,110.00	82.10	4,192.10
Size 2 15 HP, 480V	L1@2.00	Ea	4,110.00	82.10	4,192.10
Size 2 25 HP, 480V	L1@2.00	Ea	4,110.00	82.10	4,192.10
Size 3 20 HP, 208V	L1@3.00	Ea	7,010.00	123.00	7,133.00
Size 3 25 HP, 230V	L1@3.00	Ea	7,010.00	123.00	7,133.00
Size 3 50 HP, 480V	L1@3.00	Ea	7,070.00	123.00	7,193.00

NEMA 12 combination three pole AC magnetic starters with class R fuse disconnect

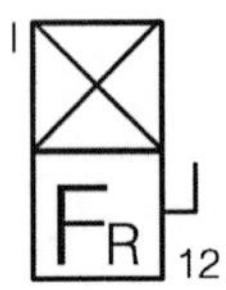

Material	Craft@Hrs	Unit	Material Cost	Labor Cost	Installed Cost
Size 0 3 HP, 208V	L1@1.50	Ea	1,510.00	61.50	1,571.50
Size 0 3 HP, 230V	L1@1.50	Ea	1,510.00	61.50	1,571.50
Size 0 5 HP, 480V	L1@1.50	Ea	1,540.00	61.50	1,601.50
Size 1 5 HP, 208V	L1@1.50	Ea	1,560.00	61.50	1,621.50
Size 1 7.5 HP, 208V	L1@1.50	Ea	1,570.00	61.50	1,631.50
Size 1 5 HP, 230V	L1@1.50	Ea	1,560.00	61.50	1,621.50
Size 1 7.5 HP, 230V	L1@1.50	Ea	1,570.00	61.50	1,631.50
Size 1 10 HP, 480V	L1@1.50	Ea	1,570.00	61.50	1,631.50
Size 2 10 HP, 208V	L1@2.00	Ea	2,380.00	82.10	2,462.10
Size 2 15 HP, 230V	L1@2.00	Ea	2,380.00	82.10	2,462.10
Size 2 15 HP, 480V	L1@2.00	Ea	2,390.00	82.10	2,472.10
Size 2 25 HP, 480V	L1@2.00	Ea	2,390.00	82.10	2,472.10
Size 3 20 HP, 208V	L1@3.00	Ea	3,790.00	123.00	3,913.00
Size 3 25 HP, 208V	L1@3.00	Ea	4,080.00	123.00	4,203.00
Size 3 25 HP, 230V	L1@3.00	Ea	3,790.00	123.00	3,913.00
Size 3 50 HP, 480V	L1@3.00	Ea	3,860.00	123.00	3,983.00
Size 4 40 HP, 208V	L1@4.00	Ea	7,670.00	164.00	7,834.00
Size 4 50 HP, 230V	L1@4.00	Ea	7,670.00	164.00	7,834.00
Size 4 100 HP, 480V	L1@4.00	Ea	7,670.00	164.00	7,834.00

Use these figures to estimate the cost of combination starters installed in buildings under the conditions described on pages 5 and 6. Costs listed are for each starter installed. The crew size is one electrician working at a labor cost of $41.03. These costs include layout, material handling, and normal waste. Add for overload relay heater elements, sales tax, delivery, supervision, mobilization, demobilization, cleanup, overhead and profit. Note: AC magnetic starters are non-reversing with three melting alloy overload relays for motor protection. The heating elements must be appropriate for the full load amp rating of the connected motor.

Fuses are on pages 283 to 305.

Combination AC Magnetic Three Pole Starters with Fusible & Non-fusible Disconnect and Overload Relays

NEMA 12 combination three pole AC magnetic starters with class R fuse disconnect

Material	Craft@Hrs	Unit	Material Cost	Labor Cost	Installed Cost
Size 5 75 HP, 208V	L2@7.00	Ea	17,300.00	287.00	17,587.00
Size 5 100 HP, 230V	L2@7.00	Ea	17,300.00	287.00	17,587.00
Size 5 200 HP, 480V	L2@7.00	Ea	17,300.00	287.00	17,587.00
Size 6 150 HP, 208V	L2@10.0	Ea	40,400.00	410.00	40,810.00
Size 6 200 HP, 230V	L2@10.0	Ea	40,400.00	410.00	40,810.00
Size 6 400 HP, 480V	L2@10.0	Ea	40,400.00	410.00	40,810.00

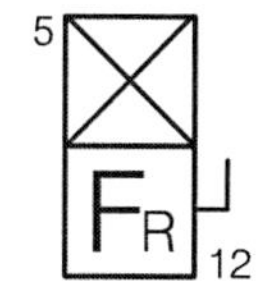

NEMA 1 combination three pole AC magnetic starters, oversized enclosure, non-fused

Material	Craft@Hrs	Unit	Material Cost	Labor Cost	Installed Cost
Size 0 3 HP, 208V	L1@1.30	Ea	1,860.00	53.30	1,913.30
Size 0 3 HP, 230V	L1@1.30	Ea	1,860.00	53.30	1,913.30
Size 0 5 HP, 480V	L1@1.30	Ea	1,860.00	53.30	1,913.30
Size 1 7.5 HP, 208V	L1@1.30	Ea	1,540.00	53.30	1,593.30
Size 1 7.5 HP, 230V	L1@1.30	Ea	1,540.00	53.30	1,593.30
Size 1 10 HP, 480V	L1@1.30	Ea	1,540.00	53.30	1,593.30
Size 2 10 HP, 208V	L1@1.60	Ea	2,220.00	65.60	2,285.60
Size 2 15 HP, 230V	L1@1.60	Ea	2,220.00	65.60	2,285.60
Size 2 25 HP, 480V	L1@1.60	Ea	2,220.00	65.60	2,285.60

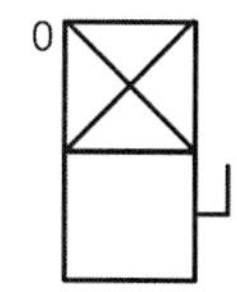

NEMA 4 combination three pole AC magnetic starters, oversized enclosure, non-fused

Material	Craft@Hrs	Unit	Material Cost	Labor Cost	Installed Cost
Size 0 3 HP, 208V	L1@1.60	Ea	3,210.00	65.60	3,275.60
Size 0 3 HP, 230V	L1@1.60	Ea	3,210.00	65.60	3,275.60
Size 0 5 HP, 480V	L1@1.60	Ea	3,210.00	65.60	3,275.60
Size 1 7.5 HP, 208V	L1@1.60	Ea	3,280.00	65.60	3,345.60
Size 1 7.5 HP, 230V	L1@1.60	Ea	3,280.00	65.60	3,345.60
Size 1 10 HP, 480V	L1@1.60	Ea	3,280.00	65.60	3,345.60
Size 2 10 HP, 208V	L1@2.10	Ea	4,590.00	86.20	4,676.20
Size 2 15 HP, 230V	L1@2.10	Ea	4,590.00	86.20	4,676.20
Size 2 25 HP, 480V	L1@2.10	Ea	4,590.00	86.20	4,676.20

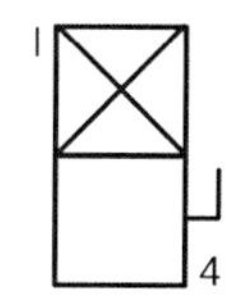

NEMA 12 combination three pole AC magnetic starters, oversized enclosure, non-fused

Material	Craft@Hrs	Unit	Material Cost	Labor Cost	Installed Cost
Size 0 3 HP, 208V	L1@1.60	Ea	1,990.00	65.60	2,055.60
Size 0 3 HP, 230V	L1@1.60	Ea	1,990.00	65.60	2,055.60
Size 0 5 HP, 480V	L1@1.60	Ea	1,990.00	65.60	2,055.60
Size 1 7.5 HP, 208V	L1@1.60	Ea	2,030.00	65.60	2,095.60
Size 1 7.5 HP, 230V	L1@1.60	Ea	2,030.00	65.60	2,095.60
Size 1 10 HP, 480V	L1@1.60	Ea	2,030.00	65.60	2,095.60
Size 2 10 HP, 208V	L1@2.10	Ea	2,800.00	86.20	2,886.20
Size 2 15 HP, 230V	L1@2.10	Ea	2,800.00	86.20	2,886.20
Size 2 25 HP, 480V	L1@2.10	Ea	2,800.00	86.20	2,886.20

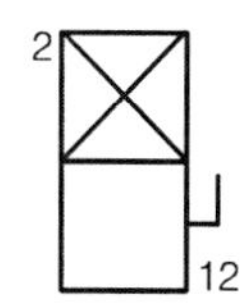

Use these figures to estimate the cost of combination starters installed in buildings under the conditions described on pages 5 and 6. Costs listed are for each starter installed. The crew size is one electrician for starters up to size 4 and two electricians for over size 4. Cost per manhour is $41.03. These costs include layout, material handling, and normal waste. Add for overload relay heater elements, sales tax, delivery, supervision, mobilization, demobilization, cleanup, overhead and profit. Note: AC magnetic starters are non-reversing with three melting alloy overload relays for motor protection. The heating elements must be appropriate for the full load amp rating of the connected motor.

Fuses are on pages 283 to 305.

Combination AC Magnetic Three Pole Starters in Oversized Enclosure with Fusible Disconnect and Overload Relays

NEMA 1 combination three pole AC magnetic starters, oversized enclosure, fusible disconnect

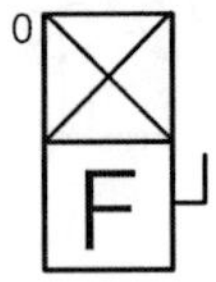

Material	Craft@Hrs	Unit	Material Cost	Labor Cost	Installed Cost
Size 0 3 HP, 208V	L1@1.30	Ea	1,510.00	53.30	1,563.30
Size 0 3 HP, 230V	L1@1.30	Ea	1,510.00	53.30	1,563.30
Size 0 5 HP, 480V	L1@1.30	Ea	1,540.00	53.30	1,593.30
Size 1 5 HP, 208V	L1@1.30	Ea	1,540.00	53.30	1,593.30
Size 1 7.5 HP, 208V	L1@1.30	Ea	1,550.00	53.30	1,603.30
Size 1 5 HP, 230V	L1@1.30	Ea	1,540.00	53.30	1,593.30
Size 1 7.5 HP, 230V	L1@1.30	Ea	1,570.00	53.30	1,623.30
Size 1 10 HP, 480V	L1@1.30	Ea	1,570.00	53.30	1,623.30
Size 2 10 HP, 208V	L1@1.60	Ea	2,240.00	65.60	2,305.60
Size 2 15 HP, 230V	L1@1.60	Ea	2,160.00	65.60	2,225.60
Size 2 15 HP, 480V	L1@1.60	Ea	2,250.00	65.60	2,315.60
Size 2 25 HP, 480V	L1@1.60	Ea	2,270.00	65.60	2,335.60

NEMA 4 combination three pole AC magnetic starters, oversized enclosure, fusible disconnect

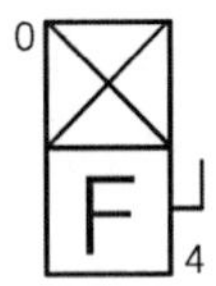

Material	Craft@Hrs	Unit	Material Cost	Labor Cost	Installed Cost
Size 0 3 HP, 208V	L1@1.60	Ea	3,270.00	65.60	3,335.60
Size 0 3 HP, 230V	L1@1.60	Ea	3,270.00	65.60	3,335.60
Size 0 5 HP, 480V	L1@1.60	Ea	3,280.00	65.60	3,345.60
Size 1 5 HP, 208V	L1@1.60	Ea	3,300.00	65.60	3,365.60
Size 1 7.5 HP, 208V	L1@1.60	Ea	3,340.00	65.60	3,405.60
Size 1 5 HP, 230V	L1@1.60	Ea	3,330.00	65.60	3,395.60
Size 1 7.5 HP, 230V	L1@1.60	Ea	3,330.00	65.60	3,395.60
Size 1 10 HP, 480V	L1@1.60	Ea	3,330.00	65.60	3,395.60
Size 2 10 HP, 208V	L1@2.10	Ea	4,630.00	86.20	4,716.20
Size 2 15 HP, 230V	L1@2.10	Ea	4,630.00	86.20	4,716.20
Size 2 15 HP, 480V	L1@2.10	Ea	4,650.00	86.20	4,736.20
Size 2 25 HP, 480V	L1@2.10	Ea	4,660.00	86.20	4,746.20

NEMA 12 combination three pole AC magnetic starters, oversized enclosure, fusible disconnect

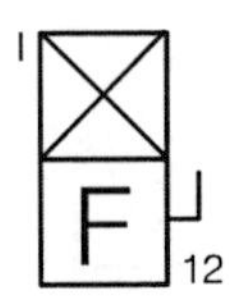

Material	Craft@Hrs	Unit	Material Cost	Labor Cost	Installed Cost
Size 0 3 HP, 208V	L1@1.60	Ea	2,030.00	65.60	2,095.60
Size 0 3 HP, 230V	L1@1.60	Ea	2,030.00	65.60	2,095.60
Size 0 5 HP, 480V	L1@1.60	Ea	2,040.00	65.60	2,105.60
Size 1 5 HP, 208V	L1@1.60	Ea	2,030.00	65.60	2,095.60
Size 1 7.5 HP, 208V	L1@1.60	Ea	2,100.00	65.60	2,165.60
Size 1 5 HP, 230V	L1@1.60	Ea	2,050.00	65.60	2,115.60
Size 1 7.5 HP, 230V	L1@1.60	Ea	2,100.00	65.60	2,165.60
Size 1 10 HP, 480V	L1@1.60	Ea	2,100.00	65.60	2,165.60

Use these figures to estimate the cost of combination starters installed in buildings under the conditions described on pages 5 and 6. Costs listed are for each starter installed. The crew size is one electrician for starters up to size 4 and two electricians for over size 4. Cost per manhour is $41.03. These costs include layout, material handling, and normal waste. Add for overload relay heater elements, sales tax, delivery, supervision, mobilization, demobilization, cleanup, overhead and profit. Note: AC magnetic starters are non-reversing with three melting alloy overload relays for motor protection. The heating elements must be appropriate for the full load amp rating of the connected motor.

Fuses are on pages 283 to 305.

NEMA 12 combination three pole AC magnetic starters, oversized enclosure, fusible disconnect

Material	Craft@Hrs	Unit	Material Cost	Labor Cost	Installed Cost
Size 2 10 HP, 208V	L1@2.10	Ea	2,800.00	86.20	2,886.20
Size 2 15 HP, 230V	L1@2.10	Ea	2,800.00	86.20	2,886.20
Size 2 25 HP, 480V	L1@2.10	Ea	2,910.00	86.20	2,996.20

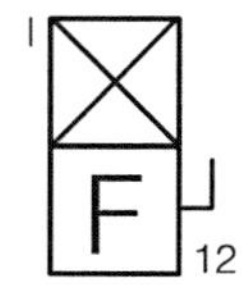

NEMA 1 combination three pole AC magnetic starters, circuit breaker disconnect

Material	Craft@Hrs	Unit	Material Cost	Labor Cost	Installed Cost
Size 0 3 HP, 208V	L1@1.25	Ea	1,580.00	51.30	1,631.30
Size 0 3 HP, 230V	L1@1.25	Ea	1,580.00	51.30	1,631.30
Size 0 5 HP, 480V	L1@1.25	Ea	1,580.00	51.30	1,631.30
Size 0 5 HP, 600V	L1@1.25	Ea	1,580.00	51.30	1,631.30
Size 1 7.5 HP, 208V	L1@1.25	Ea	1,660.00	51.30	1,711.30
Size 1 7.5 HP, 230V	L1@1.25	Ea	1,660.00	51.30	1,711.30
Size 1 10 HP, 480V	L1@1.25	Ea	1,660.00	51.30	1,711.30
Size 1 10 HP, 600V	L1@1.25	Ea	1,660.00	51.30	1,711.30
Size 2 10 HP, 208V	L1@1.50	Ea	2,310.00	61.50	2,371.50
Size 2 15 HP, 230V	L1@1.50	Ea	2,310.00	61.50	2,371.50
Size 2 25 HP, 480V	L1@1.50	Ea	2,310.00	61.50	2,371.50
Size 2 25 HP, 600V	L1@1.50	Ea	2,310.00	61.50	2,371.50
Size 3 25 HP, 208V	L1@2.00	Ea	3,340.00	82.10	3,422.10
Size 3 30 HP, 230V	L1@2.00	Ea	3,340.00	82.10	3,422.10
Size 3 50 HP, 480V	L1@2.00	Ea	3,340.00	82.10	3,422.10
Size 3 50 HP, 600V	L1@2.00	Ea	3,340.00	82.10	3,422.10
Size 4 40 HP, 208V	L1@3.00	Ea	7,260.00	123.00	7,383.00
Size 4 50 HP, 230V	L1@3.00	Ea	7,260.00	123.00	7,383.00
Size 4 100 HP, 480V	L1@3.00	Ea	7,260.00	123.00	7,383.00
Size 4 100 HP, 600V	L1@3.00	Ea	7,260.00	123.00	7,383.00
Size 5 75 HP, 208V	L2@4.00	Ea	17,000.00	164.00	17,164.00
Size 5 100 HP, 230V	L2@4.00	Ea	17,000.00	164.00	17,164.00
Size 5 200 HP, 480V	L2@4.00	Ea	17,000.00	164.00	17,164.00
Size 5 200 HP, 600V	L2@4.00	Ea	17,000.00	164.00	17,164.00
Size 6 150 HP, 208V	L2@8.00	Ea	36,300.00	328.00	36,628.00
Size 6 200 HP, 230V	L2@8.00	Ea	36,300.00	328.00	36,628.00
Size 6 400 HP, 480V	L2@8.00	Ea	36,300.00	328.00	36,628.00
Size 6 400 HP, 600V	L2@8.00	Ea	36,300.00	328.00	36,628.00

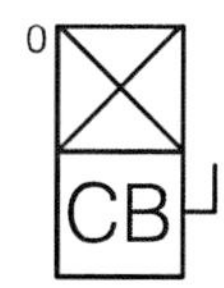

Use these figures to estimate the cost of combination starters installed in buildings under the conditions described on pages 5 and 6. Costs listed are for each starter installed. The crew size is one electrician for starters up to size 4 and two electricians for over size 4. Cost per manhour is $41.03. These costs include layout, material handling, and normal waste. Add for overload relay heater elements, sales tax, delivery, supervision, mobilization, demobilization, cleanup, overhead and profit. Note: AC magnetic starters are non-reversing with three melting alloy overload relays for motor protection. The heating elements must be appropriate for the full load amp rating of the connected motor. Circuit breakers are either 22,000 or 100,000 amp interrupt capacity.

Fuses are on pages 283 to 305.

Combination AC Magnetic Three Pole Starters with Circuit Breaker Disconnect and Overload Relays

NEMA 4 combination three pole AC magnetic starters, C.B. disconnect and overload relays

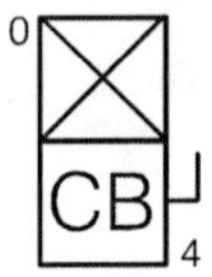

Material	Craft@Hrs	Unit	Material Cost	Labor Cost	Installed Cost
Size 0 3 HP, 208V	L1@1.50	Ea	2,760.00	61.50	2,821.50
Size 0 3 HP, 230V	L1@1.50	Ea	2,760.00	61.50	2,821.50
Size 0 5 HP, 480V	L1@1.50	Ea	2,760.00	61.50	2,821.50
Size 0 5 HP, 600V	L1@1.50	Ea	2,760.00	61.50	2,821.50
Size 1 7.5 HP, 208V	L1@1.50	Ea	2,790.00	61.50	2,851.50
Size 1 7.5 HP, 230V	L1@1.50	Ea	2,790.00	61.50	2,851.50
Size 1 10 HP, 480V	L1@1.50	Ea	2,790.00	61.50	2,851.50
Size 1 10 HP, 600V	L1@1.50	Ea	2,790.00	61.50	2,851.50
Size 2 10 HP, 208V	L1@2.00	Ea	4,110.00	82.10	4,192.10
Size 2 15 HP, 230V	L1@2.00	Ea	4,110.00	82.10	4,192.10
Size 2 25 HP, 480V	L1@2.00	Ea	4,110.00	82.10	4,192.10
Size 2 25 HP, 600V	L1@2.00	Ea	4,110.00	82.10	4,192.10
Size 3 25 HP, 208V	L1@3.00	Ea	6,440.00	123.00	6,563.00
Size 3 30 HP, 230V	L1@3.00	Ea	6,440.00	123.00	6,563.00
Size 3 50 HP, 480V	L1@3.00	Ea	6,440.00	123.00	6,563.00
Size 3 50 HP, 600V	L1@3.00	Ea	6,440.00	123.00	6,563.00
Size 4 40 HP, 208V	L1@4.00	Ea	11,200.00	164.00	11,364.00
Size 4 50 HP, 230V	L1@4.00	Ea	11,200.00	164.00	11,364.00
Size 4 100 HP, 480V	L1@4.00	Ea	11,200.00	164.00	11,364.00
Size 4 100 HP, 600V	L1@4.00	Ea	11,200.00	164.00	11,364.00
Size 5 75 HP, 208V	L2@7.00	Ea	27,800.00	287.00	28,087.00
Size 5 100 HP, 230V	L2@7.00	Ea	27,800.00	287.00	28,087.00
Size 5 200 HP, 480V	L2@7.00	Ea	27,800.00	287.00	28,087.00
Size 5 200 HP, 600V	L2@7.00	Ea	27,800.00	287.00	28,087.00
Size 6 150 HP, 208V	L2@10.0	Ea	42,200.00	410.00	42,610.00
Size 6 200 HP, 230V	L2@10.0	Ea	42,200.00	410.00	42,610.00
Size 6 400 HP, 480V	L2@10.0	Ea	42,200.00	410.00	42,610.00
Size 6 400 HP, 600V	L2@10.0	Ea	42,200.00	410.00	42,610.00

NEMA 4X combination three pole AC magnetic starters, C.B. disconnect and overload relays

Material	Craft@Hrs	Unit	Material Cost	Labor Cost	Installed Cost
Size 0 3 HP, 208V	L1@1.75	Ea	3,140.00	71.80	3,211.80
Size 0 3 HP, 230V	L1@1.75	Ea	3,140.00	71.80	3,211.80
Size 0 5 HP, 480V	L1@1.75	Ea	3,140.00	71.80	3,211.80
Size 0 5 HP, 600V	L1@1.75	Ea	3,140.00	71.80	3,211.80
Size 1 7.5 HP, 208V	L1@1.75	Ea	3,210.00	71.80	3,281.80
Size 1 7.5 HP, 230V	L1@1.75	Ea	3,210.00	71.80	3,281.80
Size 1 10 HP, 480V	L1@1.75	Ea	3,210.00	71.80	3,281.80
Size 1 10 HP, 600V	L1@1.75	Ea	3,210.00	71.80	3,281.80

Use these figures to estimate the cost of combination starters installed in buildings under the conditions described on pages 5 and 6. Costs listed are for each starter installed. The crew size is one electrician for starters up to size 4 and two electricians for over size 4. Cost per manhour is $41.03. These costs include layout, material handling, and normal waste. Add for overload relay heater elements, sales tax, delivery, supervision, mobilization, demobilization, cleanup, overhead and profit. Note: AC magnetic starters are non-reversing with three melting alloy overload relays for motor protection. The heating elements must be appropriate for the full load amp rating of the connected motor. Circuit breakers are either 22,000 or 100,000 amp interrupt capacity.

Combination AC Magnetic Three Pole Starters with Circuit Breaker Disconnect and Overload Relays

NEMA 4X combination three pole AC magnetic starters, C.B. disconnect and overload relays

Material	Craft@Hrs	Unit	Material Cost	Labor Cost	Installed Cost
Size 2 10 HP, 208V	L1@2.00	Ea	4,510.00	82.10	4,592.10
Size 2 15 HP, 230V	L1@2.00	Ea	4,510.00	82.10	4,592.10
Size 2 25 HP, 480V	L1@2.00	Ea	4,510.00	82.10	4,592.10
Size 2 25 HP, 600V	L1@2.00	Ea	4,510.00	82.10	4,592.10
Size 3 25 HP, 208V	L1@3.00	Ea	7,080.00	123.00	7,203.00
Size 3 30 HP, 230V	L1@3.00	Ea	7,080.00	123.00	7,203.00
Size 3 50 HP, 480V	L1@3.00	Ea	7,080.00	123.00	7,203.00
Size 3 50 HP, 600V	L1@3.00	Ea	7,080.00	123.00	7,203.00
Size 4 40 HP, 208V	L1@4.00	Ea	12,400.00	164.00	12,564.00
Size 4 50 HP, 230V	L1@4.00	Ea	12,400.00	164.00	12,564.00
Size 4 100 HP, 480V	L1@4.00	Ea	12,400.00	164.00	12,564.00
Size 4 100 HP, 600V	L1@4.00	Ea	12,400.00	164.00	12,564.00

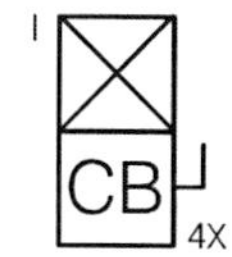

NEMA 7 & 9 combination three pole AC magnetic starters, C.B. disconnect and overload relays

Material	Craft@Hrs	Unit	Material Cost	Labor Cost	Installed Cost
Size 0 3 HP, 208V	L1@1.75	Ea	3,330.00	71.80	3,401.80
Size 0 3 HP, 230V	L1@1.75	Ea	3,330.00	71.80	3,401.80
Size 0 5 HP, 480V	L1@1.75	Ea	3,330.00	71.80	3,401.80
Size 0 5 HP, 600V	L1@1.75	Ea	3,330.00	71.80	3,401.80
Size 1 7.5 HP, 208V	L1@1.75	Ea	3,260.00	71.80	3,331.80
Size 1 7.5 HP, 230V	L1@1.75	Ea	3,380.00	71.80	3,451.80
Size 1 10 HP, 480V	L1@1.75	Ea	3,380.00	71.80	3,451.80
Size 1 10 HP, 600V	L1@1.75	Ea	3,380.00	71.80	3,451.80
Size 2 10 HP, 208V	L1@2.25	Ea	4,510.00	92.30	4,602.30
Size 2 15 HP, 230V	L1@2.25	Ea	4,510.00	92.30	4,602.30
Size 2 25 HP, 480V	L1@2.25	Ea	4,510.00	92.30	4,602.30
Size 2 25 HP, 600V	L1@2.25	Ea	4,510.00	92.30	4,602.30
Size 3 25 HP, 208V	L1@3.50	Ea	7,480.00	144.00	7,624.00
Size 3 30 HP, 230V	L1@3.50	Ea	7,480.00	144.00	7,624.00
Size 3 50 HP, 480V	L1@3.50	Ea	7,480.00	144.00	7,624.00
Size 3 50 HP, 600V	L1@3.50	Ea	7,480.00	144.00	7,624.00
Size 4 40 HP, 208V	L1@4.50	Ea	11,400.00	185.00	11,585.00
Size 4 50 HP, 230V	L1@4.50	Ea	11,400.00	185.00	11,585.00
Size 4 100 HP, 480V	L1@4.50	Ea	11,400.00	185.00	11,585.00
Size 4 100 HP, 600V	L1@4.50	Ea	11,400.00	185.00	11,585.00
Size 5 75 HP, 208V	L2@8.00	Ea	24,900.00	328.00	25,228.00
Size 5 100 HP, 230V	L2@8.00	Ea	24,900.00	328.00	25,228.00
Size 5 200 HP, 480V	L2@8.00	Ea	24,900.00	328.00	25,228.00
Size 5 200 HP, 600V	L2@8.00	Ea	24,900.00	328.00	25,228.00

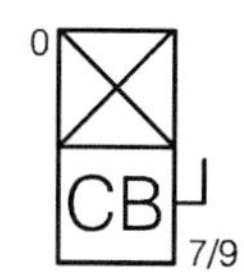

Use these figures to estimate the cost of combination starters installed in buildings under the conditions described on pages 5 and 6. Costs listed are for each starter installed. The crew size is one electrician for starters up to size 4 and two electricians for over size 4. Cost per manhour is $41.03. These costs include layout, material handling, and normal waste. Add for overload relay heater elements, sales tax, delivery, supervision, mobilization, demobilization, cleanup, overhead and profit. Note: AC magnetic starters are non-reversing with three melting alloy overload relays for motor protection. The heating elements must be appropriate for the full load amp rating of the connected motor. Circuit breakers are either 22,000 or 100,000 amp interrupt capacity.

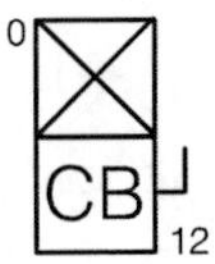

NEMA 12 combination three pole AC magnetic starters, C.B. disconnect and overload relays

Material	Craft@Hrs	Unit	Material Cost	Labor Cost	Installed Cost
Size 0 3 HP, 208V	L1@1.60	Ea	1,890.00	65.60	1,955.60
Size 0 3 HP, 230V	L1@1.60	Ea	1,890.00	65.60	1,955.60
Size 0 5 HP, 480V	L1@1.60	Ea	1,890.00	65.60	1,955.60
Size 0 5 HP, 600V	L1@1.60	Ea	1,890.00	65.60	1,955.60
Size 1 7.5 HP, 208V	L1@1.60	Ea	1,960.00	65.60	2,025.60
Size 1 7.5 HP, 230V	L1@1.60	Ea	1,960.00	65.60	2,025.60
Size 1 10 HP, 480V	L1@1.60	Ea	1,960.00	65.60	2,025.60
Size 1 10 HP, 600V	L1@1.60	Ea	1,960.00	65.60	2,025.60
Size 2 10 HP, 208V	L1@2.10	Ea	2,720.00	86.20	2,806.20
Size 2 15 HP, 230V	L1@2.10	Ea	2,720.00	86.20	2,806.20
Size 2 25 HP, 480V	L1@2.10	Ea	2,720.00	86.20	2,806.20
Size 2 25 HP, 600V	L1@2.10	Ea	2,720.00	86.20	2,806.20
Size 3 25 HP, 208V	L1@3.25	Ea	3,870.00	133.00	4,003.00
Size 3 30 HP, 230V	L1@3.25	Ea	3,870.00	133.00	4,003.00
Size 3 50 HP, 480V	L1@3.25	Ea	3,870.00	133.00	4,003.00
Size 3 50 HP, 600V	L1@3.25	Ea	3,870.00	133.00	4,003.00
Size 4 40 HP, 208V	L1@4.50	Ea	8,730.00	185.00	8,915.00
Size 4 50 HP, 230V	L1@4.50	Ea	8,730.00	185.00	8,915.00
Size 4 100 HP, 480V	L1@4.50	Ea	8,730.00	185.00	8,915.00
Size 4 100 HP, 600V	L1@4.50	Ea	8,730.00	185.00	8,915.00
Size 5 75 HP, 208V	L2@8.50	Ea	19,600.00	349.00	19,949.00
Size 5 100 HP, 230V	L2@8.50	Ea	19,600.00	349.00	19,949.00
Size 5 200 HP, 480V	L2@8.50	Ea	19,600.00	349.00	19,949.00
Size 5 200 HP, 600V	L2@8.50	Ea	19,600.00	349.00	19,949.00
Size 6 150 HP, 208V	L2@12.5	Ea	39,500.00	513.00	40,013.00
Size 6 200 HP, 230V	L2@12.5	Ea	39,500.00	513.00	40,013.00
Size 6 400 HP, 480V	L2@12.5	Ea	39,500.00	513.00	40,013.00
Size 6 400 HP, 600V	L2@12.5	Ea	39,500.00	513.00	40,013.00

Use these figures to estimate the cost of combination starters installed in buildings under the conditions described on pages 5 and 6. Costs listed are for each starter installed. The crew size is one electrician for starters up to size 4 and two electricians for over size 4. Cost per manhour is $41.03. These costs include layout, material handling, and normal waste. Add for overload relay heater elements, sales tax, delivery, supervision, mobilization, demobilization, cleanup, overhead and profit. Note: AC magnetic starters are non-reversing with three melting alloy overload relays for motor protection. The heating elements must be appropriate for the full load amp rating of the connected motor. Circuit breakers are either 22,000 or 100,000 amp interrupt capacity.

Combination AC Magnetic Three Pole Starters in Oversized Enclosure with Circuit Breaker Disconnect and Overload Relays

NEMA 1 combination three pole AC magnetic starters, oversized enclosure, C.B. disconnect

Material	Craft@Hrs	Unit	Material Cost	Labor Cost	Installed Cost
Size 0 3 HP, 208V	L1@1.30	Ea	1,900.00	53.30	1,953.30
Size 0 3 HP, 230V	L1@1.30	Ea	1,900.00	53.30	1,953.30
Size 0 5 HP, 480V	L1@1.30	Ea	1,900.00	53.30	1,953.30
Size 0 5 HP, 600V	L1@1.30	Ea	1,900.00	53.30	1,953.30
Size 1 7.5 HP, 208V	L1@1.30	Ea	1,980.00	53.30	2,033.30
Size 1 7.5 HP, 230V	L1@1.30	Ea	1,980.00	53.30	2,033.30
Size 1 10 HP, 480V	L1@1.30	Ea	1,980.00	53.30	2,033.30
Size 1 10 HP, 600V	L1@1.30	Ea	1,980.00	53.30	2,033.30
Size 2 10 HP, 208V	L1@1.60	Ea	2,620.00	65.60	2,685.60
Size 2 15 HP, 230V	L1@1.60	Ea	2,620.00	65.60	2,685.60
Size 2 25 HP, 480V	L1@1.60	Ea	2,620.00	65.60	2,685.60
Size 2 25 HP, 600V	L1@1.60	Ea	2,620.00	65.60	2,685.60

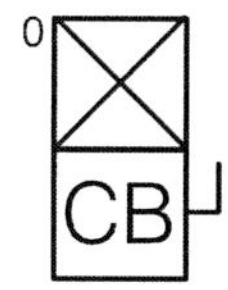

NEMA 4 combination three pole AC magnetic starters, oversized enclosure, C.B. disconnect

Material	Craft@Hrs	Unit	Material Cost	Labor Cost	Installed Cost
Size 0 3 HP, 208V	L1@1.60	Ea	3,650.00	65.60	3,715.60
Size 0 3 HP, 230V	L1@1.60	Ea	3,650.00	65.60	3,715.60
Size 0 5 HP, 480V	L1@1.60	Ea	3,650.00	65.60	3,715.60
Size 0 5 HP, 600V	L1@1.60	Ea	3,650.00	65.60	3,715.60
Size 1 7.5 HP, 208V	L1@1.60	Ea	3,700.00	65.60	3,765.60
Size 1 7.5 HP, 230V	L1@1.60	Ea	3,700.00	65.60	3,765.60
Size 1 10 HP, 480V	L1@1.60	Ea	3,700.00	65.60	3,765.60
Size 1 10 HP, 600V	L1@1.60	Ea	3,700.00	65.60	3,765.60
Size 2 10 HP, 208V	L1@2.10	Ea	5,010.00	86.20	5,096.20
Size 2 15 HP, 230V	L1@2.10	Ea	5,010.00	86.20	5,096.20
Size 2 25 HP, 480V	L1@2.10	Ea	5,010.00	86.20	5,096.20
Size 2 25 HP, 600V	L1@2.10	Ea	5,010.00	86.20	5,096.20

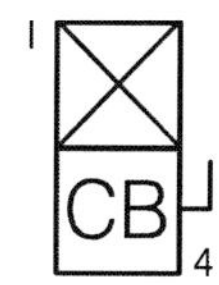

NEMA 12 combination three pole AC magnetic starters, oversized enclosure, C.B. disconnect

Material	Craft@Hrs	Unit	Material Cost	Labor Cost	Installed Cost
Size 0 3 HP, 208V	L1@1.60	Ea	2,430.00	65.60	2,495.60
Size 0 3 HP, 230V	L1@1.60	Ea	2,430.00	65.60	2,495.60
Size 0 5 HP, 480V	L1@1.60	Ea	2,430.00	65.60	2,495.60
Size 0 5 HP, 600V	L1@1.60	Ea	2,430.00	65.60	2,495.60

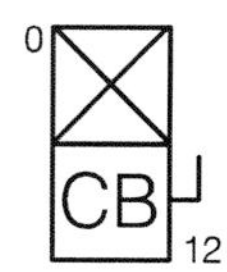

Use these figures to estimate the cost of combination starters installed in buildings under the conditions described on pages 5 and 6. Costs listed are for each starter installed. The crew size is one electrician for starters up to size 4 and two electricians for over size 4. Cost per manhour is $41.03. These costs include layout, material handling, and normal waste. Add for overload relay heater elements, sales tax, delivery, supervision, mobilization, demobilization, cleanup, overhead and profit. Note: AC magnetic starters are non-reversing with three melting alloy overload relays for motor protection. The heating elements must be appropriate for the full load amp rating of the connected motor. Circuit breakers are either 22,000 or 100,000 amp interrupt capacity.

Motor Control Equipment

NEMA 12 combination three pole AC magnetic starters in oversize enclosure with C.B. disconnect

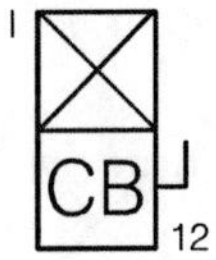

Material	Craft@Hrs	Unit	Material Cost	Labor Cost	Installed Cost
Size 1 7.5 HP, 208V	L1@1.60	Ea	2,140.00	65.60	2,205.60
Size 1 7.5 HP, 230V	L1@1.60	Ea	2,140.00	65.60	2,205.60
Size 1 10 HP, 480V	L1@1.60	Ea	2,140.00	65.60	2,205.60
Size 1 10 HP, 600V	L1@1.60	Ea	2,140.00	65.60	2,205.60
Size 2 10 HP, 208V	L1@2.10	Ea	2,820.00	86.20	2,906.20
Size 2 15 HP, 230V	L1@2.10	Ea	2,820.00	86.20	2,906.20
Size 2 25 HP, 480V	L1@2.10	Ea	2,820.00	86.20	2,906.20
Size 2 25 HP, 600V	L1@2.10	Ea	2,820.00	86.20	2,906.20

NEMA 1 two & three unit general purpose motor control stations

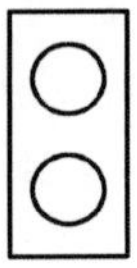

Material	Craft@Hrs	Unit	Material Cost	Labor Cost	Installed Cost
Stop-start	L1@0.40	Ea	85.20	16.40	101.60
Stop-start lockout stop	L1@0.40	Ea	97.00	16.40	113.40
Stop-start main. contact	L1@0.40	Ea	97.00	16.40	113.40
Forward-reverse	L1@0.40	Ea	97.00	16.40	113.40
Open-close	L1@0.40	Ea	97.00	16.40	113.40
Up-down	L1@0.40	Ea	97.00	16.40	113.40
On-off, main. contact	L1@0.40	Ea	97.00	16.40	113.40
Hand-auto main. contact	L1@0.40	Ea	97.00	16.40	113.40
Hand-off-auto	L1@0.50	Ea	109.00	20.50	129.50

NEMA 1 two & three unit heavy duty motor control stations

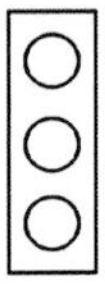

Material	Craft@Hrs	Unit	Material Cost	Labor Cost	Installed Cost
Stop-start	L1@0.50	Ea	219.00	20.50	239.50
Stop-start lockout stop	L1@0.50	Ea	219.00	20.50	239.50
Stop-start main. contact	L1@0.50	Ea	219.00	20.50	239.50
Forward-reverse	L1@0.50	Ea	219.00	20.50	239.50
Open-close	L1@0.50	Ea	219.00	20.50	239.50
Up-down	L1@0.50	Ea	219.00	20.50	239.50
On-off, main. contact	L1@0.50	Ea	219.00	20.50	239.50
Hand-auto main. contact	L1@0.50	Ea	254.00	20.50	274.50
Hand-off-auto	L1@0.60	Ea	268.00	24.60	292.60

Use these figures to estimate the cost of combination starters and control stations installed in buildings under the conditions described on pages 5 and 6. Costs listed for control stations are for each station installed. The crew size is one electrician working at a labor cost of $41.03 per manhour. These costs include the enclosure, layout, material handling, and normal waste. Add for sales tax, delivery, supervision, mobilization, demobilization, cleanup, overhead and profit. Note: Be sure to select the right switch for the control sequence. Many other general purpose and heavy duty control stations are available when other combinations and features are required. Most units can be assembled into special configurations to meet specific needs.

NEMA 1 general purpose flush mounted two & three unit motor control stations

Material	Craft@Hrs	Unit	Material Cost	Labor Cost	Installed Cost
Stop-start	L1@0.40	Ea	103.00	16.40	119.40
Stop-start lockout stop	L1@0.40	Ea	144.00	16.40	160.40
Stop-start main. contact	L1@0.40	Ea	144.00	16.40	160.40
Forward-reverse	L1@0.40	Ea	123.00	16.40	139.40
Open-close	L1@0.40	Ea	123.00	16.40	139.40
Up-down	L1@0.40	Ea	123.00	16.40	139.40
On-off	L1@0.40	Ea	123.00	16.40	139.40
On-off, main. contact	L1@0.40	Ea	143.00	16.40	159.40
Hand-auto, main. contact	L1@0.40	Ea	143.00	16.40	159.40
Hand-off auto	L1@0.60	Ea	311.00	24.60	335.60

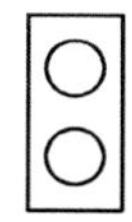

NEMA 4 general purpose two & three unit motor control stations

Material	Craft@Hrs	Unit	Material Cost	Labor Cost	Installed Cost
Stop-start	L1@0.50	Ea	212.00	20.50	232.50
Stop-start lockout stop	L1@0.50	Ea	212.00	20.50	232.50
Stop-start main. contact	L1@0.50	Ea	285.00	20.50	305.50
Forward-reverse	L1@0.50	Ea	249.00	20.50	269.50
Open-close	L1@0.50	Ea	249.00	20.50	269.50
Up-down	L1@0.50	Ea	249.00	20.50	269.50
On-off	L1@0.50	Ea	249.00	20.50	269.50
On-off, main. contact	L1@0.50	Ea	285.00	20.50	305.50
Hand-auto main. contact	L1@0.50	Ea	285.00	20.50	305.50
Stop-start main. contact	L1@0.60	Ea	285.00	24.60	309.60

NEMA 4 heavy duty two & three unit motor control stations

Material	Craft@Hrs	Unit	Material Cost	Labor Cost	Installed Cost
Stop-start	L1@0.60	Ea	212.00	24.60	236.60
Stop-start lockout stop	L1@0.60	Ea	212.00	24.60	236.60
Stop-start main. contact	L1@0.60	Ea	285.00	24.60	309.60
Forward-reverse	L1@0.60	Ea	249.00	24.60	273.60
Open-close	L1@0.60	Ea	249.00	24.60	273.60
Up-down	L1@0.60	Ea	249.00	24.60	273.60
On-off	L1@0.60	Ea	249.00	24.60	273.60
Jog	L1@0.60	Ea	249.00	24.60	273.60
Manual-auto	L1@0.60	Ea	249.00	24.60	273.60
Hand-off-auto	L1@0.75	Ea	285.00	30.80	315.80

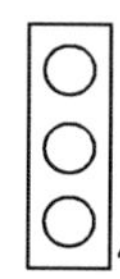

Use these figures to estimate the cost of control stations installed in buildings under the conditions described on pages 5 and 6. Costs listed are for each station installed. The crew size is one electrician working at a labor cost of $41.03 per manhour. These costs include the enclosure, layout, material handling, and normal waste. Add for sales tax, delivery, supervision, mobilization, demobilization, cleanup, overhead and profit. Note: Be sure to select the right switch for the control sequence. Many other general purpose and heavy duty control stations are available when other combinations and features are required. Most units can be assembled into special configurations to meet specific needs.

Section 14: Trenching and Excavation

Many electrical estimators feel like a fish out of water when estimating trenching and excavation. There are too many unknowns. Every soil type has unique characteristics ready to trap the unwary and unprepared. What was supposed to be a simple trench can become a financial disaster when shoring or dewatering become necessary, when you discover an outcropping of rock at mid-trench, when you cut into an unrecorded water line or unmarked septic system or when a downpour fills a newly dug trench with silt.

If you feel uncomfortable about the unknowns when estimating trenching and excavation, take heart. Many seasoned excavation estimators feel exactly the same way. It goes with the territory. And as long as trenching and excavation are included with the electrical portion of the work, electrical estimators will have to work up bids for earthwork.

Of course, the easiest way to handle trenching and excavation is to get a firm quote from a dirt contractor. The more bids you get, the better the chances of getting an attractive price. But no matter how many bids you get, my advice is to protect yourself. Know enough about estimating trenching and excavation to read and evaluate each bid received. Be able to recognize a bid that doesn't include everything that's required or includes prices for more work than is necessary.

There are ways to reduce your risk and produce consistently reliable estimates for excavation and trenching — even if you're not a licensed, card-carrying earthwork estimator. With a little care and practice, you'll feel as confident about trench estimates as any other part of an electrical estimator's work.

Electrical contractors use many types of trenchers. Wheel trenchers dig a lot of trench in a short time and adapt well to most soil types. But they're impractical on the smaller jobs most electrical contractors handle regularly. Backhoes are the industry standard for deeper and wider trenches. But chain trenchers are a better choice for narrower and shallower trench work. The best choice in equipment depends on soil conditions, size of the trench, and the amount of work necessary. There's no single piece of equipment that fits every job.

Many electrical contractors have their own trenching equipment and routinely do all or nearly all of their own excavation. Others rent what's needed on a job by job basis. Of course, there are advantages to doing your own trenching. It's done when you want and the way you want, not when your trenching sub has the time. Many smaller electrical contractors can profit by doing their own trenching ***if*** they have an experienced operator available and ***if*** they estimate trench costs accurately.

My advice is to avoid buying excavation equipment if good rental equipment is available locally and if your need for a particular piece of equipment is less than two weeks out of most months. After that, buying or leasing on a long-term basis may lower equipment costs.

No matter whether you own or rent equipment, keep track of equipment productivity rates and operating costs. There should be two profits on every job: One goes into your wallet. The other is what you learn that will make future estimates more accurate. If you're not keeping good cost records, you're missing half the profit each job offers. This is especially true in earthwork where costs can multiply very quickly.

The manhour tables in this section will help if you have no other reliable data. But no manhour figure you find in a book can be as reliable as your own cost data on work done by your crews, with your equipment and under conditions you know and understand. What's listed in this section will be either too high or too low for most contractors on most jobs. But the figures I've selected will be a good starting point for most trenching work with most types of equipment and in most types of soil. Refining these figures for the work you do is your job.

Doing the Take-Off

The first step on every trench estimate is to look for boring data logs on the plans. They're your best clue about subgrade conditions. If there are no logs, make an on-site inspection — with a shovel. You'll seldom have to trench more than 3 or 4 feet below the surface. Scrape off the top 12 inches to remove any topsoil. What you see about a foot down is probably the same soil that's 3 feet below the surface.

Sandy soil is easy trenching. But sand tends to dry out and cave in. When digging in sand, schedule work so that duct lines go in immediately after the trenching. Backfill as soon as possible.

In very hard or rocky soil, a tracked trencher is best. A backhoe probably won't be adequate. No matter what type of equipment is used, make an allowance if it isn't in good condition, if it's prone to breakdowns or if the operator isn't experienced.

Location of the trench can make a big difference in the manhours and equipment time required. Poor access always slows production and can make it impossible to use the most productive equipment.

Always consider what you're going to do with the spoil that's removed, even when it's all going to go back in the trench eventually. On some sites there won't be room to stockpile spoil beside the trench. Even if there's plenty of room, some soil or debris will have to be hauled away on many jobs. If you're breaking out pavement, hauling broken pavement to a legal dump and paying the dump fee may be major cost items.

Don't automatically think of trenching equipment every time you see a trench on the plans. Sometimes it's cheaper to put a laborer to work with a shovel for a day or two. Consider all the costs involved in renting a trencher, moving it to the site, hiring an operator to do a few hours of work, and then taking the trencher back to the yard. The fastest way to dig a trench isn't always the cheapest.

Of course, there are limits to hand digging. Trenches over about 4 feet deep are usually machine work, especially in rocky soil or hardpan. And hot or cold weather will slow hand production more than machine work.

Consider jacking or boring conduit under streets and walkways rather than breaking pavement and trenching. Pipe jacking is expensive. But it's easy to get a bid on this type of work. And that bid may cut your costs when pavement sawing, excavation, backfill, hauling debris, dump fees, and patching pavement or concrete are considered. Jacking conduit is routine for street light and signal contractors.

Examine the cost of pavement removal and patching very carefully on any job where a line has to pass under an existing pavement or concrete surface. Most specs require that pavement be saw cut before being broken out. Saw cutting can be expensive, especially on a slope or where access is restricted. Backfill usually has to include a specific sub-base material between the compacted soil and the asphalt or concrete patch. Getting a small delivery of exactly the right sub-base may be expensive.

Occasionally some earthwork other than trenching will be included in the electrical portion of the plans. Foundations for light standards are probably the most common example. Estimating work like this is similar to estimating any trench. Study the plans. Figure the volume of soil to move. Make an educated guess of labor and equipment productivity per hour. Then multiply the cost per hour by the number of hours required.

You'll find that it's cheaper to drill light pole foundations on larger jobs. Consider renting a drill rig with an experienced operator if you don't have someone who meets that description on your payroll. In many areas you can get a quote on drilling per hole or for the entire job.

If you need less than a few hour's work with a drilling rig, consider hand digging. Even if the work's done with a drill rig, there's plenty of hand work. The pit may have to be shaped by hand. You'll have to spread the piles of spoil or load them for disposal. And some backfilling may be needed.

Trenching and Excavation

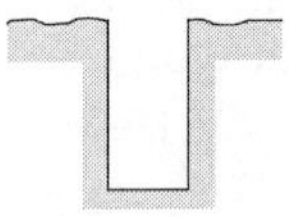

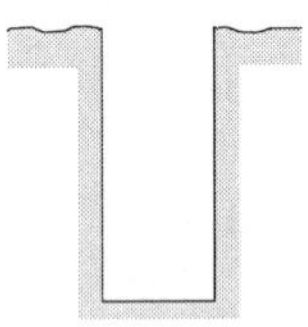

Material	Craft@Hrs	Unit	Material Cost	Equipment Cost	Labor Cost	Installed Cost
Trenching per linear foot of trench						
10" x 12"	L1@0.02	LF	—	.82	5.91	6.73
12" x 12"	L1@0.03	LF	—	1.23	6.31	7.54
18" x 12"	L1@0.04	LF	—	1.64	6.92	8.56
24" x 12"	L1@0.05	LF	—	2.05	7.28	9.33
10" x 18"	L1@0.03	LF	—	1.23	6.71	7.94
12" x 18"	L1@0.04	LF	—	1.64	7.06	8.70
18" x 18"	L1@0.05	LF	—	2.05	7.66	9.71
24" x 18"	L1@0.06	LF	—	2.46	8.05	10.51
10" x 24"	L1@0.04	LF	—	1.64	6.82	8.46
12" x 24"	L1@0.05	LF	—	2.05	7.24	9.29
18" x 24"	L1@0.06	LF	—	2.46	7.84	10.30
24" x 24"	L1@0.07	LF	—	2.87	8.25	11.12
10" x 30"	L1@0.05	LF	—	2.05	7.06	9.11
12" x 30"	L1@0.06	LF	—	2.46	7.44	9.90
18" x 30"	L1@0.07	LF	—	2.87	8.25	11.12
24" x 30"	L1@0.08	LF	—	3.28	8.61	11.89
10" x 36"	L1@0.06	LF	—	2.46	7.06	9.52
12" x 36"	L1@0.07	LF	—	2.87	7.84	10.71
18" x 36"	L1@0.08	LF	—	3.28	8.61	11.89
24" x 36"	L1@0.09	LF	—	3.69	8.99	12.68
12" x 40"	L1@0.07	LF	—	2.87	7.84	10.71
18" x 40"	L1@0.08	LF	—	3.28	8.61	11.89
24" x 40"	L1@0.09	LF	—	3.69	9.39	13.08
30" x 40"	L1@0.10	LF	—	4.10	10.20	14.30
12" x 48"	L1@0.08	LF	—	3.28	8.61	11.89
18" x 48"	L1@0.09	LF	—	3.69	9.39	13.08
24" x 48"	L1@0.10	LF	—	4.10	10.20	14.30
30" x 48"	L1@0.11	LF	—	4.51	12.60	17.11
18" x 54"	L1@0.10	LF	—	4.10	9.78	13.88
24" x 54"	L1@0.11	LF	—	4.51	10.60	15.11
30" x 54"	L1@0.12	LF	—	4.92	12.60	17.52
36" x 54"	L1@0.13	LF	—	5.33	13.70	19.03
18" x 60"	L1@0.11	LF	—	4.51	13.70	18.21
24" x 60"	L1@0.12	LF	—	4.92	15.60	20.52
30" x 60"	L1@0.13	LF	—	5.33	17.30	22.63
36" x 60"	L1@0.14	LF	—	5.74	21.50	27.24
Pit excavation by pit size						
24" x 24" x 24"	L1@0.14	Ea	—	5.74	15.60	21.34
24" x 36" x 24"	L1@0.18	Ea	—	7.39	19.70	27.09
24" x 48" x 24"	L1@0.20	Ea	—	8.21	31.30	39.51
24" x 24" x 36"	L1@0.21	Ea	—	8.62	41.10	49.72
24" x 36" x 36"	L1@0.27	Ea	—	11.10	46.60	57.70
24" x 48" x 36"	L1@0.30	Ea	—	12.30	54.60	66.90
24" x 24" x 48"	L1@0.28	Ea	—	11.50	78.40	89.90
24" x 36" x 48"	L1@0.36	Ea	—	14.80	155.00	169.80
24" x 48" x 48"	L1@0.40	Ea	—	16.40	233.00	249.40

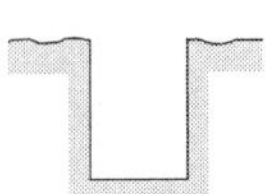

Use these figures to estimate the cost of trenching and excavation done under the conditions described on pages 5 and 6. Costs listed are for each linear foot of trench excavated or for each pit excavated. The crew is one electrician working at a labor cost of $41.03 manhour. These costs include trenching or excavation equipment and layout. Add for surface cutting, patching, encasement, warning tape, barricades, shoring, clean-up, compaction, delivery, supervision, mobilization, demobilization, overhead and profit. Note: Costs will be higher when obstructions or poor digging conditions delay production. It's important to check ground conditions before estimating excavation work.

Section 15: Surface Raceways

Surface raceways have been used since the early 1900's to cover and protect wiring that's not concealed in a wall. Originally it was used to add circuits in buildings that were erected before electrical power was widely available. Today, surface raceway can be used any time circuits have to be added or altered to meet changing needs of the occupants in the interior of offices, hospitals, stores, residences, and manufacturing plants.

Raceway and fittings are made from steel, aluminum and plastic to meet a wide variety of surface wiring requirements. Smaller sizes are made for situations where only a few conductors are needed to carry lower amperage. Larger sizes are available to carry more conductors and supply heavier electrical loads. Most manufacturers sell fittings that can be used to route raceway around nearly any obstacle on the wall surface. But it's important to use fittings appropriate for the type and size of raceway being installed.

Surface raceway can be installed on the ceiling, on the wall, in shelving, in counters or in any other dry interior location. It can be installed on drywall, plaster, wood, paneling, unit masonry or concrete surfaces. Surface raceway comes with a factory finish but can be painted after installation.

Use a hacksaw to cut surface raceway. Before assembling the cut piece and pulling wire, be sure to deburr the inside and outside of the cut end. This can be done with a small flat file or an electrician's pocket knife. The hacksaw blade should be shatterproof, high-speed steel with 40 teeth per inch. A shear can be used in place of a hacksaw when cutting the base or cover of $1\frac{9}{32}$" by ¾" raceway.

Many types of companion outlet boxes are available for each size of raceway. These boxes are designed for surface mounting. Most come in several depths, rectangular or round for junction boxes or fixtures, single gang for junction boxes or wiring devices, and multi-ganged for grouping wiring devices.

Fittings are available for connecting surface raceway to conduit, to other concealed wiring, or for connecting surface raceway of one size to surface raceway of another size.

Pulling wire through surface raceway is easier if you use a special wire pulley. The pulley is clamped to the raceway at internal elbows and guides wire around the fitting. Use a fish tape leader to pull several conductors through the raceway.

One-Piece Raceway

One-piece surface raceway comes in three sizes: ½" by $\frac{11}{32}$", ¾" by $\frac{17}{32}$", and ¾" by $\frac{21}{32}$". This raceway is shaped to look like wood molding so it isn't so obvious when installed. Fittings are made to look like the raceway. Wire size and number of conductors determine the size of raceway to use.

All surface raceway and fittings should carry the Underwriter's Laboratory label. The *National Electrical Code*, Articles 352 and 353 are your primary guide when installing surface raceway. But many other sections of the *NEC* will apply too.

Surface raceway is manufactured in 10 foot lengths with 10 pieces to the bundle. Each length comes with a coupling.

A hand bender is available for bending ¾" surface raceway. It's used like an EMT hand bender. The bender can make simple offsets or short bends in a flat position where the raceway is installed against a ceiling or wall. An adapter can be inserted into the bender for bending ½" surface raceway.

Two-Piece Raceway

Two-piece raceway comes without couplings and in the following sizes:

$1\frac{9}{32}$" by ¾" raceway is made of .040" galvanized steel and packed in bundles of 10. Bundles of 10-foot lengths have 100 feet to the bundle. Bundles of 5-foot lengths have 50 feet to the bundle. The cover is scored every 3 inches for easy breakoff.

$\frac{19}{32}$" by ¾" type 304 stainless steel raceway is available in 5-foot lengths.

1¼" by $\frac{7}{8}$" .040" steel raceway comes in 10-foot lengths, 100 feet to a carton. The cover comes in 5-foot lengths, 100 feet to a carton, and is scored every 3 inches. It can be ordered unscored if necessary.

$2\frac{3}{8}$" by ¾" .040" galvanized steel raceway comes in 5- or 10-foot lengths. Base and cover are packed separately.

$2^3/_4$" by $1^{17}/_{32}$" .040" steel raceway comes 50 feet to a carton. Bases are 10 feet long and covers are 5 feet long.

Bases for 4¾" by 1¾" are made of .050" galvanized steel. Covers and dividers are .040" galvanized steel. Base comes in 10-foot lengths, 50 feet per carton. Cover and divider come in 5-foot lengths. Base, cover and divider are packed separately.

Bases for 4¾" by $3^9/_{16}$" surface raceway are .060" galvanized steel. Covers are .040" galvanized steel. Base and cover are packed 20 feet per carton. Base comes in 5- or 10-foot lengths, cover comes in 5-foot lengths.

Larger surface raceway is used in laboratories, TV studios, communication centers, computer rooms, control rooms, plants, and indoor areas that are dry and exposed. Base for two-piece raceway can be installed on any of the surfaces mentioned above for one-piece raceways.

Overfloor Raceway

Overfloor raceway comes in two sizes. The smaller size is $1^9/_{16}$" by $^{11}/_{32}$", .040" galvanized steel and comes in 5- or 10-foot lengths with 50 feet per carton. The raceway base is scored every 3 inches for easy breakoff. It can be cut with the same hacksaw as other surface raceway. It's recommended that a reinforcing member be inserted in overfloor raceway where heavy traffic is likely. The larger overfloor raceway size is $2^7/_{32}$" by $^{23}/_{32}$". The base is .040" galvanized steel and the cover is .050" steel. The raceway comes in 5- or 10-foot lengths with 5 pieces per carton. The base is scored every 3 inches for easy breakoff.

Many fittings can be used with overfloor raceway. Adapters are available for joining overfloor raceway to wall-mounted surface raceway. Of course, junction boxes and outlet boxes are made for distribution of power on the floor surface. Overfloor raceway is used to protect power, telephone, computer or signal wiring. It's tapered at the top edges so it's less of an obstruction when laid on the floor. The raceway comes with a factory finish and can be field painted.

Multi-Outlet Systems

Multi-outlet strips are a good choice in laboratories, office work stations, shops, stores, hobby shops, over work benches, in schools, hospitals, and clean rooms where many electrical appliances may be in use in a small area. These outlet strips can be used indoors in any non-hazardous location not subject to mechanical injury or abuse.

Multi-outlet strip is specified by the outlet spacing, the type of outlet, color and number of circuits. A wide range of fittings are available. The base and covers are cut to fit and deburred the same as other surface raceway. The covers are factory punched for outlet spacing. The outlets are prewired at the desired spacing as a single or multiple circuit.

Factory preassembled multi-outlet strip is available complete with cord and plug. The units come in a strip length of 2, 3, 5 and 6 feet, with a duplex receptacle or 6 single receptacles. The cords are usually 2 to 6 feet long. These units are handy when temporary displays are being installed. Colors and finishes available include antique bronze, ivory, brown, and stainless steel. Strips are available with a button-reset mini-breaker or lighted on-off switch.

Telepower Poles

A simple way to provide electrical, telephone, or computer wiring to a work station or desk located away from a wall is to use a telepower pole. The pole is attached to the drop ceiling grid system and secured to the work station or to the floor next to the desk. A junction box on the top of the pole above the grid system provides the connection point for electrical, telephone and computer system wiring. The telepower pole has divided raceway for each of the conductors. Receptacles for electrical service are factory installed in the assembled pole.

A range of colors and styles is available. The pole can be attached to the floor or merely set on a carpeted floor. A universal base plate is made for either application.

The telepower poles are about 10'5" inches long. When used in a room with a ceiling grid higher than that, use a section of two-piece raceway to extend the telepower pole through the ceiling grid. First remove the top junction box. Cut at least 1 foot from one side of the telepower pole, leaving one side longer than the other. Cut the two-piece raceway to the desired length. Tie the sections together with stove bolts. The section added should be painted to match the telepower pole.

Channel Wire Systems

Channel wire systems are popular in manufacturing plants, warehouses, storage areas, parts rooms, assembly areas and other large work areas. The channel is a simple $1^5/_8$" by $1^5/_8$" steel strut through which a flat conductor assembly has been drawn. The flat conductor is rated for 30 amps at 277/480 volts.

A special junction box is attached to the strut at the feed-in end. At the other end of the strut an insulated end cap is installed. Splice plates are used where the strut is coupled together.

Color coded caps can be inserted anywhere along the strut to tap into the flat conductor assembly. This offers access to power along the entire length of the channel. The color-coded caps provide connections to the phase conductors. For 277 volt power, the black colored cap connects phase A and neutral. The red cap connects phase B and neutral, and the blue cap connects phase C and neutral. For 480 volt power, the black over red cap connects to phase A and phase B, the red over blue cap connects to phase B and phase C, the blue over black cap connects to phase B and phase A.

A hanger assembly is used with each color coded cap to complete the tap to the flat conductors. Later the tap assembly can be relocated because the flat conductor is self healing.

For a high bay lighting job, the strut is simply installed in rows where the lighting will be placed. The feed-in box is installed on each run of strut, and an insulated end cap is installed on the end of each strut run. The strut is supported with standard strut hanger assemblies spaced close enough to provide adequate support for the lighting fixtures planned. After the strut is installed, the flat wire is drawn into the strut and connected to the feed-in junction boxes, and wired back through conduit to the panel. The color coded cap is installed with the fixture feed hanger. The fixture is then attached to the hanger assembly and lamped.

When fluorescent fixtures are installed, one color coded cap with hanger and one fixture hanger is used to secure each 4 foot or 8 foot fixture to the channel. Only one feed-in assembly is required for each fluorescent fixture.

Dual feed-in junction boxes are available. Two sets of lighting loads can be connected to the electrical system from the same junction box. On longer struts more dual feed-in junction boxes may be used.

System Engineering

Leading manufacturers of surface raceway and channel wiring can usually assist you with design and layout of a channel wiring system. An indoor lighting system can be designed for fluorescent, mercury vapor, metal halide, high pressure sodium or low pressure sodium lighting, depending on the light density needed. Your material supplier will probably be able to suggest the name of a manufacturer that can provide this design service.

Steel Surface Raceway

One-piece steel surface raceway

Material	Craft@Hrs	Unit	Material Cost	Labor Cost	Installed Cost
1/2" x 11/32"	L1@3.00	CLF	82.60	123.00	205.60
3/4" x 17/32"	L1@3.50	CLF	89.80	144.00	233.80
3/4" x 21/32"	L1@3.75	CLF	101.00	154.00	255.00

Two-piece steel surface raceway, base only

Material	Craft@Hrs	Unit	Material Cost	Labor Cost	Installed Cost
1-9/32" x 3/4"	L1@3.50	CLF	94.80	144.00	238.80
1-1/4" x 7/8"	L1@3.75	CLF	142.00	154.00	296.00
2-3/8" x 3/4"	L1@4.00	CLF	142.00	164.00	306.00
2-3/4" x 1-17/32"	L1@5.00	CLF	232.00	205.00	437.00
4-3/4" x 1-3/4"	L1@6.00	CLF	396.00	246.00	642.00
4-3/4" x 3-9/16"	L1@6.50	CLF	734.00	267.00	1,001.00

Two-piece steel surface raceway, cover only

Material	Craft@Hrs	Unit	Material Cost	Labor Cost	Installed Cost
1-1/4" x 7/8"	L1@2.00	CLF	89.80	82.10	171.90
2-3/8" x 3/4"	L1@2.25	CLF	85.20	92.30	177.50
2-3/4" x 1-17/32"	L1@2.50	CLF	122.00	103.00	225.00
4-3/4" x 1-3/4"	L1@3.00	CLF	229.00	123.00	352.00
4-3/4" x 3-9/16"	L1@3.25	CLF	251.00	133.00	384.00

Overfloor steel surface raceway

Material	Craft@Hrs	Unit	Material Cost	Labor Cost	Installed Cost
1-9/16" x 11/32"	L1@4.50	CLF	189.00	185.00	374.00
2-7/32" x 23/32"	L1@6.00	CLF	291.00	246.00	537.00

Overfloor steel surface raceway fittings, 1-9/16" x 11/32"

Material	Craft@Hrs	Unit	Material Cost	Labor Cost	Installed Cost
Reinforcing member	L1@2.75	CLF	469.00	113.00	582.00
Wire clip	L1@0.05	Ea	.30	2.05	2.35
Fiber bushing	L1@0.05	Ea	.62	2.05	2.67
Two-hole strap	L1@0.05	Ea	.56	2.05	2.61
Station tubing bracket	L1@0.10	Ea	1.47	4.10	5.57
Flat elbow, 90 degree	L1@0.20	Ea	4.94	8.21	13.15
Internal elbow, 90 degree	L1@0.25	Ea	4.94	10.30	15.24
External elbow, 90 degree	L1@0.25	Ea	6.23	10.30	16.53
Adapter fitting	L1@0.15	Ea	6.35	6.15	12.50
Telephone outlet	L1@0.25	Ea	10.40	10.30	20.70
Narrow tel outlet	L1@0.25	Ea	10.00	10.30	20.30
Utility box	L1@0.25	Ea	9.22	10.30	19.52
Narrow junction box	L1@0.25	Ea	12.80	10.30	23.10
Combination connector	L1@0.20	Ea	6.24	8.21	14.45

Use these figures to estimate the cost of surface metal raceway installed in a building under the conditions described in the chapter text. Costs listed are for each 100 linear feet installed. The crew size is one electrician working at a labor cost of $41.03 per manhour. These costs include typical surface installation and layout for walls, ceiling, laboratory counters, work stations and overfloor, material handling and normal waste. Add for various fittings, boxes, sales tax, delivery, supervision, mobilization, demobilization, cleanup, overhead and profit. Note: Raceway runs are 100' long. Shorter runs will take more labor and longer runs will take less labor per linear foot.

Fittings for one-piece steel 1/2" x 11/32" surface raceway

Material	Craft@Hrs	Unit	Material Cost	Labor Cost	Installed Cost
Coupling	L1@0.05	Ea	.39	2.05	2.44
Bushing	L1@0.05	Ea	.31	2.05	2.36
Supporting clip	L1@0.05	Ea	.34	2.05	2.39
One hole strap	L1@0.05	Ea	.27	2.05	2.32
Connection cover	L1@0.05	Ea	.43	2.05	2.48
Flat elbow, 90 degree	L1@0.15	Ea	1.69	6.15	7.84
Internal elbow, 90 degree	L1@0.15	Ea	3.22	6.15	9.37
Internal twisted 90 degree	L1@0.15	Ea	4.24	6.15	10.39
External elbow, 90 degree	L1@0.15	Ea	2.71	6.15	8.86
Adjustable J-box	L1@0.20	Ea	8.18	8.21	16.39
Extension adapter	L1@0.15	Ea	6.06	6.15	12.21
Reducing adapter	L1@0.15	Ea	3.34	6.15	9.49

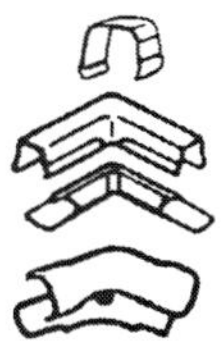

Fittings for one-piece steel 3/4" x 17/32" surface raceway

Material	Craft@Hrs	Unit	Material Cost	Labor Cost	Installed Cost
Bushing	L1@0.05	Ea	.24	2.05	2.29
One-hole strap	L1@0.05	Ea	.27	2.05	2.32
Connector cover	L1@0.05	Ea	.34	2.05	2.39
Flat elbow, 90 degree	L1@0.15	Ea	1.40	6.15	7.55
Flat elbow, 45 degree	L1@0.15	Ea	4.08	6.15	10.23
Internal elbow, 90 degree	L1@0.15	Ea	1.75	6.15	7.90
External elbow, 90 degree	L1@0.15	Ea	1.64	6.15	7.79
Internal corner coupling	L1@0.15	Ea	1.98	6.15	8.13

Fittings for one-piece steel 3/4" x 17/32" or 3/4" x 21/32" surface raceway

Material	Craft@Hrs	Unit	Material Cost	Labor Cost	Installed Cost
Flexible section, 18"	L1@0.20	Ea	16.10	8.21	24.31
Coupling	L1@0.05	Ea	.34	2.05	2.39
Supporting clip	L1@0.05	Ea	.38	2.05	2.43
Ground clamp	L1@0.05	Ea	4.26	2.05	6.31
Internal twist. elbow	L1@0.15	Ea	4.05	6.15	10.20
Tee	L1@0.20	Ea	3.36	8.21	11.57
Internal pull elbow	L1@0.20	Ea	11.20	8.21	19.41
Internal elbow, 90 degree	L1@0.15	Ea	2.74	6.15	8.89
Corner box	L1@0.20	Ea	11.60	8.21	19.81

Overfloor steel boxes for 1-9/16" x 11/32" surface raceway

Material	Craft@Hrs	Unit	Material Cost	Labor Cost	Installed Cost
4" dia junction box	L1@0.25	Ea	8.15	10.30	18.45
Duplex recept & box	L1@0.35	Ea	15.50	14.40	29.90
Box for single recept	L1@0.25	Ea	9.76	10.30	20.06
Box for duplex recept	L1@0.25	Ea	12.40	10.30	22.70
Telephone outlet	L1@0.25	Ea	12.10	10.30	22.40

Use these figures to estimate the cost of surface metal raceway and fittings installed in a building under the conditions described in the chapter text. The crew size is one electrician working at a labor cost of $41.03 per manhour. These costs include layout, material handling and normal waste. Add for sales tax, delivery, supervision, mobilization, demobilization, cleanup, overhead and profit.

Steel Surface Raceway Fittings

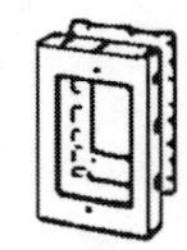

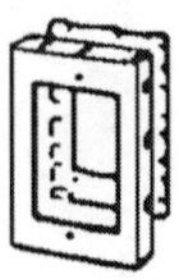

Material	Craft@Hrs	Unit	Material Cost	Labor Cost	Installed Cost
Steel boxes for 1/2" x 11/32" or 3/4" x 17/32" surface raceway					
Single pole sw & box	L1@0.35	Ea	11.00	14.40	25.40
Duplex recept. & box	L1@0.25	Ea	12.90	10.30	23.20
Utility box	L1@0.20	Ea	7.29	8.21	15.50
Steel boxes for 3/4" x 17/32" or 3/4" x 21/32" surface raceway					
Corner box, small	L1@0.25	Ea	8.70	10.30	19.00
Corner box, large	L1@0.25	Ea	12.40	10.30	22.70
Utility box, 3" dia.	L1@0.25	Ea	7.29	10.30	17.59
Blank cover	L1@0.05	Ea	3.90	2.05	5.95
Extension box, 3"	L1@0.20	Ea	8.84	8.21	17.05
Utility box, 4"	L1@0.25	Ea	7.85	10.30	18.15
Distribution box, 4"	L1@0.25	Ea	8.84	10.30	19.14
Blank cover, 4"	L1@0.05	Ea	3.90	2.05	5.95
Extension box, 4"	L1@0.20	Ea	9.58	8.21	17.79
Extension box, 5-1/2"	L1@0.25	Ea	9.76	10.30	20.06
Fixture box, 4-3/4"	L1@0.25	Ea	8.88	10.30	19.18
Fixture box, 5-1/2"	L1@0.30	Ea	9.02	12.30	21.32
Fixture box, 6-3/8"	L1@0.30	Ea	10.60	12.30	22.90
Device box, 1-3/8" deep	L1@0.25	Ea	5.99	10.30	16.29
Device box, 2 gang, 2-3/4"	L1@0.30	Ea	10.60	12.30	22.90
Device box, 3 gang, 2-3/4"	L1@0.40	Ea	26.90	16.40	43.30
Device box, 4 gang, 2-3/4"	L1@0.50	Ea	43.50	20.50	64.00
Device box, 5 gang, 2-3/4"	L1@0.65	Ea	74.30	26.70	101.00
Device box, 6 gang, 2-3/4"	L1@0.75	Ea	78.10	30.80	108.90
Steel fittings for 3/4" x 17/32" or 3/4" x 21/32" surface raceway					
1/2" male box conn.	L1@0.05	Ea	2.58	2.05	4.63
3/4" male box conn.	L1@0.06	Ea	7.25	2.46	9.71
1/2" female box conn.	L1@0.05	Ea	2.66	2.05	4.71
3/4" female box conn.	L1@0.06	Ea	8.41	2.46	10.87
1/2" male elbow box	L1@0.20	Ea	8.36	8.21	16.57
1/2" female elbow box	L1@0.20	Ea	8.36	8.21	16.57
Combination connector	L1@0.10	Ea	2.97	4.10	7.07
Adjust. offset conn.	L1@0.10	Ea	8.12	4.10	12.22
Kick plate	L1@0.10	Ea	12.00	4.10	16.10
Armored cable conn.	L1@0.05	Ea	2.74	2.05	4.79

Use these figures to estimate the cost of surface metal raceway and fittings installed in a building under the conditions described in the chapter text. The crew size is one electrician working at a labor cost of $41.03 per manhour. These costs include layout, material handling and normal waste. Add for sales tax, delivery, supervision, mobilization, demobilization, cleanup, overhead and profit.

Surface Metal Raceway, Fittings, Tools and Assemblies

Overfloor steel fittings for 2-7/32" x 23/32" surface raceway

Material	Craft@Hrs	Unit	Material Cost	Labor Cost	Installed Cost
Wire clip	L1@0.05	Ea	.60	2.05	2.65
Fiber bushing	L1@0.05	Ea	.94	2.05	2.99
Flat elbow, 90 degree	L1@0.25	Ea	7.81	10.30	18.11
Telephone elbow	L1@0.25	Ea	17.10	10.30	27.40
Tel service fit small	L1@0.25	Ea	14.20	10.30	24.50
Telephone service fit	L1@0.25	Ea	17.10	10.30	27.40
Junction box	L1@0.25	Ea	12.00	10.30	22.30

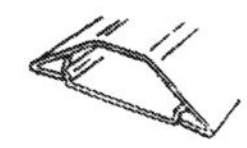

Tools for 1/2" and 3/4" steel surface raceway

Material	Craft@Hrs	Unit	Material Cost	Labor Cost	Installed Cost
Bender	--	Ea	76.30	--	76.30
Bender adapter	--	Ea	15.00	--	15.00
Cutting box	--	Ea	80.60	--	80.60
Wire pulley	--	Ea	29.20	--	29.20
Cover removal tool	--	Ea	15.00	--	15.00
Canopy cutter	--	Ea	401.00	--	401.00
Shear for 2-3/8"	--	Ea	264.00	--	264.00

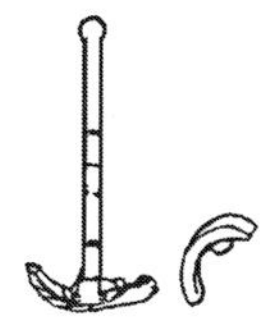

Enamel spray paint for steel surface raceway

Material	Craft@Hrs	Unit	Material Cost	Labor Cost	Installed Cost
Buff	--	Ea	14.80	--	14.80
Gray	--	Ea	14.80	--	14.80
Putty	--	Ea	14.80	--	14.80

Two-piece 1-9/32" x 3/4" surface metal raceway assembly, with cord

Material	Craft@Hrs	Unit	Material Cost	Labor Cost	Installed Cost
24" with 3 outlets	L1@0.15	Ea	67.40	6.15	73.55
36" with 6 outlets	L1@0.20	Ea	75.60	8.21	83.81
60" with 6 outlets	L1@0.25	Ea	84.30	10.30	94.60
72" with 6 outlets	L1@0.30	Ea	88.70	12.30	101.00

Two-piece surface metal raceway, 1-9/32" x 3/4", prewired, single circuit, grounding

Material	Craft@Hrs	Unit	Material Cost	Labor Cost	Installed Cost
12" long, 2 outlets	L1@0.30	Ea	30.70	12.30	43.00
36" long, 6 outlets	L1@0.40	Ea	29.30	16.40	45.70
60" long, 5 outlets	L1@0.50	Ea	33.80	20.50	54.30
60" long, 10 outlets	L1@0.60	Ea	45.50	24.60	70.10
72" long, 4 outlets	L1@0.50	Ea	31.90	20.50	52.40
72" long, 6 outlets	L1@0.60	Ea	36.30	24.60	60.90
72" long, 8 outlets	L1@0.70	Ea	43.20	28.70	71.90
72" long, 12 outlets	L1@0.80	Ea	46.60	32.80	79.40

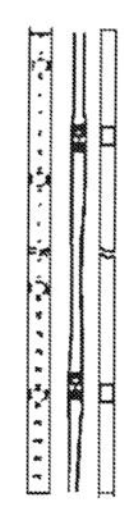

Use these figures to estimate the cost of surface metal raceway, assemblies and fittings installed in a building under the conditions described in the chapter text. The crew size is one electrician working at a labor cost of $41.03 per manhour. These costs include layout, material handling and normal waste. Add for sales tax, delivery, supervision, mobilization, demobilization, cleanup, overhead and profit.

Surface Metal Raceway Assemblies and Fittings

Two-piece surface metal raceway, 1-9/32" x 3/4", prewired, 1 circuit, 3-wire

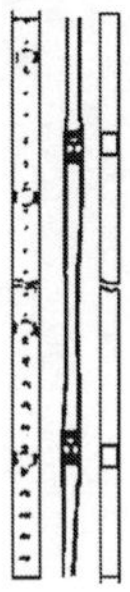

Material	Craft@Hrs	Unit	Material Cost	Labor Cost	Installed Cost
12" long, 2 outlets	L1@0.35	Ea	32.90	14.40	47.30
36" long, 6 outlets	L1@0.45	Ea	68.60	18.50	87.10
60" long, 5 outlets	L1@0.55	Ea	35.90	22.60	58.50
60" long, 10 outlets	L1@0.65	Ea	91.20	26.70	117.90
72" long, 4 outlets	L1@0.55	Ea	36.10	22.60	58.70
72" long, 6 outlets	L1@0.65	Ea	42.40	26.70	69.10
72" long, 8 outlets	L1@0.75	Ea	46.00	30.80	76.80
72" long, 12 outlets	L1@0.85	Ea	56.50	34.90	91.40

Two-piece surface metal raceway, 1-9/32" x 3/4" prewired, 2 circuit, 3-wire

Material	Craft@Hrs	Unit	Material Cost	Labor Cost	Installed Cost
60" long, 5 outlets	L1@0.60	Ea	39.60	24.60	64.20
72" long, 4 outlets	L1@0.60	Ea	44.10	24.60	68.70
72" long, 6 outlets	L1@0.70	Ea	53.10	28.70	81.80
72" long, 8 outlets	L1@0.80	Ea	68.90	32.80	101.70

Surface metal raceway, 2 circuit, 4-wire, grounding, outlets alternately wired

Material	Craft@Hrs	Unit	Material Cost	Labor Cost	Installed Cost
60" long, 5 outlets	L1@0.65	Ea	42.40	26.70	69.10
72" long, 4 outlets	L1@0.65	Ea	43.50	26.70	70.20
72" long, 6 outlets	L1@0.75	Ea	50.50	30.80	81.30
72" long, 8 outlets	L1@0.85	Ea	53.10	34.90	88.00

Two-piece 1-9/32" x 3/4" surface metal raceway assembly fittings

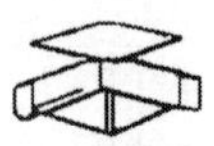
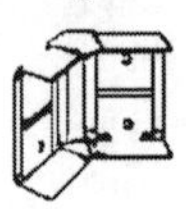

Material	Craft@Hrs	Unit	Material Cost	Labor Cost	Installed Cost
Flat elbow, 90 degree	L1@0.30	Ea	4.01	12.30	16.31
Splice cover	L1@0.05	Ea	2.52	2.05	4.57
Tee	L1@0.35	Ea	10.40	14.40	24.80
Internal corner coupling	L1@0.10	Ea	1.31	4.10	5.41
External corner coupling	L1@0.10	Ea	2.97	4.10	7.07
Ivory receptacle cap	L1@0.05	Ea	.48	2.05	2.53
Single pole switch	L1@0.35	Ea	14.60	14.40	29.00
Flush plate adapter	L1@0.20	Ea	7.93	8.21	16.14
Side reducing conn.	L1@0.10	Ea	9.66	4.10	13.76
End reducing conn.	L1@0.05	Ea	3.10	2.05	5.15
Wire clip	L1@0.05	Ea	.35	2.05	2.40
Coupling	L1@0.05	Ea	.49	2.05	2.54
Supporting clip	L1@0.05	Ea	1.69	2.05	3.74
Cover clip	L1@0.05	Ea	.93	2.05	2.98
T-bar clip	L1@0.10	Ea	4.08	4.10	8.18
Ground clip	L1@0.05	Ea	3.74	2.05	5.79
End entrance fitting	L1@0.15	Ea	4.40	6.15	10.55
End entr. fit. long	L1@0.20	Ea	14.70	8.21	22.91
Blank end fitting	L1@0.05	Ea	.69	2.05	2.74

Use these figures to estimate the cost of surface metal raceway assemblies and fittings installed in a building under the conditions described in the chapter text. The crew size is one electrician working at a labor cost of $41.03 per manhour. These costs include typical layout material handling and normal waste. Add for sales tax, delivery, supervision, mobilization, demobilization, cleanup, overhead and profit.

Material	Craft@Hrs	Unit	Material Cost	Labor Cost	Installed Cost

Stainless steel surface raceway, 1-9/32" x 3/4", prewired, 1 circuit, 3-wire, with ends & couplings

Material	Craft@Hrs	Unit	Material Cost	Labor Cost	Installed Cost
36" long, 6 outlets	L1@0.50	Ea	68.60	20.50	89.10
60" long, 5 outlets	L1@0.60	Ea	82.40	24.60	107.00
60" long, 10 outlets	L1@0.70	Ea	91.20	28.70	119.90
72" long, 6 outlets	L1@0.70	Ea	103.00	28.70	131.70

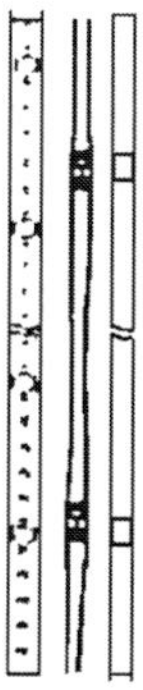

Two-piece steel surface raceway, 2-3/8" x 3/4", prewired receptacles, single circuit, 3-wire

Material	Craft@Hrs	Unit	Material Cost	Labor Cost	Installed Cost
60" long, 1 outlet	L1@0.55	Ea	30.10	22.60	52.70
60" long, 2 outlets	L1@0.60	Ea	30.30	24.60	54.90
72" long, 4 outlets	L1@0.65	Ea	33.80	26.70	60.50

Fittings for 2-3/8" x 3/4" steel surface raceway with prewired receptacles

Material	Craft@Hrs	Unit	Material Cost	Labor Cost	Installed Cost
Coupling	L1@0.05	Ea	2.21	2.05	4.26
Wire clip	L1@0.05	Ea	.88	2.05	2.93
Cover clip	L1@0.05	Ea	1.28	2.05	3.33
End blank fitting	L1@0.05	Ea	1.37	2.05	3.42
End trim fitting	L1@0.05	Ea	11.60	2.05	13.65
End entrance fitting	L1@0.30	Ea	13.80	12.30	26.10
End reducing fitting	L1@0.10	Ea	6.43	4.10	10.53
Internal coupling	L1@0.25	Ea	6.38	10.30	16.68
External coupling	L1@0.25	Ea	13.30	10.30	23.60
Flat elbow, 90 degree	L1@0.30	Ea	27.60	12.30	39.90
Wall box connector	L1@0.30	Ea	9.58	12.30	21.88
Single pole switch 15A	L1@0.25	Ea	16.20	10.30	26.50
Three-way switch 15A	L1@0.30	Ea	22.30	12.30	34.60
Single recept. cover	L1@0.15	Ea	9.76	6.15	15.91

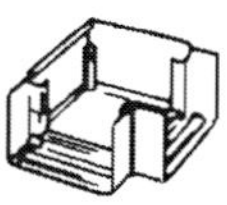

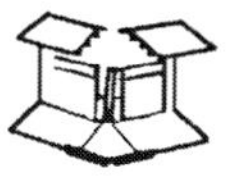

Use these figures to estimate the cost of surface metal raceway assemblies and fittings installed in a building under the conditions described in the chapter text. The crew size is one electrician working at a labor cost of $41.03 per manhour. These costs include typical layout material handling and normal waste. Add for sales tax, delivery, supervision, mobilization, demobilization, cleanup, overhead and profit.

Overhead Steel Channel Wiring System and Fittings

Overhead steel channel system, 1-5/8" x 1-5/8"

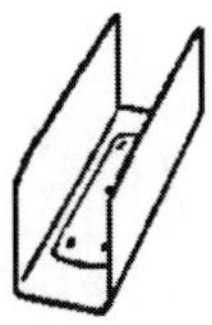

Material	Craft@Hrs	Unit	Material Cost	Labor Cost	Installed Cost
10' lengths	L1@4.00	CLF	551.00	164.00	715.00
20' lengths	L1@3.75	CLF	551.00	154.00	705.00

Overhead steel channel flat wire, 600 volt

Material	Craft@Hrs	Unit	Material Cost	Labor Cost	Installed Cost
#10 stranded	L1@8.00	MLF	467.00	328.00	795.00

Overhead steel channel system fittings

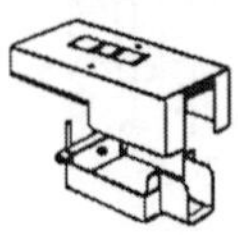

Material	Craft@Hrs	Unit	Material Cost	Labor Cost	Installed Cost
Fixture hanger	L1@0.15	Ea	8.20	6.15	14.35
End cap, insulated	L1@0.15	Ea	11.60	6.15	17.75
Coupling	L1@0.15	Ea	11.40	6.15	17.55
Channel hanger	L1@0.10	Ea	4.05	4.10	8.15
Splice connector	L1@0.25	Ea	5.46	10.30	15.76
Junction box	L1@0.30	Ea	31.90	12.30	44.20
Dual feed J-box	L1@0.30	Ea	50.00	12.30	62.30

Overhead steel channel system power taps and fixture hangers

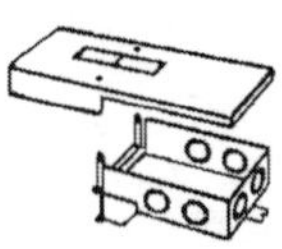

Material	Craft@Hrs	Unit	Material Cost	Labor Cost	Installed Cost
Phase A, black	L1@0.20	Ea	18.70	8.21	26.91
Phase A-B, bk & red	L1@0.20	Ea	18.70	8.21	26.91
Phase B, red	L1@0.20	Ea	18.70	8.21	26.91
Phase B-C, red & blue	L1@0.20	Ea	18.70	8.21	26.91
Phase C, blue	L1@0.20	Ea	18.70	8.21	26.91
Phase A-C, bk & blue	L1@0.20	Ea	18.70	8.21	26.91

Overhead steel channel system power taps with box & locking receptacle

Material	Craft@Hrs	Unit	Material Cost	Labor Cost	Installed Cost
Phase A, black	L1@0.20	Ea	79.30	8.21	87.51
Phase B, red	L1@0.20	Ea	79.30	8.21	87.51
Phase C, blue	L1@0.20	Ea	79.30	8.21	87.51

Overhead steel channel system tools

Material	Craft@Hrs	Unit	Material Cost	Labor Cost	Installed Cost
Cable pulling leader	--	Ea	50.90	--	50.90
Cable pull-in guide	--	Ea	78.60	--	78.60

Use these figures to estimate the cost of surface metal raceway and fittings and installed in a building under the conditions described in the chapter text. The crew size is one electrician working at a labor cost of $41.03 per manhour. These costs include layout, material handling and normal waste. Add sales tax, delivery, supervision, mobilization, demobilization, cleanup, overhead and profit.

Two-piece overhead steel raceway lateral 1-9/32" x 3/4", wired with one 125 volt circuit

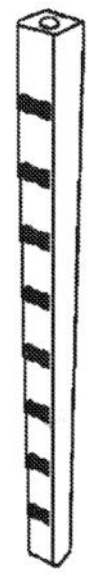

Material	Craft@Hrs	Unit	Material Cost	Labor Cost	Installed Cost
24" outlet spacing	L1@6.00	CLF	606.00	246.00	852.00
30" outlet spacing	L1@6.00	CLF	525.00	246.00	771.00

Two-piece overhead steel raceway lateral 1-9/32" x 3/4", wired with one 277 volt circuit

Material	Craft@Hrs	Unit	Material Cost	Labor Cost	Installed Cost
24" outlet spacing	L1@6.00	CLF	606.00	246.00	852.00
30" outlet spacing	L1@6.00	CLF	525.00	246.00	771.00

Two-piece overhead steel raceway lateral 1-9/32" x 3/4", wired with two 125 volt circuits

Material	Craft@Hrs	Unit	Material Cost	Labor Cost	Installed Cost
24" outlet spacing	L1@6.00	CLF	635.00	246.00	881.00
30" outlet spacing	L1@6.00	CLF	553.00	246.00	799.00

Two-piece lateral 2-3/4" x 1-17/32", wired with one 125 volt circuit

Material	Craft@Hrs	Unit	Material Cost	Labor Cost	Installed Cost
24" outlet spacing	L1@8.00	CLF	939.00	328.00	1,267.00
30" outlet spacing	L1@8.00	CLF	842.00	328.00	1,170.00

Two-piece lateral 2-3/4" x 1-17/32", wired with one 277 volt circuit

Material	Craft@Hrs	Unit	Material Cost	Labor Cost	Installed Cost
24" outlet spacing	L1@8.00	CLF	939.00	328.00	1,267.00
30" outlet spacing	L1@8.00	CLF	842.00	328.00	1,170.00

Two-piece lateral 2-3/4" x 1-17/32", wired with two 125 volt circuits

Material	Craft@Hrs	Unit	Material Cost	Labor Cost	Installed Cost
24" outlet spacing	L1@8.00	CLF	976.00	328.00	1,304.00
30" outlet spacing	L1@8.00	CLF	930.00	328.00	1,258.00

Two-piece lateral 2-3/4" x 1-17/32", wired with two 277 volt circuits

Material	Craft@Hrs	Unit	Material Cost	Labor Cost	Installed Cost
24" outlet spacing	L1@8.00	CLF	976.00	328.00	1,304.00

Two-piece lateral 2-3/4" x 1-17/32", wired with two 125/277 volt circuits alternating

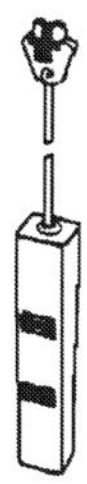

Material	Craft@Hrs	Unit	Material Cost	Labor Cost	Installed Cost
24" outlet spacing	L1@8.25	CLF	1,000.00	338.00	1,338.00

Two-piece lateral 2-3/4" x 1-17/32", wired with two 125 and two 277 volt circuits alternating

Material	Craft@Hrs	Unit	Material Cost	Labor Cost	Installed Cost
24" outlet spacing	L1@8.50	CLF	1,370.00	349.00	1,719.00

Use these figures to estimate the cost of surface metal raceway and fittings installed in a building under the conditions described in the chapter text. The crew size is one electrician working at a labor cost of $41.03 per manhour. These costs include layout, material handling and normal waste. Add for sales tax, delivery, supervision, mobilization, demobilization, cleanup, overhead and profit.

Telephone-Power Poles

Two-piece steel raceway lateral 4-3/4" x 3-9/16", wired with four 277 volt circuits

Material	Craft@Hrs	Unit	Material Cost	Labor Cost	Installed Cost
24" outlet spacing	L1@10.0	CLF	1,110.00	410.00	1,520.00

Two-piece lateral 4-3/4" x 3-9/16", wired with three 125 volt and four 277 volt circuits

Material	Craft@Hrs	Unit	Material Cost	Labor Cost	Installed Cost
24" outlet spacing	L1@10.3	CLF	1,320.00	423.00	1,743.00

Two-piece lateral 4-3/4" x 3-9/16", wired with five 125 volt and six 277 volt circuits

Material	Craft@Hrs	Unit	Material Cost	Labor Cost	Installed Cost
24" outlet spacing	L1@10.5	CLF	1,450.00	431.00	1,881.00

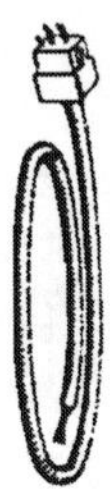

Fixture whips for overhead steel raceway

Material	Craft@Hrs	Unit	Material Cost	Labor Cost	Installed Cost
15A, 125V, 6' long	L1@0.20	Ea	63.00	8.21	71.21
15A, 125V, 15' long	L1@0.25	Ea	93.10	10.30	103.40
15A, 277V, 6' long	L1@0.20	Ea	55.30	8.21	63.51
15A, 277V, 15' long	L1@0.25	Ea	107.00	10.30	117.30
20A, 125V, 6' long	L1@0.20	Ea	63.00	8.21	71.21
20A, 125V, 15' long	L1@0.25	Ea	93.10	10.30	103.40

Jumper whips for overhead steel raceway

Material	Craft@Hrs	Unit	Material Cost	Labor Cost	Installed Cost
20A, 125V, 18' long	L1@0.25	Ea	130.00	10.30	140.30
Flat elbow, 90 degree	L1@0.25	Ea	103.00	10.30	113.30

Switching whips for overhead steel raceway

Material	Craft@Hrs	Unit	Material Cost	Labor Cost	Installed Cost
20A, 125V, 18' long	L1@0.25	Ea	177.00	10.30	187.30
20A, 277V, 18' long	L1@0.50	Ea	132.00	20.50	152.50

Cable adapter assemblies for overhead steel raceway

Material	Craft@Hrs	Unit	Material Cost	Labor Cost	Installed Cost
20A, 125V, 6' long	L1@0.15	Ea	63.00	6.15	69.15
20A, 125V, 12' long	L1@0.20	Ea	93.10	8.21	101.31
20A, 125V, 18' long	L1@0.25	Ea	132.00	10.30	142.30
20A, 277V, 6' long	L1@0.15	Ea	59.30	6.15	65.45

Two-piece overhead steel raceway lateral extension with wiring harness, 4-3/4" x 3-9/16"

Material	Craft@Hrs	Unit	Material Cost	Labor Cost	Installed Cost
15A, 277V, 14" long	L1@0.15	Ea	169.00	6.15	175.15
20A, 277V, 22" long	L1@0.20	Ea	175.00	8.21	183.21
20A, 277V, 36" long	L1@0.30	Ea	274.00	12.30	286.30

Use these figures to estimate the cost of surface metal raceway and fittings installed in a building under the conditions described in the chapter text. The crew size is one electrician working at a labor cost of $41.03 per manhour. These costs include layout, material handling and normal waste. Add sales tax, delivery, supervision, mobilization, demobilization, cleanup, overhead and profit.

Material	Craft@Hrs	Unit	Material Cost	Labor Cost	Installed Cost
Fittings for overhead steel raceway 2-3/4" x 1-17/32"					
Switch cover	L1@0.10	Ea	6.15	4.10	10.15
Sign recept cover	L1@0.10	Ea	6.15	4.10	10.15
Fixture/drop cover	L1@0.10	Ea	6.15	4.10	10.15
Tap-off fitting	L1@0.10	Ea	6.44	4.10	10.54
Single recept cover	L1@0.10	Ea	6.15	4.10	10.15
Circuit breaker box	L1@0.20	Ea	26.90	8.21	35.11
Extension adapter	L1@0.15	Ea	8.42	6.15	14.57
Take-off connector	L1@0.10	Ea	22.30	4.10	26.40
Conduit connector	L1@0.10	Ea	36.10	4.10	40.20
Panel connector	L1@0.30	Ea	8.41	12.30	20.71
End reducing conn.	L1@0.10	Ea	7.44	4.10	11.54
External corner coupl	L1@0.25	Ea	15.40	10.30	25.70
Sign recept. plate	L1@0.15	Ea	6.15	6.15	12.30
Single recept. plate	L1@0.15	Ea	6.15	6.15	12.30
Switch plate	L1@0.15	Ea	6.15	6.15	12.30
Duplex plate	L1@0.15	Ea	6.15	6.15	12.30
Utility box	L1@0.25	Ea	29.60	10.30	39.90
Deep sw box, 2-gang	L1@0.30	Ea	54.70	12.30	67.11
Fittings for overhead steel raceway 4-3/4" x 3-9/16"					
Coupling	L1@0.10	Ea	6.38	4.10	10.48
"C" hanger	L1@0.20	Ea	12.00	8.21	20.21
Blank end fitting	L1@0.15	Ea	8.18	6.15	14.33
Tap-off fitting	L1@0.25	Ea	8.45	10.30	18.75
Take-off fitting	L1@0.30	Ea	57.30	12.30	69.60
Tel conn housing	L1@0.30	Ea	77.90	12.30	90.20
Cable adapt assembly	L1@0.25	Ea	28.30	10.30	38.60
Receptacle cap	L1@0.05	Ea	.56	2.05	2.61

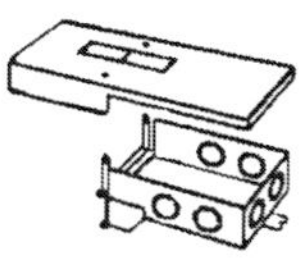

Use these figures to estimate the cost of surface metal raceway fittings installed in a building under the conditions described in the chapter text. The crew size is one electrician working at a labor cost of $41.03 per manhour. These costs include layout, material handling and normal waste. Add for sales tax, delivery, supervision, mobilization, demobilization, cleanup, overhead and profit.

Telephone-Power Poles

Material	Craft@Hrs	Unit	Material Cost	Labor Cost	Installed Cost
Telephone-power pole assemblies, 1-1/4" x 1-3/4", 2 power outlets					
10' pole, flush boot	L1@1.25	Ea	166.00	51.30	217.30
10'-4" pole, 2" boot	L1@1.25	Ea	157.00	51.30	208.30
12'-2" pole, 2" boot	L1@1.25	Ea	209.00	51.30	260.30
T-bar hanger clamp	L1@0.30	Ea	7.08	12.30	19.38
Telephone-power pole assemblies, 2-1/8" x 2-1/8", 4 power outlets					
10'-5" pole, flush boot	L1@1.30	Ea	137.00	53.30	190.30
12'-5" pole, flush boot	L1@1.35	Ea	179.00	55.40	234.40
15'-5" pole, flush boot	L1@1.40	Ea	228.00	57.40	285.40
10'-5" pole isol gr.	L1@1.35	Ea	161.00	55.40	216.40
Cabling network adapter	L1@0.50	Ea	21.10	20.50	41.60
Telephone-power pole assemblies, 2-3/4" x 1-7/16", 2 duplex outlets					
10'-5" pole, 2" foot	L1@1.30	Ea	214.00	53.30	267.30
12'-5" pole, 2" foot	L1@1.35	Ea	235.00	55.40	290.40
Cabling network adapter	L1@0.50	Ea	23.80	20.50	44.30
Telephone-power pole assemblies, 2-3/4" x 1-7/16", 4 power outlets					
10'-5" pole, 2" foot	L1@1.40	Ea	195.00	57.40	252.40
12'-5" pole, 2" foot	L1@1.45	Ea	235.00	59.50	294.50
Cabling network adapter	L1@0.50	Ea	23.80	20.50	44.30
Telephone-power pole assemblies, 2-3/4" x 2-7/8", 2 duplex outlets					
10'-5" pole, flush boot	L1@1.45	Ea	302.00	59.50	361.50
Telephone-power pole assemblies, 2" x 2", 4 power outlets					
10'-5" pole, adj foot	L1@1.30	Ea	308.00	53.30	361.30
10'-5" w/ lt woodgrain	L1@1.30	Ea	329.00	53.30	382.30
10'-5" w/ dk woodgrain	L1@1.30	Ea	329.00	53.30	382.30
12'-5" pole, adj foot	L1@1.40	Ea	279.00	57.40	336.40
15'-5" pole, adj foot	L1@1.50	Ea	356.00	61.50	417.50
Tel entrance fitting	L1@0.30	Ea	10.70	12.30	23.00
Cabling network adapter	L1@0.50	Ea	21.10	20.50	41.60

Use these figures to estimate the cost of telephone-power poles installed in a building under good conditions as described in the chapter text. Costs listed are for each telephone-power pole or listed fitting. The crew size is one electrician working at a labor cost of $41.03 per manhour. These costs include typical installation and layout for office furniture, counters and work stations, material handling and normal waste. Add for various fittings, boxes, sales tax, delivery, supervision, mobilization, demobilization, cleanup, overhead and profit.

Material	Craft@Hrs	Unit	Material Cost	Labor Cost	Installed Cost

Telephone-power pole assemblies, 1-5/16" x 1-5/8", 4 power outlets

Material	Craft@Hrs	Unit	Material Cost	Labor Cost	Installed Cost
10'-4" pole, flush boot	L1@1.30	Ea	174.00	53.30	227.30
10'-4" 3 unit pole	L1@1.40	Ea	203.00	57.40	260.40
Tel entrance fitting	L1@0.30	Ea	10.70	12.30	23.00
Cabling network adapter	L1@0.50	Ea	21.10	20.50	41.60
Add-on compartment	L1@0.50	Ea	83.20	20.50	103.70

Telephone-power pole assemblies, 2" x 2", 4 power outlets

Material	Craft@Hrs	Unit	Material Cost	Labor Cost	Installed Cost
10'-4" pole, flush boot	L1@1.40	Ea	238.00	57.40	295.40
10'-4" pole, w/lt woodgrain	L1@1.40	Ea	255.00	57.40	312.40
10'-4" pole, w/dk woodgrain	L1@1.40	Ea	255.00	57.40	312.40
12'-4" pole, flush boot	L1@1.50	Ea	279.00	61.50	340.50
15'-4" pole, flush boot	L1@1.70	Ea	356.00	69.80	425.80
Tel entrance fittings	L1@0.30	Ea	10.70	12.30	23.00

Telephone-power pole assemblies, 2" x 2", dedicated/isolated ground

Material	Craft@Hrs	Unit	Material Cost	Labor Cost	Installed Cost
10'-4", dedicated	L1@1.40	Ea	264.00	57.40	321.40
10'-4", isolated	L1@1.40	Ea	257.00	57.40	314.40
Tel entrance fitting	L1@0.30	Ea	10.70	12.30	23.00

Telephone-communications pole, 2" x 2"

Material	Craft@Hrs	Unit	Material Cost	Labor Cost	Installed Cost
10'-4" pole, flush boot	L1@1.00	Ea	121.00	41.00	162.00

Telephone-communications pole, 2-1/8" x 2-1/8"

Material	Craft@Hrs	Unit	Material Cost	Labor Cost	Installed Cost
10'-4" pole, flush boot	L1@1.00	Ea	209.00	41.00	250.00

Use these figures to estimate the cost of telephone-power poles installed in a building under good conditions as described in the chapter text. Costs listed are for each telephone-power pole or listed fitting. The crew size is one electrician working at a labor cost of $41.03 per manhour. These costs include typical installation and layout for office furniture, counters and work stations, material handling and normal waste. Add for various fittings, boxes, sales tax, delivery, supervision, mobilization, demobilization, cleanup, overhead and profit.

Section 16: Grounding

Article 250 of the *National Electrical Code* sets the standard for grounding and bonding AC electrical systems from 50 volts to 1,000 volts when used to supply power to a premise's wiring systems.

Grounding is intended to provide a path of least resistance from the power source to the earth so voltage is drained away before it can do any damage to people or equipment. The *NEC* considers an electrical system grounded when it's "connected to earth or to some conducting body that serves in place of the earth."

The *NEC* describes the purpose of grounding as follows: "Systems and circuits conductors are grounded to limit voltages due to lightning, line surges or unintentional contact with higher voltage lines, and to stabilize the voltage to ground during normal operation. System and circuit conductors are solidly grounded to facilitate overcurrent device operation in case of ground faults."

Most building inspectors consider grounding to be an important part of every electrical system. They're likely to check the grounding system very carefully.

Seven Common Grounding Methods

There are seven common ways of grounding an electrical system. I'll describe each of these in the paragraphs that follow. It's important to understand that the grounding method you select has to be appropriate for the electrical system you're estimating. Your building inspector may approve some types of grounding and not others for any particular job. It's a good idea to check with the building official if you're not sure of the type of grounding system required.

Here are the seven common grounding methods:

1. Connect to the incoming water supply metallic piping.
2. Connect to a driven ground rod at the point of electrical service.
3. Connect to a driven ground pipe at the point of electrical service.
4. Connect to a Ufer ground system.
5. Connect to a buried copper ring.
6. Connect to the building frame.
7. Connect to a plate electrode.

Connect to Underground Water Supply

This is probably the most common type of grounding system. The underground metallic water pipe must be in direct contact with the earth for a distance of at least 10 feet. Connection to the water pipe is made with an approved water pipe clamp, usually above ground, and in an area protected from mechanical injury. If the grounding conductor is installed in conduit to prevent damage, the water pipe clamp should be equipped with a ground clamp hub sized for the conduit.

In most areas, the inspector will approve a ground conductor with an armor jacket installed from the service equipment to the water pipe clamp that's attached to the underground metallic water pipe.

Connect to Driven Electrode

A driven ground rod or electrode has to be installed so that at least 8 feet of electrode is in contact with the earth. The electrode has to be made of steel or iron and must be at least $^5/_8$" in diameter. Nonferrous or stainless steel rods have to be at least ½" in diameter. The most common type of nonferrous electrode is made of steel and has a copper jacket. These are known as copper-clad ground rods. The exposed upper end of the ground rod must be protected against physical damage. If rock or hard soil below the surface makes it hard to drive the electrode, you can drive the rod at an angle not more than 45 degrees from the vertical. You can also bury it in a trench that's at least 2½ feet deep.

If you use metallic ground pipe or conduit for grounding, it has to be at least ¾" in diameter. The outer surface has to be galvanized or otherwise metal-coated for corrosion protection. In many areas the soil is highly corrosive. Even galvanized pipe will deteriorate in a few years. In any case, the pipe should be installed the same way as a ground rod would be installed.

It's not always easy to drive an 8-foot ground rod or pipe into the soil. If a sledge hammer won't drive a rod, a compressor or pneumatic drill may be required. A pneumatic drill can cut through hard earth and even some rock. When the hole has been cut to the required depth, insert the ground rod or pipe. Then wash backfill into the hole with water. The water helps increase the effectiveness of the grounding.

Connect to Ufer System

The Ufer ground system is installed in a building foundation when the foundation is poured. The

Ufer electrode has to be surrounded by at least 2" of concrete poured near the bottom of a foundation or footing that's in direct contact with the earth. Two types of material are acceptable for a Ufer ground:

1. At least 20 feet of one or more ½" steel reinforcing bars or rods.
2. At least 20 feet of bare copper conductor not smaller than number 4 AWG.

Connect to a Bare Copper Ring

Use a ground ring that encircles the building in direct contact with the earth at a depth of not less than 2½ feet. The bare copper conductor has to be at least 20 feet long and not smaller than number 2 AWG.

Connect to Building Frame

The metal frame of the building has to be grounded to the earth. This connection should have a sufficiently low impedance and enough current-carrying capacity to prevent an accumulation of voltage within the frame.

Connect to Plate Electrode

Plate electrodes are acceptable for grounding if the electrode has at least 2 square feet of surface exposed to the earth. There are two types of plates:

1. Iron or steel plate electrodes have to be at least ¼" thick.
2. Nonferrous electrodes have to be at least .06" thick.

The grounding electrode conductor has to be one continuous piece. No splices are permitted. Use Table 16-1 (*NEC* Table 250-66) to size the grounding electrode conductor.

Other Grounding Requirements

Certain types of electrical equipment need a separate grounding system. Metallic raceway serving the equipment and the metallic enclosure may have to be grounded. Electrically connected equipment housings must also be grounded. Table 16-2 (*NEC* Table 250-122) identifies the size of these grounding conductors.

Here are some definitions the *NEC* uses to describe grounding requirements.

Grounded

Connected to earth or to some conducting body that serves in place of earth.

Grounded Conductor

A system of circuit conductors that is intentionally grounded.

Grounding Conductor

A conductor used to connect equipment or the grounded circuit of a wiring system to a grounding electrode or electrodes.

Table 250.66
Grounding Electrode Conductor for Alternating-Current Systems

Size of Largest Ungrounded Service-Entrance Conductor or Equivalent Area for Parallel Conductors[a] (AWG/kcmil)		Size of Grounding Electrode Conductor (AWG/kcmil)	
Copper	**Aluminum or Copper-Clad Aluminum**	**Copper**	**Aluminum or Copper-Clad Aluminum[b]**
2 or smaller	1/0 or smaller	8	6
1 or 1/0	2/0 or 3/0	6	4
2/0 or 3/0	4/0 or 250	4	2
Over 3/0 through 350	Over 250 through 500	2	1/0
Over 350 through 600	Over 500 through 900	1/0	3/0
Over 600 through 1100	Over 900 through 1750	2/0	4/0
Over 1100	Over 1750	3/0	250

Notes:
1. If multiple sets of service-entrance conductors connect directly to a service drop, set of overhead service conductors, set of underground service conductors, or service lateral, the equivalent size of the largest service-entrance conductor shall be determined by the largest sum of the areas of the corresponding conductors of each set.
2. Where there are no service-entrance conductors, the grounding electrode conductor size shall be determined by the equivalent size of the largest service-entrance conductor required for the load to be served.

[a]This table also applies to the derived conductors of separately derived ac systems.
[b]See installation restrictions in Section 250.64(A).

Table 16-1

Grounding Conductor, Equipment

The conductor used to connect the noncurrent-carrying metal parts of equipment, raceways, and other enclosures to the system grounded conductor and/or to the grounding electrode conductor at the service equipment or at the source of a separately derived system.

Grounding Electrode Conductor

The conductor used to connect the grounding electrode to the equipment grounding conductor and/or to the grounded conductor of the circuit at the service equipment or at the source of a separately derived system.

Ground Fault Circuit Interrupter

A device intended for the protection of personnel that functions to deenergize a circuit or portion thereof within an established period of time when a current to ground exceeds some predetermined value that is less than that required to operate the overcurrent protective device of the supply circuit.

Ground Fault Protection of Equipment

A system intended to provide protection of equipment from damaging line-to-ground fault currents by operating to cause a disconnecting means to open all ungrounded conductors of the faulted circuit. This protection is provided at current levels less than those required to protect conductors from damage through the operation of a supply circuit overcurrent device.

Ground Testing

Some contracts require that the grounding system be tested to see if it works as intended. Resistance is measured in ohms. The device used for this test is called a ***ground megger***. It checks the resistance of the earth.

To test resistance of soil, drive two ½" by 36" steel rods into the earth between 1 and 2 feet deep at a specified distance from the ground electrode. Then connect a #12 copper wire to each rod, to the ground electrode and to the megger. The megger indicates the resistance of the earth between electrodes.

Table 250.122
Minimum Size Equipment Grounding Conductors for Grounding Raceway and Equipment

Rating or Setting of Automatic Overcurrent Device in Circuit Ahead of Equipment, Conduit, etc., Not Exceeding (Amperes)	Size (AWG or kcmil)	
	Copper	Aluminum or Copper-Clad Aluminum.*
15	14	12
20	12	10
60	10	8
100	8	6
200	6	4
300	4	2
400	3	1
500	2	1/0
600	1	2/0
800	1/0	3/0
1000	2/0	4/0
1200	3/0	250
1600	4/0	350
2000	250	400
2500	350	600
3000	400	600
4000	500	750
5000	700	1200
6000	800	1200

Note: Where necessary to comply with Section 250.4(A)(5) or (B)(4), the equipment grounding conductor shall be sized larger than given in this table.
*See installation restrictions in 250.120.
From the ***National Electrical Code*** *©2016, National Fire Protection Association.*

Table 16-2

The *NEC* assumes ground resistance of 25 ohms or less. If resistance is higher than that, additional rod or rods will be required. Don't install these rods until a change order has been signed. When a megger reading is required, keep a record of those readings in your job file.

The earth's resistance will change as moisture levels in the soil change. Wetter soil is a better conductor. Dryer soil has higher resistance. When resistance is high and the climate is normally dry, you may have to increase the conducting capacity of the ground. Try drilling an oversized hole about 2" to 3" in diameter. Install a ½" to 1" ground rod. Mix copper sulfate with the backfill soil and wash it into the hole with water. After a few days, the copper sulfate should provide a much lower resistance reading.

Grounding Problems

Good grounding isn't always easy. Think about the type of grounding that will provide economical yet durable protection for the electric system. Some soils are very corrosive and limit the life of an electrode or even the underground water service pipe. As the metal is eaten away, the conducting capacity drops, reducing the margin of safety.

The Ufer ground system provides many advantages if the soil is corrosive. The main limitation is that a Ufer ground has to be installed in the building's foundation in contact with the earth when the foundation is poured.

Another choice would be dual grounding: Provide two or more grounds. That may help ensure an adequate grounding system for many years. For example, you might install a driven electrode and also connect to the underground water supply system. Any combination of two or more connections would be ideal. But be sure that both grounds meet code requirements.

Connecting the Ground

The connection to either a Ufer ground, a ground rod or a building frame should be made with an exothermic weld. The weld is made by fusing the conductor to the electrode.

Start by selecting a carbon mold the right size to accept the ground conductor and the electrode. The fusing process is quick and simple. Make sure both surfaces on the electrode and conductor are clean and bright. Then place the mold on the electrode and conductor. Select a powder charge that's the right size for the mold. Place the fusing slug and charge. Then set the powder off with a sparking device.

Let the weld cool before removing the mold. When the connection is cool, chip off the slag and check the weld for blistering. If the weld is puffy and weak, start over with a new weld.

Every conductor connection should be cleaned carefully. Remove any corrosion, insulation or dirt with a wire brush. A small piece of sandpaper is excellent for cleaning the connections.

Grounding Electrical Devices

Metallic outlet boxes have a threaded screw hole in the back of the box. Use this hole for attaching a ground conductor from the wiring device's ground terminal. Ground clips have been approved for clipping bonding conductors to the side of metallic boxes.

If you use nonmetallic conduit and boxes, install a ground or bonding conductor along with the circuit conductors. Splice grounding and bonding conductors and the device grounding jumpers in nonmetallic boxes.

Conduit will provide a continuous ground from each load all the way back to the service entrance equipment or distribution panel.

Wire comes with either a bare ground wire or with the ground wire insulated and colored green. Green insulation is used on wire sizes 8 and smaller. On heavier gauge wire, you may have to color the ground green or wrap green marking tape around the green conductor.

Metallic conduit that's stubbed up under electrical equipment should be bonded or grounded to the service equipment with a jumper. Install a threaded fitting called a ground bushing on the conduit or conduit fitting. The ground bushing has either a ground lug or a ground terminal for attaching the jumper wire.

No ground bushing is needed when nonmetallic conduit is used to connect electrical equipment. But a ground wire has to be attached to the electrical equipment ground. Ream the nonmetallic conduit to removed burrs before attaching the ground wire.

Aluminum Conductors

The *NEC* doesn't permit the use of aluminum electrodes for grounding purposes. When aluminum wire is used in a grounding system, be sure the points of contact are thoroughly cleaned of oxidation before making the connection. Clean the contacts with a brush and solvent. Then apply an anti-oxidation preparation to the joint. Make the connection with compression connectors rated for aluminum.

A few days after making the connection, retighten the connector. An aluminum connector tends to loosen when first installed. As it loosens, resistance increases. That weakens the connection and may increase resistance even more. It's a good idea to recheck connections to aluminum conductors a year after the work is completed.

Grounding

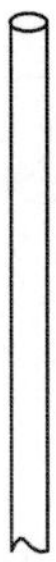

Soft drawn solid copper ground wire

Material	Craft@Hrs	Unit	Material Cost	Labor Cost	Installed Cost
#14	L1@4.00	KLF	132.00	164.00	296.00
#12	L1@4.25	KLF	208.00	174.00	382.00
#10	L1@4.50	KLF	302.00	185.00	487.00
#8	L1@4.75	KLF	538.00	195.00	733.00
#6	L1@5.00	KLF	1,020.00	205.00	1,225.00
#4	L1@6.00	KLF	1,630.00	246.00	1,876.00
#2	L1@7.00	KLF	2,310.00	287.00	2,597.00

Soft drawn stranded bare copper ground wire

Material	Craft@Hrs	Unit	Material Cost	Labor Cost	Installed Cost
#8	L1@4.75	KLF	580.00	195.00	775.00
#6	L1@5.00	KLF	1,020.00	205.00	1,225.00
#4	L1@6.00	KLF	1,630.00	246.00	1,876.00
#2	L1@8.00	KLF	2,400.00	328.00	2,728.00
#1	L1@10.0	KLF	3,240.00	410.00	3,650.00
#1/0	L1@11.0	KLF	3,610.00	451.00	4,061.00
#2/0	L1@12.0	KLF	4,220.00	492.00	4,712.00
#3/0	L1@13.0	KLF	5,360.00	533.00	5,893.00
#4/0	L1@14.0	KLF	6,710.00	574.00	7,284.00

Wiring device grounding jumpers, insulated

Material	Craft@Hrs	Unit	Material Cost	Labor Cost	Installed Cost
#12	L1@0.05	Ea	.92	2.05	2.97

Ground clip

Material	Craft@Hrs	Unit	Material Cost	Labor Cost	Installed Cost
Box grounding clip	L1@0.05	Ea	.49	2.05	2.54

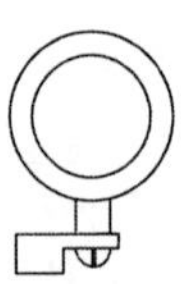

Insulated ground bushing

Material	Craft@Hrs	Unit	Material Cost	Labor Cost	Installed Cost
1/2"	L1@0.10	Ea	7.26	4.10	11.36
3/4"	L1@0.10	Ea	9.30	4.10	13.40
1"	L1@0.10	Ea	10.30	4.10	14.40
1-1/4"	L1@0.15	Ea	14.30	6.15	20.45
1-1/2"	L1@0.15	Ea	15.70	6.15	21.85
2"	L1@0.20	Ea	21.00	8.21	29.21
2-1/2"	L1@0.20	Ea	37.00	8.21	45.21
3"	L1@0.25	Ea	47.90	10.30	58.20
3-1/2"	L1@0.25	Ea	53.70	10.30	64.00
4"	L1@0.30	Ea	67.20	12.30	79.50
5"	L1@0.40	Ea	259.00	16.40	275.40
6"	L1@0.50	Ea	397.00	20.50	417.50

Use these figures to estimate the cost of bare copper wire, grounding jumpers, ground clips and insulated ground bushings installed in buildings, under the conditions described on pages 5 and 6. Costs listed are for each 1000 linear feet installed and each part installed. The crew size is one electrician working at a labor cost of $41.03 per manhour. These costs include wire make-up, continuity testing, reel set-up, layout, material handling, and normal waste. Add for connectors, sales tax, delivery, supervision, mobilization, demobilization, cleanup, overhead and profit. Note: The *NEC* permits installation of bare copper wire in conduit with conductors. But bare copper wire installed outside conduit must be protected against damage. Some dealers sell bare copper wire by weight rather than length. The table on page 102 shows weights per 1000 linear feet.

Material	Craft@Hrs	Unit	Material Cost	Labor Cost	Installed Cost

One hole solder type grounding lugs

#10	L1@0.10	Ea	1.24	4.10	5.34
#8	L1@0.15	Ea	1.57	6.15	7.72
#6	L1@0.15	Ea	1.73	6.15	7.88
#4	L1@0.20	Ea	1.87	8.21	10.08
#2	L1@0.20	Ea	2.50	8.21	10.71
#1/0	L1@0.25	Ea	3.31	10.30	13.61
#2/0	L1@0.25	Ea	4.07	10.30	14.37
#3/0	L1@0.30	Ea	5.54	12.30	17.84
#4/0	L1@0.30	Ea	7.09	12.30	19.39
# 250 KCMIL	L1@0.40	Ea	13.40	16.40	29.80
# 400 KCMIL	L1@0.50	Ea	20.30	20.50	40.80
# 500 KCMIL	L1@0.50	Ea	31.30	20.50	51.80
# 600 KCMIL	L1@0.50	Ea	33.60	20.50	54.10
# 800 KCMIL	L1@0.60	Ea	56.20	24.60	80.80
#1000 KCMIL	L1@0.70	Ea	61.60	28.70	90.30
#1500 KCMIL	L1@0.75	Ea	110.00	30.80	140.80

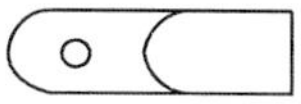

One hole solderless type grounding lugs

#10	L1@0.05	Ea	1.70	2.05	3.75
#6	L1@0.08	Ea	1.73	3.28	5.01
#4	L1@0.10	Ea	2.46	4.10	6.56
#2	L1@0.10	Ea	2.46	4.10	6.56
#1/0	L1@0.15	Ea	4.79	6.15	10.94
#3/0	L1@0.20	Ea	9.25	8.21	17.46
#4/0	L1@0.20	Ea	10.90	8.21	19.11
#350 KCMIL	L1@0.30	Ea	20.60	12.30	32.90
#500 KCMIL	L1@0.40	Ea	29.60	16.40	46.00
#1000 KCMIL	L1@0.60	Ea	74.00	24.60	98.60

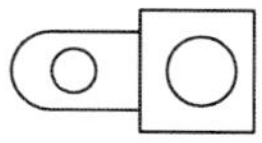

Water pipe ground clamps

1"	L1@0.10	Ea	4.70	4.10	8.80
2"	L1@0.15	Ea	6.47	6.15	12.62
2-1/2"	L1@0.20	Ea	27.90	8.21	36.11
5"	L1@0.30	Ea	42.50	12.30	54.80
6"	L1@0.40	Ea	64.30	16.40	80.70

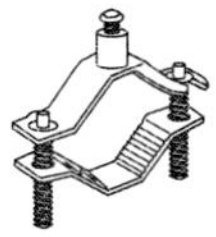

Ground clamp conduit hubs

Armored cable	L1@0.15	Ea	9.71	6.15	15.86
1/2"	L1@0.20	Ea	4.70	8.21	12.91
3/4"	L1@0.25	Ea	8.43	10.30	18.73
1"	L1@0.30	Ea	16.40	12.30	28.70
1-1/4"	L1@0.35	Ea	27.20	14.40	41.60

Use these figures to estimate the cost of conductor lugs, ground clamps and ground hubs installed under conditions described on pages 5 and 6. Costs are for each lug installed. The crew size is one electrician working at a labor cost of $41.03 per manhour. The costs include lug or ground fitting (and no other material) wire cutting, insulation stripping for termination, layout, material handling and normal waste. Add for insulating the lug (if needed), sales tax, delivery, supervision, mobilization, demobilization, cleanup, overhead and profit. Note: Always recheck solderless lugs for tightness.

Grounding

Material	Craft@Hrs	Unit	Material Cost	Labor Cost	Installed Cost

Copper-clad ground rods

Material	Craft@Hrs	Unit	Material Cost	Labor Cost	Installed Cost
1/2" dia x 8' long	L1@0.70	Ea	13.90	28.70	42.60
5/8" dia x 8' long	L1@0.75	Ea	15.80	30.80	46.60
3/4" dia x 8' long	L1@0.90	Ea	31.40	36.90	68.30
1/2" dia x 10' long	L1@0.75	Ea	19.60	30.80	50.40
5/8" dia x 10' long	L1@0.90	Ea	21.10	36.90	58.00
3/4" dia x 10' long	L1@1.00	Ea	32.80	41.00	73.80
1" dia x 10' long	L1@1.25	Ea	100.00	51.30	151.30
1" dia x 20' long	L1@2.00	Ea	211.00	82.10	293.10

Copper-clad sectional ground rods

Material	Craft@Hrs	Unit	Material Cost	Labor Cost	Installed Cost
5/8" dia x 8' long	L1@1.70	Ea	21.10	69.80	90.90
1/2" dia x 10' long	L1@1.70	Ea	21.80	69.80	91.60
5/8" dia x 10' long	L1@1.75	Ea	24.40	71.80	96.20
3/4" dia x 10' long	L1@2.00	Ea	35.00	82.10	117.10
1" dia x 10' long	L1@2.25	Ea	105.00	92.30	197.30

Ground rod couplings

Material	Craft@Hrs	Unit	Material Cost	Labor Cost	Installed Cost
1/2"	L1@0.10	Ea	12.20	4.10	16.30
5/8"	L1@0.10	Ea	15.80	4.10	19.90
3/4"	L1@0.15	Ea	21.80	6.15	27.95
1"	L1@0.20	Ea	26.20	8.21	34.41

Ground rod clamps

Material	Craft@Hrs	Unit	Material Cost	Labor Cost	Installed Cost
1/2"	L1@0.10	Ea	3.14	4.10	7.24
5/8"	L1@0.10	Ea	2.78	4.10	6.88
3/4"	L1@0.15	Ea	3.97	6.15	10.12
1"	L1@0.20	Ea	9.02	8.21	17.23

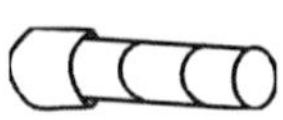

Ground rod stud bolts for sectional rod

Material	Craft@Hrs	Unit	Material Cost	Labor Cost	Installed Cost
1/2"	L1@0.25	Ea	13.70	10.30	24.00
5/8"	L1@0.25	Ea	12.00	10.30	22.30
3/4"	L1@0.30	Ea	13.30	12.30	25.60
1"	L1@0.40	Ea	23.50	16.40	39.90

Use these figures to estimate the cost of ground rods installed under the conditions described on pages 5 and 6. Costs listed are for each piece installed. The crew size is one electrician working at a labor cost of $41.03 per manhour. These costs include the ground rod or clamp only (and no other material), wire cutting, stripping for termination, layout, material handling and normal waste. Add for sales tax, delivery, supervision, mobilization, demobilization, cleanup, overhead and profit. Note: Always recheck mechanical connections for tightness.

Grounding locknuts

Material	Craft@Hrs	Unit	Material Cost	Labor Cost	Installed Cost
1/2"	L1@0.05	Ea	1.88	2.05	3.93
3/4"	L1@0.06	Ea	2.40	2.46	4.86
1"	L1@0.08	Ea	3.27	3.28	6.55
1-1/4"	L1@0.10	Ea	4.38	4.10	8.48
1-1/2"	L1@0.10	Ea	5.91	4.10	10.01
2"	L1@0.15	Ea	8.01	6.15	14.16
2-1/2"	L1@0.20	Ea	16.20	8.21	24.41
3"	L1@0.20	Ea	20.60	8.21	28.81
3-1/2"	L1@0.25	Ea	33.30	10.30	43.60
4"	L1@0.30	Ea	42.70	12.30	55.00

Exothermic copper grounding connections

Material	Craft@Hrs	Unit	Material Cost	Labor Cost	Installed Cost
2/0 - 1/2" rod	L1@0.40	Ea	16.20	16.40	32.60
2/0 - 5/8" rod	L1@0.50	Ea	16.20	20.50	36.70
2/0 - 3/4" rod	L1@0.50	Ea	16.20	20.50	36.70
2/0 - 1" rod	L1@0.60	Ea	16.20	24.60	40.80
4/0 - 1/2" rod	L1@0.50	Ea	17.90	20.50	38.40
4/0 - 5/8" rod	L1@0.50	Ea	17.90	20.50	38.40
4/0 - 3/4" rod	L1@0.60	Ea	17.90	24.60	42.50
4/0 - 1" rod	L1@0.75	Ea	17.90	30.80	48.70
2/0 - 2/0 splice	L1@0.40	Ea	15.50	16.40	31.90
2/0 - 4/0 splice	L1@0.50	Ea	15.50	20.50	36.00
4/0 - 4/0 splice	L1@0.50	Ea	16.20	20.50	36.70
2/0 - 2/0 x 2/0	L1@0.70	Ea	16.20	28.70	44.90
2/0 - 4/0 x 4/0	L1@0.80	Ea	17.90	32.80	50.70
4/0 - 4/0 x 4/0	L1@1.00	Ea	18.70	41.00	59.70
2/0 - lug	L1@0.40	Ea	14.40	16.40	30.80
4/0 - lug	L1@0.50	Ea	15.50	20.50	36.00
2/0 - steel beam	L1@0.40	Ea	15.50	16.40	31.90
4/0 - steel beam	L1@0.50	Ea	16.20	20.50	36.70
Exothermic mold	--	Ea	154.00	--	154.00

Use these figures to estimate the cost of grounding connections installed under the conditions described on pages 5 and 6. Costs listed are for each piece installed. The crew size is one electrician working at a labor cost of $41.03 per manhour. These costs include the grounding locknut or exothermic connection only (and no other material), wire cutting, stripping for termination, layout, material handling and normal waste. Add for sales tax, delivery, supervision, mobilization, demobilization, cleanup, overhead and profit. Note: Always recheck mechanical connections for tightness. Be sure exothermic powder charges are stored in a dry place. Exothermic molds are designed for the connection required.

Section 17: Assemblies

The first sixteen chapters in this manual assume that your take-off form will list every box, plate, strap and plaster ring needed to do the work. That's the only way to make perfectly accurate electrical estimates: Measure and count everything that goes into the job. But it's also slow, tedious, demanding work.

Many experienced electrical estimators use shortcuts that save time without sacrificing accuracy. They group associated materials into assemblies that can be priced as a unit. For example, why price wire and couplings and conduit and connectors and straps separately when they're used together and in about the same proportion on nearly every circuit? Grouping materials into assemblies saves counting, measuring and calculating time and can produce results nearly as accurate as more exhaustive detailed estimates.

The remainder of this section has assembly cost estimates for the most common electrical work. Labor and material prices are the same as in the prior sections. The only difference here is that materials are grouped to speed and simplify your estimates. If you decide to use costs from this section, all you have to do is count symbols on the E sheets and measure circuit lengths. List each assembly on your take-off sheet. Then use the figures in this section to find the estimated cost per assembly.

Unfortunately, there's one major disadvantage to estimating by assembly: The number of possible assemblies is almost infinite. I couldn't possibly list every electrical assembly you're likely to need on the next job. There are far too many combinations. Even if you prefer assembly pricing, at least occasionally you're going to have to figure the wire, conduit, couplings and straps separately.

The assemblies listed here are the most common combinations, using standard commercial grade materials that meet code requirements. But you won't be able to use these costs on jobs that require oversize outlet boxes, set minimum conduit sizes, specify unusual wire insulation or identify lighting fixtures by catalog number. None of the assemblies in this section cover situations like that. But I believe you'll find that most switch and receptacle circuits on most jobs can be priced from the pages that follow.

Conduit and Wire

Cost tables for conduit and wire begin with assemblies for empty conduit with 2 set screw connectors, 9 set screw couplings and 9 one-hole straps per 100 linear feet. Simply measure conduit length for each size and type and multiply the length (in hundreds of feet) by the cost per 100 feet. The next set of cost tables include conduit, connectors, couplings, straps, a pull line, plus insulated copper wire. Note that costs will be slightly higher on short runs because more than two set screw connectors are required for each 100 linear feet.

Outlet Boxes and Wiring Devices

Assemblies on these pages include an outlet box, a wiring device, a cover plate and a plaster ring (when needed). Some outlet boxes can be assembled into multi-ganged outlets when more than one device is required. When using costs in this section, begin by deciding what type of assembly is required. Then count on the E sheets the symbols that represent each assembly. Finally, multiply the cost per assembly by the number of assemblies.

Switches

Assemblies are included for 15 amp and 20 amp switches, whether single pole, double pole, three way or four way and for both 120 volts and 277 volt circuits.

Receptacles

Assemblies are included for both 15 amp and 20 amp circuits and for both single and duplex receptacles. The 15 amp receptacles are NEMA 5-15R. The 20 amp receptacles are NEMA 5-20R.

Plastic Plates

Switch and receptacle plates are smooth plastic and are listed by color. All plates are good quality as produced by the major manufacturers.

Lighting Fixtures

Costs for lighting fixtures assume factory-assembled units with lamps and a 5' flexible conduit whip. The whip has copper wire and flex connectors. Using assembled fluorescent fixtures like these can reduce installation costs, especially in a T-bar grid ceiling.

Material	Craft@Hrs	Unit	Material Cost	Labor Cost	Installed Cost
100' 1/2" EMT conduit, 2 set screw connectors, 9 set screw couplings and 9 one-hole straps					
Empty conduit	L1@4.07	CLF	39.50	167.00	206.50
1 plastic pull line	L1@4.38	CLF	40.00	180.00	220.00
1 #14THHN, solid	L1@4.67	CLF	51.70	192.00	243.70
2 #14THHN, solid	L1@5.27	CLF	63.40	216.00	279.40
3 #14THHN, solid	L1@5.87	CLF	75.10	241.00	316.10
4 #14THHN, solid	L1@6.47	CLF	86.80	265.00	351.80
5 #14THHN, solid	L1@7.07	CLF	98.50	290.00	388.50
6 #14THHN, solid	L1@7.67	CLF	110.00	315.00	425.00
1 #14THHN, stranded	L1@4.67	CLF	49.50	192.00	241.50
2 #14THHN, stranded	L1@5.27	CLF	59.00	216.00	275.00
3 #14THHN, stranded	L1@5.87	CLF	68.50	241.00	309.50
4 #14THHN, stranded	L1@6.47	CLF	77.80	265.00	342.80
5 #14THHN, stranded	L1@7.07	CLF	87.30	290.00	377.30
6 #14THHN, stranded	L1@7.67	CLF	96.80	315.00	411.80
1 #12THHN, solid	L1@4.77	CLF	56.90	196.00	252.90
2 #12THHN, solid	L1@5.47	CLF	73.70	224.00	297.70
3 #12THHN, solid	L1@6.17	CLF	90.50	253.00	343.50
4 #12THHN, solid	L1@6.87	CLF	107.00	282.00	389.00
5 #12THHN, solid	L1@7.57	CLF	124.00	311.00	435.00
6 #12THHN, solid	L1@8.27	CLF	141.00	339.00	480.00
1 #12THHN, stranded	L1@4.77	CLF	53.50	196.00	249.50
2 #12THHN, stranded	L1@5.47	CLF	67.10	224.00	291.10
3 #12THHN, stranded	L1@6.17	CLF	80.70	253.00	333.70
4 #12THHN, stranded	L1@6.87	CLF	94.30	282.00	376.30
5 #12THHN, stranded	L1@7.57	CLF	108.00	311.00	419.00
6 #12THHN, stranded	L1@8.27	CLF	122.00	339.00	461.00
1 #10THHN, solid	L1@4.87	CLF	65.60	200.00	265.60
2 #10THHN, solid	L1@5.67	CLF	91.20	233.00	324.20
3 #10THHN, solid	L1@6.47	CLF	117.00	265.00	382.00
4 #10THHN, solid	L1@7.27	CLF	142.00	298.00	440.00
5 #10THHN, solid	L1@8.07	CLF	168.00	331.00	499.00
6 #10THHN, solid	L1@8.87	CLF	193.00	364.00	557.00
1 #10THHN, stranded	L1@4.87	CLF	60.40	200.00	260.40
2 #10THHN, stranded	L1@5.67	CLF	80.70	233.00	313.70
3 #10THHN, stranded	L1@6.47	CLF	101.00	265.00	366.00
4 #10THHN, stranded	L1@7.27	CLF	122.00	298.00	420.00
5 #10THHN, stranded	L1@8.07	CLF	141.00	331.00	472.00
6 #10THHN, stranded	L1@8.87	CLF	162.00	364.00	526.00

Use these figures to estimate the cost of installing assemblies under the conditions described on pages 5 and 6. Costs listed are for each assembly installed. The crew is one electrician working at a labor cost of $41.03 per manhour. These costs include layout, material handling and normal waste. Add for the sales tax, delivery, supervision, mobilization, demobilization, cleanup, overhead and profit.

3/4" EMT Conduit Assemblies

Material	Craft@Hrs	Unit	Material Cost	Labor Cost	Installed Cost
100' 3/4" EMT conduit, 2 set screw connectors, 9 set screw couplings and 9 one-hole straps					
Empty conduit	L1@4.52	CLF	73.30	185.00	258.30
1 plastic pull line	L1@4.82	CLF	73.80	198.00	271.80
1 #14THHN, solid	L1@5.12	CLF	85.50	210.00	295.50
2 #14THHN, solid	L1@5.72	CLF	97.20	235.00	332.20
3 #14THHN, solid	L1@6.32	CLF	109.00	259.00	368.00
4 #14THHN, solid	L1@6.92	CLF	120.00	284.00	404.00
5 #14THHN, solid	L1@7.52	CLF	132.00	309.00	441.00
6 #14THHN, solid	L1@8.12	CLF	144.00	333.00	477.00
1 #14THHN, stranded	L1@5.12	CLF	83.30	210.00	293.30
2 #14THHN, stranded	L1@5.72	CLF	92.80	235.00	327.80
3 #14THHN, stranded	L1@6.32	CLF	102.00	259.00	361.00
4 #14THHN, stranded	L1@6.92	CLF	112.00	284.00	396.00
5 #14THHN, stranded	L1@7.52	CLF	122.00	309.00	431.00
6 #14THHN, stranded	L1@8.12	CLF	131.00	333.00	464.00
1 #12THHN, solid	L1@5.22	CLF	90.70	214.00	304.70
2 #12THHN, solid	L1@5.92	CLF	107.00	243.00	350.00
3 #12THHN, solid	L1@6.62	CLF	124.00	272.00	396.00
4 #12THHN, solid	L1@7.32	CLF	141.00	300.00	441.00
5 #12THHN, solid	L1@8.02	CLF	158.00	329.00	487.00
6 #12THHN, solid	L1@8.72	CLF	175.00	358.00	533.00
1 #12THHN, stranded	L1@5.22	CLF	87.50	214.00	301.50
2 #12THHN, stranded	L1@5.92	CLF	101.00	243.00	344.00
3 #12THHN, stranded	L1@6.62	CLF	115.00	272.00	387.00
4 #12THHN, stranded	L1@7.32	CLF	129.00	300.00	429.00
5 #12THHN, stranded	L1@8.02	CLF	141.00	329.00	470.00
6 #12THHN, stranded	L1@8.72	CLF	155.00	358.00	513.00
1 #10THHN, solid	L1@5.32	CLF	99.50	218.00	317.50
2 #10THHN, solid	L1@6.12	CLF	125.00	251.00	376.00
3 #10THHN, solid	L1@6.92	CLF	151.00	284.00	435.00
4 #10THHN, solid	L1@7.72	CLF	176.00	317.00	493.00
5 #10THHN, solid	L1@8.52	CLF	202.00	350.00	552.00
6 #10THHN, solid	L1@9.32	CLF	227.00	382.00	609.00
1 #10THHN, stranded	L1@5.32	CLF	94.20	218.00	312.20
2 #10THHN, stranded	L1@6.12	CLF	115.00	251.00	366.00
3 #10THHN, stranded	L1@6.92	CLF	134.00	284.00	418.00
4 #10THHN, stranded	L1@7.72	CLF	155.00	317.00	472.00
5 #10THHN, stranded	L1@8.52	CLF	176.00	350.00	526.00
6 #10THHN, stranded	L1@9.32	CLF	196.00	382.00	578.00

Use these figures to estimate the cost of installing assemblies under the conditions described on pages 5 and 6. Costs listed are for each assembly installed. The crew is one electrician working at a labor cost of $41.03 per manhour. These costs include layout, material handling and normal waste. Add for the sales tax, delivery, supervision, mobilization, demobilization, cleanup, overhead and profit.

100' 1" EMT conduit, 2 set screw connectors, 9 set screw couplings and 9 one-hole straps

Material	Craft@Hrs	Unit	Material Cost	Labor Cost	Installed Cost
Empty conduit	L1@5.33	CLF	123.00	219.00	342.00
1 plastic pull line	L1@5.63	CLF	124.00	231.00	355.00
1 #14THHN, solid	L1@5.93	CLF	136.00	243.00	379.00
2 #14THHN, solid	L1@6.53	CLF	147.00	268.00	415.00
3 #14THHN, solid	L1@7.13	CLF	159.00	293.00	452.00
4 #14THHN, solid	L1@7.73	CLF	170.00	317.00	487.00
5 #14THHN, solid	L1@8.33	CLF	182.00	342.00	524.00
6 #14THHN, solid	L1@8.93	CLF	193.00	366.00	559.00
1 #14THHN, stranded	L1@5.93	CLF	133.00	243.00	376.00
2 #14THHN, stranded	L1@6.53	CLF	142.00	268.00	410.00
3 #14THHN, stranded	L1@7.13	CLF	152.00	293.00	445.00
4 #14THHN, stranded	L1@7.73	CLF	161.00	317.00	478.00
5 #14THHN, stranded	L1@8.33	CLF	171.00	342.00	513.00
6 #14THHN, stranded	L1@8.93	CLF	181.00	366.00	547.00
1 #12THHN, solid	L1@6.03	CLF	140.00	247.00	387.00
2 #12THHN, solid	L1@6.73	CLF	158.00	276.00	434.00
3 #12THHN, solid	L1@7.43	CLF	174.00	305.00	479.00
4 #12THHN, solid	L1@8.13	CLF	191.00	334.00	525.00
5 #12THHN, solid	L1@8.83	CLF	207.00	362.00	569.00
6 #12THHN, solid	L1@9.53	CLF	225.00	391.00	616.00
1 #12THHN, stranded	L1@6.03	CLF	137.00	247.00	384.00
2 #12THHN, stranded	L1@6.73	CLF	151.00	276.00	427.00
3 #12THHN, stranded	L1@7.43	CLF	164.00	305.00	469.00
4 #12THHN, stranded	L1@8.13	CLF	178.00	334.00	512.00
5 #12THHN, stranded	L1@8.83	CLF	191.00	362.00	553.00
6 #12THHN, stranded	L1@9.53	CLF	205.00	391.00	596.00
1 #10THHN, solid	L1@6.13	CLF	149.00	252.00	401.00
2 #10THHN, solid	L1@6.93	CLF	175.00	284.00	459.00
3 #10THHN, solid	L1@7.73	CLF	200.00	317.00	517.00
4 #10THHN, solid	L1@8.53	CLF	226.00	350.00	576.00
5 #10THHN, solid	L1@9.33	CLF	251.00	383.00	634.00
6 #10THHN, solid	L1@10.1	CLF	277.00	414.00	691.00
1 #10THHN, stranded	L1@6.13	CLF	144.00	252.00	396.00
2 #10THHN, stranded	L1@6.93	CLF	164.00	284.00	448.00
3 #10THHN, stranded	L1@7.73	CLF	184.00	317.00	501.00
4 #10THHN, stranded	L1@8.53	CLF	205.00	350.00	555.00
5 #10THHN, stranded	L1@9.33	CLF	226.00	383.00	609.00
6 #10THHN, stranded	L1@10.1	CLF	246.00	414.00	660.00

Use these figures to estimate the cost of installing assemblies under the conditions described on pages 5 and 6. Costs listed are for each assembly installed. The crew is one electrician working at a labor cost of $41.03 per manhour. These costs include layout, material handling and normal waste. Add for the sales tax, delivery, supervision, mobilization, demobilization, cleanup, overhead and profit.

1-1/4" EMT Conduit Assemblies

Material	Craft@Hrs	Unit	Material Cost	Labor Cost	Installed Cost
100' 1-1/4" EMT conduit, 2 set screw connectors, 9 set screw couplings and 9 one-hole straps					
Empty conduit	L1@6.14	CLF	191.00	252.00	443.00
1 plastic pull line	L1@6.44	CLF	191.00	264.00	455.00
1 #14THHN, solid	L1@6.74	CLF	203.00	277.00	480.00
2 #14THHN, solid	L1@7.34	CLF	214.00	301.00	515.00
3 #14THHN, solid	L1@7.94	CLF	226.00	326.00	552.00
4 #14THHN, solid	L1@8.54	CLF	237.00	350.00	587.00
5 #14THHN, solid	L1@9.14	CLF	249.00	375.00	624.00
6 #14THHN, solid	L1@9.74	CLF	262.00	400.00	662.00
1 #14THHN, stranded	L1@6.74	CLF	200.00	277.00	477.00
2 #14THHN, stranded	L1@7.34	CLF	210.00	301.00	511.00
3 #14THHN, stranded	L1@7.94	CLF	220.00	326.00	546.00
4 #14THHN, stranded	L1@8.54	CLF	229.00	350.00	579.00
5 #14THHN, stranded	L1@9.14	CLF	239.00	375.00	614.00
6 #14THHN, stranded	L1@9.74	CLF	248.00	400.00	648.00
1 #12THHN, solid	L1@6.84	CLF	209.00	281.00	490.00
2 #12THHN, solid	L1@7.54	CLF	225.00	309.00	534.00
3 #12THHN, solid	L1@8.24	CLF	242.00	338.00	580.00
4 #12THHN, solid	L1@8.94	CLF	258.00	367.00	625.00
5 #12THHN, solid	L1@9.64	CLF	276.00	396.00	672.00
6 #12THHN, solid	L1@10.3	CLF	292.00	423.00	715.00
1 #12THHN, stranded	L1@6.84	CLF	205.00	281.00	486.00
2 #12THHN, stranded	L1@7.54	CLF	218.00	309.00	527.00
3 #12THHN, stranded	L1@8.24	CLF	232.00	338.00	570.00
4 #12THHN, stranded	L1@8.94	CLF	246.00	367.00	613.00
5 #12THHN, stranded	L1@9.64	CLF	259.00	396.00	655.00
6 #12THHN, stranded	L1@10.3	CLF	272.00	423.00	695.00
1 #10THHN, solid	L1@6.94	CLF	217.00	285.00	502.00
2 #10THHN, solid	L1@7.74	CLF	242.00	318.00	560.00
3 #10THHN, solid	L1@8.54	CLF	268.00	350.00	618.00
4 #10THHN, solid	L1@9.34	CLF	293.00	383.00	676.00
5 #10THHN, solid	L1@10.1	CLF	320.00	414.00	734.00
6 #10THHN, solid	L1@10.9	CLF	345.00	447.00	792.00
1 #10THHN, stranded	L1@6.94	CLF	212.00	285.00	497.00
2 #10THHN, stranded	L1@7.74	CLF	232.00	318.00	550.00
3 #10THHN, stranded	L1@8.54	CLF	253.00	350.00	603.00
4 #10THHN, stranded	L1@9.34	CLF	272.00	383.00	655.00
5 #10THHN, stranded	L1@10.1	CLF	293.00	414.00	707.00
6 #10THHN, stranded	L1@10.9	CLF	313.00	447.00	760.00

Use these figures to estimate the cost of installing assemblies under the conditions described on pages 5 and 6. Costs listed are for each assembly installed. The crew is one electrician working at a labor cost of $41.03 per manhour. These costs include layout, material handling and normal waste. Add for the sales tax, delivery, supervision, mobilization, demobilization, cleanup, overhead and profit.

Material	Craft@Hrs	Unit	Material Cost	Labor Cost	Installed Cost
100' 3/8" aluminum flexible conduit and 2 screw-in connectors					
Empty conduit	L1@2.53	CLF	21.50	104.00	125.50
1 plastic pull line	L1@2.83	CLF	22.00	116.00	138.00
1 #14THHN, solid	L1@3.13	CLF	33.70	128.00	161.70
2 #14THHN, solid	L1@3.73	CLF	45.40	153.00	198.40
3 #14THHN, solid	L1@4.33	CLF	57.10	178.00	235.10
4 #14THHN, solid	L1@4.93	CLF	68.80	202.00	270.80
5 #14THHN, solid	L1@5.53	CLF	80.50	227.00	307.50
6 #14THHN, solid	L1@6.13	CLF	92.20	252.00	344.20
1 #14THHN, stranded	L1@3.13	CLF	31.50	128.00	159.50
2 #14THHN, stranded	L1@3.73	CLF	41.00	153.00	194.00
3 #14THHN, stranded	L1@4.33	CLF	50.40	178.00	228.40
4 #14THHN, stranded	L1@4.93	CLF	59.90	202.00	261.90
5 #14THHN, stranded	L1@5.53	CLF	69.40	227.00	296.40
6 #14THHN, stranded	L1@6.13	CLF	78.80	252.00	330.80
1 #12THHN, solid	L1@3.23	CLF	38.80	133.00	171.80
2 #12THHN, solid	L1@3.93	CLF	55.70	161.00	216.70
3 #12THHN, solid	L1@4.63	CLF	72.50	190.00	262.50
4 #12THHN, solid	L1@5.33	CLF	89.30	219.00	308.30
5 #12THHN, solid	L1@6.03	CLF	106.00	247.00	353.00
6 #12THHN, solid	L1@6.73	CLF	123.00	276.00	399.00
1 #12THHN, stranded	L1@3.23	CLF	35.60	133.00	168.60
2 #12THHN, stranded	L1@3.93	CLF	49.10	161.00	210.10
3 #12THHN, stranded	L1@4.63	CLF	62.70	190.00	252.70
4 #12THHN, stranded	L1@5.33	CLF	76.20	219.00	295.20
5 #12THHN, stranded	L1@6.03	CLF	89.80	247.00	336.80
6 #12THHN, stranded	L1@6.73	CLF	103.00	276.00	379.00
1 #10THHN, solid	L1@3.33	CLF	47.60	137.00	184.60
2 #10THHN, solid	L1@4.13	CLF	73.20	169.00	242.20
3 #10THHN, solid	L1@4.93	CLF	98.80	202.00	300.80
4 #10THHN, solid	L1@5.73	CLF	124.00	235.00	359.00
5 #10THHN, solid	L1@6.53	CLF	151.00	268.00	419.00
6 #10THHN, solid	L1@7.33	CLF	176.00	301.00	477.00
1 #10THHN, stranded	L1@3.33	CLF	42.40	137.00	179.40
2 #10THHN, stranded	L1@4.13	CLF	62.70	169.00	231.70
3 #10THHN, stranded	L1@4.93	CLF	83.10	202.00	285.10
4 #10THHN, stranded	L1@5.73	CLF	103.00	235.00	338.00
5 #10THHN, stranded	L1@6.53	CLF	124.00	268.00	392.00
6 #10THHN, stranded	L1@7.33	CLF	144.00	301.00	445.00

Use these figures to estimate the cost of installing assemblies under the conditions described on pages 5 and 6. Costs listed are for each assembly installed. The crew is one electrician working at a labor cost of $41.03 per manhour. These costs include layout, material handling and normal waste. Add for the sales tax, delivery, supervision, mobilization, demobilization, cleanup, overhead and profit.

1/2" Aluminum Flex Conduit Assemblies

Material	Craft@Hrs	Unit	Material Cost	Labor Cost	Installed Cost
100' 1/2" aluminum flexible conduit and 2 screw-in connectors					
Empty conduit	L1@2.73	CLF	18.00	112.00	130.00
1 plastic pull line	L1@3.03	CLF	18.40	124.00	142.40
1 #14THHN, solid	L1@3.33	CLF	30.10	137.00	167.10
2 #14THHN, solid	L1@3.93	CLF	41.80	161.00	202.80
3 #14THHN, solid	L1@4.53	CLF	53.50	186.00	239.50
4 #14THHN, solid	L1@5.13	CLF	65.20	210.00	275.20
5 #14THHN, solid	L1@5.73	CLF	76.90	235.00	311.90
6 #14THHN, solid	L1@6.33	CLF	88.60	260.00	348.60
1 #14THHN, stranded	L1@3.33	CLF	27.90	137.00	164.90
2 #14THHN, stranded	L1@3.93	CLF	37.40	161.00	198.40
3 #14THHN, stranded	L1@4.53	CLF	46.90	186.00	232.90
4 #14THHN, stranded	L1@5.13	CLF	56.30	210.00	266.30
5 #14THHN, stranded	L1@5.73	CLF	65.80	235.00	300.80
6 #14THHN, stranded	L1@6.33	CLF	75.30	260.00	335.30
1 #12THHN, solid	L1@3.43	CLF	35.30	141.00	176.30
2 #12THHN, solid	L1@4.13	CLF	52.10	169.00	221.10
3 #12THHN, solid	L1@4.83	CLF	68.90	198.00	266.90
4 #12THHN, solid	L1@5.53	CLF	85.80	227.00	312.80
5 #12THHN, solid	L1@6.23	CLF	103.00	256.00	359.00
6 #12THHN, solid	L1@6.93	CLF	119.00	284.00	403.00
1 #12THHN, stranded	L1@3.43	CLF	32.00	141.00	173.00
2 #12THHN, stranded	L1@4.13	CLF	45.60	169.00	214.60
3 #12THHN, stranded	L1@4.83	CLF	59.20	198.00	257.20
4 #12THHN, stranded	L1@5.53	CLF	72.70	227.00	299.70
5 #12THHN, stranded	L1@6.23	CLF	86.30	256.00	342.30
6 #12THHN, stranded	L1@6.93	CLF	99.90	284.00	383.90
1 #10THHN, solid	L1@3.53	CLF	44.00	145.00	189.00
2 #10THHN, solid	L1@4.33	CLF	69.60	178.00	247.60
3 #10THHN, solid	L1@5.13	CLF	95.20	210.00	305.20
4 #10THHN, solid	L1@5.93	CLF	120.00	243.00	363.00
5 #10THHN, solid	L1@6.73	CLF	146.00	276.00	422.00
6 #10THHN, solid	L1@7.53	CLF	173.00	309.00	482.00
1 #10THHN, stranded	L1@3.53	CLF	38.80	145.00	183.80
2 #10THHN, stranded	L1@4.33	CLF	59.20	178.00	237.20
3 #10THHN, stranded	L1@5.13	CLF	79.50	210.00	289.50
4 #10THHN, stranded	L1@5.93	CLF	99.90	243.00	342.90
5 #10THHN, stranded	L1@6.73	CLF	120.00	276.00	396.00
6 #10THHN, stranded	L1@7.53	CLF	140.00	309.00	449.00

Use these figures to estimate the cost of installing assemblies under the conditions described on pages 5 and 6. Costs listed are for each assembly installed. The crew is one electrician working at a labor cost of $41.03 per manhour. These costs include layout, material handling and normal waste. Add for the sales tax, delivery, supervision, mobilization, demobilization, cleanup, overhead and profit.

3/4" Aluminum Flex Conduit Assemblies

Material	Craft@Hrs	Unit	Material Cost	Labor Cost	Installed Cost
100' 3/4" aluminum flexible conduit and 2 screw-in connectors					
Empty conduit	L1@3.05	CLF	25.00	125.00	150.00
1 plastic pull line	L1@3.35	CLF	25.50	137.00	162.50
1 #14THHN, solid	L1@3.65	CLF	37.20	150.00	187.20
2 #14THHN, solid	L1@4.25	CLF	48.80	174.00	222.80
3 #14THHN, solid	L1@4.85	CLF	60.50	199.00	259.50
4 #14THHN, solid	L1@5.45	CLF	72.20	224.00	296.20
5 #14THHN, solid	L1@6.05	CLF	83.90	248.00	331.90
6 #14THHN, solid	L1@6.65	CLF	95.60	273.00	368.60
1 #14THHN, stranded	L1@3.65	CLF	34.90	150.00	184.90
2 #14THHN, stranded	L1@4.25	CLF	44.40	174.00	218.40
3 #14THHN, stranded	L1@4.85	CLF	53.90	199.00	252.90
4 #14THHN, stranded	L1@5.45	CLF	63.40	224.00	287.40
5 #14THHN, stranded	L1@6.05	CLF	72.70	248.00	320.70
6 #14THHN, stranded	L1@6.65	CLF	82.20	273.00	355.20
1 #12THHN, solid	L1@3.75	CLF	42.30	154.00	196.30
2 #12THHN, solid	L1@4.45	CLF	59.10	183.00	242.10
3 #12THHN, solid	L1@5.15	CLF	76.00	211.00	287.00
4 #12THHN, solid	L1@5.85	CLF	92.80	240.00	332.80
5 #12THHN, solid	L1@6.55	CLF	110.00	269.00	379.00
6 #12THHN, solid	L1@7.25	CLF	126.00	297.00	423.00
1 #12THHN, stranded	L1@3.75	CLF	39.00	154.00	193.00
2 #12THHN, stranded	L1@4.45	CLF	52.60	183.00	235.60
3 #12THHN, stranded	L1@5.15	CLF	66.10	211.00	277.10
4 #12THHN, stranded	L1@5.85	CLF	79.70	240.00	319.70
5 #12THHN, stranded	L1@6.55	CLF	93.30	269.00	362.30
6 #12THHN, stranded	L1@7.25	CLF	107.00	297.00	404.00
1 #10THHN, solid	L1@3.85	CLF	51.10	158.00	209.10
2 #10THHN, solid	L1@4.65	CLF	76.70	191.00	267.70
3 #10THHN, solid	L1@5.45	CLF	102.00	224.00	326.00
4 #10THHN, solid	L1@6.25	CLF	127.00	256.00	383.00
5 #10THHN, solid	L1@7.05	CLF	153.00	289.00	442.00
6 #10THHN, solid	L1@7.85	CLF	180.00	322.00	502.00
1 #10THHN, stranded	L1@3.85	CLF	80.50	158.00	238.50
2 #10THHN, stranded	L1@4.65	CLF	66.10	191.00	257.10
3 #10THHN, stranded	L1@5.45	CLF	86.40	224.00	310.40
4 #10THHN, stranded	L1@6.25	CLF	107.00	256.00	363.00
5 #10THHN, stranded	L1@7.05	CLF	127.00	289.00	416.00
6 #10THHN, stranded	L1@7.85	CLF	147.00	322.00	469.00

Use these figures to estimate the cost of installing assemblies under the conditions described on pages 5 and 6. Costs listed are for each assembly installed. The crew is one electrician working at a labor cost of $41.03 per manhour. These costs include layout, material handling and normal waste. Add for the sales tax, delivery, supervision, mobilization, demobilization, cleanup, overhead and profit.

1" Aluminum Flex Conduit Assemblies

Material	Craft@Hrs	Unit	Material Cost	Labor Cost	Installed Cost
100' 1" aluminum flexible conduit and 2 screw-in connectors					
Empty conduit	L1@3.36	CLF	47.10	138.00	185.10
1 plastic pull line	L1@3.66	CLF	47.60	150.00	197.60
1 #14THHN, solid	L1@3.96	CLF	59.30	162.00	221.30
2 #14THHN, solid	L1@4.56	CLF	71.00	187.00	258.00
3 #14THHN, solid	L1@5.16	CLF	82.70	212.00	294.70
4 #14THHN, solid	L1@5.76	CLF	94.40	236.00	330.40
5 #14THHN, solid	L1@6.36	CLF	106.00	261.00	367.00
6 #14THHN, solid	L1@6.96	CLF	118.00	286.00	404.00
1 #14THHN, stranded	L1@3.96	CLF	57.10	162.00	219.10
2 #14THHN, stranded	L1@4.56	CLF	66.60	187.00	253.60
3 #14THHN, stranded	L1@5.16	CLF	76.00	212.00	288.00
4 #14THHN, stranded	L1@5.76	CLF	85.50	236.00	321.50
5 #14THHN, stranded	L1@6.36	CLF	95.00	261.00	356.00
6 #14THHN, stranded	L1@6.96	CLF	104.00	286.00	390.00
1 #12THHN, solid	L1@4.06	CLF	64.50	167.00	231.50
2 #12THHN, solid	L1@4.76	CLF	81.30	195.00	276.30
3 #12THHN, solid	L1@5.46	CLF	98.10	224.00	322.10
4 #12THHN, solid	L1@6.16	CLF	115.00	253.00	368.00
5 #12THHN, solid	L1@6.86	CLF	132.00	281.00	413.00
6 #12THHN, solid	L1@7.56	CLF	148.00	310.00	458.00
1 #12THHN, stranded	L1@4.06	CLF	61.20	167.00	228.20
2 #12THHN, stranded	L1@4.76	CLF	74.70	195.00	269.70
3 #12THHN, stranded	L1@5.46	CLF	88.30	224.00	312.30
4 #12THHN, stranded	L1@6.16	CLF	102.00	253.00	355.00
5 #12THHN, stranded	L1@6.86	CLF	115.00	281.00	396.00
6 #12THHN, stranded	L1@7.56	CLF	129.00	310.00	439.00
1 #10THHN, solid	L1@4.16	CLF	73.20	171.00	244.20
2 #10THHN, solid	L1@4.96	CLF	98.80	204.00	302.80
3 #10THHN, solid	L1@5.76	CLF	124.00	236.00	360.00
4 #10THHN, solid	L1@6.56	CLF	151.00	269.00	420.00
5 #10THHN, solid	L1@7.36	CLF	176.00	302.00	478.00
6 #10THHN, solid	L1@8.16	CLF	202.00	335.00	537.00
1 #10THHN, stranded	L1@4.16	CLF	68.00	171.00	239.00
2 #10THHN, stranded	L1@4.96	CLF	88.30	204.00	292.30
3 #10THHN, stranded	L1@5.76	CLF	109.00	236.00	345.00
4 #10THHN, stranded	L1@6.56	CLF	129.00	269.00	398.00
5 #10THHN, stranded	L1@7.36	CLF	149.00	302.00	451.00
6 #10THHN, stranded	L1@8.16	CLF	169.00	335.00	504.00

Use these figures to estimate the cost of installing assemblies under the conditions described on pages 5 and 6. Costs listed are for each assembly installed. The crew is one electrician working at a labor cost of $41.03 per manhour. These costs include layout, material handling and normal waste. Add for the sales tax, delivery, supervision, mobilization, demobilization, cleanup, overhead and profit.

3/8" Steel Flex Conduit Assemblies

Material	Craft@Hrs	Unit	Material Cost	Labor Cost	Installed Cost
100' 3/8" steel flexible conduit and 2 screw-in connectors					
Empty conduit	L1@2.73	CLF	19.70	112.00	131.70
1 plastic pull line	L1@3.03	CLF	20.20	124.00	144.20
1 #14THHN, solid	L1@3.33	CLF	31.90	137.00	168.90
2 #14THHN, solid	L1@3.93	CLF	43.60	161.00	204.60
3 #14THHN, solid	L1@4.53	CLF	55.30	186.00	241.30
4 #14THHN, solid	L1@5.13	CLF	67.00	210.00	277.00
5 #14THHN, solid	L1@5.73	CLF	78.70	235.00	313.70
6 #14THHN, solid	L1@6.33	CLF	90.40	260.00	350.40
1 #14THHN, stranded	L1@3.33	CLF	29.70	137.00	166.70
2 #14THHN, stranded	L1@3.93	CLF	39.20	161.00	200.20
3 #14THHN, stranded	L1@4.53	CLF	48.70	186.00	234.70
4 #14THHN, stranded	L1@5.13	CLF	58.00	210.00	268.00
5 #14THHN, stranded	L1@5.73	CLF	67.50	235.00	302.50
6 #14THHN, stranded	L1@6.33	CLF	77.00	260.00	337.00
1 #12THHN, solid	L1@3.43	CLF	37.10	141.00	178.10
2 #12THHN, solid	L1@4.13	CLF	53.90	169.00	222.90
3 #12THHN, solid	L1@4.83	CLF	70.70	198.00	268.70
4 #12THHN, solid	L1@5.53	CLF	87.60	227.00	314.60
5 #12THHN, solid	L1@6.23	CLF	104.00	256.00	360.00
6 #12THHN, solid	L1@6.93	CLF	122.00	284.00	406.00
1 #12THHN, stranded	L1@3.43	CLF	33.70	141.00	174.70
2 #12THHN, stranded	L1@4.13	CLF	47.30	169.00	216.30
3 #12THHN, stranded	L1@4.83	CLF	60.80	198.00	258.80
4 #12THHN, stranded	L1@5.53	CLF	74.50	227.00	301.50
5 #12THHN, stranded	L1@6.23	CLF	88.00	256.00	344.00
6 #12THHN, stranded	L1@6.93	CLF	102.00	284.00	386.00
1 #10THHN, solid	L1@3.53	CLF	45.80	145.00	190.80
2 #10THHN, solid	L1@4.33	CLF	71.40	178.00	249.40
3 #10THHN, solid	L1@5.13	CLF	97.00	210.00	307.00
4 #10THHN, solid	L1@5.93	CLF	123.00	243.00	366.00
5 #10THHN, solid	L1@6.73	CLF	148.00	276.00	424.00
6 #10THHN, solid	L1@7.53	CLF	174.00	309.00	483.00
1 #10THHN, stranded	L1@3.53	CLF	40.50	145.00	185.50
2 #10THHN, stranded	L1@4.33	CLF	60.80	178.00	238.80
3 #10THHN, stranded	L1@5.13	CLF	81.20	210.00	291.20
4 #10THHN, stranded	L1@5.93	CLF	102.00	243.00	345.00
5 #10THHN, stranded	L1@6.73	CLF	122.00	276.00	398.00
6 #10THHN, stranded	L1@7.53	CLF	142.00	309.00	451.00

Use these figures to estimate the cost of installing assemblies under the conditions described on pages 5 and 6. Costs listed are for each assembly installed. The crew is one electrician working at a labor cost of $41.03 per manhour. These costs include layout, material handling and normal waste. Add for the sales tax, delivery, supervision, mobilization, demobilization, cleanup, overhead and profit.

1/2" Steel Flex Conduit Assemblies

Material	Craft@Hrs	Unit	Material Cost	Labor Cost	Installed Cost
100' 1/2" steel flexible conduit and 2 screw-in connectors					
Empty conduit	L1@3.03	CLF	17.60	124.00	141.60
1 plastic pull line	L1@3.33	CLF	18.10	137.00	155.10
1 #14THHN, solid	L1@3.63	CLF	29.80	149.00	178.80
2 #14THHN, solid	L1@4.23	CLF	41.50	174.00	215.50
3 #14THHN, solid	L1@4.83	CLF	53.20	198.00	251.20
4 #14THHN, solid	L1@5.43	CLF	64.90	223.00	287.90
5 #14THHN, solid	L1@6.03	CLF	76.60	247.00	323.60
6 #14THHN, solid	L1@6.63	CLF	88.30	272.00	360.30
1 #14THHN, stranded	L1@3.63	CLF	27.60	149.00	176.60
2 #14THHN, stranded	L1@4.23	CLF	37.10	174.00	211.10
3 #14THHN, stranded	L1@4.83	CLF	46.50	198.00	244.50
4 #14THHN, stranded	L1@5.43	CLF	56.00	223.00	279.00
5 #14THHN, stranded	L1@6.03	CLF	65.40	247.00	312.40
6 #14THHN, stranded	L1@6.63	CLF	74.90	272.00	346.90
1 #12THHN, solid	L1@3.73	CLF	35.00	153.00	188.00
2 #12THHN, solid	L1@4.43	CLF	51.80	182.00	233.80
3 #12THHN, solid	L1@5.13	CLF	68.60	210.00	278.60
4 #12THHN, solid	L1@5.83	CLF	85.50	239.00	324.50
5 #12THHN, solid	L1@6.53	CLF	102.00	268.00	370.00
6 #12THHN, solid	L1@7.23	CLF	119.00	297.00	416.00
1 #12THHN, stranded	L1@3.73	CLF	31.60	153.00	184.60
2 #12THHN, stranded	L1@4.43	CLF	45.20	182.00	227.20
3 #12THHN, stranded	L1@5.13	CLF	58.70	210.00	268.70
4 #12THHN, stranded	L1@5.83	CLF	72.40	239.00	311.40
5 #12THHN, stranded	L1@6.53	CLF	86.00	268.00	354.00
6 #12THHN, stranded	L1@7.23	CLF	99.50	297.00	396.50
1 #10THHN, solid	L1@3.83	CLF	43.70	157.00	200.70
2 #10THHN, solid	L1@4.63	CLF	69.30	190.00	259.30
3 #10THHN, solid	L1@5.43	CLF	94.90	223.00	317.90
4 #10THHN, solid	L1@6.23	CLF	120.00	256.00	376.00
5 #10THHN, solid	L1@7.03	CLF	146.00	288.00	434.00
6 #10THHN, solid	L1@7.83	CLF	171.00	321.00	492.00
1 #10THHN, stranded	L1@3.83	CLF	38.50	157.00	195.50
2 #10THHN, stranded	L1@4.63	CLF	58.70	190.00	248.70
3 #10THHN, stranded	L1@5.43	CLF	79.10	223.00	302.10
4 #10THHN, stranded	L1@6.23	CLF	99.50	256.00	355.50
5 #10THHN, stranded	L1@7.03	CLF	119.00	288.00	407.00
6 #10THHN, stranded	L1@7.83	CLF	140.00	321.00	461.00

Use these figures to estimate the cost of installing assemblies under the conditions described on pages 5 and 6. Costs listed are for each assembly installed. The crew is one electrician working at a labor cost of $41.03 per manhour. These costs include layout, material handling and normal waste. Add for the sales tax, delivery, supervision, mobilization, demobilization, cleanup, overhead and profit.

Material	Craft@Hrs	Unit	Material Cost	Labor Cost	Installed Cost
100' 3/4" steel flexible conduit and 2 screw-in connectors					
Empty conduit	L1@3.30	CLF	24.20	135.00	159.20
1 plastic pull line	L1@3.60	CLF	24.70	148.00	172.70
1 #14THHN, solid	L1@3.90	CLF	36.40	160.00	196.40
2 #14THHN, solid	L1@4.50	CLF	48.10	185.00	233.10
3 #14THHN, solid	L1@5.10	CLF	59.80	209.00	268.80
4 #14THHN, solid	L1@5.70	CLF	71.50	234.00	305.50
5 #14THHN, solid	L1@6.30	CLF	83.20	258.00	341.20
6 #14THHN, solid	L1@6.90	CLF	94.90	283.00	377.90
1 #14THHN, stranded	L1@3.90	CLF	34.20	160.00	194.20
2 #14THHN, stranded	L1@4.50	CLF	43.70	185.00	228.70
3 #14THHN, stranded	L1@5.10	CLF	53.20	209.00	262.20
4 #14THHN, stranded	L1@5.70	CLF	62.60	234.00	296.60
5 #14THHN, stranded	L1@6.30	CLF	72.10	258.00	330.10
6 #14THHN, stranded	L1@6.90	CLF	81.60	283.00	364.60
1 #12THHN, solid	L1@4.00	CLF	41.60	164.00	205.60
2 #12THHN, solid	L1@4.70	CLF	58.40	193.00	251.40
3 #12THHN, solid	L1@5.40	CLF	75.20	222.00	297.20
4 #12THHN, solid	L1@6.10	CLF	92.10	250.00	342.10
5 #12THHN, solid	L1@6.80	CLF	109.00	279.00	388.00
6 #12THHN, solid	L1@7.50	CLF	126.00	308.00	434.00
1 #12THHN, stranded	L1@4.00	CLF	38.20	164.00	202.20
2 #12THHN, stranded	L1@4.70	CLF	51.80	193.00	244.80
3 #12THHN, stranded	L1@5.40	CLF	65.40	222.00	287.40
4 #12THHN, stranded	L1@6.10	CLF	79.00	250.00	329.00
5 #12THHN, stranded	L1@6.80	CLF	92.60	279.00	371.60
6 #12THHN, stranded	L1@7.50	CLF	106.00	308.00	414.00
1 #10THHN, solid	L1@4.10	CLF	50.30	168.00	218.30
2 #10THHN, solid	L1@4.90	CLF	75.90	201.00	276.90
3 #10THHN, solid	L1@5.70	CLF	101.00	234.00	335.00
4 #10THHN, solid	L1@6.50	CLF	127.00	267.00	394.00
5 #10THHN, solid	L1@7.30	CLF	153.00	300.00	453.00
6 #10THHN, solid	L1@8.10	CLF	178.00	332.00	510.00
1 #10THHN, stranded	L1@4.10	CLF	45.10	168.00	213.10
2 #10THHN, stranded	L1@4.90	CLF	65.40	201.00	266.40
3 #10THHN, stranded	L1@5.70	CLF	85.70	234.00	319.70
4 #10THHN, stranded	L1@6.50	CLF	106.00	267.00	373.00
5 #10THHN, stranded	L1@7.30	CLF	126.00	300.00	426.00
6 #10THHN, stranded	L1@8.10	CLF	147.00	332.00	479.00

Use these figures to estimate the cost of installing assemblies under the conditions described on pages 5 and 6. Costs listed are for each assembly installed. The crew is one electrician working at a labor cost of $41.03 per manhour. These costs include layout, material handling and normal waste. Add for the sales tax, delivery, supervision, mobilization, demobilization, cleanup, overhead and profit.

1" Steel Flex Conduit Assemblies

Material	Craft@Hrs	Unit	Material Cost	Labor Cost	Installed Cost
100' 1" steel flexible conduit and 2 screw-in connectors					
Empty conduit	L1@3.56	CLF	44.50	146.00	190.50
1 plastic pull line	L1@3.86	CLF	44.90	158.00	202.90
1 #14THHN, solid	L1@4.16	CLF	56.60	171.00	227.60
2 #14THHN, solid	L1@4.76	CLF	68.30	195.00	263.30
3 #14THHN, solid	L1@5.36	CLF	80.00	220.00	300.00
4 #14THHN, solid	L1@5.96	CLF	91.70	245.00	336.70
5 #14THHN, solid	L1@6.56	CLF	103.00	269.00	372.00
6 #14THHN, solid	L1@7.16	CLF	115.00	294.00	409.00
1 #14THHN, stranded	L1@4.16	CLF	54.40	171.00	225.40
2 #14THHN, stranded	L1@4.76	CLF	63.90	195.00	258.90
3 #14THHN, stranded	L1@5.36	CLF	73.30	220.00	293.30
4 #14THHN, stranded	L1@5.96	CLF	82.80	245.00	327.80
5 #14THHN, stranded	L1@6.56	CLF	92.30	269.00	361.30
6 #14THHN, stranded	L1@7.16	CLF	102.00	294.00	396.00
1 #12THHN, solid	L1@4.26	CLF	61.90	175.00	236.90
2 #12THHN, solid	L1@4.96	CLF	78.70	204.00	282.70
3 #12THHN, solid	L1@5.66	CLF	95.50	232.00	327.50
4 #12THHN, solid	L1@6.36	CLF	112.00	261.00	373.00
5 #12THHN, solid	L1@7.06	CLF	129.00	290.00	419.00
6 #12THHN, solid	L1@7.76	CLF	146.00	318.00	464.00
1 #12THHN, stranded	L1@4.26	CLF	58.50	175.00	233.50
2 #12THHN, stranded	L1@4.96	CLF	72.10	204.00	276.10
3 #12THHN, stranded	L1@5.66	CLF	85.60	232.00	317.60
4 #12THHN, stranded	L1@6.36	CLF	99.20	261.00	360.20
5 #12THHN, stranded	L1@7.06	CLF	113.00	290.00	403.00
6 #12THHN, stranded	L1@7.76	CLF	126.00	318.00	444.00
1 #10THHN, solid	L1@4.36	CLF	70.50	179.00	249.50
2 #10THHN, solid	L1@5.16	CLF	96.10	212.00	308.10
3 #10THHN, solid	L1@5.96	CLF	122.00	245.00	367.00
4 #10THHN, solid	L1@6.76	CLF	147.00	277.00	424.00
5 #10THHN, solid	L1@7.56	CLF	173.00	310.00	483.00
6 #10THHN, solid	L1@8.36	CLF	198.00	343.00	541.00
1 #10THHN, stranded	L1@4.36	CLF	65.30	179.00	244.30
2 #10THHN, stranded	L1@5.16	CLF	85.60	212.00	297.60
3 #10THHN, stranded	L1@5.96	CLF	106.00	245.00	351.00
4 #10THHN, stranded	L1@6.76	CLF	126.00	277.00	403.00
5 #10THHN, stranded	L1@7.56	CLF	147.00	310.00	457.00
6 #10THHN, stranded	L1@8.36	CLF	167.00	343.00	510.00

Use these figures to estimate the cost of installing assemblies under the conditions described on pages 5 and 6. Costs listed are for each assembly installed. The crew is one electrician working at a labor cost of $41.03 per manhour. These costs include layout, material handling and normal waste. Add for the sales tax, delivery, supervision, mobilization, demobilization, cleanup, overhead and profit.

Material	Craft@Hrs	Unit	Material Cost	Labor Cost	Installed Cost
100' 1/2" PVC conduit, 2 terminal adapters and 2 elbows					
Empty conduit	L1@3.30	CLF	36.40	135.00	171.40
1 plastic pull line	L1@3.60	CLF	36.80	148.00	184.80
1 #14THHN, solid	L1@4.20	CLF	48.50	172.00	220.50
2 #14THHN, solid	L1@4.80	CLF	60.20	197.00	257.20
3 #14THHN, solid	L1@5.40	CLF	71.90	222.00	293.90
4 #14THHN, solid	L1@6.00	CLF	83.60	246.00	329.60
5 #14THHN, solid	L1@6.60	CLF	95.30	271.00	366.30
6 #14THHN, solid	L1@7.20	CLF	107.00	295.00	402.00
1 #14THHN, stranded	L1@4.20	CLF	46.30	172.00	218.30
2 #14THHN, stranded	L1@4.80	CLF	55.80	197.00	252.80
3 #14THHN, stranded	L1@5.40	CLF	65.20	222.00	287.20
4 #14THHN, stranded	L1@6.00	CLF	74.70	246.00	320.70
5 #14THHN, stranded	L1@6.60	CLF	84.20	271.00	355.20
6 #14THHN, stranded	L1@7.20	CLF	93.70	295.00	388.70
1 #12THHN, solid	L1@4.00	CLF	53.60	164.00	217.60
2 #12THHN, solid	L1@4.70	CLF	70.50	193.00	263.50
3 #12THHN, solid	L1@5.40	CLF	87.30	222.00	309.30
4 #12THHN, solid	L1@6.10	CLF	104.00	250.00	354.00
5 #12THHN, solid	L1@6.80	CLF	120.00	279.00	399.00
6 #12THHN, solid	L1@7.50	CLF	138.00	308.00	446.00
1 #12THHN, stranded	L1@4.00	CLF	50.40	164.00	214.40
2 #12THHN, stranded	L1@4.70	CLF	63.90	193.00	256.90
3 #12THHN, stranded	L1@5.40	CLF	77.50	222.00	299.50
4 #12THHN, stranded	L1@6.10	CLF	91.10	250.00	341.10
5 #12THHN, stranded	L1@6.80	CLF	105.00	279.00	384.00
6 #12THHN, stranded	L1@7.50	CLF	118.00	308.00	426.00
1 #10THHN, solid	L1@4.10	CLF	62.40	168.00	230.40
2 #10THHN, solid	L1@4.90	CLF	88.00	201.00	289.00
3 #10THHN, solid	L1@5.70	CLF	114.00	234.00	348.00
4 #10THHN, solid	L1@6.50	CLF	139.00	267.00	406.00
5 #10THHN, solid	L1@7.30	CLF	164.00	300.00	464.00
6 #10THHN, solid	L1@8.10	CLF	190.00	332.00	522.00
1 #10THHN, stranded	L1@4.10	CLF	57.20	168.00	225.20
2 #10THHN, stranded	L1@4.90	CLF	77.50	201.00	278.50
3 #10THHN, stranded	L1@5.70	CLF	97.90	234.00	331.90
4 #10THHN, stranded	L1@6.50	CLF	118.00	267.00	385.00
5 #10THHN, stranded	L1@7.30	CLF	139.00	300.00	439.00
6 #10THHN, stranded	L1@8.10	CLF	159.00	332.00	491.00

Use these figures to estimate the cost of installing assemblies under the conditions described on pages 5 and 6. Costs listed are for each assembly installed. The crew is one electrician working at a labor cost of $41.03 per manhour. These costs include layout, material handling and normal waste. Add for the sales tax, delivery, supervision, mobilization, demobilization, cleanup, overhead and profit.

3/4" PVC Conduit Assemblies

Material	Craft@Hrs	Unit	Material Cost	Labor Cost	Installed Cost
100' 3/4" PVC conduit, 2 terminal adapters and 2 elbows					
Empty conduit	L1@3.32	CLF	43.40	136.00	179.40
1 plastic pull line	L1@3.63	CLF	43.90	149.00	192.90
1 #14THHN, solid	L1@3.92	CLF	55.60	161.00	216.60
2 #14THHN, solid	L1@4.52	CLF	67.30	185.00	252.30
3 #14THHN, solid	L1@5.12	CLF	79.00	210.00	289.00
4 #14THHN, solid	L1@5.72	CLF	90.70	235.00	325.70
5 #14THHN, solid	L1@6.32	CLF	102.00	259.00	361.00
6 #14THHN, solid	L1@6.92	CLF	114.00	284.00	398.00
1 #14THHN, stranded	L1@3.92	CLF	53.40	161.00	214.40
2 #14THHN, stranded	L1@4.52	CLF	62.80	185.00	247.80
3 #14THHN, stranded	L1@5.12	CLF	72.30	210.00	282.30
4 #14THHN, stranded	L1@5.72	CLF	81.80	235.00	316.80
5 #14THHN, stranded	L1@6.32	CLF	91.30	259.00	350.30
6 #14THHN, stranded	L1@6.92	CLF	101.00	284.00	385.00
1 #12THHN, solid	L1@4.02	CLF	60.70	165.00	225.70
2 #12THHN, solid	L1@4.72	CLF	77.60	194.00	271.60
3 #12THHN, solid	L1@5.42	CLF	94.40	222.00	316.40
4 #12THHN, solid	L1@6.12	CLF	111.00	251.00	362.00
5 #12THHN, solid	L1@6.82	CLF	129.00	280.00	409.00
6 #12THHN, solid	L1@7.52	CLF	145.00	309.00	454.00
1 #12THHN, stranded	L1@4.02	CLF	57.50	165.00	222.50
2 #12THHN, stranded	L1@4.72	CLF	71.00	194.00	265.00
3 #12THHN, stranded	L1@5.42	CLF	84.60	222.00	306.60
4 #12THHN, stranded	L1@6.12	CLF	98.10	251.00	349.10
5 #12THHN, stranded	L1@6.82	CLF	112.00	280.00	392.00
6 #12THHN, stranded	L1@7.52	CLF	125.00	309.00	434.00
1 #10THHN, solid	L1@4.12	CLF	69.50	169.00	238.50
2 #10THHN, solid	L1@4.92	CLF	95.10	202.00	297.10
3 #10THHN, solid	L1@5.72	CLF	120.00	235.00	355.00
4 #10THHN, solid	L1@6.52	CLF	146.00	268.00	414.00
5 #10THHN, solid	L1@7.32	CLF	171.00	300.00	471.00
6 #10THHN, solid	L1@8.12	CLF	198.00	333.00	531.00
1 #10THHN, stranded	L1@4.12	CLF	64.30	169.00	233.30
2 #10THHN, stranded	L1@4.92	CLF	84.60	202.00	286.60
3 #10THHN, stranded	L1@5.72	CLF	105.00	235.00	340.00
4 #10THHN, stranded	L1@6.52	CLF	125.00	268.00	393.00
5 #10THHN, stranded	L1@7.32	CLF	146.00	300.00	446.00
6 #10THHN, stranded	L1@8.12	CLF	166.00	333.00	499.00

Use these figures to estimate the cost of installing assemblies under the conditions described on pages 5 and 6. Costs listed are for each assembly installed. The crew is one electrician working at a labor cost of $41.03 per manhour. These costs include layout, material handling and normal waste. Add for the sales tax, delivery, supervision, mobilization, demobilization, cleanup, overhead and profit.

1" PVC Conduit Assemblies

Material	Craft@Hrs	Unit	Material Cost	Labor Cost	Installed Cost
100' 1" PVC conduit, 2 terminal adapters and 2 elbows					
Empty conduit	L1@3.46	CLF	66.30	142.00	208.30
1 plastic pull line	L1@3.76	CLF	66.70	154.00	220.70
1 #14THHN, solid	L1@4.06	CLF	78.40	167.00	245.40
2 #14THHN, solid	L1@4.66	CLF	90.10	191.00	281.10
3 #14THHN, solid	L1@5.26	CLF	102.00	216.00	318.00
4 #14THHN, solid	L1@5.86	CLF	114.00	240.00	354.00
5 #14THHN, solid	L1@6.46	CLF	125.00	265.00	390.00
6 #14THHN, solid	L1@7.06	CLF	137.00	290.00	427.00
1 #14THHN, stranded	L1@4.06	CLF	76.20	167.00	243.20
2 #14THHN, stranded	L1@4.66	CLF	85.60	191.00	276.60
3 #14THHN, stranded	L1@5.26	CLF	95.10	216.00	311.10
4 #14THHN, stranded	L1@5.86	CLF	105.00	240.00	345.00
5 #14THHN, stranded	L1@6.46	CLF	114.00	265.00	379.00
6 #14THHN, stranded	L1@7.06	CLF	124.00	290.00	414.00
1 #12THHN, solid	L1@4.16	CLF	83.50	171.00	254.50
2 #12THHN, solid	L1@4.86	CLF	100.00	199.00	299.00
3 #12THHN, solid	L1@5.56	CLF	117.00	228.00	345.00
4 #12THHN, solid	L1@6.26	CLF	134.00	257.00	391.00
5 #12THHN, solid	L1@6.96	CLF	151.00	286.00	437.00
6 #12THHN, solid	L1@7.66	CLF	168.00	314.00	482.00
1 #12THHN, stranded	L1@4.16	CLF	80.30	171.00	251.30
2 #12THHN, stranded	L1@4.86	CLF	93.80	199.00	292.80
3 #12THHN, stranded	L1@5.56	CLF	107.00	228.00	335.00
4 #12THHN, stranded	L1@6.26	CLF	120.00	257.00	377.00
5 #12THHN, stranded	L1@6.96	CLF	134.00	286.00	420.00
6 #12THHN, stranded	L1@7.66	CLF	148.00	314.00	462.00
1 #10THHN, solid	L1@4.26	CLF	92.30	175.00	267.30
2 #10THHN, solid	L1@5.06	CLF	118.00	208.00	326.00
3 #10THHN, solid	L1@5.86	CLF	144.00	240.00	384.00
4 #10THHN, solid	L1@6.66	CLF	169.00	273.00	442.00
5 #10THHN, solid	L1@7.46	CLF	195.00	306.00	501.00
6 #10THHN, solid	L1@8.26	CLF	220.00	339.00	559.00
1 #10THHN, stranded	L1@4.26	CLF	87.10	175.00	262.10
2 #10THHN, stranded	L1@5.06	CLF	107.00	208.00	315.00
3 #10THHN, stranded	L1@5.86	CLF	127.00	240.00	367.00
4 #10THHN, stranded	L1@6.66	CLF	148.00	273.00	421.00
5 #10THHN, stranded	L1@7.46	CLF	168.00	306.00	474.00
6 #10THHN, stranded	L1@8.26	CLF	189.00	339.00	528.00

Use these figures to estimate the cost of installing assemblies under the conditions described on pages 5 and 6. Costs listed are for each assembly installed. The crew is one electrician working at a labor cost of $41.03 per manhour. These costs include layout, material handling and normal waste. Add for the sales tax, delivery, supervision, mobilization, demobilization, cleanup, overhead and profit.

1-1/4" PVC Conduit Assemblies

100' 1-1/4" PVC conduit, 2 terminal adapters and 2 elbows

Material	Craft@Hrs	Unit	Material Cost	Labor Cost	Installed Cost
Empty conduit	L1@3.60	CLF	93.70	148.00	241.70
1 plastic pull line	L1@3.90	CLF	94.20	160.00	254.20
1 #14THHN, solid	L1@4.20	CLF	106.00	172.00	278.00
2 #14THHN, solid	L1@4.80	CLF	117.00	197.00	314.00
3 #14THHN, solid	L1@5.40	CLF	130.00	222.00	352.00
4 #14THHN, solid	L1@6.00	CLF	141.00	246.00	387.00
5 #14THHN, solid	L1@6.60	CLF	153.00	271.00	424.00
6 #14THHN, solid	L1@7.20	CLF	164.00	295.00	459.00
1 #14THHN, stranded	L1@4.20	CLF	104.00	172.00	276.00
2 #14THHN, stranded	L1@4.80	CLF	113.00	197.00	310.00
3 #14THHN, stranded	L1@5.40	CLF	123.00	222.00	345.00
4 #14THHN, stranded	L1@6.00	CLF	132.00	246.00	378.00
5 #14THHN, stranded	L1@6.60	CLF	141.00	271.00	412.00
6 #14THHN, stranded	L1@7.20	CLF	151.00	295.00	446.00
1 #12THHN, solid	L1@4.30	CLF	111.00	176.00	287.00
2 #12THHN, solid	L1@5.00	CLF	127.00	205.00	332.00
3 #12THHN, solid	L1@5.70	CLF	145.00	234.00	379.00
4 #12THHN, solid	L1@6.40	CLF	161.00	263.00	424.00
5 #12THHN, solid	L1@7.10	CLF	178.00	291.00	469.00
6 #12THHN, solid	L1@7.80	CLF	195.00	320.00	515.00
1 #12THHN, stranded	L1@4.30	CLF	108.00	176.00	284.00
2 #12THHN, stranded	L1@5.00	CLF	122.00	205.00	327.00
3 #12THHN, stranded	L1@5.70	CLF	134.00	234.00	368.00
4 #12THHN, stranded	L1@6.40	CLF	148.00	263.00	411.00
5 #12THHN, stranded	L1@7.10	CLF	162.00	291.00	453.00
6 #12THHN, stranded	L1@7.80	CLF	176.00	320.00	496.00
1 #10THHN, solid	L1@4.40	CLF	119.00	181.00	300.00
2 #10THHN, solid	L1@5.20	CLF	145.00	213.00	358.00
3 #10THHN, solid	L1@6.00	CLF	171.00	246.00	417.00
4 #10THHN, solid	L1@6.80	CLF	197.00	279.00	476.00
5 #10THHN, solid	L1@7.60	CLF	222.00	312.00	534.00
6 #10THHN, solid	L1@8.40	CLF	248.00	345.00	593.00
1 #10THHN, stranded	L1@4.40	CLF	114.00	181.00	295.00
2 #10THHN, stranded	L1@5.20	CLF	134.00	213.00	347.00
3 #10THHN, stranded	L1@6.00	CLF	155.00	246.00	401.00
4 #10THHN, stranded	L1@6.80	CLF	176.00	279.00	455.00
5 #10THHN, stranded	L1@7.60	CLF	196.00	312.00	508.00
6 #10THHN, stranded	L1@8.40	CLF	217.00	345.00	562.00

Use these figures to estimate the cost of installing assemblies under the conditions described on pages 5 and 6. Costs listed are for each assembly installed. The crew is one electrician working at a labor cost of $41.03 per manhour. These costs include layout, material handling and normal waste. Add for the sales tax, delivery, supervision, mobilization, demobilization, cleanup, overhead and profit.

Material	Craft@Hrs	Unit	Material Cost	Labor Cost	Installed Cost
100' 1/2" galvanized rigid conduit, 2 terminations and 9 one-hole straps					
Empty conduit	L1@4.55	CLF	241.00	187.00	428.00
1 plastic pull line	L1@4.85	CLF	241.00	199.00	440.00
1 #14THHN, solid	L1@5.15	CLF	253.00	211.00	464.00
2 #14THHN, solid	L1@5.75	CLF	264.00	236.00	500.00
3 #14THHN, solid	L1@6.35	CLF	277.00	261.00	538.00
4 #14THHN, solid	L1@6.95	CLF	288.00	285.00	573.00
5 #14THHN, solid	L1@7.55	CLF	300.00	310.00	610.00
6 #14THHN, solid	L1@8.15	CLF	312.00	334.00	646.00
1 #14THHN, stranded	L1@5.15	CLF	250.00	211.00	461.00
2 #14THHN, stranded	L1@5.75	CLF	261.00	236.00	497.00
3 #14THHN, stranded	L1@6.35	CLF	270.00	261.00	531.00
4 #14THHN, stranded	L1@6.95	CLF	279.00	285.00	564.00
5 #14THHN, stranded	L1@7.55	CLF	288.00	310.00	598.00
6 #14THHN, stranded	L1@8.15	CLF	298.00	334.00	632.00
1 #12THHN, solid	L1@5.25	CLF	258.00	215.00	473.00
2 #12THHN, solid	L1@5.95	CLF	275.00	244.00	519.00
3 #12THHN, solid	L1@6.65	CLF	292.00	273.00	565.00
4 #12THHN, solid	L1@7.35	CLF	308.00	302.00	610.00
5 #12THHN, solid	L1@8.05	CLF	326.00	330.00	656.00
6 #12THHN, solid	L1@8.75	CLF	342.00	359.00	701.00
1 #12THHN, stranded	L1@5.25	CLF	255.00	215.00	470.00
2 #12THHN, stranded	L1@5.95	CLF	269.00	244.00	513.00
3 #12THHN, stranded	L1@6.65	CLF	281.00	273.00	554.00
4 #12THHN, stranded	L1@7.35	CLF	295.00	302.00	597.00
5 #12THHN, stranded	L1@8.05	CLF	309.00	330.00	639.00
6 #12THHN, stranded	L1@8.75	CLF	322.00	359.00	681.00
1 #10THHN, solid	L1@5.35	CLF	266.00	220.00	486.00
2 #10THHN, solid	L1@6.15	CLF	292.00	252.00	544.00
3 #10THHN, solid	L1@6.95	CLF	319.00	285.00	604.00
4 #10THHN, solid	L1@7.75	CLF	344.00	318.00	662.00
5 #10THHN, solid	L1@8.55	CLF	370.00	351.00	721.00
6 #10THHN, solid	L1@9.35	CLF	395.00	384.00	779.00
1 #10THHN, stranded	L1@5.35	CLF	262.00	220.00	482.00
2 #10THHN, stranded	L1@6.15	CLF	281.00	252.00	533.00
3 #10THHN, stranded	L1@6.95	CLF	302.00	285.00	587.00
4 #10THHN, stranded	L1@7.75	CLF	322.00	318.00	640.00
5 #10THHN, stranded	L1@8.55	CLF	343.00	351.00	694.00
6 #10THHN, stranded	L1@9.35	CLF	364.00	384.00	748.00

Use these figures to estimate the cost of installing assemblies under the conditions described on pages 5 and 6. Costs listed are for each assembly installed. The crew is one electrician working at a labor cost of $41.03 per manhour. These costs include layout, material handling and normal waste. Add for the sales tax, delivery, supervision, mobilization, demobilization, cleanup, overhead and profit.

3/4" Galvanized Rigid Conduit Assemblies

Material	Craft@Hrs	Unit	Material Cost	Labor Cost	Installed Cost
100' 3/4" galvanized rigid conduit, 2 terminations and 9 one-hole straps					
Empty conduit	L1@5.16	CLF	253.00	212.00	465.00
1 plastic pull line	L1@5.46	CLF	253.00	224.00	477.00
1 #14THHN, solid	L1@5.76	CLF	264.00	236.00	500.00
2 #14THHN, solid	L1@6.36	CLF	276.00	261.00	537.00
3 #14THHN, solid	L1@6.96	CLF	288.00	286.00	574.00
4 #14THHN, solid	L1@7.56	CLF	300.00	310.00	610.00
5 #14THHN, solid	L1@8.16	CLF	312.00	335.00	647.00
6 #14THHN, solid	L1@8.76	CLF	323.00	359.00	682.00
1 #14THHN, stranded	L1@5.76	CLF	262.00	236.00	498.00
2 #14THHN, stranded	L1@6.36	CLF	272.00	261.00	533.00
3 #14THHN, stranded	L1@6.96	CLF	281.00	286.00	567.00
4 #14THHN, stranded	L1@7.56	CLF	291.00	310.00	601.00
5 #14THHN, stranded	L1@8.16	CLF	300.00	335.00	635.00
6 #14THHN, stranded	L1@8.76	CLF	309.00	359.00	668.00
1 #12THHN, solid	L1@5.86	CLF	270.00	240.00	510.00
2 #12THHN, solid	L1@6.56	CLF	286.00	269.00	555.00
3 #12THHN, solid	L1@7.26	CLF	304.00	298.00	602.00
4 #12THHN, solid	L1@7.96	CLF	320.00	327.00	647.00
5 #12THHN, solid	L1@8.66	CLF	337.00	355.00	692.00
6 #12THHN, solid	L1@9.36	CLF	353.00	384.00	737.00
1 #12THHN, stranded	L1@5.86	CLF	266.00	240.00	506.00
2 #12THHN, stranded	L1@6.56	CLF	280.00	269.00	549.00
3 #12THHN, stranded	L1@7.26	CLF	293.00	298.00	591.00
4 #12THHN, stranded	L1@7.96	CLF	307.00	327.00	634.00
5 #12THHN, stranded	L1@8.66	CLF	321.00	355.00	676.00
6 #12THHN, stranded	L1@9.36	CLF	335.00	384.00	719.00
1 #10THHN, solid	L1@5.96	CLF	278.00	245.00	523.00
2 #10THHN, solid	L1@6.76	CLF	304.00	277.00	581.00
3 #10THHN, solid	L1@7.56	CLF	330.00	310.00	640.00
4 #10THHN, solid	L1@8.36	CLF	356.00	343.00	699.00
5 #10THHN, solid	L1@9.16	CLF	381.00	376.00	757.00
6 #10THHN, solid	L1@9.96	CLF	407.00	409.00	816.00
1 #10THHN, stranded	L1@5.96	CLF	273.00	245.00	518.00
2 #10THHN, stranded	L1@6.76	CLF	293.00	277.00	570.00
3 #10THHN, stranded	L1@7.56	CLF	314.00	310.00	624.00
4 #10THHN, stranded	L1@8.36	CLF	335.00	343.00	678.00
5 #10THHN, stranded	L1@9.16	CLF	354.00	376.00	730.00
6 #10THHN, stranded	L1@9.96	CLF	375.00	409.00	784.00

Use these figures to estimate the cost of installing assemblies under the conditions described on pages 5 and 6. Costs listed are for each assembly installed. The crew is one electrician working at a labor cost of $41.03 per manhour. These costs include layout, material handling and normal waste. Add for the sales tax, delivery, supervision, mobilization, demobilization, cleanup, overhead and profit.

1" Galvanized Rigid Conduit Assemblies

Material	Craft@Hrs	Unit	Material Cost	Labor Cost	Installed Cost
100' 1" galvanized rigid conduit, 2 terminations and 9 one-hole straps					
Empty conduit	L1@5.88	CLF	396.00	241.00	637.00
1 plastic pull line	L1@6.18	CLF	396.00	254.00	650.00
1 #14THHN, solid	L1@6.48	CLF	408.00	266.00	674.00
2 #14THHN, solid	L1@7.08	CLF	419.00	290.00	709.00
3 #14THHN, solid	L1@7.68	CLF	431.00	315.00	746.00
4 #14THHN, solid	L1@8.28	CLF	444.00	340.00	784.00
5 #14THHN, solid	L1@8.88	CLF	455.00	364.00	819.00
6 #14THHN, solid	L1@9.48	CLF	467.00	389.00	856.00
1 #14THHN, stranded	L1@6.48	CLF	405.00	266.00	671.00
2 #14THHN, stranded	L1@7.08	CLF	416.00	290.00	706.00
3 #14THHN, stranded	L1@7.68	CLF	425.00	315.00	740.00
4 #14THHN, stranded	L1@8.28	CLF	434.00	340.00	774.00
5 #14THHN, stranded	L1@8.88	CLF	444.00	364.00	808.00
6 #14THHN, stranded	L1@9.48	CLF	453.00	389.00	842.00
1 #12THHN, solid	L1@6.58	CLF	414.00	270.00	684.00
2 #12THHN, solid	L1@7.28	CLF	430.00	299.00	729.00
3 #12THHN, solid	L1@7.98	CLF	447.00	327.00	774.00
4 #12THHN, solid	L1@8.68	CLF	463.00	356.00	819.00
5 #12THHN, solid	L1@9.38	CLF	481.00	385.00	866.00
6 #12THHN, solid	L1@10.1	CLF	497.00	414.00	911.00
1 #12THHN, stranded	L1@6.58	CLF	410.00	270.00	680.00
2 #12THHN, stranded	L1@7.28	CLF	424.00	299.00	723.00
3 #12THHN, stranded	L1@7.98	CLF	437.00	327.00	764.00
4 #12THHN, stranded	L1@8.68	CLF	451.00	356.00	807.00
5 #12THHN, stranded	L1@9.38	CLF	465.00	385.00	850.00
6 #12THHN, stranded	L1@10.1	CLF	477.00	414.00	891.00
1 #10THHN, solid	L1@6.68	CLF	422.00	274.00	696.00
2 #10THHN, solid	L1@7.48	CLF	447.00	307.00	754.00
3 #10THHN, solid	L1@8.28	CLF	474.00	340.00	814.00
4 #10THHN, solid	L1@9.08	CLF	499.00	373.00	872.00
5 #10THHN, solid	L1@9.88	CLF	525.00	405.00	930.00
6 #10THHN, solid	L1@10.7	CLF	550.00	439.00	989.00
1 #10THHN, stranded	L1@6.68	CLF	417.00	274.00	691.00
2 #10THHN, stranded	L1@7.48	CLF	437.00	307.00	744.00
3 #10THHN, stranded	L1@8.28	CLF	458.00	340.00	798.00
4 #10THHN, stranded	L1@9.08	CLF	477.00	373.00	850.00
5 #10THHN, stranded	L1@9.88	CLF	498.00	405.00	903.00
6 #10THHN, stranded	L1@10.7	CLF	519.00	439.00	958.00

Use these figures to estimate the cost of installing assemblies under the conditions described on pages 5 and 6. Costs listed are for each assembly installed. The crew is one electrician working at a labor cost of $41.03 per manhour. These costs include layout, material handling and normal waste. Add for the sales tax, delivery, supervision, mobilization, demobilization, cleanup, overhead and profit.

1-1/4" Galvanized Rigid Conduit Assemblies

Material	Craft@Hrs	Unit	Material Cost	Labor Cost	Installed Cost
100' 1-1/4" galvanized rigid conduit, 2 terminations and 9 one-hole straps					
Empty conduit	L1@8.10	CLF	414.00	332.00	746.00
1 plastic pull line	L1@8.40	CLF	415.00	345.00	760.00
1 #14THHN, solid	L1@8.70	CLF	426.00	357.00	783.00
2 #14THHN, solid	L1@9.30	CLF	438.00	382.00	820.00
3 #14THHN, solid	L1@9.90	CLF	449.00	406.00	855.00
4 #14THHN, solid	L1@10.5	CLF	461.00	431.00	892.00
5 #14THHN, solid	L1@11.1	CLF	473.00	455.00	928.00
6 #14THHN, solid	L1@11.7	CLF	484.00	480.00	964.00
1 #14THHN, stranded	L1@8.70	CLF	424.00	357.00	781.00
2 #14THHN, stranded	L1@9.30	CLF	433.00	382.00	815.00
3 #14THHN, stranded	L1@9.90	CLF	443.00	406.00	849.00
4 #14THHN, stranded	L1@10.5	CLF	453.00	431.00	884.00
5 #14THHN, stranded	L1@11.1	CLF	462.00	455.00	917.00
6 #14THHN, stranded	L1@11.7	CLF	474.00	480.00	954.00
1 #12THHN, solid	L1@8.80	CLF	431.00	361.00	792.00
2 #12THHN, solid	L1@9.50	CLF	448.00	390.00	838.00
3 #12THHN, solid	L1@10.2	CLF	465.00	419.00	884.00
4 #12THHN, solid	L1@10.9	CLF	482.00	447.00	929.00
5 #12THHN, solid	L1@11.6	CLF	499.00	476.00	975.00
6 #12THHN, solid	L1@12.3	CLF	515.00	505.00	1,020.00
1 #12THHN, stranded	L1@8.80	CLF	429.00	361.00	790.00
2 #12THHN, stranded	L1@9.50	CLF	441.00	390.00	831.00
3 #12THHN, stranded	L1@10.2	CLF	455.00	419.00	874.00
4 #12THHN, stranded	L1@10.9	CLF	469.00	447.00	916.00
5 #12THHN, stranded	L1@11.6	CLF	482.00	476.00	958.00
6 #12THHN, stranded	L1@12.3	CLF	496.00	505.00	1,001.00
1 #10THHN, solid	L1@8.90	CLF	440.00	365.00	805.00
2 #10THHN, solid	L1@9.70	CLF	466.00	398.00	864.00
3 #10THHN, solid	L1@10.5	CLF	491.00	431.00	922.00
4 #10THHN, solid	L1@11.3	CLF	517.00	464.00	981.00
5 #10THHN, solid	L1@12.1	CLF	542.00	496.00	1,038.00
6 #10THHN, solid	L1@12.9	CLF	568.00	529.00	1,097.00
1 #10THHN, stranded	L1@8.90	CLF	434.00	365.00	799.00
2 #10THHN, stranded	L1@9.70	CLF	455.00	398.00	853.00
3 #10THHN, stranded	L1@10.5	CLF	476.00	431.00	907.00
4 #10THHN, stranded	L1@11.3	CLF	496.00	464.00	960.00
5 #10THHN, stranded	L1@12.1	CLF	517.00	496.00	1,013.00
6 #10THHN, stranded	L1@12.9	CLF	536.00	529.00	1,065.00

Use these figures to estimate the cost of installing assemblies under the conditions described on pages 5 and 6. Costs listed are for each assembly installed. The crew is one electrician working at a labor cost of $41.03 per manhour. These costs include layout, material handling and normal waste. Add for the sales tax, delivery, supervision, mobilization, demobilization, cleanup, overhead and profit.

Material	Craft@Hrs	Unit	Material Cost	Labor Cost	Installed Cost
Switches in 1-1/2" deep handy boxes with handy box covers					
1-pole 15 amp brown	L1@0.38	Ea	8.33	15.60	23.93
2-pole 15 amp brown	L1@0.43	Ea	16.20	17.60	33.80
3-way 15 amp brown	L1@0.43	Ea	9.66	17.60	27.26
4-way 15 amp brown	L1@0.48	Ea	22.40	19.70	42.10
1-pole 15 amp brown 277V	L1@0.38	Ea	8.35	15.60	23.95
2-pole 15 amp brown 277V	L1@0.43	Ea	16.20	17.60	33.80
3-way 15 amp brown 277V	L1@0.43	Ea	9.92	17.60	27.52
4-way 15 amp brown 277V	L1@0.48	Ea	22.40	19.70	42.10
1-pole 15 amp ivory	L1@0.38	Ea	8.33	15.60	23.93
2-pole 15 amp ivory	L1@0.43	Ea	16.20	17.60	33.80
3-way 15 amp ivory	L1@0.43	Ea	9.66	17.60	27.26
4-way 15 amp ivory	L1@0.48	Ea	22.40	19.70	42.10
1-pole 15 amp ivory 277V	L1@0.38	Ea	5.22	15.60	20.82
2-pole 15 amp ivory 277V	L1@0.43	Ea	16.20	17.60	33.80
3-way 15 amp ivory 277V	L1@0.43	Ea	9.92	17.60	27.52
4-way 15 amp ivory 277V	L1@0.48	Ea	22.40	19.70	42.10
1-pole 20 amp brown	L1@0.38	Ea	8.95	15.60	24.55
2-pole 20 amp brown	L1@0.43	Ea	18.70	17.60	36.30
3-way 20 amp brown	L1@0.43	Ea	10.10	17.60	27.70
4-way 20 amp brown	L1@0.48	Ea	32.80	19.70	52.50
1-pole 20 amp brown 277V	L1@0.38	Ea	14.60	15.60	30.20
2-pole 20 amp brown 277V	L1@0.43	Ea	18.00	17.60	35.60
3-way 20 amp brown 277V	L1@0.43	Ea	16.40	17.60	34.00
4-way 20 amp brown 277V	L1@0.48	Ea	35.90	19.70	55.60
1-pole 20 amp ivory	L1@0.38	Ea	8.95	15.60	24.55
2-pole 20 amp ivory	L1@0.43	Ea	18.70	17.60	36.30
3-way 20 amp ivory	L1@0.43	Ea	10.10	17.60	27.70
4-way 20 amp ivory	L1@0.48	Ea	32.80	19.70	52.50
1-pole 20 amp ivory 277V	L1@0.38	Ea	14.60	15.60	30.20
2-pole 20 amp ivory 277V	L1@0.43	Ea	18.00	17.60	35.60
3-way 20 amp ivory 277V	L1@0.43	Ea	16.40	17.60	34.00
4-way 20 amp ivory 277V	L1@0.48	Ea	35.90	19.70	55.60

S

Key switches in 1-1/2" deep handy boxes with handy box covers

Material	Craft@Hrs	Unit	Material Cost	Labor Cost	Installed Cost
1-pole 15 amp	L1@0.38	Ea	17.70	15.60	33.30
2-pole 15 amp	L1@0.43	Ea	23.80	17.60	41.40
3-way 15 amp	L1@0.43	Ea	22.00	17.60	39.60
4-way 15 amp	L1@0.48	Ea	39.80	19.70	59.50
1-pole 20 amp	L1@0.38	Ea	24.80	15.60	40.40
2-pole 20 amp	L1@0.43	Ea	28.20	17.60	45.80
3-way 20 amp	L1@0.43	Ea	26.50	17.60	44.10
4-way 20 amp	L1@0.48	Ea	47.00	19.70	66.70

S_K

Use these figures to estimate the cost of installing assemblies under the conditions described on pages 5 and 6. Costs listed are for each assembly installed. The crew is one electrician working at a labor cost of $41.03 per manhour. These costs include layout, material handling and normal waste. Add for the sales tax, delivery, supervision, mobilization, demobilization, cleanup, overhead and profit.

1-7/8" Deep Handy Box Switch Assemblies

Material	Craft@Hrs	Unit	Material Cost	Labor Cost	Installed Cost
Switches in 1-7/8" deep handy boxes with handy box covers					
1-pole 15 amp brown	L1@0.40	Ea	8.52	16.40	24.92
2-pole 15 amp brown	L1@0.45	Ea	16.40	18.50	34.90
3-way 15 amp brown	L1@0.45	Ea	9.86	18.50	28.36
4-way 15 amp brown	L1@0.50	Ea	22.60	20.50	43.10
1-pole 15 amp brown 277V	L1@0.40	Ea	8.55	16.40	24.95
2-pole 15 amp brown 277V	L1@0.45	Ea	16.40	18.50	34.90
3-way 15 amp brown 277V	L1@0.45	Ea	10.10	18.50	28.60
4-way 15 amp brown 277V	L1@0.50	Ea	22.60	20.50	43.10
1-pole 15 amp ivory	L1@0.40	Ea	8.52	16.40	24.92
2-pole 15 amp ivory	L1@0.45	Ea	16.40	18.50	34.90
3-way 15 amp ivory	L1@0.45	Ea	9.86	18.50	28.36
4-way 15 amp ivory	L1@0.50	Ea	22.60	20.50	43.10
1-pole 15 amp ivory 277V	L1@0.40	Ea	8.55	16.40	24.95
2-pole 15 amp ivory 277V	L1@0.45	Ea	16.40	18.50	34.90
3-way 15 amp ivory 277V	L1@0.45	Ea	10.10	18.50	28.60
4-way 15 amp ivory 277V	L1@0.50	Ea	22.60	20.50	43.10
1-pole 20 amp brown	L1@0.40	Ea	9.15	16.40	25.55
2-pole 20 amp brown	L1@0.45	Ea	18.90	18.50	37.40
3-way 20 amp brown	L1@0.45	Ea	10.30	18.50	28.80
4-way 20 amp brown	L1@0.50	Ea	33.00	20.50	53.50
1-pole 20 amp brown 277V	L1@0.40	Ea	14.80	16.40	31.20
2-pole 20 amp brown 277V	L1@0.45	Ea	18.20	18.50	36.70
3-way 20 amp brown 277V	L1@0.45	Ea	16.70	18.50	35.20
4-way 20 amp brown 277V	L1@0.50	Ea	36.10	20.50	56.60
1-pole 20 amp ivory	L1@0.40	Ea	9.15	16.40	25.55
2-pole 20 amp ivory	L1@0.45	Ea	18.90	18.50	37.40
3-way 20 amp ivory	L1@0.45	Ea	10.30	18.50	28.80
4-way 20 amp ivory	L1@0.50	Ea	33.00	20.50	53.50
1-pole 20 amp ivory 277V	L1@0.40	Ea	14.80	16.40	31.20
2-pole 20 amp ivory 277V	L1@0.45	Ea	18.20	18.50	36.70
3-way 20 amp ivory 277V	L1@0.45	Ea	16.70	18.50	35.20
4-way 20 amp ivory 277V	L1@0.50	Ea	36.10	20.50	56.60
Key switches in 1-7/8" deep handy boxes with handy box covers					
1-pole 15 amp	L1@0.40	Ea	17.90	16.40	34.30
2-pole 15 amp	L1@0.45	Ea	24.00	18.50	42.50
3-way 15 amp	L1@0.45	Ea	22.20	18.50	40.70
4-way 15 amp	L1@0.50	Ea	40.00	20.50	60.50
1-pole 20 amp	L1@0.40	Ea	25.00	16.40	41.40
2-pole 20 amp	L1@0.45	Ea	28.40	18.50	46.90
3-way 20 amp	L1@0.45	Ea	26.70	18.50	45.20
4-way 20 amp	L1@0.50	Ea	47.30	20.50	67.80

S

S_K

Use these figures to estimate the cost of installing assemblies under the conditions described on pages 5 and 6. Costs listed are for each assembly installed. The crew is one electrician working at a labor cost of $41.03 per manhour. These costs include layout, material handling and normal waste. Add for the sales tax, delivery, supervision, mobilization, demobilization, cleanup, overhead and profit.

2-1/8" Deep Handy Box Switch Assemblies

Material	Craft@Hrs	Unit	Material Cost	Labor Cost	Installed Cost
Switches in 2-1/8" deep handy boxes with handy box covers					
1-pole 15 amp brown	L1@0.43	Ea	12.20	17.60	29.80
2-pole 15 amp brown	L1@0.48	Ea	20.00	19.70	39.70
3-way 15 amp brown	L1@0.48	Ea	13.40	19.70	33.10
4-way 15 amp brown	L1@0.53	Ea	26.20	21.70	47.90
1-pole 15 amp brown 277V	L1@0.43	Ea	12.20	17.60	29.80
2-pole 15 amp brown 277V	L1@0.48	Ea	20.00	19.70	39.70
3-way 15 amp brown 277V	L1@0.48	Ea	13.70	19.70	33.40
4-way 15 amp brown 277V	L1@0.53	Ea	26.20	21.70	47.90
1-pole 15 amp ivory	L1@0.43	Ea	12.20	17.60	29.80
2-pole 15 amp ivory	L1@0.48	Ea	20.00	19.70	39.70
3-way 15 amp ivory	L1@0.48	Ea	13.40	19.70	33.10
4-way 15 amp ivory	L1@0.53	Ea	26.20	21.70	47.90
1-pole 15 amp ivory 277V	L1@0.43	Ea	12.20	17.60	29.80
2-pole 15 amp ivory 277V	L1@0.48	Ea	20.00	19.70	39.70
3-way 15 amp ivory 277V	L1@0.48	Ea	13.70	19.70	33.40
4-way 15 amp ivory 277V	L1@0.53	Ea	26.20	21.70	47.90
1-pole 20 amp brown	L1@0.43	Ea	12.80	17.60	30.40
2-pole 20 amp brown	L1@0.48	Ea	22.50	19.70	42.20
3-way 20 amp brown	L1@0.48	Ea	13.90	19.70	33.60
4-way 20 amp brown	L1@0.53	Ea	36.60	21.70	58.30
1-pole 20 amp brown 277V	L1@0.43	Ea	18.30	17.60	35.90
2-pole 20 amp brown 277V	L1@0.48	Ea	21.90	19.70	41.60
3-way 20 amp brown 277V	L1@0.48	Ea	20.30	19.70	40.00
4-way 20 amp brown 277V	L1@0.53	Ea	39.80	21.70	61.50
1-pole 20 amp ivory	L1@0.43	Ea	12.80	17.60	30.40
2-pole 20 amp ivory	L1@0.48	Ea	22.50	19.70	42.20
3-way 20 amp ivory	L1@0.48	Ea	13.90	19.70	33.60
4-way 20 amp ivory	L1@0.53	Ea	36.60	21.70	58.30
1-pole 20 amp ivory 277V	L1@0.43	Ea	18.30	17.60	35.90
2-pole 20 amp ivory 277V	L1@0.48	Ea	21.90	19.70	41.60
3-way 20 amp ivory 277V	L1@0.48	Ea	20.30	19.70	40.00
4-way 20 amp ivory 277V	L1@0.53	Ea	39.80	21.70	61.50
Key switches in 2-1/8" deep handy boxes with handy box covers					
1-pole 15 amp	L1@0.43	Ea	21.50	17.60	39.10
2-pole 15 amp	L1@0.48	Ea	27.60	19.70	47.30
3-way 15 amp	L1@0.48	Ea	25.80	19.70	45.50
4-way 15 amp	L1@0.53	Ea	43.60	21.70	65.30
1-pole 20 amp	L1@0.43	Ea	28.60	17.60	46.20
2-pole 20 amp	L1@0.48	Ea	32.10	19.70	51.80
3-way 20 amp	L1@0.48	Ea	30.30	19.70	50.00
4-way 20 amp	L1@0.53	Ea	50.90	21.70	72.60

S

S_K

Use these figures to estimate the cost of installing assemblies under the conditions described on pages 5 and 6. Costs listed are for each assembly installed. The crew is one electrician working at a labor cost of $41.03 per manhour. These costs include layout, material handling and normal waste. Add for the sales tax, delivery, supervision, mobilization, demobilization, cleanup, overhead and profit.

2-1/2" Deep Handy Box Switch Assemblies

Material	Craft@Hrs	Unit	Material Cost	Labor Cost	Installed Cost
Switches in 2-1/2" deep handy boxes with handy box covers					
1-pole 15 amp brown	L1@0.43	Ea	8.98	17.60	26.58
2-pole 15 amp brown	L1@0.48	Ea	16.90	19.70	36.60
3-way 15 amp brown	L1@0.48	Ea	10.30	19.70	30.00
4-way 15 amp brown	L1@0.53	Ea	23.00	21.70	44.70
1-pole 15 amp brown 277V	L1@0.43	Ea	9.00	17.60	26.60
2-pole 15 amp brown 277V	L1@0.48	Ea	16.90	19.70	36.60
3-way 15 amp brown 277V	L1@0.48	Ea	10.60	19.70	30.30
4-way 15 amp brown 277V	L1@0.53	Ea	23.00	21.70	44.70
1-pole 15 amp ivory	L1@0.43	Ea	8.98	17.60	26.58
2-pole 15 amp ivory	L1@0.48	Ea	16.90	19.70	36.60
3-way 15 amp ivory	L1@0.48	Ea	10.30	19.70	30.00
4-way 15 amp ivory	L1@0.53	Ea	23.00	21.70	44.70
1-pole 15 amp ivory 277V	L1@0.43	Ea	9.00	17.60	26.60
2-pole 15 amp ivory 277V	L1@0.48	Ea	16.90	19.70	36.60
3-way 15 amp ivory 277V	L1@0.48	Ea	10.60	19.70	30.30
4-way 15 amp ivory 277V	L1@0.53	Ea	23.00	21.70	44.70
1-pole 20 amp brown	L1@0.43	Ea	9.60	17.60	27.20
2-pole 20 amp brown	L1@0.48	Ea	19.40	19.70	39.10
3-way 20 amp brown	L1@0.48	Ea	10.70	19.70	30.40
4-way 20 amp brown	L1@0.53	Ea	33.40	21.70	55.10
1-pole 20 amp brown 277V	L1@0.43	Ea	9.00	17.60	26.60
2-pole 20 amp brown 277V	L1@0.48	Ea	16.90	19.70	36.60
3-way 20 amp brown 277V	L1@0.48	Ea	10.60	19.70	30.30
4-way 20 amp brown 277V	L1@0.53	Ea	23.00	21.70	44.70
1-pole 20 amp ivory	L1@0.43	Ea	9.60	17.60	27.20
2-pole 20 amp ivory	L1@0.48	Ea	19.40	19.70	39.10
3-way 20 amp ivory	L1@0.48	Ea	10.70	19.70	30.40
4-way 20 amp ivory	L1@0.53	Ea	33.40	21.70	55.10
1-pole 20 amp ivory 277V	L1@0.43	Ea	9.00	17.60	26.60
2-pole 20 amp ivory 277V	L1@0.48	Ea	16.90	19.70	36.60
3-way 20 amp ivory 277V	L1@0.48	Ea	10.60	19.70	30.30
4-way 20 amp ivory 277V	L1@0.53	Ea	23.00	21.70	44.70
Key switches in 2-1/2" deep handy boxes with handy box covers					
1-pole 15 amp	L1@0.43	Ea	18.40	17.60	36.00
2-pole 15 amp	L1@0.48	Ea	24.50	19.70	44.20
3-way 15 amp	L1@0.48	Ea	22.70	19.70	42.40
4-way 15 amp	L1@0.53	Ea	40.40	21.70	62.10
1-pole 20 amp	L1@0.43	Ea	25.40	17.60	43.00
2-pole 20 amp	L1@0.48	Ea	28.90	19.70	48.60
3-way 20 amp	L1@0.48	Ea	27.20	19.70	46.90
4-way 20 amp	L1@0.53	Ea	47.70	21.70	69.40

S

S_K

Use these figures to estimate the cost of installing assemblies under the conditions described on pages 5 and 6. Costs listed are for each assembly installed. The crew is one electrician working at a labor cost of $41.03 per manhour. These costs include layout, material handling and normal waste. Add for the sales tax, delivery, supervision, mobilization, demobilization, cleanup, overhead and profit.

2" Deep 15 Amp Sectional Box Switch Assemblies

Switches, 15 amp 1-pole in 2" deep sectional box and plastic plate

S

Material	Craft@Hrs	Unit	Material Cost	Labor Cost	Installed Cost
1 gang, brown or ivory	L1@0.38	Ea	10.60	15.60	26.20
2 gang, brown or ivory	L1@0.76	Ea	21.20	31.20	52.40
3 gang, brown or ivory	L1@1.14	Ea	31.80	46.80	78.60
4 gang, brown or ivory	L1@1.52	Ea	42.40	62.40	104.80
5 gang, brown or ivory	L1@1.90	Ea	53.00	78.00	131.00
6 gang, brown or ivory	L1@2.28	Ea	63.60	93.50	157.10

Switches, 15 amp 2-pole in 2" deep sectional box and plastic plate

S_2

Material	Craft@Hrs	Unit	Material Cost	Labor Cost	Installed Cost
1 gang, brown or ivory	L1@0.43	Ea	18.40	17.60	36.00
2 gang, brown or ivory	L1@0.86	Ea	36.90	35.30	72.20
3 gang, brown or ivory	L1@1.29	Ea	55.40	52.90	108.30
4 gang, brown or ivory	L1@1.72	Ea	73.90	70.60	144.50
5 gang, brown or ivory	L1@2.15	Ea	92.30	88.20	180.50
6 gang, brown or ivory	L1@2.58	Ea	110.00	106.00	216.00

Switches, 15 amp 3-way in 2" deep sectional box and plastic plate

S_3

Material	Craft@Hrs	Unit	Material Cost	Labor Cost	Installed Cost
1 gang, brown or ivory	L1@0.43	Ea	12.20	17.60	29.80
2 gang, brown or ivory	L1@0.86	Ea	24.40	35.30	59.70
3 gang, brown or ivory	L1@1.29	Ea	36.50	52.90	89.40
4 gang, brown or ivory	L1@1.72	Ea	48.70	70.60	119.30
5 gang, brown or ivory	L1@2.15	Ea	60.90	88.20	149.10
6 gang, brown or ivory	L1@2.58	Ea	73.10	106.00	179.10

Switches, 15 amp 4-way in 2" deep sectional box and plastic plate

S_4

Material	Craft@Hrs	Unit	Material Cost	Labor Cost	Installed Cost
1 gang, brown or ivory	L1@0.48	Ea	24.70	19.70	44.40
2 gang, brown or ivory	L1@0.96	Ea	49.30	39.40	88.70
3 gang, brown or ivory	L1@1.44	Ea	73.90	59.10	133.00
4 gang, brown or ivory	L1@1.92	Ea	98.60	78.80	177.40
5 gang, brown or ivory	L1@2.40	Ea	123.00	98.50	221.50
6 gang, brown or ivory	L1@2.88	Ea	148.00	118.00	266.00

Switches, 15 amp key in 2" deep sectional box and plastic plate

S_K

Material	Craft@Hrs	Unit	Material Cost	Labor Cost	Installed Cost
1 gang 1-pole	L1@0.38	Ea	20.00	15.60	35.60
1 gang 2-pole	L1@0.43	Ea	26.00	17.60	43.60
1 gang 3-way	L1@0.43	Ea	24.30	17.60	41.90
1 gang 4-way	L1@0.48	Ea	42.00	19.70	61.70

Use these figures to estimate the cost of installing assemblies under the conditions described on pages 5 and 6. Costs listed are for each assembly installed. The crew is one electrician working at a labor cost of $41.03 per manhour. These costs include layout, material handling and normal waste. Add for the sales tax, delivery, supervision, mobilization, demobilization, cleanup, overhead and profit.

2-1/2" Deep 15 Amp Sectional Box Switch Assemblies

	Material	Craft@Hrs	Unit	Material Cost	Labor Cost	Installed Cost
	Switches, 15 amp 1-pole in 2-1/2" deep sectional box and plastic plate					
	1 gang, brown or ivory	L1@0.40	Ea	9.78	16.40	26.18
	2 gang, brown or ivory	L1@0.80	Ea	19.60	32.80	52.40
S	3 gang, brown or ivory	L1@1.20	Ea	29.40	49.20	78.60
	4 gang, brown or ivory	L1@1.60	Ea	39.10	65.60	104.70
	5 gang, brown or ivory	L1@2.00	Ea	48.90	82.10	131.00
	6 gang, brown or ivory	L1@2.40	Ea	58.70	98.50	157.20
	Switches, 15 amp 2-pole in 2-1/2" deep sectional box and plastic plate					
	1 gang, brown or ivory	L1@0.45	Ea	17.70	18.50	36.20
	2 gang, brown or ivory	L1@0.90	Ea	35.30	36.90	72.20
S_2	3 gang, brown or ivory	L1@1.35	Ea	53.00	55.40	108.40
	4 gang, brown or ivory	L1@1.80	Ea	70.60	73.90	144.50
	5 gang, brown or ivory	L1@2.25	Ea	88.30	92.30	180.60
	6 gang, brown or ivory	L1@2.70	Ea	106.00	111.00	217.00
	Switches, 15 amp 3-way in 2-1/2" deep sectional box and plastic plate					
	1 gang, brown or ivory	L1@0.45	Ea	11.30	18.50	29.80
	2 gang, brown or ivory	L1@0.90	Ea	22.70	36.90	59.60
S_3	3 gang, brown or ivory	L1@1.35	Ea	34.00	55.40	89.40
	4 gang, brown or ivory	L1@1.80	Ea	45.40	73.90	119.30
	5 gang, brown or ivory	L1@2.25	Ea	56.80	92.30	149.10
	6 gang, brown or ivory	L1@2.70	Ea	68.20	111.00	179.20
	Switches, 15 amp 4-way in 2-1/2" deep sectional box and plastic plate					
	1 gang, brown or ivory	L1@0.50	Ea	23.80	20.50	44.30
	2 gang, brown or ivory	L1@1.00	Ea	47.70	41.00	88.70
S_4	3 gang, brown or ivory	L1@1.50	Ea	71.50	61.50	133.00
	4 gang, brown or ivory	L1@2.00	Ea	95.30	82.10	177.40
	5 gang, brown or ivory	L1@2.50	Ea	119.00	103.00	222.00
	6 gang, brown or ivory	L1@3.00	Ea	143.00	123.00	266.00
	Switches, 15 amp key in 2-1/2" deep sectional box and plastic plate					
	1 gang 1-pole	L1@0.40	Ea	19.20	16.40	35.60
S_K	1 gang 2-pole	L1@0.45	Ea	25.20	18.50	43.70
	1 gang 3-way	L1@0.45	Ea	23.40	18.50	41.90
	1 gang 4-way	L1@0.50	Ea	41.20	20.50	61.70

Use these figures to estimate the cost of installing assemblies under the conditions described on pages 5 and 6. Costs listed are for each assembly installed. The crew is one electrician working at a labor cost of $41.03 per manhour. These costs include layout, material handling and normal waste. Add for the sales tax, delivery, supervision, mobilization, demobilization, cleanup, overhead and profit.

2-3/4" Deep 15 Amp Sectional Box Switch Assemblies

Switches, 15 amp 1-pole in 2-3/4" deep sectional box and plastic plate

Material	Craft@Hrs	Unit	Material Cost	Labor Cost	Installed Cost	
1 gang, brown or ivory	L1@0.43	Ea	11.20	17.60	28.80	
2 gang, brown or ivory	L1@0.86	Ea	22.50	35.30	57.80	
3 gang, brown or ivory	L1@1.29	Ea	33.70	52.90	86.60	
4 gang, brown or ivory	L1@1.72	Ea	44.90	70.60	115.50	S
5 gang, brown or ivory	L1@2.15	Ea	56.10	88.20	144.30	
6 gang, brown or ivory	L1@2.58	Ea	67.30	106.00	173.30	

Switches, 15 amp 2-pole in 2-3/4" deep sectional box and plastic plate

Material	Craft@Hrs	Unit	Material Cost	Labor Cost	Installed Cost	
1 gang, brown or ivory	L1@0.48	Ea	19.00	19.70	38.70	
2 gang, brown or ivory	L1@0.96	Ea	38.20	39.40	77.60	
3 gang, brown or ivory	L1@1.44	Ea	57.20	59.10	116.30	S_2
4 gang, brown or ivory	L1@1.92	Ea	76.40	78.80	155.20	
5 gang, brown or ivory	L1@2.40	Ea	95.40	98.50	193.90	
6 gang, brown or ivory	L1@2.88	Ea	114.00	118.00	232.00	

Switches, 15 amp 3-way in 2-3/4" deep sectional box and plastic plate

Material	Craft@Hrs	Unit	Material Cost	Labor Cost	Installed Cost	
1 gang, brown or ivory	L1@0.48	Ea	12.80	19.70	32.50	
2 gang, brown or ivory	L1@0.96	Ea	25.60	39.40	65.00	
3 gang, brown or ivory	L1@1.44	Ea	38.40	59.10	97.50	S_3
4 gang, brown or ivory	L1@1.92	Ea	51.20	78.80	130.00	
5 gang, brown or ivory	L1@2.40	Ea	64.00	98.50	162.50	
6 gang, brown or ivory	L1@2.88	Ea	76.80	118.00	194.80	

Switches, 15 amp 4-way in 2-3/4" deep sectional box and plastic plate

Material	Craft@Hrs	Unit	Material Cost	Labor Cost	Installed Cost	
1 gang, brown or ivory	L1@0.53	Ea	25.30	21.70	47.00	
2 gang, brown or ivory	L1@1.06	Ea	50.60	43.50	94.10	
3 gang, brown or ivory	L1@1.59	Ea	75.80	65.20	141.00	S_4
4 gang, brown or ivory	L1@2.12	Ea	101.00	87.00	188.00	
5 gang, brown or ivory	L1@2.65	Ea	126.00	109.00	235.00	
6 gang, brown or ivory	L1@3.18	Ea	152.00	130.00	282.00	

Switches, 15 amp key in 2-3/4" deep sectional box and plastic plate

Material	Craft@Hrs	Unit	Material Cost	Labor Cost	Installed Cost	
1 gang 1-pole	L1@0.43	Ea	20.60	17.60	38.20	
1 gang 2-pole	L1@0.48	Ea	26.60	19.70	46.30	
1 gang 3-way	L1@0.48	Ea	24.90	19.70	44.60	S_K
1 gang 4-way	L1@0.53	Ea	42.70	21.70	64.40	

Use these figures to estimate the cost of installing assemblies under the conditions described on pages 5 and 6. Costs listed are for each assembly installed. The crew is one electrician working at a labor cost of $41.03 per manhour. These costs include layout, material handling and normal waste. Add for the sales tax, delivery, supervision, mobilization, demobilization, cleanup, overhead and profit.

3-1/2" Deep 15 Amp Sectional Box Switch Assemblies

	Material	Craft@Hrs	Unit	Material Cost	Labor Cost	Installed Cost
	Switches, 15 amp 1-pole in 3-1/2" deep sectional box and plastic plate					
	1 gang, brown or ivory	L1@0.48	Ea	11.40	19.70	31.10
	2 gang, brown or ivory	L1@0.96	Ea	22.50	39.40	61.90
S	3 gang, brown or ivory	L1@1.44	Ea	33.70	59.10	92.80
	4 gang, brown or ivory	L1@1.92	Ea	44.90	78.80	123.70
	5 gang, brown or ivory	L1@2.40	Ea	56.10	98.50	154.60
	6 gang, brown or ivory	L1@2.88	Ea	67.30	118.00	185.30
	Switches, 15 amp 2-pole in 3-1/2" deep sectional box and plastic plate					
	1 gang, brown or ivory	L1@0.53	Ea	19.00	21.70	40.70
	2 gang, brown or ivory	L1@1.06	Ea	38.20	43.50	81.70
S_2	3 gang, brown or ivory	L1@1.59	Ea	57.20	65.20	122.40
	4 gang, brown or ivory	L1@2.12	Ea	76.40	87.00	163.40
	5 gang, brown or ivory	L1@2.65	Ea	95.40	109.00	204.40
	6 gang, brown or ivory	L1@3.18	Ea	114.00	130.00	244.00
	Switches, 15 amp 3-way in 3-1/2" deep sectional box and plastic plate					
	1 gang, brown or ivory	L1@0.53	Ea	12.80	21.70	34.50
	2 gang, brown or ivory	L1@1.06	Ea	25.60	43.50	69.10
S_3	3 gang, brown or ivory	L1@1.59	Ea	38.40	65.20	103.60
	4 gang, brown or ivory	L1@2.12	Ea	51.20	87.00	138.20
	5 gang, brown or ivory	L1@2.65	Ea	64.00	109.00	173.00
	6 gang, brown or ivory	L1@3.18	Ea	76.80	130.00	206.80
	Switches, 15 amp 4-way in 3-1/2" deep sectional box and plastic plate					
	1 gang, brown or ivory	L1@0.58	Ea	25.30	23.80	49.10
	2 gang, brown or ivory	L1@1.16	Ea	50.60	47.60	98.20
S_4	3 gang, brown or ivory	L1@1.74	Ea	75.80	71.40	147.20
	4 gang, brown or ivory	L1@2.32	Ea	101.00	95.20	196.20
	5 gang, brown or ivory	L1@2.90	Ea	126.00	119.00	245.00
	6 gang, brown or ivory	L1@3.48	Ea	152.00	143.00	295.00
	Switches, 15 amp key in 3-1/2" deep sectional box and plastic plate					
	1 gang 1-pole	L1@0.48	Ea	20.60	19.70	40.30
S_K	1 gang 2-pole	L1@0.53	Ea	26.60	21.70	48.30
	1 gang 3-way	L1@0.53	Ea	24.90	21.70	46.60
	1 gang 4-way	L1@0.58	Ea	42.70	23.80	66.50

Use these figures to estimate the cost of installing assemblies under the conditions described on pages 5 and 6. Costs listed are for each assembly installed. The crew is one electrician working at a labor cost of $41.03 per manhour. These costs include layout, material handling and normal waste. Add for the sales tax, delivery, supervision, mobilization, demobilization, cleanup, overhead and profit.

2" Deep 15 Amp Sectional Box Switch Assemblies

Material	Craft@Hrs	Unit	Material Cost	Labor Cost	Installed Cost	
Switches, 15 amp 1-pole in 2" deep sectional box and plastic plate						
1 gang, white	L1@0.38	Ea	10.60	15.60	26.20	
2 gang, white	L1@0.76	Ea	21.20	31.20	52.40	
3 gang, white	L1@1.14	Ea	31.80	46.80	78.60	S
4 gang, white	L1@1.52	Ea	42.40	62.40	104.80	
5 gang, white	L1@1.90	Ea	53.00	78.00	131.00	
6 gang, white	L1@2.28	Ea	63.60	93.50	157.10	
Switches, 15 amp 2-pole in 2" deep sectional box and plastic plate						
1 gang, white	L1@0.43	Ea	18.40	17.60	36.00	
2 gang, white	L1@0.86	Ea	36.90	35.30	72.20	
3 gang, white	L1@1.29	Ea	55.40	52.90	108.30	S_2
4 gang, white	L1@1.72	Ea	73.90	70.60	144.50	
5 gang, white	L1@2.15	Ea	92.30	88.20	180.50	
6 gang, white	L1@2.58	Ea	110.00	106.00	216.00	
Switches, 15 amp 3-way in 2" deep sectional box and plastic plate						
1 gang, white	L1@0.43	Ea	12.20	17.60	29.80	
2 gang, white	L1@0.86	Ea	24.40	35.30	59.70	
3 gang, white	L1@1.29	Ea	36.50	52.90	89.40	S_3
4 gang, white	L1@1.72	Ea	50.60	70.60	121.20	
5 gang, white	L1@2.15	Ea	60.90	88.20	149.10	
6 gang, white	L1@2.58	Ea	73.10	106.00	179.10	
Switches, 15 amp 4-way in 2" deep sectional box and plastic plate						
1 gang, white	L1@0.48	Ea	24.70	19.70	44.40	
2 gang, white	L1@0.96	Ea	49.30	39.40	88.70	
3 gang, white	L1@1.44	Ea	73.90	59.10	133.00	S_4
4 gang, white	L1@1.92	Ea	98.60	78.80	177.40	
5 gang, white	L1@2.40	Ea	123.00	98.50	221.50	
6 gang, white	L1@2.88	Ea	148.00	118.00	266.00	

Use these figures to estimate the cost of installing assemblies under the conditions described on pages 5 and 6. Costs listed are for each assembly installed. The crew is one electrician working at a labor cost of $41.03 per manhour. These costs include layout, material handling and normal waste. Add for the sales tax, delivery, supervision, mobilization, demobilization, cleanup, overhead and profit.

2-1/2" Deep 15 Amp Sectional Box Switch Assemblies

Switches, 15 amp 1-pole in 2-1/2" deep sectional box and plastic plate

S

Material	Craft@Hrs	Unit	Material Cost	Labor Cost	Installed Cost
1 gang ivory	L1@0.40	Ea	9.78	16.40	26.18
2 gang ivory	L1@0.80	Ea	19.60	32.80	52.40
3 gang ivory	L1@1.20	Ea	29.40	49.20	78.60
4 gang ivory	L1@1.60	Ea	39.10	65.60	104.70
5 gang ivory	L1@2.00	Ea	48.90	82.10	131.00
6 gang ivory	L1@2.40	Ea	58.70	98.50	157.20

Switches, 15 amp 2-pole in 2-1/2" deep sectional box and plastic plate

S_2

Material	Craft@Hrs	Unit	Material Cost	Labor Cost	Installed Cost
1 gang ivory	L1@0.45	Ea	17.70	18.50	36.20
2 gang ivory	L1@0.90	Ea	35.30	36.90	72.20
3 gang ivory	L1@1.35	Ea	53.00	55.40	108.40
4 gang ivory	L1@1.80	Ea	70.60	73.90	144.50
5 gang ivory	L1@2.25	Ea	88.30	92.30	180.60
6 gang ivory	L1@2.70	Ea	106.00	111.00	217.00

Switches, 15 amp 3-way in 2-1/2" deep sectional box and plastic plate

S_3

Material	Craft@Hrs	Unit	Material Cost	Labor Cost	Installed Cost
1 gang ivory	L1@0.45	Ea	11.30	18.50	29.80
2 gang ivory	L1@0.90	Ea	22.70	36.90	59.60
3 gang ivory	L1@1.35	Ea	34.00	55.40	89.40
4 gang ivory	L1@1.80	Ea	45.40	73.90	119.30
5 gang ivory	L1@2.25	Ea	56.80	92.30	149.10
6 gang ivory	L1@2.70	Ea	68.20	111.00	179.20

Switches, 15 amp 4-way in 2-1/2" deep sectional box and plastic plate

S_4

Material	Craft@Hrs	Unit	Material Cost	Labor Cost	Installed Cost
1 gang ivory	L1@0.50	Ea	23.80	20.50	44.30
2 gang ivory	L1@1.00	Ea	47.70	41.00	88.70
3 gang ivory	L1@1.50	Ea	71.50	61.50	133.00
4 gang ivory	L1@2.00	Ea	95.30	82.10	177.40
5 gang ivory	L1@2.50	Ea	119.00	103.00	222.00
6 gang ivory	L1@3.00	Ea	143.00	123.00	266.00

Use these figures to estimate the cost of installing assemblies under the conditions described on pages 5 and 6. Costs listed are for each assembly installed. The crew is one electrician working at a labor cost of $41.03 per manhour. These costs include layout, material handling and normal waste. Add for the sales tax, delivery, supervision, mobilization, demobilization, cleanup, overhead and profit.

2-3/4" Deep 15 Amp Sectional Box Switch Assemblies

Switches, 15 amp 1-pole in 2-3/4" deep sectional box and plastic plate

Material	Craft@Hrs	Unit	Material Cost	Labor Cost	Installed Cost	
1 gang ivory	L1@0.43	Ea	11.20	17.60	28.80	
2 gang ivory	L1@0.86	Ea	22.50	35.30	57.80	
3 gang ivory	L1@1.29	Ea	33.70	52.90	86.60	S
4 gang ivory	L1@1.72	Ea	44.90	70.60	115.50	
5 gang ivory	L1@2.15	Ea	56.10	88.20	144.30	
6 gang ivory	L1@2.58	Ea	67.30	106.00	173.30	

Switches, 15 amp 2-pole in 2-3/4" deep sectional box and plastic plate

Material	Craft@Hrs	Unit	Material Cost	Labor Cost	Installed Cost	
1 gang ivory	L1@0.48	Ea	19.00	19.70	38.70	
2 gang ivory	L1@0.96	Ea	38.20	39.40	77.60	
3 gang ivory	L1@1.44	Ea	57.20	59.10	116.30	S_2
4 gang ivory	L1@1.92	Ea	76.40	78.80	155.20	
5 gang ivory	L1@2.40	Ea	95.40	98.50	193.90	
6 gang ivory	L1@2.88	Ea	114.00	118.00	232.00	

Switches, 15 amp 3-way in 2-3/4" deep sectional box and plastic plate

Material	Craft@Hrs	Unit	Material Cost	Labor Cost	Installed Cost	
1 gang ivory	L1@0.48	Ea	12.80	19.70	32.50	
2 gang ivory	L1@0.96	Ea	25.60	39.40	65.00	
3 gang ivory	L1@1.44	Ea	38.40	59.10	97.50	S_3
4 gang ivory	L1@1.92	Ea	51.20	78.80	130.00	
5 gang ivory	L1@2.40	Ea	64.00	98.50	162.50	
6 gang ivory	L1@2.88	Ea	76.80	118.00	194.80	

Switches, 15 amp 4-way in 2-3/4" deep sectional box and plastic plate

Material	Craft@Hrs	Unit	Material Cost	Labor Cost	Installed Cost	
1 gang ivory	L1@0.53	Ea	25.30	21.70	47.00	
2 gang ivory	L1@1.06	Ea	48.70	43.50	92.20	
3 gang ivory	L1@1.59	Ea	75.80	65.20	141.00	S_4
4 gang ivory	L1@2.12	Ea	101.00	87.00	188.00	
5 gang ivory	L1@2.65	Ea	126.00	109.00	235.00	
6 gang ivory	L1@3.18	Ea	152.00	130.00	282.00	

Use these figures to estimate the cost of installing assemblies under the conditions described on pages 5 and 6. Costs listed are for each assembly installed. The crew is one electrician working at a labor cost of $41.03 per manhour. These costs include layout, material handling and normal waste. Add for the sales tax, delivery, supervision, mobilization, demobilization, cleanup, overhead and profit.

3-1/2" Deep 15 Amp Sectional Box Switch Assemblies

	Material	Craft@Hrs	Unit	Material Cost	Labor Cost	Installed Cost
	Switches, 15 amp 1-pole in 3-1/2" deep sectional box and plastic plate					
	1 gang, white	L1@0.48	Ea	11.40	19.70	31.10
	2 gang, white	L1@0.96	Ea	22.50	39.40	61.90
S	3 gang, white	L1@1.44	Ea	33.70	59.10	92.80
	4 gang, white	L1@1.92	Ea	44.90	78.80	123.70
	5 gang, white	L1@2.40	Ea	56.10	98.50	154.60
	6 gang, white	L1@2.88	Ea	67.30	118.00	185.30
	Switches, 15 amp 2-pole in 3-1/2" deep sectional box and plastic plate					
	1 gang, white	L1@0.53	Ea	19.00	21.70	40.70
	2 gang, white	L1@1.06	Ea	38.20	43.50	81.70
S_2	3 gang, white	L1@1.59	Ea	57.20	65.20	122.40
	4 gang, white	L1@2.12	Ea	76.40	87.00	163.40
	5 gang, white	L1@2.65	Ea	95.40	109.00	204.40
	6 gang, white	L1@3.18	Ea	114.00	130.00	244.00
	Switches, 15 amp 3-way in 3-1/2" deep sectional box and plastic plate					
	1 gang, white	L1@0.53	Ea	12.80	21.70	34.50
	2 gang, white	L1@1.06	Ea	25.60	43.50	69.10
S_3	3 gang, white	L1@1.59	Ea	38.40	65.20	103.60
	4 gang, white	L1@2.12	Ea	51.20	87.00	138.20
	5 gang, white	L1@2.65	Ea	64.00	109.00	173.00
	6 gang, white	L1@3.18	Ea	76.80	130.00	206.80
	Switches, 15 amp 4-way in 3-1/2" deep sectional box and plastic plate					
	1 gang, white	L1@0.58	Ea	25.30	23.80	49.10
S_4	2 gang, white	L1@1.16	Ea	50.60	47.60	98.20
	3 gang, white	L1@1.74	Ea	75.80	71.40	147.20
	4 gang, white	L1@2.32	Ea	101.00	95.20	196.20
	5 gang, white	L1@2.90	Ea	126.00	119.00	245.00
	6 gang, white	L1@3.48	Ea	152.00	143.00	295.00

Use these figures to estimate the cost of installing assemblies under the conditions described on pages 5 and 6. Costs listed are for each assembly installed. The crew is one electrician working at a labor cost of $41.03 per manhour. These costs include layout, material handling and normal waste. Add for the sales tax, delivery, supervision, mobilization, demobilization, cleanup, overhead and profit.

2" Deep 20 Amp Sectional Box Switch Assemblies

Material	Craft@Hrs	Unit	Material Cost	Labor Cost	Installed Cost	
Switches, 20 amp 1-pole in 2" deep sectional box and plastic plate						
1 gang, brown or ivory	L1@0.38	Ea	11.20	15.60	26.80	
2 gang, brown or ivory	L1@0.76	Ea	22.40	31.20	53.60	
3 gang, brown or ivory	L1@1.14	Ea	33.60	46.80	80.40	S
4 gang, brown or ivory	L1@1.52	Ea	44.80	62.40	107.20	
5 gang, brown or ivory	L1@1.90	Ea	56.00	78.00	134.00	
6 gang, brown or ivory	L1@2.28	Ea	67.20	93.50	160.70	
Switches, 20 amp 2-pole in 2" deep sectional box and plastic plate						
1 gang, brown or ivory	L1@0.43	Ea	20.90	17.60	38.50	
2 gang, brown or ivory	L1@0.86	Ea	41.90	35.30	77.20	
3 gang, brown or ivory	L1@1.29	Ea	62.90	52.90	115.80	S_2
4 gang, brown or ivory	L1@1.72	Ea	83.90	70.60	154.50	
5 gang, brown or ivory	L1@2.15	Ea	105.00	88.20	193.20	
6 gang, brown or ivory	L1@2.58	Ea	126.00	106.00	232.00	
Switches, 20 amp 3-way in 2" deep sectional box and plastic plate						
1 gang, brown or ivory	L1@0.43	Ea	12.40	17.60	30.00	
2 gang, brown or ivory	L1@0.86	Ea	24.80	35.30	60.10	
3 gang, brown or ivory	L1@1.29	Ea	38.00	52.90	90.90	S_3
4 gang, brown or ivory	L1@1.72	Ea	49.40	70.60	120.00	
5 gang, brown or ivory	L1@2.15	Ea	61.80	88.20	150.00	
6 gang, brown or ivory	L1@2.58	Ea	74.20	106.00	180.20	
Switches, 20 amp 4-way in 2" deep sectional box and plastic plate						
1 gang, brown or ivory	L1@0.48	Ea	35.10	19.70	54.80	
2 gang, brown or ivory	L1@0.96	Ea	70.00	39.40	109.40	
3 gang, brown or ivory	L1@1.44	Ea	105.00	59.10	164.10	S_4
4 gang, brown or ivory	L1@1.92	Ea	141.00	78.80	219.80	
5 gang, brown or ivory	L1@2.40	Ea	175.00	98.50	273.50	
6 gang, brown or ivory	L1@2.88	Ea	210.00	118.00	328.00	
Switches, 20 amp key in 2" deep sectional box and plastic plate						
1-pole 20 amp	L1@0.38	Ea	27.10	15.60	42.70	
2-pole 20 amp	L1@0.43	Ea	30.50	17.60	48.10	S_K
3-way 20 amp	L1@0.43	Ea	28.70	17.60	46.30	
4-way 20 amp	L1@0.48	Ea	49.20	19.70	68.90	

Use these figures to estimate the cost of installing assemblies under the conditions described on pages 5 and 6. Costs listed are for each assembly installed. The crew is one electrician working at a labor cost of $41.03 per manhour. These costs include layout, material handling and normal waste. Add for the sales tax, delivery, supervision, mobilization, demobilization, cleanup, overhead and profit.

2-1/2" Deep 20 Amp Sectional Box Switch Assemblies

	Material	Craft@Hrs	Unit	Material Cost	Labor Cost	Installed Cost
	Switches, 20 amp 1-pole in 2-1/2" deep sectional box and plastic plate					
	1 gang, brown or ivory	L1@0.40	Ea	10.40	16.40	26.80
	2 gang, brown or ivory	L1@0.80	Ea	20.80	32.80	53.60
S	3 gang, brown or ivory	L1@1.20	Ea	31.10	49.20	80.30
	4 gang, brown or ivory	L1@1.60	Ea	41.50	65.60	107.10
	5 gang, brown or ivory	L1@2.00	Ea	51.90	82.10	134.00
	6 gang, brown or ivory	L1@2.40	Ea	62.30	98.50	160.80
	Switches, 20 amp 2-pole in 2-1/2" deep sectional box and plastic plate					
	1 gang, brown or ivory	L1@0.45	Ea	20.20	18.50	38.70
	2 gang, brown or ivory	L1@0.90	Ea	40.30	36.90	77.20
S_2	3 gang, brown or ivory	L1@1.35	Ea	60.50	55.40	115.90
	4 gang, brown or ivory	L1@1.80	Ea	80.60	73.90	154.50
	5 gang, brown or ivory	L1@2.25	Ea	101.00	92.30	193.30
	6 gang, brown or ivory	L1@2.70	Ea	121.00	111.00	232.00
	Switches, 20 amp 3-way in 2-1/2" deep sectional box and plastic plate					
	1 gang, brown or ivory	L1@0.45	Ea	11.60	18.50	30.10
	2 gang, brown or ivory	L1@0.90	Ea	23.10	36.90	60.00
S_3	3 gang, brown or ivory	L1@1.35	Ea	34.70	55.40	90.10
	4 gang, brown or ivory	L1@1.80	Ea	46.20	73.90	120.10
	5 gang, brown or ivory	L1@2.25	Ea	57.80	92.30	150.10
	6 gang, brown or ivory	L1@2.70	Ea	69.30	111.00	180.30
	Switches, 20 amp 4-way in 2-1/2" deep sectional box and plastic plate					
	1 gang, brown or ivory	L1@0.50	Ea	34.20	20.50	54.70
	2 gang, brown or ivory	L1@1.00	Ea	68.50	41.00	109.50
S_4	3 gang, brown or ivory	L1@1.50	Ea	103.00	61.50	164.50
	4 gang, brown or ivory	L1@2.00	Ea	137.00	82.10	219.10
	5 gang, brown or ivory	L1@2.50	Ea	171.00	103.00	274.00
	6 gang, brown or ivory	L1@3.00	Ea	205.00	123.00	328.00
	Switches, 20 amp key in 2-1/2" deep sectional box and plastic plate					
	1 gang 1-pole	L1@0.40	Ea	26.20	16.40	42.60
	1 gang 2-pole	L1@0.45	Ea	29.70	18.50	48.20
S_K	1 gang 3-way	L1@0.45	Ea	27.90	18.50	46.40
	1 gang 4-way	L1@0.50	Ea	48.50	20.50	69.00

Use these figures to estimate the cost of installing assemblies under the conditions described on pages 5 and 6. Costs listed are for each assembly installed. The crew is one electrician working at a labor cost of $41.03 per manhour. These costs include layout, material handling and normal waste. Add for the sales tax, delivery, supervision, mobilization, demobilization, cleanup, overhead and profit.

2-3/4" Deep 20 Amp Sectional Box Switch Assemblies

Switches, 20 amp 1-pole in 2-3/4" deep sectional box and plastic plate

Material	Craft@Hrs	Unit	Material Cost	Labor Cost	Installed Cost
1 gang, brown or ivory	L1@0.43	Ea	11.90	17.60	29.50
2 gang, brown or ivory	L1@0.86	Ea	23.60	35.30	58.90
3 gang, brown or ivory	L1@1.29	Ea	35.50	52.90	88.40
4 gang, brown or ivory	L1@1.72	Ea	47.30	70.60	117.90
5 gang, brown or ivory	L1@2.15	Ea	59.10	88.20	147.30
6 gang, brown or ivory	L1@2.58	Ea	71.00	106.00	177.00

S

Switches, 20 amp 2-pole in 2-3/4" deep sectional box and plastic plate

Material	Craft@Hrs	Unit	Material Cost	Labor Cost	Installed Cost
1 gang, brown or ivory	L1@0.48	Ea	21.50	19.70	41.20
2 gang, brown or ivory	L1@0.96	Ea	43.20	39.40	82.60
3 gang, brown or ivory	L1@1.44	Ea	64.70	59.10	123.80
4 gang, brown or ivory	L1@1.92	Ea	86.40	78.80	165.20
5 gang, brown or ivory	L1@2.40	Ea	108.00	98.50	206.50
6 gang, brown or ivory	L1@2.88	Ea	129.00	118.00	247.00

S_2

Switches, 20 amp 3-way in 2-3/4" deep sectional box and plastic plate

Material	Craft@Hrs	Unit	Material Cost	Labor Cost	Installed Cost
1 gang, brown or ivory	L1@0.48	Ea	13.00	19.70	32.70
2 gang, brown or ivory	L1@0.96	Ea	26.00	39.40	65.40
3 gang, brown or ivory	L1@1.44	Ea	38.90	59.10	98.00
4 gang, brown or ivory	L1@1.92	Ea	51.90	78.80	130.70
5 gang, brown or ivory	L1@2.40	Ea	64.90	98.50	163.40
6 gang, brown or ivory	L1@2.88	Ea	78.00	118.00	196.00

S_3

Switches, 20 amp 4-way in 2-3/4" deep sectional box and plastic plate

Material	Craft@Hrs	Unit	Material Cost	Labor Cost	Installed Cost
1 gang, brown or ivory	L1@0.53	Ea	35.70	21.70	57.40
2 gang, brown or ivory	L1@1.06	Ea	71.30	43.50	114.80
3 gang, brown or ivory	L1@1.59	Ea	107.00	65.20	172.20
4 gang, brown or ivory	L1@2.12	Ea	143.00	87.00	230.00
5 gang, brown or ivory	L1@2.65	Ea	178.00	109.00	287.00
6 gang, brown or ivory	L1@3.18	Ea	214.00	130.00	344.00

S_4

Switches, 20 amp key in 2-3/4" deep sectional box and plastic plate

Material	Craft@Hrs	Unit	Material Cost	Labor Cost	Installed Cost
1 gang 1-pole	L1@0.43	Ea	27.70	17.60	45.30
1 gang 2-pole	L1@0.48	Ea	31.10	19.70	50.80
1 gang 3-way	L1@0.48	Ea	29.40	19.70	49.10
1 gang 4-way	L1@0.53	Ea	50.00	21.70	71.70

S_K

Use these figures to estimate the cost of installing assemblies under the conditions described on pages 5 and 6. Costs listed are for each assembly installed. The crew is one electrician working at a labor cost of $41.03 per manhour. These costs include layout, material handling and normal waste. Add for the sales tax, delivery, supervision, mobilization, demobilization, cleanup, overhead and profit.

3-1/2" Deep 20 Amp Sectional Box Switch Assemblies

Switches, 20 amp 1-pole in 3-1/2" deep sectional box and plastic plate

S

Material	Craft@Hrs	Unit	Material Cost	Labor Cost	Installed Cost
1 gang, brown or ivory	L1@0.48	Ea	12.10	19.70	31.80
2 gang, brown or ivory	L1@0.96	Ea	24.10	39.40	63.50
3 gang, brown or ivory	L1@1.44	Ea	36.20	59.10	95.30
4 gang, brown or ivory	L1@1.92	Ea	48.30	78.80	127.10
5 gang, brown or ivory	L1@2.40	Ea	60.50	98.50	159.00
6 gang, brown or ivory	L1@2.88	Ea	72.50	118.00	190.50

Switches, 20 amp 2-pole in 3-1/2" deep sectional box and plastic plate

S_2

Material	Craft@Hrs	Unit	Material Cost	Labor Cost	Installed Cost
1 gang, brown or ivory	L1@0.53	Ea	21.50	21.70	43.20
2 gang, brown or ivory	L1@1.06	Ea	43.20	43.50	86.70
3 gang, brown or ivory	L1@1.59	Ea	64.70	65.20	129.90
4 gang, brown or ivory	L1@2.12	Ea	86.40	87.00	173.40
5 gang, brown or ivory	L1@2.65	Ea	108.00	109.00	217.00
6 gang, brown or ivory	L1@3.18	Ea	129.00	130.00	259.00

Switches, 20 amp 3-way in 3-1/2" deep sectional box and plastic plate

S_3

Material	Craft@Hrs	Unit	Material Cost	Labor Cost	Installed Cost
1 gang, brown or ivory	L1@0.53	Ea	13.00	21.70	34.70
2 gang, brown or ivory	L1@1.06	Ea	26.00	43.50	69.50
3 gang, brown or ivory	L1@1.59	Ea	38.90	65.20	104.10
4 gang, brown or ivory	L1@2.12	Ea	51.90	87.00	138.90
5 gang, brown or ivory	L1@2.65	Ea	64.90	109.00	173.90
6 gang, brown or ivory	L1@3.18	Ea	78.00	130.00	208.00

Switches, 20 amp 4-way in 3-1/2" deep sectional box and plastic plate

S_4

Material	Craft@Hrs	Unit	Material Cost	Labor Cost	Installed Cost
1 gang, brown or ivory	L1@0.58	Ea	35.70	23.80	59.50
2 gang, brown or ivory	L1@1.16	Ea	71.30	47.60	118.90
3 gang, brown or ivory	L1@1.74	Ea	107.00	71.40	178.40
4 gang, brown or ivory	L1@2.32	Ea	143.00	95.20	238.20
5 gang, brown or ivory	L1@2.90	Ea	178.00	119.00	297.00
6 gang, brown or ivory	L1@3.48	Ea	214.00	143.00	357.00

Switches, 20 amp key in 3-1/2" deep sectional box and plastic plate

S_K

Material	Craft@Hrs	Unit	Material Cost	Labor Cost	Installed Cost
1 gang 1-pole	L1@0.48	Ea	27.70	19.70	47.40
1 gang 2-pole	L1@0.53	Ea	31.10	21.70	52.80
1 gang 3-way	L1@0.53	Ea	29.40	21.70	51.10
1 gang 4-way	L1@0.58	Ea	50.00	23.80	73.80

Use these figures to estimate the cost of installing assemblies under the conditions described on pages 5 and 6. Costs listed are for each assembly installed. The crew is one electrician working at a labor cost of $41.03 per manhour. These costs include layout, material handling and normal waste. Add for the sales tax, delivery, supervision, mobilization, demobilization, cleanup, overhead and profit.

2" Deep 20 Amp Sectional Box Switch Assemblies

Material	Craft@Hrs	Unit	Material Cost	Labor Cost	Installed Cost	
Switches, 20 amp 1-pole in 2" deep sectional box and plastic plate						
1 gang, white	L1@0.38	Ea	11.20	15.60	26.80	
2 gang, white	L1@0.76	Ea	22.40	31.20	53.60	
3 gang, white	L1@1.14	Ea	33.60	46.80	80.40	S
4 gang, white	L1@1.52	Ea	44.80	62.40	107.20	
5 gang, white	L1@1.90	Ea	56.00	78.00	134.00	
6 gang, white	L1@2.28	Ea	67.20	93.50	160.70	
Switches, 20 amp 2-pole in 2" deep sectional box and plastic plate						
1 gang, white	L1@0.43	Ea	20.90	17.60	38.50	
2 gang, white	L1@0.86	Ea	41.90	35.30	77.20	
3 gang, white	L1@1.29	Ea	62.90	52.90	115.80	S_2
4 gang, white	L1@1.72	Ea	83.90	70.60	154.50	
5 gang, white	L1@2.15	Ea	105.00	88.20	193.20	
6 gang, white	L1@2.58	Ea	126.00	106.00	232.00	
Switches, 20 amp 3-way in 2" deep sectional box and plastic plate						
1 gang, white	L1@0.43	Ea	12.40	17.60	30.00	
2 gang, white	L1@0.86	Ea	24.80	35.30	60.10	
3 gang, white	L1@1.29	Ea	37.10	52.90	90.00	S_3
4 gang, white	L1@1.72	Ea	49.40	70.60	120.00	
5 gang, white	L1@2.15	Ea	61.80	88.20	150.00	
6 gang, white	L1@2.58	Ea	74.20	106.00	180.20	
Switches, 20 amp 4-way in 2" deep sectional box and plastic plate						
1 gang, white	L1@0.48	Ea	35.10	19.70	54.80	
2 gang, white	L1@0.96	Ea	70.00	39.40	109.40	
3 gang, white	L1@1.44	Ea	105.00	59.10	164.10	S_4
4 gang, white	L1@1.92	Ea	141.00	78.80	219.80	
5 gang, white	L1@2.40	Ea	175.00	98.50	273.50	
6 gang, white	L1@2.88	Ea	210.00	118.00	328.00	

Use these figures to estimate the cost of installing assemblies under the conditions described on pages 5 and 6. Costs listed are for each assembly installed. The crew is one electrician working at a labor cost of $41.03 per manhour. These costs include layout, material handling and normal waste. Add for the sales tax, delivery, supervision, mobilization, demobilization, cleanup, overhead and profit.

2-1/2" Deep 20 Amp Sectional Box Switch Assemblies

Switches, 20 amp 1-pole in 2-1/2" deep sectional box and plastic plate

S

Material	Craft@Hrs	Unit	Material Cost	Labor Cost	Installed Cost
1 gang, white	L1@0.40	Ea	10.40	16.40	26.80
2 gang, white	L1@0.80	Ea	20.80	32.80	53.60
3 gang, white	L1@1.20	Ea	31.10	49.20	80.30
4 gang, white	L1@1.60	Ea	41.50	65.60	107.10
5 gang, white	L1@2.00	Ea	51.90	82.10	134.00
6 gang, white	L1@2.40	Ea	62.30	98.50	160.80

Switches, 20 amp 2-pole in 2-1/2" deep sectional box and plastic plate

S_2

Material	Craft@Hrs	Unit	Material Cost	Labor Cost	Installed Cost
1 gang, white	L1@0.45	Ea	20.20	18.50	38.70
2 gang, white	L1@0.90	Ea	40.30	36.90	77.20
3 gang, white	L1@1.35	Ea	60.50	55.40	115.90
4 gang, white	L1@1.80	Ea	80.60	73.90	154.50
5 gang, white	L1@2.25	Ea	101.00	92.30	193.30
6 gang, white	L1@2.70	Ea	121.00	111.00	232.00

Switches, 20 amp 3-way in 2-1/2" deep sectional box and plastic plate

S_3

Material	Craft@Hrs	Unit	Material Cost	Labor Cost	Installed Cost
1 gang, white	L1@0.45	Ea	11.60	18.50	30.10
2 gang, white	L1@0.90	Ea	23.10	36.90	60.00
3 gang, white	L1@1.35	Ea	34.70	55.40	90.10
4 gang, white	L1@1.80	Ea	46.20	73.90	120.10
5 gang, white	L1@2.25	Ea	57.80	92.30	150.10
6 gang, white	L1@2.70	Ea	69.30	111.00	180.30

Switches, 20 amp 4-way in 2-1/2" deep sectional box and plastic plate

S_4

Material	Craft@Hrs	Unit	Material Cost	Labor Cost	Installed Cost
1 gang, white	L1@0.50	Ea	34.20	20.50	54.70
2 gang, white	L1@1.00	Ea	68.50	41.00	109.50
3 gang, white	L1@1.50	Ea	103.00	61.50	164.50
4 gang, white	L1@2.00	Ea	137.00	82.10	219.10
5 gang, white	L1@2.50	Ea	171.00	103.00	274.00
6 gang, white	L1@3.00	Ea	205.00	123.00	328.00

Use these figures to estimate the cost of installing assemblies under the conditions described on pages 5 and 6. Costs listed are for each assembly installed. The crew is one electrician working at a labor cost of $41.03 per manhour. These costs include layout, material handling and normal waste. Add for the sales tax, delivery, supervision, mobilization, demobilization, cleanup, overhead and profit.

2-3/4" Deep 20 Amp Sectional Box Switch Assemblies

Switches, 20 amp 1-pole in 2-3/4" deep sectional box and plastic plate

Material	Craft@Hrs	Unit	Material Cost	Labor Cost	Installed Cost
1 gang, white	L1@0.43	Ea	11.90	17.60	29.50
2 gang, white	L1@0.86	Ea	23.60	35.30	58.90
3 gang, white	L1@1.29	Ea	35.50	52.90	88.40
4 gang, white	L1@1.72	Ea	47.30	70.60	117.90
5 gang, white	L1@2.15	Ea	59.10	88.20	147.30
6 gang, white	L1@2.58	Ea	71.00	106.00	177.00

S

Switches, 20 amp 2-pole in 2-3/4" deep sectional box and plastic plate

Material	Craft@Hrs	Unit	Material Cost	Labor Cost	Installed Cost
1 gang, white	L1@0.48	Ea	21.50	19.70	41.20
2 gang, white	L1@0.96	Ea	43.20	39.40	82.60
3 gang, white	L1@1.44	Ea	64.70	59.10	123.80
4 gang, white	L1@1.92	Ea	86.40	78.80	165.20
5 gang, white	L1@2.40	Ea	108.00	98.50	206.50
6 gang, white	L1@2.88	Ea	129.00	118.00	247.00

S_2

Switches, 20 amp 3-way in 2-3/4" deep sectional box and plastic plate

Material	Craft@Hrs	Unit	Material Cost	Labor Cost	Installed Cost
1 gang, white	L1@0.48	Ea	13.00	19.70	32.70
2 gang, white	L1@0.96	Ea	26.00	39.40	65.40
3 gang, white	L1@1.44	Ea	38.90	59.10	98.00
4 gang, white	L1@1.92	Ea	51.90	78.80	130.70
5 gang, white	L1@2.40	Ea	64.90	98.50	163.40
6 gang, white	L1@2.88	Ea	78.00	118.00	196.00

S_3

Switches, 20 amp 4-way in 2-3/4" deep sectional box and plastic plate

Material	Craft@Hrs	Unit	Material Cost	Labor Cost	Installed Cost
1 gang, white	L1@0.53	Ea	35.70	21.70	57.40
2 gang, white	L1@1.06	Ea	71.30	43.50	114.80
3 gang, white	L1@1.59	Ea	107.00	65.20	172.20
4 gang, white	L1@2.12	Ea	143.00	87.00	230.00
5 gang, white	L1@2.65	Ea	178.00	109.00	287.00
6 gang, white	L1@3.18	Ea	214.00	130.00	344.00

S_4

Use these figures to estimate the cost of installing assemblies under the conditions described on pages 5 and 6. Costs listed are for each assembly installed. The crew is one electrician working at a labor cost of $41.03 per manhour. These costs include layout, material handling and normal waste. Add for the sales tax, delivery, supervision, mobilization, demobilization, cleanup, overhead and profit.

3-1/2" Deep 20 Amp Sectional Box Switch Assemblies

Switches, 20 amp 1-pole in 3-1/2" deep sectional box and plastic plate

S

Material	Craft@Hrs	Unit	Material Cost	Labor Cost	Installed Cost
1 gang, white	L1@0.48	Ea	12.10	19.70	31.80
2 gang, white	L1@0.96	Ea	24.10	39.40	63.50
3 gang, white	L1@1.44	Ea	36.20	59.10	95.30
4 gang, white	L1@1.92	Ea	48.30	78.80	127.10
5 gang, white	L1@2.40	Ea	60.50	98.50	159.00
6 gang, white	L1@2.88	Ea	72.50	118.00	190.50

Switches, 20 amp 2-pole in 3-1/2" deep sectional box and plastic plate

S_2

Material	Craft@Hrs	Unit	Material Cost	Labor Cost	Installed Cost
1 gang, white	L1@0.53	Ea	21.50	21.70	43.20
2 gang, white	L1@1.06	Ea	43.20	43.50	86.70
3 gang, white	L1@1.59	Ea	64.70	65.20	129.90
4 gang, white	L1@2.12	Ea	86.40	87.00	173.40
5 gang, white	L1@2.65	Ea	108.00	109.00	217.00
6 gang, white	L1@3.18	Ea	129.00	130.00	259.00

Switches, 20 amp 3-way in 3-1/2" deep sectional box and plastic plate

S_3

Material	Craft@Hrs	Unit	Material Cost	Labor Cost	Installed Cost
1 gang, white	L1@0.53	Ea	13.00	21.70	34.70
2 gang, white	L1@1.06	Ea	26.00	43.50	69.50
3 gang, white	L1@1.59	Ea	38.90	65.20	104.10
4 gang, white	L1@2.12	Ea	51.90	87.00	138.90
5 gang, white	L1@2.65	Ea	64.90	109.00	173.90
6 gang, white	L1@3.18	Ea	78.00	130.00	208.00

Switches, 20 amp 4-way in 3-1/2" deep sectional box and plastic plate

S_4

Material	Craft@Hrs	Unit	Material Cost	Labor Cost	Installed Cost
1 gang, white	L1@0.58	Ea	35.70	23.80	59.50
2 gang, white	L1@1.16	Ea	71.30	47.60	118.90
3 gang, white	L1@1.74	Ea	107.00	71.40	178.40
4 gang, white	L1@2.32	Ea	143.00	95.20	238.20
5 gang, white	L1@2.90	Ea	178.00	119.00	297.00
6 gang, white	L1@3.48	Ea	214.00	143.00	357.00

Use these figures to estimate the cost of installing assemblies under the conditions described on pages 5 and 6. Costs listed are for each assembly installed. The crew is one electrician working at a labor cost of $41.03 per manhour. These costs include layout, material handling and normal waste. Add for the sales tax, delivery, supervision, mobilization, demobilization, cleanup, overhead and profit.

4/S x 1-1/4" Deep, 15 Amp, 1 Gang Switch Assemblies

Material	Craft@Hrs	Unit	Material Cost	Labor Cost	Installed Cost	
1 gang brown switches, 15 amp, 1-pole, in 4/S x 1-1/4" deep box, ring and plastic plate						
4-S-ring flat	L1@0.53	Ea	13.10	21.70	34.80	
4-S-ring 1/4"	L1@0.53	Ea	13.50	21.70	35.20	
4-S-ring 1/2"	L1@0.53	Ea	12.30	21.70	34.00	
4-S-ring 5/8"	L1@0.53	Ea	11.00	21.70	32.70	
4-S-ring 3/4"	L1@0.53	Ea	11.30	21.70	33.00	
4-S-ring 1"	L1@0.53	Ea	11.90	21.70	33.60	
4-S-ring 1-1/4"	L1@0.53	Ea	13.20	21.70	34.90	S
4-S-tile ring 1/2"	L1@0.53	Ea	12.60	21.70	34.30	
4-S-tile ring 3/4"	L1@0.53	Ea	12.70	21.70	34.40	
4-S-tile ring 1"	L1@0.53	Ea	12.80	21.70	34.50	
4-S-tile ring 1-1/4"	L1@0.54	Ea	13.20	22.20	35.40	
4-S-tile ring 1-1/2"	L1@0.54	Ea	15.60	22.20	37.80	
4-S-tile ring 2"	L1@0.54	Ea	15.90	22.20	38.10	
1 gang brown switches, 15 amp, 2-pole, in 4/S x 1-1/4" deep box, ring and plastic plate						
4-S-ring flat	L1@0.58	Ea	21.00	23.80	44.80	
4-S-ring 1/4"	L1@0.58	Ea	21.40	23.80	45.20	
4-S-ring 1/2"	L1@0.58	Ea	20.10	23.80	43.90	
4-S-ring 5/8"	L1@0.58	Ea	18.90	23.80	42.70	
4-S-ring 3/4"	L1@0.58	Ea	19.30	23.80	43.10	
4-S-ring 1"	L1@0.58	Ea	19.80	23.80	43.60	
4-S-ring 1-1/4"	L1@0.58	Ea	21.00	23.80	44.80	S_2
4-S-tile ring 1/2"	L1@0.58	Ea	20.40	23.80	44.20	
4-S-tile ring 3/4"	L1@0.58	Ea	20.50	23.80	44.30	
4-S-tile ring 1"	L1@0.58	Ea	20.70	23.80	44.50	
4-S-tile ring 1-1/4"	L1@0.59	Ea	21.00	24.20	45.20	
4-S-tile ring 1-1/2"	L1@0.59	Ea	23.50	24.20	47.70	
4-S-tile ring 2"	L1@0.59	Ea	23.70	24.20	47.90	
1 gang brown switches, 15 amp, 3-way, in 4/S x 1-1/4" deep box, ring and plastic plate						
4-S-ring flat	L1@0.58	Ea	14.70	23.80	38.50	
4-S-ring 1/4"	L1@0.58	Ea	15.10	23.80	38.90	
4-S-ring 1/2"	L1@0.58	Ea	13.80	23.80	37.60	
4-S-ring 5/8"	L1@0.58	Ea	12.70	23.80	36.50	
4-S-ring 3/4"	L1@0.58	Ea	12.90	23.80	36.70	
4-S-ring 1"	L1@0.58	Ea	13.40	23.80	37.20	
4-S-ring 1-1/4"	L1@0.58	Ea	14.80	23.80	38.60	S_3
4-S-tile ring 1/2"	L1@0.58	Ea	14.20	23.80	38.00	
4-S-tile ring 3/4"	L1@0.58	Ea	14.30	23.80	38.10	
4-S-tile ring 1"	L1@0.58	Ea	14.40	23.80	38.20	
4-S-tile ring 1-1/4"	L1@0.59	Ea	14.80	24.20	39.00	
4-S-tile ring 1-1/2"	L1@0.59	Ea	17.20	24.20	41.40	
4-S-tile ring 2"	L1@0.59	Ea	17.50	24.20	41.70	

Use these figures to estimate the cost of installing assemblies under the conditions described on pages 5 and 6. Costs listed are for each assembly installed. The crew is one electrician working at a labor cost of $41.03 per manhour. These costs include layout, material handling and normal waste. Add for the sales tax, delivery, supervision, mobilization, demobilization, cleanup, overhead and profit.

4/S x 1-1/4" Deep, 15 Amp, 1 Gang Switch Assemblies

1 gang brown switches, 15 amp, 4-way, in 4/S x 1-1/4" deep box, ring and plastic plate

S_4

Material	Craft@Hrs	Unit	Material Cost	Labor Cost	Installed Cost
4-S-ring flat	L1@0.63	Ea	27.20	25.80	53.00
4-S-ring 1/4"	L1@0.63	Ea	27.60	25.80	53.40
4-S-ring 1/2"	L1@0.63	Ea	26.30	25.80	52.10
4-S-ring 5/8"	L1@0.63	Ea	25.10	25.80	50.90
4-S-ring 3/4"	L1@0.63	Ea	25.40	25.80	51.20
4-S-ring 1"	L1@0.63	Ea	25.90	25.80	51.70
4-S-ring 1-1/4"	L1@0.63	Ea	27.30	25.80	53.10
4-S-tile ring 1/2"	L1@0.63	Ea	26.60	25.80	52.40
4-S-tile ring 3/4"	L1@0.63	Ea	26.70	25.80	52.50
4-S-tile ring 1"	L1@0.63	Ea	26.90	25.80	52.70
4-S-tile ring 1-1/4"	L1@0.64	Ea	27.30	26.30	53.60
4-S-tile ring 1-1/2"	L1@0.64	Ea	29.70	26.30	56.00
4-S-tile ring 2"	L1@0.64	Ea	30.00	26.30	56.30

1 gang key switches, 15 amp, 1-pole, in 4/S x 1-1/4" deep box, ring and plastic plate

S_K

Material	Craft@Hrs	Unit	Material Cost	Labor Cost	Installed Cost
4-S-ring flat	L1@0.53	Ea	22.50	21.70	44.20
4-S-ring 1/4"	L1@0.53	Ea	22.90	21.70	44.60
4-S-ring 1/2"	L1@0.53	Ea	21.60	21.70	43.30
4-S-ring 5/8"	L1@0.53	Ea	20.50	21.70	42.20
4-S-ring 3/4"	L1@0.53	Ea	20.70	21.70	42.40
4-S-ring 1"	L1@0.53	Ea	21.20	21.70	42.90
4-S-ring 1-1/4"	L1@0.53	Ea	22.60	21.70	44.30
4-S-tile ring 1/2"	L1@0.53	Ea	22.00	21.70	43.70
4-S-tile ring 3/4"	L1@0.53	Ea	22.10	21.70	43.80
4-S-tile ring 1"	L1@0.53	Ea	22.30	21.70	44.00
4-S-tile ring 1-1/4"	L1@0.54	Ea	22.60	22.20	44.80
4-S-tile ring 1-1/2"	L1@0.54	Ea	25.00	22.20	47.20
4-S-tile ring 2"	L1@0.54	Ea	25.30	22.20	47.50

1 gang key switches, 15 amp, 2-pole, in 4/S x 1-1/4" deep box, ring and plastic plate

S_{K2}

Material	Craft@Hrs	Unit	Material Cost	Labor Cost	Installed Cost
4-S-ring flat	L1@0.58	Ea	28.60	23.80	52.40
4-S-ring 1/4"	L1@0.58	Ea	29.00	23.80	52.80
4-S-ring 1/2"	L1@0.58	Ea	27.70	23.80	51.50
4-S-ring 5/8"	L1@0.58	Ea	26.50	23.80	50.30
4-S-ring 3/4"	L1@0.58	Ea	26.90	23.80	50.70
4-S-ring 1"	L1@0.58	Ea	27.70	23.80	51.50
4-S-ring 1-1/4"	L1@0.58	Ea	28.60	23.80	52.40
4-S-tile ring 1/2"	L1@0.58	Ea	28.00	23.80	51.80
4-S-tile ring 3/4"	L1@0.58	Ea	28.10	23.80	51.90
4-S-tile ring 1"	L1@0.58	Ea	28.30	23.80	52.10
4-S-tile ring 1-1/4"	L1@0.59	Ea	28.60	24.20	52.80
4-S-tile ring 1-1/2"	L1@0.59	Ea	31.10	24.20	55.30
4-S-tile ring 2"	L1@0.59	Ea	31.30	24.20	55.50

Use these figures to estimate the cost of installing assemblies under the conditions described on pages 5 and 6. Costs listed are for each assembly installed. The crew is one electrician working at a labor cost of $41.03 per manhour. These costs include layout, material handling and normal waste. Add for the sales tax, delivery, supervision, mobilization, demobilization, cleanup, overhead and profit.

4/S x 1-1/4" Deep, 15 Amp, 1 and 2 Gang Switch Assemblies

1 gang key switches, 15 amp, 3-way, in 4/S x 1-1/4" deep box, ring and plastic plate

Material	Craft@Hrs	Unit	Material Cost	Labor Cost	Installed Cost	
4-S-ring flat	L1@0.58	Ea	26.90	23.80	50.70	
4-S-ring 1/4"	L1@0.58	Ea	27.20	23.80	51.00	
4-S-ring 1/2"	L1@0.58	Ea	25.90	23.80	49.70	
4-S-ring 5/8"	L1@0.58	Ea	24.80	23.80	48.60	
4-S-ring 3/4"	L1@0.58	Ea	25.00	23.80	48.80	
4-S-ring 1"	L1@0.58	Ea	25.50	23.80	49.30	
4-S-ring 1-1/4"	L1@0.58	Ea	26.90	23.80	50.70	S_{K3}
4-S-tile ring 1/2"	L1@0.58	Ea	26.20	23.80	50.00	
4-S-tile ring 3/4"	L1@0.58	Ea	26.30	23.80	50.10	
4-S-tile ring 1"	L1@0.58	Ea	26.50	23.80	50.30	
4-S-tile ring 1-1/4"	L1@0.59	Ea	26.90	24.20	51.10	
4-S-tile ring 1-1/2"	L1@0.59	Ea	29.20	24.20	53.40	
4-S-tile ring 2"	L1@0.59	Ea	29.60	24.20	53.80	

1 gang key switches, 15 amp, 4-way, in 4/S x 1-1/4" deep box, ring and plastic plate

Material	Craft@Hrs	Unit	Material Cost	Labor Cost	Installed Cost	
4-S-ring flat	L1@0.63	Ea	44.50	25.80	70.30	
4-S-ring 1/4	L1@0.63	Ea	45.00	25.80	70.80	
4-S-ring 1/2"	L1@0.63	Ea	43.70	25.80	69.50	
4-S-ring 5/8"	L1@0.63	Ea	42.50	25.80	68.30	
4-S-ring 3/4"	L1@0.63	Ea	42.80	25.80	68.60	
4-S-ring 1"	L1@0.63	Ea	43.30	25.80	69.10	
4-S-ring 1-1/4"	L1@0.63	Ea	44.70	25.80	70.50	S_{K4}
4-S-tile ring 1/2"	L1@0.63	Ea	44.00	25.80	69.80	
4-S-tile ring 3/4"	L1@0.63	Ea	44.10	25.80	69.90	
4-S-tile ring 1"	L1@0.63	Ea	44.20	25.80	70.00	
4-S-tile ring 1-1/4"	L1@0.64	Ea	44.70	26.30	71.00	
4-S-tile ring 1-1/2"	L1@0.64	Ea	47.00	26.30	73.30	
4-S-tile ring 2"	L1@0.64	Ea	47.40	26.30	73.70	

2 gang brown switches, 15 amp, 1-pole, 4/S x 1-1/4" deep box, ring and plastic plate

Material	Craft@Hrs	Unit	Material Cost	Labor Cost	Installed Cost	
4-S-ring flat	L1@0.77	Ea	18.90	31.60	50.50	
4-S-ring 1/4	L1@0.77	Ea	20.00	31.60	51.60	
4-S-ring 1/2"	L1@0.77	Ea	16.70	31.60	48.30	
4-S-ring 5/8"	L1@0.77	Ea	17.00	31.60	48.60	
4-S-ring 3/4"	L1@0.77	Ea	17.40	31.60	49.00	
4-S-ring 1"	L1@0.77	Ea	19.40	31.60	51.00	
4-S-ring 1-1/4"	L1@0.77	Ea	19.90	31.60	51.50	SS
4-S-tile ring 1/2"	L1@0.77	Ea	16.90	31.60	48.50	
4-S-tile ring 3/4"	L1@0.77	Ea	17.90	31.60	49.50	
4-S-tile ring 1"	L1@0.77	Ea	19.40	31.60	51.00	
4-S-tile ring 1-1/4"	L1@0.79	Ea	19.90	32.40	52.30	
4-S-tile ring 1-1/2"	L1@0.79	Ea	21.50	32.40	53.90	
4-S-tile ring 2"	L1@0.79	Ea	23.50	32.40	55.90	

Use these figures to estimate the cost of installing assemblies under the conditions described on pages 5 and 6. Costs listed are for each assembly installed. The crew is one electrician working at a labor cost of $41.03 per manhour. These costs include layout, material handling and normal waste. Add for the sales tax, delivery, supervision, mobilization, demobilization, cleanup, overhead and profit.

4/S x 1-1/4" Deep, 15 Amp, 2 Gang Switch Assemblies

2 gang brown switches, 15 amp, 2-pole, in 4/S x 1-1/4" deep box, ring and plastic plate

S_2S_2

Material	Craft@Hrs	Unit	Material Cost	Labor Cost	Installed Cost
4-S-ring flat	L1@0.87	Ea	34.70	35.70	70.40
4-S-ring 1/4"	L1@0.87	Ea	35.80	35.70	71.50
4-S-ring 1/2"	L1@0.87	Ea	32.40	35.70	68.10
4-S-ring 5/8"	L1@0.87	Ea	32.70	35.70	68.40
4-S-ring 3/4"	L1@0.87	Ea	33.10	35.70	68.80
4-S-ring 1"	L1@0.87	Ea	35.10	35.70	70.80
4-S-ring 1-1/4"	L1@0.87	Ea	35.60	35.70	71.30
4-S-tile ring 1/2"	L1@0.87	Ea	32.60	35.70	68.30
4-S-tile ring 3/4"	L1@0.87	Ea	33.60	35.70	69.30
4-S-tile ring 1"	L1@0.87	Ea	35.10	35.70	70.80
4-S-tile ring 1-1/4"	L1@0.89	Ea	35.60	36.50	72.10
4-S-tile ring 1-1/2"	L1@0.89	Ea	37.30	36.50	73.80
4-S-tile ring 2"	L1@0.89	Ea	34.00	36.50	70.50

2 gang brown switches, 15 amp, 3-way, in 4/S x 1-1/4" deep box, ring and plastic plate

S_3S_3

4-S-ring flat	L1@0.87	Ea	22.10	35.70	57.80
4-S-ring 1/4"	L1@0.87	Ea	23.10	35.70	58.80
4-S-ring 1/2"	L1@0.87	Ea	19.80	35.70	55.50
4-S-ring 5/8"	L1@0.87	Ea	20.20	35.70	55.90
4-S-ring 3/4"	L1@0.87	Ea	20.50	35.70	56.20
4-S-ring 1"	L1@0.87	Ea	22.50	35.70	58.20
4-S-ring 1-1/4"	L1@0.87	Ea	23.10	35.70	58.80
4-S-tile ring 1/2"	L1@0.87	Ea	20.00	35.70	55.70
4-S-tile ring 3/4"	L1@0.87	Ea	21.00	35.70	56.70
4-S-tile ring 1"	L1@0.87	Ea	22.50	35.70	58.20
4-S-tile ring 1-1/4"	L1@0.89	Ea	23.00	36.50	59.50
4-S-tile ring 1-1/2"	L1@0.89	Ea	24.70	36.50	61.20
4-S-tile ring 2"	L1@0.89	Ea	26.60	36.50	63.10

2 gang brown switches, 15 amp, 4-way, in 4/S x 1-1/4" deep box, ring and plastic plate

S_4S_4

4-S-ring flat	L1@0.97	Ea	47.00	39.80	86.80
4-S-ring 1/4"	L1@0.97	Ea	48.10	39.80	87.90
4-S-ring 1/2"	L1@0.97	Ea	44.70	39.80	84.50
4-S-ring 5/8"	L1@0.97	Ea	45.10	39.80	84.90
4-S-ring 3/4"	L1@0.97	Ea	45.40	39.80	85.20
4-S-ring 1"	L1@0.97	Ea	47.50	39.80	87.30
4-S-ring 1-1/4"	L1@0.97	Ea	48.00	39.80	87.80
4-S-tile ring 1/2"	L1@0.97	Ea	45.00	39.80	84.80
4-S-tile ring 3/4"	L1@0.97	Ea	46.00	39.80	85.80
4-S-tile ring 1"	L1@0.97	Ea	47.50	39.80	87.30
4-S-tile ring 1-1/4"	L1@0.99	Ea	48.00	40.60	88.60
4-S-tile ring 1-1/2"	L1@0.99	Ea	49.60	40.60	90.20
4-S-tile ring 2"	L1@0.99	Ea	51.60	40.60	92.20

Use these figures to estimate the cost of installing assemblies under the conditions described on pages 5 and 6. Costs listed are for each assembly installed. The crew is one electrician working at a labor cost of $41.03 per manhour. These costs include layout, material handling and normal waste. Add for the sales tax, delivery, supervision, mobilization, demobilization, cleanup, overhead and profit.

4/S x 1-1/4" Deep, 15 Amp, 1 Gang Switch Assemblies

1 gang ivory switches, 15 amp, 1-pole, in 4/S x 1-1/4" deep box, ring and plastic plate

Material	Craft@Hrs	Unit	Material Cost	Labor Cost	Installed Cost
4-S-ring flat	L1@0.53	Ea	13.10	21.70	34.80
4-S-ring 1/4"	L1@0.53	Ea	13.50	21.70	35.20
4-S-ring 1/2"	L1@0.53	Ea	12.30	21.70	34.00
4-S-ring 5/8"	L1@0.53	Ea	11.00	21.70	32.70
4-S-ring 3/4"	L1@0.53	Ea	11.30	21.70	33.00
4-S-ring 1"	L1@0.53	Ea	11.90	21.70	33.60
4-S-ring 1-1/4"	L1@0.53	Ea	13.20	21.70	34.90
4-S-tile ring 1/2"	L1@0.53	Ea	12.60	21.70	34.30
4-S-tile ring 3/4"	L1@0.53	Ea	12.70	21.70	34.40
4-S-tile ring 1"	L1@0.53	Ea	12.80	21.70	34.50
4-S-tile ring 1-1/4"	L1@0.54	Ea	13.20	22.20	35.40
4-S-tile ring 1-1/2"	L1@0.54	Ea	15.60	22.20	37.80
4-S-tile ring 2"	L1@0.54	Ea	15.90	22.20	38.10

S

1 gang ivory switches, 15 amp, 2-pole, in 4/S x 1-1/4" deep box, ring and plastic plate

Material	Craft@Hrs	Unit	Material Cost	Labor Cost	Installed Cost
4-S-ring flat	L1@0.58	Ea	21.00	23.80	44.80
4-S-ring 1/4"	L1@0.58	Ea	21.40	23.80	45.20
4-S-ring 1/2"	L1@0.58	Ea	20.10	23.80	43.90
4-S-ring 5/8"	L1@0.58	Ea	18.90	23.80	42.70
4-S-ring 3/4"	L1@0.58	Ea	19.30	23.80	43.10
4-S-ring 1"	L1@0.58	Ea	19.80	23.80	43.60
4-S-ring 1-1/4"	L1@0.58	Ea	21.00	23.80	44.80
4-S-tile ring 1/2"	L1@0.58	Ea	20.40	23.80	44.20
4-S-tile ring 3/4"	L1@0.58	Ea	20.50	23.80	44.30
4-S-tile ring 1"	L1@0.58	Ea	20.70	23.80	44.50
4-S-tile ring 1-1/4"	L1@0.59	Ea	21.00	24.20	45.20
4-S-tile ring 1-1/2"	L1@0.59	Ea	23.50	24.20	47.70
4-S-tile ring 2"	L1@0.59	Ea	23.70	24.20	47.90

S_2

1 gang ivory switches, 15 amp, 3-way, in 4/S x 1-1/4" deep box, ring and plastic plate

Material	Craft@Hrs	Unit	Material Cost	Labor Cost	Installed Cost
4-S-ring flat	L1@0.58	Ea	14.70	23.80	38.50
4-S-ring 1/4"	L1@0.58	Ea	15.10	23.80	38.90
4-S-ring 1/2"	L1@0.58	Ea	13.80	23.80	37.60
4-S-ring 5/8"	L1@0.58	Ea	12.70	23.80	36.50
4-S-ring 3/4"	L1@0.58	Ea	12.90	23.80	36.70
4-S-ring 1"	L1@0.58	Ea	13.40	23.80	37.20
4-S-ring 1-1/4"	L1@0.58	Ea	14.80	23.80	38.60
4-S-tile ring 1/2"	L1@0.58	Ea	14.20	23.80	38.00
4-S-tile ring 3/4"	L1@0.58	Ea	14.30	23.80	38.10
4-S-tile ring 1"	L1@0.58	Ea	14.40	23.80	38.20
4-S-tile ring 1-1/4"	L1@0.59	Ea	14.80	24.20	39.00
4-S-tile ring 1-1/2"	L1@0.59	Ea	17.20	24.20	41.40
4-S-tile ring 2"	L1@0.59	Ea	17.50	24.20	41.70

S_3

Use these figures to estimate the cost of installing assemblies under the conditions described on pages 5 and 6. Costs listed are for each assembly installed. The crew is one electrician working at a labor cost of $41.03 per manhour. These costs include layout, material handling and normal waste. Add for the sales tax, delivery, supervision, mobilization, demobilization, cleanup, overhead and profit.

4/S x 1-1/4" Deep, 15 Amp, 1 and 2 Gang Switch Assemblies

Material	Craft@Hrs	Unit	Material Cost	Labor Cost	Installed Cost
1 gang ivory switches, 15 amp, 4-way, in 4/S x 1-1/4" deep box, ring and plastic plate					
4-S-ring flat	L1@0.63	Ea	27.20	25.80	53.00
4-S-ring 1/4"	L1@0.63	Ea	13.50	25.80	39.30
4-S-ring 1/2"	L1@0.63	Ea	27.60	25.80	53.40
4-S-ring 5/8"	L1@0.63	Ea	26.30	25.80	52.10
4-S-ring 3/4"	L1@0.63	Ea	25.10	25.80	50.90
4-S-ring 1"	L1@0.63	Ea	25.40	25.80	51.20
4-S-ring 1-1/4"	L1@0.63	Ea	25.90	25.80	51.70
4-S-tile ring 1/2"	L1@0.63	Ea	27.30	25.80	53.10
4-S-tile ring 3/4"	L1@0.63	Ea	26.60	25.80	52.40
4-S-tile ring 1"	L1@0.63	Ea	26.70	25.80	52.50
4-S-tile ring 1-1/4"	L1@0.64	Ea	27.30	26.30	53.60
4-S-tile ring 1-1/2"	L1@0.64	Ea	29.70	26.30	56.00
4-S-tile ring 2"	L1@0.64	Ea	30.00	26.30	56.30
2 gang ivory switches, 15 amp, 1-pole, in 4/S x 1-1/4" deep box, ring and plastic plate					
4-S-ring flat	L1@0.77	Ea	18.90	31.60	50.50
4-S-ring 1/4"	L1@0.77	Ea	20.00	31.60	51.60
4-S-ring 1/2"	L1@0.77	Ea	16.70	31.60	48.30
4-S-ring 5/8"	L1@0.77	Ea	17.00	31.60	48.60
4-S-ring 3/4"	L1@0.77	Ea	17.40	31.60	49.00
4-S-ring 1"	L1@0.77	Ea	19.40	31.60	51.00
4-S-ring 1-1/4"	L1@0.77	Ea	19.90	31.60	51.50
4-S-tile ring 1/2"	L1@0.77	Ea	16.90	31.60	48.50
4-S-tile ring 3/4"	L1@0.77	Ea	17.90	31.60	49.50
4-S-tile ring 1"	L1@0.77	Ea	19.40	31.60	51.00
4-S-tile ring 1-1/4"	L1@0.79	Ea	19.90	32.40	52.30
4-S-tile ring 1-1/2"	L1@0.79	Ea	21.50	32.40	53.90
4-S-tile ring 2"	L1@0.79	Ea	23.50	32.40	55.90

S_4

SS

Use these figures to estimate the cost of installing assemblies under the conditions described on pages 5 and 6. Costs listed are for each assembly installed. The crew is one electrician working at a labor cost of $41.03 per manhour. These costs include layout, material handling and normal waste. Add for the sales tax, delivery, supervision, mobilization, demobilization, cleanup, overhead and profit.

4/S x 1-1/4" Deep, 15 Amp, 2 Gang Switch Assemblies

2 gang ivory switches, 15 amp, 2-pole, in 4/S x 1-1/4" deep box, ring and plastic plate

Material	Craft@Hrs	Unit	Material Cost	Labor Cost	Installed Cost
4-S-ring flat	L1@0.87	Ea	34.70	35.70	70.40
4-S-ring 1/4"	L1@0.87	Ea	35.70	35.70	71.40
4-S-ring 1/2"	L1@0.87	Ea	32.40	35.70	68.10
4-S-ring 5/8"	L1@0.87	Ea	32.70	35.70	68.40
4-S-ring 3/4"	L1@0.87	Ea	33.10	35.70	68.80
4-S-ring 1"	L1@0.87	Ea	35.10	35.70	70.80
4-S-ring 1-1/4"	L1@0.87	Ea	35.60	35.70	71.30
4-S-tile ring 1/2"	L1@0.87	Ea	32.60	35.70	68.30
4-S-tile ring 3/4"	L1@0.87	Ea	33.60	35.70	69.30
4-S-tile ring 1"	L1@0.87	Ea	35.10	35.70	70.80
4-S-tile ring 1-1/4"	L1@0.89	Ea	35.60	36.50	72.10
4-S-tile ring 1-1/2"	L1@0.89	Ea	37.30	36.50	73.80
4-S-tile ring 2"	L1@0.89	Ea	39.20	36.50	75.70

S_2S_2

2 gang ivory switches, 15 amp, 3-way, in 4/S x 1-1/4" deep box, ring and plastic plate

Material	Craft@Hrs	Unit	Material Cost	Labor Cost	Installed Cost
4-S-ring flat	L1@0.87	Ea	22.10	35.70	57.80
4-S-ring 1/4"	L1@0.87	Ea	23.10	35.70	58.80
4-S-ring 1/2"	L1@0.87	Ea	19.80	35.70	55.50
4-S-ring 5/8"	L1@0.87	Ea	20.20	35.70	55.90
4-S-ring 3/4"	L1@0.87	Ea	20.50	35.70	56.20
4-S-ring 1"	L1@0.87	Ea	22.50	35.70	58.20
4-S-ring 1-1/4"	L1@0.87	Ea	23.10	35.70	58.80
4-S-tile ring 1/2"	L1@0.87	Ea	20.00	35.70	55.70
4-S-tile ring 3/4"	L1@0.87	Ea	21.00	35.70	56.70
4-S-tile ring 1"	L1@0.87	Ea	22.50	35.70	58.20
4-S-tile ring 1-1/4"	L1@0.89	Ea	23.00	36.50	59.50
4-S-tile ring 1-1/2"	L1@0.89	Ea	24.70	36.50	61.20
4-S-tile ring 2"	L1@0.89	Ea	26.60	36.50	63.10

S_3S_3

2 gang ivory switches, 15 amp, 4-way, in 4/S x 1-1/4" deep box, ring and plastic plate

Material	Craft@Hrs	Unit	Material Cost	Labor Cost	Installed Cost
4-S-ring flat	L1@0.97	Ea	47.00	39.80	86.80
4-S-ring 1/4"	L1@0.97	Ea	48.10	39.80	87.90
4-S-ring 1/2"	L1@0.97	Ea	44.70	39.80	84.50
4-S-ring 5/8"	L1@0.97	Ea	45.10	39.80	84.90
4-S-ring 3/4"	L1@0.97	Ea	45.40	39.80	85.20
4-S-ring 1"	L1@0.97	Ea	47.50	39.80	87.30
4-S-ring 1-1/4"	L1@0.97	Ea	48.00	39.80	87.80
4-S-tile ring 1/2"	L1@0.97	Ea	45.00	39.80	84.80
4-S-tile ring 3/4"	L1@0.97	Ea	46.00	39.80	85.80
4-S-tile ring 1"	L1@0.97	Ea	47.50	39.80	87.30
4-S-tile ring 1-1/4"	L1@0.99	Ea	48.00	40.60	88.60
4-S-tile ring 1-1/2"	L1@0.99	Ea	49.60	40.60	90.20
4-S-tile ring 2"	L1@0.99	Ea	51.60	40.60	92.20

S_4S_4

Use these figures to estimate the cost of installing assemblies under the conditions described on pages 5 and 6. Costs listed are for each assembly installed. The crew is one electrician working at a labor cost of $41.03 per manhour. These costs include layout, material handling and normal waste. Add for the sales tax, delivery, supervision, mobilization, demobilization, cleanup, overhead and profit.

4/S x 1-1/2" Deep, 15 Amp, 1 Gang Switch Assemblies

Material	Craft@Hrs	Unit	Material Cost	Labor Cost	Installed Cost
1 gang brown switches, 15 amp, 1-pole, in 4/S x 1-1/2" deep box, ring and plastic plate					
4-S-ring flat	L1@0.53	Ea	12.70	21.70	34.40
4-S-ring 1/4"	L1@0.53	Ea	13.10	21.70	34.80
4-S-ring 1/2"	L1@0.53	Ea	11.80	21.70	33.50
4-S-ring 5/8"	L1@0.53	Ea	10.60	21.70	32.30
4-S-ring 3/4"	L1@0.53	Ea	10.90	21.70	32.60
4-S-ring 1"	L1@0.53	Ea	11.40	21.70	33.10
4-S-ring 1-1/4"	L1@0.53	Ea	12.80	21.70	34.50
4-S-tile ring 1/2"	L1@0.53	Ea	12.10	21.70	33.80
4-S-tile ring 3/4"	L1@0.53	Ea	12.20	21.70	33.90
4-S-tile ring 1"	L1@0.53	Ea	12.40	21.70	34.10
4-S-tile ring 1-1/4"	L1@0.54	Ea	12.70	22.20	34.90
4-S-tile ring 1-1/2"	L1@0.54	Ea	15.20	22.20	37.40
4-S-tile ring 2"	L1@0.54	Ea	15.40	22.20	37.60
1 gang brown switches, 15 amp, 2-pole, in 4/S x 1-1/2" deep box, ring and plastic plate					
4-S-ring flat	L1@0.58	Ea	20.60	23.80	44.40
4-S-ring 1/4"	L1@0.58	Ea	20.90	23.80	44.70
4-S-ring 1/2"	L1@0.58	Ea	19.70	23.80	43.50
4-S-ring 5/8"	L1@0.58	Ea	18.50	23.80	42.30
4-S-ring 3/4"	L1@0.58	Ea	18.70	23.80	42.50
4-S-ring 1"	L1@0.58	Ea	19.30	23.80	43.10
4-S-ring 1-1/4"	L1@0.58	Ea	20.60	23.80	44.40
4-S-tile ring 1/2"	L1@0.58	Ea	20.00	23.80	43.80
4-S-tile ring 3/4"	L1@0.58	Ea	20.10	23.80	43.90
4-S-tile ring 1"	L1@0.58	Ea	20.30	23.80	44.10
4-S-tile ring 1-1/4"	L1@0.59	Ea	20.60	24.20	44.80
4-S-tile ring 1-1/2"	L1@0.59	Ea	23.00	24.20	47.20
4-S-tile ring 2"	L1@0.59	Ea	23.30	24.20	47.50
1 gang brown switches, 15 amp, 3-way, in 4/S x 1-1/2" deep box, ring and plastic plate					
4-S-ring flat	L1@0.58	Ea	14.30	23.80	38.10
4-S-ring 1/4"	L1@0.58	Ea	14.70	23.80	38.50
4-S-ring 1/2"	L1@0.58	Ea	13.30	23.80	37.10
4-S-ring 5/8"	L1@0.58	Ea	12.20	23.80	36.00
4-S-ring 3/4"	L1@0.58	Ea	12.50	23.80	36.30
4-S-ring 1"	L1@0.58	Ea	13.00	23.80	36.80
4-S-ring 1-1/4"	L1@0.58	Ea	14.40	23.80	38.20
4-S-tile ring 1/2"	L1@0.58	Ea	13.70	23.80	37.50
4-S-tile ring 3/4"	L1@0.58	Ea	13.70	23.80	37.50
4-S-tile ring 1"	L1@0.58	Ea	13.90	23.80	37.70
4-S-tile ring 1-1/4"	L1@0.59	Ea	14.30	24.20	38.50
4-S-tile ring 1-1/2"	L1@0.59	Ea	16.80	24.20	41.00
4-S-tile ring 2"	L1@0.59	Ea	17.00	24.20	41.20

S

S_2

S_3

Use these figures to estimate the cost of installing assemblies under the conditions described on pages 5 and 6. Costs listed are for each assembly installed. The crew is one electrician working at a labor cost of $41.03 per manhour. These costs include layout, material handling and normal waste. Add for the sales tax, delivery, supervision, mobilization, demobilization, cleanup, overhead and profit.

1 gang brown switches, 15 amp, 4-way, in 4/S x 1-1/2" deep box, ring and plastic plate

Material	Craft@Hrs	Unit	Material Cost	Labor Cost	Installed Cost	
4-S-ring flat	L1@0.63	Ea	26.70	25.80	52.50	
4-S-ring 1/4"	L1@0.63	Ea	27.20	25.80	53.00	
4-S-ring 1/2"	L1@0.63	Ea	25.80	25.80	51.60	
4-S-ring 5/8"	L1@0.63	Ea	24.70	25.80	50.50	
4-S-ring 3/4"	L1@0.63	Ea	25.00	25.80	50.80	
4-S-ring 1"	L1@0.63	Ea	25.50	25.80	51.30	
4-S-ring 1-1/4"	L1@0.63	Ea	26.70	25.80	52.50	S_4
4-S-tile ring 1/2"	L1@0.63	Ea	26.10	25.80	51.90	
4-S-tile ring 3/4"	L1@0.63	Ea	26.20	25.80	52.00	
4-S-tile ring 1"	L1@0.63	Ea	26.40	25.80	52.20	
4-S-tile ring 1-1/4"	L1@0.64	Ea	26.70	26.30	53.00	
4-S-tile ring 1-1/2"	L1@0.64	Ea	29.20	26.30	55.50	
4-S-tile ring 2"	L1@0.64	Ea	29.50	26.30	55.80	

1 gang key switches, 15 amp, 1-pole, in 4/S x 1-1/2" deep box, ring and plastic plate

4-S-ring flat	L1@0.53	Ea	22.10	21.70	43.80	
4-S-ring 1/4"	L1@0.53	Ea	22.50	21.70	44.20	
4-S-ring 1/2"	L1@0.53	Ea	21.10	21.70	42.80	
4-S-ring 5/8"	L1@0.53	Ea	20.00	21.70	41.70	
4-S-ring 3/4"	L1@0.53	Ea	20.30	21.70	42.00	
4-S-ring 1"	L1@0.53	Ea	20.80	21.70	42.50	
4-S-ring 1-1/4"	L1@0.53	Ea	22.20	21.70	43.90	S_K
4-S-tile ring 1/2"	L1@0.53	Ea	21.50	21.70	43.20	
4-S-tile ring 3/4"	L1@0.53	Ea	21.50	21.70	43.20	
4-S-tile ring 1"	L1@0.53	Ea	21.80	21.70	43.50	
4-S-tile ring 1-1/4"	L1@0.54	Ea	22.10	22.20	44.30	
4-S-tile ring 1-1/2"	L1@0.54	Ea	24.60	22.20	46.80	
4-S-tile ring 2"	L1@0.54	Ea	24.80	22.20	47.00	

1 gang key switches, 15 amp, 2-pole, in 4/S x 1-1/2" deep box, ring and plastic plate

4-S-ring flat	L1@0.58	Ea	28.10	23.80	51.90	
4-S-ring 1/4"	L1@0.58	Ea	28.50	23.80	52.30	
4-S-ring 1/2"	L1@0.58	Ea	27.30	23.80	51.10	
4-S-ring 5/8"	L1@0.58	Ea	26.10	23.80	49.90	
4-S-ring 3/4"	L1@0.58	Ea	26.30	23.80	50.10	
4-S-ring 1"	L1@0.58	Ea	26.90	23.80	50.70	
4-S-ring 1-1/4"	L1@0.58	Ea	28.20	23.80	52.00	S_{K2}
4-S-tile ring 1/2"	L1@0.58	Ea	27.60	23.80	51.40	
4-S-tile ring 3/4"	L1@0.58	Ea	27.70	23.80	51.50	
4-S-tile ring 1"	L1@0.58	Ea	27.80	23.80	51.60	
4-S-tile ring 1-1/4"	L1@0.59	Ea	28.20	24.20	52.40	
4-S-tile ring 1-1/2"	L1@0.59	Ea	30.60	24.20	54.80	
4-S-tile ring 2"	L1@0.59	Ea	30.90	24.20	55.10	

Use these figures to estimate the cost of installing assemblies under the conditions described on pages 5 and 6. Costs listed are for each assembly installed. The crew is one electrician working at a labor cost of $41.03 per manhour. These costs include layout, material handling and normal waste. Add for the sales tax, delivery, supervision, mobilization, demobilization, cleanup, overhead and profit.

4/S x 1-1/2" Deep, 15 Amp, 1 and 2 Gang Switch Assemblies

	Material	Craft@Hrs	Unit	Material Cost	Labor Cost	Installed Cost
	1 gang key switches, 15 amp, 3-way, in 4/S x 1-1/2" deep box, ring and plastic plate					
	4-S-ring flat	L1@0.58	Ea	26.30	23.80	50.10
	4-S-ring 1/4"	L1@0.58	Ea	26.70	23.80	50.50
	4-S-ring 1/2"	L1@0.58	Ea	25.50	23.80	49.30
	4-S-ring 5/8"	L1@0.58	Ea	24.30	23.80	48.10
	4-S-ring 3/4"	L1@0.58	Ea	24.60	23.80	48.40
S_{K3}	4-S-ring 1"	L1@0.58	Ea	25.10	23.80	48.90
	4-S-ring 1-1/4"	L1@0.58	Ea	26.40	23.80	50.20
	4-S-tile ring 1/2"	L1@0.58	Ea	25.80	23.80	49.60
	4-S-tile ring 3/4"	L1@0.58	Ea	25.90	23.80	49.70
	4-S-tile ring 1"	L1@0.58	Ea	26.00	23.80	49.80
	4-S-tile ring 1-1/4"	L1@0.59	Ea	26.30	24.20	50.50
	4-S-tile ring 1-1/2"	L1@0.59	Ea	28.80	24.20	53.00
	4-S-tile ring 2"	L1@0.59	Ea	29.10	24.20	53.30
	1 gang key switches, 15 amp, 4-way, in 4/S x 1-1/2" deep box, ring and plastic plate					
	4-S-ring flat	L1@0.63	Ea	44.10	25.80	69.90
	4-S-ring 1/4"	L1@0.63	Ea	44.50	25.80	70.30
	4-S-ring 1/2"	L1@0.63	Ea	43.20	25.80	69.00
	4-S-ring 5/8"	L1@0.63	Ea	42.00	25.80	67.80
	4-S-ring 3/4"	L1@0.63	Ea	42.40	25.80	68.20
S_{K4}	4-S-ring 1"	L1@0.63	Ea	42.90	25.80	68.70
	4-S-ring 1-1/4"	L1@0.63	Ea	44.10	25.80	69.90
	4-S-tile ring 1/2"	L1@0.63	Ea	43.50	25.80	69.30
	4-S-tile ring 3/4"	L1@0.63	Ea	43.60	25.80	69.40
	4-S-tile ring 1"	L1@0.63	Ea	43.80	25.80	69.60
	4-S-tile ring 1-1/4"	L1@0.64	Ea	44.10	26.30	70.40
	4-S-tile ring 1-1/2"	L1@0.64	Ea	46.60	26.30	72.90
	4-S-tile ring 2"	L1@0.64	Ea	46.80	26.30	73.10
	2 gang brown switches, 15 amp, 1-pole, in 4/S x 1-1/2" deep box, ring and plastic plate					
	4-S-ring flat	L1@0.77	Ea	18.50	31.60	50.10
	4-S-ring 1/4"	L1@0.77	Ea	19.50	31.60	51.10
	4-S-ring 1/2"	L1@0.77	Ea	16.10	31.60	47.70
	4-S-ring 5/8"	L1@0.77	Ea	16.50	31.60	48.10
	4-S-ring 3/4"	L1@0.77	Ea	16.90	31.60	48.50
	4-S-ring 1"	L1@0.77	Ea	18.80	31.60	50.40
SS	4-S-ring 1-1/4"	L1@0.77	Ea	19.50	31.60	51.10
	4-S-tile ring 1/2"	L1@0.77	Ea	16.40	31.60	48.00
	4-S-tile ring 3/4"	L1@0.77	Ea	17.40	31.60	49.00
	4-S-tile ring 1"	L1@0.77	Ea	18.80	31.60	50.40
	4-S-tile ring 1-1/4"	L1@0.79	Ea	19.40	32.40	51.80
	4-S-tile ring 1-1/2"	L1@0.79	Ea	21.10	32.40	53.50
	4-S-tile ring 2"	L1@0.79	Ea	23.00	32.40	55.40

Use these figures to estimate the cost of installing assemblies under the conditions described on pages 5 and 6. Costs listed are for each assembly installed. The crew is one electrician working at a labor cost of $41.03 per manhour. These costs include layout, material handling and normal waste. Add for the sales tax, delivery, supervision, mobilization, demobilization, cleanup, overhead and profit.

4/S x 1-1/2" Deep, 15 Amp, 2 Gang Switch Assemblies

Material	Craft@Hrs	Unit	Material Cost	Labor Cost	Installed Cost	
2 gang brown switches, 15 amp, 2-pole, in 4/S x 1-1/2" deep box, ring and plastic plate						
4-S-ring flat	L1@0.87	Ea	34.20	35.70	69.90	
4-S-ring 1/4"	L1@0.87	Ea	35.30	35.70	71.00	
4-S-ring 1/2"	L1@0.87	Ea	31.80	35.70	67.50	
4-S-ring 5/8"	L1@0.87	Ea	32.30	35.70	68.00	
4-S-ring 3/4"	L1@0.87	Ea	32.60	35.70	68.30	
4-S-ring 1"	L1@0.87	Ea	34.60	35.70	70.30	
4-S-ring 1-1/4"	L1@0.87	Ea	35.20	35.70	70.90	S_2S_2
4-S-tile ring 1/2"	L1@0.87	Ea	32.20	35.70	67.90	
4-S-tile ring 3/4"	L1@0.87	Ea	33.20	35.70	68.90	
4-S-tile ring 1"	L1@0.87	Ea	34.70	35.70	70.40	
4-S-tile ring 1-1/4"	L1@0.89	Ea	35.10	36.50	71.60	
4-S-tile ring 1-1/2"	L1@0.89	Ea	36.80	36.50	73.30	
4-S-tile ring 2"	L1@0.89	Ea	38.70	36.50	75.20	
2 gang brown switches, 15 amp, 3-way, in 4/S x 1-1/2" deep box, ring and plastic plate						
4-S-ring flat	L1@0.87	Ea	21.60	35.70	57.30	
4-S-ring 1/4"	L1@0.87	Ea	22.70	35.70	58.40	
4-S-ring 1/2"	L1@0.87	Ea	19.30	35.70	55.00	
4-S-ring 5/8"	L1@0.87	Ea	19.70	35.70	55.40	
4-S-ring 3/4"	L1@0.87	Ea	20.00	35.70	55.70	
4-S-ring 1"	L1@0.87	Ea	22.00	35.70	57.70	
4-S-ring 1-1/4"	L1@0.87	Ea	22.60	35.70	58.30	S_3S_3
4-S-tile ring 1/2"	L1@0.87	Ea	19.60	35.70	55.30	
4-S-tile ring 3/4"	L1@0.87	Ea	20.60	35.70	56.30	
4-S-tile ring 1"	L1@0.87	Ea	22.10	35.70	57.80	
4-S-tile ring 1-1/4"	L1@0.89	Ea	22.60	36.50	59.10	
4-S-tile ring 1-1/2"	L1@0.89	Ea	24.30	36.50	60.80	
4-S-tile ring 2"	L1@0.89	Ea	26.10	36.50	62.60	
2 gang brown switches, 15 amp, 4-way, in 4/S x 1-1/2" deep box, ring and plastic plate						
4-S-ring flat	L1@0.97	Ea	46.60	39.80	86.40	
4-S-ring 1/4"	L1@0.97	Ea	47.60	39.80	87.40	
4-S-ring 1/2"	L1@0.97	Ea	44.20	39.80	84.00	
4-S-ring 5/8"	L1@0.97	Ea	44.70	39.80	84.50	
4-S-ring 3/4"	L1@0.97	Ea	45.00	39.80	84.80	
4-S-ring 1"	L1@0.97	Ea	46.90	39.80	86.70	
4-S-ring 1-1/4"	L1@0.97	Ea	47.60	39.80	87.40	S_4S_4
4-S-tile ring 1/2"	L1@0.97	Ea	44.50	39.80	84.30	
4-S-tile ring 3/4"	L1@0.97	Ea	45.50	39.80	85.30	
4-S-tile ring 1"	L1@0.97	Ea	46.90	39.80	86.70	
4-S-tile ring 1-1/4"	L1@0.99	Ea	47.50	40.60	88.10	
4-S-tile ring 1-1/2"	L1@0.99	Ea	49.20	40.60	89.80	
4-S-tile ring 2"	L1@0.99	Ea	51.10	40.60	91.70	

Use these figures to estimate the cost of installing assemblies under the conditions described on pages 5 and 6. Costs listed are for each assembly installed. The crew is one electrician working at a labor cost of $41.03 per manhour. These costs include layout, material handling and normal waste. Add for the sales tax, delivery, supervision, mobilization, demobilization, cleanup, overhead and profit.

4/S x 1-1/2" Deep, 15 Amp, 1 Gang Switch Assemblies

1 gang ivory switches, 15 amp, 1-pole, in 4/S x 1-1/2" deep box, ring and plastic plate

S

Material	Craft@Hrs	Unit	Material Cost	Labor Cost	Installed Cost
4-S-ring flat	L1@0.53	Ea	12.70	21.70	34.40
4-S-ring 1/4"	L1@0.53	Ea	13.10	21.70	34.80
4-S-ring 1/2"	L1@0.53	Ea	11.80	21.70	33.50
4-S-ring 5/8"	L1@0.53	Ea	10.60	21.70	32.30
4-S-ring 3/4"	L1@0.53	Ea	10.90	21.70	32.60
4-S-ring 1"	L1@0.53	Ea	11.40	21.70	33.10
4-S-ring 1-1/4"	L1@0.53	Ea	12.80	21.70	34.50
4-S-tile ring 1/2"	L1@0.53	Ea	12.10	21.70	33.80
4-S-tile ring 3/4"	L1@0.53	Ea	12.20	21.70	33.90
4-S-tile ring 1"	L1@0.53	Ea	12.40	21.70	34.10
4-S-tile ring 1-1/4"	L1@0.54	Ea	12.70	22.20	34.90
4-S-tile ring 1-1/2"	L1@0.54	Ea	15.20	22.20	37.40
4-S-tile ring 2"	L1@0.54	Ea	15.40	22.20	37.60

1 gang ivory switches, 15 amp, 2-pole, in 4/S x 1-1/2" deep box, ring and plastic plate

S_2

Material	Craft@Hrs	Unit	Material Cost	Labor Cost	Installed Cost
4-S-ring flat	L1@0.58	Ea	20.60	23.80	44.40
4-S-ring 1/4"	L1@0.58	Ea	20.90	23.80	44.70
4-S-ring 1/2"	L1@0.58	Ea	19.70	23.80	43.50
4-S-ring 5/8"	L1@0.58	Ea	18.50	23.80	42.30
4-S-ring 3/4"	L1@0.58	Ea	18.70	23.80	42.50
4-S-ring 1"	L1@0.58	Ea	19.30	23.80	43.10
4-S-ring 1-1/4"	L1@0.58	Ea	20.60	23.80	44.40
4-S-tile ring 1/2"	L1@0.58	Ea	20.00	23.80	43.80
4-S-tile ring 3/4"	L1@0.58	Ea	20.10	23.80	43.90
4-S-tile ring 1"	L1@0.58	Ea	20.30	23.80	44.10
4-S-tile ring 1-1/4"	L1@0.59	Ea	20.60	24.20	44.80
4-S-tile ring 1-1/2"	L1@0.59	Ea	23.00	24.20	47.20
4-S-tile ring 2"	L1@0.59	Ea	23.30	24.20	47.50

1 gang ivory switches, 15 amp, 3-way, in 4/S x 1-1/2" deep box, ring and plastic plate

S_3

Material	Craft@Hrs	Unit	Material Cost	Labor Cost	Installed Cost
4-S-ring flat	L1@0.58	Ea	14.30	23.80	38.10
4-S-ring 1/4"	L1@0.58	Ea	14.70	23.80	38.50
4-S-ring 1/2"	L1@0.58	Ea	13.30	23.80	37.10
4-S-ring 5/8"	L1@0.58	Ea	12.20	23.80	36.00
4-S-ring 3/4"	L1@0.58	Ea	12.50	23.80	36.30
4-S-ring 1"	L1@0.58	Ea	13.00	23.80	36.80
4-S-ring 1-1/4"	L1@0.58	Ea	14.40	23.80	38.20
4-S-tile ring 1/2"	L1@0.58	Ea	13.70	23.80	37.50
4-S-tile ring 3/4"	L1@0.58	Ea	13.70	23.80	37.50
4-S-tile ring 1"	L1@0.58	Ea	13.90	23.80	37.70
4-S-tile ring 1-1/4"	L1@0.59	Ea	14.30	24.20	38.50
4-S-tile ring 1-1/2"	L1@0.59	Ea	16.80	24.20	41.00
4-S-tile ring 2"	L1@0.59	Ea	17.00	24.20	41.20

Use these figures to estimate the cost of installing assemblies under the conditions described on pages 5 and 6. Costs listed are for each assembly installed. The crew is one electrician working at a labor cost of $41.03 per manhour. These costs include layout, material handling and normal waste. Add for the sales tax, delivery, supervision, mobilization, demobilization, cleanup, overhead and profit.

4/S x 1-1/2" Deep, 15 Amp, 1 and 2 Gang Switch Assemblies

1 gang ivory switches, 15 amp, 4-way, in 4/S x 1-1/2" deep box, ring and plastic plate

Material	Craft@Hrs	Unit	Material Cost	Labor Cost	Installed Cost
4-S-ring flat	L1@0.63	Ea	26.70	25.80	52.50
4-S-ring 1/4"	L1@0.63	Ea	27.20	25.80	53.00
4-S-ring 1/2"	L1@0.63	Ea	25.80	25.80	51.60
4-S-ring 5/8"	L1@0.63	Ea	24.70	25.80	50.50
4-S-ring 3/4"	L1@0.63	Ea	25.00	25.80	50.80
4-S-ring 1"	L1@0.63	Ea	25.50	25.80	51.30
4-S-ring 1-1/4"	L1@0.63	Ea	26.70	25.80	52.50
4-S-tile ring 1/2"	L1@0.63	Ea	26.10	25.80	51.90
4-S-tile ring 3/4"	L1@0.63	Ea	26.20	25.80	52.00
4-S-tile ring 1"	L1@0.63	Ea	26.40	25.80	52.20
4-S-tile ring 1-1/4"	L1@0.64	Ea	26.70	26.30	53.00
4-S-tile ring 1-1/2"	L1@0.64	Ea	29.20	26.30	55.50
4-S-tile ring 2"	L1@0.64	Ea	29.50	26.30	55.80

S_4

2 gang ivory switches, 15 amp, 1-pole, in 4/S x 1-1/2" deep box, ring and plastic plate

Material	Craft@Hrs	Unit	Material Cost	Labor Cost	Installed Cost
4-S-ring flat	L1@0.77	Ea	18.50	31.60	50.10
4-S-ring 1/4"	L1@0.77	Ea	19.50	31.60	51.10
4-S-ring 1/2"	L1@0.77	Ea	16.10	31.60	47.70
4-S-ring 5/8"	L1@0.77	Ea	16.50	31.60	48.10
4-S-ring 3/4"	L1@0.77	Ea	16.90	31.60	48.50
4-S-ring 1"	L1@0.77	Ea	18.80	31.60	50.40
4-S-ring 1-1/4"	L1@0.77	Ea	19.50	31.60	51.10
4-S-tile ring 1/2"	L1@0.77	Ea	16.40	31.60	48.00
4-S-tile ring 3/4"	L1@0.77	Ea	17.40	31.60	49.00
4-S-tile ring 1"	L1@0.77	Ea	18.80	31.60	50.40
4-S-tile ring 1-1/4"	L1@0.79	Ea	19.40	32.40	51.80
4-S-tile ring 1-1/2"	L1@0.79	Ea	21.10	32.40	53.50
4-S-tile ring 2"	L1@0.79	Ea	23.00	32.40	55.40

SS

Use these figures to estimate the cost of installing assemblies under the conditions described on pages 5 and 6. Costs listed are for each assembly installed. The crew is one electrician working at a labor cost of $41.03 per manhour. These costs include layout, material handling and normal waste. Add for the sales tax, delivery, supervision, mobilization, demobilization, cleanup, overhead and profit.

4/S x 1-1/2" Deep, 15 Amp, 2 Gang Switch Assemblies

	Material	Craft@Hrs	Unit	Material Cost	Labor Cost	Installed Cost
	2 gang ivory switches, 15 amp, 2-pole, in 4/S x 1-1/2" deep box, ring and plastic plate					
	4-S-ring flat	L1@0.87	Ea	34.20	35.70	69.90
	4-S-ring 1/4"	L1@0.87	Ea	35.30	35.70	71.00
	4-S-ring 1/2"	L1@0.87	Ea	31.80	35.70	67.50
	4-S-ring 5/8"	L1@0.87	Ea	32.30	35.70	68.00
	4-S-ring 3/4"	L1@0.87	Ea	32.60	35.70	68.30
	4-S-ring 1"	L1@0.87	Ea	34.60	35.70	70.30
S_2S_2	4-S-ring 1-1/4"	L1@0.87	Ea	35.20	35.70	70.90
	4-S-tile ring 1/2"	L1@0.87	Ea	32.20	35.70	67.90
	4-S-tile ring 3/4"	L1@0.87	Ea	33.20	35.70	68.90
	4-S-tile ring 1"	L1@0.87	Ea	34.70	35.70	70.40
	4-S-tile ring 1-1/4"	L1@0.89	Ea	35.10	36.50	71.60
	4-S-tile ring 1-1/2"	L1@0.89	Ea	36.80	36.50	73.30
	4-S-tile ring 2"	L1@0.89	Ea	38.70	36.50	75.20
	2 gang ivory switches, 15 amp, 3-way, in 4/S x 1-1/2" deep box, ring and plastic plate					
	4-S-ring flat	L1@0.87	Ea	21.60	35.70	57.30
	4-S-ring 1/4"	L1@0.87	Ea	22.70	35.70	58.40
	4-S-ring 1/2"	L1@0.87	Ea	19.30	35.70	55.00
	4-S-ring 5/8"	L1@0.87	Ea	19.70	35.70	55.40
	4-S-ring 3/4"	L1@0.87	Ea	20.00	35.70	55.70
S_3S_3	4-S-ring 1"	L1@0.87	Ea	22.00	35.70	57.70
	4-S-ring 1-1/4"	L1@0.87	Ea	22.60	35.70	58.30
	4-S-tile ring 1/2"	L1@0.87	Ea	19.60	35.70	55.30
	4-S-tile ring 3/4"	L1@0.87	Ea	20.60	35.70	56.30
	4-S-tile ring 1"	L1@0.87	Ea	22.10	35.70	57.80
	4-S-tile ring 1-1/4"	L1@0.89	Ea	22.60	36.50	59.10
	4-S-tile ring 1-1/2"	L1@0.89	Ea	24.30	36.50	60.80
	4-S-tile ring 2"	L1@0.89	Ea	26.10	36.50	62.60
	2 gang ivory switches, 15 amp, 4-way, in 4/S x 1-1/2" deep box, ring and plastic plate					
	4-S-ring flat	L1@0.97	Ea	46.60	39.80	86.40
	4-S-ring 1/4"	L1@0.97	Ea	47.60	39.80	87.40
	4-S-ring 1/2"	L1@0.97	Ea	44.20	39.80	84.00
	4-S-ring 5/8"	L1@0.97	Ea	44.70	39.80	84.50
	4-S-ring 3/4"	L1@0.97	Ea	45.00	39.80	84.80
S_4S_4	4-S-ring 1"	L1@0.97	Ea	46.90	39.80	86.70
	4-S-ring 1-1/4"	L1@0.97	Ea	47.60	39.80	87.40
	4-S-tile ring 1/2"	L1@0.97	Ea	44.50	39.80	84.30
	4-S-tile ring 3/4"	L1@0.97	Ea	45.50	39.80	85.30
	4-S-tile ring 1"	L1@0.97	Ea	46.90	39.80	86.70
	4-S-tile ring 1-1/4"	L1@0.99	Ea	47.50	40.60	88.10
	4-S-tile ring 1-1/2"	L1@0.99	Ea	49.20	40.60	89.80
	4-S-tile ring 2"	L1@0.99	Ea	51.10	40.60	91.70

Use these figures to estimate the cost of installing assemblies under the conditions described on pages 5 and 6. Costs listed are for each assembly installed. The crew is one electrician working at a labor cost of $41.03 per manhour. These costs include layout, material handling and normal waste. Add for the sales tax, delivery, supervision, mobilization, demobilization, cleanup, overhead and profit.

4/S x 2-1/8" Deep, 15 Amp, 1 Gang Switch Assemblies

1 gang brown switches, 15 amp, 1-pole, in 4/S x 2-1/8" deep box, ring and plastic plate

Material	Craft@Hrs	Unit	Material Cost	Labor Cost	Installed Cost
4-S-ring flat	L1@0.55	Ea	15.00	22.60	37.60
4-S-ring 1/4"	L1@0.55	Ea	15.40	22.60	38.00
4-S-ring 1/2"	L1@0.55	Ea	14.20	22.60	36.80
4-S-ring 5/8"	L1@0.55	Ea	12.90	22.60	35.50
4-S-ring 3/4"	L1@0.55	Ea	13.20	22.60	35.80
4-S-ring 1"	L1@0.55	Ea	13.70	22.60	36.30
4-S-ring 1-1/4"	L1@0.55	Ea	15.10	22.60	37.70
4-S-tile ring 1/2"	L1@0.55	Ea	14.50	22.60	37.10
4-S-tile ring 3/4"	L1@0.55	Ea	14.60	22.60	37.20
4-S-tile ring 1"	L1@0.55	Ea	14.70	22.60	37.30
4-S-tile ring 1-1/4"	L1@0.56	Ea	15.10	23.00	38.10
4-S-tile ring 1-1/2"	L1@0.56	Ea	17.50	23.00	40.50
4-S-tile ring 2"	L1@0.56	Ea	17.80	23.00	40.80

S

1 gang brown switches, 15 amp, 2-pole, in 4/S x 2-1/8" deep box, ring and plastic plate

Material	Craft@Hrs	Unit	Material Cost	Labor Cost	Installed Cost
4-S-ring flat	L1@0.60	Ea	22.90	24.60	47.50
4-S-ring 1/4"	L1@0.60	Ea	23.30	24.60	47.90
4-S-ring 1/2"	L1@0.60	Ea	22.00	24.60	46.60
4-S-ring 5/8"	L1@0.60	Ea	20.80	24.60	45.40
4-S-ring 3/4"	L1@0.60	Ea	21.10	24.60	45.70
4-S-ring 1"	L1@0.60	Ea	21.60	24.60	46.20
4-S-ring 1-1/4"	L1@0.60	Ea	22.90	24.60	47.50
4-S-tile ring 1/2"	L1@0.60	Ea	22.30	24.60	46.90
4-S-tile ring 3/4"	L1@0.60	Ea	22.40	24.60	47.00
4-S-tile ring 1"	L1@0.60	Ea	22.60	24.60	47.20
4-S-tile ring 1-1/4"	L1@0.61	Ea	22.90	25.00	47.90
4-S-tile ring 1-1/2"	L1@0.61	Ea	25.40	25.00	50.40
4-S-tile ring 2"	L1@0.61	Ea	25.60	25.00	50.60

S_2

1 gang brown switches, 15 amp, 3-way, in 4/S x 2-1/8" deep box, ring and plastic plate

Material	Craft@Hrs	Unit	Material Cost	Labor Cost	Installed Cost
4-S-ring flat	L1@0.60	Ea	16.50	24.60	41.10
4-S-ring 1/4	L1@0.60	Ea	17.00	24.60	41.60
4-S-ring 1/2"	L1@0.60	Ea	15.70	24.60	40.30
4-S-ring 5/8"	L1@0.60	Ea	14.60	24.60	39.20
4-S-ring 3/4"	L1@0.60	Ea	14.80	24.60	39.40
4-S-ring 1"	L1@0.60	Ea	15.30	24.60	39.90
4-S-ring 1-1/4"	L1@0.60	Ea	16.70	24.60	41.30
4-S-tile ring 1/2"	L1@0.60	Ea	16.00	24.60	40.60
4-S-tile ring 3/4"	L1@0.60	Ea	16.10	24.60	40.70
4-S-tile ring 1"	L1@0.60	Ea	16.20	24.60	40.80
4-S-tile ring 1-1/4"	L1@0.61	Ea	16.70	25.00	41.70
4-S-tile ring 1-1/2"	L1@0.61	Ea	19.00	25.00	44.00
4-S-tile ring 2"	L1@0.61	Ea	19.40	25.00	44.40

S_3

Use these figures to estimate the cost of installing assemblies under the conditions described on pages 5 and 6. Costs listed are for each assembly installed. The crew is one electrician working at a labor cost of $41.03 per manhour. These costs include layout, material handling and normal waste. Add for the sales tax, delivery, supervision, mobilization, demobilization, cleanup, overhead and profit.

4/S x 2-1/8" Deep, 15 Amp, 1 Gang Switch Assemblies

	Material	Craft@Hrs	Unit	Material Cost	Labor Cost	Installed Cost
	1 gang brown switches, 15 amp, 4-way, in 4/S x 2-1/8" deep box, ring and plastic plate					
	4-S-ring flat	L1@0.65	Ea	29.00	26.70	55.70
	4-S-ring 1/4"	L1@0.65	Ea	29.50	26.70	56.20
	4-S-ring 1/2"	L1@0.65	Ea	28.20	26.70	54.90
	4-S-ring 5/8"	L1@0.65	Ea	27.00	26.70	53.70
S_4	4-S-ring 3/4"	L1@0.65	Ea	27.30	26.70	54.00
	4-S-ring 1"	L1@0.65	Ea	27.80	26.70	54.50
	4-S-ring 1-1/4"	L1@0.65	Ea	29.10	26.70	55.80
	4-S-tile ring 1/2"	L1@0.65	Ea	28.50	26.70	55.20
	4-S-tile ring 3/4"	L1@0.65	Ea	28.60	26.70	55.30
	4-S-tile ring 1"	L1@0.65	Ea	28.70	26.70	55.40
	4-S-tile ring 1-1/4"	L1@0.66	Ea	29.10	27.10	56.20
	4-S-tile ring 1-1/2"	L1@0.66	Ea	31.50	27.10	58.60
	4-S-tile ring 2"	L1@0.66	Ea	31.80	27.10	58.90
	1 gang key switches, 15 amp, 1-pole, in 4/S x 2-1/8" deep box, ring and plastic plate					
	4-S-ring flat	L1@0.55	Ea	24.50	22.60	47.10
	4-S-ring 1/4"	L1@0.55	Ea	24.80	22.60	47.40
	4-S-ring 1/2"	L1@0.55	Ea	23.50	22.60	46.10
	4-S-ring 5/8"	L1@0.55	Ea	22.40	22.60	45.00
	4-S-ring 3/4"	L1@0.55	Ea	22.60	22.60	45.20
S_K	4-S-ring 1"	L1@0.55	Ea	23.10	22.60	45.70
	4-S-ring 1-1/4"	L1@0.55	Ea	24.50	22.60	47.10
	4-S-tile ring 1/2"	L1@0.55	Ea	23.80	22.60	46.40
	4-S-tile ring 3/4"	L1@0.55	Ea	23.90	22.60	46.50
	4-S-tile ring 1"	L1@0.55	Ea	24.10	22.60	46.70
	4-S-tile ring 1-1/4"	L1@0.56	Ea	24.50	23.00	47.50
	4-S-tile ring 1-1/2"	L1@0.56	Ea	26.90	23.00	49.90
	4-S-tile ring 2"	L1@0.56	Ea	27.20	23.00	50.20
	1 gang key switches, 15 amp, 2-pole, in 4/S x 2-1/8" deep box, ring and plastic plate					
	4-S-ring flat	L1@0.60	Ea	30.50	24.60	55.10
	4-S-ring 1/4"	L1@0.60	Ea	30.90	24.60	55.50
	4-S-ring 1/2"	L1@0.60	Ea	29.60	24.60	54.20
	4-S-ring 5/8"	L1@0.60	Ea	28.40	24.60	53.00
	4-S-ring 3/4"	L1@0.60	Ea	28.70	24.60	53.30
S_{K2}	4-S-ring 1"	L1@0.60	Ea	29.10	24.60	53.70
	4-S-ring 1-1/4"	L1@0.60	Ea	30.50	24.60	55.10
	4-S-tile ring 1/2"	L1@0.60	Ea	29.90	24.60	54.50
	4-S-tile ring 3/4"	L1@0.60	Ea	30.00	24.60	54.60
	4-S-tile ring 1"	L1@0.60	Ea	30.20	24.60	54.80
	4-S-tile ring 1-1/4"	L1@0.61	Ea	30.50	25.00	55.50
	4-S-tile ring 1-1/2"	L1@0.61	Ea	33.00	25.00	58.00
	4-S-tile ring 2"	L1@0.61	Ea	33.20	25.00	58.20

Use these figures to estimate the cost of installing assemblies under the conditions described on pages 5 and 6. Costs listed are for each assembly installed. The crew is one electrician working at a labor cost of $41.03 per manhour. These costs include layout, material handling and normal waste. Add for the sales tax, delivery, supervision, mobilization, demobilization, cleanup, overhead and profit.

4/S x 2-1/8" Deep, 15 Amp, 1 and 2 Gang Switch Assemblies

1 gang key switches, 15 amp, 3-way, in 4/S x 2-1/8" deep box, ring and plastic plate

Material	Craft@Hrs	Unit	Material Cost	Labor Cost	Installed Cost
4-S-ring flat	L1@0.60	Ea	28.70	24.60	53.30
4-S-ring 1/4"	L1@0.60	Ea	29.00	24.60	53.60
4-S-ring 1/2"	L1@0.60	Ea	27.80	24.60	52.40
4-S-ring 5/8"	L1@0.60	Ea	26.60	24.60	51.20
4-S-ring 3/4"	L1@0.60	Ea	26.90	24.60	51.50
4-S-ring 1"	L1@0.60	Ea	27.40	24.60	52.00
4-S-ring 1-1/4"	L1@0.60	Ea	28.70	24.60	53.30
4-S-tile ring 1/2"	L1@0.60	Ea	28.10	24.60	52.70
4-S-tile ring 3/4"	L1@0.60	Ea	28.20	24.60	52.80
4-S-tile ring 1"	L1@0.60	Ea	28.40	24.60	53.00
4-S-tile ring 1-1/4"	L1@0.61	Ea	28.70	25.00	53.70
4-S-tile ring 1-1/2"	L1@0.61	Ea	31.10	25.00	56.10
4-S-tile ring 2"	L1@0.61	Ea	31.40	25.00	56.40

S_{K3}

1 gang key switches, 15 amp, 4-way, in 4/S x 2-1/8" deep box, ring and plastic plate

Material	Craft@Hrs	Unit	Material Cost	Labor Cost	Installed Cost
4-S-ring flat	L1@0.65	Ea	46.40	26.70	73.10
4-S-ring 1/4"	L1@0.65	Ea	46.80	26.70	73.50
4-S-ring 1/2"	L1@0.65	Ea	45.60	26.70	72.30
4-S-ring 5/8"	L1@0.65	Ea	44.30	26.70	71.00
4-S-ring 3/4"	L1@0.65	Ea	44.70	26.70	71.40
4-S-ring 1"	L1@0.65	Ea	45.20	26.70	71.90
4-S-ring 1-1/4"	L1@0.65	Ea	46.50	26.70	73.20
4-S-tile ring 1/2"	L1@0.65	Ea	45.90	26.70	72.60
4-S-tile ring 3/4"	L1@0.65	Ea	46.00	26.70	72.70
4-S-tile ring 1"	L1@0.65	Ea	46.10	26.70	72.80
4-S-tile ring 1-1/4"	L1@0.66	Ea	46.50	27.10	73.60
4-S-tile ring 1-1/2"	L1@0.66	Ea	48.90	27.10	76.00
4-S-tile ring 2"	L1@0.66	Ea	49.20	27.10	76.30

S_{K4}

2 gang brown switches, 15 amp, 1-pole, in 4/S x 2-1/8" deep box, ring and plastic plate

Material	Craft@Hrs	Unit	Material Cost	Labor Cost	Installed Cost
4-S-ring flat	L1@0.79	Ea	20.80	32.40	53.20
4-S-ring 1/4"	L1@0.79	Ea	21.90	32.40	54.30
4-S-ring 1/2"	L1@0.79	Ea	18.50	32.40	50.90
4-S-ring 5/8"	L1@0.79	Ea	18.80	32.40	51.20
4-S-ring 3/4"	L1@0.79	Ea	19.30	32.40	51.70
4-S-ring 1"	L1@0.79	Ea	21.20	32.40	53.60
4-S-ring 1-1/4"	L1@0.79	Ea	21.80	32.40	54.20
4-S-tile ring 1/2"	L1@0.79	Ea	18.70	32.40	51.10
4-S-tile ring 3/4"	L1@0.79	Ea	19.80	32.40	52.20
4-S-tile ring 1"	L1@0.79	Ea	21.20	32.40	53.60
4-S-tile ring 1-1/4"	L1@0.81	Ea	21.80	33.20	55.00
4-S-tile ring 1-1/2"	L1@0.81	Ea	23.40	33.20	56.60
4-S-tile ring 2"	L1@0.81	Ea	25.40	33.20	58.60

SS

Use these figures to estimate the cost of installing assemblies under the conditions described on pages 5 and 6. Costs listed are for each assembly installed. The crew is one electrician working at a labor cost of $41.03 per manhour. These costs include layout, material handling and normal waste. Add for the sales tax, delivery, supervision, mobilization, demobilization, cleanup, overhead and profit.

4/S x 2-1/8" Deep, 15 Amp, 2 Gang Switch Assemblies

2 gang brown switches, 15 amp, 2-pole, in 4/S x 2-1/8" deep box, ring and plastic plate

S_2S_2

Material	Craft@Hrs	Unit	Material Cost	Labor Cost	Installed Cost
4-S-ring flat	L1@0.89	Ea	36.50	36.50	73.00
4-S-ring 1/4"	L1@0.89	Ea	37.60	36.50	74.10
4-S-ring 1/2"	L1@0.89	Ea	34.20	36.50	70.70
4-S-ring 5/8"	L1@0.89	Ea	34.70	36.50	71.20
4-S-ring 3/4"	L1@0.89	Ea	35.00	36.50	71.50
4-S-ring 1"	L1@0.89	Ea	36.90	36.50	73.40
4-S-ring 1-1/4"	L1@0.89	Ea	37.50	36.50	74.00
4-S-tile ring 1/2"	L1@0.89	Ea	34.50	36.50	71.00
4-S-tile ring 3/4"	L1@0.89	Ea	35.50	36.50	72.00
4-S-tile ring 1"	L1@0.89	Ea	36.90	36.50	73.40
4-S-tile ring 1-1/4"	L1@0.91	Ea	37.50	37.30	74.80
4-S-tile ring 1-1/2"	L1@0.91	Ea	39.10	37.30	76.40
4-S-tile ring 2"	L1@0.91	Ea	41.10	37.30	78.40

2 gang brown switches, 15 amp, 3-way, in 4/S x 2-1/8" deep box, ring and plastic plate

S_3S_3

Material	Craft@Hrs	Unit	Material Cost	Labor Cost	Installed Cost
4-S-ring flat	L1@0.89	Ea	23.90	36.50	60.40
4-S-ring 1/4"	L1@0.89	Ea	25.00	36.50	61.50
4-S-ring 1/2"	L1@0.89	Ea	21.60	36.50	58.10
4-S-ring 5/8"	L1@0.89	Ea	22.10	36.50	58.60
4-S-ring 3/4"	L1@0.89	Ea	22.40	36.50	58.90
4-S-ring 1"	L1@0.89	Ea	24.40	36.50	60.90
4-S-ring 1-1/4"	L1@0.89	Ea	25.00	36.50	61.50
4-S-tile ring 1/2"	L1@0.89	Ea	21.90	36.50	58.40
4-S-tile ring 3/4"	L1@0.89	Ea	22.90	36.50	59.40
4-S-tile ring 1"	L1@0.89	Ea	24.40	36.50	60.90
4-S-tile ring 1-1/4"	L1@0.91	Ea	24.90	37.30	62.20
4-S-tile ring 1-1/2"	L1@0.91	Ea	26.50	37.30	63.80
4-S-tile ring 2"	L1@0.91	Ea	28.50	37.30	65.80

2 gang brown switches, 15 amp, 4-way, in 4/S x 2-1/8" deep box, ring and plastic plate

S_4S_4

Material	Craft@Hrs	Unit	Material Cost	Labor Cost	Installed Cost
4-S-ring flat	L1@0.99	Ea	48.90	40.60	89.50
4-S-ring 1/4"	L1@0.99	Ea	50.00	40.60	90.60
4-S-ring 1/2"	L1@0.99	Ea	46.50	40.60	87.10
4-S-ring 5/8"	L1@0.99	Ea	46.90	40.60	87.50
4-S-ring 3/4"	L1@0.99	Ea	47.30	40.60	87.90
4-S-ring 1"	L1@0.99	Ea	49.30	40.60	89.90
4-S-ring 1-1/4"	L1@0.99	Ea	49.90	40.60	90.50
4-S-tile ring 1/2"	L1@0.99	Ea	46.80	40.60	87.40
4-S-tile ring 3/4"	L1@0.99	Ea	47.90	40.60	88.50
4-S-tile ring 1"	L1@0.99	Ea	49.30	40.60	89.90
4-S-tile ring 1-1/4"	L1@1.01	Ea	49.90	41.40	91.30
4-S-tile ring 1-1/2"	L1@1.01	Ea	51.50	41.40	92.90
4-S-tile ring 2"	L1@1.01	Ea	53.50	41.40	94.90

Use these figures to estimate the cost of installing assemblies under the conditions described on pages 5 and 6. Costs listed are for each assembly installed. The crew is one electrician working at a labor cost of $41.03 per manhour. These costs include layout, material handling and normal waste. Add for the sales tax, delivery, supervision, mobilization, demobilization, cleanup, overhead and profit.

4/S x 2-1/8" Deep, 15 Amp, 1 and 2 Gang Switch Assemblies

Material	Craft@Hrs	Unit	Material Cost	Labor Cost	Installed Cost	
1 gang ivory switches, 15 amp, 1-pole, in 4/S x 2-1/8" deep box, ring and plastic plate						
4-S-ring flat	L1@0.55	Ea	15.00	22.60	37.60	
4-S-ring 1/4"	L1@0.55	Ea	15.40	22.60	38.00	
4-S-ring 1/2"	L1@0.55	Ea	14.20	22.60	36.80	
4-S-ring 5/8"	L1@0.55	Ea	12.90	22.60	35.50	
4-S-ring 3/4"	L1@0.55	Ea	13.20	22.60	35.80	
4-S-ring 1"	L1@0.55	Ea	13.70	22.60	36.30	
4-S-ring 1-1/4"	L1@0.55	Ea	15.10	22.60	37.70	S
4-S-tile ring 1/2"	L1@0.55	Ea	14.50	22.60	37.10	
4-S-tile ring 3/4"	L1@0.55	Ea	14.60	22.60	37.20	
4-S-tile ring 1"	L1@0.55	Ea	14.70	22.60	37.30	
4-S-tile ring 1-1/4"	L1@0.56	Ea	15.10	23.00	38.10	
4-S-tile ring 1-1/2"	L1@0.56	Ea	17.50	23.00	40.50	
4-S-tile ring 2"	L1@0.56	Ea	17.80	23.00	40.80	
1 gang ivory switches, 15 amp, 2-pole, in 4/S x 2-1/8" deep box, ring and plastic plate						
4-S-ring flat	L1@0.60	Ea	22.90	24.60	47.50	
4-S-ring 1/4"	L1@0.60	Ea	23.30	24.60	47.90	
4-S-ring 1/2"	L1@0.60	Ea	22.00	24.60	46.60	
4-S-ring 5/8"	L1@0.60	Ea	20.80	24.60	45.40	
4-S-ring 3/4"	L1@0.60	Ea	21.10	24.60	45.70	
4-S-ring 1"	L1@0.60	Ea	21.60	24.60	46.20	
4-S-ring 1-1/4"	L1@0.60	Ea	22.90	24.60	47.50	S_2
4-S-tile ring 1/2"	L1@0.60	Ea	22.30	24.60	46.90	
4-S-tile ring 3/4"	L1@0.60	Ea	22.40	24.60	47.00	
4-S-tile ring 1"	L1@0.60	Ea	22.60	24.60	47.20	
4-S-tile ring 1-1/4"	L1@0.61	Ea	22.90	25.00	47.90	
4-S-tile ring 1-1/2"	L1@0.61	Ea	25.40	25.00	50.40	
1 gang ivory switches, 15 amp, 3-way, in 4/S x 2-1/8" deep box, ring and plastic plate						
4-S-ring flat	L1@0.60	Ea	16.50	24.60	41.10	
4-S-ring 1/4"	L1@0.60	Ea	17.00	24.60	41.60	
4-S-ring 1/2"	L1@0.60	Ea	15.70	24.60	40.30	
4-S-ring 5/8"	L1@0.60	Ea	14.60	24.60	39.20	
4-S-ring 3/4"	L1@0.60	Ea	14.80	24.60	39.40	
4-S-ring 1"	L1@0.60	Ea	15.30	24.60	39.90	
4-S-ring 1-1/4"	L1@0.60	Ea	16.70	24.60	41.30	S_3
4-S-tile ring 1/2"	L1@0.60	Ea	16.00	24.60	40.60	
4-S-tile ring 3/4"	L1@0.60	Ea	16.10	24.60	40.70	
4-S-tile ring 1"	L1@0.60	Ea	16.20	24.60	40.80	
4-S-tile ring 1-1/4"	L1@0.61	Ea	16.70	25.00	41.70	
4-S-tile ring 1-1/2"	L1@0.61	Ea	19.00	25.00	44.00	
4-S-tile ring 2"	L1@0.61	Ea	19.40	25.00	44.40	

Use these figures to estimate the cost of installing assemblies under the conditions described on pages 5 and 6. Costs listed are for each assembly installed. The crew is one electrician working at a labor cost of $41.03 per manhour. These costs include layout, material handling and normal waste. Add for the sales tax, delivery, supervision, mobilization, demobilization, cleanup, overhead and profit.

4/S x 2-1/8" Deep, 15 Amp, 1 and 2 Gang Switch Assemblies

1 gang ivory switches, 15 amp, 4-way, in 4/S x 2-1/8" deep box, ring and plastic plate

	Material	Craft@Hrs	Unit	Material Cost	Labor Cost	Installed Cost
	4-S-ring flat	L1@0.65	Ea	29.00	26.70	55.70
	4-S-ring 1/4"	L1@0.65	Ea	29.50	26.70	56.20
	4-S-ring 1/2"	L1@0.65	Ea	28.20	26.70	54.90
	4-S-ring 5/8"	L1@0.65	Ea	27.00	26.70	53.70
	4-S-ring 3/4"	L1@0.65	Ea	27.30	26.70	54.00
S_4	4-S-ring 1"	L1@0.65	Ea	27.80	26.70	54.50
	4-S-ring 1-1/4"	L1@0.65	Ea	29.10	26.70	55.80
	4-S-tile ring 1/2"	L1@0.65	Ea	28.50	26.70	55.20
	4-S-tile ring 3/4"	L1@0.65	Ea	28.60	26.70	55.30
	4-S-tile ring 1"	L1@0.65	Ea	28.70	26.70	55.40
	4-S-tile ring 1-1/4"	L1@0.66	Ea	29.10	27.10	56.20
	4-S-tile ring 1-1/2"	L1@0.66	Ea	31.50	27.10	58.60
	4-S-tile ring 2"	L1@0.66	Ea	31.80	27.10	58.90

2 gang ivory switches, 15 amp, 1-pole, in 4/S x 2-1/8" deep box, ring and plastic plate

	4-S-ring flat	L1@0.79	Ea	20.80	32.40	53.20
	4-S-ring 1/4"	L1@0.79	Ea	21.90	32.40	54.30
	4-S-ring 1/2"	L1@0.79	Ea	18.50	32.40	50.90
	4-S-ring 5/8"	L1@0.79	Ea	18.80	32.40	51.20
	4-S-ring 3/4"	L1@0.79	Ea	19.30	32.40	51.70
	4-S-ring 1"	L1@0.79	Ea	21.20	32.40	53.60
SS	4-S-ring 1-1/4"	L1@0.79	Ea	21.80	32.40	54.20
	4-S-tile ring 1/2"	L1@0.79	Ea	18.70	32.40	51.10
	4-S-tile ring 3/4"	L1@0.79	Ea	19.80	32.40	52.20
	4-S-tile ring 1"	L1@0.79	Ea	21.20	32.40	53.60
	4-S-tile ring 1-1/4"	L1@0.81	Ea	21.80	33.20	55.00
	4-S-tile ring 1-1/2"	L1@0.81	Ea	23.40	33.20	56.60
	4-S-tile ring 2"	L1@0.81	Ea	25.40	33.20	58.60

Use these figures to estimate the cost of installing assemblies under the conditions described on pages 5 and 6. Costs listed are for each assembly installed. The crew is one electrician working at a labor cost of $41.03 per manhour. These costs include layout, material handling and normal waste. Add for the sales tax, delivery, supervision, mobilization, demobilization, cleanup, overhead and profit.

4/S x 2-1/8" Deep, 15 Amp, 2 Gang Switch Assemblies

2 gang ivory switches, 15 amp, 2-pole, in 4/S x 2-1/8" deep box, ring and plastic plate

Material	Craft@Hrs	Unit	Material Cost	Labor Cost	Installed Cost
4-S-ring flat	L1@0.89	Ea	36.50	36.50	73.00
4-S-ring 1/4"	L1@0.89	Ea	38.20	36.50	74.70
4-S-ring 1/2"	L1@0.89	Ea	34.20	36.50	70.70
4-S-ring 5/8"	L1@0.89	Ea	34.70	36.50	71.20
4-S-ring 3/4"	L1@0.89	Ea	35.00	36.50	71.50
4-S-ring 1"	L1@0.89	Ea	36.90	36.50	73.40
4-S-ring 1-1/4"	L1@0.89	Ea	37.50	36.50	74.00
4-S-tile ring 1/2"	L1@0.89	Ea	34.50	36.50	71.00
4-S-tile ring 3/4"	L1@0.89	Ea	35.50	36.50	72.00
4-S-tile ring 1"	L1@0.89	Ea	36.90	36.50	73.40
4-S-tile ring 1-1/4"	L1@0.91	Ea	37.50	37.30	74.80
4-S-tile ring 1-1/2"	L1@0.91	Ea	39.10	37.30	76.40
4-S-tile ring 2"	L1@0.91	Ea	41.10	37.30	78.40

S_2S_2

2 gang ivory switches, 15 amp, 3-way, in 4/S x 2-1/8" deep box, ring and plastic plate

Material	Craft@Hrs	Unit	Material Cost	Labor Cost	Installed Cost
4-S-ring flat	L1@0.89	Ea	23.90	36.50	60.40
4-S-ring 1/4"	L1@0.89	Ea	25.00	36.50	61.50
4-S-ring 1/2"	L1@0.89	Ea	21.60	36.50	58.10
4-S-ring 5/8"	L1@0.89	Ea	22.10	36.50	58.60
4-S-ring 3/4"	L1@0.89	Ea	22.40	36.50	58.90
4-S-ring 1"	L1@0.89	Ea	24.40	36.50	60.90
4-S-ring 1-1/4"	L1@0.89	Ea	25.00	36.50	61.50
4-S-tile ring 1/2"	L1@0.89	Ea	21.90	36.50	58.40
4-S-tile ring 3/4"	L1@0.89	Ea	22.90	36.50	59.40
4-S-tile ring 1"	L1@0.89	Ea	24.40	36.50	60.90
4-S-tile ring 1-1/4"	L1@0.91	Ea	24.90	37.30	62.20
4-S-tile ring 1-1/2"	L1@0.91	Ea	26.50	37.30	63.80
4-S-tile ring 2"	L1@0.91	Ea	28.50	37.30	65.80

S_3S_3

2 gang ivory switches, 15 amp, 4-way, in 4/S x 2-1/8" deep box, ring and plastic plate

Material	Craft@Hrs	Unit	Material Cost	Labor Cost	Installed Cost
4-S-ring flat	L1@0.99	Ea	48.90	40.60	89.50
4-S-ring 1/4"	L1@0.99	Ea	50.00	40.60	90.60
4-S-ring 1/2"	L1@0.99	Ea	46.50	40.60	87.10
4-S-ring 5/8"	L1@0.99	Ea	46.90	40.60	87.50
4-S-ring 3/4"	L1@0.99	Ea	47.30	40.60	87.90
4-S-ring 1"	L1@0.99	Ea	49.30	40.60	89.90
4-S-ring 1-1/4"	L1@0.99	Ea	49.90	40.60	90.50
4-S-tile ring 1/2"	L1@0.99	Ea	46.80	40.60	87.40
4-S-tile ring 3/4"	L1@0.99	Ea	47.90	40.60	88.50
4-S-tile ring 1"	L1@0.99	Ea	49.30	40.60	89.90
4-S-tile ring 1-1/4"	L1@1.01	Ea	49.90	41.40	91.30
4-S-tile ring 1-1/2"	L1@1.01	Ea	51.50	41.40	92.90
4-S-tile ring 2"	L1@1.01	Ea	53.50	41.40	94.90

S_4S_4

Use these figures to estimate the cost of installing assemblies under the conditions described on pages 5 and 6. Costs listed are for each assembly installed. The crew is one electrician working at a labor cost of $41.03 per manhour. These costs include layout, material handling and normal waste. Add for the sales tax, delivery, supervision, mobilization, demobilization, cleanup, overhead and profit.

Handy Box and Receptacle Assemblies

Receptacles in 1-1/2" deep handy boxes with handy box cover

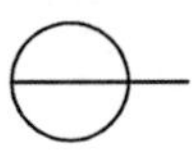

Material	Craft@Hrs	Unit	Material Cost	Labor Cost	Installed Cost
Single 15 amp brown	L1@0.39	Ea	19.40	16.00	35.40
Single 15 amp ivory	L1@0.39	Ea	19.40	16.00	35.40
Single 15 amp white	L1@0.39	Ea	19.40	16.00	35.40
Duplex 15 amp brown	L1@0.38	Ea	17.40	15.60	33.00
Duplex 15 amp ivory	L1@0.38	Ea	17.40	15.60	33.00
Duplex 15 amp white	L1@0.38	Ea	17.40	15.60	33.00
Single 20 amp brown	L1@0.38	Ea	20.10	15.60	35.70
Single 20 amp ivory	L1@0.38	Ea	20.10	15.60	35.70
Single 20 amp white	L1@0.38	Ea	20.10	15.60	35.70
Duplex 20 amp brown	L1@0.38	Ea	21.00	15.60	36.60
Duplex 20 amp ivory	L1@0.38	Ea	21.00	15.60	36.60
Duplex 20 amp white	L1@0.38	Ea	21.00	15.60	36.60

Receptacles in 1-7/8" deep handy boxes with handy box cover

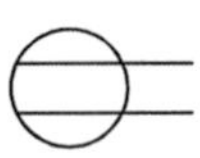

Material	Craft@Hrs	Unit	Material Cost	Labor Cost	Installed Cost
Single 15 amp brown	L1@0.41	Ea	19.60	16.80	36.40
Single 15 amp ivory	L1@0.41	Ea	19.60	16.80	36.40
Single 15 amp white	L1@0.41	Ea	19.60	16.80	36.40
Duplex 15 amp brown	L1@0.40	Ea	17.70	16.40	34.10
Duplex 15 amp ivory	L1@0.40	Ea	17.70	16.40	34.10
Duplex 15 amp white	L1@0.40	Ea	17.70	16.40	34.10
Single 20 amp brown	L1@0.40	Ea	20.40	16.40	36.80
Single 20 amp ivory	L1@0.40	Ea	20.40	16.40	36.80
Single 20 amp white	L1@0.40	Ea	20.40	16.40	36.80
Duplex 20 amp brown	L1@0.40	Ea	21.20	16.40	37.60
Duplex 20 amp ivory	L1@0.40	Ea	21.20	16.40	37.60
Duplex 20 amp white	L1@0.40	Ea	21.20	16.40	37.60

Receptacles in 2-1/8" deep handy boxes with handy box cover

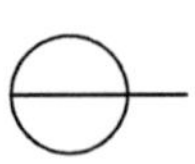

Material	Craft@Hrs	Unit	Material Cost	Labor Cost	Installed Cost
Single 15 amp brown	L1@0.44	Ea	23.30	18.10	41.40
Single 15 amp ivory	L1@0.44	Ea	23.30	18.10	41.40
Single 15 amp white	L1@0.44	Ea	23.30	18.10	41.40
Duplex 15 amp brown	L1@0.43	Ea	21.30	17.60	38.90
Duplex 15 amp ivory	L1@0.43	Ea	21.30	17.60	38.90
Duplex 15 amp white	L1@0.43	Ea	21.30	17.60	38.90

Use these figures to estimate the cost of installing assemblies under the conditions described on pages 5 and 6. Costs listed are for each assembly installed. The crew is one electrician working at a labor cost of $41.03 per manhour. These costs include layout, material handling and normal waste. Add for the sales tax, delivery, supervision, mobilization, demobilization, cleanup, overhead and profit.

Receptacles in 2-1/8" deep handy boxes with handy box cover

Material	Craft@Hrs	Unit	Material Cost	Labor Cost	Installed Cost
Single 20 amp brown	L1@0.43	Ea	24.00	17.60	41.60
Single 20 amp ivory	L1@0.43	Ea	24.00	17.60	41.60
Single 20 amp white	L1@0.43	Ea	24.00	17.60	41.60
Duplex 20 amp brown	L1@0.43	Ea	24.70	17.60	42.30
Duplex 20 amp ivory	L1@0.43	Ea	24.70	17.60	42.30
Duplex 20 amp white	L1@0.43	Ea	24.70	17.60	42.30

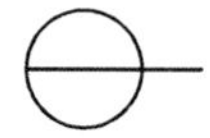

Receptacles in 1-7/8" deep handy boxes with handy box cover and flat bracket

Material	Craft@Hrs	Unit	Material Cost	Labor Cost	Installed Cost
Single 15 amp brown	L1@0.41	Ea	23.10	16.80	39.90
Single 15 amp ivory	L1@0.41	Ea	23.10	16.80	39.90
Single 15 amp white	L1@0.41	Ea	23.10	16.80	39.90
Duplex 15 amp brown	L1@0.40	Ea	21.10	16.40	37.50
Duplex 15 amp ivory	L1@0.40	Ea	21.10	16.40	37.50
Duplex 15 amp white	L1@0.40	Ea	21.10	16.40	37.50
Single 20 amp brown	L1@0.40	Ea	23.80	16.40	40.20
Single 20 amp ivory	L1@0.40	Ea	23.80	16.40	40.20
Single 20 amp white	L1@0.40	Ea	23.80	16.40	40.20
Duplex 20 amp brown	L1@0.40	Ea	24.60	16.40	41.00
Duplex 20 amp ivory	L1@0.40	Ea	24.60	16.40	41.00
Duplex 20 amp white	L1@0.40	Ea	24.60	16.40	41.00

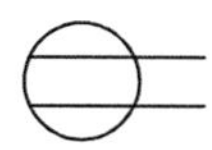

Receptacles in 2-1/8" deep handy boxes with handy box cover and flat bracket

Material	Craft@Hrs	Unit	Material Cost	Labor Cost	Installed Cost
Single 15 amp brown	L1@0.44	Ea	23.30	18.10	41.40
Single 15 amp ivory	L1@0.44	Ea	23.30	18.10	41.40
Single 15 amp white	L1@0.44	Ea	23.30	18.10	41.40
Duplex 15 amp brown	L1@0.43	Ea	21.30	17.60	38.90
Duplex 15 amp ivory	L1@0.43	Ea	21.30	17.60	38.90
Duplex 15 amp white	L1@0.43	Ea	21.30	17.60	38.90
Single 20 amp brown	L1@0.43	Ea	24.00	17.60	41.60
Single 20 amp ivory	L1@0.43	Ea	24.00	17.60	41.60
Single 20 amp white	L1@0.43	Ea	24.00	17.60	41.60
Duplex 20 amp brown	L1@0.43	Ea	24.70	17.60	42.30
Duplex 20 amp ivory	L1@0.43	Ea	24.70	17.60	42.30
Duplex 20 amp white	L1@0.43	Ea	24.70	17.60	42.30

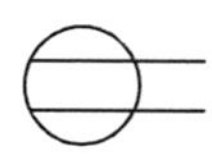

Use these figures to estimate the cost of installing assemblies under the conditions described on pages 5 and 6. Costs listed are for each assembly installed. The crew is one electrician working at a labor cost of $41.03 per manhour. These costs include layout, material handling and normal waste. Add for the sales tax, delivery, supervision, mobilization, demobilization, cleanup, overhead and profit.

Sectional Box and Receptacle Assemblies

Material	Craft@Hrs	Unit	Material Cost	Labor Cost	Installed Cost
Receptacles in 2" deep sectional box and plastic plate					
Single 15 amp brown	L1@0.39	Ea	21.70	16.00	37.70
Single 15 amp ivory	L1@0.39	Ea	21.70	16.00	37.70
Single 15 amp white	L1@0.39	Ea	21.70	16.00	37.70
Duplex 15 amp brown	L1@0.38	Ea	19.70	15.60	35.30
Duplex 15 amp ivory	L1@0.38	Ea	19.70	15.60	35.30
Duplex 15 amp white	L1@0.38	Ea	19.70	15.60	35.30
Single 20 amp brown	L1@0.38	Ea	22.40	15.60	38.00
Single 20 amp ivory	L1@0.38	Ea	22.40	15.60	38.00
Single 20 amp white	L1@0.38	Ea	22.40	15.60	38.00
Duplex 20 amp brown	L1@0.38	Ea	23.20	15.60	38.80
Duplex 20 amp ivory	L1@0.38	Ea	23.20	15.60	38.80
Duplex 20 amp white	L1@0.38	Ea	23.20	15.60	38.80
Receptacles in 2" deep 2 gang device					
Single 15 amp brown	L1@0.78	Ea	43.30	32.00	75.30
Single 15 amp ivory	L1@0.78	Ea	43.30	32.00	75.30
Single 15 amp white	L1@0.78	Ea	43.30	32.00	75.30
Duplex 15 amp brown	L1@0.76	Ea	39.40	31.20	70.60
Duplex 15 amp ivory	L1@0.76	Ea	39.40	31.20	70.60
Duplex 15 amp white	L1@0.76	Ea	39.40	31.20	70.60
Single 20 amp brown	L1@0.76	Ea	44.80	31.20	76.00
Single 20 amp ivory	L1@0.76	Ea	44.80	31.20	76.00
Single 20 amp white	L1@0.76	Ea	44.80	31.20	76.00
Duplex 20 amp brown	L1@0.76	Ea	46.40	31.20	77.60
Duplex 20 amp ivory	L1@0.76	Ea	46.40	31.20	77.60
Duplex 20 amp white	L1@0.76	Ea	46.40	31.20	77.60

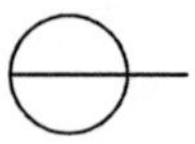

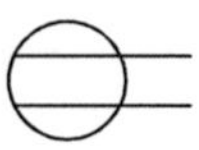

Use these figures to estimate the cost of installing assemblies under the conditions described on pages 5 and 6. Costs listed are for each assembly installed. The crew is one electrician working at a labor cost of $41.03 per manhour. These costs include layout, material handling and normal waste. Add for the sales tax, delivery, supervision, mobilization, demobilization, cleanup, overhead and profit.

Receptacles in 2-1/2" deep sectional box and plastic plate

Material	Craft@Hrs	Unit	Material Cost	Labor Cost	Installed Cost
Single 15 amp brown	L1@0.41	Ea	20.90	16.80	37.70
Single 15 amp ivory	L1@0.41	Ea	20.90	16.80	37.70
Single 15 amp white	L1@0.41	Ea	20.90	16.80	37.70
Duplex 15 amp brown	L1@0.40	Ea	18.90	16.40	35.30
Duplex 15 amp ivory	L1@0.40	Ea	18.90	16.40	35.30
Duplex 15 amp white	L1@0.40	Ea	18.90	16.40	35.30
Single 20 amp brown	L1@0.40	Ea	21.60	16.40	38.00
Single 20 amp ivory	L1@0.40	Ea	21.60	16.40	38.00
Single 20 amp white	L1@0.40	Ea	21.60	16.40	38.00
Duplex 20 amp brown	L1@0.40	Ea	22.40	16.40	38.80
Duplex 20 amp ivory	L1@0.40	Ea	22.40	16.40	38.80
Duplex 20 amp white	L1@0.40	Ea	22.40	16.40	38.80

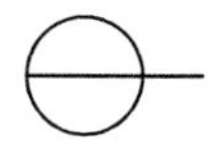

Receptacles in 2-1/2" deep 2 gang device

Material	Craft@Hrs	Unit	Material Cost	Labor Cost	Installed Cost
Single 15 amp brown	L1@0.82	Ea	41.60	33.60	75.20
Single 15 amp ivory	L1@0.82	Ea	41.60	33.60	75.20
Single 15 amp white	L1@0.82	Ea	41.60	33.60	75.20
Duplex 15 amp brown	L1@0.80	Ea	37.70	32.80	70.50
Duplex 15 amp ivory	L1@0.80	Ea	37.70	32.80	70.50
Duplex 15 amp white	L1@0.80	Ea	37.70	32.80	70.50
Single 20 amp brown	L1@0.80	Ea	43.20	32.80	76.00
Single 20 amp ivory	L1@0.80	Ea	43.20	32.80	76.00
Single 20 amp white	L1@0.80	Ea	43.20	32.80	76.00
Duplex 20 amp brown	L1@0.80	Ea	44.80	32.80	77.60
Duplex 20 amp ivory	L1@0.80	Ea	44.80	32.80	77.60
Duplex 20 amp white	L1@0.80	Ea	44.80	32.80	77.60

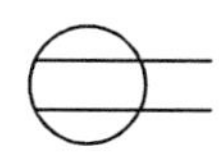

Use these figures to estimate the cost of installing assemblies under the conditions described on pages 5 and 6. Costs listed are for each assembly installed. The crew is one electrician working at a labor cost of $41.03 per manhour. These costs include layout, material handling and normal waste. Add for the sales tax, delivery, supervision, mobilization, demobilization, cleanup, overhead and profit.

Sectional Box and Receptacle Assemblies

Receptacles in 2-3/4" deep sectional box and plastic plate

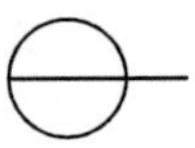

Material	Craft@Hrs	Unit	Material Cost	Labor Cost	Installed Cost
Single 15 amp brown	L1@0.44	Ea	22.30	18.10	40.40
Single 15 amp ivory	L1@0.44	Ea	22.30	18.10	40.40
Single 15 amp white	L1@0.44	Ea	22.30	18.10	40.40
Duplex 15 amp brown	L1@0.43	Ea	20.40	17.60	38.00
Duplex 15 amp ivory	L1@0.43	Ea	20.40	17.60	38.00
Duplex 15 amp white	L1@0.43	Ea	20.40	17.60	38.00
Single 20 amp brown	L1@0.43	Ea	23.10	17.60	40.70
Single 20 amp ivory	L1@0.43	Ea	23.10	17.60	40.70
Single 20 amp white	L1@0.43	Ea	23.10	17.60	40.70
Duplex 20 amp brown	L1@0.43	Ea	23.80	17.60	41.40
Duplex 20 amp ivory	L1@0.43	Ea	23.80	17.60	41.40
Duplex 20 amp white	L1@0.43	Ea	23.80	17.60	41.40

Receptacles in 2-3/4" deep 2 gang device

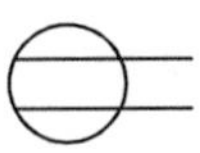

Material	Craft@Hrs	Unit	Material Cost	Labor Cost	Installed Cost
Single 15 amp brown	L1@0.88	Ea	44.60	36.10	80.70
Single 15 amp ivory	L1@0.88	Ea	44.60	36.10	80.70
Single 15 amp white	L1@0.88	Ea	44.60	36.10	80.70
Duplex 15 amp brown	L1@0.86	Ea	40.60	35.30	75.90
Duplex 15 amp ivory	L1@0.86	Ea	40.60	35.30	75.90
Duplex 15 amp white	L1@0.86	Ea	40.60	35.30	75.90
Single 20 amp brown	L1@0.86	Ea	46.00	35.30	81.30
Single 20 amp ivory	L1@0.86	Ea	46.00	35.30	81.30
Single 20 amp white	L1@0.86	Ea	46.00	35.30	81.30
Duplex 20 amp brown	L1@0.86	Ea	47.70	35.30	83.00
Duplex 20 amp ivory	L1@0.86	Ea	47.70	35.30	83.00
Duplex 20 amp white	L1@0.86	Ea	47.70	35.30	83.00

Use these figures to estimate the cost of installing assemblies under the conditions described on pages 5 and 6. Costs listed are for each assembly installed. The crew is one electrician working at a labor cost of $41.03 per manhour. These costs include layout, material handling and normal waste. Add for the sales tax, delivery, supervision, mobilization, demobilization, cleanup, overhead and profit.

Sectional Box and Receptacle Assemblies

Material	Craft@Hrs	Unit	Material Cost	Labor Cost	Installed Cost
Receptacles in 3-1/2" deep sectional box and plastic plate					
Single 15 amp brown	L1@0.49	Ea	22.50	20.10	42.60
Single 15 amp ivory	L1@0.49	Ea	22.50	20.10	42.60
Single 15 amp white	L1@0.49	Ea	22.50	20.10	42.60
Duplex 15 amp brown	L1@0.48	Ea	20.60	19.70	40.30
Duplex 15 amp ivory	L1@0.48	Ea	20.60	19.70	40.30
Duplex 15 amp white	L1@0.48	Ea	20.60	19.70	40.30
Single 20 amp brown	L1@0.48	Ea	23.30	19.70	43.00
Single 20 amp ivory	L1@0.48	Ea	23.30	19.70	43.00
Single 20 amp white	L1@0.48	Ea	23.30	19.70	43.00
Duplex 20 amp brown	L1@0.48	Ea	22.40	19.70	42.10
Duplex 20 amp ivory	L1@0.48	Ea	22.40	19.70	42.10
Duplex 20 amp white	L1@0.48	Ea	22.40	19.70	42.10
Receptacles in 3-1/2" deep 2 gang device					
Single 15 amp brown	L1@0.98	Ea	45.10	40.20	85.30
Single 15 amp ivory	L1@0.98	Ea	45.10	40.20	85.30
Single 15 amp white	L1@0.98	Ea	45.10	40.20	85.30
Duplex 15 amp brown	L1@0.96	Ea	41.10	39.40	80.50
Duplex 15 amp ivory	L1@0.96	Ea	41.10	39.40	80.50
Duplex 15 amp white	L1@0.96	Ea	41.10	39.40	80.50
Single 20 amp brown	L1@0.96	Ea	46.50	39.40	85.90
Single 20 amp ivory	L1@0.96	Ea	46.50	39.40	85.90
Single 20 amp white	L1@0.96	Ea	46.50	39.40	85.90
Duplex 20 amp brown	L1@0.96	Ea	44.80	39.40	84.20
Duplex 20 amp ivory	L1@0.96	Ea	44.80	39.40	84.20
Duplex 20 amp white	L1@0.96	Ea	44.80	39.40	84.20

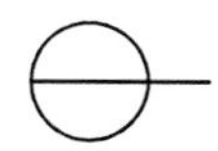

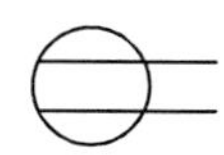

Use these figures to estimate the cost of installing assemblies under the conditions described on pages 5 and 6. Costs listed are for each assembly installed. The crew is one electrician working at a labor cost of $41.03 per manhour. These costs include layout, material handling and normal waste. Add for the sales tax, delivery, supervision, mobilization, demobilization, cleanup, overhead and profit.

4/S x 1-1/2" Deep, 15 Amp, 1 Gang
Single Receptacle Assemblies

1 gang brown single receptacle, 15 amp, in 4/S x 1-1/2" deep box, ring and plastic plate

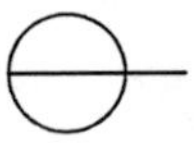

Material	Craft@Hrs	Unit	Material Cost	Labor Cost	Installed Cost
4-S-ring flat	L1@0.54	Ea	23.80	22.20	46.00
4-S-ring 1/4"	L1@0.54	Ea	24.20	22.20	46.40
4-S-ring 1/2"	L1@0.54	Ea	22.80	22.20	45.00
4-S-ring 5/8"	L1@0.54	Ea	21.70	22.20	43.90
4-S-ring 3/4"	L1@0.54	Ea	22.00	22.20	44.20
4-S-ring 1"	L1@0.54	Ea	22.50	22.20	44.70
4-S-ring 1-1/4"	L1@0.54	Ea	23.80	22.20	46.00
4-S-tile ring 1/2"	L1@0.54	Ea	23.90	22.20	46.10
4-S-tile ring 3/4"	L1@0.54	Ea	23.30	22.20	45.50
4-S-tile ring 1"	L1@0.54	Ea	23.50	22.20	45.70
4-S-tile ring 1-1/4"	L1@0.55	Ea	23.80	22.60	46.40
4-S-tile ring 1-1/2"	L1@0.55	Ea	26.30	22.60	48.90
4-S-tile ring 2"	L1@0.55	Ea	26.50	22.60	49.10

1 gang ivory single receptacle, 15 amp, in 4/S x 1-1/2" deep box, ring and plastic plate

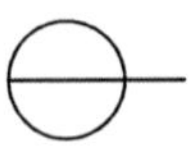

Material	Craft@Hrs	Unit	Material Cost	Labor Cost	Installed Cost
4-S-ring flat	L1@0.54	Ea	23.80	22.20	46.00
4-S-ring 1/4"	L1@0.54	Ea	24.20	22.20	46.40
4-S-ring 1/2"	L1@0.54	Ea	22.80	22.20	45.00
4-S-ring 5/8"	L1@0.54	Ea	21.70	22.20	43.90
4-S-ring 3/4"	L1@0.54	Ea	22.00	22.20	44.20
4-S-ring 1"	L1@0.54	Ea	22.50	22.20	44.70
4-S-ring 1-1/4"	L1@0.54	Ea	23.80	22.20	46.00
4-S-tile ring 1/2"	L1@0.54	Ea	23.20	22.20	45.40
4-S-tile ring 3/4"	L1@0.54	Ea	23.30	22.20	45.50
4-S-tile ring 1"	L1@0.54	Ea	23.50	22.20	45.70
4-S-tile ring 1-1/4"	L1@0.55	Ea	23.80	22.60	46.40
4-S-tile ring 1-1/2"	L1@0.55	Ea	26.30	22.60	48.90
4-S-tile ring 2"	L1@0.55	Ea	26.50	22.60	49.10

1 gang white single receptacle, 15 amp, in 4/S x 1-1/2" deep box, ring and plastic plate

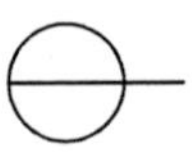

Material	Craft@Hrs	Unit	Material Cost	Labor Cost	Installed Cost
4-S-ring flat	L1@0.54	Ea	23.80	22.20	46.00
4-S-ring 1/4"	L1@0.54	Ea	24.20	22.20	46.40
4-S-ring 1/2"	L1@0.54	Ea	22.80	22.20	45.00
4-S-ring 5/8"	L1@0.54	Ea	21.70	22.20	43.90
4-S-ring 3/4"	L1@0.54	Ea	22.00	22.20	44.20
4-S-ring 1"	L1@0.54	Ea	22.50	22.20	44.70
4-S-ring 1-1/4"	L1@0.54	Ea	23.80	22.20	46.00
4-S-tile ring 1/2"	L1@0.54	Ea	23.20	22.20	45.40
4-S-tile ring 3/4"	L1@0.54	Ea	23.30	22.20	45.50
4-S-tile ring 1"	L1@0.54	Ea	23.50	22.20	45.70
4-S-tile ring 1-1/4"	L1@0.55	Ea	23.80	22.60	46.40
4-S-tile ring 1-1/2"	L1@0.55	Ea	26.30	22.60	48.90
4-S-tile ring 2"	L1@0.55	Ea	26.50	22.60	49.10

Use these figures to estimate the cost of installing assemblies under the conditions described on pages 5 and 6. Costs listed are for each assembly installed. The crew is one electrician working at a labor cost of $41.03 per manhour. These costs include layout, material handling and normal waste. Add for the sales tax, delivery, supervision, mobilization, demobilization, cleanup, overhead and profit.

4/S x 1-1/2" Deep, 15 Amp, 1 Gang Duplex Receptacle Assemblies

1 gang brown duplex receptacle, 15 amp, in 4/S x 1-1/2" deep box, ring and plastic plate

Material	Craft@Hrs	Unit	Material Cost	Labor Cost	Installed Cost
4-S-ring flat	L1@0.53	Ea	21.80	21.70	43.50
4-S-ring 1/4"	L1@0.53	Ea	22.20	21.70	43.90
4-S-ring 1/2"	L1@0.53	Ea	20.90	21.70	42.60
4-S-ring 5/8"	L1@0.53	Ea	19.70	21.70	41.40
4-S-ring 3/4"	L1@0.53	Ea	20.00	21.70	41.70
4-S-ring 1"	L1@0.53	Ea	20.60	21.70	42.30
4-S-ring 1-1/4"	L1@0.53	Ea	21.80	21.70	43.50
4-S-tile ring 1/2"	L1@0.53	Ea	21.20	21.70	42.90
4-S-tile ring 3/4"	L1@0.53	Ea	21.30	21.70	43.00
4-S-tile ring 1"	L1@0.53	Ea	21.50	21.70	43.20
4-S-tile ring 1-1/4"	L1@0.54	Ea	21.80	22.20	44.00
4-S-tile ring 1-1/2"	L1@0.54	Ea	24.30	22.20	46.50
4-S-tile ring 2"	L1@0.54	Ea	24.50	22.20	46.70

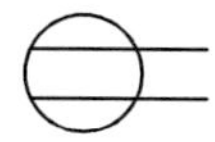

1 gang ivory duplex receptacle, 15 amp, in 4/S x 1-1/2" deep box, ring and plastic plate

Material	Craft@Hrs	Unit	Material Cost	Labor Cost	Installed Cost
4-S-ring flat	L1@0.53	Ea	21.80	21.70	43.50
4-S-ring 1/4"	L1@0.53	Ea	22.20	21.70	43.90
4-S-ring 1/2"	L1@0.53	Ea	20.90	21.70	42.60
4-S-ring 5/8"	L1@0.53	Ea	19.70	21.70	41.40
4-S-ring 3/4"	L1@0.53	Ea	20.00	21.70	41.70
4-S-ring 1"	L1@0.53	Ea	20.60	21.70	42.30
4-S-ring 1-1/4"	L1@0.53	Ea	21.80	21.70	43.50
4-S-tile ring 1/2"	L1@0.53	Ea	21.20	21.70	42.90
4-S-tile ring 3/4"	L1@0.53	Ea	21.30	21.70	43.00
4-S-tile ring 1"	L1@0.53	Ea	21.50	21.70	43.20
4-S-tile ring 1-1/4"	L1@0.54	Ea	21.80	22.20	44.00
4-S-tile ring 1-1/2"	L1@0.54	Ea	24.30	22.20	46.50
4-S-tile ring 2"	L1@0.54	Ea	24.50	22.20	46.70

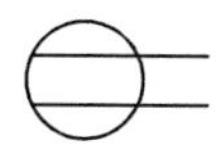

1 gang white duplex receptacle, 15 amp, in 4/S x 1-1/2" deep box, ring and plastic plate

Material	Craft@Hrs	Unit	Material Cost	Labor Cost	Installed Cost
4-S-ring flat	L1@0.53	Ea	21.80	21.70	43.50
4-S-ring 1/4"	L1@0.53	Ea	22.20	21.70	43.90
4-S-ring 1/2"	L1@0.53	Ea	20.90	21.70	42.60
4-S-ring 5/8"	L1@0.53	Ea	19.70	21.70	41.40
4-S-ring 3/4"	L1@0.53	Ea	20.00	21.70	41.70
4-S-ring 1"	L1@0.53	Ea	20.60	21.70	42.30
4-S-ring 1-1/4"	L1@0.53	Ea	21.80	21.70	43.50
4-S-tile ring 1/2"	L1@0.53	Ea	21.20	21.70	42.90
4-S-tile ring 3/4"	L1@0.53	Ea	21.30	21.70	43.00
4-S-tile ring 1"	L1@0.53	Ea	21.50	21.70	43.20
4-S-tile ring 1-1/4"	L1@0.54	Ea	21.80	22.20	44.00
4-S-tile ring 1-1/2"	L1@0.54	Ea	24.30	22.20	46.50
4-S-tile ring 2"	L1@0.54	Ea	24.50	22.20	46.70

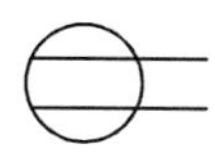

Use these figures to estimate the cost of installing assemblies under the conditions described on pages 5 and 6. Costs listed are for each assembly installed. The crew is one electrician working at a labor cost of $41.03 per manhour. These costs include layout, material handling and normal waste. Add for the sales tax, delivery, supervision, mobilization, demobilization, cleanup, overhead and profit.

4/S x 1-1/2" Deep, 20 Amp, 1 Gang Single Receptacle Assemblies

Material	Craft@Hrs	Unit	Material Cost	Labor Cost	Installed Cost
1 gang brown single receptacle, 20 amp, in 4/S x 1-1/2" deep box, ring and plastic plate					
4-S-ring flat	L1@0.53	Ea	24.50	21.70	46.20
4-S-ring 1/4"	L1@0.53	Ea	24.90	21.70	46.60
4-S-ring 1/2"	L1@0.53	Ea	23.60	21.70	45.30
4-S-ring 5/8"	L1@0.53	Ea	22.40	21.70	44.10
4-S-ring 3/4"	L1@0.53	Ea	22.70	21.70	44.40
4-S-ring 1"	L1@0.53	Ea	23.30	21.70	45.00
4-S-ring 1-1/4"	L1@0.53	Ea	24.60	21.70	46.30
4-S-tile ring 1/2"	L1@0.53	Ea	23.90	21.70	45.60
4-S-tile ring 3/4"	L1@0.53	Ea	24.00	21.70	45.70
4-S-ring 1"	L1@0.53	Ea	24.20	21.70	45.90
4-S-ring 1-1/4"	L1@0.54	Ea	24.50	22.20	46.70
4-S-tile ring 1-1/2"	L1@0.54	Ea	27.00	22.20	49.20
4-S-tile ring 2"	L1@0.54	Ea	27.20	22.20	49.40
1 gang ivory single receptacle, 20 amp, in 4/S x 1-1/2" deep box, ring and plastic plate					
4-S-ring flat	L1@0.53	Ea	24.50	21.70	46.20
4-S-ring 1/4"	L1@0.53	Ea	24.90	21.70	46.60
4-S-ring 1/2"	L1@0.53	Ea	23.60	21.70	45.30
4-S-ring 5/8"	L1@0.53	Ea	22.40	21.70	44.10
4-S-ring 3/4"	L1@0.53	Ea	22.70	21.70	44.40
4-S-ring 1"	L1@0.53	Ea	23.30	21.70	45.00
4-S-ring 1-1/4"	L1@0.53	Ea	24.60	21.70	46.30
4-S-tile ring 1/2"	L1@0.53	Ea	23.90	21.70	45.60
4-S-tile ring 3/4"	L1@0.53	Ea	24.00	21.70	45.70
4-S-tile ring 1"	L1@0.53	Ea	24.20	21.70	45.90
4-S-tile ring 1-1/4"	L1@0.54	Ea	24.50	22.20	46.70
4-S-tile ring 1-1/2"	L1@0.54	Ea	27.00	22.20	49.20
4-S-tile ring 2"	L1@0.54	Ea	27.20	22.20	49.40
1 gang white single receptacle, 20 amp, in 4/S x 1-1/2" deep box, ring and plastic plate					
4-S-ring flat	L1@0.53	Ea	24.50	21.70	46.20
4-S-ring 1/4"	L1@0.53	Ea	24.90	21.70	46.60
4-S-ring 1/2"	L1@0.53	Ea	23.60	21.70	45.30
4-S-ring 5/8"	L1@0.53	Ea	22.40	21.70	44.10
4-S-ring 3/4"	L1@0.53	Ea	22.70	21.70	44.40
4-S-ring 1"	L1@0.53	Ea	23.30	21.70	45.00
4-S-ring 1-1/4"	L1@0.53	Ea	24.60	21.70	46.30
4-S-tile ring 1/2"	L1@0.53	Ea	23.90	21.70	45.60
4-S-tile ring 3/4"	L1@0.53	Ea	24.00	21.70	45.70
4-S-tile ring 1"	L1@0.53	Ea	24.20	21.70	45.90
4-S-tile ring 1-1/4"	L1@0.54	Ea	24.50	22.20	46.70
4-S-tile ring 1-1/2"	L1@0.54	Ea	27.00	22.20	49.20
4-S-tile ring 2"	L1@0.54	Ea	27.20	22.20	49.40

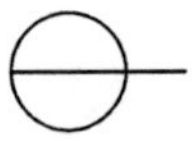
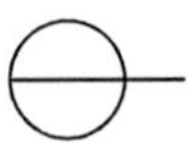
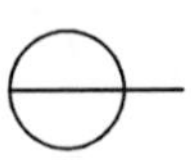

Use these figures to estimate the cost of installing assemblies under the conditions described on pages 5 and 6. Costs listed are for each assembly installed. The crew is one electrician working at a labor cost of $41.03 per manhour. These costs include layout, material handling and normal waste. Add for the sales tax, delivery, supervision, mobilization, demobilization, cleanup, overhead and profit.

1 gang brown duplex receptacle, 20 amp, in 4/S x 1-1/2" deep box, ring and plastic plate

Material	Craft@Hrs	Unit	Material Cost	Labor Cost	Installed Cost
4-S-ring flat	L1@0.53	Ea	25.30	21.70	47.00
4-S-ring 1/4"	L1@0.53	Ea	25.70	21.70	47.40
4-S-ring 1/2"	L1@0.53	Ea	24.40	21.70	46.10
4-S-ring 5/8"	L1@0.53	Ea	23.30	21.70	45.00
4-S-ring 3/4"	L1@0.53	Ea	23.60	21.70	45.30
4-S-ring 1"	L1@0.53	Ea	24.00	21.70	45.70
4-S-ring 1-1/4"	L1@0.53	Ea	25.30	21.70	47.00
4-S-tile ring 1/2"	L1@0.53	Ea	24.70	21.70	46.40
4-S-tile ring 3/4"	L1@0.53	Ea	24.80	21.70	46.50
4-S-tile ring 1"	L1@0.53	Ea	25.00	21.70	46.70
4-S-tile ring 1-1/4"	L1@0.54	Ea	25.30	22.20	47.50
4-S-tile ring 1-1/2"	L1@0.54	Ea	27.80	22.20	50.00
4-S-tile ring 2"	L1@0.54	Ea	28.00	22.20	50.20

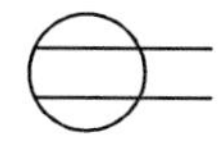

1 gang ivory duplex receptacle, 20 amp, in 4/S x 1-1/2" deep box, ring and plastic plate

Material	Craft@Hrs	Unit	Material Cost	Labor Cost	Installed Cost
4-S-ring flat	L1@0.53	Ea	25.30	21.70	47.00
4-S-ring 1/4"	L1@0.53	Ea	25.70	21.70	47.40
4-S-ring 1/2"	L1@0.53	Ea	24.40	21.70	46.10
4-S-ring 5/8"	L1@0.53	Ea	23.30	21.70	45.00
4-S-ring 3/4"	L1@0.53	Ea	23.60	21.70	45.30
4-S-ring 1"	L1@0.53	Ea	24.00	21.70	45.70
4-S-ring 1-1/4"	L1@0.53	Ea	25.30	21.70	47.00
4-S-tile ring 1/2"	L1@0.53	Ea	24.70	21.70	46.40
4-S-tile ring 3/4"	L1@0.53	Ea	24.80	21.70	46.50
4-S-tile ring 1"	L1@0.53	Ea	25.00	21.70	46.70
4-S-tile ring 1-1/4"	L1@0.54	Ea	25.30	22.20	47.50
4-S-tile ring 1-1/2"	L1@0.54	Ea	27.80	22.20	50.00
4-S-tile ring 2"	L1@0.54	Ea	28.00	22.20	50.20

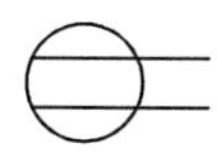

1 gang white duplex receptacle, 20 amp, in 4/S x 1-1/2" deep box, ring and plastic plate

Material	Craft@Hrs	Unit	Material Cost	Labor Cost	Installed Cost
4-S-ring flat	L1@0.53	Ea	25.30	21.70	47.00
4-S-ring 1/4"	L1@0.53	Ea	25.70	21.70	47.40
4-S-ring 1/2"	L1@0.53	Ea	24.40	21.70	46.10
4-S-ring 5/8"	L1@0.53	Ea	23.30	21.70	45.00
4-S-ring 3/4"	L1@0.53	Ea	23.60	21.70	45.30
4-S-ring 1"	L1@0.53	Ea	24.00	21.70	45.70
4-S-ring 1-1/4"	L1@0.53	Ea	25.30	21.70	47.00
4-S-tile ring 1/2"	L1@0.53	Ea	24.70	21.70	46.40
4-S-tile ring 3/4"	L1@0.53	Ea	24.80	21.70	46.50
4-S-tile ring 1"	L1@0.53	Ea	25.00	21.70	46.70
4-S-tile ring 1-1/4"	L1@0.54	Ea	25.30	22.20	47.50
4-S-tile ring 1-1/2"	L1@0.54	Ea	27.80	22.20	50.00
4-S-tile ring 2"	L1@0.54	Ea	27.00	22.20	49.20

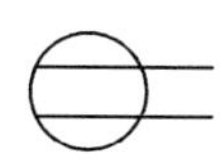

Use these figures to estimate the cost of installing assemblies under the conditions described on pages 5 and 6. Costs listed are for each assembly installed. The crew is one electrician working at a labor cost of $41.03 per manhour. These costs include layout, material handling and normal waste. Add for the sales tax, delivery, supervision, mobilization, demobilization, cleanup, overhead and profit.

Troffer Fluorescent Assemblies

Framed troffer lay-in T-bar fluorescent fixtures with lamps, 5' aluminum flex whip, 3-#12 THHN solid copper wire and wire connectors

24" wide steel frame troffer lay-in T-bar fixtures, 5' flex whip, 3-#12 THHN solid copper wire

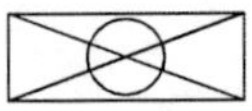

Material	Craft@Hrs	Unit	Material Cost	Labor Cost	Installed Cost
48" 2 lamp	L1@1.06	Ea	211.00	43.50	254.50
48" 3 lamp	L1@1.16	Ea	259.00	47.60	306.60
48" 4 lamp	L1@1.26	Ea	286.00	51.70	337.70
48" 2 lamp energy saver	L1@1.06	Ea	217.00	43.50	260.50
48" 3 lamp energy saver	L1@1.16	Ea	284.00	47.60	331.60
48" 4 lamp energy saver	L1@1.26	Ea	306.00	51.70	357.70

Recessed 24" wide steel frame troffer T-bar fixtures, 5' flex whip, 3-#12 THHN solid copper wire

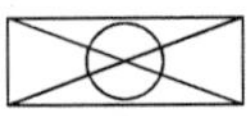

Material	Craft@Hrs	Unit	Material Cost	Labor Cost	Installed Cost
48" 2 lamp	L1@1.06	Ea	219.00	43.50	262.50
48" 3 lamp	L1@1.16	Ea	271.00	47.60	318.60
48" 4 lamp	L1@1.26	Ea	298.00	51.70	349.70
48" 2 lamp energy saver	L1@1.06	Ea	230.00	43.50	273.50
48" 3 lamp energy saver	L1@1.16	Ea	301.00	47.60	348.60
48" 4 lamp energy saver	L1@1.26	Ea	319.00	51.70	370.70

24" wide aluminum frame troffer T-bar fixtures, 5' flex whip, 3-#12 THHN solid copper wire

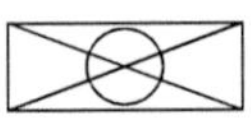

Material	Craft@Hrs	Unit	Material Cost	Labor Cost	Installed Cost
48" 2 lamp	L1@1.06	Ea	217.00	43.50	260.50
48" 3 lamp	L1@1.16	Ea	266.00	47.60	313.60
48" 4 lamp	L1@1.26	Ea	303.00	51.70	354.70
48" 2 lamp energy saver	L1@1.06	Ea	240.00	43.50	283.50
48" 3 lamp energy saver	L1@1.16	Ea	304.00	47.60	351.60
48" 4 lamp energy saver	L1@1.26	Ea	337.00	51.70	388.70

Recessed 24" wide aluminum frame troffer T-bar fixtures, 5' flex whip, 3-#12 THHN solid copper wire

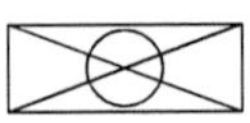

Material	Craft@Hrs	Unit	Material Cost	Labor Cost	Installed Cost
48" 2 lamp	L1@1.06	Ea	227.00	43.50	270.50
48" 3 lamp	L1@1.16	Ea	277.00	47.60	324.60
48" 4 lamp	L1@1.26	Ea	309.00	51.70	360.70
48" 2 lamp energy saver	L1@1.06	Ea	246.00	43.50	289.50
48" 3 lamp energy saver	L1@1.16	Ea	317.00	47.60	364.60
48" 4 lamp energy saver	L1@1.26	Ea	341.00	51.70	392.70

Use these figures to estimate the cost of installing assemblies under the conditions described on pages 5 and 6. Costs listed are for each assembly installed. The crew is one electrician working at a labor cost of $41.03 per manhour. These costs include layout, material handling and normal waste. Add for the sales tax, delivery, supervision, mobilization, demobilization, cleanup, overhead and profit.

Section 18: Communications

This section lists many types and sizes of communications wire and cable, as well as connectors and terminators commonly used in this work. The material tables are designed to make it easy to identify the components you need to estimate costs for your project.

As usual in estimating, there are many variables. It's up to the estimator to make the necessary adjustments to fit the task at hand. A small compact job site with short runs of cabling will require more labor hours per thousand feet than a job that's spread out over many thousand square feet, with long cable runs. This section uses an average between the short and long runs of cabling, so some adjustment in manhours may be necessary.

Every estimator has studied the time it takes to do certain tasks — and they'll all arrive at slightly different manhour values. If you believe your company's crew can install the work in more or less time than is listed here, make an adjustment by splitting the difference between what's listed in this section and what you believe it should be. If you get the job, pay close attention to the performance of the installation crews. Check their actual performance against the estimated time, so you can make adjustments for future estimates. But don't change the estimated manhours too much. Just because your crew did one particular job at a better rate doesn't mean that'll happen on all your jobs. There are unique conditions on every project. The manhours in this book are averages of multiple jobs, and include the "surprises."

Cable Distribution Systems

It may take longer to install communications cable in a conduit system than it would to install the cable in an accessible ceiling space. The specifications may require at least one of the following cable distribution systems.

Surface Cabling

The communications cable is installed directly to the surface with staples or straps.

Conduit System (see Section 1)

The specifications may require the electrical contractor to install a conduit system complete with boxes for the communications devices. Sometimes the electricians are required to install a pull string in the empty conduits. This makes it easier for the electronics technicians to pull in their cables in the conduit systems. If there's no requirement for the electrician to install a pull string, the communications tech must either install the pull string or use a *fish tape* in order to pull in the cable.

Metal Wireway (see Section 6)

The specifications may require the electrical contractor to install a metal wireway (*gutter*) system for the installation of the communications cable. The wireway could be either *screw cover* or *hinged cover*. The gutter comes in many sizes with an assortment of fittings. The communications cable is usually bundled together in the gutter and tagged for identification.

Underfloor Raceway (see Section 7)

The underfloor raceway system is usually installed by the electrical contractor during construction of the concrete floor. The raceway is connected to flush floor junction boxes for easy access. The underfloor *duct* usually has threaded hubs just below the floor surface on 2-foot centers for mounting an above-floor device or junction box. There may be at least two ducts running in parallel, one for electrical power, the other for cable. After the devices and junction boxes are mounted on the low voltage duct, the cable is then fished from the access junction box, through the duct low voltage system and to the location of the device.

Cable Tray System (see Section 9)

The cable tray system is installed by the electrical contractor during the building construction. The tray system is installed in the ceiling space or exposed overhead and may be a simple *ladder* tray, an *expanded metal* tray or *sheet metal* tray. Cable trays are made of galvanized sheet metal or aluminum. Some trays have covers to protect the cable, while others are open. The cable is bundled and tagged for identification, then laid in the cable tray.

Surface Raceway (see Section 15)

Surface raceway is usually installed by the electrical contractor during the building construction. The raceway is installed as a finish product on walls and ceilings, with junction boxes for access. Surface raceway boxes are installed for devices and may require plates to accommodate cable connections to equipment. The raceway fittings, such as elbows and tees, have removable covers for pulling in the cable. The raceway is made of either sheet metal or plastic, and is painted.

Cellular Flooring

Some buildings have cellular flooring that's installed by the iron workers as the building is being constructed. The cellular flooring, made of sheet metal, is usually installed in large office areas. The cells span the selected floor area and are connected with *header* ducts installed by the electrical contractor. Usually, a lightweight concrete floor is poured over the top of the cellular floor.

A 2-inch hole is drilled through the lightweight concrete floor and into the selected cell. Cells are designated for electrical power, communications and electronic uses. The cable is pulled from a terminal board or cabinet, through the header duct, to the cell and to the device outlet.

The header duct has removable covers for easy access to the cells.

Retrofitting

The biggest challenge is to install a communications system in an existing building without doing too much demolition. Some of the methods listed above may be the answer — or you may have to use some creativity to come up with a workable plan that will keep the owner happy.

The manufacturers seem to come up with newer and better ways to meet the needs of the retrofitter. Your first step should be to contact your supplier; they're willing to help you find the right answers.

Material	Craft@Hrs	Unit	Material Cost	Labor Cost	Installed Cost

22 AWG stranded communications cable, unshielded twisted pair, 80 degree, PVC jacket, 150 volt

Material	Craft@Hrs	Unit	Material Cost	Labor Cost	Installed Cost
2 pairs	L2@5.50	KLF	430.00	226.00	656.00
3 pairs	L2@5.75	KLF	597.00	236.00	833.00
4 pairs	L2@6.00	KLF	768.00	246.00	1,014.00
5 pairs	L2@6.25	KLF	1,150.00	256.00	1,406.00
6 pairs	L2@6.50	KLF	1,540.00	267.00	1,807.00
9 pairs	L2@7.25	KLF	1,640.00	297.00	1,937.00
11 pairs	L2@7.75	KLF	3,580.00	318.00	3,898.00
12 pairs	L2@8.00	KLF	3,720.00	328.00	4,048.00
15 pairs	L2@8.75	KLF	3,850.00	359.00	4,209.00
19 pairs	L2@9.75	KLF	5,540.00	400.00	5,940.00

22 AWG solid communications cable, unshielded twisted pair, 80 degree, PVC jacket, 150 volt

Material	Craft@Hrs	Unit	Material Cost	Labor Cost	Installed Cost
1 pair	L2@5.25	KLF	199.00	215.00	414.00
2 pairs	L2@5.50	KLF	381.00	226.00	607.00
3 pairs	L2@5.75	KLF	599.00	236.00	835.00
4 pairs	L2@6.00	KLF	794.00	246.00	1,040.00
5 pairs	L2@6.25	KLF	1,010.00	256.00	1,266.00
6 pairs	L2@6.50	KLF	1,250.00	267.00	1,517.00
11 pairs	L2@7.75	KLF	2,950.00	318.00	3,268.00
13 pairs	L2@8.25	KLF	3,360.00	338.00	3,698.00
16 pairs	L2@9.00	KLF	4,470.00	369.00	4,839.00
27 pairs	L2@11.8	KLF	4,570.00	484.00	5,054.00

18 AWG stranded communications cable, unshielded twisted pair, 80 degree, PVC jacket, 300 volt

Material	Craft@Hrs	Unit	Material Cost	Labor Cost	Installed Cost
1 pair	L2@5.50	KLF	685.00	226.00	911.00
2 pairs	L2@5.75	KLF	875.00	236.00	1,111.00
3 pairs	L2@6.00	KLF	1,250.00	246.00	1,496.00
4 pairs	L2@6.25	KLF	1,620.00	256.00	1,876.00
5 pairs	L2@6.50	KLF	1,970.00	267.00	2,237.00
6 pairs	L2@6.75	KLF	2,920.00	277.00	3,197.00
8 pairs	L2@7.25	KLF	3,720.00	297.00	4,017.00
9 pairs	L2@7.50	KLF	4,620.00	308.00	4,928.00
15 pairs	L2@8.75	KLF	6.500.00	359.00	6,859.00
19 pairs	L2@10.0	KLF	9,660.00	410.00	10,070.00

Use these figures to estimate the cost of communications cable installed in conduit under the conditions described on pages 5 and 6. Costs listed are for each 1,000 linear feet installed. The crew is two technicians. The labor cost per manhour is $41.03. These costs include terminations, reel setup, cable identification, cable testing, layout, material handling, and normal waste. Add for sales tax, delivery, supervision, mobilization, demobilization, cleanup, overhead and profit.

Communications Cable

Material	Craft@Hrs	Unit	Material Cost	Labor Cost	Installed Cost
24 AWG stranded communications cable, individually shielded twisted pair, 80 degree, PVC jacket, 300 volt, with drain wire					
1 pair	L2@5.00	KLF	283.00	205.00	488.00
2 pairs	L2@5.25	KLF	376.00	215.00	591.00
3 pairs	L2@5.50	KLF	512.00	226.00	738.00
4 pairs	L2@5.75	KLF	680.00	236.00	916.00
5 pairs	L2@6.00	KLF	821.00	246.00	1,067.00
6 pairs	L2@6.25	KLF	1,030.00	256.00	1,286.00
7 pairs	L2@6.50	KLF	1,250.00	267.00	1,517.00
8 pairs	L2@6.75	KLF	1,380.00	277.00	1,657.00
9 pairs	L2@7.00	KLF	1,510.00	287.00	1,797.00
10 pairs	L2@7.25	KLF	1,550.00	297.00	1,847.00
15 pairs	L2@8.50	KLF	2,450.00	349.00	2,799.00
19 pairs	L2@9.50	KLF	3,770.00	390.00	4,160.00
25 pairs	L2@11.0	KLF	4,290.00	451.00	4,741.00
22 AWG stranded communications cable, individually shielded twisted pair, 80 degree, PVC jacket, 300 volt, with drain wire					
2 pairs	L2@5.75	KLF	879.00	236.00	1,115.00
3 pairs	L2@6.00	KLF	1,250.00	246.00	1,496.00
4 pairs	L2@6.25	KLF	1,660.00	256.00	1,916.00
6 pairs	L2@6.75	KLF	2,600.00	277.00	2,877.00
9 pairs	L2@7.50	KLF	3,810.00	308.00	4,118.00
19 pairs	L2@10.0	KLF	7,940.00	410.00	8,350.00
27 pairs	L2@12.0	KLF	11,600.00	492.00	12,092.00
18 AWG stranded communications cable, individually shielded twisted pair, 80 degree, PVC jacket, 300 volt, with drain wire					
2 pairs	L2@6.00	KLF	1,440.00	246.00	1,686.00
3 pairs	L2@6.25	KLF	1,920.00	256.00	2,176.00
4 pairs	L2@6.50	KLF	2,500.00	267.00	2,767.00
6 pairs	L2@7.00	KLF	3,910.00	287.00	4,197.00
9 pairs	L2@7.25	KLF	5,900.00	297.00	6,197.00
15 pairs	L2@9.25	KLF	9,800.00	380.00	10,180.00

Use these figures to estimate the cost of communications cable installed in conduit under the conditions described on pages 5 and 6. Costs listed are for each 1,000 linear feet installed. The crew is two technicians. The labor cost per manhour is $41.03. These costs include terminations, reel setup, cable identification, cable testing, layout, material handling, and normal waste. Add for sales tax, delivery, supervision, mobilization, demobilization, cleanup, overhead and profit.

Material	Craft@Hrs	Unit	Material Cost	Labor Cost	Installed Cost
22 AWG stranded communications cable, individually shielded twisted pair, 60 degree, 300 volt					
3 pairs	L2@5.50	KLF	692.00	226.00	918.00
6 pairs	L2@6.25	KLF	1,330.00	256.00	1,586.00
9 pairs	L2@7.00	KLF	3,210.00	287.00	3,497.00
11 pairs	L2@7.50	KLF	3,590.00	308.00	3,898.00
12 pairs	L2@7.75	KLF	4,930.00	318.00	5,248.00
15 pairs	L2@8.50	KLF	4,900.00	349.00	5,249.00
17 pairs	L2@9.00	KLF	6,360.00	369.00	6,729.00
19 pairs	L2@9.50	KLF	6,260.00	390.00	6,650.00
27 pairs	L2@11.5	KLF	11,200.00	472.00	11,672.00
20 AWG stranded communications cable, individually shielded twisted pair, 60 degree, 350 volt, with drain wire					
2 pairs	L2@5.75	KLF	893.00	236.00	1,129.00
3 pairs	L2@6.00	KLF	1,100.00	246.00	1,346.00
6 pairs	L2@6.75	KLF	2,600.00	277.00	2,877.00
9 pairs	L2@7.50	KLF	4,200.00	308.00	4,508.00
12 pairs	L2@8.25	KLF	5,450.00	338.00	5,788.00
15 pairs	L2@9.00	KLF	8,250.00	369.00	8,619.00
18 AWG stranded communications cable, individually shielded twisted pair, 60 degree, 400 volt, with drain wire					
3 pairs	L2@6.25	KLF	1,550.00	256.00	1,806.00
6 pairs	L2@7.00	KLF	3,210.00	287.00	3,497.00
9 pairs	L2@7.75	KLF	5,040.00	318.00	5,358.00
22 AWG stranded multi-conductor unshielded communications cable, 80 degree, PVC jacket, 150 volt					
2 pairs	L2@5.00	KLF	202.00	205.00	407.00
3 pairs	L2@5.25	KLF	268.00	215.00	483.00
4 pairs	L2@5.50	KLF	338.00	226.00	564.00
5 pairs	L2@5.75	KLF	437.00	236.00	673.00
6 pairs	L2@6.00	KLF	502.00	246.00	748.00
7 pairs	L2@6.25	KLF	547.00	256.00	803.00

Use these figures to estimate the cost of communications cable installed in conduit under the conditions described on pages 5 and 6. Costs listed are for each 1,000 linear feet installed. The crew is two technicians. The labor cost per manhour is $41.03. These costs include terminations, reel setup, cable identification, cable testing, layout, material handling, and normal waste. Add for sales tax, delivery, supervision, mobilization, demobilization, cleanup, overhead and profit.

Communications Cable

Material	Craft@Hrs	Unit	Material Cost	Labor Cost	Installed Cost
22 AWG stranded multi-conductor unshielded communications cable, 80 degree, PVC jacket, 150 volt					
8 pairs	L2@6.50	KLF	655.00	267.00	922.00
9 pairs	L2@6.75	KLF	655.00	277.00	932.00
10 pairs	L2@7.00	KLF	767.00	287.00	1,054.00
12 pairs	L2@7.50	KLF	1,040.00	308.00	1,348.00
15 pairs	L2@8.00	KLF	1,440.00	328.00	1,768.00
20 pairs	L2@10.0	KLF	1,990.00	410.00	2,400.00
25 pairs	L2@11.3	KLF	2,270.00	464.00	2,734.00
30 pairs	L2@12.5	KLF	3,770.00	513.00	4,283.00
40 pairs	L2@15.0	KLF	4,720.00	615.00	5,335.00
50 pairs	L2@17.5	KLF	5,920.00	718.00	6,638.00
60 pairs	L2@20.0	KLF	6,500.00	821.00	7,321.00
20 AWG stranded multi-conductor unshielded communications cable, 80 degree, PVC jacket, 300 volt					
2 pairs	L2@5.50	KLF	358.00	226.00	540.00
3 pairs	L2@5.75	KLF	406.00	236.00	642.00
4 pairs	L2@6.00	KLF	456.00	246.00	702.00
5 pairs	L2@6.25	KLF	636.00	256.00	892.00
6 pairs	L2@6.50	KLF	705.00	267.00	972.00
7 pairs	L2@6.75	KLF	790.00	277.00	1,067.00
8 pairs	L2@7.00	KLF	1,150.00	287.00	1,437.00
9 pairs	L2@7.25	KLF	1,070.00	297.00	1,367.00
10 pairs	L2@7.50	KLF	1,280.00	308.00	1,588.00
12 pairs	L2@8.00	KLF	1,510.00	328.00	1,838.00
15 pairs	L2@8.75	KLF	1,920.00	359.00	2,279.00

Use these figures to estimate the cost of communications cable installed in conduit under the conditions described on pages 5 and 6. Costs listed are for each 1,000 linear feet installed. The crew is two technicians. The labor cost per manhour is $41.03. These costs include terminations, reel setup, cable identification, cable testing, layout, material handling, and normal waste. Add for sales tax, delivery, supervision, mobilization, demobilization, cleanup, overhead and profit.

Material	Craft@Hrs	Unit	Material Cost	Labor Cost	Installed Cost
24 AWG stranded multi-conductor foil/braid shielded communications cable, 80 degree, PVC jacket, 300 volt					
2 pairs	L2@6.00	KLF	2,450.00	246.00	2,696.00
3 pairs	L2@6.25	KLF	2,780.00	256.00	3,036.00
4 pairs	L2@6.50	KLF	2,960.00	267.00	3,227.00
6 pairs	L2@7.25	KLF	3,470.00	297.00	3,767.00
8 pairs	L2@7.75	KLF	3,910.00	318.00	4,228.00
10 pairs	L2@8.25	KLF	4,330.00	338.00	4,668.00
15 pairs	L2@9.50	KLF	5,990.00	390.00	6,380.00
20 pairs	L2@10.8	KLF	7,850.00	443.00	8,293.00
25 pairs	L2@12.0	KLF	8,840.00	492.00	9,332.00
30 pairs	L2@13.3	KLF	10,600.00	546.00	11,146.00
40 pairs	L2@15.8	KLF	14,400.00	648.00	15,048.00
50 pairs	L2@17.8	KLF	17,500.00	730.00	18,230.00
24 AWG telephone cable, 60 degree, 90 volt, solid twisted pairs, tinned copper					
2 pairs	L2@5.25	KLF	253.00	215.00	468.00
6 pairs	L2@6.50	KLF	637.00	267.00	904.00
10 pairs	L2@7.50	KLF	971.00	308.00	1,279.00
25 pairs	L2@11.3	KLF	2,330.00	464.00	2,794.00
24 AWG telephone cable, 60 degree, 90 volt, solid bare copper					
6 pairs	L2@6.50	KLF	148.00	267.00	415.00
22 AWG telephone cable, 60 degree, 90 volt, solid bare copper					
2 pairs	L2@5.25	KLF	64.40	215.00	279.40
3 pairs	L2@5.50	KLF	84.90	226.00	310.90
4 pairs	L2@5.75	KLF	88.00	236.00	324.00
6 pairs	L2@6.25	KLF	194.00	256.00	450.00

Use these figures to estimate the cost of communications cable installed in conduit under the conditions described on pages 5 and 6. Costs listed are for each 1,000 linear feet installed. The crew is two technicians. The labor cost per manhour is $41.03. These costs include terminations, reel setup, cable identification, cable testing, layout, material handling, and normal waste. Add for sales tax, delivery, supervision, mobilization, demobilization, cleanup, overhead and profit.

Communications Cable

Coaxial communications cable

Material	Craft@Hrs	Unit	Material Cost	Labor Cost	Installed Cost
6/U 18 AWG, solid	L2@7.00	KLF	577.00	287.00	864.00
6A/U 21 AWG, solid	L2@7.00	KLF	3,060.00	287.00	3,347.00
8/U 13 AWG, stranded	L2@8.00	KLF	1,440.00	328.00	1,768.00
8A/U 13 AWG, stranded	L2@8.00	KLF	1,920.00	328.00	2,248.00
11/U 18 AWG, stranded	L2@8.50	KLF	1,610.00	349.00	1,959.00
11/U 14 AWG, solid	L2@9.00	KLF	1,440.00	369.00	1,809.00
11A/U 18 AWG, stranded	L2@8.50	KLF	2,500.00	349.00	2,849.00
58/U 20 AWG, solid	L2@8.00	KLF	393.00	328.00	721.00
58A/U 20 AWG, stranded	L2@8.50	KLF	485.00	349.00	834.00
58C/U 20 AWG, stranded	L2@8.50	KLF	1,250.00	349.00	1,599.00
59/U 22 AWG, solid	L2@8.00	KLF	433.00	328.00	761.00
59/U 22 AWG, stranded	L2@8.50	KLF	799.00	349.00	1,148.00
59B/U 23 AWG, solid	L2@8.00	KLF	1,010.00	328.00	1,338.00
62/U 22 AWG, solid	L2@8.00	KLF	584.00	328.00	912.00
62A/U 22 AWG, solid	L2@8.00	KLF	814.00	328.00	1,142.00

Twinaxial communications cable

Material	Craft@Hrs	Unit	Material Cost	Labor Cost	Installed Cost
95 ohm 18 AWG, str	L2@9.00	KLF	4,280.00	369.00	4,649.00
100 ohm 18 AWG, str	L2@9.00	KLF	1,260.00	369.00	1,629.00
100 ohm 20 AWG, str	L2@9.00	KLF	1,210.00	369.00	1,579.00
78 ohm 20 AWG, str, data	L2@9.50	KLF	1,200.00	390.00	1,590.00
100 ohm 22 AWG, sol, data	L2@9.50	KLF	1,310.00	390.00	1,700.00
124 ohm 16 AWG, sol, data	L2@9.50	KLF	2,910.00	390.00	3,300.00
124 ohm 25 AWG, str, data	L2@9.50	KLF	650.00	390.00	1,040.00

Plenum-rated coaxial communications cable

Material	Craft@Hrs	Unit	Material Cost	Labor Cost	Installed Cost
11/U 14 AWG, solid	L2@8.50	KLF	5,820.00	349.00	6,169.00
58C/U 20 AWG, solid	L2@8.00	KLF	2,200.00	328.00	2,528.00
58/U 20 AWG, solid	L2@8.00	KLF	1,740.00	328.00	2,068.00
59/U 20 AWG, solid	L2@8.25	KLF	2,230.00	338.00	2,568.00
59B/U 23 AWG, solid	L2@8.25	KLF	4,540.00	338.00	4,878.00
62A/U 22 AWG, solid	L2@8.25	KLF	1,920.00	338.00	2,258.00

Plenum-rated twinaxial communications cable

Material	Craft@Hrs	Unit	Material Cost	Labor Cost	Installed Cost
78 ohm 20 AWG, str	L2@8.00	KLF	2,780.00	328.00	3,108.00

Use these figures to estimate the cost of communications cable installed in conduit under the conditions described on pages 5 and 6. Costs listed are for each 1,000 linear feet installed. The crew is two technicians. The labor cost per manhour is $41.03. These costs include terminations, reel setup, cable identification, cable testing, layout, material handling, and normal waste. Add for sales tax, delivery, supervision, mobilization, demobilization, cleanup, overhead and profit.

Local Area Network (LAN) trunk coaxial cable

Material		Craft@Hrs	Unit	Material Cost	Labor Cost	Installed Cost
6/U	18 AWG, solid	L2@9.00	KLF	772.00	369.00	1,141.00
58C/U	20 AWG, str	L2@9.50	KLF	1,160.00	390.00	1,550.00

Local Area Network (LAN) trunk - coaxial plenum cable

Material		Craft@Hrs	Unit	Material Cost	Labor Cost	Installed Cost
6/U	18 AWG, solid	L2@9.50	KLF	3,050.00	390.00	3,440.00
58C/U	20 AWG, str	L2@9.75	KLF	1,740.00	400.00	2,140.00

Transceiver/drop twisted pair shielded cable

Material		Craft@Hrs	Unit	Material Cost	Labor Cost	Installed Cost
4-3 pairs	22 AWG, str plus					
1 pair	20 AWG, str	L2@12.0	KLF	3,310.00	492.00	3,802.00
4 pairs	20 AWG, str	L2@7.00	KLF	3,310.00	287.00	3,597.00

Use these figures to estimate the cost of communications cable installed in conduit under the conditions described on pages 5 and 6. Costs listed are for each 1,000 linear feet installed. The crew is two technicians. The labor cost per manhour is $41.03. These costs include terminations, reel setup, cable identification, cable testing, layout, material handling, and normal waste. Add for sales tax, delivery, supervision, mobilization, demobilization, cleanup, overhead and profit.

Cable Fittings

Material	Craft@Hrs	Unit	Material Cost	Labor Cost	Installed Cost
Cable contacts, pins for use with plugs					
30 - 26 AWG crimp snap	L1@.100	Ea	.21	4.10	4.31
26 - 22 AWG crimp snap	L1@.100	Ea	.21	4.10	4.31
24 - 20 AWG crimp snap	L1@.100	Ea	.21	4.10	4.31
30 - 26 AWG HDE insl.	L1@.100	Ea	.21	4.10	4.31
26 - 22 AWG HDE Insl.	L2@.100	Ea	.21	4.10	4.31
30 - 26 AWG solder cup	L1@.150	Ea	.24	6.15	6.39
28 - 24 AWG solder cup	L1@.150	Ea	.24	6.15	6.39
26 - 22 AWG solder cup	L1@.150	Ea	.24	6.15	6.39
24 - 20 AWG solder cup	L1@.150	Ea	.24	6.15	6.39
22 - 20 AWG solder cup	L1@.150	Ea	.24	6.15	6.39
Cable contacts, sockets for use with receptacles					
30 - 26 AWG crimp snap	L1@.100	Ea	.21	4.10	4.31
26 - 22 AWG crimp snap	L1@.100	Ea	.21	4.10	4.31
24 - 20 AWG crimp snap	L1@.100	Ea	.21	4.10	4.31
30 - 26 AWG HDE insl.	L1@.100	Ea	.23	4.10	4.33
26 - 22 AWG HDE insl.	L1@.100	Ea	.23	4.10	4.33
22 - 20 AWG HDE insl.	L1@.100	Ea	.23	4.10	4.33
30 - 26 AWG solder cup	L1@.150	Ea	.25	6.15	6.40
28 - 24 AWG solder cup	L1@.150	Ea	.25	6.15	6.40
26 - 22 AWG solder cup	L1@.150	Ea	.25	6.15	6.40
24 - 20 AWG solder cup	L1@.150	Ea	.25	6.15	6.40
22 - 20 AWG solder cup	L1@.150	Ea	.25	6.15	6.40
Cable hardware for crimp snap - HDE and solder cup contacts					
Male screw retainer, yellow	L1@.100	Ea	.51	4.10	4.61
Male screw retainer, clear	L1@.100	Ea	.51	4.10	4.61
Fem. screw retainer, yellow	L1@.100	Ea	1.69	4.10	5.79
Fem. screw retainer, clear	L1@.100	Ea	1.69	4.10	5.79

Use these figures to estimate the cost of cable fittings installed under the conditions described on pages 5 and 6. The crew size is one technician. The labor cost per manhour is $41.03. These costs include cable stripping, layout, material handling, and normal waste. Add for sales tax, delivery, supervision, mobilization, demobilization, cleanup, overhead and profit.

Material	Craft@Hrs	Unit	Material Cost	Labor Cost	Installed Cost

Cable hardware for crimp snap, HDE and solder clamps

Material	Craft@Hrs	Unit	Material Cost	Labor Cost	Installed Cost
4 - 40 slide lock post, yellow	L1@.150	Ea	2.01	6.15	8.16
2 - 56 slide lock post, clear	L1@.150	Ea	2.01	6.15	8.16
15 position slide latch	L1@.150	Ea	3.93	6.15	10.08
25 position slide latch	L1@.150	Ea	4.63	6.15	10.78

Subminiature D cable connectors, plugs - shell with contacts, zinc, crimp snap-in

Material	Craft@Hrs	Unit	Material Cost	Labor Cost	Installed Cost
24 - 20 AWG, 9 position	L1@.250	Ea	2.90	10.30	13.20
24 - 20 AWG, 15 position	L1@.300	Ea	3.77	12.30	16.07
24 - 20 AWG, 25 position	L1@.400	Ea	5.56	16.40	21.96

Subminiature D cable connectors, plugs - shell with contacts, tin with ground indents, HDE insulation displacement

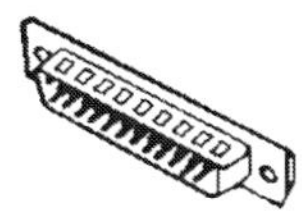

Material	Craft@Hrs	Unit	Material Cost	Labor Cost	Installed Cost
26 - 22 AWG, 9 position	L1@.250	Ea	3.33	10.30	13.63
26 - 22 AWG, 15 position	L1@.300	Ea	4.22	12.30	16.52
26 - 22 AWG, 25 position	L1@.400	Ea	5.40	16.40	21.80

Subminiature D cable connectors, plugs - shell with contacts, solder cup

Material	Craft@Hrs	Unit	Material Cost	Labor Cost	Installed Cost
30 - 18 AWG, 9 position	L1@.300	Ea	1.69	12.30	13.99
30 - 18 AWG, 15 position	L1@.400	Ea	2.31	16.40	18.71
30 - 18 AWG, 25 position	L1@.500	Ea	3.20	20.50	23.70

Subminiature D cable connectors, plugs - shell with contacts, plastic, HDE insulation displacement

Material	Craft@Hrs	Unit	Material Cost	Labor Cost	Installed Cost
26 - 22 AWG, 9 position	L1@.250	Ea	2.43	10.30	12.73
26 - 22 AWG, 15 position	L1@.300	Ea	3.13	12.30	15.43
26 - 22 AWG, 25 position	L1@.400	Ea	3.93	16.40	20.33

Subminiature D cable connectors, plug kits - shell with contacts - backshell & hardware, crimp snap-in

Material	Craft@Hrs	Unit	Material Cost	Labor Cost	Installed Cost
24 - 20 AWG, 9 position	L1@.250	Ea	5.05	10.30	15.35
24 - 20 AWG, 15 position	L1@.300	Ea	6.98	12.30	19.28
24 - 20 AWG, 25 position	L1@.400	Ea	7.49	16.40	23.89

Use these figures to estimate the cost of cable fittings installed under the conditions described on pages 5 and 6. The crew size is one technician. The labor cost per manhour is $41.03. These costs include cable stripping, layout, material handling, and normal waste. Add for sales tax, delivery, supervision, mobilization, demobilization, cleanup, overhead and profit.

Cable Fittings

Material	Craft@Hrs	Unit	Material Cost	Labor Cost	Installed Cost

Subminiature D cable connectors, plug kits - shell with contacts - backshell & hardware, crimp snap-in

26 - 20 AWG, 9 position	L1@.250	Ea	7.72	10.30	18.02
24 - 20 AWG, 15 position	L1@.300	Ea	8.53	12.30	20.83
24 - 20 AWG, 25 position	L1@.400	Ea	10.50	16.40	26.90

Subminiature D cable connectors, plug kits, shell with contacts, backshell & hardware, tin with ground indents, HDE insulation displacement

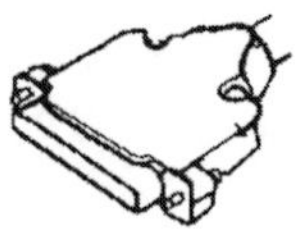

26 - 22 AWG, 9 position	L1@.250	Ea	7.10	10.30	17.40
24 - 20 AWG, 15 position	L1@.300	Ea	8.18	12.30	20.48
24 - 20 AWG, 25 position	L1@.400	Ea	9.35	16.40	25.75

Subminiature D cable connectors, plug kits, shell with contacts, backshell & hardware, tin with ground indents, solder cup

30 - 18 AWG, 9 position	L1@.300	Ea	6.30	12.30	18.60
30 - 18 AWG, 15 position	L1@.400	Ea	6.73	16.40	23.13
30 - 18 AWG, 25 position	L1@.500	Ea	7.72	20.50	28.22

Subminiature D cable connectors, plug kits, shell with contacts, backshell & hardware, tin with ground indents, plastic

26 - 22 AWG, 9 position	L1@.250	Ea	4.94	10.30	15.24
26 - 22 AWG, 15 position	L1@.300	Ea	6.30	12.30	18.60
26 - 22 AWG, 25 position	L1@.400	Ea	6.54	16.40	22.94

Subminiature D cable connectors, receptacles - shell with contacts, zinc, crimp snap-in

24 - 20 AWG, 9 position	L1@.250	Ea	3.14	10.30	13.44
24 - 20 AWG, 15 position	L1@.300	Ea	3.93	12.30	16.23
24 - 20 AWG, 25 position	L1@.400	Ea	5.57	16.40	21.97

Subminiature D cable connectors, receptacles - shell with contacts, tin with ground indents, crimp snap-in

24 - 20 AWG, 9 position	L1@.250	Ea	2.89	10.30	13.19
24 - 20 AWG, 15 position	L1@.300	Ea	3.93	12.30	16.23
24 - 20 AWG, 25 position	L1@.400	Ea	5.57	16.40	21.97

Subminiature D cable connectors, receptacles - shell with contacts, tin with ground indents, plastic

26 - 22 AWG, 9 position	L1@.250	Ea	2.61	10.30	12.91
26 - 22 AWG, 15 position	L1@.300	Ea	3.33	12.30	15.63
26 - 22 AWG, 25 position	L1@.400	Ea	4.79	16.40	21.19

Use these figures to estimate the cost of cable fittings installed under the conditions described on pages 5 and 6. The crew size is one technician. The labor cost per manhour is $41.03. These costs include cable stripping, layout, material handling, and normal waste. Add for sales tax, delivery, supervision, mobilization, demobilization, cleanup, overhead and profit.

Material	Craft@Hrs	Unit	Material Cost	Labor Cost	Installed Cost

Subminiature D cable connectors, receptacle kits - shell with contacts, backshell & hardware, zinc, crimp snap-in

24 - 20 AWG, 9 position	L1@.250	Ea	5.40	10.30	15.70
24 - 20 AWG, 15 position	L1@.300	Ea	7.33	12.30	19.63
24 - 20 AWG, 25 position	L1@.400	Ea	7.94	16.40	24.34

Subminiature D cable connectors, receptacle kits - shell with contacts, backshell & hardware, HDE insulation displacement

26 - 22 AWG, 9 position	L1@.250	Ea	7.62	10.30	17.92
26 - 22 AWG, 15 position	L1@.300	Ea	8.35	12.30	20.65
26 - 22 AWG, 25 position	L1@.400	Ea	10.00	16.40	26.40

Subminiature D cable connectors, receptacle kits - shell with contacts, backshell & hardware, solder cap

30 - 18 AWG, 9 position	L1@.300	Ea	6.43	12.30	18.73
30 - 18 AWG, 15 position	L1@.400	Ea	7.33	16.40	23.73
30 - 18 AWG, 25 position	L1@.500	Ea	8.35	20.50	28.85

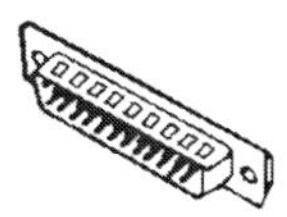

Subminiature D cable connectors, receptacle kits - shell with contacts, backshell & hardware, plastic

26 - 22 AWG, 9 position	L1@.250	Ea	4.97	10.30	15.27
26 - 22 AWG, 15 position	L1@.300	Ea	6.43	12.30	18.73
26 - 22 AWG, 25 position	L1@.400	Ea	7.62	16.40	24.02

Subminiature D cable connectors, straight exit backshells, plastic

9 position, .240 ferrules	L1@.150	Ea	1.86	6.15	8.01
15 position, .430 ferrules	L1@.150	Ea	3.14	6.15	9.29
25 position, .430 ferrules	L1@.150	Ea	3.62	6.15	9.77

Subminiature D cable connectors, straight exit backshells, metal plated plastic

9 position, .240 ferrules	L1@.150	Ea	3.62	6.15	9.77
15 position, .430 ferrules	L1@.150	Ea	3.77	6.15	9.92
25 position, .430 ferrules	L1@.150	Ea	3.93	6.15	10.08

Subminiature D cable connectors, straight exit backshells, metal

9 position, .240 ferrules	L1@.150	Ea	6.59	6.15	12.74
15 position, .430 ferrules	L1@.150	Ea	6.73	6.15	12.88
25 position, .430 ferrules	L1@.150	Ea	6.89	6.15	13.04

Use these figures to estimate the cost of cable fittings installed under the conditions described on pages 5 and 6. The crew size is one technician. The labor cost per manhour is $41.03. These costs include cable stripping, layout, material handling, and normal waste. Add for sales tax, delivery, supervision, mobilization, demobilization, cleanup, overhead and profit.

Cable Fittings

Material	Craft@Hrs	Unit	Material Cost	Labor Cost	Installed Cost
Data cable connectors, 36 position, unshielded					
28 - 26 AWG, stranded wire	L1@.150	Ea	3.52	6.15	9.67
26 - 24 AWG, sol or str wire	L1@.150	Ea	3.52	6.15	9.67
Data cable connectors, 50 position, shielded					
28 - 26 AWG, stranded wire	L1@.150	Ea	6.33	6.15	12.48
26 - 24 AWG, sol or str wire	L1@.150	Ea	6.33	6.15	12.48
Subminiature to modular jack adapters					
25 pos. plug - 6 pos.	L1@.250	Ea	15.90	10.30	26.20
25 pos. recept - 6 pos.	L1@.250	Ea	15.90	10.30	26.20
25 pos. plug - 8 pos.	L1@.250	Ea	16.10	10.30	26.40
25 pos. recept - 8 pos.	L1@.250	Ea	16.10	10.30	26.40
Field programmable RS-232 cable connectors					
Plug	L1@.150	Ea	11.90	6.15	18.05
Receptacle	L1@.150	Ea	11.90	6.15	18.05
Centerline clamp cable connectors .085, plug type					
With snap-on cable clamp	L1@.050	Ea	2.92	2.05	4.97
W/ ratcheted cable clamp	L1@.100	Ea	3.34	4.10	7.44
Cable connector savers - 25 position					
Shielded with cover	L1@.200	Ea	55.60	8.21	63.81
Plastic without cover	L1@.200	Ea	41.00	8.21	49.21
Plastic with cover	L1@.200	Ea	70.40	8.21	78.61
Cable connector gender changer - 25 position assemblies					
Shielded-cover, plug	L1@.250	Ea	83.80	10.30	94.10
Plastic-cover, plug	L1@.250	Ea	49.70	10.30	60.00
Shielded-no cover, recept.	L1@.300	Ea	51.10	12.30	63.40
Shielded-cover, receptacle	L1@.300	Ea	81.80	12.30	94.10
Plastic-no cover, recept.	L1@.300	Ea	50.10	12.30	62.40
Plastic-cover, receptacle	L1@.300	Ea	77.40	12.30	89.70

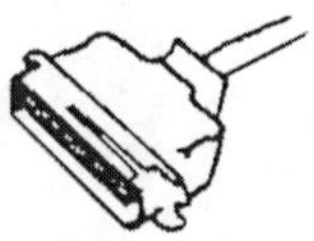

Use these figures to estimate the cost of cable fittings installed under the conditions described on pages 5 and 6. The crew size is one technician. The labor cost per manhour is $41.03. These costs include cable stripping, layout, material handling, and normal waste. Add for sales tax, delivery, supervision, mobilization, demobilization, cleanup, overhead and profit.

Material	Craft@Hrs	Unit	Material Cost	Labor Cost	Installed Cost
Strain relief cable cover, 180 degree					
.500 O.D. & smaller cable	L1@.100	Ea	3.47	4.10	7.57
.500 O.D & larger cable	L1@.150	Ea	3.21	6.15	9.36
.290 - .360 cable O.D.	L1@.100	Ea	1.73	4.10	5.83
.360 - .430 cable O.D.	L1@.150	Ea	1.59	6.15	7.74
Baluns, RJ11 to BNC plug, twisted pair cable					
14' twisted pair, plugs	L1@.150	Ea	27.00	6.15	33.15
30' twisted pair, plugs	L1@.200	Ea	33.30	8.21	41.51
8" coaxial pigtail, plugs	L1@.100	Ea	23.10	4.10	27.20
RJ11 jack to BNC plug	L1@.100	Ea	23.10	4.10	27.20
RJ11 to DPC plug filtered	L1@.100	Ea	35.60	4.10	39.70
RJ45 plug to BNC plug					
14' twisted pair, plugs	L1@.150	Ea	27.00	6.15	33.15
30' twisted pair, plugs	L1@.200	Ea	33.30	8.21	41.51
Keyed RJ45 plug to BNC plug with coaxial pigtail plugs					
8"	L1@.100	Ea	23.10	4.10	27.20
Cable fittings, twisted pair to BNC plugs					
Screw terminal to BNC plug	L1@.250	Ea	23.10	10.30	33.40
Screw terminal to panel	L1@.250	Ea	23.10	10.30	33.40
RJ11 jack to BNC jack	L1@.100	Ea	22.60	4.10	26.70
RJ11 jack to BNC jack	L1@.100	Ea	22.50	4.10	26.60
RJ11 jack to pan. BNC jack	L1@.100	Ea	29.10	4.10	33.20
RJ45 jack to BNC jack	L1@.100	Ea	24.00	4.10	28.10
RJ45 jack to pan. BNC jack	L1@.100	Ea	22.60	4.10	26.70
RJ11 to twinaxial plug	L1@.150	Ea	38.20	6.15	44.35
RJ11 with 14' twist-pr. cable	L1@.150	Ea	41.50	6.15	47.65

Use these figures to estimate the cost of cable fittings installed under the conditions described on pages 5 and 6. The crew size is one technician. The labor cost per manhour is $41.03. These costs include cable stripping, layout, material handling, and normal waste. Add for sales tax, delivery, supervision, mobilization, demobilization, cleanup, overhead and profit.

Cable Fittings

Material	Craft@Hrs	Unit	Material Cost	Labor Cost	Installed Cost

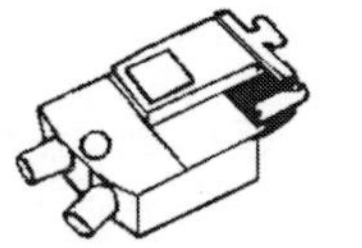

Baluns, 4-position data connectors, 93 ohm to 150 ohm coaxial shielded twisted pair

Material	Craft@Hrs	Unit	Material Cost	Labor Cost	Installed Cost
Conn. to dual BNC jack	L1@.250	Ea	56.30	10.30	66.60
Conn. to single BNC jack	L1@.250	Ea	32.60	10.30	42.90

Connector to modular jack IMP, matching, 105 ohm to 150 ohm twisted pair cable

Material	Craft@Hrs	Unit	Material Cost	Labor Cost	Installed Cost
6 pos. unshielded to shield	L1@.200	Ea	37.80	8.21	46.01
8 pos. unshielded to shield	L1@.200	Ea	37.80	8.21	46.01

Modular couplers, in-line

Material	Craft@Hrs	Unit	Material Cost	Labor Cost	Installed Cost
6 position	L1@.050	Ea	4.09	2.05	6.14
8 position	L1@.050	Ea	4.40	2.05	6.45
8 position, keyed	L1@.050	Ea	4.71	2.05	6.76

Modular couplers, right angle

Material	Craft@Hrs	Unit	Material Cost	Labor Cost	Installed Cost
6 position	L1@.050	Ea	4.25	2.05	6.30
8 position	L1@.050	Ea	4.40	2.05	6.45
8 position, keyed	L1@.050	Ea	3.96	2.05	6.01

Modular keystone jacks, telephone cable

Material	Craft@Hrs	Unit	Material Cost	Labor Cost	Installed Cost
22 - 24 AWG, 3 pair	L1@.050	Ea	4.97	2.05	7.02
24 - 28 AWG, 3 pair	L1@.050	Ea	4.97	2.05	7.02
19 - 22 AWG, 3 pair	L1@.050	Ea	5.27	2.05	7.32
22 - 24 AWG, 4 pair	L1@.050	Ea	5.13	2.05	7.18
24 - 28 AWG, 4 pair	L1@.050	Ea	5.44	2.05	7.49
22 - 24 AWG, 4 pr, keyed	L1@.050	Ea	5.27	2.05	7.32
24 - 28 AWG, 4 pr, keyed	L1@.050	Ea	5.61	2.05	7.66

Telephone splice connectors

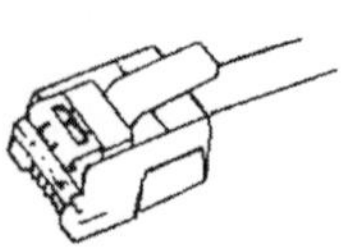

Material	Craft@Hrs	Unit	Material Cost	Labor Cost	Installed Cost
2-wire, translucent	L1@.250	Ea	93.20	10.30	103.50
2-wire, yellow	L1@.250	Ea	67.50	10.30	77.80
2-wire, clear	L1@.250	Ea	80.00	10.30	90.30
3-wire, translucent	L1@.300	Ea	111.00	12.30	123.30
3-wire, yellow	L1@.300	Ea	78.70	12.30	91.00
3-wire, clear	L1@.300	Ea	98.10	12.30	110.40

Use these figures to estimate the cost of cable fittings installed under the conditions described on pages 5 and 6. The crew size is one technician. The labor cost per manhour is $41.03. These costs include cable stripping, layout, material handling, and normal waste. Add for sales tax, delivery, supervision, mobilization, demobilization, cleanup, overhead and profit.

Material	Craft@Hrs	Unit	Material Cost	Labor Cost	Installed Cost
BNC crimp-on plug connector for RG/U cable					
Commercial silver, 58	L1@.100	Ea	1.66	4.10	5.76
Commercial gold, 58	L1@.100	Ea	2.53	4.10	6.63
Standard silver, 58	L1@.100	Ea	3.29	4.10	7.39
Standard gold, 58	L1@.100	Ea	9.33	4.10	13.43
Comm. silver, hex, 58	L1@.100	Ea	1.83	4.10	5.93
Comm. gold, hex, 58	L1@.100	Ea	3.16	4.10	7.26
Comm. silver, plenum, 58	L1@.100	Ea	1.88	4.10	5.98
Comm. gold, plenum, 58	L1@.100	Ea	2.17	4.10	6.27
Std. silver, plenum, 58	L1@.100	Ea	3.30	4.10	7.40
Std. gold, plenum, 58	L1@.100	Ea	4.83	4.10	8.93
Std. gold, 90 degree, 58	L1@.100	Ea	19.80	4.10	23.90
Comm. silver, 59/62	L1@.100	Ea	1.69	4.10	5.79
Comm. gold, 59/62	L1@.100	Ea	2.63	4.10	6.73
Std. silver, 59/62	L1@.100	Ea	3.66	4.10	7.76
Std. gold, 59/62	L1@.100	Ea	4.09	4.10	8.19
Comm. silver, hex, 59/62	L1@.100	Ea	3.34	4.10	7.44
Comm. gold, hex, 59/62	L1@.100	Ea	3.96	4.10	8.06
Comm. silv, plenum, 59/62	L1@.100	Ea	1.86	4.10	5.96
Comm. gold, plenum, 59/62	L1@.100	Ea	2.17	4.10	6.27
Std. silver, plenum, 59/62	L1@.100	Ea	3.34	4.10	7.44
Std. gold, plenum, 59/62	L1@.100	Ea	3.96	4.10	8.06
Std. gold, 90 degree, 59/62	L1@.100	Ea	19.80	4.10	23.90
Comm. silver, hi temp, 59	L1@.100	Ea	2.91	4.10	7.01
Comm. gold, hi temp, 59	L1@.100	Ea	19.10	4.10	23.20
BNC crimp-on connector, jack type for RG/U cable					
Comm. silver, 58	L1@.150	Ea	3.34	6.15	9.49
Comm. gold, 58	L1@.150	Ea	3.66	6.15	9.81
Comm. silver, plenum, 58	L1@.150	Ea	4.40	6.15	10.55
Comm. gold, plenum, 58	L1@.150	Ea	5.13	6.15	11.28
Std. gold plenum, 58	L1@.150	Ea	5.13	6.15	11.28
Comm. gold, bulkhead, 58	L1@.150	Ea	5.44	6.15	11.59
Std. gold, bulkhead, 58	L1@.150	Ea	6.35	6.15	12.50

Use these figures to estimate the cost of cable fittings installed under the conditions described on pages 5 and 6. The crew size is one technician. The labor cost per manhour is $41.03. These costs include cable stripping, layout, material handling, and normal waste. Add for sales tax, delivery, supervision, mobilization, demobilization, cleanup, overhead and profit.

Cable Fittings

BNC twist-on plug connector for RG/U cable

Material	Craft@Hrs	Unit	Material Cost	Labor Cost	Installed Cost
Straight plug, 58	L1@.200	Ea	2.76	8.21	10.97
Straight plug, 59/62	L1@.200	Ea	2.17	8.21	10.38
Straight plug, plenum, 59/62	L1@.200	Ea	2.76	8.21	10.97
Straight jack, 58	L1@.200	Ea	3.31	8.21	11.52
Straight jack, 59/62	L1@.200	Ea	3.66	8.21	11.87
Straight jack, 59	L1@.200	Ea	10.70	8.21	18.91
90 degree plug, 58	L1@.250	Ea	24.10	10.30	34.40
90 degree plug, 59/62	L1@.250	Ea	24.10	10.30	34.40

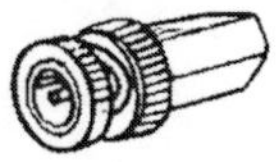

THC threaded crimp-on connector for RG/U cable

Material	Craft@Hrs	Unit	Material Cost	Labor Cost	Installed Cost
Straight plug, 58	L1@.200	Ea	3.34	8.21	11.55
Straight plug, 59/62	L1@.200	Ea	3.34	8.21	11.55
Straight jack, 58	L1@.200	Ea	26.60	8.21	34.81
Straight jack, 59/62	L1@.200	Ea	23.10	8.21	31.31

F type connectors for RG/U cable

Material	Craft@Hrs	Unit	Material Cost	Labor Cost	Installed Cost
Twist-on plug, 59/62	L1@.250	Ea	.66	10.30	10.96
Crimp-on, 59	L1@.250	Ea	.61	10.30	10.91
Crimp-on, 6	L1@.250	Ea	.61	10.30	10.91
2 pc, crimp-on, 6	L1@.250	Ea	1.11	10.30	11.41

N type connectors for RG/U cable

Material	Craft@Hrs	Unit	Material Cost	Labor Cost	Installed Cost
Comm. plug, 8	L1@.200	Ea	6.64	8.21	14.85
Std. plug, plenum	L1@.200	Ea	10.70	8.21	18.91
Jack to jack, plenum	L1@.250	Ea	16.10	10.30	26.40
Plug terminator	L1@.200	Ea	12.40	8.21	20.61
Jack terminator	L1@.200	Ea	12.40	8.21	20.61
Crimp plug, RGS/11	L1@.200	Ea	8.11	8.21	16.32

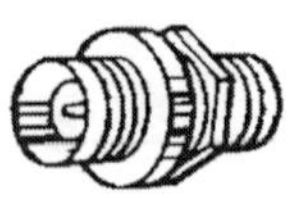

Use these figures to estimate the cost of cable fittings installed under the conditions described on pages 5 and 6. The crew size is one technician. The labor cost per manhour is $41.03. These costs include cable stripping, layout, material handling, and normal waste. Add for sales tax, delivery, supervision, mobilization, demobilization, cleanup, overhead and profit.

Material	Craft@Hrs	Unit	Material Cost	Labor Cost	Installed Cost
Twinaxial cable connectors					
Solder plug	L1@.300	Ea	5.13	12.30	17.43
Crimp-on plug	L1@.200	Ea	6.49	8.21	14.70
Solder jack	L1@.300	Ea	10.40	12.30	22.70
In-line splice, jack to jack	L1@.250	Ea	8.11	10.30	18.41
Bulkhead jack, solder	L1@.300	Ea	23.70	12.30	36.00
Coaxial cable adapters					
In-line splice, jack to jack	L1@.300	Ea	6.20	12.30	18.50
Bulk. splice, jack to jack	L1@.250	Ea	7.08	10.30	17.38
Right angle plug, 58	L1@.300	Ea	19.80	12.30	32.10
Right angle plug, 59/62	L1@.300	Ea	19.80	12.30	32.10
Bulkhead jack, 58	L1@.250	Ea	6.64	10.30	16.94
Bulkhead jack, 59/62	L1@.250	Ea	6.64	10.30	16.94
T-connector	L1@.300	Ea	13.70	12.30	26.00

Use these figures to estimate the cost of cable fittings installed under the conditions described on pages 5 and 6. The crew size is one technician. The labor cost per manhour is $41.03. These costs include cable stripping, layout, material handling, and normal waste. Add for sales tax, delivery, supervision, mobilization, demobilization, cleanup, overhead and profit.

Wire Conversion Table

For insulated conductors rated 0-2000 volts 60 degrees C (140 degrees F), not more than three conductors in a raceway or cable or earth (direct burial) on ambient temperatures of 30 degrees C (86 degrees F)

60 degrees C		90 degrees C	
Copper	**Aluminum**	**Copper**	**Aluminum**
# 10	# 8	# 10	# 8
# 8	# 6	# 8	# 6
# 6	# 4	# 6	# 4
# 4	# 2	# 4	# 2
# 2	# 1/0	# 2	# 1/0
# 1	# 2/0	# 1	# 2/0
# 1/0	# 3/0	# 1/0	# 3/0
# 2/0	# 4/0	# 2/0	# 4/0
# 3/0	# 250 kcmil	# 3/0	# 250 kcmil
# 4/0	# 350 kcmil	# 4/0	# 350 kcmil
# 250 kcmil	# 400 kcmil	# 250 kcmil	# 400 kcmil
# 300 kcmil	# 500 kcmil	# 300 kcmil	# 500 kcmil
# 350 kcmil	# 500 kcmil	# 350 kcmil	# 500 kcmil
# 400 kcmil	# 600 kcmil	# 400 kcmil	# 600 kcmil
# 500 kcmil	# 750 kcmil	# 500 kcmil	# 750 kcmil
# 600 kcmil	# 900 kcmil	# 600 kcmil	# 900 kcmil
# 750 kcmil	# 1250 kcmil	# 750 kcmil	# 1250 kcmil

See the tables in NEC Section 310 for conductor ampacities for general wiring

Section 19: Undercarpet Wiring Systems

In a large office open space area being prepared for either tenant improvements or rehab to accommodate desk and furniture layout, ***undercarpet wiring systems*** can be the easiest solution. In this case the structure is existing and the planned locations for the work station desks and other office furniture will require wiring systems.

Installing these systems in an existing structure can be very expensive and probably disruptive to existing tenants sharing space in the building. If the tenant improvements are on the ground floor, the systems will probably require cutting and patching the floor for the conduit systems.

When the improvements are on floors above the ground floor, conduit systems might be installed in the ceiling cavity of the floor below. In order to gain access to the floor above, the floor is core drilled for conduit stubups at the location of the service outlets.

The local building code may require that penetrations between floors be fireproofed in an approved manner. An oversized core penetration should be made to allow for adequate fireproofing material to stop the spread of fire.

Designers and manufacturers have engineered the undercarpet wiring system concept for a low cost, quick and clean alternative in maintaining an open office landscape. The wiring system can be designed to fulfill the user's needs and installed just before the carpeting is laid.

When the need arises to relocate a desk or office furniture, the carpet can be cut to extend the wiring systems to additional outlets and can be easily patched by a carpet layer. Then when more work stations are needed, the carpet can be cut, the wiring systems extended and the carpet patched. This wiring system is very flexible and economical for a large open space office environment.

Care should be exercised in making the office furniture planned layout. The service outlets should be accurately located to fit the required need.

As the need for office work stations increases or adjustments are needed, the wiring services can be modified. A well-planned layout should take this into consideration during the planning of the services.

The installation of the actual material is simple. The location of the service outlets are marked on the bare clean floor. The straight path for the cable from the wall junction boxes to the service outlet can be marked with a chalk line, the floor swept clean and the cable attached to the floor with the adhesive on the cable. The same procedure is used for runs between outlet locations.

Care must be used to protect the wiring systems from damage by other trades, construction equipment and furniture handling. The wiring system cables are tough and for added protection, a 7"- wide top shield is installed over the cable.

If your company is not familiar with undercarpet wiring systems, ask your supplier for help in planning the layout. The supplier will probably contact a manufacturer's representative for assistance on product information.

Material	Craft@Hrs	Unit	Material Cost	Labor Cost	Installed Cost
20 amp, 300 volt, 3-conductor #12 AWG					
2" x 50'	L2@0.30	Ea	41.20	12.30	53.50
2" x 100'	L2@0.60	Ea	80.20	24.60	104.80
2" x 250'	L2@1.50	Ea	176.00	61.50	237.50
20 amp, 300 volt, 4-conductor #12 AWG					
2-1/2" x 50'	L2@0.40	Ea	54.90	16.40	71.30
2-1/2" x 100'	L2@0.80	Ea	110.00	32.80	142.80
2-1/2" x 250'	L2@2.00	Ea	275.00	82.10	357.10
20 amp, 300 volt, 4-conductor #12 AWG with isolated ground					
2-1/2" x 100'	L2@0.80	Ea	111.00	32.80	143.80
30 amp, 300 volt, 3-conductor #12 AWG, 100' reel					
White, green, black	L2@0.35	Ea	16.40	14.40	30.80
White, green, red	L2@0.35	Ea	17.60	14.40	32.00
White, green, blue	L2@0.35	Ea	17.60	14.40	32.00
30 amp, 300 volt, 3-conductor #10 AWG, 100' reel					
White, green, black	L2@0.40	Ea	21.50	16.40	37.90
White, green, red	L2@0.40	Ea	19.10	16.40	35.50
White, green, blue	L2@0.40	Ea	19.10	16.40	35.50
20 amp, 300 volt, 5-conductor #12 AWG, 100' reel					
White, green, black, blue & red	L2@0.40	Ea	27.50	16.40	43.90
White, green, green/yellow, red & blue, with isolated ground	L2@0.40	Ea	31.10	16.40	47.50
30 amp, 300 volt, 5-conductor #10 AWG, 100' reel					
White, green, black, red & blue	L2@0.45	Ea	31.10	18.50	49.60
White, green, green/yellow, red & blue with isolated ground	L2@0.45	Ea	31.80	18.50	50.30

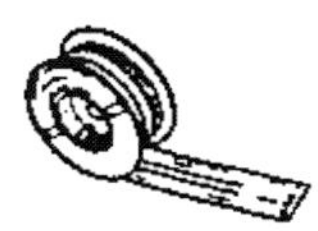

Use these figures to estimate the cost of undercarpet wiring systems installed in a building under the conditions described on pages 5 and 6. Costs listed are for each item installed. The crew is one or two electricians, as indicated, working at a labor cost of $41.03 per manhour. These costs include layout, material handling, and normal waste. Add for sales tax, delivery, supervision, mobilization, demobilization, cleanup, overhead and profit.

Undercarpet Wiring Systems

Material	Craft@Hrs	Unit	Material Cost	Labor Cost	Installed Cost
Integrated voice & data cable, 100' reel					
2 pair shielded	L2@0.40	Ea	23.10	16.40	39.50
3 pair shielded	L2@0.40	Ea	19.70	16.40	36.10
4 pair shielded	L2@0.50	Ea	23.80	20.50	44.30
Color coded cable, 100' reel					
2 pair twisted	L2@0.40	Ea	110.00	16.40	126.40
4 pair twisted	L2@0.40	Ea	90.30	16.40	106.70
Top shield tape					
7" x 50'	L1@0.25	Ea	15.70	10.30	26.00
7" x 100'	L1@0.50	Ea	31.80	20.50	52.30
7" x 250'	L1@1.25	Ea	79.10	51.30	130.40
Station wire connectors					
2 pair jack	L1@0.15	Ea	4.74	6.15	10.89
3 pair jack	L1@0.15	Ea	6.30	6.15	12.45
3 pair jack	L1@0.20	Ea	7.10	8.21	15.31
3 pair plug	L1@0.15	Ea	5.53	6.15	11.68
4 pair plug	L1@0.20	Ea	6.30	8.21	14.51
Tap & splice adapter					
3-conductor adapter	L1@0.10	Ea	34.80	4.10	38.90
Insulation kit					
Tap & splice	L1@0.15	Ea	38.70	6.15	44.85
Transition boxes					
Tel/data access	L1@0.30	Ea	30.10	12.30	42.40
Transition partitions					
Hi/Lo separator	L1@0.35	Ea	25.10	14.40	39.50

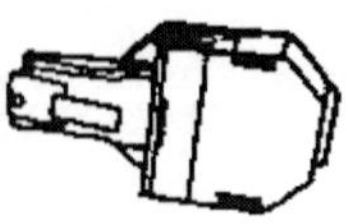

Use these figures to estimate the cost of undercarpet wiring systems installed in a building under the conditions described on pages 5 and 6. Costs listed are for each item installed. The crew is one or two electricians, as indicated, working at a labor cost of $41.03 per manhour. These costs include layout, material handling, and normal waste. Add for sales tax, delivery, supervision, mobilization, demobilization, cleanup, overhead and profit.

Material	Craft@Hrs	Unit	Material Cost	Labor Cost	Installed Cost
Transition junction boxes					
Surface mounted	L1@0.30	Ea	232.00	12.30	244.30
Flush floor mounted	L1@0.50	Ea	226.00	20.50	246.50
Stud mounted	L1@0.35	Ea	72.80	14.40	87.20
Thin stud mounted	L1@0.35	Ea	65.00	14.40	79.40
Stud mounted, DLB	L1@0.35	Ea	97.20	14.40	111.60
Single terminal box	L1@0.30	Ea	57.50	12.30	69.80
Terminal blocks					
Round to flat cable	L1@0.35	Ea	25.90	14.40	40.30
3-wire underflow duct	L1@0.35	Ea	30.10	14.40	44.50
3-wire insul. displace.	L1@0.25	Ea	55.30	10.30	65.60
5-wire insul. displace.	L1@0.30	Ea	79.10	12.30	91.40
Four way power intrafacer					
Transition assembly	L1@0.25	Ea	88.10	10.30	98.40
Insulated ground transition assembly	L1@0.35	Ea	91.60	14.40	106.00
Five way power intrafacer					
Transition assembly	L1@0.35	Ea	72.80	14.40	87.20
Insulated ground transition assembly	L1@0.40	Ea	79.60	16.40	96.00
Pedestals for power, less duplex receptacles					
For one duplex outlet	L1@0.30	Ea	122.00	12.30	134.30
For two duplex outlets	L1@0.40	Ea	137.00	16.40	153.40
Pedestals					
Communications	L1@0.30	Ea	58.50	12.30	70.80

Use these figures to estimate the cost of undercarpet wiring systems installed in a building under the conditions described on pages 5 and 6. Costs listed are for each item installed. The crew is one electrician working at a labor cost of $41.03 per manhour. These costs include layout, material handling, and normal waste. Add for sales tax, delivery, supervision, mobilization, demobilization, cleanup, overhead and profit.

Index

A

B

D

E

F

G

H

I

J

K

L

M

N

O

P

Q

R

S

T

U

V

W

X

Z

Practical References for Builders

National Estimator Cloud

Generate professional construction estimates for all residential and commercial construction from your internet browser. Includes 10 Craftsman construction cost databases, over 40,000 labor and material costs for construction, in an easy-to-use format. Cost estimates are well-organized and thoroughly indexed to speed and simplify writing estimates for nearly any residential or light commercial construction project – new construction, improvement or repair. Convert the bid to an invoice – in either QuickBooks Desktop or QuickBooks Online. Access your estimates from anywhere and on any device with a Web browser. **Just $9.49 a month.** Visit https://craftsman-book.com/national-estimator-cloud for more details.

Electrical Blueprint Reading Revised eBook

Shows how to read and interpret electrical drawings, wiring diagrams, and specifications for constructing electrical systems. Shows how a typical lighting and power layout would appear on a plan, and explains what to do to execute the plan. Describes how to use a panelboard or heating schedule, and includes typical electrical specifications.
208 pages, $14.88 at www.craftsman-book.com

Construction Contract Writer

Relying on a "one-size-fits-all" boilerplate construction contract to fit your jobs can be dangerous — almost as dangerous as a handshake agreement. *Construction Contract Writer* lets you draft a contract in minutes that precisely fits your needs and the particular job, and meets both state and federal requirements. You just answer a series of questions — like an interview — to construct a legal contract for each project you take on. Anticipate where disputes could arise and settle them in the contract before they happen. Include the warranty protection you intend, the payment schedule, and create subcontracts from the prime contract by just clicking a box. Includes a feedback button to an attorney on the Craftsman staff to help should you get stumped — *No extra charge.* **$149.95.** Download the *Construction Contract Writer* at: http://www.constructioncontractwriter.com

Insurance Replacement Estimator

Insurance underwriters demand detailed, accurate valuation data. There's no better authority on replacement cost for single-family homes than the Insurance Replacement Estimator. In minutes you get an insurance to value report showing the cost of re-construction based on your specification. You can generate and save unlimited reports.

For more information, visit
www.craftsman-book.com/insurance-replacement-estimator-online-software

The Complete Book of Home Inspections

This comprehensive manual covers every aspect of home inspection, from the tools required through the inspection of roofs, walls, interior rooms, windows and doors; garages, attics, basements and crawl spaces; paved areas around the structure, landscaping, insect damage and rot; electrical systems, HVAC systems, plumbing systems, and swimming pools. Covers energy considerations and environmental concerns such as radon and mold. Includes hundreds of photos and illustrations to help you understand, check, and identify potential problems. **290 pages, 8 x 10, $19.95**

Electrical Solar Essentials Quick-Card

This 8-page guide covers the growing field of solar energy. Focusing primarily on typical residential installations and the basics of what goes into them, this essential reference is filled with tables, tips, how-to's, why's and wherefore's. This is truly a must-have for anyone interested in solar: contractors, homeowners, etc. **8 pages, 8½ x 11, $7.95**

Electrical Inspection Notes

In this pocket-sized flip chart, you'll find code compliance information to help you make sure that every part of your electrical work is up to code. Here you'll find checklists, calculations, diagrams, plain-English code explanations, tables and charts, and who is responsible for what task during each step of the project. It lists everything to check for in the design stage, what to check for in interior electrical work, conductors, grounding, wiring methods, conduits, outlets, circuit panels, lighting, testing methods, exterior lighting, electrical service, heating, low voltage and more. **236 pages, 3 x 6, $24.95**

Estimating Electrical Construction Revised

Estimating the cost of electrical work can be a very detailed and exacting discipline. It takes specialized skills and knowledge to create reliable estimates for electrical work. See how an expert estimates materials and labor for residential and commercial electrical construction. Learn how to use labor units, the plan take-off, and the bid summary to make an accurate estimate, how to deal with suppliers, use pricing sheets, and modify labor units. This book provides extensive labor unit tables and blank forms on a CD for estimating your next electrical job. **272 pages, 8½ x 11, $59.00**
***eBook (PDF) also available;* $29.50 at** www.craftsman-book.com

Blueprint Reading for the Building Trades eBook

How to read and understand construction documents, blueprints, and schedules. Includes layouts of structural, mechanical, HVAC and electrical drawings. Shows how to interpret sectional views, follow diagrams and schematics, and covers common problems with construction specifications. **$8.38 at** www.craftsman-book.com

Ugly's Electrical Book 2017

UGLY'S Electrical References is designed to be used as an on-the-job reference for electricians, engineers, contractors, designers, maintenance workers, instructors, and the military; UGLY'S contains the most commonly required electrical information in an easy-to-read and easy-to-access format. **198 pages, 4 x 6, $24.95**

Craftsman eLibrary

Craftsman's eLibrary license gives you immediate access to 60+ PDF eBooks in our bookstore for 12 full months!
You pay only one low price. $129.99

Visit www.craftsman-book.com for more details.

Contractor's Guide to QuickBooks by Online Accounting

This book is designed to help a contractor, bookkeeper and their accountant set up and use QuickBooks Desktop specifically for the construction industry. No use re-inventing the wheel, we have used this system with contractors for over 30 years. It works and is now the national standard. By following the steps we outlined in the book you, too, can set up a good system for job costing as well as financial reporting.
156 pages, 8½ x 11, $68.50